Eleventh Edition

Digital Fundamentals

Thomas L. Floyd

PEARSON

Boston   Columbus   Indianapolis   New York   San Francisco   Hoboken
Amsterdam   Cape Town   Dubai   London   Madrid   Milan   Munich   Paris   Montreal   Toronto
Delhi   Mexico City   São Paulo   Sydney   Hong Kong   Seoul   Singapore   Taipei   Tokyo

Product Manager: Lindsey Prudhomme Gill
Program Manager: Maren Beckman
Project Manager: Rex Davidson
Editorial Assistant: Nancy Kesterson
Team Lead Program Manager: Laura Weaver
Team Lead Project Manager: JoEllen Gohr
Director of Marketing: David Gesell
Senior Marketing Coordinator: Stacey Martinez
Senior Marketing Assistant: Les Roberts
Procurement Specialist: Deidra M. Skahill
Media Project Manager: Noelle Chun
Media Project Coordinator: April Cleland
Creative Director: Andrea Nix
Art Director: Diane Y. Ernsberger
Cover Designer: Cenveo Publisher Services
Cover Image: Hamara/Shutterstock.com
Full-Service Project Management: Sherrill Redd/iEnergizer Aptara®, Inc.
Composition: iEnergizer Aptara®, Inc.
Printer/Binder: R.R. Donnelley & Sons/Willard
Cover Printer: Phoenix Color Hagerstown
Text Font: Times Roman

Credits and acknowledgments for materials borrowed from other sources and reproduced, with permission, in this textbook appear on the appropriate page within text.

Many of the designations by manufacturers and sellers to distinguish their products are claimed as trademarks. Where those designations appear in this book, and the publisher was aware of a trademark claim, the designations have been printed in initial caps or all caps.

Library of Congress Cataloging-in-Publication Data
Floyd, Thomas L.
  Digital fundamentals / Thomas L. Floyd.—Eleventh edition.
     pages cm
  ISBN 978-0-13-273796-8 (0-13-273796-5)   1. Digital electronics.   2. Logic circuits.   I. Title.
TK7868.D5F53 2015
621.381—dc23

                                                                 2014017765

10 9 8 7 6 5 4 3 2 1

ISBN 10:   0-13-273796-5
ISBN 13: 978-0-13-273796-8

This eleventh edition of *Digital Fundamentals* continues a long tradition of presenting a strong foundation in the core fundamentals of digital technology. This text provides basic concepts reinforced by plentiful illustrations, examples, exercises, and applications. Applied Logic features, Implementation features, troubleshooting sections, programmable logic and PLD programming, integrated circuit technologies, and the special topics of signal conversion and processing, data transmission, and data processing and control are included in addition to the core fundamentals. New topics and features have been added to this edition, and many other topics have been enhanced.

The approach used in *Digital Fundamentals* allows students to master the all-important fundamental concepts before getting into more advanced or optional topics. The range of topics provides the flexibility to accommodate a variety of program requirements. For example, some of the design-oriented or application-oriented topics may not be appropriate in some courses. Some programs may not cover programmable logic and PLD programming, while others may not have time to include data transmission or data processing. Also, some programs may not cover the details of "inside-the-chip" circuitry. These and other areas can be omitted or lightly covered without affecting the coverage of the fundamental topics. A background in transistor circuits is not a prerequisite for this textbook, and the coverage of integrated circuit technology (inside-the-chip circuits) is optionally presented.

New in This Edition

- New page layout and design for better visual appearance and ease of use
- Revised and improved topics
- Obsolete devices have been deleted.
- The *Applied Logic* features (formerly *System Applications*) have been revised and new topics added. Also, the VHDL code for PLD implementation is introduced and illustrated.
- A new boxed feature, entitled *Implementation*, shows how various logic functions can be implemented using fixed-function devices or by writing a VHDL program for PLD implementation.
- Boolean simplification coverage now includes the Quine-McCluskey method and the Espresso method is introduced.
- A discussion of Moore and Mealy state machines has been added.
- The chapter on programmable logic has been modified and improved.
- A discussion of memory hierarchy has been added.
- A new chapter on data transmission, including an extensive coverage of standard busses has been added.
- The chapter on computers has been completely revised and is now entitled "Data Processing and Control."
- A more extensive coverage and use of VHDL. There is a tutorial on the website at www.pearsonhighered.com/careersresources.com.
- More emphasis on D flip-flops

Standard Features

- Full-color format
- Core fundamentals are presented without being intermingled with advanced or peripheral topics.
- *InfoNotes* are sidebar features that provide interesting information in a condensed form.
- A chapter outline, chapter objectives, introduction, and key terms list appear on the opening page of each chapter.
- Within the chapter, the key terms are highlighted in color boldface. Each key term is defined at the end of the chapter as well as in the comprehensive glossary at the end of the book. Glossary terms are indicated by black boldface in the text.
- Reminders inform students where to find the answers to the various exercises and problems throughout each chapter.
- Section introduction and objectives are at the beginning of each section within a chapter.
- Checkup exercises conclude each section in a chapter with answers at the end of the chapter.
- Each worked example has a *Related Problem* with an answer at the end of the chapter.
- *Hands-On Tips* interspersed throughout provide useful and practical information.
- Multisim files (newer versions) on the website provide circuits that are referenced in the text for optional simulation and troubleshooting.
- The operation and application of test instruments, including the oscilloscope, logic analyzer, function generator, and DMM, are covered.
- Troubleshooting sections in many chapters
- Introduction to programmable logic
- Chapter summary
- True/False quiz at end of each chapter
- Multiple-choice self-test at the end of each chapter
- Extensive sectionalized problem sets at the end of each chapter with answers to odd-numbered problems at the end of the book.
- Troubleshooting, applied logic, and special design problems are provided in many chapters.
- Coverage of bipolar and CMOS IC technologies. Chapter 15 is designed as a "floating chapter" to provide optional coverage of IC technology (inside-the-chip circuitry) at any point in the course. Chapter 15 is online at www.pearsonhighered.com/careersresources.

Accompanying Student Resources

- *Experiments in Digital Fundamentals*, eleventh edition: lab manual by Dave Buchla and Doug Joksch.
- *Multisim Circuits*. The MultiSim files on the website includes selected circuits from the text that are indicated by the icon in Figure P-1.

MultiSim

FIGURE P-1

Other student resources available on the website:

1. Chapter 15, "Integrated Circuit Technologies"
2. VHDL tutorial

3. Verilog tutorial

4. MultiSim tutorial

5. Altera Quartus II tutorial

6. Xilinx ISE tutorial

7. Five-variable Karnaugh map tutorial

8. Hamming code tutorial

9. Quine-McCluskey method tutorial

10. Espresso algorithm tutorial

11. Selected VHDL programs for downloading

12. Programming the elevator controller using Altera Quartus II

Using Website VHDL Programs

VHDL programs in the text that have a corresponding VHDL file on the website are indicated by the icon in Figure P-2. These website VHDL files can be downloaded and used in conjunction with the PLD development software (Altera Quartus II or Xilinx ISE) to implement a circuit in a programmable logic device.

FIGURE P-2

Instructor Resources

- *Image Bank (0132738295)* This is a download of all the images in the text.

- *Online Course Support* If your program is offered in a distance learning format, please contact your local Pearson sales representative for a list of product solutions.

- *Instructor's Resource Manual (0132737957)* Includes worked-out solutions to chapter problems, solutions to Applied Logic Exercises, a summary of Multisim simulation results, and worked-out lab results for the lab manual by Dave Buchla and Doug Joksch.

- *TestGen (0132738287)* This computerized test bank contains over 650 questions.

- **Download Instructor Resources from the Instructor Resource Center**
  To access supplementary materials online, instructors need to request an instructor access code. Go to www.pearsonhighered.com/irc to register for an instructor access code. Within 48 hours of registering, you will receive a confirming e-mail including an instructor access code. Once you have received your code, locate your text in the online catalog and click on the Instructor Resources button on the left side of the catalog product page. Select a supplement, and a login page will appear. Once you have logged in, you can access instructor material for all Pearson textbooks. If you have any difficulties accessing the site or downloading a supplement, please contact Customer Service at http://247pearsoned.custhelp.com/.

Illustration of Book Features

Chapter Opener   Each chapter begins with an opener, which includes a list of the sections in the chapter, chapter objectives, introduction, a list of key terms, and a website reference for chapter study aids. A typical chapter opener is shown in Figure P-3.

Section Opener   Each section in a chapter begins with a brief introduction that includes a general overview and section objectives. An illustration is shown in Figure P-4.

Section Checkup Each section ends with a review consisting of questions or exercises that emphasize the main concepts presented in the section. This feature is shown in Figure P-4. Answers to the Section Checkups are at the end of the chapter.

CHAPTER 3

Logic Gates

CHAPTER OUTLINE

CHAPTER OBJECTIVES

- Describe the operation of the inverter, the AND gate, and the OR gate
- Describe the operation of the NAND gate and the NOR gate
- Express the operation of NOT, AND, OR, NAND, and NOR gates with Boolean algebra
- Describe the operation of the exclusive-OR and exclusive-NOR gates
- Use logic gates in simple applications
- Recognize and use both the distinctive shape logic gate symbols and the rectangular outline logic gate symbols of ANSI/IEEE Standard 91-1984/Std. 91a-1991
- Construct timing diagrams showing the proper time relationships of inputs and outputs for the various logic gates
- Discuss the basic concepts of programmable logic
- Make basic comparisons between the major IC technologies—CMOS and bipolar (TTL)
- Explain how the different series within the CMOS and bipolar (TTL) families differ from each other
- Define *propagation delay time, power dissipation, speed-power product,* and *fan-out* in relation to logic gates
- List specific fixed-function integrated circuit devices that contain the various logic gates
- Troubleshoot logic gates for opens and shorts by using the oscilloscope

KEY TERMS

Key terms are in order of appearance in the chapter.

- Inverter
- Truth table
- Boolean algebra
- Complement
- AND gate
- OR gate
- NAND gate
- NOR gate
- Exclusive-OR gate
- Exclusive-NOR gate
- AND array
- Fuse
- Antifuse
- EPROM
- EEPROM
- Flash
- SRAM
- Target device
- JTAG
- VHDL
- CMOS
- Bipolar
- Propagation delay time
- Fan-out
- Unit load

VISIT THE WEBSITE

Study aids for this chapter are available at http://www.pearsonhighered.com/careersresources/

INTRODUCTION

The emphasis in this chapter is on the operation, application, and troubleshooting of logic gates. The relationship of input and output waveforms of a gate using timing diagrams is thoroughly covered.

Logic symbols used to represent the logic gates are in accordance with ANSI/IEEE Standard 91-1984/Std. 91a-1991. This standard has been adopted by private industry and the military for use in internal documentation as well as published literature.

FIGURE P-3

SECTION 5–1 CHECKUP

Answers are at the end of the chapter.

1. Determine the output (1 or 0) of a 4-variable AND-OR-Invert circuit for each of the following input conditions:
   (a) $A = 1, B = 0, C = 1, D = 0$ (b) $A = 1, B = 1, C = 0, D = 1$
   (c) $A = 0, B = 1, C = 1, D = 1$
2. Determine the output (1 or 0) of an exclusive-OR gate for each of the following input conditions:
   (a) $A = 1, B = 0$ (b) $A = 1, B = 1$
   (c) $A = 0, B = 1$ (d) $A = 0, B = 0$
3. Develop the truth table for a certain 3-input logic circuit with the output expression $X = A\overline{B}C + \overline{A}BC + \overline{A}\,\overline{B}\,\overline{C} + AB\overline{C} + ABC$.
4. Draw the logic diagram for an exclusive-NOR circuit.

5–2 Implementing Combinational Logic

In this section, examples are used to illustrate how to implement a logic circuit from a Boolean expression or a truth table. Minimization of a logic circuit using the methods covered in Chapter 4 is also included.

After completing this section, you should be able to

- Implement a logic circuit from a Boolean expression
- Implement a logic circuit from a truth table
- Minimize a logic circuit

For every Boolean expression there is a logic circuit, and for every logic circuit there is a Boolean expression.

From a Boolean Expression to a Logic Circuit

Let's examine the following Boolean expression:

$$X = AB + CDE$$

A brief inspection shows that this expression is composed of two terms, AB and CDE, with a domain of five variables. The first term is formed by ANDing A with B, and the second term is formed by ANDing C, D, and E. The two terms are then ORed to form the output X. These operations are indicated in the structure of the expression as follows:

$$X = \underbrace{AB}_{\text{AND}} \overset{\text{OR}}{+} \underbrace{CDE}_{\text{AND}}$$

Note that in this particular expression, the AND operations forming the two individual terms, AB and CDE, must be performed *before* the terms can be ORed.

To implement this Boolean expression, a 2-input AND gate is required to form the term AB, and a 3-input AND gate is needed to form the term CDE. A 2-input OR gate is then required to combine the two AND terms. The resulting logic circuit is shown in Figure 5–9.

As another example, let's implement the following expression:

$$X = AB(C\overline{D} + EF)$$

InfoNote

Many control programs require logic operations to be performed by a computer. A driver program is a control program that is used with computer peripherals. For example, a mouse driver requires logic tests to determine if a button has been pressed and further logic operations to determine if it has moved, either horizontally or vertically. Within the heart of a microprocessor is the arithmetic logic unit (ALU), which performs these logic operations as directed by program instructions. All of the logic described in this chapter can also be performed by the ALU, given the proper instructions.

FIGURE P-4

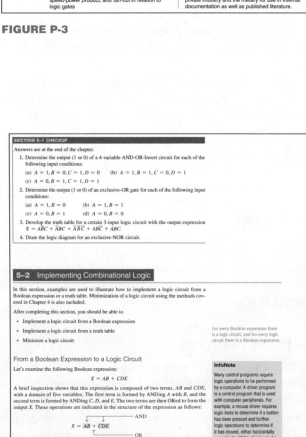

Worked Examples and Related Problems   There is an abundance of worked out examples that help to illustrate and clarify basic concepts or specific procedures. Each example ends with a *Related Problem* that reinforces or expands on the example by requiring the student to work through a problem similar to the example. A typical worked example with *Related Problem* is shown in Figure P-5.

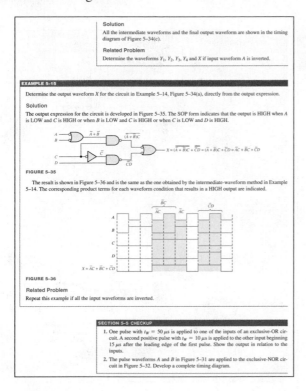

FIGURE P-5

Troubleshooting Section   Many chapters include a troubleshooting section that relates to the topics covered in the chapter and that emphasizes troubleshooting techniques and the use of test instruments and circuit simulation. A portion of a typical troubleshooting section is illustrated in Figure P-6.

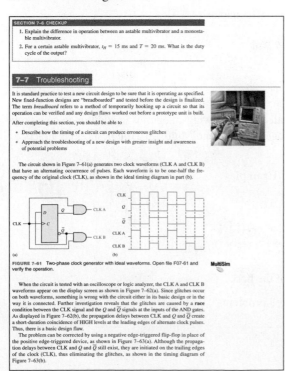

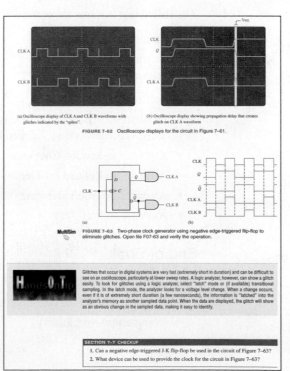

FIGURE P-6

Applied Logic   Appearing at the end of many chapters, this feature presents a practical application of the concepts and procedures covered in the chapter. In most chapters, this feature presents a "real-world" application in which analysis, troubleshooting, design, VHDL programming, and simulation are implemented. Figure P-7 shows a portion of a typical Applied Logic feature.

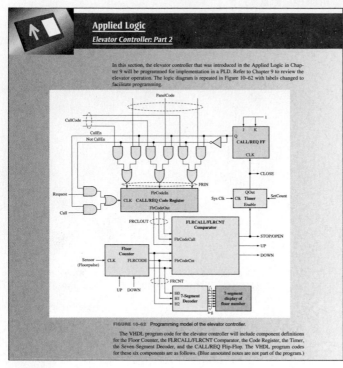

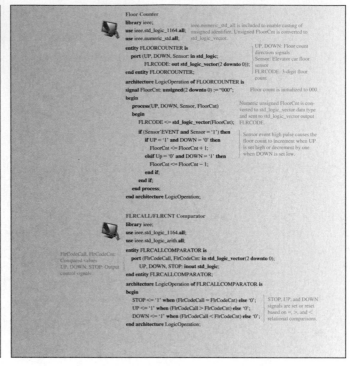

FIGURE P-7

End of Chapter

The following features are at the end of each chapter:

- Summary
- Key term glossary
- True/false quiz
- Self-test
- Problem set that includes some or all of the following categories in addition to core problems: Troubleshooting, Applied Logic, Design, and Multisim Troubleshooting Practice.
- Answers to Section Checkups
- Answers to Related Problems for Examples
- Answers to True/False quiz
- Answers to Self-Test

End of Book

The following features are at the end of the book.

- Answers to selected odd-numbered problems
- Comprehensive glossary
- Index

To the Student

Digital technology pervades almost everything in our daily lives. For example, cell phones and other types of wireless communications, television, radio, process controls, automotive electronics, consumer electronics, aircraft navigation— to name only a few applications— depend heavily on digital electronics.

A strong grounding in the fundamentals of digital technology will prepare you for the highly skilled jobs of the future. The single most important thing you can do is to understand the core fundamentals. From there you can go anywhere.

In addition, programmable logic is important in many applications and that topic in introduced in this book and example programs are given along with an online tutorial. Of course, efficient troubleshooting is a skill that is also widely sought after by potential employers. Troubleshooting and testing methods from traditional prototype testing to more advanced techniques such as boundary scan are covered.

To the Instructor

Generally, time limitations or program emphasis determines the topics to be covered in a course. It is not uncommon to omit or condense topics or to alter the sequence of certain topics in order to customize the material for a particular course. This textbook is specifically designed to provide great flexibility in topic coverage.

Certain topics are organized in separate chapters, sections, or features such that if they are omitted the rest of the coverage is not affected. Also, if these topics are included, they flow seamlessly with the rest of the coverage. The book is organized around a core of fundamental topics that are, for the most part, essential in any digital course. Around this core, there are other topics that can be included or omitted, depending on the course emphasis and/or other factors. Even within the core, selected topics can be omitted. Figure P-8 illustrates this concept.

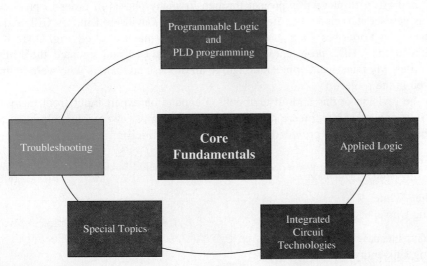

FIGURE P-8

◆ *Core Fundamentals*   The fundamental topics of digital technology should be covered in all programs. Linked to the core are several "satellite" topics that may be considered for omission or inclusion, depending on your course goals. All topics presented in this text are important in digital technology, but each block surrounding the core can be omitted, depending on your particular goals, without affecting the core fundamentals.

◆ *Programmable Logic and PLD Programming*   Although they are important topics, programmable logic and VHDL can be omitted; however, it is highly recommended that you cover this topic if at all possible. You can cover as little or as much as you consider appropriate for your program.

◆ *Troubleshooting*   Troubleshooting sections appear in many chapters and include the application and operation of laboratory instruments.

◆ *Applied Logic*   Selected real-world applications appear in many chapters.

◆ *Integrated Circuit Technologies*   Chapter 15 is an online chapter. Some or all of the topics in Chapter 15 can be covered at selected points if you wish to discuss details of the circuitry that make up digital integrated circuits. Chapter 15 can be omitted without any impact on the rest of the book.

◆ *Special Topics*   These topics are *Signal Interfacing and Processing, Data Transmission,* and *Data Processing and Control* in Chapters 12, 13, and 14 respectively, as well as selected topics in other chapters. These are topics that may not be essential for your course or are covered in another course. Also, within each block in Figure P-8 you can choose to omit or deemphasize some topics because of time constraints or other priorities in your particular program. For example in the core fundamentals, the Quine-McCluskey method, cyclic redundancy code, carry look-ahead adders, or sequential logic design could possibly be omitted. Additionally, any or all of Multisim features throughout the book can be treated as optional. Other topics may also be candidates for omission or light coverage. Whether you choose a minimal coverage of only core fundamentals, a full-blown coverage of all the topics, or anything in between, this book can be adapted to your needs.

Acknowledgments

This revision of *Digital Fundamentals* has been made possible by the work and skills of many people. I think that we have accomplished what we set out to do, and that was to further improve an already very successful textbook and make it even more useful to the student and instructor by presenting not only basics but also up-to-date and leading-edge technology.

Those at Pearson Education who have, as always, contributed a great amount of time, talent, and effort to move this project through its many phases in order to produce the book as you see it, include, but are not limited to, Rex Davidson, Lindsey Gill, and Vern Anthony. Lois Porter has done another excellent job of manuscript editing. Doug Joksch contributed the VHDL programming. Gary Snyder revised and updated the Multisim circuit files. My thanks and appreciation go to all of these and others who were indirectly involved in the project.

In the revision of this and all textbooks, I depend on expert input from many users as well as nonusers. My sincere thanks to the following reviewers who submitted many valuable suggestions and provided lots of constructive criticism:

Dr. Cuiling Gong,
Texas Christian University;

Jonathan White,
Harding University;

Zane Gastineau,
Harding University; and

Dr. Eric Bothur,
Midlands Technical College.

I also want to thank all of the members of the Pearson sales force whose efforts have helped make this text available to a large number of users. In addition, I am grateful to all of you who have adopted this text for your classes or for your own use. Without you we would not be in business. I hope that you find this eleventh edition of *Digital Fundamentals* to be even better than earlier editions and that it will continue to be a valuable learning tool and reference for the student.

Tom Floyd

CONTENTS

Introductory Concepts

CHAPTER OBJECTIVES

■ Explain the basic differences between digital and analog quantities

■ Show how voltage levels are used to represent digital quantities

■ Describe various parameters of a pulse waveform such as rise time, fall time, pulse width, frequency, period, and duty cycle

■ Explain the basic logic functions of NOT, AND, and OR

■ Describe several types of logic operations and explain their application in an example system

■ Describe programmable logic, discuss the various types, and describe how PLDs are programmed

■ Identify fixed-function digital integrated circuits according to their complexity and the type of circuit packaging

■ Identify pin numbers on integrated circuit packages

■ Recognize various instruments and understand how they are used in measurement and troubleshooting digital circuits and systems

■ Describe basic troubleshooting methods

KEY TERMS

Key terms are in order of appearance in the chapter.

■ Analog
■ Digital
■ Binary
■ Bit
■ Pulse
■ Duty cycle
■ Clock
■ Timing diagram
■ Data
■ Serial
■ Parallel
■ Logic
■ Input
■ Output
■ Gate

■ NOT
■ Inverter
■ AND
■ OR
■ Programmable logic
■ SPLD
■ CPLD
■ FPGA
■ Microcontroller
■ Embedded system
■ Compiler
■ Integrated circuit (IC)
■ Fixed-function logic
■ Troubleshooting

VISIT THE WEBSITE

Study aids for this chapter are available at
http://www.pearsonhighered.com/careersresources/

INTRODUCTION

The term *digital* is derived from the way operations are performed, by counting digits. For many years, applications of digital electronics were confined to computer systems. Today, digital technology is applied in a wide range of areas in addition to computers. Such applications as television, communications systems, radar, navigation and guidance systems, military systems, medical instrumentation, industrial process control, and consumer electronics use digital techniques. Over the years digital technology has progressed from vacuum-tube circuits to discrete transistors to complex integrated circuits, many of which contain millions of transistors, and many of which are programmable.

This chapter introduces you to digital electronics and provides a broad overview of many important concepts, components, and tools.

1–1 Digital and Analog Quantities

Electronic circuits can be divided into two broad categories, digital and analog. Digital electronics involves quantities with discrete values, and analog electronics involves quantities with continuous values. Although you will be studying digital fundamentals in this book, you should also know something about analog because many applications require both; and interfacing between analog and digital is important.

After completing this section, you should be able to

- Define *analog*
- Define *digital*
- Explain the difference between digital and analog quantities
- State the advantages of digital over analog
- Give examples of how digital and analog quantities are used in electronics

An **analog*** quantity is one having continuous values. A **digital** quantity is one having a discrete set of values. Most things that can be measured quantitatively occur in nature in analog form. For example, the air temperature changes over a continuous range of values. During a given day, the temperature does not go from, say, 70° to 71° instantaneously; it takes on all the infinite values in between. If you graphed the temperature on a typical summer day, you would have a smooth, continuous curve similar to the curve in Figure 1–1. Other examples of analog quantities are time, pressure, distance, and sound.

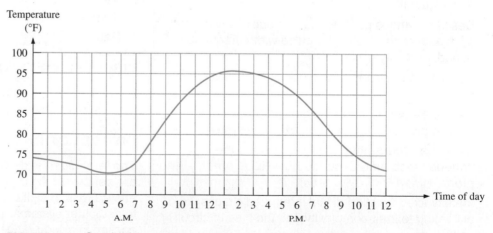

FIGURE 1–1   Graph of an analog quantity (temperature versus time).

Rather than graphing the temperature on a continuous basis, suppose you just take a temperature reading every hour. Now you have sampled values representing the temperature at discrete points in time (every hour) over a 24-hour period, as indicated in Figure 1–2.

---

*All bold terms are important and are defined in the end-of-book glossary. The blue bold terms are key terms and are included in a Key Term glossary at the end of each chapter.

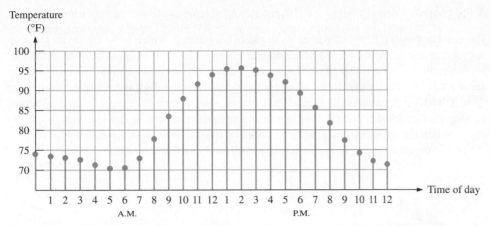

FIGURE 1–2 Sampled-value representation (quantization) of the analog quantity in Figure 1–1. Each value represented by a dot can be digitized by representing it as a digital code that consists of a series of 1s and 0s.

You have effectively converted an analog quantity to a form that can now be digitized by representing each sampled value by a digital code. It is important to realize that Figure 1–2 itself is not the digital representation of the analog quantity.

The Digital Advantage

Digital representation has certain advantages over analog representation in electronics applications. For one thing, digital data can be processed and transmitted more efficiently and reliably than analog data. Also, digital data has a great advantage when storage is necessary. For example, music when converted to digital form can be stored more compactly and reproduced with greater accuracy and clarity than is possible when it is in analog form. Noise (unwanted voltage fluctuations) does not affect digital data nearly as much as it does analog signals.

An Analog System

A public address system, used to amplify sound so that it can be heard by a large audience, is one simple example of an application of analog electronics. The basic diagram in Figure 1–3 illustrates that sound waves, which are analog in nature, are picked up by a microphone and converted to a small analog voltage called the audio signal. This voltage varies continuously as the volume and frequency of the sound changes and is applied to the input of a linear amplifier. The output of the amplifier, which is an increased reproduction of input voltage, goes to the speaker(s). The speaker changes the amplified audio signal back to sound waves that have a much greater volume than the original sound waves picked up by the microphone.

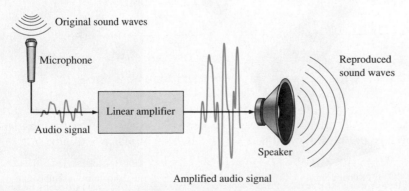

FIGURE 1–3 A basic audio public address system.

A System Using Digital and Analog Methods

The compact disk (CD) player is an example of a system in which both digital and analog circuits are used. The simplified block diagram in Figure 1–4 illustrates the basic principle. Music in digital form is stored on the compact disk. A laser diode optical system picks up the digital data from the rotating disk and transfers it to the **digital-to-analog converter (DAC)**. The DAC changes the digital data into an analog signal that is an electrical reproduction of the original music. This signal is amplified and sent to the speaker for you to enjoy. When the music was originally recorded on the CD, a process, essentially the reverse of the one described here, using an **analog-to-digital converter (ADC)** was used.

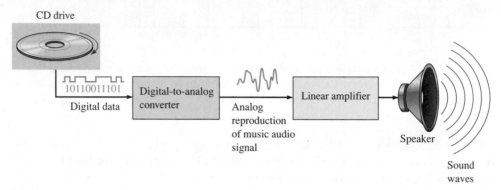

FIGURE 1–4   Basic block diagram of a CD player. Only one channel is shown.

Mechatronics

Both digital and analog electronics are used in the control of various mechanical systems. The interdisciplinary field that comprises both mechanical and electronic components is known as **mechatronics**.

Mechatronic systems are found in homes, industry, and transportation. Most home appliances consist of both mechanical and electronic components. Electronics controls the operation of a washing machine in terms of water flow, temperature, and type of cycle. Manufacturing industries rely heavily on mechatronics for process control and assembly. In automotive and other types of manufacturing, robotic arms perform precision welding, painting, and other functions on the assembly line. Automobiles themselves are mechatronic machines; a digital computer controls functions such as braking, engine parameters, fuel flow, safety features, and monitoring.

Figure 1–5(a) is a basic block diagram of a mechatronic system. A simple robotic arm is shown in Figure 1–5(b), and robotic arms on an automotive assembly line are shown in part (c).

(a) Mechatronic system block diagram   (b) Robotic arm   (c) Automotive assembly line

FIGURE 1–5   Example of a mechatronic system and application.   Part (b) Beawolf/Fotolia; Part (c) Small Town Studio/Fotolia.

The movement of the arm in any quadrant and to any specified position is accomplished with some type of digital control such as a microcontroller.

SECTION 1–1 CHECKUP

Answers are at the end of the chapter.

1. Define *analog*.

2. Define *digital*.

3. Explain the difference between a digital quantity and an analog quantity.

4. Give an example of a system that is analog and one that is a combination of both digital and analog. Name a system that is entirely digital.

5. What does a mechatronic system consist of?

1–2 Binary Digits, Logic Levels, and Digital Waveforms

Digital electronics involves circuits and systems in which there are only two possible states. These states are represented by two different voltage levels: A HIGH and a LOW. The two states can also be represented by current levels, bits and bumps on a CD or DVD, etc. In digital systems such as computers, combinations of the two states, called *codes,* are used to represent numbers, symbols, alphabetic characters, and other types of information. The two-state number system is called *binary,* and its two digits are 0 and 1. A binary digit is called a *bit*.

After completing this section, you should be able to

◆ Define *binary*

◆ Define *bit*

◆ Name the bits in a binary system

◆ Explain how voltage levels are used to represent bits

◆ Explain how voltage levels are interpreted by a digital circuit

◆ Describe the general characteristics of a pulse

◆ Determine the amplitude, rise time, fall time, and width of a pulse

◆ Identify and describe the characteristics of a digital waveform

◆ Determine the amplitude, period, frequency, and duty cycle of a digital waveform

◆ Explain what a timing diagram is and state its purpose

◆ Explain serial and parallel data transfer and state the advantage and disadvantage of each

Binary Digits

Each of the two digits in the **binary** system, 1 and 0, is called a **bit**, which is a contraction of the words *binary digit*. In digital circuits, two different voltage levels are used to represent the two bits. Generally, 1 is represented by the higher voltage, which we will refer to as a HIGH, and a 0 is represented by the lower voltage level, which we will refer to as a LOW. This is called **positive logic** and will be used throughout the book.

$$\textbf{HIGH} = 1 \quad \textbf{and} \quad \textbf{LOW} = 0$$

InfoNote

The concept of a digital computer can be traced back to Charles Babbage, who developed a crude mechanical computation device in the 1830s. John Atanasoff was the first to apply electronic processing to digital computing in 1939. In 1946, an electronic digital computer called ENIAC was implemented with vacuum-tube circuits. Even though it took up an entire room, ENIAC didn't have the computing power of your handheld calculator.

Another system in which a 1 is represented by a LOW and a 0 is represented by a HIGH is called *negative logic.*

Groups of bits (combinations of 1s and 0s), called *codes,* are used to represent numbers, letters, symbols, instructions, and anything else required in a given application.

Logic Levels

The voltages used to represent a 1 and a 0 are called *logic levels.* Ideally, one voltage level represents a HIGH and another voltage level represents a LOW. In a practical digital circuit, however, a HIGH can be any voltage between a specified minimum value and a specified maximum value. Likewise, a LOW can be any voltage between a specified minimum and a specified maximum. There can be no overlap between the accepted range of HIGH levels and the accepted range of LOW levels.

Figure 1–6 illustrates the general range of LOWs and HIGHs for a digital circuit. The variable $V_{H(max)}$ represents the maximum HIGH voltage value, and $V_{H(min)}$ represents the minimum HIGH voltage value. The maximum LOW voltage value is represented by $V_{L(max)}$, and the minimum LOW voltage value is represented by $V_{L(min)}$. The voltage values between $V_{L(max)}$ and $V_{H(min)}$ are unacceptable for proper operation. A voltage in the unacceptable range can appear as either a HIGH or a LOW to a given circuit. For example, the HIGH input values for a certain type of digital circuit technology called CMOS may range from 2 V to 3.3 V and the LOW input values may range from 0 V to 0.8 V. If a voltage of 2.5 V is applied, the circuit will accept it as a HIGH or binary 1. If a voltage of 0.5 V is applied, the circuit will accept it as a LOW or binary 0. For this type of circuit, voltages between 0.8 V and 2 V are unacceptable.

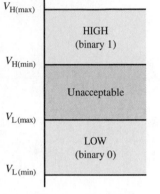

FIGURE 1–6 Logic level ranges of voltage for a digital circuit.

Digital Waveforms

Digital waveforms consist of voltage levels that are changing back and forth between the HIGH and LOW levels or states. Figure 1–7(a) shows that a single positive-going **pulse** is generated when the voltage (or current) goes from its normally LOW level to its HIGH level and then back to its LOW level. The negative-going pulse in Figure 1–7(b) is generated when the voltage goes from its normally HIGH level to its LOW level and back to its HIGH level. A digital waveform is made up of a series of pulses.

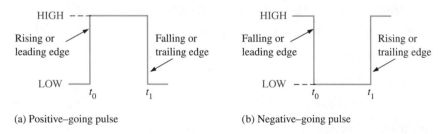

(a) Positive–going pulse   (b) Negative–going pulse

FIGURE 1–7 Ideal pulses.

The Pulse

As indicated in Figure 1–7, a pulse has two edges: a **leading edge** that occurs first at time t_0 and a **trailing edge** that occurs last at time t_1. For a positive-going pulse, the leading edge is a rising edge, and the trailing edge is a falling edge. The pulses in Figure 1–7 are ideal because the rising and falling edges are assumed to change in zero time (instantaneously). In practice, these transitions never occur instantaneously, although for most digital work you can assume ideal pulses.

Figure 1–8 shows a nonideal pulse. In reality, all pulses exhibit some or all of these characteristics. The overshoot and ringing are sometimes produced by stray inductive and

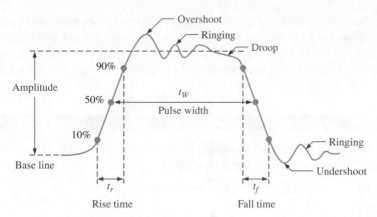

FIGURE 1–8 Nonideal pulse characteristics.

capacitive effects. The droop can be caused by stray capacitive and circuit resistance, forming an *RC* circuit with a low time constant.

The time required for a pulse to go from its LOW level to its HIGH level is called the **rise time** (t_r), and the time required for the transition from the HIGH level to the LOW level is called the **fall time** (t_f). In practice, it is common to measure rise time from 10% of the pulse **amplitude** (height from baseline) to 90% of the pulse amplitude and to measure the fall time from 90% to 10% of the pulse amplitude, as indicated in Figure 1–8. The bottom 10% and the top 10% of the pulse are not included in the rise and fall times because of the nonlinearities in the waveform in these areas. The **pulse width** (t_W) is a measure of the duration of the pulse and is often defined as the time interval between the 50% points on the rising and falling edges, as indicated in Figure 1–8.

Waveform Characteristics

Most waveforms encountered in digital systems are composed of series of pulses, sometimes called *pulse trains,* and can be classified as either periodic or nonperiodic. A **periodic** pulse waveform is one that repeats itself at a fixed interval, called a **period** (**T**). The **frequency** (f) is the rate at which it repeats itself and is measured in hertz (Hz). A nonperiodic pulse waveform, of course, does not repeat itself at fixed intervals and may be composed of pulses of randomly differing pulse widths and/or randomly differing time intervals between the pulses. An example of each type is shown in Figure 1–9.

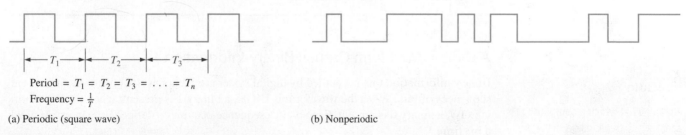

Period $= T_1 = T_2 = T_3 = \ldots = T_n$
Frequency $= \frac{1}{T}$

(a) Periodic (square wave)   (b) Nonperiodic

FIGURE 1–9 Examples of digital waveforms.

The frequency (f) of a pulse (digital) waveform is the reciprocal of the period. The relationship between frequency and period is expressed as follows:

$$f = \frac{1}{T}$$ Equation 1–1

$$T = \frac{1}{f}$$ Equation 1–2

An important characteristic of a periodic digital waveform is its **duty cycle**, which is the ratio of the pulse width (t_W) to the period (T). It can be expressed as a percentage.

$$\text{Duty cycle} = \left(\frac{t_W}{T}\right)100\% \qquad\qquad \text{Equation 1–3}$$

EXAMPLE 1–1

A portion of a periodic digital waveform is shown in Figure 1–10. The measurements are in milliseconds. Determine the following:

(a) period   (b) frequency   (c) duty cycle

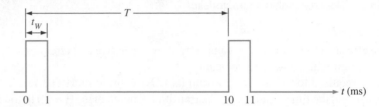

FIGURE 1–10

Solution

(a) The period (T) is measured from the edge of one pulse to the corresponding edge of the next pulse. In this case T is measured from leading edge to leading edge, as indicated. T equals **10 ms**.

(b) $f = \dfrac{1}{T} = \dfrac{1}{10 \text{ ms}} = \textbf{100 Hz}$

(c) $\text{Duty cycle} = \left(\dfrac{t_W}{T}\right)100\% = \left(\dfrac{1 \text{ ms}}{10 \text{ ms}}\right)100\% = \textbf{10\%}$

Related Problem*

A periodic digital waveform has a pulse width of 25 μs and a period of 150 μs. Determine the frequency and the duty cycle.

---

*Answers are at the end of the chapter.

A Digital Waveform Carries Binary Information

InfoNote

The speed at which a computer can operate depends on the type of microprocessor used in the system. The speed specification, for example 3.5 GHz, of a computer is the maximum clock frequency at which the microprocessor can run.

Binary information that is handled by digital systems appears as waveforms that represent sequences of bits. When the waveform is HIGH, a binary 1 is present; when the waveform is LOW, a binary 0 is present. Each bit in a sequence occupies a defined time interval called a **bit time**.

The Clock

In digital systems, all waveforms are synchronized with a basic timing waveform called the **clock.** The clock is a periodic waveform in which each interval between pulses (the period) equals the time for one bit.

An example of a clock waveform is shown in Figure 1–11. Notice that, in this case, each change in level of waveform A occurs at the leading edge of the clock waveform. In other cases, level changes occur at the trailing edge of the clock. During each bit time of the clock, waveform A is either HIGH or LOW. These HIGHs and LOWs represent a sequence

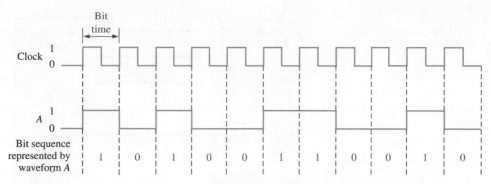

FIGURE 1–11 Example of a clock waveform synchronized with a waveform representation of a sequence of bits.

of bits as indicated. A group of several bits can contain binary information, such as a number or a letter. The clock waveform itself does not carry information.

Timing Diagrams

A **timing diagram** is a graph of digital waveforms showing the actual time relationship of two or more waveforms and how each waveform changes in relation to the others. By looking at a timing diagram, you can determine the states (HIGH or LOW) of all the waveforms at any specified point in time and the exact time that a waveform changes state relative to the other waveforms. Figure 1–12 is an example of a timing diagram made up of four waveforms. From this timing diagram you can see, for example, that the three waveforms A, B, and C are HIGH only during bit time 7 (shaded area) and they all change back LOW at the end of bit time 7.

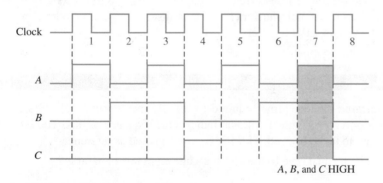

FIGURE 1–12 Example of a timing diagram.

Data Transfer

Data refers to groups of bits that convey some type of information. Binary data, which are represented by digital waveforms, must be transferred from one device to another within a digital system or from one system to another in order to accomplish a given purpose. For example, numbers stored in binary form in the memory of a computer must be transferred to the computer's central processing unit in order to be added. The sum of the addition must then be transferred to a monitor for display and/or transferred back to the memory. As illustrated in Figure 1–13, binary data are transferred in two ways: serial and parallel.

When bits are transferred in **serial** form from one point to another, they are sent one bit at a time along a single line, as illustrated in Figure 1–13(a). During the time interval from t_0 to t_1, the first bit is transferred. During the time interval from t_1 to t_2, the second bit is transferred, and so on. To transfer eight bits in series, it takes eight time intervals.

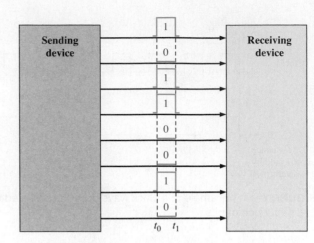

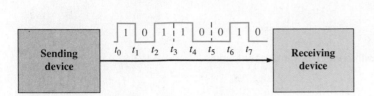

(a) Serial transfer of 8 bits of binary data. Interval t_0 to t_1 is first.

(b) Parallel transfer of 8 bits of binary data. The beginning time is t_0.

FIGURE 1–13 Illustration of serial and parallel transfer of binary data. Only the data lines are shown.

When bits are transferred in **parallel** form, all the bits in a group are sent out on separate lines at the same time. There is one line for each bit, as shown in Figure 1–13(b) for the example of eight bits being transferred. To transfer eight bits in parallel, it takes one time interval compared to eight time intervals for the serial transfer.

To summarize, an advantage of serial transfer of binary data is that a minimum of only one line is required. In parallel transfer, a number of lines equal to the number of bits to be transferred at one time is required. A disadvantage of serial transfer is that it takes longer to transfer a given number of bits than with parallel transfer at the same clock frequency. For example, if one bit can be transferred in 1 μs, then it takes 8 μs to serially transfer eight bits but only 1 μs to parallel transfer eight bits. A disadvantage of parallel transfer is that it takes more lines than serial transfer.

EXAMPLE 1–2

(a) Determine the total time required to serially transfer the eight bits contained in waveform A of Figure 1–14, and indicate the sequence of bits. The left-most bit is the first to be transferred. The 1 MHz clock is used as reference.

(b) What is the total time to transfer the same eight bits in parallel?

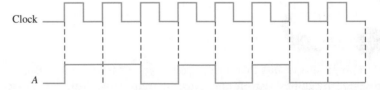

FIGURE 1–14

Solution

(a) Since the frequency of the clock is 1 MHz, the period is

$$T = \frac{1}{f} = \frac{1}{1\text{ MHz}} = 1\ \mu\text{s}$$

It takes 1 μs to transfer each bit in the waveform. The total transfer time for 8 bits is

$$8 \times 1\ \mu\text{s} = \mathbf{8\ \mu s}$$

To determine the sequence of bits, examine the waveform in Figure 1–14 during each bit time. If waveform *A* is HIGH during the bit time, a 1 is transferred. If waveform *A* is LOW during the bit time, a 0 is transferred. The bit sequence is illustrated in Figure 1–15. The left-most bit is the first to be transferred.

FIGURE 1–15

(b) A parallel transfer would take **1 μs** for all eight bits.

Related Problem

If binary data are transferred on a USB at the rate of 480 million bits per second (480 Mbps), how long will it take to serially transfer 16 bits?

SECTION 1–2 CHECKUP

1. Define *binary*.
2. What does *bit* mean?
3. What are the bits in a binary system?
4. How are the rise time and fall time of a pulse measured?
5. Knowing the period of a waveform, how do you find the frequency?
6. Explain what a clock waveform is.
7. What is the purpose of a timing diagram?
8. What is the main advantage of parallel transfer over serial transfer of binary data?

1–3   Basic Logic Functions

In its basic form, logic is the realm of human reasoning that tells you a certain proposition (declarative statement) is true if certain conditions are true. Propositions can be classified as true or false. Many situations and processes that you encounter in your daily life can be expressed in the form of propositional, or logic, functions. Since such functions are true/false or yes/no statements, digital circuits with their two-state characteristics are applicable.

After completing this section, you should be able to

* List three basic logic functions
* Define the NOT function
* Define the AND function
* Define the OR function

Several propositions, when combined, form propositional, or logic, functions. For example, the propositional statement "The light is on" will be true if "The bulb is not burned out" is true and if "The switch is on" is true. Therefore, this logical statement can be made: *The light is on only if the bulb is not burned out and the switch is on.* In this example the first statement is true only if the last two statements are true. The first statement ("The light is on") is then the basic proposition, and the other two statements are the conditions on which the proposition depends.

In the 1850s, the Irish logician and mathematician George Boole developed a mathematical system for formulating logic statements with symbols so that problems can be written and solved in a manner similar to ordinary algebra. Boolean algebra, as it is known today, is applied in the design and analysis of digital systems and will be covered in detail in Chapter 4.

The term **logic** is applied to digital circuits used to implement logic functions. Several kinds of digital logic **circuits** are the basic elements that form the building blocks for such complex digital systems as the computer. We will now look at these elements and discuss their functions in a very general way. Later chapters will cover these circuits in detail.

Three basic logic functions (NOT, AND, and OR) are indicated by standard distinctive shape symbols in Figure 1–16. Alternate standard symbols for these logic functions will be introduced in Chapter 3. The lines connected to each symbol are the **inputs** and **outputs**. The inputs are on the left of each symbol and the output is on the right. A circuit that performs a specified logic function (AND, OR) is called a logic **gate**. AND and OR gates can have any number of inputs, as indicated by the dashes in the figure.

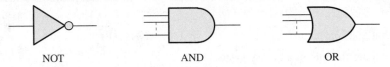

NOT
AND
OR

FIGURE 1–16 The basic logic functions and symbols.

In logic functions, the true/false conditions mentioned earlier are represented by a HIGH (true) and a LOW (false). Each of the three basic logic functions produces a unique response to a given set of conditions.

NOT

The **NOT** function changes one logic level to the opposite logic level, as indicated in Figure 1–17. When the input is HIGH (1), the output is LOW (0). When the input is LOW, the output is HIGH. In either case, the output is *not* the same as the input. The NOT function is implemented by a logic circuit known as an **inverter**.

HIGH (1)
LOW (0)
LOW (0)
HIGH (1)

FIGURE 1–17 The NOT function.

AND

The **AND** function produces a HIGH output only when all the inputs are HIGH, as indicated in Figure 1–18 for the case of two inputs. When one input is HIGH *and* the other input is HIGH, the output is HIGH. When any or all inputs are LOW, the output is LOW. The AND function is implemented by a logic circuit known as an *AND gate.*

HIGH (1)
HIGH (1)
HIGH (1)
LOW (0)
HIGH (1)
LOW (0)
HIGH (1)
LOW (0)
LOW (0)
LOW (0)
LOW (0)
LOW (0)

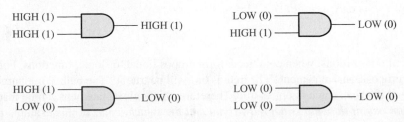

FIGURE 1–18 The AND function.

OR

The **OR** function produces a HIGH output when one or more inputs are HIGH, as indicated in Figure 1–19 for the case of two inputs. When one input is HIGH *or* the other input is HIGH *or* both inputs are HIGH, the output is HIGH. When both inputs are LOW, the output is LOW. The OR function is implemented by a logic circuit known as an *OR gate*.

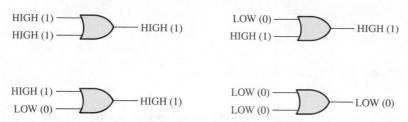

FIGURE 1–19 The OR function.

SECTION 1–3 CHECKUP

1. When does the NOT function produce a HIGH output?
2. When does the AND function produce a HIGH output?
3. When does the OR function produce a HIGH output?
4. What is an inverter?
5. What is a logic gate?

1–4   Combinational and Sequential Logic Functions

The three basic logic functions AND, OR, and NOT can be combined to form various other types of more complex logic functions, such as comparison, arithmetic, code conversion, encoding, decoding, data selection, counting, and storage. A digital system is an arrangement of the individual logic functions connected to perform a specified operation or produce a defined output. This section provides an overview of important logic functions and illustrates how they can be used in a specific system.

After completing this section, you should be able to

♦ List several types of logic functions

♦ Describe comparison and list the four arithmetic functions

♦ Describe code conversion, encoding, and decoding

♦ Describe multiplexing and demultiplexing

♦ Describe the counting function

♦ Describe the storage function

♦ Explain the operation of the tablet-bottling system

The Comparison Function

Magnitude comparison is performed by a logic circuit called a **comparator**, covered in Chapter 6. A comparator compares two quantities and indicates whether or not they are equal. For example, suppose you have two numbers and wish to know if they are equal or not equal and, if not equal, which is greater. The comparison function is represented in

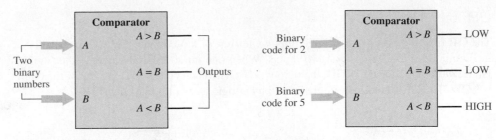

FIGURE 1–20 The comparison function.

Figure 1–20. One number in binary form (represented by logic levels) is applied to input A, and the other number in binary form (represented by logic levels) is applied to input B. The outputs indicate the relationship of the two numbers by producing a HIGH level on the proper output line. Suppose that a binary representation of the number 2 is applied to input A and a binary representation of the number 5 is applied to input B. (The binary representation of numbers and symbols is discussed in Chapter 2.) A HIGH level will appear on the $A < B$ (A is less than B) output, indicating the relationship between the two numbers (2 is less than 5). The wide arrows represent a group of parallel lines on which the bits are transferred.

InfoNote

In a microprocessor, the arithmetic logic unit (ALU) performs the operations of add, subtract, multiply, and divide as well as the logic operations on digital data as directed by a series of instructions. A typical ALU is constructed of many thousands of logic gates.

The Arithmetic Functions

Addition

Addition is performed by a logic circuit called an **adder**, covered in Chapter 6. An adder adds two binary numbers (on inputs A and B with a carry input C_{in}) and generates a sum (Σ) and a carry output (C_{out}), as shown in Figure 1–21(a). Figure 1–21(b) illustrates the addition of 3 and 9. You know that the sum is 12; the adder indicates this result by producing 2 on the sum output and 1 on the carry output. Assume that the carry input in this example is 0.

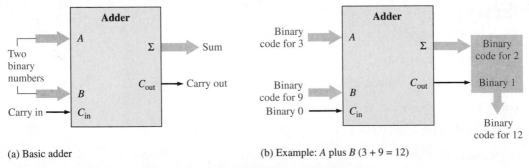

FIGURE 1–21 The addition function.

Subtraction

Subtraction is also performed by a logic circuit. A **subtracter** requires three inputs: the two numbers that are to be subtracted and a borrow input. The two outputs are the difference and the borrow output. When, for instance, 5 is subtracted from 8 with no borrow input, the difference is 3 with no borrow output. You will see in Chapter 2 how subtraction can actually be performed by an adder because subtraction is simply a special case of addition.

Multiplication

Multiplication is performed by a logic circuit called a *multiplier.* Numbers are always multiplied two at a time, so two inputs are required. The output of the multiplier is the product. Because multiplication is simply a series of additions with shifts in the positions of the partial products, it can be performed by using an adder in conjunction with other circuits.

Division

Division can be performed with a series of subtractions, comparisons, and shifts, and thus it can also be done using an adder in conjunction with other circuits. Two inputs to the divider are required, and the outputs generated are the quotient and the remainder.

The Code Conversion Function

A **code** is a set of bits arranged in a unique pattern and used to represent specified information. A code converter changes one form of coded information into another coded form. Examples are conversion between binary and other codes such as the binary coded decimal (BCD) and the Gray code. Various types of codes are covered in Chapter 2, and code converters are covered in Chapter 6.

The Encoding Function

The encoding function is performed by a logic circuit called an **encoder**, covered in Chapter 6. The encoder converts information, such as a decimal number or an alphabetic character, into some coded form. For example, one certain type of encoder converts each of the decimal digits, 0 through 9, to a binary code. A HIGH level on the input corresponding to a specific decimal digit produces logic levels that represent the proper binary code on the output lines.

Figure 1–22 is a simple illustration of an encoder used to convert (encode) a calculator keystroke into a binary code that can be processed by the calculator circuits.

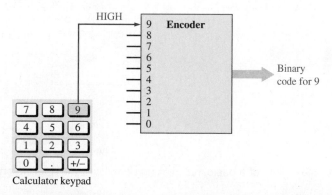

FIGURE 1–22   An encoder used to encode a calculator keystroke into a binary code for storage or for calculation.

The Decoding Function

The decoding function is performed by a logic circuit called a **decoder**, covered in Chapter 6. The decoder converts coded information, such as a binary number, into a noncoded form, such as a decimal form. For example, one particular type of decoder converts a 4-bit binary code into the appropriate decimal digit.

Figure 1–23 is a simple illustration of one type of decoder that is used to activate a 7-segment display. Each of the seven segments of the display is connected to an output line from the decoder. When a particular binary code appears on the decoder inputs, the appropriate output lines are activated and light the proper segments to display the decimal digit corresponding to the binary code.

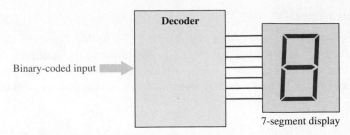

FIGURE 1–23 A decoder used to convert a special binary code into a 7-segment decimal readout.

The Data Selection Function

Two types of circuits that select data are the multiplexer and the demultiplexer. The **multiplexer**, or mux for short, is a logic circuit that switches digital data from several input lines onto a single output line in a specified time sequence. Functionally, a multiplexer can be represented by an electronic switch operation that sequentially connects each of the input lines to the output line. The **demultiplexer** (demux) is a logic circuit that switches digital data from one input line to several output lines in a specified time sequence. Essentially, the demux is a mux in reverse.

Multiplexing and demultiplexing are used when data from several sources are to be transmitted over one line to a distant location and redistributed to several destinations. Figure 1–24 illustrates this type of application where digital data from three sources are sent out along a single line to three terminals at another location.

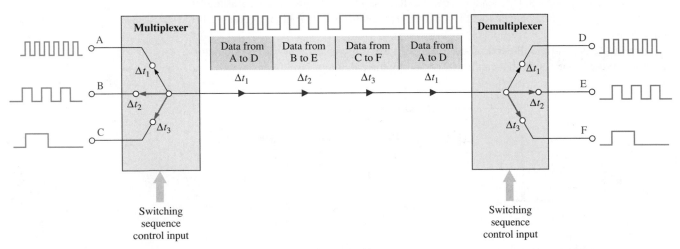

FIGURE 1–24 Illustration of a basic multiplexing/demultiplexing application.

In Figure 1–24, data from input A are connected to the output line during time interval Δt_1 and transmitted to the demultiplexer that connects them to output D. Then, during interval Δt_2, the multiplexer switches to input B and the demultiplexer switches to output E. During interval Δt_3, the multiplexer switches to input C and the demultiplexer switches to output F.

To summarize, during the first time interval, input A data go to output D. During the second time interval, input B data go to output E. During the third time interval, input C data go to output F. After this, the sequence repeats. Because the time is divided up among several sources and destinations where each has its turn to send and receive data, this process is called *time division multiplexing* (TDM).

The Storage Function

Storage is a function that is required in most digital systems, and its purpose is to retain binary data for a period of time. Some storage devices are used for short-term storage and some

InfoNote

The internal computer memories, RAM and ROM, as well as the smaller caches are semiconductor memories. The registers in a microprocessor are constructed of semiconductor flip-flops. Opto-magnetic disk memories are used in the internal hard drive and for the CD-ROM.

are used for long-term storage. A storage device can "memorize" a bit or a group of bits and retain the information as long as necessary. Common types of storage devices are flip-flops, registers, semiconductor memories, magnetic disks, magnetic tape, and optical disks (CDs).

Flip-flops

A **flip-flop** is a bistable (two stable states) logic circuit that can store only one bit at a time, either a 1 or a 0. The output of a flip-flop indicates which bit it is storing. A HIGH output indicates that a 1 is stored and a LOW output indicates that a 0 is stored. Flip-flops are implemented with logic gates and are covered in Chapter 7.

Registers

A **register** is formed by combining several flip-flops so that groups of bits can be stored. For example, an 8-bit register is constructed from eight flip-flops. In addition to storing bits, registers can be used to shift the bits from one position to another within the register or out of the register to another circuit; therefore, these devices are known as *shift registers*. Shift registers are covered in Chapter 8.

The two basic types of shift registers are serial and parallel. The bits are stored in a serial shift register one at a time, as illustrated in Figure 1–25. A good analogy to the serial shift register is loading passengers onto a bus single file through the door. They also exit the bus single file.

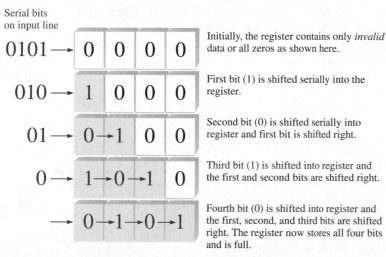

FIGURE 1–25 Example of the operation of a 4-bit serial shift register. Each block represents one storage "cell" or flip-flop.

The bits are stored in a parallel register simultaneously from parallel lines, as shown in Figure 1–26. For this case, a good analogy is loading and unloading passengers on a roller coaster where they enter all of the cars in parallel and exit in parallel.

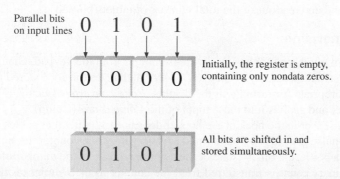

FIGURE 1–26 Example of the operation of a 4-bit parallel shift register.

Semiconductor Memories

Semiconductor memories are devices typically used for storing large numbers of bits. In one type of memory, called the *read-only memory* or ROM, the binary data are permanently or semipermanently stored and cannot be readily changed. In the *random-access memory* or RAM, the binary data are temporarily stored and can be easily changed. Memories are covered in Chapter 11.

Magnetic Memories

Magnetic disk memories are used for mass storage of binary data. An example is a computer's internal hard disk. Magnetic tape is still used to some extent in memory applications and for backing up data from other storage devices.

Optical Memories

CDs, DVDs, and Blu-ray Discs are storage devices based on laser technology. Data are represented by pits and lands on concentric tracks. A laser beam is used to store the data on the disc and to read the data from the disc.

The Counting Function

The counting function is important in digital systems. There are many types of digital **counters**, but their basic purpose is to count events represented by changing levels or pulses. To count, the counter must "remember" the present number so that it can go to the next proper number in sequence. Therefore, storage capability is an important characteristic of all counters, and flip-flops are generally used to implement them. Figure 1–27 illustrates the basic idea of counter operation. Counters are covered in Chapter 9.

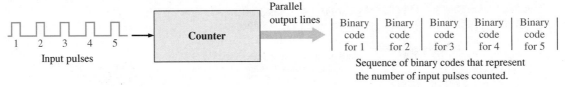

FIGURE 1–27   Illustration of basic counter operation.

A Process Control System

A system for bottling vitamin tablets is shown in the block diagram of Figure 1–28. This example system shows how the various logic functions that have been introduced can be used together to form a total system. To begin, the tablets are fed into a large funnel-type hopper. The narrow neck of the hopper creates a serial flow of tablets into a bottle on the conveyor belt below. Only one tablet at a time passes the sensor, so the tablets can be counted. The system controls the number of tablets into each bottle and displays a continually updated readout of the total number of tablets bottled.

General Operation

The maximum number of tablets per bottle is entered from the keypad, changed to a code by the *Encoder,* and stored in *Register A. Decoder A* changes the code stored in the register to a form appropriate for turning on the display. *Code converter A* changes the code to a binary number and applies it to the *A* input of the *Comparator* (Comp).

An optical sensor in the neck of the hopper detects each tablet that passes and produces a pulse. This pulse goes to the *Counter* and advances it by one count; thus, any time during the filling of a bottle, the binary state of the counter represents the number of tablets in the bottle. The binary count is transferred from the counter to the *B* input of the comparator (Comp). The *A* input of the comparator is the binary number for the maximum tablets per bottle. Now, let's say that the present number of tablets per bottle is 50. When the binary

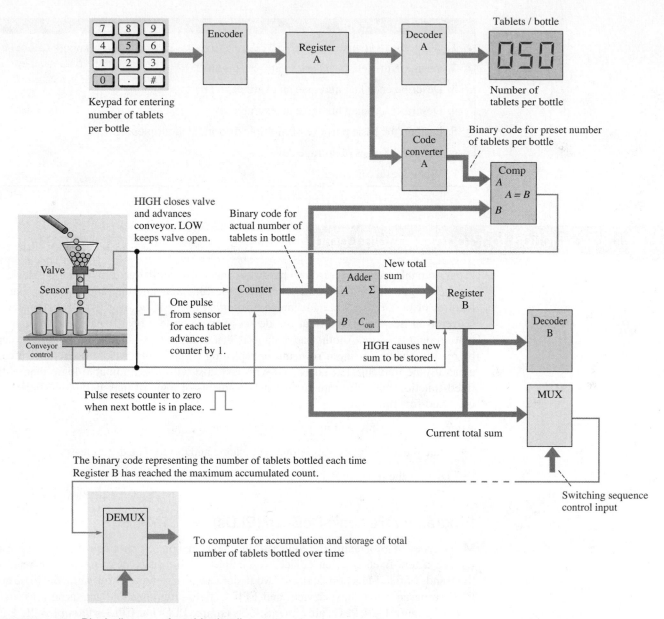

FIGURE 1–28   Block diagram of a tablet-bottling system.

number in the counter reaches 50, the $A = B$ output of the comparator goes HIGH, indicating that the bottle is full.

The HIGH output of the comparator causes the valve in the neck of the hopper to close and stop the flow of tablets. At the same time, the HIGH output of the comparator activates the conveyor, which moves the next empty bottle into place under the hopper. When the bottle is in place, the conveyor control issues a pulse that resets the counter to zero. As a result, the output of the comparator goes back LOW and causes the hopper valve to restart the flow of tablets.

For each bottle filled, the maximum binary number in the counter is transferred to the A input of the *Adder*. The B input of the adder comes from *Register B* that stores the total number of tablets bottled up through the last bottle filled. The adder produces a new cumulative sum that is then stored in register B, replacing the previous sum. This keeps a running total of the tablets bottled during a given run.

The cumulative sum stored in register B goes to *Decoder B*, which detects when *Register B* has reached its maximum capacity and enables the *MUX*, which converts the binary from parallel to serial form for transmission to the remote DEMUX. The *DEMUX* converts the data back to parallel form for storage.

SECTION 1–4 CHECKUP

1. What does a comparator do?

2. What are the four basic arithmetic operations?

3. Describe encoding and give an example.

4. Describe decoding and give an example.

5. Explain the basic purpose of multiplexing and demultiplexing.

6. Name four types of storage devices.

7. What does a counter do?

1–5 Introduction to Programmable Logic

Programmable logic requires both hardware and software. **Programmable logic** devices can be programmed to perform specified logic functions and operations by the manufacturer or by the user. One advantage of programmable logic over fixed-function logic (covered in Section 1–6) is that the devices use much less board space for an equivalent amount of logic. Another advantage is that, with programmable logic, designs can be readily changed without rewiring or replacing components. Also, a logic design can generally be implemented faster and with less cost with programmable logic than with fixed-function logic. To implement small segments of logic, it may be more efficient to use fixed-function logic.

After completing this section, you should be able to

◆ State the major types of programmable logic and discuss the differences

◆ Discuss the programmable logic design process

Programmable Logic Devices (PLDs)

Many types of programmable logic are available, ranging from small devices that can replace a few fixed-function devices to complex high-density devices that can replace thousands of fixed-function devices. Two major categories of user-programmable logic are **PLD** (programmable logic device) and **FPGA** (field-programmable gate array), as indicated in Figure 1–29. PLDs are either SPLDs (simple PLDs) or CPLDs (complex PLDs).

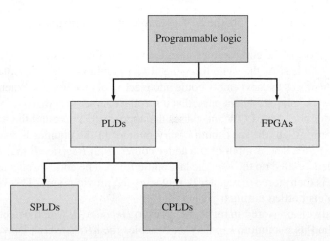

FIGURE 1–29   Programmable logic hierarchy.

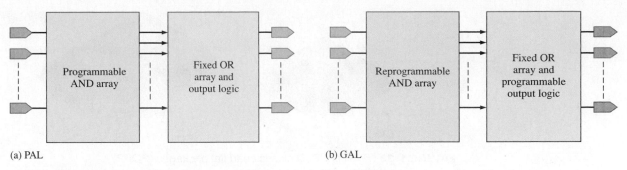

(a) PAL   (b) GAL

FIGURE 1–30 Block diagrams of simple programmable logic devices (SPLDs).

Simple Programmable Logic Device (SPLD)

The SPLD was the original PLD and is still available for small-scale applications. Generally, an **SPLD** can replace up to ten fixed-function ICs and their interconnections, depending on the type of functions and the specific SPLD. Most SPLDs are in one of two categories: PAL and GAL. A **PAL** (programmable array logic) is a device that can be programmed one time. It consists of a programmable array of AND gates and a fixed array of OR gates, as shown in Figure 1–30(a). A **GAL** (generic array logic) is a device that is basically a PAL that can be reprogrammed many times. It consists of a reprogrammable array of AND gates and a fixed array of OR gates with programmable ouputs, as shown in Figure 1–30(b). A typical SPLD package is shown in Figure 1–31 and generally has from 24 to 28 pins.

FIGURE 1–31 A typical SPLD package.

Complex Programmable Logic Device (CPLD)

As technology progressed and the amount of circuitry that could be put on a chip (chip density) increased, manufacturers were able to put more than one SPLD on a single chip and the CPLD was born. Essentially, the **CPLD** is a device containing multiple SPLDs and can replace many fixed-function ICs. Figure 1–32 shows a basic CPLD block diagram with four logic array blocks (LABs) and a programmable interconnection array (PIA). Depending on the specific CPLD, there can be from two to sixty-four LABs. Each logic array block is roughly equivalent to one SPLD.

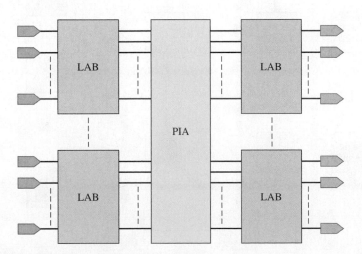

FIGURE 1–32 General block diagram of a CPLD.

Generally, CPLDs can be used to implement any of the logic functions discussed earlier, for example, decoders, encoders, multiplexers, demultiplexers, and adders. They are available in a variety of configurations, typically ranging from 44 to 160 pin packages. Examples of CPLD packages are shown in Figure 1–33.

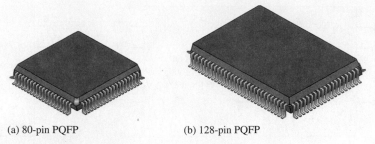

(a) 80-pin PQFP   (b) 128-pin PQFP

FIGURE 1–33   Typical CPLD plastic quad flat packages (PQFP).

Field-Programmable Gate Array (FPGA)

An **FPGA** is generally more complex and has a much higher density than a CPLD, although their applications can sometimes overlap. As mentioned, the SPLD and the CPLD are closely related because the CPLD basically contains a number of SPLDs. The FPGA, however, has a different internal structure (architecture), as illustrated in Figure 1–34. The three basic elements in an FPGA are the logic block, the programmable interconnections, and the input/output (I/O) blocks.

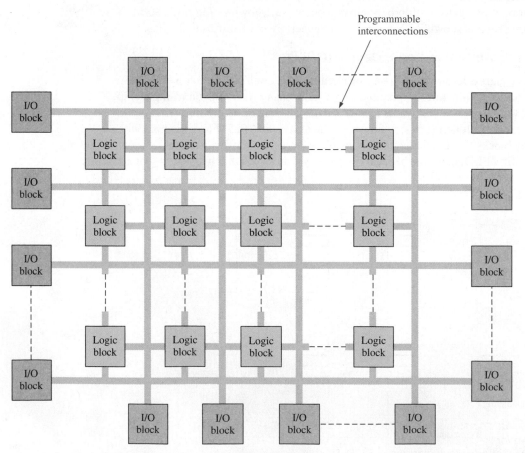

FIGURE 1–34   Basic structure of an FPGA.

The logic blocks in an FPGA are not as complex as the logic array blocks (LABs) in a CPLD, but generally there are many more of them. When the logic blocks are relatively simple, the FPGA architecture is called *fine-grained*. When the logic blocks are larger and

(a) Top view   (b) Bottom view

FIGURE 1–35   A typical ball-grid array (BGA) package.

more complex, the architecture is called *coarse-grained.* The I/O blocks are on the outer edges of the structure and provide individually selectable input, output, or bidirectional access to the outside world. The distributed programmable interconnection matrix provides for interconnection of the logic blocks and connection to inputs and outputs. Large FPGAs can have tens of thousands of logic blocks in addition to memory and other resources. A typical FPGA ball-grid array package is shown in Figure 1–35. These types of packages can have over 1000 input and output pins.

The Programming Process

An SPLD, CPLD, or FPGA can be thought of as a "blank slate" on which you implement a specified circuit or system design using a certain process. This process requires a software development package installed on a computer to implement a circuit design in the programmable chip. The computer must be interfaced with a development board or programming fixture containing the device, as illustrated in Figure 1–36.

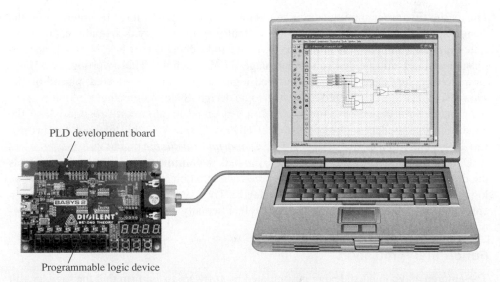

PLD development board

Programmable logic device

FIGURE 1–36   Basic setup for programming a PLD or FPGA. Graphic entry of a logic circuit is shown for illustration. Text entry such as VHDL can also be used.   (Photo courtesy of Digilent, Inc.)

Several steps, called the *design flow,* are involved in the process of implementing a digital logic design in a programmable logic device. A block diagram of a typical programming process is shown in Figure 1–37. As indicated, the design flow has access to development software.

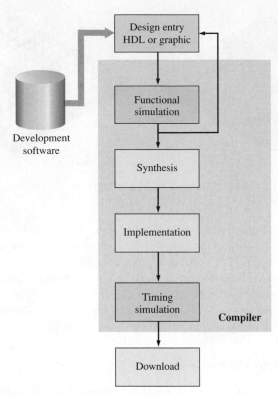

FIGURE 1–37   Basic programmable logic design flow block diagram.

Design Entry

This is the first programming step. The circuit or system design must be entered into the design application software using text-based entry, graphic entry (schematic capture), or state diagram description. Design entry is device independent. Text-based entry is accomplished with a hardware description language (**HDL**) such as VHDL, Verilog, or AHDL. Graphic (schematic) entry allows prestored logic functions to be selected, placed on the screen, and then interconnected to create a logic design. State-diagram entry requires specification of both the states through which a sequential logic circuit progresses and the conditions that produce each state change. VHDL will be used in this textbook to illustrate text-based entry of a digital design. *A VHDL tutorial is available on the website.*

Once a design has been entered, it is compiled. A **compiler** is a program that controls the design flow process and translates source code into object code in a format that can be logically tested or downloaded to a target device. The source code is created during design entry, and the object code is the final code that actually causes the design to be implemented in the programmable device.

Functional Simulation

The entered and compiled design is simulated by software to confirm that the logic circuit functions as expected. The simulation will verify that correct outputs are produced for a specified set of inputs. A device-independent software tool for doing this is generally called a *waveform editor.* Any flaws demonstrated by the simulation would be corrected by going back to design entry and making appropriate changes.

Synthesis

Synthesis is where the design is translated into a netlist, which has a standard form and is device independent.

Implementation

Implementation is where the logic structures described by the netlist are mapped into the actual structure of the specific device being programmed. The implementation process is called *fitting* or *place and route* and results in an output called a bitstream, which is device dependent.

Timing Simulation

This step comes after the design is mapped into the specific device. The timing simulation is basically used to confirm that there are no design flaws or timing problems due to propagation delays.

Download

Once a bitstream has been generated for a specific programmable device, it has to be downloaded to the device to implement the software design in hardware. Some programmable devices have to be installed in a special piece of equipment called a *device programmer* or on a development board. Other types of devices can be programmed while in a system—called in-system programming (ISP)—using a standard JTAG (Joint Test Action Group) interface. Some devices are volatile, which means they lose their contents when reset or when power is turned off. In this case, the bitstream data must be stored in a memory and reloaded into the device after each reset or power-off. Also, the contents of an ISP device can be manipulated or upgraded while it is operating in a system. This is called "on-the-fly" reconfiguration.

The Microcontroller

A microcontroller is different than a PLD. The internal circuits of a microcontroller are fixed, and a program (series of instructions) directs the microcontroller operation in order to achieve a specific outcome. The internal circuitry of a PLD is programmed into it, and once programmed, the circuitry performs required operations. Thus, a program determines microcontroller operation, but in a PLD a program determines the logic function. Microcontrollers are generally programmed with either the C language or the BASIC language.

A **microcontroller** is basically a special-purpose small computer. Microcontrollers are generally used for embedded system applications. An **embedded system** is a system that is designed to perform one or a few dedicated functions within a larger system. By contrast, a general-purpose computer, such as a laptop, is designed to perform a wide range of functions and applications.

Embedded microcontrollers are used in many common applications. The embedded microcontroller is part of a complete system, which may include additional electronics and mechanical parts. For example, a microcontroller in a television set displays the input from the remote unit on the screen and controls the channel selection, audio, and various menu adjustments like brightness and contrast. In an automobile a microcontroller takes engine sensor inputs and controls spark timing and fuel mixture. Other applications include home appliances, thermostats, cell phones, and toys.

SECTION 1–5 CHECKUP

1. List three major categories of programmable logic devices and specify their acronyms.

2. How does a CPLD differ from an SPLD?

3. Name the steps in the programming process.

4. Briefly explain each step named in question 3.

5. What are the two main functional characteristics of a microcontroller?

1–6   Fixed-Function Logic Devices

All the logic elements and functions that have been discussed are generally available in integrated circuit (IC) form. Digital systems have incorporated ICs for many years because of their small size, high reliability, low cost, and low power consumption. Despite the trend toward programmable logic, fixed-function logic continues to be used although on a more limited basis in specific applications. It is important to be able to recognize the IC packages and to know how the pin connections are numbered, as well as to be familiar with the way in which circuit complexities and circuit technologies determine the various IC classifications.

After completing this section, you should be able to

 * Recognize the difference between through-hole devices and surface-mount fixed-function devices

 * Identify dual in-line packages (DIP)

 * Identify small-outline integrated circuit packages (SOIC)

 * Identify plastic leaded chip carrier packages (PLCC)

 * Identify leadless ceramic chip carrier packages (LCC)

 * Determine pin numbers on various types of IC packages

 * Explain the complexity classifications for fixed-function ICs

A monolithic **integrated circuit (IC)** is an electronic circuit that is constructed entirely on a single small chip of silicon. All the components that make up the circuit—transistors, diodes, resistors, and capacitors—are an integral part of that single chip. Fixed-function logic and programmable logic are two broad categories of digital ICs. In **fixed-function logic** devices, the logic functions are set by the manufacturer and cannot be altered.

Figure 1–38 shows a cutaway view of one type of fixed-function IC package with the circuit chip shown within the package. Points on the chip are connected to the package pins to allow input and output connections to the outside world.

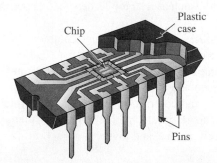

FIGURE 1–38   Cutaway view of one type of fixed-function IC package (dual in-line package) showing the chip mounted inside, with connections to input and output pins.

IC Packages

Integrated circuit (IC) packages are classified according to the way they are mounted on printed circuit boards (PCBs) as either through-hole mounted or surface mounted. The through-hole type packages have pins (leads) that are inserted through holes in the PCB and can be soldered to conductors on the opposite side. The most common type of through-hole package is the dual in-line package (**DIP**) shown in Figure 1–39(a).

(a) Dual in-line package (DIP)   (b) Small-outline IC (SOIC)

FIGURE 1–39 Examples of through-hole and surface-mounted devices. The DIP is larger than the SOIC with the same number of leads. This particular DIP is approximately 0.785 in. long, and the SOIC is approximately 0.385 in. long.

Another type of IC package uses surface-mount technology (**SMT**). Surface mounting is a space-saving alternative to through-hole mounting. The holes through the PCB are unnecessary for SMT. The pins of surface-mounted packages are soldered directly to conductors on one side of the board, leaving the other side free for additional circuits. Also, for a circuit with the same number of pins, a surface-mounted package is much smaller than a dual in-line package because the pins are placed closer together. An example of a surface-mounted package is the small-outline integrated circuit (**SOIC**) shown in Figure 1–39(b).

Various types of SMT packages are available in a range of sizes, depending on the number of leads (more leads are required for more complex circuits and lead configurations). Examples of several types are shown in Figure 1–40. As you can see, the leads of the **SSOP** (shrink small-outline package) are formed into a "gull-wing" shape. The leads of the **PLCC** (plastic-leaded chip carrier) are turned under the package in a J-type shape. Instead of leads, the **LCC** (leadless ceramic chip) has metal contacts molded into its ceramic body. The LQFP (low-profile quad flat package) also has gull-wing leads. Both the CSP (chip scale package) and the FBGA (fine-pitch ball grid array) have contacts embedded in the bottom of the package.

(a) SSOP (153 × 193 mils)   (b) PLCC (350 × 350 mils)   (c) LCC (350 × 350 mils)

(d) LQFP (7 × 7 mm)   (e) Laminate CSP bottom view   (f) FBGA bottom view
                                  (3.5 × 3.5 mm)   (4 × 4 mm)

FIGURE 1–40 Examples of SMT package configurations.   Parts (e) and (f) show bottom views.

Pin Numbering

All IC packages have a standard format for numbering the pins (leads). The dual in-line packages (DIPs) and the shrink small-outline packages (SSOP) have the numbering arrangement illustrated in Figure 1–41(a) for a 16-pin package. Looking at the top of the package, pin 1 is indicated by an identifier that can be either a small dot, a notch, or a beveled edge. The dot is always next to pin 1. Also, with the notch oriented upward, pin 1 is always the top left pin, as indicated. Starting with pin 1, the pin numbers increase as you go down, then across and up. The highest pin number is always to the right of the notch or opposite the dot.

The PLCC and LCC packages have leads arranged on all four sides. Pin 1 is indicated by a dot or other index mark and is located at the center of one set of leads. The pin numbers increase going counterclockwise as viewed from the top of the package. The highest pin number is always to the right of pin 1. Figure 1–41(b) illustrates this format for a 20-pin PLCC package.

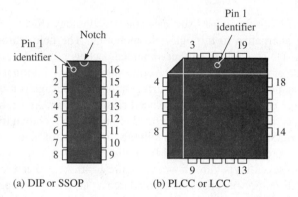

(a) DIP or SSOP   (b) PLCC or LCC

FIGURE 1–41   Pin numbering for two examples of standard types of IC packages. Top views are shown.

Complexity Classifications for Fixed-Function ICs

Fixed-function digital ICs are classified according to their complexity. They are listed here from the least complex to the most complex. The complexity figures stated here for SSI, MSI, LSI, VLSI, and ULSI are generally accepted, but definitions may vary from one source to another.

- **Small-scale integration (SSI)** describes fixed-function ICs that have up to ten equivalent gate circuits on a single chip, and they include basic gates and flip-flops.

- **Medium-scale integration (MSI)** describes integrated circuits that have from 10 to 100 equivalent gates on a chip. They include logic functions such as encoders, decoders, counters, registers, multiplexers, arithmetic circuits, small memories, and others.

- **Large-scale integration (LSI)** is a classification of ICs with complexities of from more than 100 to 10,000 equivalent gates per chip, including memories.

- **Very large-scale integration (VLSI)** describes integrated circuits with complexities of from more than 10,000 to 100,000 equivalent gates per chip.

- **Ultra large-scale integration (ULSI)** describes very large memories, larger **microprocessors**, and larger single-chip computers. Complexities of more than 100,000 equivalent gates per chip are classified as ULSI.

Integrated Circuit Technologies

The types of transistors with which all integrated circuits are implemented are either MOSFETs (metal-oxide semiconductor field-effect transistors) or bipolar junction transistors. A circuit technology that uses MOSFETs is CMOS (complementary MOS). One type of fixed-function digital circuit technology uses bipolar junction transistors and is sometimes called TTL (transistor-transistor logic). BiCMOS uses a combination of both CMOS and bipolar.

All gates and other functions can be implemented with either type of circuit technology. SSI and MSI circuits are generally available in both CMOS and bipolar. LSI, VLSI, and ULSI are generally implemented with CMOS because it requires less area on a chip and consumes less power. There is more on these integrated technologies in Chapter 3. Refer to Chapter 15 Integrated Circuit Technologies on the website for a thorough coverage.

SECTION 1–6 CHECKUP

1. What is an integrated circuit?

2. Define the terms DIP, SMT, SOIC, SSI, MSI, LSI, VLSI and ULSI.

3. Generally, in what classification does a fixed-function IC with the following number of equivalent gates fall?

   (a) 10

   (b) 75

   (c) 500

   (d) 15,000

   (e) 200,000

1–7   Test and Measurement Instruments

A variety of instruments are available for use in troubleshooting and testing. Some common types of instruments are introduced and discussed in this section.

After completing this section, you should be able to

- Distinguish between an analog and a digital oscilloscope

- Recognize common oscilloscope controls

- Determine amplitude, period, and frequency of a pulse waveform with an oscilloscope

- Discuss the logic analyzer and some common formats

- Describe the purpose of the digital multimeter (DMM), the dc power supply, the logic probe, and the logic pulser

The Oscilloscope

The oscilloscope (scope for short) is one of the most widely used instruments for general testing and troubleshooting. The scope is basically a graph-displaying device that traces the graph of a measured electrical signal on its screen. In most applications, the graph shows how signals change over time. The vertical axis of the display screen represents voltage, and the horizontal axis represents time. Amplitude, period, and frequency of a signal can be measured using the oscilloscope. Also, the pulse width, duty cycle, rise time, and fall time of a pulse waveform can be determined. Most scopes can display at least two signals on the screen at one time, enabling their time relationship to be observed. A typical digital oscilloscopes with a voltage probe connected is shown in Figure 1–42.

InfoNote

The analog scope was the earliest type of oscilloscope, but it has largely been replaced by the digital scope although analog scopes may still occasionally be found. The analog scope used a cathode ray tube (CRT) to display waveforms by sweeping an electron beam across the screen and controlling its up and down motion according to the measured waveform. Analog scopes were more limited in features than digital scopes in terms of storing and displaying waveform details.

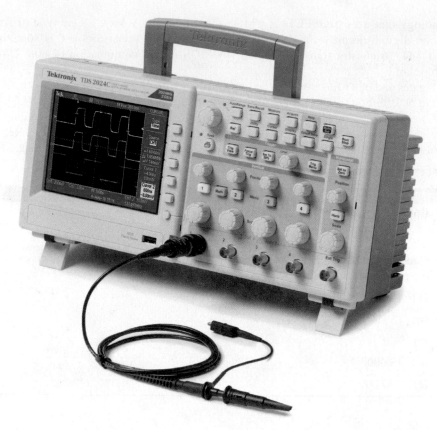

FIGURE 1–42 Typical digital oscilloscope with voltage probe.   Used with permission from Tektronix, Inc.

A digital scope converts the measured waveform to digital information by a sampling process in an analog-to-digital converter (ADC). The digital information is then used to reconstruct the waveform on the screen. Figure 1–43 shows a basic block diagram for a digital oscilloscope.

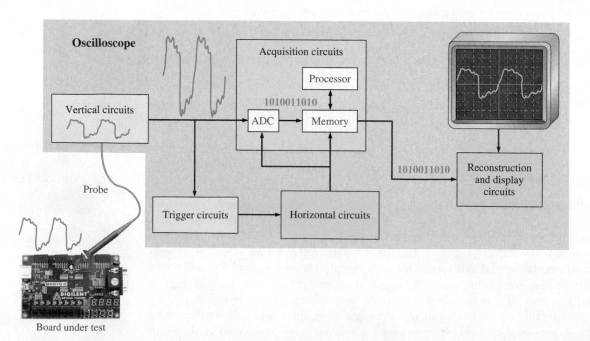

FIGURE 1–43 Block diagram of a digital oscilloscope.   (Photo courtesy of Digilent, Inc.)

Oscilloscope Controls

A front panel view of a typical four-channel digital oscilloscope is shown in Figure 1–44 (Some scopes have only two channels). Instruments vary depending on model and manufacturer, but most have certain common features. For example, each of the four vertical sections contain a Position control, a channel menu button, and a scale (volts/div) control. The horizontal section also contains a scale (sec/div) control.

Some of the main oscilloscope controls are now discussed. Refer to the user manual for complete details of your particular scope.

Vertical Controls

In the vertical section of the scope in Figure 1–44, there are identical controls for each of the four channels (1, 2, 3, and 4). The Position control lets you position a displayed waveform up or down vertically on the screen. The buttons on the right side of the screen provide for the selection of several items that appear on the screen, such as the coupling modes (ac, dc, or ground), coarse or fine adjustment for the scale (volts/div), signal inversion, and other parameters. The volts/div control adjusts the number of volts represented by each vertical division on the screen. The volts/div setting for each channel is displayed on the bottom of the screen.

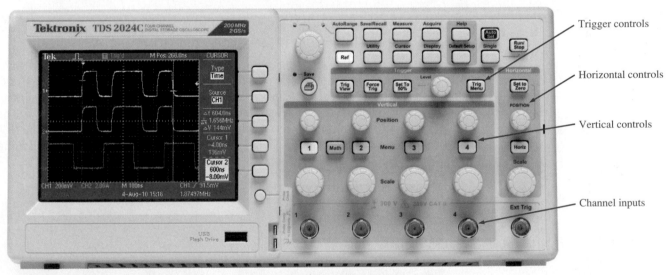

FIGURE 1–44 A typical digital oscilloscope front panel. Numbers below screen indicate the values for each division on the vertical (voltage) and horizontal (time) scales and can be varied using the vertical and horizontal controls on the scope.   Used with permission from Tektronix, Inc.

Horizontal Controls

In the horizontal section, the controls apply to all channels. The Position control lets you move a displayed waveform left or right horizontally on the screen. The Menu buttons provide for the selection of several items that appear on the screen such as the main time base, expanded view of a portion of a waveform, and other parameters. The sec/div control adjusts the time represented by each horizontal division or main time base. The sec/div setting is displayed at the bottom of the screen.

Trigger Controls

In the Trigger control section, the Level control determines the point on the triggering waveform where triggering occurs to initiate the sweep to display input waveforms. The

Trig Menu button provides for the selection of several items that appear on the screen, including edge or slope triggering, trigger source, trigger mode, and other parameters. There is also an input for an external trigger signal.

Triggering stabilizes a waveform on the screen or properly triggers on a pulse that occurs only one time or randomly. Also, it allows you to observe time delays between two waveforms. Figure 1–45 compares a triggered to an untriggered signal. The untriggered signal tends to drift across the screen, producing what appears to be multiple waveforms.

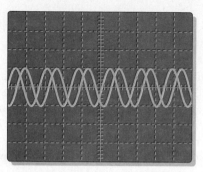   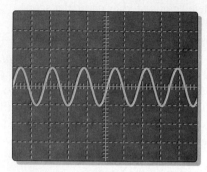

(a) Untriggered waveform display   (b) Triggered waveform display

FIGURE 1–45   Comparison of an untriggered and a triggered waveform on an oscilloscope.

Coupling a Signal into the Scope

Coupling is the method used to connect a signal voltage to be measured into the oscilloscope. DC and AC coupling are usually selected from the Vertical menu on a scope. DC coupling allows a waveform including its dc component to be displayed. AC coupling blocks the dc component of a signal so that you see the waveform centered at 0 V. The Ground mode allows you to connect the channel input to ground to see where the 0 V reference is on the screen. Figure 1–46 illustrates the result of DC and AC coupling using a pulse waveform that has a dc component.

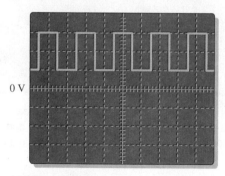   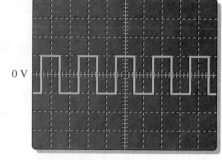

(a) DC coupled waveform   (b) AC coupled waveform

FIGURE 1–46   Displays of the same waveform having a dc component.

The voltage **probe**, shown connected to the oscilloscope in Figure 1–42, is essential for connecting a signal to the scope. Since all instruments tend to affect the circuit being measured due to loading, most scope probes provide a high series resistance to minimize loading effects. Probes that have a series resistance ten times larger than the input resistance of the scope are called $\times 10$ probes. Probes with no series resistance are called $\times 1$ probes. The oscilloscope adjusts its calibration for the attenuation of the type of probe being used. For most measurements, the $\times 10$ probe should be used. However, if you are measuring very small signals, a $\times 1$ may be the best choice.

The probe has an adjustment that allows you to compensate for the input capacitance of the scope. Most scopes have a probe compensation output that provides a calibrated square

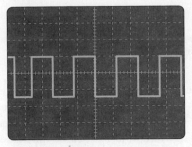
Properly compensated

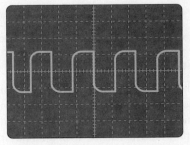
Undercompensated

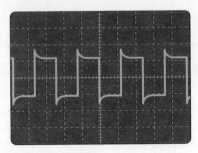
Overcompensated

FIGURE 1–47 Probe compensation conditions.

wave for probe compensation. Before making a measurement, you should make sure that the probe is properly compensated to eliminate any distortion introduced. Typically, there is a screw or other means of adjusting compensation on a probe. Figure 1–47 shows scope waveforms for three probe conditions: properly compensated, undercompensated, and overcompensated. If the waveform appears either over- or undercompensated, adjust the probe until the properly compensated square wave is achieved.

EXAMPLE 1–3

Based on the readouts, determine the amplitude and the period of the pulse waveform on the screen of a digital oscilloscope as shown in Figure 1–48. Also, calculate the frequency.

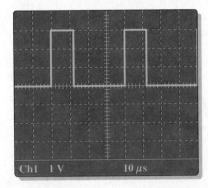

FIGURE 1–48

Solution

The volts/div setting is 1 V. The pulses are three divisions high. Since each division represents 1 V, the pulse amplitude is

$$\text{Amplitude} = (3\text{ div})(1\text{ V/div}) = \mathbf{3\ V}$$

The sec/div setting is 10 μs. A full cycle of the waveform (from beginning of one pulse to the beginning of the next) covers four divisions; therefore, the period is

$$\text{Period} = (4\text{ div})(10\ \mu\text{s/div}) = \mathbf{40\ \mu s}$$

The frequency is calculated as

$$f = \frac{1}{T} = \frac{1}{40\ \mu s} = \mathbf{25\ kHz}$$

Related Problem

For a volts/div setting of 4 V and sec/div setting of 2 ms, determine the amplitude and period of the pulse shown on the screen in Figure 1–48.

Oscilloscope Specifications

Several key specifications define the performance of a digital oscilloscope.

Bandwidth

The bandwidth describes the frequency range of an input signal that can be processed by the oscilloscope without being significantly distorted. **Bandwidth** is the frequency at which a sinusoidal input signal is attenuated to 70.7 percent of its original amplitude. As a rule of thumb, use a scope with a minimum bandwidth of at least twice the highest frequency component in the input signal.

Pulse signals have sharp rising and falling edges and are composed of high-frequency harmonics. For example, a 10 MHz pulse waveform such as a square wave contains a 10 MHz sine wave (fundamental) and a large number of significant higher-frequency sine waves called *harmonics*. In order to accurately capture the shape of the signal, the oscilloscope must have a bandwidth to capture several of these harmonics. If a sufficient number of harmonics are not captured, the resulting signal will be distorted and an incorrect measurement will result.

Sampling Rate

The **sampling rate** is the rate at which the analog-to-digital converter (ADC) in the oscilloscope is clocked to digitize the incoming signal. The sampling rate and bandwidth are not directly related, but the sampling rate should be at least five times the bandwidth. Figure 1–49 illustrates the difference between a low sampling rate and a much higher sampling rate. Part (a) shows how a sampling rate that is too low distorts the shape of the rising edge. In part (b), the higher sampling rate results in a much more accurate representation of the rising edge. When the sampling rate is sufficiently high, the signal can be precisely reproduced.

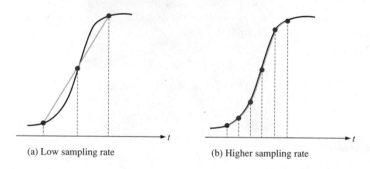

(a) Low sampling rate   (b) Higher sampling rate

FIGURE 1–49   Example of sampling a waveform. The dashed lines represent the clock (sampling) rate. The incoming signal is black and the resulting representation is blue. The red dots are the points at which the waveform values are sampled.

Record Length

The **record length** is the number of samples (data points) that the oscilloscope can capture and store. The capacity of acquisition memory determines the maximum record length. The memory must be able to store all the data points that are sampled during a certain time interval. The relationship between acquisition time, sampling rate, and record length is

$$\text{Acquisition time} = \frac{\text{Record length}}{\text{Sampling rate}}$$

Both the acquisition time (length of time that samples are taken) and/or sampling rate are limited by the record length of the oscilloscope. For example, if the record length is 1 Msample (1 million samples) and the sampling rate is 200 Msample/s, the oscilloscope acquisition time is 1 Msample ÷ 200 Msample/s = 5 ms. Therefore, one 5 ms segment of the sampled signal can be captured and stored at a time.

Resolution

The **resolution** is the number of bits used to digitally represent a sampled value. The number of discrete voltage levels used to represent a signal is defined as 2^x, where x is the resolution in bits. For example, if the resolution is four bits, $2^4 = 16$ levels can be represented. If the resolution is eight bits, $2^8 = 256$ levels can be represented. The more levels that are used to represent a signal, the higher the resolution and thus a more accurate representation is obtained. Also, the higher the resolution, the smaller the signal that can be measured.

Vertical Sensitivity

The vertical sensitivity indicates how much the oscilloscope's vertical amplifier can amplify a signal. Vertical sensitivity is usually given in volts, millivolts (mV), or microvolts (μV) per vertical division on the screen.

Horizontal Accuracy

The horizontal accuracy or time base indicates how accurately the horizontal system can display the timing of a signal, usually expressed as a percentage. The time base is shown on the horizontal axis of the screen in units of seconds per division.

The Logic Analyzer

Logic analyzers are used for measurements of multiple digital signals and measurement situations with difficult trigger requirements. Basically, the logic analyzer came about as a result of microprocessors in which troubleshooting or debugging required many more inputs than an oscilloscope offered. Many oscilloscopes have two input channels and some are available with four. Logic analyzers are typically available with from 16 to 136 input channels. Generally, an oscilloscope is used either when amplitude, frequency, and other timing parameters of a few signals at a time or when parameters such an rise and fall times, overshoot, and delay times need to be measured. The logic analyzer is used when the logic levels of a large number of signals need to be determined and for the correlation of simultaneous signals based on their timing relationships. A typical logic analyzer is shown in Figure 1–50, and a simplified block diagram is in Figure 1–51.

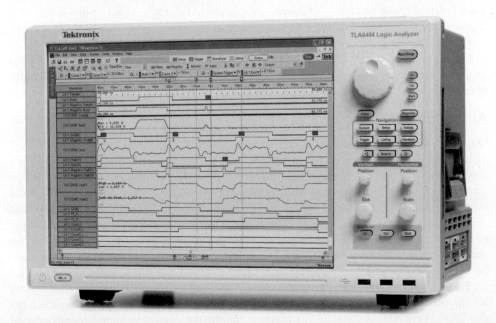

FIGURE 1–50   Typical logic analyzer.   Used with permission from Tektronix, Inc.

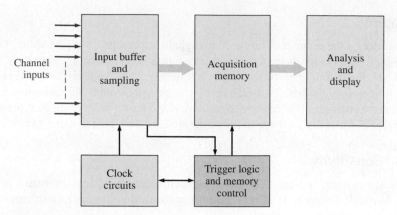

FIGURE 1–51   Simplified block diagram of a logic analyzer.

Data Acquisition

The large number of signals that can be acquired at one time is a major factor that distinguishes a logic analyzer from an oscilloscope. Generally, the two types of data acquisition in a logic analyzer are the timing acquisition and the state acquisition. Timing acquisition is used primarily when the timing relationships among the various signals need to be determined. State acquisition is used when you need to view the sequence of states as they appear in a system under test.

It is often helpful to have correlated timing and state data, and most logic analyzers can simultaneously acquire that data. For example, a problem may initially be detected as an invalid state. However, the invalid condition may be caused by a timing violation in the system under test. Without both types of information available at the same time, isolating the problem could be very difficult.

Channel Count and Memory Depth

Logic analyzers contain a real-time acquisition memory in which sampled data from all the channels are stored as they occur. Two features that are of primary importance are the channel count and the memory depth. The acquisition memory can be thought of as having a width equal to the number of channels and a depth that is the number of bits that can be captured by each channel during a certain time interval.

Channel count determines the number of signals that can be acquired simultaneously. In certain types of systems, a large number of signals are present, such as on the data bus in a microprocessor-based system. The depth of the acquisition memory (record length) determines the amount of data from a given channel that you can view at any given time.

Analysis and Display

Once data has been sampled and stored in the acquisition memory, it can typically be used in several different display and analysis modes. The waveform display is much like the display on an oscilloscope where you can view the time relationship of multiple signals. The listing display indicates the state of the system under test by showing the values of the input waveforms (1s and 0s) at various points in time (sample points). Typically, this data can be displayed in hexadecimal or other formats. Figure 1–52 shows simplified versions of these two display modes. The listing display samples correspond to the sampled points shown in red on the waveform display. You will study binary and hexadecimal (hex) numbers in the next chapter.

Two more modes that are useful in computer and microprocessor-based system testing are the instruction trace and the source code debug. The instruction trace determines and displays instructions that occur, for example, on the data bus in a microprocessor-based system. In this mode the op-codes and the mnemonics (English-like names) of instructions

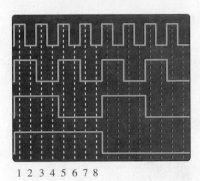

Sample	Binary	Hex	Time
1	1111	F	1 ns
2	1110	E	10 ns
3	1101	D	20 ns
4	1100	C	30 ns
5	1011	B	40 ns
6	1010	A	50 ns
7	1001	9	60 ns
8	1000	8	70 ns

1 2 3 4 5 6 7 8

(a) Waveform display   (b) Listing display

FIGURE 1–52   Two logic analyzer display modes.

are generally displayed as well as their corresponding memory address. Many logic analyzers also include a source code debug mode, which essentially allows you to see what is actually going on in the system under test when a program instruction is executed.

Probes

Three basic types of probes are used with logic analyzers. One is a multichannel probe, as shown in Figure 1–53, that can be attached to points on a circuit board under test. Another type of multichannel probe, similar to the one shown, plugs into dedicated sockets mounted on a circuit board. A third type is a single-channel clip-on probe.

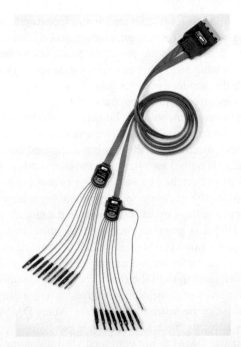

FIGURE 1–53   A typical multichannel logic analyzer probe.   Used with permission from Tektronix, Inc.

Signal Generators

Logic Signal Source

These instruments are also known as pulse generators and function generators. They are specifically designed to generate digital signals with precise edge placement and amplitudes and to produce the streams of 1s and 0s needed to test computer buses, microprocessors, and other digital systems.

Arbitrary Waveform Generators and Function Generators

The arbitrary waveform generator can be used to generate standard signals like sine waves, triangular waves, and pulses as well as signals with various shapes and characteristics. Waveforms can be defined by mathematical or graphical input. A typical arbitrary waveform generator is shown in Figure 1–54(a).

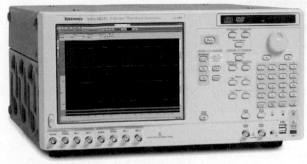

(a) Arbitrary waveform generator

(b) Function generator

FIGURE 1–54 Typical signal generators.   Used with permission from Tektronix, Inc.

The function generator, shown in part (b), provides pulse, sine, and triangular waveforms, often with programmable capability. Signal generators have logic-compatible outputs to provide the proper level and drive for inputs to digital circuits.

The Digital Multimeter (DMM)

The digital multimeter (DMM) is a versatile instrument found on virtually all workbenches. All DMMs can make basic ac and dc voltage, current, and resistance measurements. Voltage and resistance measurements are the principal quantities measured with DMMs. For current measurements, the leads are switched to a separate set of jacks and placed in series with the current path. In this mode, the meter acts like a short circuit, so serious problems can occur if the meter is incorrectly placed in parallel.

In addition to the basic measurements, most DMMs can also test diodes and capacitors and frequently will have other capabilities such as frequency measurements. Most new DMMs have an autoranging feature, meaning that the user is not required to select a range for making a measurement. If the range is not set automatically, the user needs to set the range switch for voltage measurements *higher* than the expected reading to avoid damage to the meter.

In digital circuits, DMMs are the preferred instrument for setting dc power supply voltages or checking the supply voltage on various points in the circuit. Because digital signals are nonsinusoidal, the DMM is generally *not* used for measurements of digital signals (although the average or rms value can be determined in some cases). For signal measurements, the oscilloscope is the preferred instrument.

In addition, DMMs are used in digital circuits for testing continuity between points in a circuit and checking resistors with the ohmmeter function. For checking a circuit path or looking for a short, DMMs are the instrument of choice. Many DMMs sound a beep or tone when there is continuity between the leads, making it handy to trace paths without having to look at the display. If the DMM is not equipped with a continuity test, the ohmmeter function can be used instead. Measurements of continuity or resistance are never done in "live" circuits, as any circuit voltage will disrupt the readings and can be dangerous.

Typical test bench and handheld DMMs are shown in Figure 1–55.

The DC Power Supply

This instrument is an indispensable instrument on any test bench. The power supply converts ac power from the standard wall outlet into regulated dc voltage. All digital circuits require dc voltage. Most logic circuits require from 1.2 V to 5 V to operate. The power supply is used to power circuits during design, development, and troubleshooting when in-system power is not available. A typical test bench dc power supply is shown in Figure 1–56.

(a) Bench-type DMM   (b) Handheld DMM

FIGURE 1–55   Typical DMMs.   Used with permission from (a) B+K Precision®; (b) Fluke

FIGURE 1–56   Typical bench-type dc power supply.   Used with permission from Tektronix, Inc.

The Logic Probe and Logic Pulser

The logic probe is a convenient, inexpensive handheld tool that provides a means of troubleshooting a digital circuit by sensing various conditions at a point in a circuit. A probe can detect high-level voltage, low-level voltage, single pulses, repetitive pulses, and opens on a PCB. The probe lamp indicates the condition that exists at a certain point, as indicated in the figure.

The logic pulser produces a repetitive pulse waveform that can be applied to any point in a circuit. You can apply pulses at one point in a circuit with the pulser and check another point for resulting pulses with a logic probe.

SECTION 1–7 CHECKUP

   1. What is the basic function of an oscilloscope?

   2. Name two main differences between an oscilloscope and a logic analyzer?

   3. What does the volts/div control on an oscilloscope do?

   4. What does the sec/div control on an oscilloscope do?

   5. What is *record length* in relation to a digital oscilloscope?

   6. What is the purpose of a function generator?

1–8   Introduction to Troubleshooting

Troubleshooting is the process of recognizing, isolating, and correcting a fault or failure in a system. To be an effective troubleshooter, you must understand how the system works and be able to recognize incorrect performance. Troubleshooting can be at the system level, the circuit board level, or the component level. Today, troubleshooting down to the board level is usually sufficient. Once a board is determined to be defective, it is usually replaced with a new one. However, if the circuit board is to be saved, component-level troubleshooting may be necessary.

After completing this section, you should be able to

- ◆ Describe the steps in a troubleshooting procedure
- ◆ Discuss the half-splitting method
- ◆ Discuss the signal-tracing method

Basic Hardware Troubleshooting Methods

Troubleshooting at a system level requires good detective work. When a problem occurs, the list of potential causes is usually quite large. You must gather a sufficient amount of detailed information and systematically narrow the list of potential causes to determine the problem. As a general guide to troubleshooting a system, the following steps should be followed:

1. Gather information on the problem.
2. Identify the symptoms and possible failures.
3. Isolate point(s) of failure.
4. Apply proper tools to determine the cause of the problem.
5. Fix the problem.

Check the Obvious

After collecting information on the problem, make sure to first check for obvious faults: absence of DC power, blown fuses, tripped circuit breakers, faulty burned out indicators such as lamps, loose connectors, broken or loose wires, switches in the wrong position, physical damages, boards not properly inserted, wire fragments or solder splashes shorting components, and poor quality contacts on printed circuit boards. For any troubleshooting task, you must have a system/circuit diagram. Other useful documents are a table of signal characteristics and a prewritten troubleshooting guide for the specific system.

Proper grounding is important when you set up to take measurements or work on a system. Properly grounding the oscilloscope protects you from shock, and grounding yourself protects circuits from damage. Grounding the oscilloscope means to connect it to earth ground by plugging the three-prong power cord into a grounded outlet. Grounding yourself means using a wrist-type grounding strap, particularly when you are working with CMOS logic. The wrist strap must have a high-value resistor between the strap and ground for protection against accidental contact with a voltage source.

For accurate measurements, make sure that the ground in the circuit you are testing is the same as the scope ground. This can be done by connecting the ground lead on the scope probe to a known ground point in the circuit, such as the metal chassis or a ground point on the circuit.

Replacement

Assume that a given system has multiple circuit boards. The simplest and quickest way to fix a problem is by replacing the circuit boards one by one with a known good board until the problem is corrected. This approach, of course, requires that duplicate boards be available. Another drawback to this approach is that an outside source may be causing the fault, such as a short in a connector; and by replacing the board, the fault is transferred to the new board.

Reproducing the Symptoms

Once the symptoms of a faulty system are identified, find a way to reproduce the problem. If the problem can be reproduced, it can be isolated and resolved. In some systems, the symptom may be self-evident, but in others it may have to be induced by application of a level or signal at a given point. Once this is done, then a systematic approach can be used to isolate the cause or causes of a problem. You should always consider the possibility that there is more than one fault.

If the symptoms are intermittent, the task of troubleshooting becomes more difficult. For example, in some cases a component may be temperature sensitive and fail only when the temperature is too high or too low. In these cases, the temperature can be varied by the simple process of blowing cool air on the component of concern to lower the temperature or using a heat gun to raise it, while monitoring the operation of the system.

Half-Splitting Method

In this procedure, you check for the presence or absence of a signal at a point halfway between input and output. If the signal is present, you know the fault is in the second half. If the signal is absent, you know the fault is in the first half. Then you split the defective half in half and check for a signal. The process is continued until a certain area of the system has been isolated. This may be a single circuit board in a system with many circuit boards or a component on a given circuit board. In a large system, this procedure can save a lot of time over moving down the line checking each block or stage as you go. This method is usually best applied in large complex systems. Figure 1–57 is a simple illustration of this method. The system is represented with the four green blocks. Additional steps are added to left or right for additional blocks.

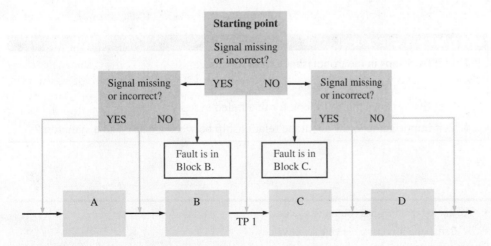

FIGURE 1–57   Concept of the half-splitting method. The blue arrows indicate the test points.

Signal-Tracing Method

Signal tracing is the procedure of tracking signals as they progress through a system from input to output. Signal tracing can be used with half-splitting, where you check for a signal at each point from where the absence of a signal was detected. Signal tracing can also begin

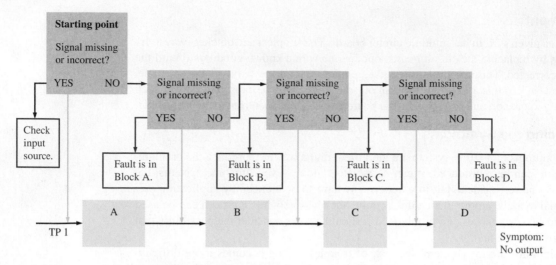

FIGURE 1–58 Concept of the signal-tracing method. Input to output is shown. The same applies if you start at the output and go toward the input.

at the output where there is an incorrect or absent signal and go back toward the input from point to point until a correct signal is found. Also, you can begin at the input and check the signal and move toward the output from point to point until the correct signal is lost. In both cases, the fault would be between the point and the output. Of course, you must know what the signal is supposed to look like in order to know if anything is wrong. Figure 1–58 illustrates the concept of signal tracing.

Signal Substitution and Injection

Signal substitution is used when the system being tested has been separated from its signal source. A generator signal is used to replace the normal signal that comes from the source when the system or portion of a system is recombined with the part that normally produces the input signal. Signal injection can be used to insert a signal at certain points in the system using the half-splitting approach.

SECTION 1–8 CHECKUP

1. List five steps in the troubleshooting procedure.
2. Name two troubleshooting methods.
3. List five obvious things to look for in a failed system.
4. Is it important to know about the relationship between a cause and a symptom?

SUMMARY

- An analog quantity has continuous values.
- A digital quantity has a discrete set of values.
- A binary digit is called a bit.
- A pulse is characterized by rise time, fall time, pulse width, and amplitude.
- The frequency of a periodic waveform is the reciprocal of the period. The formulas relating frequency and period are

$$f = \frac{1}{T} \quad \text{and} \quad T = \frac{1}{f}$$

- The duty cycle of a pulse waveform is the ratio of the pulse width to the period, expressed by the following formula as a percentage:

$$\text{Duty cycle} = \left(\frac{t_W}{T}\right)100\%$$

- A timing diagram is an arrangement of two or more waveforms showing their relationship with respect to time.

- Three basic logic operations are NOT, AND, and OR. The standard symbols for these are given in Figure 1–59.

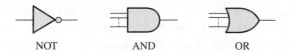

NOT   AND   OR

FIGURE 1–59

- The basic logic functions are comparison, arithmetic, code conversion, decoding, encoding, data selection, storage, and counting.

- Two types of SPLDs (simple programmable logic devices) are PAL (programmable array logic) and GAL (generic array logic).

- The CPLD (complex programmable logic device) contains multiple SPLDs with programmable interconnections.

- The FPGA (field-programmable gate array) has a different internal structure than the CPLD and is generally used for more complex circuits and systems.

- The two broad physical categories of IC packages are through-hole mounted and surface mounted.

- Three families of fixed-function integrated circuits are CMOS, bipolar, and BiCMOS.

- Bipolar is also known as TTL (transistor-transistor logic).

- The categories of ICs in terms of circuit complexity are SSI (small-scale integration), MSI (medium-scale integration), LSI, VLSI, and ULSI (large-scale, very large-scale, and ultra large-scale integration).

- Common instruments used in testing and troubleshooting digital circuits are the oscilloscope, logic analyzer, arbitrary waveform generator, data pattern generator, function generator, dc power supply, digital multimeter, logic probe, and logic pulser.

- Two basic methods of troubleshooting are the half-splitting method and the signal-tracing method.

KEY TERMS

Key terms and other bold terms in the chapter are defined in the end-of-book glossary.

Analog   Being continuous or having continuous values.

AND   A basic logic operation in which a true (HIGH) output occurs only when all the input conditions are true (HIGH).

Binary   Having two values or states; describes a number system that has a base of two and utilizes 1 and 0 as its digits.

Bit   A binary digit, which can be either a 1 or a 0.

Clock   The basic timing signal in a digital system; a periodic waveform used to synchronize operation.

Compiler   A program that controls the design flow process and translates source code into object code in a format that can be logically tested or downloaded to a target device.

CPLD   A complex programmable logic device that consists basically of multiple SPLD arrays with programmable interconnections.

Data   Information in numeric, alphabetic, or other form.

Digital   Related to digits or discrete quantities; having a set of discrete values.

Duty cycle   The ratio of the pulse width to the period of a digital waveform, expressed as a percentage.

Embedded system Generally, a single-purpose system, such as a processor, built into a larger system for the purpose of controlling the system.

Fixed-function logic A category of digital integrated circuits having functions that cannot be altered.

FPGA Field-programmable gate array.

Gate A logic circuit that performs a basic logic operation such as AND or OR.

Input The signal or line going into a circuit.

Integrated circuit (IC) A type of circuit in which all of the components are integrated on a single chip of semiconductive material of extremely small size.

Inverter A NOT circuit; a circuit that changes a HIGH to a LOW or vice versa.

Logic In digital electronics, the decision-making capability of gate circuits, in which a HIGH represents a true statement and a LOW represents a false one.

Microcontroller An integrated circuit consisting of a complete computer on a single chip and used for specified control functions.

NOT A basic logic operation that performs inversions.

OR A basic logic operation in which a true (HIGH) output occurs when one or more of the input conditions are true (HIGH).

Output The signal or line coming out of a circuit.

Parallel In digital systems, data occurring simultaneously on several lines; the transfer or processing of several bits simultaneously.

Programmable logic A category of digital integrated circuits capable of being programmed to perform specified functions.

Pulse A sudden change from one level to another, followed after a time, called the pulse width, by a sudden change back to the original level.

Serial Having one element following another, as in a serial transfer of bits; occurring in sequence rather than simultaneously.

SPLD Simple programmable logic device.

Timing diagram A graph of digital waveforms showing the time relationship of two or more waveforms.

Troubleshooting The technique or process of systematically identifying, isolating, and correcting a fault in a circuit or system.

TRUE/FALSE QUIZ

Answers are at the end of the chapter.

1. An analog quantity is one having continuous values.
2. A digital quantity has ten discrete values.
3. There are two digits in the binary system.
4. The term *bit* is short for binary digit.
5. In positive logic, a LOW level represents a binary 1.
6. If the period of a pulse waveform increases, the frequency also increases.
7. A timing diagram shows the timing relationship of two or more digital waveforms.
8. The basic logic operations are AND, OR, and MAYBE.
9. If the input to an inverter is a 1, the output is a 0.
10. Two broad types of digital integrated circuits are fixed-function and programmable.

SELF-TEST

Answers are at the end of the chapter.

1. A quantity having continuous values is
   (a) a digital quantity   (b) an analog quantity
   (c) a binary quantity   (d) a natural quantity

2. The term *bit* means
   (a) a small amount of data   (b) a 1 or a 0
   (c) binary digit   (d) both answers (b) and (c)

3. The time interval on the leading edge of a pulse between 10% and 90% of the amplitude is the
   (a) rise time   (b) fall time
   (c) pulse width   (d) period

4. A pulse in a certain waveform occurs every 10 ms. The frequency is
   (a) 1 kHz   (b) 1 Hz   (c) 100 Hz   (d) 10 Hz

5. In a certain digital waveform, the period is twice the pulse width. The duty cycle is
   (a) 100%   (b) 200%   (c) 50%

6. An inverter
   (a) performs the NOT operation   (b) changes a HIGH to a LOW
   (c) changes a LOW to a HIGH   (d) does all of the above

7. The output of an AND gate is HIGH when
   (a) any input is HIGH   (b) all inputs are HIGH
   (c) no inputs are HIGH   (d) both answers (a) and (b)

8. The output of an OR gate is HIGH when
   (a) any input is HIGH   (b) all inputs are HIGH
   (c) no inputs are HIGH   (d) both answers (a) and (b)

9. The device used to convert a binary number to a 7-segment display format is the
   (a) multiplexer   (b) encoder
   (c) decoder   (d) register

10. An example of a data storage device is
    (a) the logic gate   (b) the flip-flop   (c) the comparator
    (d) the register   (e) both answers (b) and (d)

11. VHDL is a
    (a) logic device   (b) PLD programming language
    (c) computer language   (d) very high density logic

12. A CPLD is a
    (a) controlled program logic device   (b) complex programmable logic driver
    (c) complex programmable logic device   (d) central processing logic device

13. An FPGA is a
    (a) field-programmable gate array   (b) fast programmable gate array
    (c) field-programmable generic array   (d) flash process gate application

14. A fixed-function IC package containing four AND gates is an example of
    (a) MSI   (b) SMT   (c) SOIC   (d) SSI

15. An LSI device has a circuit complexity of from
    (a) 10 to 100 equivalent gates   (b) more than 100 to 10,000 equivalent gates
    (c) 2000 to 5000 equivalent gates   (d) more than 10,000 to 100,000 equivalent gates

PROBLEMS

Answers to odd-numbered problems are at the end of the book.

Section 1–1 Digital and Analog Quantities

1. Name two advantages of digital data as compared to analog data.

2. Name an analog quantity other than temperature and sound.

3. List three common products that can have either a digital or analog output.

Section 1–2 Binary Digits, Logic Levels, and Digital Waveforms

4. Explain the difference between positive and negative logic.

5. Define the sequence of bits (1s and 0s) represented by each of the following sequences of levels:
   (a) HIGH, HIGH, LOW, HIGH, LOW, LOW, LOW, HIGH
   (b) LOW, LOW, LOW, HIGH, LOW, HIGH, LOW, HIGH, LOW

6. List the sequence of levels (HIGH and LOW) that represent each of the following bit sequences:

    (a) 1 0 1 1 1 0 1   (b) 1 1 1 0 1 0 0 1

7. For the pulse shown in Figure 1–60, graphically determine the following:

    (a) rise time   (b) fall time   (c) pulse width   (d) amplitude

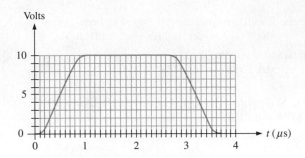

FIGURE 1–60

8. Determine the period of the digital waveform in Figure 1–61.

9. What is the frequency of the waveform in Figure 1–61?

10. Is the pulse waveform in Figure 1–61 periodic or nonperiodic?

11. Determine the duty cycle of the waveform in Figure 1–61.

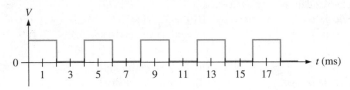

FIGURE 1–61

12. Determine the bit sequence represented by the waveform in Figure 1–62. A bit time is 1 μs in this case.

13. What is the total serial transfer time for the eight bits in Figure 1–62? What is the total parallel transfer time?

14. What is the period if the clock frequency is 3.5 GHz?

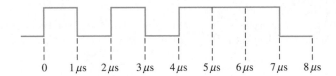

FIGURE 1–62

Section 1–3 Basic Logic Functions

15. Form a single logical statement from the following information:

    (a) The light is ON if SW1 is closed.
    (b) The light is ON if SW2 is closed.
    (c) The light is OFF if both SW1 and SW2 are open.

16. A logic circuit requires HIGHs on all its inputs to make the output HIGH. What type of logic circuit is it?

17. A basic 2-input logic circuit has a HIGH on one input and a LOW on the other input, and the output is LOW. Identify the circuit.

18. A basic 2-input logic circuit has a HIGH on one input and a LOW on the other input, and the output is HIGH. What type of logic circuit is it?

Section 1–4 Combinational and Sequential Logic Functions

19. Name the logic function of each block in Figure 1–63 based on your observation of the inputs and outputs.

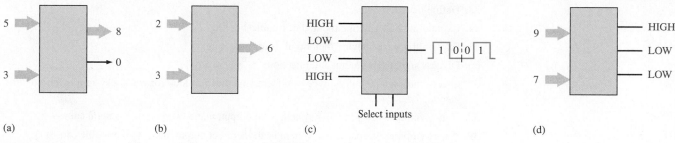

FIGURE 1–63

20. A pulse waveform with a frequency of 10 kHz is applied to the input of a counter. During 100 ms, how many pulses are counted?

21. Consider a register that can store eight bits. Assume that it has been reset so that it contains zeros in all positions. If you transfer four alternating bits (0101) serially into the register, beginning with a 1 and shifting to the right, what will the total content of the register be as soon as the fourth bit is stored?

Section 1–5 Introduction to Programmable Logic

22. Which of the following acronyms do not describe a type of programmable logic?

    PAL, GAL, SPLD, VHDL, CPLD, AHDL, FPGA

23. What do each of the following stand for?

    (a) SPLD   **(b)** CPLD   **(c)** HDL   **(d)** FPGA   **(e)** GAL

24. Define each of the following PLD programming terms:

    (a) design entry   **(b)** simulation   **(c)** compilation   **(d)** download

25. Describe the process of place-and-route.

Section 1–6 Fixed-Function Logic Devices

26. A fixed-function digital IC chip has a complexity of 200 equivalent gates. How is it classified?

27. Explain the main difference between the DIP and SMT packages.

28. Label the pin numbers on the packages in Figure 1–64. Top views are shown.

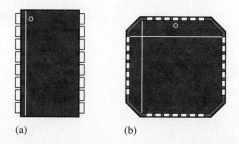

FIGURE 1–64

Section 1–7 Test and Measurement Instruments

29. A pulse is displayed on the screen of an oscilloscope, and you measure the base line as 1 V and the top of the pulse as 8 V. What is the amplitude?

30. A waveform is measured on the oscilloscope and its amplitude covers three vertical divisions. If the vertical control is set at 2 V/div, what is the total amplitude of the waveform?

31. The period of a pulse waveform measures four horizontal divisions on an oscilloscope. If the time base is set at 2 ms/div, what is the frequency of the waveform?

32. What record length is required if an oscilloscope has a sampling rate of 12 Msamples/s and the input waveform is sampled for 2 ms?

Section 1–8 Introduction to Troubleshooting

33. Define *troubleshooting*.

34. Explain the half-splitting method of troubleshooting.

35. Explain the signal-tracing method of troubleshooting.

36. Discuss signal substitution and injection.

37. Give some examples of the type of information that you look for when a system is reported to have failed.

38. If the symptom in a particular system is no output, name two possible general causes.

39. If the symptom of a particular system is an incorrect output, name two possible causes.

40. What obvious things should you look for before starting the troubleshooting process?

41. How would you isolate a fault in a system?

42. Name two common instruments used in troubleshooting.

43. Assume that you have isolated the problem down to a specific circuit board. What are your options at this point?

ANSWERS

SECTION CHECKUPS

Section 1–1 Digital and Analog Quantities

1. *Analog* means continuous.

2. *Digital* means discrete.

3. A digital quantity has a discrete set of values and an analog quantity has continuous values.

4. A public address system is analog. A CD player is analog and digital. A computer is all digital.

5. A mechatronic system consists of both mechanical and electronic components.

Section 1–2 Binary Digits, Logic Levels, and Digital Waveforms

1. Binary means having two states or values.

2. A bit is a binary digit.

3. The bits are 1 and 0.

4. Rise time: from 10% to 90% of amplitude. Fall time: from 90% to 10% of amplitude.

5. Frequency is the reciprocal of the period.

6. A clock waveform is a basic timing waveform from which other waveforms are derived.

7. A timing diagram shows the time relationship of two or more waveforms.

8. Parallel transfer is faster than serial transfer.

Section 1–3 Basic Logic Functions

1. When the input is LOW

2. When all inputs are HIGH

3. When any or all inputs are HIGH

4. An inverter is a NOT circuit.

5. A logic gate is a circuit that performs a logic operation (AND, OR).

Section 1–4 Combinational and Sequential Logic Functions

1. A comparator compares the magnitudes of two input numbers.

2. Add, subtract, multiply, and divide

3. Encoding is changing a familiar form such as decimal to a coded form such as binary.

4. Decoding is changing a code to a familiar form such as binary to decimal.

5. Multiplexing puts data from many sources onto one line. Demultiplexing takes data from one line and distributes it to many destinations.

6. Flip-flops, registers, semiconductor memories, magnetic disks

7. A counter counts events with a sequence of binary states.

Section 1–5 Introduction to Programmable Logic

1. Simple programmable logic device (SPLD), complex programmable logic device (CPLD), and field-programmable gate array (FPGA)

2. A CPLD is made up of multiple SPLDs.

3. Design entry, functional simulation, synthesis, implementation, timing simulation, and download

4. *Design entry:* The logic design is entered using development software. *Functional simulation:* The design is software simulated to make sure it works logically. *Synthesis:* The design is translated into a netlist. *Implementation:* The logic developed by the netlist is mapped into the programmable device. *Timing simulation:* The design is software simulated to confirm that there are no timing problems. *Download:* The design is placed into the programmable device.

5. The microcontroller has fixed internal circuits and its operation is directed by a program.

Section 1–6 Fixed-Function Logic Devices

1. An IC is an electronic circuit with all components integrated on a single silicon chip.

2. DIP—dual in-line package; SMT—surface-mount technology; SOIC—small-outline integrated circuit; SSI—small-scale integration; MSI—medium-scale integration; LSI—large-scale integration; VLSI—very large-scale integration; ULSI—ultra large-scale integration

3. **(a)** SSI

   (b) MSI

   (c) LSI

   (d) VLSI

   (e) ULSI

Section 1–7 Test and Measurement Instruments

1. The oscilloscope measures, processes, and displays electrical waveforms.

2. The logic analyzer has more channels than the oscillosope and has more than one data display format.

3. The volts/div control sets the voltage for each division on the screen.

4. The sec/div control sets the time for each division on the screen.

5. The function generator produces various types of waveforms.

6. The record length is the maximum number of samples that can be acquired during a given time interval.

Section 1–8 Introduction to Troubleshooting

1. Gather information, identify symptoms and possible causes, isolate point(s) of failure, apply proper tools to determine cause, and fix problem.

2. Half-splitting and signal tracing

3. Blown fuse, absence of DC power, loose connections, broken wires, loosely connected circuit board

4. Yes

RELATED PROBLEMS FOR EXAMPLES

1–1 $f = 6.67$ kHz; Duty cycle $= 16.7\%$

1–2 Serial transfer: 3.33 ns

1–3 Amplitude $= 12$ V; $T = 8$ ms

TRUE/FALSE QUIZ

1. T   **2.** F   **3.** T   **4.** T   **5.** F   **6.** F   **7.** T   **8.** F   **9.** T   **10.** T

SELF-TEST

1. (b)   **2.** (d)   **3.** (a)   **4.** (c)   **5.** (c)   **6.** (d)   **7.** (b)   **8.** (d)   **9.** (c)

10. (e)   **11.** (c)   **12.** (a)   **13.** (d)   **14.** (d)   **15.** (b)

Number Systems, Operations, and Codes

CHAPTER OUTLINE

CHAPTER OBJECTIVES

- Review the decimal number system
- Count in the binary number system
- Convert from decimal to binary and from binary to decimal
- Apply arithmetic operations to binary numbers
- Determine the 1's and 2's complements of a binary number
- Express signed binary numbers in sign-magnitude, 1's complement, 2's complement, and floating-point format
- Carry out arithmetic operations with signed binary numbers
- Convert between the binary and hexadecimal number systems
- Add numbers in hexadecimal form
- Convert between the binary and octal number systems
- Express decimal numbers in binary coded decimal (BCD) form
- Add BCD numbers
- Convert between the binary system and the Gray code
- Interpret the American Standard Code for Information Interchange (ASCII)
- Explain how to detect code errors
- Discuss the cyclic redundancy check (CRC)

KEY TERMS

Key terms are in order of appearance in the chapter.

- LSB
- MSB
- Byte
- Floating-point number
- Hexadecimal
- Octal
- BCD
- Alphanumeric
- ASCII
- Parity
- Cyclic redundancy check (CRC)

VISIT THE WEBSITE

Study aids for this chapter are available at
http://www.pearsonhighered.com/careersresources/

INTRODUCTION

The binary number system and digital codes are fundamental to computers and to digital electronics in general. In this chapter, the binary number system and its relationship to other number systems such as decimal, hexadecimal, and octal are presented. Arithmetic operations with binary numbers are covered to provide a basis for understanding how computers and many other types of digital systems work. Also, digital codes such as binary coded decimal (BCD), the Gray code, and the ASCII are covered. The parity method for detecting errors in codes is introduced. The TI-36X calculator is used to illustrate certain operations. The procedures shown may vary on other types.

2–1 Decimal Numbers

You are familiar with the decimal number system because you use decimal numbers every day. Although decimal numbers are commonplace, their weighted structure is often not understood. In this section, the structure of decimal numbers is reviewed. This review will help you more easily understand the structure of the binary number system, which is important in computers and digital electronics.

After completing this section, you should be able to

◆ Explain why the decimal number system is a weighted system

◆ Explain how powers of ten are used in the decimal system

◆ Determine the weight of each digit in a decimal number

The decimal number system has ten digits.

In the **decimal** number system each of the ten digits, 0 through 9, represents a certain quantity. As you know, the ten symbols (**digits**) do not limit you to expressing only ten different quantities because you use the various digits in appropriate positions within a number to indicate the magnitude of the quantity. You can express quantities up through nine before running out of digits; if you wish to express a quantity greater than nine, you use two or more digits, and the position of each digit within the number tells you the magnitude it represents. If, for example, you wish to express the quantity twenty-three, you use (by their respective positions in the number) the digit 2 to represent the quantity twenty and the digit 3 to represent the quantity three, as illustrated below.

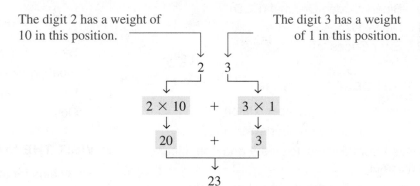

The decimal number system has a base of 10.

The position of each digit in a decimal number indicates the magnitude of the quantity represented and can be assigned a **weight**. The weights for whole numbers are positive powers of ten that increase from right to left, beginning with $10^0 = 1$.

$$\ldots 10^5 \ 10^4 \ 10^3 \ 10^2 \ 10^1 \ 10^0$$

For fractional numbers, the weights are negative powers of ten that decrease from left to right beginning with 10^{-1}.

$$10^2 \ 10^1 \ 10^0.10^{-1} \ 10^{-2} \ 10^{-3} \ldots$$

$\uparrow$ —— Decimal point

The value of a digit is determined by its position in the number.

The value of a decimal number is the sum of the digits after each digit has been multiplied by its weight, as Examples 2–1 and 2–2 illustrate.

EXAMPLE 2-1

Express the decimal number 47 as a sum of the values of each digit.

Solution

The digit 4 has a weight of 10, which is 10^1, as indicated by its position. The digit 7 has a weight of 1, which is 10^0, as indicated by its position.

$$47 = (4 \times 10^1) + (7 \times 10^0)$$
$$= (4 \times 10) + (7 \times 1) = \mathbf{40 + 7}$$

Related Problem*

Determine the value of each digit in 939.

_____

*Answers are at the end of the chapter.

EXAMPLE 2-2

Express the decimal number 568.23 as a sum of the values of each digit.

Solution

The whole number digit 5 has a weight of 100, which is 10^2, the digit 6 has a weight of 10, which is 10^1, the digit 8 has a weight of 1, which is 10^0, the fractional digit 2 has a weight of 0.1, which is 10^{-1}, and the fractional digit 3 has a weight of 0.01, which is 10^{-2}.

$$568.23 = (5 \times 10^2) + (6 \times 10^1) + (8 \times 10^0) + (2 \times 10^{-1}) + (3 \times 10^{-2})$$
$$= (5 \times 100) + (6 \times 10) + (8 \times 1) + (2 \times 0.1) + (3 \times 0.01)$$
$$= \mathbf{500} + \mathbf{60} + \mathbf{8} + \mathbf{0.2} + \mathbf{0.03}$$

Related Problem

Determine the value of each digit in 67.924.

CALCULATOR SESSION

Powers of Ten
Find the value of 10^3.

TI-36X   Step 1: [1] [0] [y^x]

Step 2: [3] [=]

1000

SECTION 2-1 CHECKUP

Answers are at the end of the chapter.

1. What weight does the digit 7 have in each of the following numbers?

   (a) 1370   (b) 6725   (c) 7051   (d) 58.72

2. Express each of the following decimal numbers as a sum of the products obtained by multiplying each digit by its appropriate weight:

   (a) 51   (b) 137   (c) 1492   (d) 106.58

2-2   Binary Numbers

The binary number system is another way to represent quantities. It is less complicated than the decimal system because the binary system has only two digits. The decimal system with its ten digits is a base-ten system; the binary system with its two digits is a base-two system. The two binary digits (bits) are 1 and 0. The position of a 1 or 0 in a binary number indicates its weight, or value within the number, just as the position of a decimal digit determines the value of that digit. The weights in a binary number are based on powers of two.

After completing this section, you should be able to

*   Count in binary

*   Determine the largest decimal number that can be represented by a given number of bits

*   Convert a binary number to a decimal number

Counting in Binary

The binary number system has two digits (bits).

To learn to count in the binary system, first look at how you count in the decimal system. You start at zero and count up to nine before you run out of digits. You then start another digit position (to the left) and continue counting 10 through 99. At this point you have exhausted all two-digit combinations, so a third digit position is needed to count from 100 through 999.

A comparable situation occurs when you count in binary, except that you have only two digits, called *bits*. Begin counting: 0, 1. At this point you have used both digits, so include another digit position and continue: 10, 11. You have now exhausted all combinations of two digits, so a third position is required. With three digit positions you can continue to count: 100, 101, 110, and 111. Now you need a fourth digit position to continue, and so on. A binary count of zero through fifteen is shown in Table 2–1. Notice the patterns with which the 1s and 0s alternate in each column.

The binary number system has a base of 2.

InfoNote

In processor operations, there are many cases where adding or subtracting 1 to a number stored in a counter is necessary. Processors have special instructions that use less time and generate less machine code than the ADD or SUB instructions. For the Intel processors, the INC (increment) instruction adds 1 to a number. For subtraction, the corresponding instruction is DEC (decrement), which subtracts 1 from a number.

TABLE 2–1

Decimal Number	Binary Number			
0	0	0	0	0
1	0	0	0	1
2	0	0	1	0
3	0	0	1	1
4	0	1	0	0
5	0	1	0	1
6	0	1	1	0
7	0	1	1	1
8	1	0	0	0
9	1	0	0	1
10	1	0	1	0
11	1	0	1	1
12	1	1	0	0
13	1	1	0	1
14	1	1	1	0
15	1	1	1	1

The value of a bit is determined by its position in the number.

CALCULATOR SESSION

Powers of Two
Find the value of 2^5.

TI-36X Step 1: [2] [y^x]

Step 2: [5] [=]

32

As you have seen in Table 2–1, four bits are required to count from zero to 15. In general, with n bits you can count up to a number equal to $2^n - 1$.

$$\text{Largest decimal number} = 2^n - 1$$

For example, with five bits ($n = 5$) you can count from zero to thirty-one.

$$2^5 - 1 = 32 - 1 = 31$$

With six bits ($n = 6$) you can count from zero to sixty-three.

$$2^6 - 1 = 64 - 1 = 63$$

An Application

Learning to count in binary will help you to basically understand how digital circuits can be used to count events. Let's take a simple example of counting tennis balls going into a box from a conveyor belt. Assume that nine balls are to go into each box.

The counter in Figure 2–1 counts the pulses from a sensor that detects the passing of a ball and produces a sequence of logic levels (digital waveforms) on each of its four parallel outputs. Each set of logic levels represents a 4-bit binary number (HIGH = 1 and LOW = 0), as indicated. As the decoder receives these waveforms, it decodes each set of four bits and converts it to the corresponding decimal number in the 7-segment display. When the counter gets to the binary state of 1001, it has counted nine tennis balls, the display shows decimal 9, and a new box is moved under the conveyor belt. Then the counter goes back to its zero state (0000), and the process starts over. (The number 9 was used only in the interest of single-digit simplicity.)

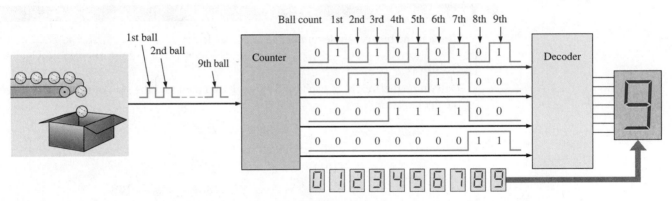

FIGURE 2–1 Illustration of a simple binary counting application.

The Weighting Structure of Binary Numbers

A binary number is a weighted number. The right-most bit is the **LSB** (least significant bit) in a binary whole number and has a weight of $2^0 = 1$. The weights increase from right to left by a power of two for each bit. The left-most bit is the **MSB** (most significant bit); its weight depends on the size of the binary number.

The weight or value of a bit increases from right to left in a binary number.

Fractional numbers can also be represented in binary by placing bits to the right of the binary point, just as fractional decimal digits are placed to the right of the decimal point. The left-most bit is the MSB in a binary fractional number and has a weight of $2^{-1} = 0.5$. The fractional weights decrease from left to right by a negative power of two for each bit.

The weight structure of a binary number is

$$2^{n-1} \ldots 2^3\, 2^2\, 2^1\, 2^0\, .\, 2^{-1}\, 2^{-2} \ldots 2^{-n}$$

$$\underset{\text{Binary point}}{\uparrow}$$

where n is the number of bits from the binary point. Thus, all the bits to the left of the binary point have weights that are positive powers of two, as previously discussed for whole numbers. All bits to the right of the binary point have weights that are negative powers of two, or fractional weights.

The powers of two and their equivalent decimal weights for an 8-bit binary whole number and a 6-bit binary fractional number are shown in Table 2–2. Notice that the weight doubles for each positive power of two and that the weight is halved for each negative power of two. You can easily extend the table by doubling the weight of the most significant positive power of two and halving the weight of the least significant negative power of two; for example, $2^9 = 512$ and $2^{-7} = 0.0078125$.

InfoNote

Processors use binary numbers to select memory locations. Each location is assigned a unique number called an *address*. Some microprocessors, for example, have 32 address lines which can select 2^{32} (4,294,967,296) unique locations.

TABLE 2–2

Binary weights.

Positive Powers of Two (Whole Numbers)									Negative Powers of Two (Fractional Number)					
2^8	2^7	2^6	2^5	2^4	2^3	2^2	2^1	2^0	2^{-1}	2^{-2}	2^{-3}	2^{-4}	2^{-5}	2^{-6}
256	128	64	32	16	8	4	2	1	1/2 0.5	1/4 0.25	1/8 0.125	1/16 0.625	1/32 0.03125	1/64 0.015625

Binary-to-Decimal Conversion

Add the weights of all 1s in a binary number to get the decimal value.

The decimal value of any binary number can be found by adding the weights of all bits that are 1 and discarding the weights of all bits that are 0.

EXAMPLE 2–3

Convert the binary whole number 1101101 to decimal.

Solution

Determine the weight of each bit that is a 1, and then find the sum of the weights to get the decimal number.

$$\text{Weight:} \quad 2^6\ 2^5\ 2^4\ 2^3\ 2^2\ 2^1\ 2^0$$
$$\text{Binary number:} \quad 1\ 1\ 0\ 1\ 1\ 0\ 1$$
$$1101101 = 2^6 + 2^5 + 2^3 + 2^2 + 2^0$$
$$= 64 + 32 + 8 + 4 + 1 = \textbf{109}$$

Related Problem

Convert the binary number 10010001 to decimal.

EXAMPLE 2–4

Convert the fractional binary number 0.1011 to decimal.

Solution

Determine the weight of each bit that is a 1, and then sum the weights to get the decimal fraction.

$$\text{Weight:} \quad 2^{-1}\ \ 2^{-2}\ \ 2^{-3}\ \ 2^{-4}$$
$$\text{Binary number:} \quad 0\ .\ 1\ \ \ 0\ \ \ 1\ \ \ 1$$
$$0.1011 = 2^{-1} + 2^{-3} + 2^{-4}$$
$$= 0.5 + 0.125 + 0.0625 = \textbf{0.6875}$$

Related Problem

Convert the binary number 10.111 to decimal.

SECTION 2–2 CHECKUP

1. What is the largest decimal number that can be represented in binary with eight bits?
2. Determine the weight of the 1 in the binary number 10000.
3. Convert the binary number 10111101.011 to decimal.

2–3 Decimal-to-Binary Conversion

In Section 2–2 you learned how to convert a binary number to the equivalent decimal number. Now you will learn two ways of converting from a decimal number to a binary number.

After completing this section, you should be able to

- ◆ Convert a decimal number to binary using the sum-of-weights method
- ◆ Convert a decimal whole number to binary using the repeated division-by-2 method
- ◆ Convert a decimal fraction to binary using the repeated multiplication-by-2 method

Sum-of-Weights Method

One way to find the binary number that is equivalent to a given decimal number is to determine the set of binary weights whose sum is equal to the decimal number. An easy way to remember binary weights is that the lowest is 1, which is 2^0, and that by doubling any weight, you get the next higher weight; thus, a list of seven binary weights would be 64, 32, 16, 8, 4, 2, 1 as you learned in the last section. The decimal number 9, for example, can be expressed as the sum of binary weights as follows:

$$9 = 8 + 1 \quad \text{or} \quad 9 = 2^3 + 2^0$$

To get the binary number for a given decimal number, find the binary weights that add up to the decimal number.

Placing 1s in the appropriate weight positions, 2^3 and 2^0, and 0s in the 2^2 and 2^1 positions determines the binary number for decimal 9.

$$
\begin{array}{cccc}
2^3 & 2^2 & 2^1 & 2^0 \\
1 & 0 & 0 & 1
\end{array}
\qquad \text{Binary number for decimal 9}
$$

EXAMPLE 2–5

Convert the following decimal numbers to binary:

(a) 12   (b) 25

(c) 58   (d) 82

Solution

(a) $12 = 8 + 4 = 2^3 + 2^2$ $\longrightarrow$ **1100**

(b) $25 = 16 + 8 + 1 = 2^4 + 2^3 + 2^0$ $\longrightarrow$ **11001**

(c) $58 = 32 + 16 + 8 + 2 = 2^5 + 2^4 + 2^3 + 2^1$ $\longrightarrow$ **111010**

(d) $82 = 64 + 16 + 2 = 2^6 + 2^4 + 2^1$ $\longrightarrow$ **1010010**

Related Problem

Convert the decimal number 125 to binary.

Repeated Division-by-2 Method

A systematic method of converting whole numbers from decimal to binary is the *repeated division-by-2* process. For example, to convert the decimal number 12 to binary, begin by dividing 12 by 2. Then divide each resulting quotient by 2 until there is a 0 whole-number quotient. The **remainders** generated by each division form the binary number. The first remainder to be produced is the LSB (least significant bit) in the binary number, and the

To get the binary number for a given decimal number, divide the decimal number by 2 until the quotient is 0. Remainders form the binary number.

last remainder to be produced is the MSB (most significant bit). This procedure is illustrated as follows for converting the decimal number 12 to binary.

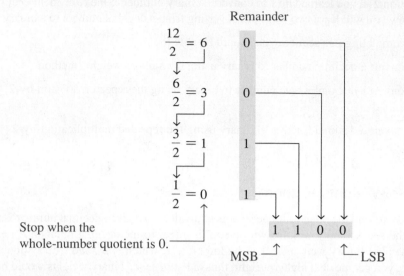

EXAMPLE 2–6

Convert the following decimal numbers to binary:

(a) 19   **(b)** 45

Solution

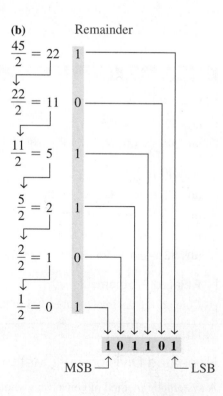

Related Problem

Convert decimal number 39 to binary.

CALCULATOR SESSION

Conversion of a Decimal Number to a Binary Number

Convert decimal 57 to binary.

DEC

TI-36X   **Step 1:**   `3rd` `EE`

Step 2:   `5` `7`

BIN

Step 3:   `3rd` `X`

111001

Converting Decimal Fractions to Binary

Examples 2–5 and 2–6 demonstrated whole-number conversions. Now let's look at fractional conversions. An easy way to remember fractional binary weights is that the most significant weight is 0.5, which is 2^{-1}, and that by halving any weight, you get the next lower weight; thus a list of four fractional binary weights would be 0.5, 0.25, 0.125, 0.0625.

Sum-of-Weights

The sum-of-weights method can be applied to fractional decimal numbers, as shown in the following example:

$$0.625 = 0.5 + 0.125 = 2^{-1} + 2^{-3} = 0.101$$

There is a 1 in the 2^{-1} position, a 0 in the 2^{-2} position, and a 1 in the 2^{-3} position.

Repeated Multiplication by 2

As you have seen, decimal whole numbers can be converted to binary by repeated division by 2. Decimal fractions can be converted to binary by repeated multiplication by 2. For example, to convert the decimal fraction 0.3125 to binary, begin by multiplying 0.3125 by 2 and then multiplying each resulting fractional part of the product by 2 until the fractional product is zero or until the desired number of decimal places is reached. The carry digits, or **carries**, generated by the multiplications produce the binary number. The first carry produced is the MSB, and the last carry is the LSB. This procedure is illustrated as follows:

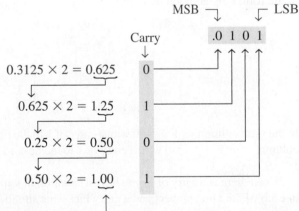

Continue to the desired number of decimal places or stop when the fractional part is all zeros.

SECTION 2–3 CHECKUP

1. Convert each decimal number to binary by using the sum-of-weights method:

   (a) 23   (b) 57   (c) 45.5

2. Convert each decimal number to binary by using the repeated division-by-2 method (repeated multiplication-by-2 for fractions):

   (a) 14   (b) 21   (c) 0.375

2–4   Binary Arithmetic

Binary arithmetic is essential in all digital computers and in many other types of digital systems. To understand digital systems, you must know the basics of binary addition, subtraction, multiplication, and division. This section provides an introduction that will be expanded in later sections.

After completing this section, you should be able to

- Add binary numbers
- Subtract binary numbers
- Multiply binary numbers
- Divide binary numbers

Binary Addition

In binary 1 + 1 = 10, not 2.

The four basic rules for adding binary digits (bits) are as follows:

$$0 + 0 = 0 \qquad \text{Sum of 0 with a carry of 0}$$
$$0 + 1 = 1 \qquad \text{Sum of 1 with a carry of 0}$$
$$1 + 0 = 1 \qquad \text{Sum of 1 with a carry of 0}$$
$$1 + 1 = 10 \qquad \text{Sum of 0 with a carry of 1}$$

Notice that the first three rules result in a single bit and in the fourth rule the addition of two 1s yields a binary two (10). When binary numbers are added, the last condition creates a sum of 0 in a given column and a carry of 1 over to the next column to the left, as illustrated in the following addition of $11 + 1$:

```
      Carry   Carry
       1  ←    1  ←
       0       1      1
      +0       0      1
      ───     ──     ──
       1      └0     └0
```

In the right column, $1 + 1 = 0$ with a carry of 1 to the next column to the left. In the middle column, $1 + 1 + 0 = 0$ with a carry of 1 to the next column to the left. In the left column, $1 + 0 + 0 = 1$.

When there is a carry of 1, you have a situation in which three bits are being added (a bit in each of the two numbers and a carry bit). This situation is illustrated as follows:

Carry bits ─────→

$$1 + 0 + 0 = 01 \qquad \text{Sum of 1 with a carry of 0}$$
$$1 + 1 + 0 = 10 \qquad \text{Sum of 0 with a carry of 1}$$
$$1 + 0 + 1 = 10 \qquad \text{Sum of 0 with a carry of 1}$$
$$1 + 1 + 1 = 11 \qquad \text{Sum of 1 with a carry of 1}$$

EXAMPLE 2–7

Add the following binary numbers:

(a) $11 + 11$ (b) $100 + 10$

(c) $111 + 11$ (d) $110 + 100$

Solution

The equivalent decimal addition is also shown for reference.

(a) $\begin{array}{r} 11 \\ +\,11 \\ \hline \mathbf{110} \end{array}$ $\begin{array}{r} 3 \\ +\,3 \\ \hline 6 \end{array}$ (b) $\begin{array}{r} 100 \\ +\,10 \\ \hline \mathbf{110} \end{array}$ $\begin{array}{r} 4 \\ +\,2 \\ \hline 6 \end{array}$

(c) $\begin{array}{r} 111 \\ +\,11 \\ \hline \mathbf{1010} \end{array}$ $\begin{array}{r} 7 \\ +\,3 \\ \hline 10 \end{array}$ (d) $\begin{array}{r} 110 \\ +\,100 \\ \hline \mathbf{1010} \end{array}$ $\begin{array}{r} 6 \\ +\,4 \\ \hline 10 \end{array}$

Related Problem

Add 1111 and 1100.

Binary Subtraction

The four basic rules for subtracting bits are as follows:

In binary $10 - 1 = 1$, not 9.

$$0 - 0 = 0$$
$$1 - 1 = 0$$
$$1 - 0 = 1$$
$$10 - 1 = 1 \qquad 0 - 1 \text{ with a borrow of } 1$$

When subtracting numbers, you sometimes have to borrow from the next column to the left. A borrow is required in binary only when you try to subtract a 1 from a 0. In this case, when a 1 is borrowed from the next column to the left, a 10 is created in the column being subtracted, and the last of the four basic rules just listed must be applied. Examples 2–8 and 2–9 illustrate binary subtraction; the equivalent decimal subtractions are also shown.

EXAMPLE 2–8

Perform the following binary subtractions:

(a) $11 - 01$ (b) $11 - 10$

Solution

(a) $\begin{array}{r} 11 \\ -\,01 \\ \hline \mathbf{10} \end{array}$ $\begin{array}{r} 3 \\ -\,1 \\ \hline 2 \end{array}$ (b) $\begin{array}{r} 11 \\ -\,10 \\ \hline \mathbf{01} \end{array}$ $\begin{array}{r} 3 \\ -\,2 \\ \hline 1 \end{array}$

No borrows were required in this example. The binary number 01 is the same as 1.

Related Problem

Subtract 100 from 111.

EXAMPLE 2–9

Subtract 011 from 101.

Solution

$\begin{array}{r} 101 \\ -\,011 \\ \hline \mathbf{010} \end{array}$ $\begin{array}{r} 5 \\ -\,3 \\ \hline 2 \end{array}$

Let's examine exactly what was done to subtract the two binary numbers since a borrow is required. Begin with the right column.

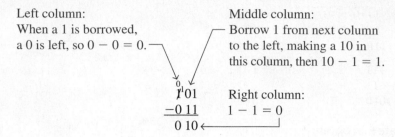

Left column:
When a 1 is borrowed,
a 0 is left, so $0 - 0 = 0$.

Middle column:
Borrow 1 from next column
to the left, making a 10 in
this column, then $10 - 1 = 1$.

$$\begin{array}{r} \overset{0}{\cancel{1}}01 \\ -0\,11 \\ \hline 0\,10 \end{array}$$

Right column:
$1 - 1 = 0$

Related Problem
Subtract 101 from 110.

Binary Multiplication

Binary multiplication of two bits is the same as multiplication of the decimal digits 0 and 1.

The four basic rules for multiplying bits are as follows:

$$0 \times 0 = 0$$
$$0 \times 1 = 0$$
$$1 \times 0 = 0$$
$$1 \times 1 = 1$$

Multiplication is performed with binary numbers in the same manner as with decimal numbers. It involves forming partial products, shifting each successive partial product left one place, and then adding all the partial products. Example 2–10 illustrates the procedure; the equivalent decimal multiplications are shown for reference.

EXAMPLE 2–10

Perform the following binary multiplications:

(a) 11×11 **(b)** 101×111

Solution

(a)

$$\begin{array}{r} 11 \\ \times\,11 \\ \hline 11 \\ +11 \\ \hline \mathbf{1001} \end{array} \qquad \begin{array}{r} 3 \\ \times\,3 \\ \hline 9 \end{array}$$

Partial products $\begin{cases} \end{cases}$

(b)

$$\begin{array}{r} 111 \\ \times\,101 \\ \hline 111 \\ 000 \\ +111 \\ \hline \mathbf{100011} \end{array} \qquad \begin{array}{r} 7 \\ \times\,5 \\ \hline 35 \end{array}$$

Partial products $\begin{cases} \end{cases}$

Related Problem
Multiply 1101×1010.

Binary Division

A calculator can be used to perform arithmetic operations with binary numbers as long as the capacity of the calculator is not exceeded.

Division in binary follows the same procedure as division in decimal, as Example 2–11 illustrates. The equivalent decimal divisions are also given.

EXAMPLE 2–11

Perform the following binary divisions:

(a) $110 \div 11$ **(b)** $110 \div 10$

Solution

10	2	**11**	3

(a) 11)110   3)6   (b) 10)110   2)6

            11   6   10   6
           ―――   ―   ―――   ―
           000   0   10   0
                                   10
                                  ―――
                                   00

Related Problem

Divide 1100 by 100.

---

1. Perform the following binary additions:

   (a) 1101 + 1010   (b) 10111 + 01101

2. Perform the following binary subtractions:

   (a) 1101 − 0100   (b) 1001 − 0111

3. Perform the indicated binary operations:

   (a) 110 × 111   (b) 1100 ÷ 011

2–5 Complements of Binary Numbers

The 1's complement and the 2's complement of a binary number are important because they permit the representation of negative numbers. The method of 2's complement arithmetic is commonly used in computers to handle negative numbers.

After completing this section, you should be able to

- ◆ Convert a binary number to its 1's complement
- ◆ Convert a binary number to its 2's complement using either of two methods

Finding the 1's Complement

The 1's **complement** of a binary number is found by changing all 1s to 0s and all 0s to 1s, as illustrated below:

Change each bit in a number to get the 1's complement.

$$1\ 0\ 1\ 1\ 0\ 0\ 1\ 0 \qquad \text{Binary number}$$
$$\downarrow\downarrow\downarrow\downarrow\downarrow\downarrow\downarrow\downarrow$$
$$0\ 1\ 0\ 0\ 1\ 1\ 0\ 1 \qquad \text{1's complement}$$

The simplest way to obtain the 1's complement of a binary number with a digital circuit is to use parallel inverters (NOT circuits), as shown in Figure 2–2 for an 8-bit binary number.

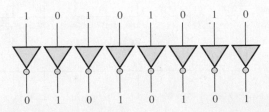

FIGURE 2–2   Example of inverters used to obtain the 1's complement of a binary number.

Finding the 2's Complement

Add 1 to the 1's complement to get the 2's complement.

The 2's complement of a binary number is found by adding 1 to the LSB of the 1's complement.

$$\text{2's complement} = \text{(1's complement)} + 1$$

EXAMPLE 2–12

Find the 2's complement of 10110010.

Solution

$$
\begin{array}{ll}
10110010 & \text{Binary number} \\
01001101 & \text{1's complement} \\
\underline{+1} & \text{Add 1} \\
\mathbf{01001110} & \text{2's complement}
\end{array}
$$

Related Problem

Determine the 2's complement of 11001011.

Change all bits to the left of the least significant 1 to get 2's complement.

An alternative method of finding the 2's complement of a binary number is as follows:

1. Start at the right with the LSB and write the bits as they are up to and including the first 1.

2. Take the 1's complements of the remaining bits.

EXAMPLE 2–13

Find the 2's complement of 10111000 using the alternative method.

Solution

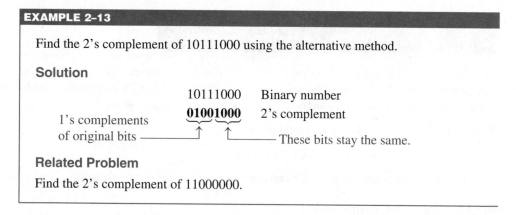

1's complements of original bits ——————↑ ↑—————— These bits stay the same.

10111000   Binary number
01001000   2's complement

Related Problem

Find the 2's complement of 11000000.

The 2's complement of a negative binary number can be realized using inverters and an adder, as indicated in Figure 2–3. This illustrates how an 8-bit number can be converted to its 2's complement by first inverting each bit (taking the 1's complement) and then adding 1 to the 1's complement with an adder circuit.

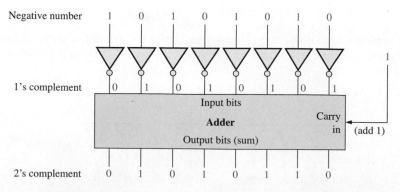

FIGURE 2–3 Example of obtaining the 2's complement of a negative binary number.

To convert from a 1's or 2's complement back to the true (uncomplemented) binary form, use the same two procedures described previously. To go from the 1's complement back to true binary, reverse all the bits. To go from the 2's complement form back to true binary, take the 1's complement of the 2's complement number and add 1 to the least significant bit.

SECTION 2–5 CHECKUP

1. Determine the 1's complement of each binary number:

   (a) 00011010　　(b) 11110111　　(c) 10001101

2. Determine the 2's complement of each binary number:

   (a) 00010110　　(b) 11111100　　(c) 10010001

2–6　Signed Numbers

Digital systems, such as the computer, must be able to handle both positive and negative numbers. A signed binary number consists of both sign and magnitude information. The sign indicates whether a number is positive or negative, and the magnitude is the value of the number. There are three forms in which signed integer (whole) numbers can be represented in binary: sign-magnitude, 1's complement, and 2's complement. Of these, the 2's complement is the most important and the sign-magnitude is the least used. Noninteger and very large or small numbers can be expressed in floating-point format.

After completing this section, you should be able to

- ◆ Express positive and negative numbers in sign-magnitude
- ◆ Express positive and negative numbers in 1's complement
- ◆ Express positive and negative numbers in 2's complement
- ◆ Determine the decimal value of signed binary numbers
- ◆ Express a binary number in floating-point format

The Sign Bit

The left-most bit in a signed binary number is the **sign bit**, which tells you whether the number is positive or negative.

A 0 sign bit indicates a positive number, and a 1 sign bit indicates a negative number.

Sign-Magnitude Form

When a signed binary number is represented in sign-magnitude, the left-most bit is the sign bit and the remaining bits are the magnitude bits. The magnitude bits are in true (uncomplemented) binary for both positive and negative numbers. For example, the decimal number +25 is expressed as an 8-bit signed binary number using the sign-magnitude form as

$$00011001$$

Sign bit ——↑　↑——Magnitude bits

The decimal number −25 is expressed as

$$10011001$$

Notice that the only difference between +25 and −25 is the sign bit because the magnitude bits are in true binary for both positive and negative numbers.

In the sign-magnitude form, a negative number has the same magnitude bits as the corresponding positive number but the sign bit is a 1 rather than a zero.

1's Complement Form

Positive numbers in 1's complement form are represented the same way as the positive sign-magnitude numbers. Negative numbers, however, are the 1's complements of the corresponding positive numbers. For example, using eight bits, the decimal number −25 is expressed as the 1's complement of +25 (00011001) as

$$11100110$$

In the 1's complement form, a negative number is the 1's complement of the corresponding positive number.

2's Complement Form

Positive numbers in 2's complement form are represented the same way as in the sign-magnitude and 1's complement forms. Negative numbers are the 2's complements of the corresponding positive numbers. Again, using eight bits, let's take decimal number −25 and express it as the 2's complement of +25 (00011001). Inverting each bit and adding 1, you get

$$-25 = 11100111$$

In the 2's complement form, a negative number is the 2's complement of the corresponding positive number.

EXAMPLE 2–14

Express the decimal number −39 as an 8-bit number in the sign-magnitude, 1's complement, and 2's complement forms.

Solution

First, write the 8-bit number for +39.

$$00100111$$

In the *sign-magnitude form*, −39 is produced by changing the sign bit to a 1 and leaving the magnitude bits as they are. The number is

$$\mathbf{10100111}$$

In the *1's complement form*, −39 is produced by taking the 1's complement of +39 (00100111).

$$\mathbf{11011000}$$

In the *2's complement form*, −39 is produced by taking the 2's complement of +39 (00100111) as follows:

```
  11011000    1's complement
+        1
  11011001    2's complement
```

Related Problem

Express +19 and −19 as 8-bit numbers in sign-magnitude, 1's complement, and 2's complement.

The Decimal Value of Signed Numbers

Sign-Magnitude

Decimal values of positive and negative numbers in the sign-magnitude form are determined by summing the weights in all the magnitude bit positions where there are 1s and ignoring those positions where there are zeros. The sign is determined by examination of the sign bit.

EXAMPLE 2–15

Determine the decimal value of this signed binary number expressed in sign-magnitude: 10010101.

Solution

The seven magnitude bits and their powers-of-two weights are as follows:

$$2^6 \quad 2^5 \quad 2^4 \quad 2^3 \quad 2^2 \quad 2^1 \quad 2^0$$
$$0 \quad\; 0 \quad\; 1 \quad\; 0 \quad\; 1 \quad\; 0 \quad\; 1$$

Summing the weights where there are 1s,

$$16 + 4 + 1 = 21$$

The sign bit is 1; therefore, the decimal number is **−21**.

Related Problem

Determine the decimal value of the sign-magnitude number 01110111.

1's Complement

Decimal values of positive numbers in the 1's complement form are determined by summing the weights in all bit positions where there are 1s and ignoring those positions where there are zeros. Decimal values of negative numbers are determined by assigning a negative value to the weight of the sign bit, summing all the weights where there are 1s, and adding 1 to the result.

EXAMPLE 2–16

Determine the decimal values of the signed binary numbers expressed in 1's complement:

(a) 00010111   **(b)** 11101000

Solution

(a) The bits and their powers-of-two weights for the positive number are as follows:

$$-2^7 \quad 2^6 \quad 2^5 \quad 2^4 \quad 2^3 \quad 2^2 \quad 2^1 \quad 2^0$$
$$0 \quad\;\; 0 \quad\; 0 \quad\; 1 \quad\; 0 \quad\; 1 \quad\; 1 \quad\; 1$$

Summing the weights where there are 1s,

$$16 + 4 + 2 + 1 = \mathbf{+23}$$

(b) The bits and their powers-of-two weights for the negative number are as follows. Notice that the negative sign bit has a weight of -2^7 or -128.

$$-2^7 \quad 2^6 \quad 2^5 \quad 2^4 \quad 2^3 \quad 2^2 \quad 2^1 \quad 2^0$$
$$1 \quad\;\; 1 \quad\; 1 \quad\; 0 \quad\; 1 \quad\; 0 \quad\; 0 \quad\; 0$$

Summing the weights where there are 1s,

$$-128 + 64 + 32 + 8 = -24$$

Adding 1 to the result, the final decimal number is

$$-24 + 1 = \mathbf{-23}$$

Related Problem

Determine the decimal value of the 1's complement number 11101011.

2's Complement

Decimal values of positive and negative numbers in the 2's complement form are determined by summing the weights in all bit positions where there are 1s and ignoring those positions where there are zeros. The weight of the sign bit in a negative number is given a negative value.

EXAMPLE 2–17

Determine the decimal values of the signed binary numbers expressed in 2's complement:

(a) 01010110   (b) 10101010

Solution

(a) The bits and their powers-of-two weights for the positive number are as follows:

$$-2^7 \quad 2^6 \quad 2^5 \quad 2^4 \quad 2^3 \quad 2^2 \quad 2^1 \quad 2^0$$
$$\;\;\;0 \quad\;\; 1 \quad\;\; 0 \quad\;\; 1 \quad\;\; 0 \quad\;\; 1 \quad\;\; 1 \quad\;\; 0$$

Summing the weights where there are 1s,

$$64 + 16 + 4 + 2 = \mathbf{+86}$$

(b) The bits and their powers-of-two weights for the negative number are as follows. Notice that the negative sign bit has a weight of $-2^7 = -128$.

$$-2^7 \quad 2^6 \quad 2^5 \quad 2^4 \quad 2^3 \quad 2^2 \quad 2^1 \quad 2^0$$
$$\;\;\;1 \quad\;\; 0 \quad\;\; 1 \quad\;\; 0 \quad\;\; 1 \quad\;\; 0 \quad\;\; 1 \quad\;\; 0$$

Summing the weights where there are 1s,

$$-128 + 32 + 8 + 2 = \mathbf{-86}$$

Related Problem

Determine the decimal value of the 2's complement number 11010111.

From these examples, you can see why the 2's complement form is preferred for representing signed integer numbers: To convert to decimal, it simply requires a summation of weights regardless of whether the number is positive or negative. The 1's complement system requires adding 1 to the summation of weights for negative numbers but not for positive numbers. Also, the 1's complement form is generally not used because two representations of zero (00000000 or 11111111) are possible.

Range of Signed Integer Numbers

The range of magnitude values represented by binary numbers depends on the number of bits (n).

We have used 8-bit numbers for illustration because the 8-bit grouping is common in most computers and has been given the special name **byte**. With one byte or eight bits, you can represent 256 different numbers. With two bytes or sixteen bits, you can represent 65,536 different numbers. With four bytes or 32 bits, you can represent 4.295×10^9 different numbers. The formula for finding the number of different combinations of n bits is

$$\text{Total combinations} = 2^n$$

For 2's complement signed numbers, the range of values for n-bit numbers is

$$\text{Range} = -(2^{n-1}) \text{ to } +(2^{n-1} - 1)$$

where in each case there is one sign bit and $n - 1$ magnitude bits. For example, with four bits you can represent numbers in 2's complement ranging from $-(2^3) = -8$ to $2^3 - 1 = +7$. Similarly, with eight bits you can go from -128 to $+127$, with sixteen bits you can go from

−32,768 to +32,767, and so on. There is one less positive number than there are negative numbers because zero is represented as a positive number (all zeros).

Floating-Point Numbers

To represent very large **integer** (whole) numbers, many bits are required. There is also a problem when numbers with both integer and fractional parts, such as 23.5618, need to be represented. The floating-point number system, based on scientific notation, is capable of representing very large and very small numbers without an increase in the number of bits and also for representing numbers that have both integer and fractional components.

A **floating-point number** (also known as a *real number*) consists of two parts plus a sign. The **mantissa** is the part of a floating-point number that represents the magnitude of the number and is between 0 and 1. The **exponent** is the part of a floating-point number that represents the number of places that the decimal point (or binary point) is to be moved.

A decimal example will be helpful in understanding the basic concept of floating-point numbers. Let's consider a decimal number which, in integer form, is 241,506,800. The mantissa is .2415068 and the exponent is 9. When the integer is expressed as a floating-point number, it is normalized by moving the decimal point to the left of all the digits so that the mantissa is a fractional number and the exponent is the power of ten. The floating-point number is written as

$$0.2415068 \times 10^9$$

For binary floating-point numbers, the format is defined by ANSI/IEEE Standard 754-1985 in three forms: *single-precision, double-precision,* and *extended-precision.* These all have the same basic formats except for the number of bits. Single-precision floating-point numbers have 32 bits, double-precision numbers have 64 bits, and extended-precision numbers have 80 bits. We will restrict our discussion to the single-precision floating-point format.

Single-Precision Floating-Point Binary Numbers

In the standard format for a single-precision binary number, the sign bit (S) is the left-most bit, the exponent (E) includes the next eight bits, and the mantissa or fractional part (F) includes the remaining 23 bits, as shown next.

←	32 bits	→
S	Exponent (E)	Mantissa (fraction, F)
1 bit	8 bits	23 bits

In the mantissa or fractional part, the binary point is understood to be to the left of the 23 bits. Effectively, there are 24 bits in the mantissa because in any binary number the left-most (most significant) bit is always a 1. Therefore, this 1 is understood to be there although it does not occupy an actual bit position.

The eight bits in the exponent represent a *biased exponent,* which is obtained by adding 127 to the actual exponent. The purpose of the bias is to allow very large or very small numbers without requiring a separate sign bit for the exponents. The biased exponent allows a range of actual exponent values from −126 to +128.

To illustrate how a binary number is expressed in floating-point format, let's use 1011010010001 as an example. First, it can be expressed as 1 plus a fractional binary number by moving the binary point 12 places to the left and then multiplying by the appropriate power of two.

$$1011010010001 = 1.011010010001 \times 2^{12}$$

Assuming that this is a positive number, the sign bit (S) is 0. The exponent, 12, is expressed as a biased exponent by adding it to 127 (12 + 127 = 139). The biased exponent (E) is expressed as the binary number 10001011. The mantissa is the fractional part (F) of the binary number, .011010010001. Because there is always a 1 to the left of the binary point

InfoNote

In addition to the CPU (central processing unit), computers use *coprocessors* to perform complicated mathematical calculations using floating-point numbers. The purpose is to increase performance by freeing up the CPU for other tasks. The mathematical coprocessor is also known as the floating-point unit (FPU).

in the power-of-two expression, it is not included in the mantissa. The complete floating-point number is

S	E	F
0	10001011	01101001000100000000000

Next, let's see how to evaluate a binary number that is already in floating-point format. The general approach to determining the value of a floating-point number is expressed by the following formula:

$$\text{Number} = (-1)^S (1 + F)(2^{E-127})$$

To illustrate, let's consider the following floating-point binary number:

S	E	F
1	10010001	10001110001000000000000

The sign bit is 1. The biased exponent is $10010001 = 145$. Applying the formula, we get

$$\text{Number} = (-1)^1 (1.10001110001)(2^{145-127})$$
$$= (-1)(1.10001110001)(2^{18}) = -1100011100010000000$$

This floating-point binary number is equivalent to $-407,688$ in decimal. Since the exponent can be any number between -126 and $+128$, extremely large and small numbers can be expressed. A 32-bit floating-point number can replace a binary integer number having 129 bits. Because the exponent determines the position of the binary point, numbers containing both integer and fractional parts can be represented.

There are two exceptions to the format for floating-point numbers: The number 0.0 is represented by all 0s, and infinity is represented by all 1s in the exponent and all 0s in the mantissa.

EXAMPLE 2–18

Convert the decimal number 3.248×10^4 to a single-precision floating-point binary number.

Solution

Convert the decimal number to binary.

$$3.248 \times 10^4 = 32480 = 111111011100000_2 = 1.11111011100000 \times 2^{14}$$

The MSB will not occupy a bit position because it is always a 1. Therefore, the mantissa is the fractional 23-bit binary number 11111011100000000000000 and the biased exponent is

$$14 + 127 = 141 = 10001101_2$$

The complete floating-point number is

0	10001101	11111011100000000000000

Related Problem

Determine the binary value of the following floating-point binary number:

0 10011000 10000100010100110000000

SECTION 2–6 CHECKUP

1. Express the decimal number $+9$ as an 8-bit binary number in the sign-magnitude system.

2. Express the decimal number -33 as an 8-bit binary number in the 1's complement system.

3. Express the decimal number -46 as an 8-bit binary number in the 2's complement system.

4. List the three parts of a signed, floating-point number.

2–7   Arithmetic Operations with Signed Numbers

In the last section, you learned how signed numbers are represented in three different forms. In this section, you will learn how signed numbers are added, subtracted, multiplied, and divided. Because the 2's complement form for representing signed numbers is the most widely used in computers and microprocessor-based systems, the coverage in this section is limited to 2's complement arithmetic. The processes covered can be extended to the other forms if necessary.

After completing this section, you should be able to

- Add signed binary numbers
- Define *overflow*
- Explain how computers add strings of numbers
- Subtract signed binary numbers
- Multiply signed binary numbers using the direct addition method
- Multiply signed binary numbers using the partial products method
- Divide signed binary numbers

Addition

The two numbers in an addition are the **addend** and the **augend**. The result is the **sum**. There are four cases that can occur when two signed binary numbers are added.

1. Both numbers positive
2. Positive number with magnitude larger than negative number
3. Negative number with magnitude larger than positive number
4. Both numbers negative

Let's take one case at a time using 8-bit signed numbers as examples. The equivalent decimal numbers are shown for reference.

Both numbers positive:

$$
\begin{array}{rr}
00000111 & 7 \\
+\ 00000100 & +\ 4 \\
\hline
00001011 & 11
\end{array}
$$

Addition of two positive numbers yields a positive number.

The sum is positive and is therefore in true (uncomplemented) binary.

Positive number with magnitude larger than negative number:

$$
\begin{array}{rr}
00001111 & 15 \\
+\ 11111010 & +\ -6 \\
\hline
\text{Discard carry} \longrightarrow 1\ \ 00001001 & 9
\end{array}
$$

Addition of a positive number and a smaller negative number yields a positive number.

The final carry bit is discarded. The sum is positive and therefore in true (uncomplemented) binary.

Negative number with magnitude larger than positive number:

$$
\begin{array}{rr}
00010000 & 16 \\
+\ 11101000 & +\ -24 \\
\hline
11111000 & -8
\end{array}
$$

Addition of a positive number and a larger negative number or two negative numbers yields a negative number in 2's complement.

The sum is negative and therefore in 2's complement form.

Both numbers negative:

$$
\begin{array}{rr}
11111011 & -5 \\
+\ 11110111 & +\ -9 \\
\hline
\text{Discard carry} \longrightarrow 1\ \ 11110010 & -14
\end{array}
$$

The final carry bit is discarded. The sum is negative and therefore in 2's complement form.

In a computer, the negative numbers are stored in 2's complement form so, as you can see, the addition process is very simple: *Add the two numbers and discard any final carry bit.*

Overflow Condition

When two numbers are added and the number of bits required to represent the sum exceeds the number of bits in the two numbers, an **overflow** results as indicated by an incorrect sign bit. An overflow can occur only when both numbers are positive or both numbers are negative. If the sign bit of the result is different than the sign bit of the numbers that are added, overflow is indicated. The following 8-bit example will illustrate this condition.

$$
\begin{array}{rr}
01111101 & 125 \\
+\ 00111010 & +\ 58 \\
\hline
10110111 & 183
\end{array}
$$

Sign incorrect ⟶
Magnitude incorrect ⟶

In this example the sum of 183 requires eight magnitude bits. Since there are seven magnitude bits in the numbers (one bit is the sign), there is a carry into the sign bit which produces the overflow indication.

Numbers Added Two at a Time

Now let's look at the addition of a string of numbers, added two at a time. This can be accomplished by adding the first two numbers, then adding the third number to the sum of the first two, then adding the fourth number to this result, and so on. This is how computers add strings of numbers. The addition of numbers taken two at a time is illustrated in Example 2–19.

EXAMPLE 2–19

Add the signed numbers: 01000100, 00011011, 00001110, and 00010010.

Solution

The equivalent decimal additions are given for reference.

$$
\begin{array}{rll}
68 & 01000100 & \\
+\ 27 & +\ 00011011 & \text{Add 1st two numbers} \\
\hline
95 & 01011111 & \text{1st sum} \\
+\ 14 & +\ 00001110 & \text{Add 3rd number} \\
\hline
109 & 01101101 & \text{2nd sum} \\
+\ 18 & +\ 00010010 & \text{Add 4th number} \\
\hline
127 & \mathbf{01111111} & \text{Final sum}
\end{array}
$$

Related Problem

Add 00110011, 10111111, and 01100011. These are signed numbers.

Subtraction

Subtraction is addition with the sign of the subtrahend changed.

Subtraction is a special case of addition. For example, subtracting +6 (the **subtrahend**) from +9 (the **minuend**) is equivalent to adding −6 to +9. Basically, *the subtraction operation changes the sign of the subtrahend and adds it to the minuend.* The result of a subtraction is called the **difference**.

The sign of a positive or negative binary number is changed by taking its 2's complement.

For example, when you take the 2's complement of the positive number 00000100 (+4), you get 11111100, which is −4 as the following sum-of-weights evaluation shows:

$$-128 + 64 + 32 + 16 + 8 + 4 = -4$$

As another example, when you take the 2's complement of the negative number 11101101 (−19), you get 00010011, which is +19 as the following sum-of-weights evaluation shows:

$$16 + 2 + 1 = 19$$

Since subtraction is simply an addition with the sign of the subtrahend changed, the process is stated as follows:

To subtract two signed numbers, take the 2's complement of the subtrahend and add. Discard any final carry bit.

Example 2–20 illustrates the subtraction process.

> When you subtract two binary numbers with the 2's complement method, it is important that both numbers have the same number of bits.

EXAMPLE 2–20

Perform each of the following subtractions of the signed numbers:

(a) 00001000 − 00000011   (b) 00001100 − 11110111

(c) 11100111 − 00010011   (d) 10001000 − 11100010

Solution

Like in other examples, the equivalent decimal subtractions are given for reference.

(a) In this case, $8 - 3 = 8 + (-3) = 5$.

	00001000	Minuend (+8)
	+ 11111101	2's complement of subtrahend (−3)
Discard carry ⟶	**1 00000101**	Difference (+5)

(b) In this case, $12 - (-9) = 12 + 9 = 21$.

00001100	Minuend (+12)
+ 00001001	2's complement of subtrahend (+9)
00010101	Difference (+21)

(c) In this case, $-25 - (+19) = -25 + (-19) = -44$.

	11100111	Minuend (−25)
	+ 11101101	2's complement of subtrahend (−19)
Discard carry	**1 11010100**	Difference (−44)

(d) In this case, $-120 - (-30) = -120 + 30 = -90$.

10001000	Minuend (−120)
+ 00011110	2's complement of subtrahend (+30)
10100110	Difference (−90)

Related Problem

Subtract 01000111 from 01011000.

Multiplication

The numbers in a multiplication are the **multiplicand**, the **multiplier**, and the **product**. These are illustrated in the following decimal multiplication:

$$
\begin{array}{rl}
8 & \text{Multiplicand} \\
\times\,3 & \text{Multiplier} \\
\hline
24 & \text{Product}
\end{array}
$$

Multiplication is equivalent to adding a number to itself a number of times equal to the multiplier.

The multiplication operation in most computers is accomplished using addition. As you have already seen, subtraction is done with an adder; now let's see how multiplication is done.

Direct addition and *partial products* are two basic methods for performing multiplication using addition. In the direct addition method, you add the multiplicand a number of times equal to the multiplier. In the previous decimal example (8×3), three multiplicands are added: $8 + 8 + 8 = 24$. The disadvantage of this approach is that it becomes very lengthy if the multiplier is a large number. For example, to multiply 350×75, you must add 350 to itself 75 times. Incidentally, this is why the term *times* is used to mean multiply.

When two binary numbers are multiplied, both numbers must be in true (uncomplemented) form. The direct addition method is illustrated in Example 2–21 adding two binary numbers at a time.

EXAMPLE 2–21

Multiply the signed binary numbers: 01001101 (multiplicand) and 00000100 (multiplier) using the direct addition method.

Solution

Since both numbers are positive, they are in true form, and the product will be positive. The decimal value of the multiplier is 4, so the multiplicand is added to itself four times as follows:

$$
\begin{array}{rl}
01001101 & \text{1st time} \\
+\,01001101 & \text{2nd time} \\
\hline
10011010 & \text{Partial sum} \\
+\,01001101 & \text{3rd time} \\
\hline
11100111 & \text{Partial sum} \\
+\,01001101 & \text{4th time} \\
\hline
\mathbf{100110100} & \text{Product}
\end{array}
$$

Since the sign bit of the multiplicand is 0, it has no effect on the outcome. All of the bits in the product are magnitude bits.

Related Problem

Multiply 01100001 by 00000110 using the direct addition method.

The partial products method is perhaps the more common one because it reflects the way you multiply longhand. The multiplicand is multiplied by each multiplier digit beginning with the least significant digit. The result of the multiplication of the multiplicand by a multiplier digit is called a *partial product*. Each successive partial product is moved (shifted) one place to the left and when all the partial products have been produced, they are added to get the final product. Here is a decimal example.

$$
\begin{array}{rl}
239 & \text{Multiplicand} \\
\times\,123 & \text{Multiplier} \\
\hline
717 & \text{1st partial product } (3 \times 239) \\
478 & \text{2nd partial product } (2 \times 239) \\
+\,239 & \text{3rd partial product } (1 \times 239) \\
\hline
29{,}397 & \text{Final product}
\end{array}
$$

The sign of the product of a multiplication depends on the signs of the multiplicand and the multiplier according to the following two rules:

- **If the signs are the same, the product is positive.**
- **If the signs are different, the product is negative.**

The basic steps in the partial products method of binary multiplication are as follows:

Step 1: Determine if the signs of the multiplicand and multiplier are the same or different. This determines what the sign of the product will be.

Step 2: Change any negative number to true (uncomplemented) form. Because most computers store negative numbers in 2's complement, a 2's complement operation is required to get the negative number into true form.

Step 3: Starting with the least significant multiplier bit, generate the partial products. When the multiplier bit is 1, the partial product is the same as the multiplicand. When the multiplier bit is 0, the partial product is zero. Shift each successive partial product one bit to the left.

Step 4: Add each successive partial product to the sum of the previous partial products to get the final product.

Step 5: If the sign bit that was determined in step 1 is negative, take the 2's complement of the product. If positive, leave the product in true form. Attach the sign bit to the product.

EXAMPLE 2–22

Multiply the signed binary numbers: 01010011 (multiplicand) and 11000101 (multiplier).

Solution

Step 1: The sign bit of the multiplicand is 0 and the sign bit of the multiplier is 1. The sign bit of the product will be 1 (negative).

Step 2: Take the 2's complement of the multiplier to put it in true form.

$$11000101 \longrightarrow 00111011$$

Step 3 and 4: The multiplication proceeds as follows. Notice that only the magnitude bits are used in these steps.

1010011	Multiplicand
× 0111011	Multiplier
1010011	1st partial product
+ 1010011	2nd partial product
11111001	Sum of 1st and 2nd
+ 0000000	3rd partial product
011111001	Sum
+ 1010011	4th partial product
1110010001	Sum
+ 1010011	5th partial product
100011000001	Sum
+ 1010011	6th partial product
1001100100001	Sum
+ 0000000	7th partial product
1001100100001	Final product

Step 5: Since the sign of the product is a 1 as determined in step 1, take the 2's complement of the product.

$$1001100100001 \longrightarrow 0110011011111$$

Attach the sign bit ———┐
                         ↓

1 0110011011111

Related Problem

Verify the multiplication is correct by converting to decimal numbers and performing the multiplication.

Division

The numbers in a division are the **dividend**, the **divisor**, and the **quotient**. These are illustrated in the following standard division format.

$$\frac{\text{dividend}}{\text{divisor}} = \text{quotient}$$

The division operation in computers is accomplished using subtraction. Since subtraction is done with an adder, division can also be accomplished with an adder.

The result of a division is called the *quotient;* the quotient is the number of times that the divisor will go into the dividend. This means that the divisor can be subtracted from the dividend a number of times equal to the quotient, as illustrated by dividing 21 by 7.

21	Dividend
− 7	1st subtraction of divisor
14	1st partial remainder
− 7	2nd subtraction of divisor
7	2nd partial remainder
− 7	3rd subtraction of divisor
0	Zero remainder

In this simple example, the divisor was subtracted from the dividend three times before a remainder of zero was obtained. Therefore, the quotient is 3.

The sign of the quotient depends on the signs of the dividend and the divisor according to the following two rules:

- **If the signs are the same, the quotient is positive.**
- **If the signs are different, the quotient is negative.**

When two binary numbers are divided, both numbers must be in true (uncomplemented) form. The basic steps in a division process are as follows:

Step 1: Determine if the signs of the dividend and divisor are the same or different. This determines what the sign of the quotient will be. Initialize the quotient to zero.

Step 2: Subtract the divisor from the dividend using 2's complement addition to get the first partial remainder and add 1 to the quotient. If this partial remainder is positive, go to step 3. If the partial remainder is zero or negative, the division is complete.

Step 3: Subtract the divisor from the partial remainder and add 1 to the quotient. If the result is positive, repeat for the next partial remainder. If the result is zero or negative, the division is complete.

Continue to subtract the divisor from the dividend and the partial remainders until there is a zero or a negative result. Count the number of times that the divisor is subtracted and you have the quotient. Example 2–23 illustrates these steps using 8-bit signed binary numbers.

EXAMPLE 2–23

Divide 01100100 by 00011001.

Solution

Step 1: The signs of both numbers are positive, so the quotient will be positive. The quotient is initially zero: 00000000.

Step 2: Subtract the divisor from the dividend using 2's complement addition (remember that final carries are discarded).

$$
\begin{array}{ll}
01100100 & \text{Dividend} \\
+\ 11100111 & \text{2's complement of divisor} \\
\hline
01001011 & \text{Positive 1st partial remainder}
\end{array}
$$

Add 1 to quotient: 00000000 + 00000001 = 00000001.

Step 3: Subtract the divisor from the 1st partial remainder using 2's complement addition.

$$
\begin{array}{ll}
01001011 & \text{1st partial remainder} \\
+\ 11100111 & \text{2's complement of divisor} \\
\hline
00110010 & \text{Positive 2nd partial remainder}
\end{array}
$$

Add 1 to quotient: 00000001 + 00000001 = 00000010.

Step 4: Subtract the divisor from the 2nd partial remainder using 2's complement addition.

$$
\begin{array}{ll}
00110010 & \text{2nd partial remainder} \\
+\ 11100111 & \text{2's complement of divisor} \\
\hline
00011001 & \text{Positive 3rd partial remainder}
\end{array}
$$

Add 1 to quotient: 00000010 + 00000001 = 00000011.

Step 5: Subtract the divisor from the 3rd partial remainder using 2's complement addition.

$$
\begin{array}{ll}
00011001 & \text{3rd partial remainder} \\
+\ 11100111 & \text{2's complement of divisor} \\
\hline
00000000 & \text{Zero remainder}
\end{array}
$$

Add 1 to quotient: 00000011 + 00000001 = **00000100** (final quotient). The process is complete.

Related Problem

Verify that the process is correct by converting to decimal numbers and performing the division.

SECTION 2–7 CHECKUP

1. List the four cases when numbers are added.

2. Add the signed numbers 00100001 and 10111100.

3. Subtract the signed numbers 00110010 from 01110111.

4. What is the sign of the product when two negative numbers are multiplied?

5. Multiply 01111111 by 00000101.

6. What is the sign of the quotient when a positive number is divided by a negative number?

7. Divide 00110000 by 00001100.

2-8   Hexadecimal Numbers

The hexadecimal number system has sixteen characters; it is used primarily as a compact way of displaying or writing binary numbers because it is very easy to convert between binary and hexadecimal. As you are probably aware, long binary numbers are difficult to read and write because it is easy to drop or transpose a bit. Since computers and microprocessors understand only 1s and 0s, it is necessary to use these digits when you program in "machine language." Imagine writing a sixteen bit instruction for a microprocessor system in 1s and 0s. It is much more efficient to use hexadecimal or octal; octal numbers are covered in Section 2–9. Hexadecimal is widely used in computer and microprocessor applications.

After completing this section, you should be able to

- ◆ List the hexadecimal characters
- ◆ Count in hexadecimal
- ◆ Convert from binary to hexadecimal
- ◆ Convert from hexadecimal to binary
- ◆ Convert from hexadecimal to decimal
- ◆ Convert from decimal to hexadecimal
- ◆ Add hexadecimal numbers
- ◆ Determine the 2's complement of a hexadecimal number
- ◆ Subtract hexadecimal numbers

The hexadecimal number system consists of digits 0–9 and letters A–F.

The **hexadecimal** number system has a base of sixteen; that is, it is composed of 16 **numeric** and alphabetic **characters**. Most digital systems process binary data in groups that are multiples of four bits, making the hexadecimal number very convenient because each hexadecimal digit represents a 4-bit binary number (as listed in Table 2–3).

TABLE 2-3

Decimal	Binary	Hexadecimal
0	0000	0
1	0001	1
2	0010	2
3	0011	3
4	0100	4
5	0101	5
6	0110	6
7	0111	7
8	1000	8
9	1001	9
10	1010	A
11	1011	B
12	1100	C
13	1101	D
14	1110	E
15	1111	F

Ten numeric digits and six alphabetic characters make up the hexadecimal number system. The use of letters A, B, C, D, E, and F to represent numbers may seem strange at first, but keep in mind that any number system is only a set of sequential symbols. If you understand what quantities these symbols represent, then the form of the symbols themselves is less important once you get accustomed to using them. We will use the subscript 16 to designate hexadecimal numbers to avoid confusion with decimal numbers. Sometimes you may see an "h" following a hexadecimal number.

Counting in Hexadecimal

How do you count in hexadecimal once you get to F? Simply start over with another column and continue as follows:

$$\ldots, E, F, 10, 11, 12, 13, 14, 15, 16, 17, 18, 19, 1A, 1B, 1C, 1D, 1E, 1F,$$
$$20, 21, 22, 23, 24, 25, 26, 27, 28, 29, 2A, 2B, 2C, 2D, 2E, 2F, 30, 31, \ldots$$

With two hexadecimal digits, you can count up to FF_{16}, which is decimal 255. To count beyond this, three hexadecimal digits are needed. For instance, 100_{16} is decimal 256, 101_{16} is decimal 257, and so forth. The maximum 3-digit hexadecimal number is FFF_{16}, or decimal 4095. The maximum 4-digit hexadecimal number is $FFFF_{16}$, which is decimal 65,535.

Binary-to-Hexadecimal Conversion

Converting a binary number to hexadecimal is a straightforward procedure. Simply break the binary number into 4-bit groups, starting at the right-most bit and replace each 4-bit group with the equivalent hexadecimal symbol.

EXAMPLE 2–24

Convert the following binary numbers to hexadecimal:

(a) 1100101001010111 **(b)** 111111000101101001

Solution

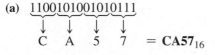 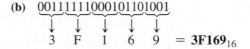

Two zeros have been added in part (b) to complete a 4-bit group at the left.

Related Problem

Convert the binary number 10011110111100011100 to hexadecimal.

Hexadecimal-to-Binary Conversion

To convert from a hexadecimal number to a binary number, reverse the process and replace each hexadecimal symbol with the appropriate four bits.

Hexadecimal is a convenient way to represent binary numbers.

EXAMPLE 2–25

Determine the binary numbers for the following hexadecimal numbers:

(a) $10A4_{16}$ **(b)** $CF8E_{16}$ **(c)** 9742_{16}

Solution

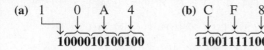 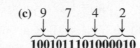

In part (a), the MSB is understood to have three zeros preceding it, thus forming a 4-bit group.

Related Problem

Convert the hexadecimal number 6BD3 to binary.

Conversion between hexadecimal and binary is direct and easy.

It should be clear that it is much easier to deal with a hexadecimal number than with the equivalent binary number. Since conversion is so easy, the hexadecimal system is widely used for representing binary numbers in programming, printouts, and displays.

Hexadecimal-to-Decimal Conversion

One way to find the decimal equivalent of a hexadecimal number is to first convert the hexadecimal number to binary and then convert from binary to decimal.

EXAMPLE 2–26

Convert the following hexadecimal numbers to decimal:

(a) $1C_{16}$ (b) $A85_{16}$

Solution

Remember, convert the hexadecimal number to binary first, then to decimal.

(a)   1   C
        ↓   ↓
$\overline{00011100} = 2^4 + 2^3 + 2^2 = 16 + 8 + 4 = \mathbf{28_{10}}$

(b)   A   8   5
       ↓   ↓   ↓
$\overline{101010000101} = 2^{11} + 2^9 + 2^7 + 2^2 + 2^0 = 2048 + 512 + 128 + 4 + 1 = \mathbf{2693_{10}}$

Related Problem

Convert the hexadecimal number 6BD to decimal.

A calculator can be used to perform arithmetic operations with hexadecimal numbers.

Another way to convert a hexadecimal number to its decimal equivalent is to multiply the decimal value of each hexadecimal digit by its weight and then take the sum of these products. The weights of a hexadecimal number are increasing powers of 16 (from right to left). For a 4-digit hexadecimal number, the weights are

$$16^3 \quad 16^2 \quad 16^1 \quad 16^0$$
$$4096 \quad 256 \quad 16 \quad 1$$

EXAMPLE 2–27

Convert the following hexadecimal numbers to decimal:

(a) $E5_{16}$ (b) $B2F8_{16}$

Solution

Recall from Table 2–3 that letters A through F represent decimal numbers 10 through 15, respectively.

(a) $E5_{16} = (E \times 16) + (5 \times 1) = (14 \times 16) + (5 \times 1) = 224 + 5 = \mathbf{229_{10}}$

(b) $B2F8_{16} = (B \times 4096) + (2 \times 256) + (F \times 16) + (8 \times 1)$
$= (11 \times 4096) + (2 \times 256) + (15 \times 16) + (8 \times 1)$
$= \quad 45,056 \quad + \quad 512 \quad + \quad 240 \quad + \quad 8 \quad = \mathbf{45,816_{10}}$

Related Problem

Convert $60A_{16}$ to decimal.

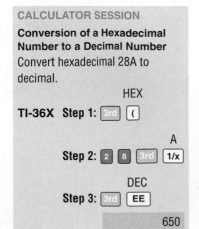

CALCULATOR SESSION

Conversion of a Hexadecimal Number to a Decimal Number
Convert hexadecimal 28A to decimal.

　　　　　　　HEX
TI-36X Step 1: [3rd] [(]

　　　　　　　　　A
Step 2: [2] [8] [3rd] [1/x]

　　　　　　　DEC
Step 3: [3rd] [EE]

　　　　　　　　650

Decimal-to-Hexadecimal Conversion

Repeated division of a decimal number by 16 will produce the equivalent hexadecimal number, formed by the remainders of the divisions. The first remainder produced is the least significant digit (LSD). Each successive division by 16 yields a remainder that becomes a digit in the equivalent hexadecimal number. This procedure is similar to repeated division by 2 for decimal-to-binary conversion that was covered in Section 2–3. Example 2–28 illustrates the procedure. Note that when a quotient has a fractional part, the fractional part is multiplied by the divisor to get the remainder.

EXAMPLE 2–28

Convert the decimal number 650 to hexadecimal by repeated division by 16.

Solution

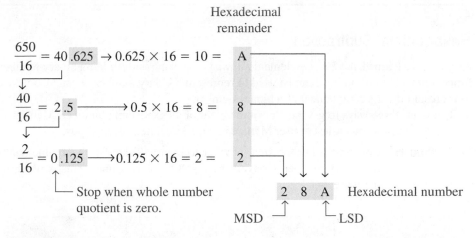

Related Problem

Convert decimal 2591 to hexadecimal.

Hexadecimal Addition

Addition can be done directly with hexadecimal numbers by remembering that the hexadecimal digits 0 through 9 are equivalent to decimal digits 0 through 9 and that hexadecimal digits A through F are equivalent to decimal numbers 10 through 15. When adding two hexadecimal numbers, use the following rules. (Decimal numbers are indicated by a subscript 10.)

1. In any given column of an addition problem, think of the two hexadecimal digits in terms of their decimal values. For instance, $5_{16} = 5_{10}$ and $C_{16} = 12_{10}$.

2. If the sum of these two digits is 15_{10} or less, bring down the corresponding hexadecimal digit.

3. If the sum of these two digits is greater than 15_{10}, bring down the amount of the sum that exceeds 16_{10} and carry a 1 to the next column.

EXAMPLE 2–29

Add the following hexadecimal numbers:

(a) $23_{16} + 16_{16}$ (b) $58_{16} + 22_{16}$ (c) $2B_{16} + 84_{16}$ (d) $DF_{16} + AC_{16}$

Solution

(a) 23_{16} right column: $3_{16} + 6_{16} = 3_{10} + 6_{10} = 9_{10} = 9_{16}$

    $+ 16_{16}$ left column: $2_{16} + 1_{16} = 2_{10} + 1_{10} = 3_{10} = 3_{16}$

     $\mathbf{39_{16}}$

(b) 58_{16}
$+ 22_{16}$
$\overline{7A_{16}}$

right column: $8_{16} + 2_{16} = 8_{10} + 2_{10} = 10_{10} = A_{16}$
left column: $5_{16} + 2_{16} = 5_{10} + 2_{10} = 7_{10} = 7_{16}$

(c) $2B_{16}$
$+ 84_{16}$
$\overline{AF_{16}}$

right column: $B_{16} + 4_{16} = 11_{10} + 4_{10} = 15_{10} = F_{16}$
left column: $2_{16} + 8_{16} = 2_{10} + 8_{10} = 10_{10} = A_{16}$

(d) DF_{16}
$+ AC_{16}$
$\overline{18B_{16}}$

right column: $F_{16} + C_{16} = 15_{10} + 12_{10} = 27_{10}$
$27_{10} - 16_{10} = 11_{10} = B_{16}$ with a 1 carry
left column: $D_{16} + A_{16} + 1_{16} = 13_{10} + 10_{10} + 1_{10} = 24_{10}$
$24_{10} - 16_{10} = 8_{10} = 8_{16}$ with a 1 carry

Related Problem

Add $4C_{16}$ and $3A_{16}$.

Hexadecimal Subtraction

As you have learned, the 2's complement allows you to subtract by adding binary numbers. Since a hexadecimal number can be used to represent a binary number, it can also be used to represent the 2's complement of a binary number.

There are three ways to get the 2's complement of a hexadecimal number. Method 1 is the most common and easiest to use. Methods 2 and 3 are alternate methods.

Method 1: Convert the hexadecimal number to binary. Take the 2's complement of the binary number. Convert the result to hexadecimal. This is illustrated in Figure 2–4.

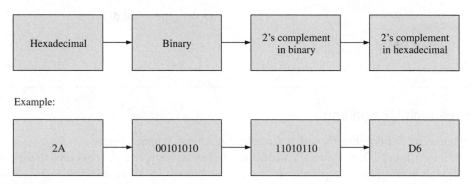

FIGURE 2–4 Getting the 2's complement of a hexadecimal number, Method 1.

Method 2: Subtract the hexadecimal number from the maximum hexadecimal number and add 1. This is illustrated in Figure 2–5.

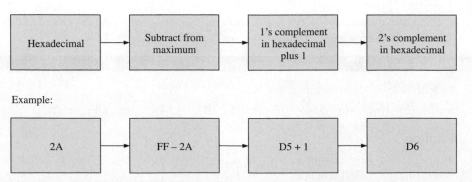

FIGURE 2–5 Getting the 2's complement of a hexadecimal number, Method 2.

Method 3: Write the sequence of single hexadecimal digits. Write the sequence in reverse below the forward sequence. The 1's complement of each hex digit is the digit directly below it. Add 1 to the resulting number to get the 2's complement. This is illustrated in Figure 2–6.

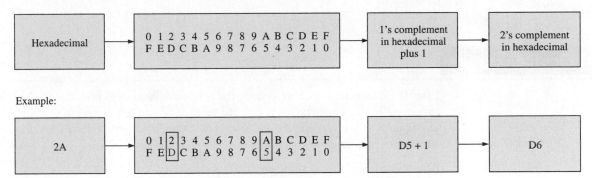

FIGURE 2–6 Getting the 2's complement of a hexadecimal number, Method 3.

EXAMPLE 2–30

Subtract the following hexadecimal numbers:

(a) $84_{16} - 2A_{16}$ (b) $C3_{16} - 0B_{16}$

Solution

(a) $2A_{16} = 00101010$

2's complement of $2A_{16} = 11010110 = D6_{16}$ (using Method 1)

$$\begin{array}{r} 84_{16} \\ + D6_{16} \\ \hline \cancel{1}5A_{16} \end{array}$$ Add
Drop carry, as in 2's complement addition

The difference is **$5A_{16}$**.

(b) $0B_{16} = 00001011$

2's complement of $0B_{16} = 11110101 = F5_{16}$ (using Method 1)

$$\begin{array}{r} C3_{16} \\ + F5_{16} \\ \hline \cancel{1}B8_{16} \end{array}$$ Add
Drop carry

The difference is **$B8_{16}$**.

Related Problem

Subtract 173_{16} from BCD_{16}.

SECTION 2–8 CHECKUP

1. Convert the following binary numbers to hexadecimal:

   (a) 10110011   (b) 110011101000

2. Convert the following hexadecimal numbers to binary:

   (a) 57_{16} (b) $3A5_{16}$ (c) $F80B_{16}$

3. Convert $9B30_{16}$ to decimal.

4. Convert the decimal number 573 to hexadecimal.

5. Add the following hexadecimal numbers directly:

    (a) $18_{16} + 34_{16}$ **(b)** $3F_{16} + 2A_{16}$

6. Subtract the following hexadecimal numbers:

    (a) $75_{16} - 21_{16}$ **(b)** $94_{16} - 5C_{16}$

2–9   Octal Numbers

Like the hexadecimal number system, the octal number system provides a convenient way to express binary numbers and codes. However, it is used less frequently than hexadecimal in conjunction with computers and microprocessors to express binary quantities for input and output purposes.

After completing this section, you should be able to

- ◆ Write the digits of the octal number system
- ◆ Convert from octal to decimal
- ◆ Convert from decimal to octal
- ◆ Convert from octal to binary
- ◆ Convert from binary to octal

The **octal** number system is composed of eight digits, which are

$$0, 1, 2, 3, 4, 5, 6, 7$$

To count above 7, begin another column and start over:

$$10, 11, 12, 13, 14, 15, 16, 17, 20, 21, \ldots$$

The octal number system has a base of 8.

Counting in octal is similar to counting in decimal, except that the digits 8 and 9 are not used. To distinguish octal numbers from decimal numbers or hexadecimal numbers, we will use the subscript 8 to indicate an octal number. For instance, 15_8 in octal is equivalent to 13_{10} in decimal and D in hexadecimal. Sometimes you may see an "o" or a "Q" following an octal number.

Octal-to-Decimal Conversion

Since the octal number system has a base of eight, each successive digit position is an increasing power of eight, beginning in the right-most column with 8^0. The evaluation of an octal number in terms of its decimal equivalent is accomplished by multiplying each digit by its weight and summing the products, as illustrated here for 2374_8.

$$\text{Weight:}\quad 8^3\ 8^2\ 8^1\ 8^0$$
$$\text{Octal number:}\quad 2\ \ 3\ 7\ 4$$

$$
\begin{aligned}
2374_8 &= (2 \times 8^3) + (3 \times 8^2) + (7 \times 8^1) + (4 \times 8^0) \\
&= (2 \times 512) + (3 \times 64) + (7 \times 8) + (4 \times 1) \\
&= \quad 1024 \quad + \quad 192 \quad + \quad 56 \quad + \quad 4 \quad = 1276_{10}
\end{aligned}
$$

Decimal-to-Octal Conversion

A method of converting a decimal number to an octal number is the repeated division-by-8 method, which is similar to the method used in the conversion of decimal numbers to binary or to hexadecimal. To show how it works, let's convert the decimal number 359 to octal. Each successive division by 8 yields a remainder that becomes a digit in the equivalent octal number. The first remainder generated is the least significant digit (LSD).

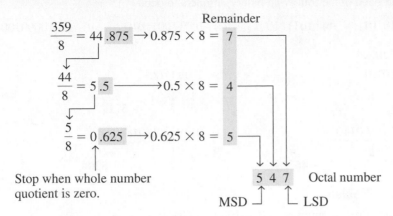

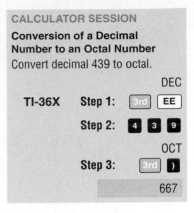

Octal-to-Binary Conversion

Because each octal digit can be represented by a 3-bit binary number, it is very easy to convert from octal to binary. Each octal digit is represented by three bits as shown in Table 2–4.

Octal is a convenient way to represent binary numbers, but it is not as commonly used as hexadecimal.

TABLE 2–4

Octal/binary conversion.

Octal Digit	0	1	2	3	4	5	6	7
Binary	000	001	010	011	100	101	110	111

To convert an octal number to a binary number, simply replace each octal digit with the appropriate three bits.

EXAMPLE 2–31

Convert each of the following octal numbers to binary:

(a) 13_8 (b) 25_8 (c) 140_8 (d) 7526_8

Solution

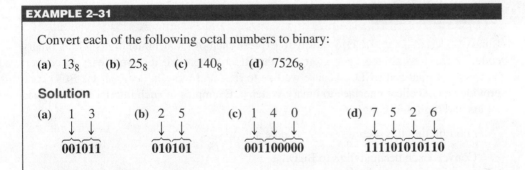

Related Problem

Convert each of the binary numbers to decimal and verify that each value agrees with the decimal value of the corresponding octal number.

Binary-to-Octal Conversion

Conversion of a binary number to an octal number is the reverse of the octal-to-binary conversion. The procedure is as follows: Start with the right-most group of three bits and, moving from right to left, convert each 3-bit group to the equivalent octal digit. If there are not three bits available for the left-most group, add either one or two zeros to make a complete group. These leading zeros do not affect the value of the binary number.

EXAMPLE 2–32

Convert each of the following binary numbers to octal:

(a) 110101   (b) 101111001   (c) 100110011010   (d) 11010000100

Solution

(a) 110101

$\underbrace{11}\underbrace{01}\underbrace{01}$
↓   ↓
6   5 = **65**$_8$

(b) 101111001

$\underbrace{101}\underbrace{111}\underbrace{001}$
↓   ↓   ↓
5   7   1 = **571**$_8$

(c) 100110011010

$\underbrace{100}\underbrace{110}\underbrace{011}\underbrace{010}$
↓   ↓   ↓   ↓
4   6   3   2 = **4632**$_8$

(d) 011010000100

$\underbrace{011}\underbrace{010}\underbrace{000}\underbrace{100}$
↓   ↓   ↓   ↓
3   2   0   4 = **3204**$_8$

Related Problem

Convert the binary number 1010101000111110010 to octal.

SECTION 2–9 CHECKUP

1. Convert the following octal numbers to decimal:
    (a) 73_8 (b) 125_8
2. Convert the following decimal numbers to octal:
    (a) 98_{10} (b) 163_{10}
3. Convert the following octal numbers to binary:
    (a) 46_8 (b) 723_8 (c) 5624_8
4. Convert the following binary numbers to octal:
    (a) 110101111   (b) 1001100010   (c) 10111111001

2–10   Binary Coded Decimal (BCD)

Binary coded decimal (BCD) is a way to express each of the decimal digits with a binary code. There are only ten code groups in the BCD system, so it is very easy to convert between decimal and BCD. Because we like to read and write in decimal, the BCD code provides an excellent interface to binary systems. Examples of such interfaces are keypad inputs and digital readouts.

After completing this section, you should be able to

- ◆ Convert each decimal digit to BCD
- ◆ Express decimal numbers in BCD
- ◆ Convert from BCD to decimal
- ◆ Add BCD numbers

The 8421 BCD Code

In BCD, 4 bits represent each decimal digit.

The 8421 code is a type of **BCD** (binary coded decimal) code. Binary coded decimal means that each decimal digit, 0 through 9, is represented by a binary code of four bits. The designation 8421 indicates the binary weights of the four bits ($2^3, 2^2, 2^1, 2^0$). The ease of conversion between 8421 code numbers and the familiar decimal numbers is the main advantage of this code. All you have to remember are the ten binary combinations that represent the ten decimal digits as shown in Table 2–5. The 8421 code is the predominant BCD code, and when we refer to BCD, we always mean the 8421 code unless otherwise stated.

TABLE 2–5

Decimal/BCD conversion.

Decimal Digit	0	1	2	3	4	5	6	7	8	9
BCD	0000	0001	0010	0011	0100	0101	0110	0111	1000	1001

Invalid Codes

You should realize that, with four bits, sixteen numbers (0000 through 1111) can be represented but that, in the 8421 code, only ten of these are used. The six code combinations that are not used—1010, 1011, 1100, 1101, 1110, and 1111—are invalid in the 8421 BCD code.

To express any decimal number in BCD, simply replace each decimal digit with the appropriate 4-bit code, as shown by Example 2–33.

EXAMPLE 2–33

Convert each of the following decimal numbers to BCD:

(a) 35   (b) 98   (c) 170   (d) 2469

Solution

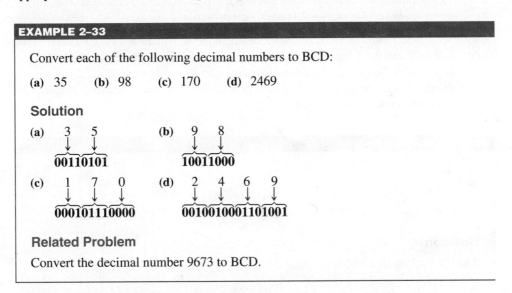

Related Problem

Convert the decimal number 9673 to BCD.

It is equally easy to determine a decimal number from a BCD number. Start at the right-most bit and break the code into groups of four bits. Then write the decimal digit represented by each 4-bit group.

EXAMPLE 2–34

Convert each of the following BCD codes to decimal:

(a) 10000110   (b) 001101010001   (c) 1001010001110000

Solution

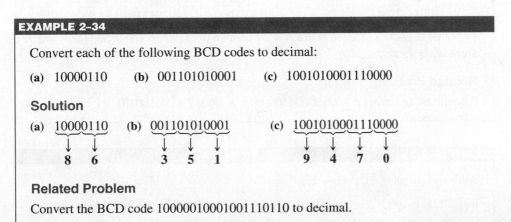

Related Problem

Convert the BCD code 10000010001001110110 to decimal.

InfoNote

BCD is sometimes used for arithmetic operations in processors. To represent BCD numbers in a processor, they usually are "packed," so that eight bits have two BCD digits. Normally, a processor will add numbers as if they were straight binary. Special instructions are available for computer programmers to correct the results when BCD numbers are added or subtracted. For example, in Assembly Language, the programmer will include a DAA (Decimal Adjust for Addition) instruction to automatically correct the answer to BCD following an addition.

Applications

Digital clocks, digital thermometers, digital meters, and other devices with seven-segment displays typically use BCD code to simplify the displaying of decimal numbers. BCD is not as efficient as straight binary for calculations, but it is particularly useful if only limited processing is required, such as in a digital thermometer.

BCD Addition

BCD is a numerical code and can be used in arithmetic operations. Addition is the most important operation because the other three operations (subtraction, multiplication, and division) can be accomplished by the use of addition. Here is how to add two BCD numbers:

Step 1: Add the two BCD numbers, using the rules for binary addition in Section 2–4.

Step 2: If a 4-bit sum is equal to or less than 9, it is a valid BCD number.

Step 3: If a 4-bit sum is greater than 9, or if a carry out of the 4-bit group is generated, it is an invalid result. Add 6 (0110) to the 4-bit sum in order to skip the six invalid states and return the code to 8421. If a carry results when 6 is added, simply add the carry to the next 4-bit group.

Example 2–35 illustrates BCD additions in which the sum in each 4-bit column is equal to or less than 9, and the 4-bit sums are therefore valid BCD numbers. Example 2–36 illustrates the procedure in the case of invalid sums (greater than 9 or a carry).

An alternative method to add BCD numbers is to convert them to decimal, perform the addition, and then convert the answer back to BCD.

EXAMPLE 2–35

Add the following BCD numbers:

(a) 0011 + 0100

(b) 00100011 + 00010101

(c) 10000110 + 00010011

(d) 010001010000 + 010000010111

Solution

The decimal number additions are shown for comparison.

(a)

0011	3
+ 0100	+ 4
0111	7

(b)

0010	0011	23
+ 0001	0101	+ 15
0011	**1000**	38

(c)

1000	0110	86
+ 0001	0011	+ 13
1001	**1001**	99

(d)

0100	0101	0000	450
+ 0100	0001	0111	+ 417
1000	**0110**	**0111**	867

Note that in each case the sum in any 4-bit column does not exceed 9, and the results are valid BCD numbers.

Related Problem

Add the BCD numbers: 1001000001000011 + 0000100100100101.

EXAMPLE 2–36

Add the following BCD numbers:

(a) 1001 + 0100

(b) 1001 + 1001

(c) 00010110 + 00010101

(d) 01100111 + 01010011

Solution

The decimal number additions are shown for comparison.

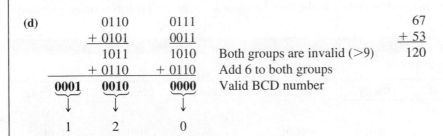

(a)
```
     1001                                              9
   + 0100                                             +4
     1101      Invalid BCD number (>9)                13
   + 0110      Add 6
0001 0011      Valid BCD number
  ↓    ↓
  1    3
```

(b)
```
     1001                                              9
   + 1001                                            + 9
   1 0010      Invalid because of carry              18
   + 0110      Add 6
0001 1000      Valid BCD number
  ↓    ↓
  1    8
```

(c)
```
0001   0110                                           16
+0001  0101                                         + 15
0010   1011      Right group is invalid (>9),         31
                 left group is valid.
     + 0110      Add 6 to invalid code. Add
                 carry, 0001, to next group.
0011   0001      Valid BCD number
  ↓      ↓
  3      1
```

(d)
```
     0110   0111                                      67
   + 0101   0011                                    + 53
     1011   1010   Both groups are invalid (>9)      120
   + 0110 + 0110   Add 6 to both groups
0001 0010   0000   Valid BCD number
  ↓    ↓      ↓
  1    2      0
```

Related Problem

Add the BCD numbers: 01001000 + 00110100.

---

1. What is the binary weight of each 1 in the following BCD numbers?

   (a) 0010   (b) 1000   (c) 0001   (d) 0100

2. Convert the following decimal numbers to BCD:

   (a) 6   (b) 15   (c) 273   (d) 849

3. What decimal numbers are represented by each BCD code?

   (a) 10001001   (b) 001001111000   (c) 000101010111

4. In BCD addition, when is a 4-bit sum invalid?

2–11   Digital Codes

Many specialized codes are used in digital systems. You have just learned about the BCD code; now let's look at a few others. Some codes are strictly numeric, like BCD, and others are alphanumeric; that is, they are used to represent numbers, letters, symbols, and instructions. The codes introduced in this section are the Gray code, the ASCII code, and the Unicode.

After completing this section, you should be able to

- ◆ Explain the advantage of the Gray code
- ◆ Convert between Gray code and binary
- ◆ Use the ASCII code
- ◆ Discuss the Unicode

The Gray Code

The single bit change characteristic of the Gray code minimizes the chance for error.

The **Gray code** is unweighted and is not an arithmetic code; that is, there are no specific weights assigned to the bit positions. The important feature of the Gray code is that *it exhibits only a single bit change from one code word to the next in sequence.* This property is important in many applications, such as shaft position encoders, where error susceptibility increases with the number of bit changes between adjacent numbers in a sequence.

Table 2–6 is a listing of the 4-bit Gray code for decimal numbers 0 through 15. Binary numbers are shown in the table for reference. Like binary numbers, *the Gray code can have any number of bits.* Notice the single-bit change between successive Gray code words. For instance, in going from decimal 3 to decimal 4, the Gray code changes from 0010 to 0110, while the binary code changes from 0011 to 0100, a change of three bits. The only bit change in the Gray code is in the third bit from the right: the other bits remain the same.

TABLE 2–6

Four-bit Gray code.

Decimal	Binary	Gray Code	Decimal	Binary	Gray Code
0	0000	0000	8	1000	1100
1	0001	0001	9	1001	1101
2	0010	0011	10	1010	1111
3	0011	0010	11	1011	1110
4	0100	0110	12	1100	1010
5	0101	0111	13	1101	1011
6	0110	0101	14	1110	1001
7	0111	0100	15	1111	1000

Binary-to-Gray Code Conversion

Conversion between binary code and Gray code is sometimes useful. The following rules explain how to convert from a binary number to a Gray code word:

1. The most significant bit (left-most) in the Gray code is the same as the corresponding MSB in the binary number.

2. Going from left to right, add each adjacent pair of binary code bits to get the next Gray code bit. Discard carries.

For example, the conversion of the binary number 10110 to Gray code is as follows:

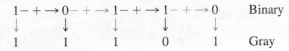

The Gray code is 11101.

Gray-to-Binary Code Conversion

To convert from Gray code to binary, use a similar method; however, there are some differences. The following rules apply:

1. The most significant bit (left-most) in the binary code is the same as the corresponding bit in the Gray code.

2. Add each binary code bit generated to the Gray code bit in the next adjacent position. Discard carries.

For example, the conversion of the Gray code word 11011 to binary is as follows:

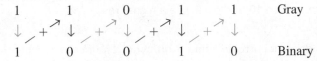

The binary number is 10010.

EXAMPLE 2–37

(a) Convert the binary number 11000110 to Gray code.

(b) Convert the Gray code 10101111 to binary.

Solution

(a) Binary to Gray code:

$$1-+\rightarrow1-+\rightarrow0-+\rightarrow0-+\rightarrow0-+\rightarrow1-+\rightarrow1-+\rightarrow0$$
$$1\quad0\quad1\quad0\quad0\quad1\quad0\quad1$$

(b) Gray code to binary:

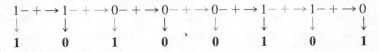

Related Problem

(a) Convert binary 101101 to Gray code.

(b) Convert Gray code 100111 to binary.

An Application

The concept of a 3-bit shaft position encoder is shown in Figure 2–7. Basically, there are three concentric rings that are segmented into eight sectors. The more sectors there are, the more accurately the position can be represented, but we are using only eight to illustrate. Each sector of each ring is either reflective or nonreflective. As the rings rotate with the shaft, they come under an IR emitter that produces three separate IR beams. A 1 is indicated where there is a reflected beam, and a 0 is indicated where there is no reflected beam. The IR detector senses the presence or absence of reflected

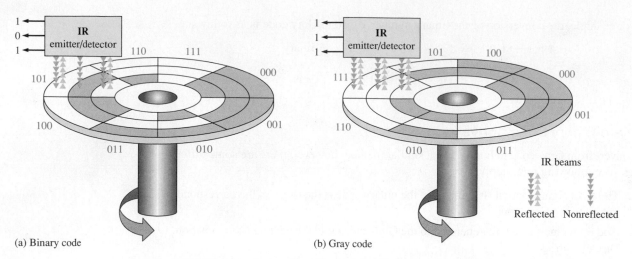

(a) Binary code

(b) Gray code

FIGURE 2–7   A simplified illustration of how the Gray code solves the error problem in shaft position encoders. Three bits are shown to illustrate the concept, although most shaft encoders use more than 10 bits to achieve a higher resolution.

beams and produces a corresponding 3-bit code. The IR emitter/detector is in a fixed position. As the shaft rotates counterclockwise through 360°, the eight sectors move under the three beams. Each beam is either reflected or absorbed by the sector surface to represent a binary or Gray code number that indicates the shaft position.

In Figure 2–7(a), the sectors are arranged in a straight binary pattern, so that the detector output goes from 000 to 001 to 010 to 011 and so on. When a beam is aligned over a reflective sector, the output is 1; when a beam is aligned over a nonreflective sector, the output is 0. If one beam is slightly ahead of the others during the transition from one sector to the next, an erroneous output can occur. Consider what happens when the beams are on the 111 sector and about to enter the 000 sector. If the MSB beam is slightly ahead, the position would be incorrectly indicated by a transitional 011 instead of a 111 or a 000. In this type of application, it is virtually impossible to maintain precise mechanical alignment of the IR emitter/detector beams; therefore, some error will usually occur at many of the transitions between sectors.

The Gray code is used to eliminate the error problem which is inherent in the binary code. As shown in Figure 2–7(b), the Gray code assures that only one bit will change between adjacent sectors. This means that even though the beams may not be in precise alignment, there will never be a transitional error. For example, let's again consider what happens when the beams are on the 111 sector and about to move into the next sector, 101. The only two possible outputs during the transition are 111 and 101, no matter how the beams are aligned. A similar situation occurs at the transitions between each of the other sectors.

Alphanumeric Codes

In order to communicate, you need not only numbers, but also letters and other symbols. In the strictest sense, **alphanumeric** codes are codes that represent numbers and alphabetic characters (letters). Most such codes, however, also represent other characters such as symbols and various instructions necessary for conveying information.

At a minimum, an alphanumeric code must represent 10 decimal digits and 26 letters of the alphabet, for a total of 36 items. This number requires six bits in each code combination because five bits are insufficient ($2^5 = 32$). There are 64 total combinations of six bits, so there are 28 unused code combinations. Obviously, in many applications, symbols other than just numbers and letters are necessary to communicate completely. You need spaces, periods, colons, semicolons, question marks, etc. You also need instructions to tell the receiving system what to do with the information. With codes that are six bits long, you can handle decimal numbers, the alphabet, and 28 other symbols. This should give you an idea of the requirements for a basic alphanumeric code. The ASCII is a common alphanumeric code and is covered next.

ASCII

ASCII is the abbreviation for American Standard Code for Information Interchange. Pronounced "askee," ASCII is a universally accepted alphanumeric code used in most computers and other electronic equipment. Most computer keyboards are standardized with the ASCII. When you enter a letter, a number, or control command, the corresponding ASCII code goes into the computer.

ASCII has 128 characters and symbols represented by a 7-bit binary code. Actually, ASCII can be considered an 8-bit code with the MSB always 0. This 8-bit code is 00 through 7F in hexadecimal. The first thirty-two ASCII characters are nongraphic commands that are never printed or displayed and are used only for control purposes. Examples of the control characters are "null," "line feed," "start of text," and "escape." The other characters are graphic symbols that can be printed or displayed and include the letters of the alphabet (lowercase and uppercase), the ten decimal digits, punctuation signs, and other commonly used symbols.

Table 2–7 is a listing of the ASCII code showing the decimal, hexadecimal, and binary representations for each character and symbol. The left section of the table lists the names of the 32 control characters (00 through 1F hexadecimal). The graphic symbols are listed in the rest of the table (20 through 7F hexadecimal).

EXAMPLE 2–38

Use Table 2–7 to determine the binary ASCII codes that are entered from the computer's keyboard when the following C language program statement is typed in. Also express each code in hexadecimal.

$$\text{if } (x > 5)$$

Solution

The ASCII code for each symbol is found in Table 2–7.

Symbol	Binary	Hexadecimal
i	1101001	69_{16}
f	1100110	66_{16}
Space	0100000	20_{16}
(	0101000	28_{16}
x	1111000	78_{16}
>	0111110	$3E_{16}$
5	0110101	35_{16}
)	0101001	29_{16}

Related Problem

Use Table 2–7 to determine the sequence of ASCII codes required for the following C program statement and express each code in hexadecimal:

$$\text{if } (y < 8)$$

The ASCII Control Characters

The first thirty-two codes in the ASCII table (Table 2–7) represent the control characters. These are used to allow devices such as a computer and printer to communicate with each other when passing information and data. The control key function allows a control character to be entered directly from an ASCII keyboard by pressing the control key (CTRL) and the corresponding symbol.

TABLE 2-7

American Standard Code for Information Interchange (ASCII).

Control Characters

Name	Dec	Binary	Hex
NUL	0	0000000	00
SOH	1	0000001	01
STX	2	0000010	02
ETX	3	0000011	03
EOT	4	0000100	04
ENQ	5	0000101	05
ACK	6	0000110	06
BEL	7	0000111	07
BS	8	0001000	08
HT	9	0001001	09
LF	10	0001010	0A
VT	11	0001011	0B
FF	12	0001100	0C
CR	13	0001101	0D
SO	14	0001110	0E
SI	15	0001111	0F
DLE	16	0010000	10
DC1	17	0010001	11
DC2	18	0010010	12
DC3	19	0010011	13
DC4	20	0010100	14
NAK	21	0010101	15
SYN	22	0010110	16
ETB	23	0010111	17
CAN	24	0011000	18
EM	25	0011001	19
SUB	26	0011010	1A
ESC	27	0011011	1B
FS	28	0011100	1C
GS	29	0011101	1D
RS	30	0011110	1E
US	31	0011111	1F

Graphic Symbols

Symbol	Dec	Binary	Hex	Symbol	Dec	Binary	Hex	Symbol	Dec	Binary	Hex
space	32	0100000	20	@	64	1000000	40	`	96	1100000	60
!	33	0100001	21	A	65	1000001	41	a	97	1100001	61
"	34	0100010	22	B	66	1000010	42	b	98	1100010	62
#	35	0100011	23	C	67	1000011	43	c	99	1100011	63
$	36	0100100	24	D	68	1000100	44	d	100	1100100	64
%	37	0100101	25	E	69	1000101	45	e	101	1100101	65
&	38	0100110	26	F	70	1000110	46	f	102	1100110	66
'	39	0100111	27	G	71	1000111	47	g	103	1100111	67
(	40	0101000	28	H	72	1001000	48	h	104	1101000	68
)	41	0101001	29	I	73	1001001	49	i	105	1101001	69
*	42	0101010	2A	J	74	1001010	4A	j	106	1101010	6A
+	43	0101011	2B	K	75	1001011	4B	k	107	1101011	6B
,	44	0101100	2C	L	76	1001100	4C	l	108	1101100	6C
-	45	0101101	2D	M	77	1001101	4D	m	109	1101101	6D
.	46	0101110	2E	N	78	1001110	4E	n	110	1101110	6E
/	47	0101111	2F	O	79	1001111	4F	o	111	1101111	6F
0	48	0110000	30	P	80	1010000	50	p	112	1110000	70
1	49	0110001	31	Q	81	1010001	51	q	113	1110001	71
2	50	0110010	32	R	82	1010010	52	r	114	1110010	72
3	51	0110011	33	S	83	1010011	53	s	115	1110011	73
4	52	0110100	34	T	84	1010100	54	t	116	1110100	74
5	53	0110101	35	U	85	1010101	55	u	117	1110101	75
6	54	0110110	36	V	86	1010110	56	v	118	1110110	76
7	55	0110111	37	W	87	1010111	57	w	119	1110111	77
8	56	0111000	38	X	88	1011000	58	x	120	1111000	78
9	57	0111001	39	Y	89	1011001	59	y	121	1111001	79
:	58	0111010	3A	Z	90	1011010	5A	z	122	1111010	7A
;	59	0111011	3B	[	91	1011011	5B	{	123	1111011	7B
<	60	0111100	3C	\	92	1011100	5C	|	124	1111100	7C
=	61	0111101	3D	]	93	1011101	5D	}	125	1111101	7D
>	62	0111110	3E	^	94	1011110	5E	~	126	1111110	7E
?	63	0111111	3F	_	95	1011111	5F	Del	127	1111111	7F

Extended ASCII Characters

In addition to the 128 standard ASCII characters, there are an additional 128 characters that were adopted by IBM for use in their PCs (personal computers). Because of the popularity of the PC, these particular extended ASCII characters are also used in applications other than PCs and have become essentially an unofficial standard.

The extended ASCII characters are represented by an 8-bit code series from hexadecimal 80 to hexadecimal FF and can be grouped into the following general categories: foreign (non-English) alphabetic characters, foreign currency symbols, Greek letters, mathematical symbols, drawing characters, bar graphing characters, and shading characters.

Unicode

Unicode provides the ability to encode all of the characters used for the written languages of the world by assigning each character a unique numeric value and name utilizing the universal character set (UCS). It is applicable in computer applications dealing with multilingual text, mathematical symbols, or other technical characters.

Unicode has a wide array of characters, and their various encoding forms are used in many environments. While ASCII basically uses 7-bit codes, Unicode uses relatively abstract "code points"—non-negative integer numbers—that map sequences of one or more bytes, using different encoding forms and schemes. To permit compatibility, Unicode assigns the first 128 code points to the same characters as ASCII. One can, therefore, think of ASCII as a 7-bit encoding scheme for a very small subset of Unicode and of the UCS.

Unicode consists of about 100,000 characters, a set of code charts for visual reference, an encoding methodology and set of standard character encodings, and an enumeration of character properties such as uppercase and lowercase. It also consists of a number of related items, such as character properties, rules for text normalization, decomposition, collation, rendering, and bidirectional display order (for the correct display of text containing both right-to-left scripts, such as Arabic or Hebrew, and left-to-right scripts).

SECTION 2–11 CHECKUP

1. Convert the following binary numbers to the Gray code:

   (a) 1100   (b) 1010   (c) 11010

2. Convert the following Gray codes to binary:

   (a) 1000   (b) 1010   (c) 11101

3. What is the ASCII representation for each of the following characters? Express each as a bit pattern and in hexadecimal notation.

   (a) K   (b) r   (c) $ (d) +

2–12   Error Codes

In this section, three methods for adding bits to codes to detect a single-bit error are discussed. The parity method of error detection is introduced, and the cyclic redundancy check is discussed. Also, the Hamming code for error detection and correction is presented.

After completing this section, you should be able to

- ◆ Determine if there is an error in a code based on the parity bit

- ◆ Assign the proper parity bit to a code

- ◆ Explain the cyclic redundancy (CRC) check

- ◆ Describe the Hamming code

Parity Method for Error Detection

A parity bit tells if the number of 1s is odd or even.

Many systems use a parity bit as a means for bit **error detection**. Any group of bits contain either an even or an odd number of 1s. A parity bit is attached to a group of bits to make the total number of 1s in a group always even or always odd. An even parity bit makes the total number of 1s even, and an odd parity bit makes the total odd.

A given system operates with even or odd **parity**, but not both. For instance, if a system operates with even parity, a check is made on each group of bits received to make sure the total number of 1s in that group is even. If there is an odd number of 1s, an error has occurred.

As an illustration of how parity bits are attached to a code, Table 2–8 lists the parity bits for each BCD number for both even and odd parity. The parity bit for each BCD number is in the *P* column.

TABLE 2–8

The BCD code with parity bits.

Even Parity		Odd Parity	
P	BCD	*P*	BCD
0	0000	1	0000
1	0001	0	0001
1	0010	0	0010
0	0011	1	0011
1	0100	0	0100
0	0101	1	0101
0	0110	1	0110
1	0111	0	0111
1	1000	0	1000
0	1001	1	1001

The parity bit can be attached to the code at either the beginning or the end, depending on system design. Notice that the total number of 1s, including the parity bit, is always even for even parity and always odd for odd parity.

Detecting an Error

A parity bit provides for the detection of a single bit error (or any odd number of errors, which is very unlikely) but cannot check for two errors in one group. For instance, let's assume that we wish to transmit the BCD code 0101. (Parity can be used with any number of bits; we are using four for illustration.) The total code transmitted, including the even parity bit, is

$$
\begin{array}{c}
\text{Even parity bit} \\
\downarrow \\
00101 \\
\uparrow \\
\text{BCD code}
\end{array}
$$

Now let's assume that an error occurs in the third bit from the left (the 1 becomes a 0).

$$
\begin{array}{c}
\text{Even parity bit} \\
\downarrow \\
00001 \\
\uparrow \\
\text{Bit error}
\end{array}
$$

When this code is received, the parity check circuitry determines that there is only a single 1 (odd number), when there should be an even number of 1s. Because an even number of 1s does not appear in the code when it is received, an error is indicated.

An odd parity bit also provides in a similar manner for the detection of a single error in a given group of bits.

EXAMPLE 2–39

Assign the proper even parity bit to the following code groups:

(a) 1010
(b) 111000
(c) 101101
(d) 1000111001001
(e) 101101011111

Solution

Make the parity bit either 1 or 0 as necessary to make the total number of 1s even. The parity bit will be the left-most bit (color).

(a) **0**1010
(b) **1**111000
(c) **0**101101
(d) **0**100011100101
(e) **1**101101011111

Related Problem

Add an even parity bit to the 7-bit ASCII code for the letter K.

EXAMPLE 2–40

An odd parity system receives the following code groups: 10110, 11010, 110011, 110101110100, and 1100010101010. Determine which groups, if any, are in error.

Solution

Since odd parity is required, any group with an even number of 1s is incorrect. The following groups are in error: **110011** and **1100010101010**.

Related Problem

The following ASCII character is received by an odd parity system: 00110111. Is it correct?

Cyclic Redundancy Check

The **cyclic redundancy check (CRC)** is a widely used code used for detecting one- and two-bit transmission errors when digital data are transferred on a communication link. The communication link can be between two computers that are connected to a network or between a digital storage device (such as a CD, DVD, or a hard drive) and a PC. If it is properly designed, the CRC can also detect multiple errors for a number of bits in sequence (burst errors). In CRC, a certain number of check bits, sometimes called a *checksum*, are appended to the data bits (added to end) that are being transmitted. The transmitted data are tested by the receiver for errors using the CRC. Not every possible error can be identified, but the CRC is much more efficient than just a simple parity check.

CRC is often described mathematically as the division of two polynomials to generate a remainder. A polynomial is a mathematical expression that is a sum of terms with positive exponents. When the coefficients are limited to 1s and 0s, it is called a *univariate polynomial*. An example of a univariate polynomial is $1x^3 + 0x^2 + 1x^1 + 1x^0$ or simply $x^3 + x^1 + x^0$, which can be fully described by the 4-bit binary number 1011. Most cyclic redundancy checks use a 16-bit or larger polynomial, but for simplicity the process is illustrated here with four bits.

Modulo-2 Operations

Simply put, CRC is based on the division of two binary numbers; and, as you know, division is just a series of subtractions and shifts. To do subtraction, a method called *modulo-2* addition can be used. Modulo-2 addition (or subtraction) is the same as binary addition with the carries discarded, as shown in the truth table in Table 2–9. **Truth tables** are widely used to describe the operation of logic circuits, as you will learn in Chapter 3. With two bits, there is a total of four possible combinations, as shown in the table. This particular table describes the modulo-2 operation also known as *exclusive-OR* and can be implemented with a logic

TABLE 2–9

Modulo-2 operation.

Input Bits	Output Bit
0 0	0
0 1	1
1 0	1
1 1	0

gate that will be introduced in Chapter 3. A simple rule for modulo-2 is that the output is 1 if the inputs are different; otherwise, it is 0.

CRC Process

The process is as follows:

1. Select a fixed generator code; it can have fewer bits than the data bits to be checked. This code is understood in advance by both the sending and receiving devices and must be the same for both.

2. Append a number of 0s equal to the number of bits in the generator code to the data bits.

3. Divide the data bits including the appended bits by the generator code bits using modulo-2.

4. If the remainder is 0, the data and appended bits are sent as is.

5. If the remainder is not 0, the appended bits are made equal to the remainder bits in order to get a 0 remainder before data are sent.

6. At the receiving end, the receiver divides the incoming appended data bit code by the same generator code as used by the sender.

7. If the remainder is 0, there is no error detected (it is possible in rare cases for multiple errors to cancel). If the remainder is not 0, an error has been detected in the transmission and a retransmission is requested by the receiver.

Figure 2–8 illustrates the CRC process.

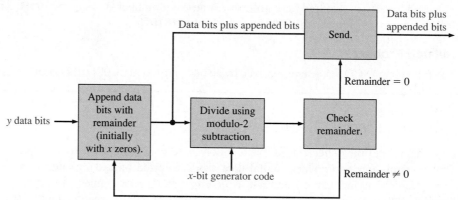

(a) Transmitting end of communication link

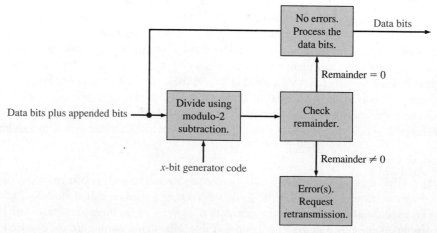

(b) Receiving end of communication link

FIGURE 2–8 The CRC process.

EXAMPLE 2–41

Determine the transmitted CRC for the following byte of data (D) and generator code (G). Verify that the remainder is 0.

$$\text{D:} \quad 11010011$$
$$\text{G:} \quad 1010$$

Solution

Since the generator code has four data bits, add four 0s (blue) to the data byte. The appended data (D′) is

$$\text{D}' = 110100110000$$

Divide the appended data by the generator code (red) using the modulo-2 operation until all bits have been used.

$$\frac{\text{D}'}{\text{G}} = \frac{110100110000}{1010}$$

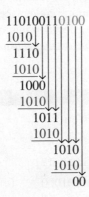

Remainder = 0100. Since the remainder is not 0, append the data with the four remainder bits (blue). Then divide by the generator code (red). The transmitted CRC is **110100110100**.

```
110100110100
1010
1110
1010
1000
1010
1011
1010
1010
1010
  00
```

Remainder = 0

Related Problem

Change the generator code to 1100 and verify that a 0 remainder results when the CRC process is applied to the data byte (11010011).

EXAMPLE 2–42

During transmission, an error occurs in the second bit from the left in the appended data byte generated in Example 2–41. The received data is

$$D' = 100100110100$$

Apply the CRC process to the received data to detect the error using the same generator code (1010).

Solution

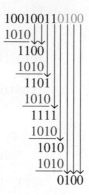

Remainder $= 0100$. Since it is not zero, an error is indicated.

Related Problem

Assume two errors in the data byte as follows: 10011011. Apply the CRC process to check for the errors using the same received data and the same generator code.

Hamming Code

The **Hamming code** is used to detect and correct a single-bit error in a transmitted code. To accomplish this, four redundancy bits are introduced in a 7-bit group of data bits. These redundancy bits are interspersed at bit positions 2^n ($n = 0, 1, 2, 3$) within the original data bits. At the end of the transmission, the redundancy bits have to be removed from the data bits. A recent version of the Hamming code places all the redundancy bits at the end of the data bits, making their removal easier than that of the interspersed bits. *A coverage of the classic Hamming code is available on the website.*

SECTION 2–12 CHECKUP

1. Which odd-parity code is in error?
   (a) 1011   (b) 1110   (c) 0101   (d) 1000
2. Which even-parity code is in error?
   (a) 11000110   (b) 00101000   (c) 10101010   (d) 11111011
3. Add an even parity bit to the end of each of the following codes.
   (a) 1010100   (b) 0100000   (c) 1110111   (d) 1000110
4. What does CRC stand for?
5. Apply modulo-2 operations to determine the following:
   (a) $1 + 1$ (b) $1 - 1$ (c) $1 - 0$ (d) $0 + 1$

SUMMARY

- A binary number is a weighted number in which the weight of each whole number digit is a positive power of two and the weight of each fractional digit is a negative power of two. The whole number weights increase from right to left—from least significant digit to most significant.
- A binary number can be converted to a decimal number by summing the decimal values of the weights of all the 1s in the binary number.
- A decimal whole number can be converted to binary by using the sum-of-weights or the repeated division-by-2 method.
- A decimal fraction can be converted to binary by using the sum-of-weights or the repeated multiplication-by-2 method.
- The basic rules for binary addition are as follows:

$$0 + 0 = 0$$
$$0 + 1 = 1$$
$$1 + 0 = 1$$
$$1 + 1 = 10$$

- The basic rules for binary subtraction are as follows:

$$0 - 0 = 0$$
$$1 - 1 = 0$$
$$1 - 0 = 1$$
$$10 - 1 = 1$$

- The 1's complement of a binary number is derived by changing 1s to 0s and 0s to 1s.
- The 2's complement of a binary number can be derived by adding 1 to the 1's complement.
- Binary subtraction can be accomplished with addition by using the 1's or 2's complement method.
- A positive binary number is represented by a 0 sign bit.
- A negative binary number is represented by a 1 sign bit.
- For arithmetic operations, negative binary numbers are represented in 1's complement or 2's complement form.
- In an addition operation, an overflow is possible when both numbers are positive or when both numbers are negative. An incorrect sign bit in the sum indicates the occurrence of an overflow.
- The hexadecimal number system consists of 16 digits and characters, 0 through 9 followed by A through F.
- One hexadecimal digit represents a 4-bit binary number, and its primary usefulness is in simplifying bit patterns and making them easier to read.
- A decimal number can be converted to hexadecimal by the repeated division-by-16 method.
- The octal number system consists of eight digits, 0 through 7.
- A decimal number can be converted to octal by using the repeated division-by-8 method.
- Octal-to-binary conversion is accomplished by simply replacing each octal digit with its 3-bit binary equivalent. The process is reversed for binary-to-octal conversion.
- A decimal number is converted to BCD by replacing each decimal digit with the appropriate 4-bit binary code.
- The ASCII is a 7-bit alphanumeric code that is used in computer systems for input and output of information.
- A parity bit is used to detect an error in a code.
- The CRC (cyclic redundancy check) is based on polynomial division using modulo-2 operations.

KEY TERMS

Key terms and other bold terms in the chapter are defined in the end-of-book glossary.

Alphanumeric   Consisting of numerals, letters, and other characters.

ASCII   American Standard Code for Information Interchange; the most widely used alphanumeric code.

BCD   Binary coded decimal; a digital code in which each of the decimal digits, 0 through 9, is represented by a group of four bits.

Byte   A group of eight bits.

Cyclic redundancy check (CRC)   A type of error detection code.

Floating-point number   A number representation based on scientific notation in which the number consists of an exponent and a mantissa.

Hexadecimal   Describes a number system with a base of 16.

LSB   Least significant bit; the right-most bit in a binary whole number or code.

MSB   Most significant bit; the left-most bit in a binary whole number or code.

Octal   Describes a number system with a base of eight.

Parity   In relation to binary codes, the condition of evenness or oddness of the number of 1s in a code group.

TRUE/FALSE QUIZ

Answers are at the end of the chapter.

1. The decimal number system is a weighted system with ten digits.
2. The binary number system is a weighted system with two digits.
3. LSB stands for lowest single bit.
4. In binary, $1 + 1 = 2$.
5. The 1's complement of the binary number 1010 is 0101.
6. The 2's complement of the binary number 0001 is 1110.
7. The right-most bit in a signed binary number is the sign bit.
8. The hexadecimal number system has 16 characters, six of which are alphabetic characters.
9. BCD stands for binary coded decimal.
10. ASCII stands for American standard code for information indication.
11. CRC stands for cyclic redundancy check.
12. The modulo-2 sum of 11 and 10 is 100.

SELF-TEST

Answers are at the end of the chapter.

1. $2 \times 10^1 + 8 \times 10^0$ is equal to
   (a) 10   (b) 280   (c) 2.8   (d) 28
2. The binary number 1101 is equal to the decimal number
   (a) 13   (b) 49   (c) 11   (d) 3
3. The binary number 11011101 is equal to the decimal number
   (a) 121   (b) 221   (c) 441   (d) 256
4. The decimal number 17 is equal to the binary number
   (a) 10010   (b) 11000   (c) 10001   (d) 01001
5. The decimal number 175 is equal to the binary number
   (a) 11001111   (b) 10101110   (c) 10101111   (d) 11101111
6. The sum of 11010 + 01111 equals
   (a) 101001   (b) 101010   (c) 110101   (d) 101000

7. The difference of 110 − 010 equals
   (a) 001   (b) 010   (c) 101   (d) 100

8. The 1's complement of 10111001 is
   (a) 01000111   (b) 01000110   (c) 11000110   (d) 10101010

9. The 2's complement of 11001000 is
   (a) 00110111   (b) 00110001   (c) 01001000   (d) 00111000

10. The decimal number +122 is expressed in the 2's complement form as
    (a) 01111010   (b) 11111010   (c) 01000101   (d) 10000101

11. The decimal number −34 is expressed in the 2's complement form as
    (a) 01011110   (b) 10100010   (c) 11011110   (d) 01011101

12. A single-precision floating-point binary number has a total of
    (a) 8 bits   (b) 16 bits   (c) 24 bits   (d) 32 bits

13. In the 2's complement form, the binary number 10010011 is equal to the decimal number
    (a) −19   (b) +109   (c) +91   (d) −109

14. The binary number 101100111001010100001 can be written in octal as
    (a) 5471230_8 (b) 5471241_8 (c) 2634521_8 (d) 23162501_8

15. The binary number 1000110101010001101111 can be written in hexadecimal as
    (a) $AD467_{16}$ (b) $8C46F_{16}$ (c) $8D46F_{16}$ (d) $AE46F_{16}$

16. The binary number for $F7A9_{16}$ is
    (a) 1111011110101001   (b) 1110111110101001
    (c) 1111111010110001   (d) 1111011010101001

17. The BCD number for decimal 473 is
    (a) 111011010   (b) 110001110011   (c) 010001110011   (d) 010011110011

18. Refer to Table 2–7. The command STOP in ASCII is
    (a) 1010011101010010011111010000   (b) 1010010100110010011101010000
    (c) 1001010110110110011101010001   (d) 1010011101010010011101100100

19. The code that has an even-parity error is
    (a) 1010011   (b) 1101000   (c) 1001000   (d) 1110111

20. In the cyclic redundancy check, the absence of errors is indicated by
    (a) Remainder = generator code   (b) Remainder = 0
    (c) Remainder = 1   (d) Quotient = 0

PROBLEMS

Answers to odd-numbered problems are at the end of the book.

Section 2–1 Decimal Numbers

1. What is the weight of the digit 6 in each of the following decimal numbers?
   (a) 1386   (b) 54,692   (c) 671,920

2. Express each of the following decimal numbers as a power of ten:
   (a) 10   (b) 100   (c) 10,000   d) 1,000,000

3. Give the value of each digit in the following decimal numbers:
   (a) 471   (b) 9356   (c) 125,000

4. How high can you count with four decimal digits?

Section 2–2 Binary Numbers

5. Convert the following binary numbers to decimal:
   (a) 11   (b) 100   (c) 111   (d) 1000
   (e) 1001   (f) 1100   (g) 1011   (h) 1111

6. Convert the following binary numbers to decimal:
   (a) 1110   (b) 1010   (c) 11100   (d) 10000
   (e) 10101   (f) 11101   (g) 10111   (h) 11111

7. Convert each binary number to decimal:

   (a) 110011.11   (b) 101010.01   (c) 1000001.111
   (d) 1111000.101   (e) 1011100.10101   (f) 1110001.0001
   (g) 1011010.1010   (h) 1111111.11111

8. What is the highest decimal number that can be represented by each of the following numbers of binary digits (bits)?

   (a) two   (b) three   (c) four   (d) five   (e) six
   (f) seven   (g) eight   (h) nine   (i) ten   (j) eleven

9. How many bits are required to represent the following decimal numbers?

   (a) 17   (b) 35   (c) 49   (d) 68
   (e) 81   (f) 114   (g) 132   (h) 205

10. Generate the binary sequence for each decimal sequence:

    (a) 0 through 7   (b) 8 through 15   (c) 16 through 31
    (d) 32 through 63   (e) 64 through 75

Section 2–3 Decimal-to-Binary Conversion

11. Convert each decimal number to binary by using the sum-of-weights method:

    (a) 10   (b) 17   (c) 24   (d) 48
    (e) 61   (f) 93   (g) 125   (h) 186

12. Convert each decimal fraction to binary using the sum-of-weights method:

    (a) 0.32   (b) 0.246   (c) 0.0981

13. Convert each decimal number to binary using repeated division by 2:

    (a) 15   (b) 21   (c) 28   (d) 34
    (e) 40   (f) 59   (g) 65   (h) 73

14. Convert each decimal fraction to binary using repeated multiplication by 2:

    (a) 0.98   (b) 0.347   (c) 0.9028

Section 2–4 Binary Arithmetic

15. Add the binary numbers:

    (a) 11 + 01   (b) 10 + 10   (c) 101 + 11
    (d) 111 + 110   (e) 1001 + 101   (f) 1101 + 1011

16. Use direct subtraction on the following binary numbers:

    (a) 11 − 1   (b) 101 − 100   (c) 110 − 101
    (d) 1110 − 11   (e) 1100 − 1001   (f) 11010 − 10111

17. Perform the following binary multiplications:

    (a) 11 × 11   (b) 100 × 10   (c) 111 × 101
    (d) 1001 × 110   (e) 1101 × 1101   (f) 1110 × 1101

18. Divide the binary numbers as indicated:

    (a) 100 ÷ 10   (b) 1001 ÷ 11   (c) 1100 ÷ 100

Section 2–5 Complements of Binary Numbers

19. What are two ways of representing zero in 1's complement form?

20. How is zero represented in 2's complement form?

21. Determine the 1's complement of each binary number:

    (a) 101   (b) 110   (c) 1010
    (d) 11010111   (e) 1110101   (f) 00001

22. Determine the 2's complement of each binary number using either method:

    (a) 10   (b) 111   (c) 1001   (d) 1101
    (e) 11100   (f) 10011   (g) 10110000   (h) 00111101

Section 2–6 Signed Numbers

23. Express each decimal number in binary as an 8-bit sign-magnitude number:

    (a) +29   (b) −85   (c) +100   (d) −123

24. Express each decimal number as an 8-bit number in the 1's complement form:

    (a) −34   (b) +57   (c) −99   (d) +115

25. Express each decimal number as an 8-bit number in the 2's complement form:

    (a) +12   (b) −68   (c) +101   (d) −125

26. Determine the decimal value of each signed binary number in the sign-magnitude form:

    (a) 10011001   (b) 01110100   (c) 10111111

27. Determine the decimal value of each signed binary number in the 1's complement form:

    (a) 10011001   (b) 01110100   (c) 10111111

28. Determine the decimal value of each signed binary number in the 2's complement form:

    (a) 10011001   (b) 01110100   (c) 10111111

29. Express each of the following sign-magnitude binary numbers in single-precision floating-point format:

    (a) 0111110000101011   (b) 100110000011000

30. Determine the values of the following single-precision floating-point numbers:

    (a) 1 10000001 01001001110001000000000
    (b) 0 11001100 10000111110100100000000

Section 2–7 Arithmetic Operations with Signed Numbers

31. Convert each pair of decimal numbers to binary and add using the 2's complement form:

    (a) 33 and 15   (b) 56 and −27   (c) −46 and 25   (d) −110 and −84

32. Perform each addition in the 2's complement form:

    (a) 00010110 + 00110011   (b) 01110000 + 10101111

33. Perform each addition in the 2's complement form:

    (a) 10001100 + 00111001   (b) 11011001 + 11100111

34. Perform each subtraction in the 2's complement form:

    (a) 00110011 − 00010000   (b) 01100101 − 11101000

35. Multiply 01101010 by 11110001 in the 2's complement form.

36. Divide 01000100 by 00011001 in the 2's complement form.

Section 2–8 Hexadecimal Numbers

37. Convert each hexadecimal number to binary:

    (a) 38_{16} (b) 59_{16} (c) $A14_{16}$ (d) $5C8_{16}$
    (e) 4100_{16} (f) $FB17_{16}$ (g) $8A9D_{16}$

38. Convert each binary number to hexadecimal:

    (a) 1110   (b) 10   (c) 10111
    (d) 10100110   (e) 111111110000 (f) 100110000010

39. Convert each hexadecimal number to decimal:

    (a) 23_{16} (b) 92_{16} (c) $1A_{16}$ (d) $8D_{16}$
    (e) $F3_{16}$ (f) EB_{16} (g) $5C2_{16}$ (h) 700_{16}

40. Convert each decimal number to hexadecimal:

    (a) 8   (b) 14   (c) 33   (d) 52
    (e) 284   (f) 2890   (g) 4019   (h) 6500

41. Perform the following additions:

    (a) $37_{16} + 29_{16}$ (b) $A0_{16} + 6B_{16}$ (c) $FF_{16} + BB_{16}$

42. Perform the following subtractions:

    (a) $51_{16} - 40_{16}$ (b) $C8_{16} - 3A_{16}$ (c) $FD_{16} - 88_{16}$

Section 2–9 Octal Numbers

43. Convert each octal number to decimal:

    (a) 12_8 **(b)** 27_8 **(c)** 56_8 **(d)** 64_8 **(e)** 103_8
    (f) 557_8 **(g)** 163_8 **(h)** 1024_8 **(i)** 7765_8

44. Convert each decimal number to octal by repeated division by 8:

    (a) 15   **(b)** 27   **(c)** 46   **(d)** 70
    (e) 100   **(f)** 142   **(g)** 219   **(h)** 435

45. Convert each octal number to binary:

    (a) 13_8 **(b)** 57_8 **(c)** 101_8 **(d)** 321_8 **(e)** 540_8
    (f) 4653_8 **(g)** 13271_8 **(h)** 45600_8 **(i)** 100213_8

46. Convert each binary number to octal:

    (a) 111   **(b)** 10   **(c)** 110111
    (d) 101010   **(e)** 1100   **(f)** 1011110
    (g) 101100011001   **(h)** 10110000011   **(i)** 111111101111000

Section 2–10 Binary Coded Decimal (BCD)

47. Convert each of the following decimal numbers to 8421 BCD:

    (a) 10   **(b)** 13   **(c)** 18   **(d)** 21   **(e)** 25   **(f)** 36
    (g) 44   **(h)** 57   **(i)** 69   **(j)** 98   **(k)** 125   **(l)** 156

48. Convert each of the decimal numbers in Problem 47 to straight binary, and compare the number of bits required with that required for BCD.

49. Convert the following decimal numbers to BCD:

    (a) 104   **(b)** 128   **(c)** 132   **(d)** 150   **(e)** 186
    (f) 210   **(g)** 359   **(h)** 547   **(i)** 1051

50. Convert each of the BCD numbers to decimal:

    (a) 0001   **(b)** 0110   **(c)** 1001
    (d) 00011000   **(e)** 00011001   **(f)** 00110010
    (g) 01000101   **(h)** 10011000   **(i)** 100001110000

51. Convert each of the BCD numbers to decimal:

    (a) 10000000   **(b)** 001000110111
    (c) 001101000110   **(d)** 010000100001
    (e) 011101010100   **(f)** 100000000000
    (g) 100101111000   **(h)** 0001011010000011
    (i) 1001000000011000   **(j)** 0110011001100111

52. Add the following BCD numbers:

    (a) 0010 + 0001   **(b)** 0101 + 0011
    (c) 0111 + 0010   **(d)** 1000 + 0001
    (e) 00011000 + 00010001   **(f)** 01100100 + 00110011
    (g) 01000000 + 01000111   **(h)** 10000101 + 00010011

53. Add the following BCD numbers:

    (a) 1000 + 0110   **(b)** 0111 + 0101
    (c) 1001 + 1000   **(d)** 1001 + 0111
    (e) 00100101 + 00100111   **(f)** 01010001 + 01011000
    (g) 10011000 + 10010111   **(h)** 010101100001 + 011100001000

54. Convert each pair of decimal numbers to BCD, and add as indicated:

    (a) 4 + 3   **(b)** 5 + 2   **(c)** 6 + 4   **(d)** 17 + 12
    (e) 28 + 23   **(f)** 65 + 58   **(g)** 113 + 101   **(h)** 295 + 157

Section 2–11 Digital Codes

55. In a certain application a 4-bit binary sequence cycles from 1111 to 0000 periodically. There are four bit changes, and because of circuit delays, these changes may not occur at the same instant. For example, if the LSB changes first, the number will appear as 1110 during the transition from 1111 to 0000 and may be misinterpreted by the system. Illustrate how the Gray code avoids this problem.

56. Convert each binary number to Gray code:

    (a) 11011   **(b)** 1001010   **(c)** 1111011101110

57. Convert each Gray code to binary:

    (a) 1010   **(b)** 00010   **(c)** 11000010001

58. Convert each of the following decimal numbers to ASCII. Refer to Table 2–7.

    (a) 1   **(b)** 3   **(c)** 6   **(d)** 10   **(e)** 18

    (f) 29   **(g)** 56   **(h)** 75   **(i)** 107

59. Determine each ASCII character. Refer to Table 2–7.

    (a) 0011000   **(b)** 1001010   **(c)** 0111101

    (d) 0100011   **(e)** 0111110   **(f)** 1000010

60. Decode the following ASCII coded message:

    1001000 1100101 1101100 1101100 1101111 0101110

    0100000 1001000 1101111 1110111 0100000 1100001

    1110010 1100101 0100000 1111001 1101111 1110101

    0111111

61. Write the message in Problem 60 in hexadecimal.

62. Convert the following statement to ASCII:

30 INPUT A, B

Section 2–12 Error Codes

63. Determine which of the following even parity codes are in error:

    (a) 100110010   **(b)** 011101010   **(c)** 10111111010001010

64. Determine which of the following odd parity codes are in error:

    (a) 11110110   **(b)** 00110001   **(c)** 01010101010101010

65. Attach the proper even parity bit to each of the following bytes of data:

    (a) 10100100   **(b)** 00001001   **(c)** 11111110

66. Apply modulo-2 to the following:

    (a) 1100 + 1011   **(b)** 1111 + 0100   **(c)** 10011001 + 100011100

67. Verify that modulo-2 subtraction is the same as modulo-2 addition by adding the result of each operation in problem 66 to either of the original numbers to get the other number. This will show that the result is the same as the difference of the two numbers.

68. Apply CRC to the data bits 10110010 using the generator code 1010 to produce the transmitted CRC code.

69. Assume that the code produced in problem 68 incurs an error in the most significant bit during transmission. Apply CRC to detect the error.

ANSWERS

SECTION CHECKUPS

Section 2–1 Decimal Numbers

  1. **(a)** 1370: 10   **(b)** 6725: 100   **(c)** 7051: 1000   **(d)** 58.72: 0.1

  2. **(a)** $51 = (5 \times 10) + (1 \times 1)$

    (b) $137 = (1 \times 100) + (3 \times 10) + (7 \times 1)$

    (c) $1492 = (1 \times 1000) + (4 \times 100) + (9 \times 10) + (2 \times 1)$

    (d) $106.58 = (1 \times 100) + (0 \times 10) + (6 \times 1) + (5 \times 0.1) + (8 \times 0.01)$

Section 2–2 Binary Numbers

1. $2^8 - 1 = 255$
2. Weight is 16.
3. $10111101.011 = 189.375$

Section 2–3 Decimal-to-Binary Conversion

1. (a) $23 = 10111$ (b) $57 = 111001$ (c) $45.5 = 101101.1$
2. (a) $14 = 1110$ (b) $21 = 10101$ (c) $0.375 = 0.011$

Section 2–4 Binary Arithmetic

1. (a) $1101 + 1010 = 10111$ (b) $10111 + 01101 = 100100$
2. (a) $1101 - 0100 = 1001$ (b) $1001 - 0111 = 0010$
3. (a) $110 \times 111 = 101010$ (b) $1100 \div 011 = 100$

Section 2–5 Complements of Binary Numbers

1. (a) 1's comp of $00011010 = 11100101$ (b) 1's comp of $11110111 = 00001000$
   (c) 1's comp of $10001101 = 01110010$
2. (a) 2's comp of $00010110 = 11101010$ (b) 2's comp of $11111100 = 00000100$
   (c) 2's comp of $10010001 = 01101111$

Section 2–6 Signed Numbers

1. Sign-magnitude: $+9 = 00001001$
2. 1's comp: $-33 = 11011110$
3. 2's comp: $-46 = 11010010$
4. Sign bit, exponent, and mantissa

Section 2–7 Arithmetic Operations with Signed Numbers

1. Cases of addition: positive number is larger, negative number is larger, both are positive, both are negative
2. $00100001 + 10111100 = 11011101$
3. $01110111 - 00110010 = 01000101$
4. Sign of product is positive.
5. $00000101 \times 01111111 = 01001111011$
6. Sign of quotient is negative.
7. $00110000 \div 00001100 = 00000100$

Section 2–8 Hexadecimal Numbers

1. (a) $10110011 = B3_{16}$ (b) $110011101000 = CE8_{16}$
2. (a) $57_{16} = 01010111$ (b) $3A5_{16} = 001110100101$
   (c) $F8OB_{16} = 1111100000001011$
3. $9B30_{16} = 39,728_{10}$
4. $573_{10} = 23D_{16}$
5. (a) $18_{16} + 34_{16} = 4C_{16}$ (b) $3F_{16} + 2A_{16} = 69_{16}$
6. (a) $75_{16} - 21_{16} = 54_{16}$ (b) $94_{16} - 5C_{16} = 38_{16}$

Section 2–9 Octal Numbers

1. (a) $73_8 = 59_{10}$ (b) $125_8 = 85_{10}$
2. (a) $98_{10} = 142_8$ (b) $163_{10} = 243_8$

3. (a) $46_8 = 100110$ **(b)** $723_8 = 111010011$ **(c)** $5624_8 = 101110010100$

4. (a) $110101111 = 657_8$ **(b)** $1001100010 = 1142_8$ **(c)** $10111111001 = 2771_8$

Section 2–10 Binary Coded Decimal (BCD)

1. (a) 0010: 2   **(b)** 1000: 8   **(c)** 0001: 1   **(d)** 0100: 4

2. (a) $6_{10} = 0110$ **(b)** $15_{10} = 00010101$ **(c)** $273_{10} = 001001110011$

    (d) $849_{10} = 100001001001$

3. (a) $10001001 = 89_{10}$ **(b)** $001001111000 = 278_{10}$ **(c)** $000101010111 = 157_{10}$

4. A 4-bit sum is invalid when it is greater than 9_{10}.

Section 2–11 Digital Codes

1. (a) $1100_2 = 1010$ Gray   **(b)** $1010_2 = 1111$ Gray   **(c)** $11010_2 = 10111$ Gray

2. (a) 1000 Gray $= 1111_2$ **(b)** 1010 Gray $= 1100_2$ **(c)** 11101 Gray $= 10110_2$

3. (a) K: $1001011 \rightarrow 4B_{16}$ **(b)** r: $1110010 \rightarrow 72_{16}$

    (c) \$: $0100100 \rightarrow 24_{16}$ **(d)** +: $0101011 \rightarrow 2B_{16}$

Section 2–12 Error Codes

1. (c) 0101 has an error.

2. (d) 11111011 has an error.

3. (a) 10101001   **(b)** 01000001   **(c)** 11101110   **(d)** 10001101

4. Cyclic redundancy check

5. (a) 0   **(b)** 0   **(c)** 1   **(d)** 1

RELATED PROBLEMS FOR EXAMPLES

2–1 9 has a value of 900, 3 has a value of 30, 9 has a value of 9.

2–2 6 has a value of 60, 7 has a value of 7, 9 has a value of 9/10 (0.9), 2 has a value of 2/100 (0.02), 4 has a value of 4/1000 (0.004).

2–3 $10010001 = 128 + 16 + 1 = 145$

2–4 $10.111 = 2 + 0.5 + 0.25 + 0.125 = 2.875$

2–5 $125 = 64 + 32 + 16 + 8 + 4 + 1 = 1111101$

2–6 $39 = 100111$

2–7 $1111 + 1100 = 11011$

2–8 $111 - 100 = 011$

2–9 $110 - 101 = 001$

2–10 $1101 \times 1010 = 10000010$

2–11 $1100 \div 100 = 11$

2–12 00110101

2–13 01000000

2–14 See Table 2–10.

TABLE 2–10			
	Sign-Magnitude	1's Comp	2's Comp
+19	00010011	00010011	00010011
−19	10010011	11101100	11101101

2–15 $01110111 = +119_{10}$

2–16 $11101011 = -20_{10}$

2–17 $11010111 = -41_{10}$

2–18 11000010001010011000000000

2–19 01010101

2–20 00010001

2–21 1001000110

2–22 $(83)(-59) = -4897$ (10110011011111 in 2's comp)

2–23 $100 \div 25 = 4$ (0100)

2–24 $4F79C_{16}$

2–25 01101011110100011_2

2–26 $6BD_{16} = 011010111101 = 2^{10} + 2^9 + 2^7 + 2^5 + 2^4 + 2^3 + 2^2 + 2^0$
$= 1024 + 512 + 128 + 32 + 16 + 8 + 4 + 1 = 1725_{10}$

2–27 $60A_{16} = (6 \times 256) + (0 \times 16) + (10 \times 1) = 1546_{10}$

2–28 $2591_{10} = A1F_{16}$

2–29 $4C_{16} + 3A_{16} = 86_{16}$

2–30 $BCD_{16} - 173_{16} = A5A_{16}$

2–31 **(a)** $001011_2 = 11_{10} = 13_8$ **(b)** $010101_2 = 21_{10} = 25_8$
    (c) $001100000_2 = 96_{10} = 140_8$ **(d)** $111101010110_2 = 3926_{10} = 7526_8$

2–32 1250762_8

2–33 1001011001110011

2–34 $82,276_{10}$

2–35 1001100101101000

2–36 10000010

2–37 **(a)** 111011 (Gray)   **(b)** 111010_2

2–38 The sequence of codes for if (y < 8) is $69_{16}66_{16}20_{16}28_{16}79_{16}3C_{16}38_{16}29_{16}$

2–39 01001011

2–40 Yes

2–41 A 0 remainder results

2–42 Errors are indicated.

TRUE/FALSE QUIZ

1. T   **2.** T   **3.** F   **4.** F   **5.** T   **6.** F   **7.** F   **8.** T   **9.** T   **10.** F
11. T   **12.** F

SELF-TEST

1. (d)   **2.** (a)   **3.** (b)   **4.** (c)   **5.** (c)   **6.** (a)   **7.** (d)   **8.** (b)
9. (d)   **10.** (a)   **11.** (c)   **12.** (d)   **13.** (d)   **14.** (b)   **15.** (c)   **16.** (a)
17. (c)   **18.** (a)   **19.** (b)   **20.** (b)

Logic Gates

CHAPTER OBJECTIVES

- Describe the operation of the inverter, the AND gate, and the OR gate

- Describe the operation of the NAND gate and the NOR gate

- Express the operation of NOT, AND, OR, NAND, and NOR gates with Boolean algebra

- Describe the operation of the exclusive-OR and exclusive-NOR gates

- Use logic gates in simple applications

- Recognize and use both the distinctive shape logic gate symbols and the rectangular outline logic gate symbols of ANSI/IEEE Standard 91-1984/Std. 91a-1991

- Construct timing diagrams showing the proper time relationships of inputs and outputs for the various logic gates

- Discuss the basic concepts of programmable logic

- Make basic comparisons between the major IC technologies—CMOS and bipolar (TTL)

- Explain how the different series within the CMOS and bipolar (TTL) families differ from each other

- Define *propagation delay time, power dissipation, speed-power product,* and *fan-out* in relation to logic gates

- List specific fixed-function integrated circuit devices that contain the various logic gates

- Troubleshoot logic gates for opens and shorts by using the oscilloscope

KEY TERMS

Key terms are in order of appearance in the chapter.

- Inverter
- Truth table
- Boolean algebra
- Complement
- AND gate
- OR gate
- NAND gate
- NOR gate
- Exclusive-OR gate
- Exclusive-NOR gate
- AND array
- Fuse
- Antifuse

- EPROM
- EEPROM
- Flash
- SRAM
- Target device
- JTAG
- VHDL
- CMOS
- Bipolar
- Propagation delay time
- Fan-out
- Unit load

VISIT THE WEBSITE

Study aids for this chapter are available at
http://www.pearsonhighered.com/careersresources/

INTRODUCTION

The emphasis in this chapter is on the operation, application, and troubleshooting of logic gates. The relationship of input and output waveforms of a gate using timing diagrams is thoroughly covered.

Logic symbols used to represent the logic gates are in accordance with ANSI/IEEE Standard 91-1984/Std. 91a-1991. This standard has been adopted by private industry and the military for use in internal documentation as well as published literature.

Both fixed-function logic and programmable logic are discussed in this chapter. Because integrated circuits (ICs) are used in all applications, the logic function of a device is generally of greater importance to the technician or technologist than the details of the component-level circuit operation within the IC package. Therefore, detailed coverage of the devices at the component level can be treated as an optional topic. Digital integrated circuit technologies are discussed in Chapter 15 on the website, all or parts of which may be introduced at appropriate points throughout the text.

Suggestion: Review Section 1–3 before you start this chapter.

3–1   The Inverter

The inverter (NOT circuit) performs the operation called *inversion* or *complementation.* The inverter changes one logic level to the opposite level. In terms of bits, it changes a 1 to a 0 and a 0 to a 1.

After completing this section, you should be able to

- ◆ Identify negation and polarity indicators

- ◆ Identify an inverter by either its distinctive shape symbol or its rectangular outline symbol

- ◆ Produce the truth table for an inverter

- ◆ Describe the logical operation of an inverter

Standard logic symbols for the **inverter** are shown in Figure 3–1. Part (a) shows the *distinctive shape* symbols, and part (b) shows the *rectangular outline* symbols. In this textbook, distinctive shape symbols are generally used; however, the rectangular outline symbols are found in many industry publications, and you should become familiar with them as well. (Logic symbols are in accordance with **ANSI/IEEE** Standard 91-1984 and its supplement Standard 91a-1991.)

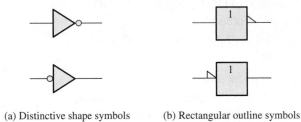

(a) Distinctive shape symbols
with negation indicators

(b) Rectangular outline symbols
with polarity indicators

FIGURE 3–1   Standard logic symbols for the inverter (ANSI/IEEE Std. 91-1984/ Std. 91a-1991).

The Negation and Polarity Indicators

The negation indicator is a "bubble" (O) that indicates **inversion** or *complementation* when it appears on the input or output of any logic element, as shown in Figure 3–1(a) for the inverter. Generally, inputs are on the left of a logic symbol and the output is on the right. When appearing on the input, the bubble means that a 0 is the active or *asserted* input state, and the input is called an active-LOW input. When appearing on the output, the bubble means that a 0 is the active or asserted output state, and the output is called an active-LOW output. The absence of a bubble on the input or output means that a 1 is the active or asserted state, and in this case, the input or output is called active-HIGH.

The polarity or level indicator is a "triangle" (◁) that indicates inversion when it appears on the input or output of a logic element, as shown in Figure 3–1(b). When appearing on the input, it means that a LOW level is the active or asserted input state. When appearing on the output, it means that a LOW level is the active or asserted output state.

Either indicator (bubble or triangle) can be used both on distinctive shape symbols and on rectangular outline symbols. Figure 3–1(a) indicates the principal inverter symbols used in this text. Note that a change in the placement of the negation or polarity indicator does not imply a change in the way an inverter operates.

Inverter Truth Table

When a HIGH level is applied to an inverter input, a LOW level will appear on its output. When a LOW level is applied to its input, a HIGH will appear on its output. This operation is summarized in Table 3–1, which shows the output for each possible input in terms of levels and corresponding bits. A table such as this is called a **truth table**.

Inverter Operation

Figure 3–2 shows the output of an inverter for a pulse input, where t_1 and t_2 indicate the corresponding points on the input and output pulse waveforms.

When the input is LOW, the output is HIGH; when the input is HIGH, the output is LOW, thereby producing an inverted output pulse.

TABLE 3–1	
Inverter truth table.	
Input	**Output**
LOW (0)	HIGH (1)
HIGH (1)	LOW (0)

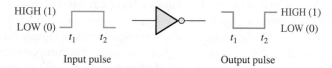

FIGURE 3–2   Inverter operation with a pulse input. Open file F03-02 to verify inverter operation. *A Multisim tutorial is available on the website.*

MultiSim

Timing Diagrams

Recall from Chapter 1 that a *timing diagram* is basically a graph that accurately displays the relationship of two or more waveforms with respect to each other on a time basis. For example, the time relationship of the output pulse to the input pulse in Figure 3–2 can be shown with a simple timing diagram by aligning the two pulses so that the occurrences of the pulse edges appear in the proper time relationship. The rising edge of the input pulse and the falling edge of the output pulse occur at the same time (ideally). Similarly, the falling edge of the input pulse and the rising edge of the output pulse occur at the same time (ideally). This timing relationship is shown in Figure 3–3. In practice, there is a very small delay from the input transition until the corresponding output transition. Timing diagrams are especially useful for illustrating the time relationship of digital waveforms with multiple pulses.

A timing diagram shows how two or more waveforms relate in time.

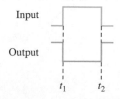

FIGURE 3–3   Timing diagram for the case in Figure 3–2.

EXAMPLE 3–1

A waveform is applied to an inverter in Figure 3–4. Determine the output waveform corresponding to the input and show the timing diagram. According to the placement of the bubble, what is the active output state?

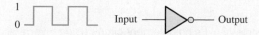

FIGURE 3–4

Solution

The output waveform is exactly opposite to the input (inverted), as shown in Figure 3–5, which is the basic timing diagram. The active or asserted output state is **0**.

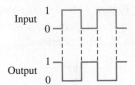

FIGURE 3–5

Related Problem*

If the inverter is shown with the negative indicator (bubble) on the input instead of the output, how is the timing diagram affected?

_____

*Answers are at the end of the chapter.

Logic Expression for an Inverter

In **Boolean algebra**, which is the mathematics of logic circuits and will be covered thoroughly in Chapter 4, a variable is generally designated by one or two letters although there can be more. Letters near the beginning of the alphabet usually designate inputs, while letters near the end of the alphabet usually designate outputs. The **complement** of a variable is designated by a bar over the letter. A variable can take on a value of either 1 or 0. If a given variable is 1, its complement is 0 and vice versa.

The operation of an inverter (NOT circuit) can be expressed as follows: If the input variable is called A and the output variable is called X, then

$$X = \overline{A}$$

This expression states that the output is the complement of the input, so if $A = 0$, then $X = 1$, and if $A = 1$, then $X = 0$. Figure 3–6 illustrates this. The complemented variable $\overline{A}$ can be read as "A bar" or "not A."

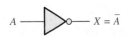

FIGURE 3–6 The inverter complements an input variable.

An Application

Figure 3–7 shows a circuit for producing the 1's complement of an 8-bit binary number. The bits of the binary number are applied to the inverter inputs and the 1's complement of the number appears on the outputs.

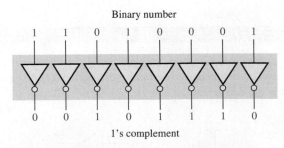

FIGURE 3–7 Example of a 1's complement circuit using inverters.

Answers are at the end of the chapter.

1. When a 1 is on the input of an inverter, what is the output?

2. An active-HIGH pulse (HIGH level when asserted, LOW level when not) is required on an inverter input.

   (a) Draw the appropriate logic symbol, using the distinctive shape and the negation indicator, for the inverter in this application.

   (b) Describe the output when a positive-going pulse is applied to the input of an inverter.

3–2 The AND Gate

The AND gate is one of the basic gates that can be combined to form any logic function. An AND gate can have two or more inputs and performs what is known as logical multiplication.

After completing this section, you should be able to

- Identify an AND gate by its distinctive shape symbol or by its rectangular outline symbol

- Describe the operation of an AND gate

- Generate the truth table for an AND gate with any number of inputs

- Produce a timing diagram for an AND gate with any specified input waveforms

- Write the logic expression for an AND gate with any number of inputs

- Discuss examples of AND gate applications

The term *gate* was introduced in Chapter 1 and is used to describe a circuit that performs a basic logic operation. The AND gate is composed of two or more inputs and a single output, as indicated by the standard logic symbols shown in Figure 3–8. Inputs are on the left, and the output is on the right in each symbol. Gates with two inputs are shown; however, an AND gate can have any number of inputs greater than one. Although examples of both distinctive shape symbols and rectangular outline symbols are shown, the distinctive shape symbol, shown in part (a), is used predominantly in this book.

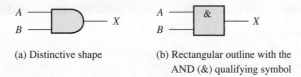

(a) Distinctive shape　　　(b) Rectangular outline with the AND (&) qualifying symbol

FIGURE 3–8 Standard logic symbols for the AND gate showing two inputs (ANSI/IEEE Std. 91-1984/Std. 91a-1991).

Operation of an AND Gate

An **AND gate** produces a HIGH output *only* when *all* of the inputs are HIGH. When any of the inputs is LOW, the output is LOW. Therefore, the basic purpose of an AND gate is to determine when certain conditions are simultaneously true, as indicated by HIGH levels on all of its inputs, and to produce a HIGH on its output to indicate that all these conditions are

InfoNote

Logic gates are one of the fundamental building blocks of digital systems. Most of the functions in a computer, with the exception of certain types of memory, are implemented with logic gates used on a very large scale. For example, a microprocessor, which is the main part of a computer, is made up of hundreds of thousands or even millions of logic gates.

An AND gate can have more than two inputs.

true. The inputs of the 2-input AND gate in Figure 3–8 are labeled A and B, and the output is labeled X. The gate operation can be stated as follows:

For a 2-input AND gate, output X is HIGH only when inputs A and B are HIGH; X is LOW when either A or B is LOW, or when both A and B are LOW.

Figure 3–9 illustrates a 2-input AND gate with all four possibilities of input combinations and the resulting output for each.

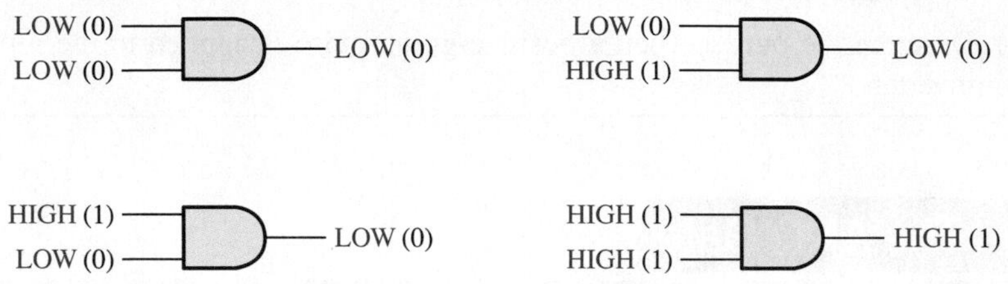

LOW (0) — LOW (0) ⟫ LOW (0)   LOW (0) — HIGH (1) ⟫ LOW (0)

HIGH (1) — LOW (0) ⟫ LOW (0)   HIGH (1) — HIGH (1) ⟫ HIGH (1)

 FIGURE 3–9   All possible logic levels for a 2-input AND gate. Open file F03-09 to verify AND gate operation.

AND Gate Truth Table

For an AND gate, all HIGH inputs produce a HIGH output.

The logical operation of a gate can be expressed with a truth table that lists all input combinations with the corresponding outputs, as illustrated in Table 3–2 for a 2-input AND gate. The truth table can be expanded to any number of inputs. Although the terms HIGH and LOW tend to give a "physical" sense to the input and output states, the truth table is shown with 1s and 0s; a HIGH is equivalent to a 1 and a LOW is equivalent to a 0 in positive logic. For any AND gate, regardless of the number of inputs, the output is HIGH *only* when *all* inputs are HIGH.

The total number of possible combinations of binary inputs to a gate is determined by the following formula:

$$N = 2^n$$

Equation 3–1

where N is the number of possible input combinations and n is the number of input variables. To illustrate,

For two input variables: $N = 2^2 = 4$ combinations
For three input variables: $N = 2^3 = 8$ combinations
For four input variables: $N = 2^4 = 16$ combinations

You can determine the number of input bit combinations for gates with any number of inputs by using Equation 3–1.

TABLE 3–2

Truth table for a 2-input AND gate.

Inputs		Output
A	B	X
0	0	0
0	1	0
1	0	0
1	1	1

1 = HIGH, 0 = LOW

EXAMPLE 3–2

TABLE 3–3

Inputs			Output
A	B	C	X
0	0	0	0
0	0	1	0
0	1	0	0
0	1	1	0
1	0	0	0
1	0	1	0
1	1	0	0
1	1	1	1

(a) Develop the truth table for a 3-input AND gate.

(b) Determine the total number of possible input combinations for a 4-input AND gate.

Solution

(a) There are eight possible input combinations ($2^3 = 8$) for a 3-input AND gate. The input side of the truth table (Table 3–3) shows all eight combinations of three bits. The output side is all 0s except when all three input bits are 1s.

(b) $N = 2^4 = \mathbf{16}$. There are 16 possible combinations of input bits for a 4-input AND gate.

Related Problem

Develop the truth table for a 4-input AND gate.

AND Gate Operation with Waveform Inputs

In most applications, the inputs to a gate are not stationary levels but are voltage waveforms that change frequently between HIGH and LOW logic levels. Now let's look at the operation of AND gates with pulse waveform inputs, keeping in mind that an AND gate obeys the truth table operation regardless of whether its inputs are constant levels or levels that change back and forth.

Let's examine the waveform operation of an AND gate by looking at the inputs with respect to each other in order to determine the output level at any given time. In Figure 3–10, inputs A and B are both HIGH (1) during the time interval, t_1, making output X HIGH (1) during this interval. During time interval t_2, input A is LOW (0) and input B is HIGH (1), so the output is LOW (0). During time interval t_3, both inputs are HIGH (1) again, and therefore the output is HIGH (1). During time interval t_4, input A is HIGH (1) and input B is LOW (0), resulting in a LOW (0) output. Finally, during time interval t_5, input A is LOW (0), input B is LOW (0), and the output is therefore LOW (0). As you know, a diagram of input and output waveforms showing time relationships is called a *timing diagram.*

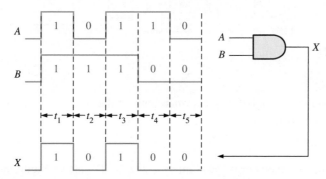

FIGURE 3–10 Example of AND gate operation with a timing diagram showing input and output relationships.

EXAMPLE 3–3

If two waveforms, A and B, are applied to the AND gate inputs as in Figure 3–11, what is the resulting output waveform?

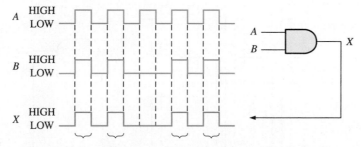

A and *B* are both HIGH during these four time intervals; therefore, *X* is HIGH.

FIGURE 3–11

Solution

The output waveform X is HIGH only when both A and B waveforms are HIGH as shown in the timing diagram in Figure 3–11.

Related Problem

Determine the output waveform and show a timing diagram if the second and fourth pulses in waveform A of Figure 3–11 are replaced by LOW levels.

Remember, when analyzing the waveform operation of logic gates, it is important to pay careful attention to the time relationships of all the inputs with respect to each other and to the output.

EXAMPLE 3–4

For the two input waveforms, A and B, in Figure 3–12, show the output waveform with its proper relation to the inputs.

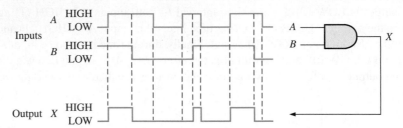

FIGURE 3–12

Solution

The output waveform is HIGH only when both of the input waveforms are HIGH as shown in the timing diagram.

Related Problem

Show the output waveform if the B input to the AND gate in Figure 3–12 is always HIGH.

EXAMPLE 3–5

For the 3-input AND gate in Figure 3–13, determine the output waveform in relation to the inputs.

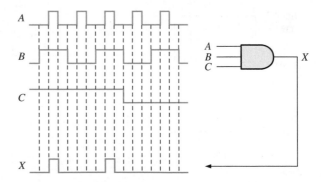

FIGURE 3–13

Solution

The output waveform X of the 3-input AND gate is HIGH only when all three input waveforms A, B, and C are HIGH.

Related Problem

What is the output waveform of the AND gate in Figure 3–13 if the C input is always HIGH?

EXAMPLE 3–6

Use Multisim to simulate a 3-input AND gate with input waveforms that cycle through binary numbers 0 through 9.

Solution

Use the Multisim word generator in the up counter mode to provide the combination of waveforms representing the binary sequence, as shown in Figure 3–14. The first three waveforms on the oscilloscope display are the inputs, and the bottom waveform is the output.

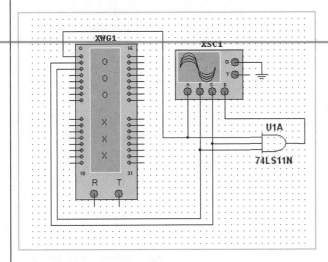

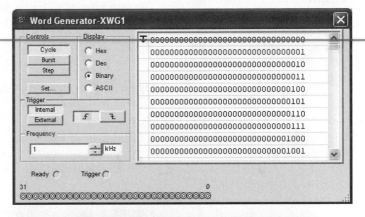

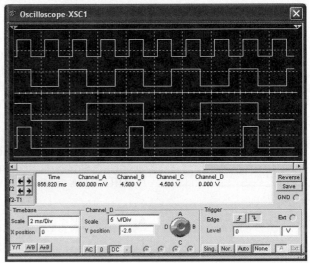

FIGURE 3–14

Related Problem

Use Multisim software to create the setup and simulate the 3-input AND gate as illustrated in this example.

MultiSim

Logic Expressions for an AND Gate

The logical AND function of two variables is represented mathematically either by placing a dot between the two variables, as $A \cdot B$, or by simply writing the adjacent letters without the dot, as AB. We will normally use the latter notation.

When variables are shown together like *ABC*, they are ANDed.

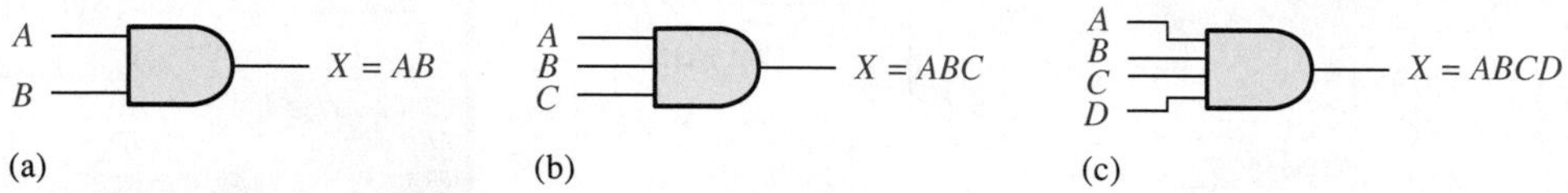

Boolean multiplication follows the same basic rules governing binary multiplication, which were discussed in Chapter 2 and are as follows:

$$0 \cdot 0 = 0$$
$$0 \cdot 1 = 0$$
$$1 \cdot 0 = 0$$
$$1 \cdot 1 = 1$$

Boolean multiplication is the same as the AND function.

The operation of a 2-input AND gate can be expressed in equation form as follows: If one input variable is A, if the other input variable is B, and if the output variable is X, then the Boolean expression is

$$X = AB$$

Figure 3–15(a) shows the AND gate logic symbol with two input variables and the output variable indicated.

(a)   (b)   (c)

FIGURE 3–15   Boolean expressions for AND gates with two, three, and four inputs.

To extend the AND expression to more than two input variables, simply use a new letter for each input variable. The function of a 3-input AND gate, for example, can be expressed as $X = ABC$, where A, B, and C are the input variables. The expression for a 4-input AND gate can be $X = ABCD$, and so on. Parts (b) and (c) of Figure 3–15 show AND gates with three and four input variables, respectively.

You can evaluate an AND gate operation by using the Boolean expressions for the output. For example, each variable on the inputs can be either a 1 or a 0; so for the 2-input AND gate, make substitutions in the equation for the output, $X = AB$, as shown in Table 3–4. This evaluation shows that the output X of an AND gate is a 1 (HIGH) only when both inputs are 1s (HIGHs). A similar analysis can be made for any number of input variables.

TABLE 3–4

A	B	AB = X
0	0	$0 \cdot 0 = 0$
0	1	$0 \cdot 1 = 0$
1	0	$1 \cdot 0 = 0$
1	1	$1 \cdot 1 = 1$

Applications

The AND Gate as an Enable/Inhibit Device

A common application of the AND gate is to **enable** (that is, to allow) the passage of a signal (pulse waveform) from one point to another at certain times and to **inhibit** (prevent) the passage at other times.

A simple example of this particular use of an AND gate is shown in Figure 3–16, where the AND gate controls the passage of a signal (waveform A) to a digital counter. The purpose of this circuit is to measure the frequency of waveform A. The enable pulse has a width of precisely 1 ms. When the enable pulse is HIGH, waveform A passes through the gate to the counter; and when the enable pulse is LOW, the signal is prevented from passing through the gate (inhibited).

During the 1 millisecond (1 ms) interval of the enable pulse, pulses in waveform A pass through the AND gate to the counter. The number of pulses passing through during the 1 ms interval is equal to the frequency of waveform A. For example, Figure 3–16 shows six pulses in one millisecond, which is a frequency of 6 kHz. If 1000 pulses pass through the gate in the 1 ms interval of the enable pulse, there are 1000 pulses/ms, or a frequency of 1 MHz.

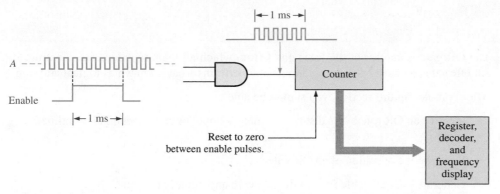

FIGURE 3–16   An AND gate performing an enable/inhibit function for a frequency counter.

The counter counts the number of pulses per second and produces a binary output that goes to a decoding and display circuit to produce a readout of the frequency. The enable pulse repeats at certain intervals and a new updated count is made so that if the frequency changes, the new value will be displayed. Between enable pulses, the counter is reset so that it starts at zero each time an enable pulse occurs. The current frequency count is stored in a register so that the display is unaffected by the resetting of the counter.

A Seat Belt Alarm System

In Figure 3–17, an AND gate is used in a simple automobile seat belt alarm system to detect when the ignition switch is on *and* the seat belt is unbuckled. If the ignition switch is on, a HIGH is produced on input A of the AND gate. If the seat belt is not properly buckled, a HIGH is produced on input B of the AND gate. Also, when the ignition switch is turned on, a timer is started that produces a HIGH on input C for 30 s. If all three conditions exist—that is, if the ignition is on *and* the seat belt is unbuckled *and* the timer is running—the output of the AND gate is HIGH, and an audible alarm is energized to remind the driver.

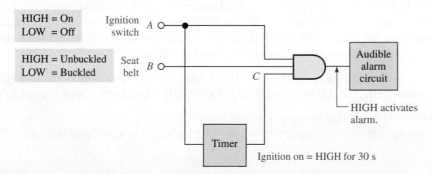

FIGURE 3–17   A simple seat belt alarm circuit using an AND gate.

SECTION 3–2 CHECKUP

1. When is the output of an AND gate HIGH?
2. When is the output of an AND gate LOW?
3. Describe the truth table for a 5-input AND gate.

3-3   The OR Gate

The OR gate is another of the basic gates from which all logic functions are constructed. An OR gate can have two or more inputs and performs what is known as logical addition.

After completing this section, you should be able to

 ◆ Identify an OR gate by its distinctive shape symbol or by its rectangular outline symbol

 ◆ Describe the operation of an OR gate

 ◆ Generate the truth table for an OR gate with any number of inputs

 ◆ Produce a timing diagram for an OR gate with any specified input waveforms

 ◆ Write the logic expression for an OR gate with any number of inputs

 ◆ Discuss an OR gate application

An OR gate can have more than two inputs.

An **OR gate** has two or more inputs and one output, as indicated by the standard logic symbols in Figure 3–18, where OR gates with two inputs are illustrated. An OR gate can have any number of inputs greater than one. Although both distinctive shape and rectangular outline symbols are shown, the distinctive shape OR gate symbol is used in this textbook.

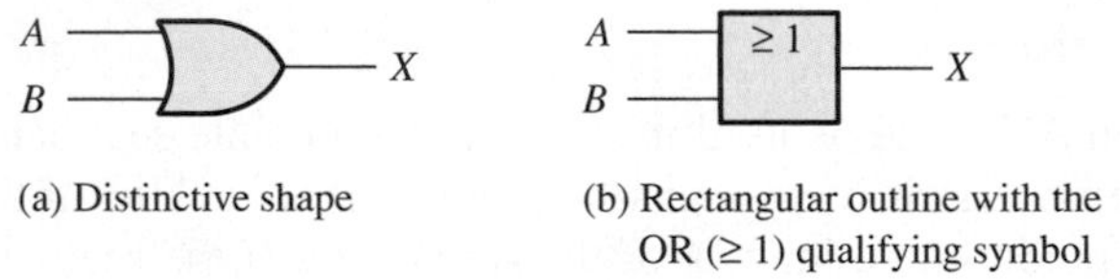

A ──╲
         ⟩── X
B ──╱

A ──▭ ≥ 1 ── X
B ──

(a) Distinctive shape

(b) Rectangular outline with the OR (≥ 1) qualifying symbol

FIGURE 3–18   Standard logic symbols for the OR gate showing two inputs (ANSI/IEEE Std. 91-1984/Std. 91a-1991).

Operation of an OR Gate

For an OR gate, at least one HIGH input produces a HIGH output.

An OR gate produces a HIGH on the output when *any* of the inputs is HIGH. The output is LOW only when all of the inputs are LOW. Therefore, an OR gate determines when one or more of its inputs are HIGH and produces a HIGH on its output to indicate this condition. The inputs of the 2-input OR gate in Figure 3–18 are labeled A and B, and the output is labeled X. The operation of the gate can be stated as follows:

> **For a 2-input OR gate, output X is HIGH when either input A or input B is HIGH, or when both A and B are HIGH; X is LOW only when both A and B are LOW.**

The HIGH level is the active or asserted output level for the OR gate. Figure 3–19 illustrates the operation for a 2-input OR gate for all four possible input combinations.

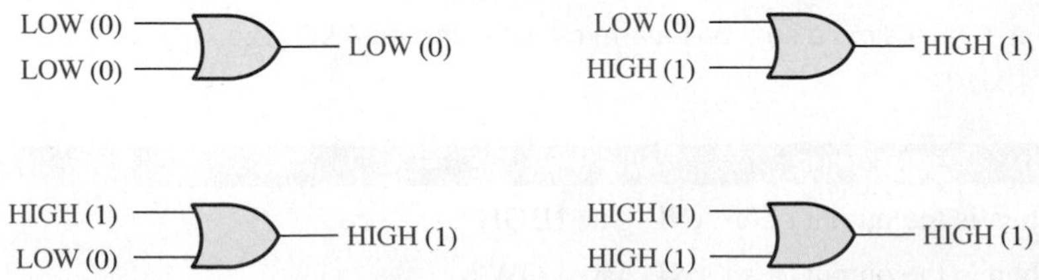

LOW (0) ──⟩── LOW (0)
LOW (0) ──

LOW (0) ──⟩── HIGH (1)
HIGH (1) ──

HIGH (1) ──⟩── HIGH (1)
LOW (0) ──

HIGH (1) ──⟩── HIGH (1)
HIGH (1) ──

MultiSim   **FIGURE 3–19**   All possible logic levels for a 2-input OR gate. Open file F03-19 to verify OR gate operation.

OR Gate Truth Table

The operation of a 2-input OR gate is described in Table 3–5. This truth table can be expanded for any number of inputs; but regardless of the number of inputs, the output is HIGH when one or more of the inputs are HIGH.

OR Gate Operation with Waveform Inputs

Now let's look at the operation of an OR gate with pulse waveform inputs, keeping in mind its logical operation. Again, the important thing in the analysis of gate operation with pulse waveforms is the time relationship of all the waveforms involved. For example, in Figure 3–20, inputs A and B are both HIGH (1) during time interval t_1, making output X HIGH (1). During time interval t_2, input A is LOW (0), but because input B is HIGH (1), the output is HIGH (1). Both inputs are LOW (0) during time interval t_3, so there is a LOW (0) output during this time. During time interval t_4, the output is HIGH (1) because input A is HIGH (1).

TABLE 3–5

Truth table for a 2-input OR gate.

Inputs		Output
A	B	X
0	0	0
0	1	1
1	0	1
1	1	1

$1 =$ HIGH, $0 =$ LOW

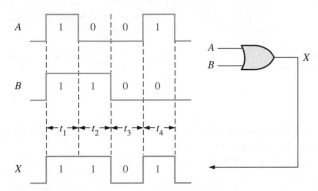

FIGURE 3–20 Example of OR gate operation with a timing diagram showing input and output time relationships.

In this illustration, we have applied the truth table operation of the OR gate to each of the time intervals during which the levels are nonchanging. Examples 3–7 through 3–9 further illustrate OR gate operation with waveforms on the inputs.

EXAMPLE 3–7

If the two input waveforms, A and B, in Figure 3–21 are applied to the OR gate, what is the resulting output waveform?

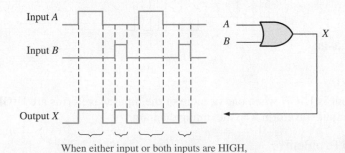

When either input or both inputs are HIGH, the output is HIGH.

FIGURE 3–21

Solution

The output waveform X of a 2-input OR gate is HIGH when either or both input waveforms are HIGH as shown in the timing diagram. In this case, both input waveforms are never HIGH at the same time.

Related Problem

Determine the output waveform and show the timing diagram if input A is changed such that it is HIGH from the beginning of the existing first pulse to the end of the existing second pulse.

EXAMPLE 3–8

For the two input waveforms, A and B, in Figure 3–22, show the output waveform with its proper relation to the inputs.

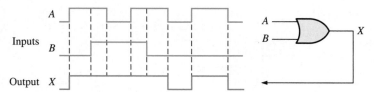

Inputs

Output X

FIGURE 3–22

Solution

When either or both input waveforms are HIGH, the output is HIGH as shown by the output waveform X in the timing diagram.

Related Problem

Determine the output waveform and show the timing diagram if the middle pulse of input A is replaced by a LOW level.

EXAMPLE 3–9

For the 3-input OR gate in Figure 3–23, determine the output waveform in proper time relation to the inputs.

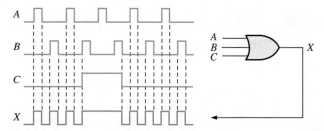

FIGURE 3–23

Solution

The output is HIGH when one or more of the input waveforms are HIGH as indicated by the output waveform X in the timing diagram.

Related Problem

Determine the output waveform and show the timing diagram if input C is always LOW.

Logic Expressions for an OR Gate

The logical OR function of two variables is represented mathematically by a $+$ between the two variables, for example, $A + B$. The plus sign is read as "OR."

When variables are separated by $+$, they are ORed.

Addition in Boolean algebra involves variables whose values are either binary 1 or binary 0. The basic rules for **Boolean addition** are as follows:

$$0 + 0 = 0$$
$$0 + 1 = 1$$
$$1 + 0 = 1$$
$$1 + 1 = 1$$

Boolean addition is the same as the OR function.

Notice that Boolean addition differs from binary addition in the case where two 1s are added. There is no carry in Boolean addition.

The operation of a 2-input OR gate can be expressed as follows: If one input variable is A, if the other input variable is B, and if the output variable is X, then the Boolean expression is

$$X = A + B$$

Figure 3–24(a) shows the OR gate logic symbol with two input variables and the output variable labeled.

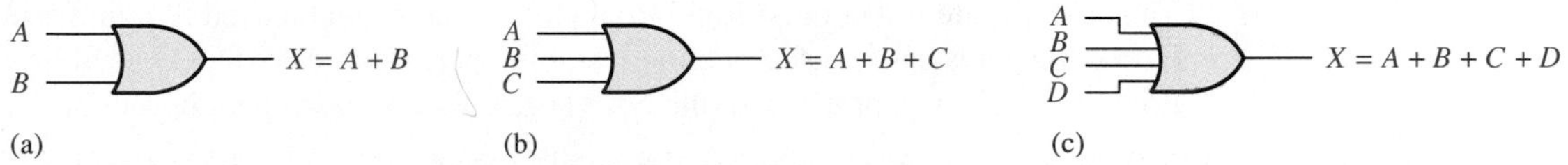

FIGURE 3–24   Boolean expressions for OR gates with two, three, and four inputs.

To extend the OR expression to more than two input variables, a new letter is used for each additional variable. For instance, the function of a 3-input OR gate can be expressed as $X = A + B + C$. The expression for a 4-input OR gate can be written as $X = A + B + C + D$, and so on. Parts (b) and (c) of Figure 3–24 show OR gates with three and four input variables, respectively.

OR gate operation can be evaluated by using the Boolean expressions for the output X by substituting all possible combinations of 1 and 0 values for the input variables, as shown in Table 3–6 for a 2-input OR gate. This evaluation shows that the output X of an OR gate is a 1 (HIGH) when any one or more of the inputs are 1 (HIGH). A similar analysis can be extended to OR gates with any number of input variables.

An Application

A simplified portion of an intrusion detection and alarm system is shown in Figure 3–25. This system could be used for one room in a home—a room with two windows and a door. The sensors are magnetic switches that produce a HIGH output when open and a LOW output when closed. As long as the windows and the door are secured, the switches are closed and all three of the OR gate inputs are LOW. When one of the windows or the door is opened, a HIGH is produced on that input to the OR gate and the gate output goes HIGH. It then activates and latches an alarm circuit to warn of the intrusion.

InfoNote

A mask operation that is used in computer programming to selectively make certain bits in a data byte equal to 1 (called setting) while not affecting any other bit is done with the OR operation. A mask is used that contains a 1 in any position where a data bit is to be set. For example, if you want to force the sign bit in an 8-bit signed number to equal 1, but leave all other bits unchanged, you can OR the data byte with the mask 10000000.

TABLE 3–6

A	B	$A + B = X$
0	0	$0 + 0 = 0$
0	1	$0 + 1 = 1$
1	0	$1 + 0 = 1$
1	1	$1 + 1 = 1$

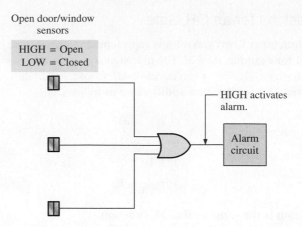

Open door/window
sensors

HIGH = Open
LOW = Closed

HIGH activates
alarm.

Alarm
circuit

FIGURE 3–25   A simplified intrusion detection system using an OR gate.

SECTION 3–3 CHECKUP

1. When is the output of an OR gate HIGH?
2. When is the output of an OR gate LOW?
3. Describe the truth table for a 3-input OR gate.

3–4   The NAND Gate

The NAND gate is a popular logic element because it can be used as a universal gate; that is, NAND gates can be used in combination to perform the AND, OR, and inverter operations. The universal property of the NAND gate will be examined thoroughly in Chapter 5.

After completing this section, you should be able to

♦ Identify a NAND gate by its distinctive shape symbol or by its rectangular outline symbol

♦ Describe the operation of a NAND gate

♦ Develop the truth table for a NAND gate with any number of inputs

♦ Produce a timing diagram for a NAND gate with any specified input waveforms

♦ Write the logic expression for a NAND gate with any number of inputs

♦ Describe NAND gate operation in terms of its negative-OR equivalent

♦ Discuss examples of NAND gate applications

The NAND gate is the same as the AND gate except the output is inverted.

The term *NAND* is a contraction of NOT-AND and implies an AND function with a complemented (inverted) output. The standard logic symbol for a 2-input NAND gate and its equivalency to an AND gate followed by an inverter are shown in Figure 3–26(a), where the symbol ≡ means equivalent to. A rectangular outline symbol is shown in part (b).

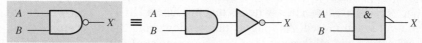

(a) Distinctive shape, 2-input NAND gate and its
    NOT/AND equivalent

(b) Rectangular outline, 2-input NAND
    gate with polarity indicator

FIGURE 3–26   Standard NAND gate logic symbols (ANSI/IEEE Std. 91-1984/Std. 91a-1991).

Operation of a NAND Gate

A **NAND gate** produces a LOW output only when all the inputs are HIGH. When any of the inputs is LOW, the output will be HIGH. For the specific case of a 2-input NAND gate, as shown in Figure 3–26 with the inputs labeled A and B and the output labeled X, the operation can be stated as follows:

> **For a 2-input NAND gate, output X is LOW only when inputs A and B are HIGH; X is HIGH when either A or B is LOW, or when both A and B are LOW.**

This operation is opposite that of the AND in terms of the output level. In a NAND gate, the LOW level (0) is the active or asserted output level, as indicated by the bubble on the output. Figure 3–27 illustrates the operation of a 2-input NAND gate for all four input combinations, and Table 3–7 is the truth table summarizing the logical operation of the 2-input NAND gate.

TABLE 3–7
Truth table for a 2-input NAND gate.

Inputs		Output
A	B	X
0	0	1
0	1	1
1	0	1
1	1	0

$1 = \text{HIGH}, 0 = \text{LOW}.$

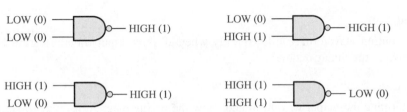

FIGURE 3–27   Operation of a 2-input NAND gate. Open file F03-27 to verify NAND gate operation.

MultiSim

NAND Gate Operation with Waveform Inputs

Now let's look at the pulse waveform operation of a NAND gate. Remember from the truth table that the only time a LOW output occurs is when all of the inputs are HIGH.

EXAMPLE 3–10

If the two waveforms A and B shown in Figure 3–28 are applied to the NAND gate inputs, determine the resulting output waveform.

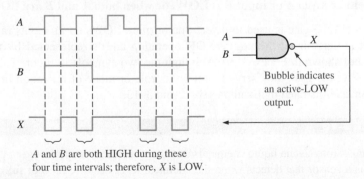

A and B are both HIGH during these four time intervals; therefore, X is LOW.

FIGURE 3–28

Solution

Output waveform X is LOW only during the four time intervals when both input waveforms A and B are HIGH as shown in the timing diagram.

Related Problem

Determine the output waveform and show the timing diagram if input waveform B is inverted.

EXAMPLE 3–11

Show the output waveform for the 3-input NAND gate in Figure 3–29 with its proper time relationship to the inputs.

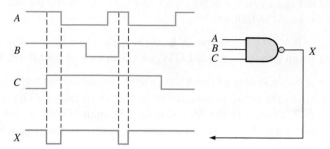

FIGURE 3–29

Solution

The output waveform X is LOW only when all three input waveforms are HIGH as shown in the timing diagram.

Related Problem

Determine the output waveform and show the timing diagram if input waveform A is inverted.

Negative-OR Equivalent Operation of a NAND Gate

Inherent in a NAND gate's operation is the fact that one or more LOW inputs produce a HIGH output. Table 3–7 shows that output X is HIGH (1) when any of the inputs, A and B, is LOW (0). From this viewpoint, a NAND gate can be used for an OR operation that requires one or more LOW inputs to produce a HIGH output. This aspect of NAND operation is referred to as **negative-OR**. The term *negative* in this context means that the inputs are defined to be in the active or asserted state when LOW.

> **For a 2-input NAND gate performing a negative-OR operation, output X is HIGH when either input A or input B is LOW, or when both A and B are LOW.**

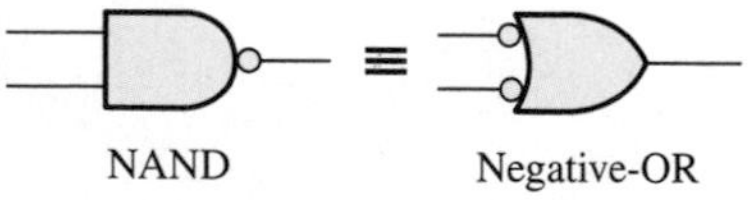

When a NAND gate is used to detect one or more LOWs on its inputs rather than all HIGHs, it is performing the negative-OR operation and is represented by the standard logic symbol shown in Figure 3–30. Although the two symbols in Figure 3–30 represent the same physical gate, they serve to define its role or mode of operation in a particular application, as illustrated by Examples 3–12 and 3–13.

NAND   Negative-OR

FIGURE 3–30 ANSI/IEEE standard symbols representing the two equivalent operations of a NAND gate.

EXAMPLE 3–12

Two tanks store certain liquid chemicals that are required in a manufacturing process. Each tank has a sensor that detects when the chemical level drops to 25% of full. The sensors produce a HIGH level of 5 V when the tanks are more than one-quarter full. When the volume of chemical in a tank drops to one-quarter full, the sensor puts out a LOW level of 0 V.

It is required that a single green light-emitting diode (LED) on an indicator panel show when both tanks are more than one-quarter full. Show how a NAND gate can be used to implement this function.

Solution

Figure 3–31 shows a NAND gate with its two inputs connected to the tank level sensors and its output connected to the indicator panel. The operation can be stated as follows: If tank A *and* tank B are above one-quarter full, the LED is on.

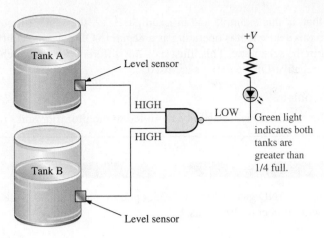

FIGURE 3–31

As long as both sensor outputs are HIGH (5 V), indicating that both tanks are more than one-quarter full, the NAND gate output is LOW (0 V). The green LED circuit is connected so that a LOW voltage turns it on. The resistor limits the LED current.

Related Problem

How can the circuit of Figure 3–31 be modified to monitor the levels in three tanks rather than two?

EXAMPLE 3–13

For the process described in Example 3–12 it has been decided to have a red LED display come on when at least one of the tanks falls to the quarter-full level rather than have the green LED display indicate when both are above one quarter. Show how this requirement can be implemented.

Solution

Figure 3–32 shows a NAND gate operating as a negative-OR gate to detect the occurrence of at least one LOW on its inputs. A sensor puts out a LOW voltage if the volume in its tank goes to one-quarter full or less. When this happens, the gate output goes HIGH. The red LED circuit in the panel is connected so that a HIGH voltage turns it on. The operation can be stated as follows: If tank *A or* tank *B or* both are below one-quarter full, the LED is on.

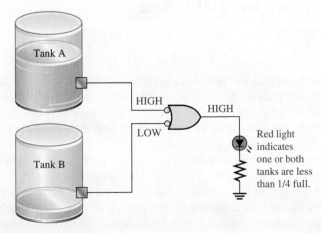

FIGURE 3–32

Notice that, in this example and in Example 3–12, the same 2-input NAND gate is used, but in this example it is operating as a negative-OR gate and a different gate symbol is used in the schematic. This illustrates the different way in which the NAND and equivalent negative-OR operations are used.

Related Problem

How can the circuit in Figure 3–32 be modified to monitor four tanks rather than two?

EXAMPLE 3–14

For the 4-input NAND gate in Figure 3–33, operating as a negative-OR gate, determine the output with respect to the inputs.

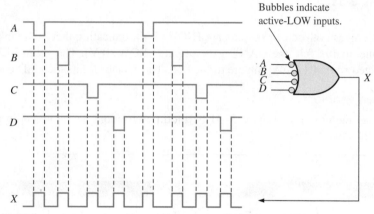

FIGURE 3–33

Solution

The output waveform X is HIGH any time an input waveform is LOW as shown in the timing diagram.

Related Problem

Determine the output waveform if input waveform A is inverted before it is applied to the gate.

Logic Expressions for a NAND Gate

A bar over a variable or variables indicates an inversion.

The Boolean expression for the output of a 2-input NAND gate is

$$X = \overline{AB}$$

This expression says that the two input variables, A and B, are first ANDed and then complemented, as indicated by the bar over the AND expression. This is a description in equation form of the operation of a NAND gate with two inputs. Evaluating this expression for all possible values of the two input variables, you get the results shown in Table 3–8.

Once an expression is determined for a given logic function, that function can be evaluated for all possible values of the variables. The evaluation tells you exactly what the output of the logic circuit is for each of the input conditions, and it therefore gives you a complete description of the circuit's logic operation. The NAND expression can be extended to more than two input variables by including additional letters to represent the other variables.

TABLE 3–8

A	B	$\overline{AB} = X$
0	0	$\overline{0 \cdot 0} = \overline{0} = 1$
0	1	$\overline{0 \cdot 1} = \overline{0} = 1$
1	0	$\overline{1 \cdot 0} = \overline{0} = 1$
1	1	$\overline{1 \cdot 1} = \overline{1} = 0$

1. When is the output of a NAND gate LOW?
2. When is the output of a NAND gate HIGH?
3. Describe the functional differences between a NAND gate and a negative-OR gate. Do they both have the same truth table?
4. Write the output expression for a NAND gate with inputs A, B, and C.

3–5   The NOR Gate

The NOR gate, like the NAND gate, is a useful logic element because it can also be used as a universal gate; that is, NOR gates can be used in combination to perform the AND, OR, and inverter operations. The universal property of the NOR gate will be examined thoroughly in Chapter 5.

After completing this section, you should be able to

* Identify a NOR gate by its distinctive shape symbol or by its rectangular outline symbol

* Describe the operation of a NOR gate

* Develop the truth table for a NOR gate with any number of inputs

* Produce a timing diagram for a NOR gate with any specified input waveforms

* Write the logic expression for a NOR gate with any number of inputs

* Describe NOR gate operation in terms of its negative-AND equivalent

* Discuss examples of NOR gate applications

The term *NOR* is a contraction of NOT-OR and implies an OR function with an inverted (complemented) output. The standard logic symbol for a 2-input NOR gate and its equivalent OR gate followed by an inverter are shown in Figure 3–34(a). A rectangular outline symbol is shown in part (b).

The NOR is the same as the OR except the output is inverted.

(a) Distinctive shape, 2-input NOR gate and its NOT/OR equivalent

(b) Rectangular outline, 2-input NOR gate with polarity indicator

FIGURE 3–34   Standard NOR gate logic symbols (ANSI/IEEE Std. 91-1984/Std. 91a-1991).

Operation of a NOR Gate

A **NOR gate** produces a LOW output when *any* of its inputs is HIGH. Only when all of its inputs are LOW is the output HIGH. For the specific case of a 2-input NOR gate, as shown in Figure 3–34 with the inputs labeled A and B and the output labeled X, the operation can be stated as follows:

For a 2-input NOR gate, output X is LOW when either input A or input B is HIGH, or when both A and B are HIGH; X is HIGH only when both A and B are LOW.

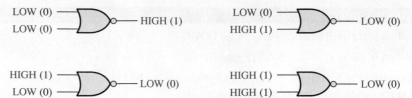

 FIGURE 3–35   Operation of a 2-input NOR gate. Open file F03-35 to verify NOR gate operation.

TABLE 3–9

Truth table for a 2-input NOR gate.

Inputs		Output
A	B	X
0	0	1
0	1	0
1	0	0
1	1	0

$1 = \text{HIGH}, 0 = \text{LOW}.$

This operation results in an output level opposite that of the OR gate. In a NOR gate, the LOW output is the active or asserted output level as indicated by the bubble on the output. Figure 3–35 illustrates the operation of a 2-input NOR gate for all four possible input combinations, and Table 3–9 is the truth table for a 2-input NOR gate.

NOR Gate Operation with Waveform Inputs

The next two examples illustrate the operation of a NOR gate with pulse waveform inputs. Again, as with the other types of gates, we will simply follow the truth table operation to determine the output waveforms in the proper time relationship to the inputs.

EXAMPLE 3–15

If the two waveforms shown in Figure 3–36 are applied to a NOR gate, what is the resulting output waveform?

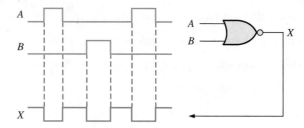

FIGURE 3–36

Solution

Whenever any input of the NOR gate is HIGH, the output is LOW as shown by the output waveform X in the timing diagram.

Related Problem

Invert input B and determine the output waveform in relation to the inputs.

EXAMPLE 3–16

Show the output waveform for the 3-input NOR gate in Figure 3–37 with the proper time relation to the inputs.

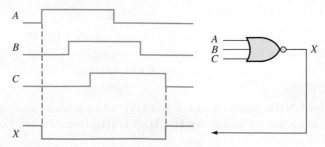

FIGURE 3–37

Solution

The output X is LOW when any input is HIGH as shown by the output waveform X in the timing diagram.

Related Problem

With the B and C inputs inverted, determine the output and show the timing diagram.

Negative-AND Equivalent Operation of the NOR Gate

A NOR gate, like the NAND, has another aspect of its operation that is inherent in the way it logically functions. Table 3–9 shows that a HIGH is produced on the gate output only when all of the inputs are LOW. From this viewpoint, a NOR gate can be used for an AND operation that requires all LOW inputs to produce a HIGH output. This aspect of NOR operation is called **negative-AND**. The term *negative* in this context means that the inputs are defined to be in the active or asserted state when LOW.

> **For a 2-input NOR gate performing a negative-AND operation, output X is HIGH only when both inputs A and B are LOW.**

When a NOR gate is used to detect all LOWs on its inputs rather than one or more HIGHs, it is performing the negative-AND operation and is represented by the standard symbol in Figure 3–38. Remember that the two symbols in Figure 3–38 represent the same physical gate and serve only to distinguish between the two modes of its operation. The following three examples illustrate this.

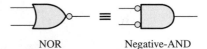

NOR   Negative-AND

FIGURE 3–38   Standard symbols representing the two equivalent operations of a NOR gate.

EXAMPLE 3–17

A device is needed to indicate when two LOW levels occur simultaneously on its inputs and to produce a HIGH output as an indication. Specify the device.

Solution

A 2-input NOR gate operating as a negative-AND gate is required to produce a HIGH output when both inputs are LOW, as shown in Figure 3–39.

LOW ⎯◦⟍
                  ⟩⎯ HIGH
LOW ⎯◦⟋

FIGURE 3–39

Related Problem

A device is needed to indicate when one or two HIGH levels occur on its inputs and to produce a LOW output as an indication. Specify the device.

EXAMPLE 3–18

As part of an aircraft's functional monitoring system, a circuit is required to indicate the status of the landing gears prior to landing. A green LED display turns on if all three gears are properly extended when the "gear down" switch has been activated in preparation for landing. A red LED display turns on if any of the gears fail to extend properly prior to landing. When a landing gear is extended, its sensor produces a LOW voltage. When a landing gear is retracted, its sensor produces a HIGH voltage. Implement a circuit to meet this requirement.

Solution

Power is applied to the circuit only when the "gear down" switch is activated. Use a NOR gate for each of the two requirements as shown in Figure 3–40. One NOR gate operates as a negative-AND to detect a LOW from each of the three landing gear sensors. When all three of the gate inputs are LOW, the three landing gears are properly extended and the resulting HIGH output from the negative-AND gate turns on the green LED display. The other NOR gate operates as a NOR to detect if one or more of the landing gears remain retracted when the "gear down" switch is activated. When one or more of the landing gears remain retracted, the resulting HIGH from the sensor is detected by the NOR gate, which produces a LOW output to turn on the red LED warning display.

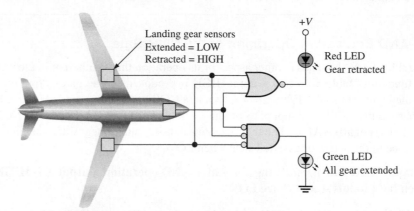

FIGURE 3–40

Related Problem

What type of gate should be used to detect if all three landing gears are retracted after takeoff, assuming a LOW output is required to activate an LED display?

When driving a load such as an LED with a logic gate, consult the manufacturer's data sheet for maximum drive capabilities (output current). A regular IC logic gate may not be capable of handling the current required by certain loads such as some LEDs. Logic gates with a buffered output, such as an open-collector (OC) or open-drain (OD) output, are available in many types of IC logic gate configurations. The output current capability of typical IC logic gates is limited to the μA or relatively low mA range. For example, standard TTL can handle output currents up to 16 mA but only when the output is LOW. Most LEDs require currents in the range of about 10 mA to 50 mA.

EXAMPLE 3–19

For the 4-input NOR gate operating as a negative-AND in Figure 3–41, determine the output relative to the inputs.

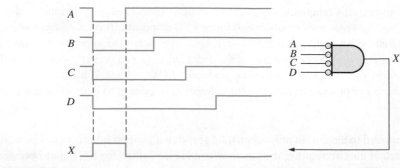

FIGURE 3–41

Solution

Any time all of the input waveforms are LOW, the output is HIGH as shown by output waveform X in the timing diagram.

Related Problem

Determine the output with input D inverted and show the timing diagram.

Logic Expressions for a NOR Gate

The Boolean expression for the output of a 2-input NOR gate can be written as

$$X = \overline{A + B}$$

This equation says that the two input variables are first ORed and then complemented, as indicated by the bar over the OR expression. Evaluating this expression, you get the results shown in Table 3–10. The NOR expression can be extended to more than two input variables by including additional letters to represent the other variables.

A	B	$\overline{A + B} = X$
0	0	$\overline{0 + 0} = \overline{0} = 1$
0	1	$\overline{0 + 1} = \overline{1} = 0$
1	0	$\overline{1 + 0} = \overline{1} = 0$
1	1	$\overline{1 + 1} = \overline{1} = 0$

TABLE 3–10

SECTION 3–5 CHECKUP

1. When is the output of a NOR gate HIGH?

2. When is the output of a NOR gate LOW?

3. Describe the functional difference between a NOR gate and a negative-AND gate. Do they both have the same truth table?

4. Write the output expression for a 3-input NOR with input variables A, B, and C.

3–6   The Exclusive-OR and Exclusive-NOR Gates

Exclusive-OR and exclusive-NOR gates are formed by a combination of other gates already discussed, as you will see in Chapter 5. However, because of their fundamental importance in many applications, these gates are often treated as basic logic elements with their own unique symbols.

After completing this section, you should be able to

◆ Identify the exclusive-OR and exclusive-NOR gates by their distinctive shape symbols or by their rectangular outline symbols

◆ Describe the operations of exclusive-OR and exclusive-NOR gates

◆ Show the truth tables for exclusive-OR and exclusive-NOR gates

◆ Produce a timing diagram for an exclusive-OR or exclusive-NOR gate with any specified input waveforms

◆ Discuss examples of exclusive-OR and exclusive-NOR gate applications

The Exclusive-OR Gate

Standard symbols for an exclusive-OR (XOR for short) gate are shown in Figure 3–42. The XOR gate has only two inputs. The **exclusive-OR gate** performs modulo-2 addition (introduced in Chapter 2). The output of an exclusive-OR gate is HIGH *only* when the two

InfoNote

Exclusive-OR gates connected to form an adder circuit allow a processor to perform addition, subtraction, multiplication, and division in its Arithmetic Logic Unit (ALU). An exclusive-OR gate combines basic AND, OR, and NOT logic.

For an exclusive-OR gate, opposite inputs make the output HIGH.

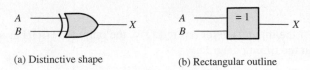

(a) Distinctive shape

(b) Rectangular outline

FIGURE 3–42　Standard logic symbols for the exclusive-OR gate.

TABLE 3–11

Truth table for an exclusive-OR gate.

Inputs		Output
A	B	X
0	0	0
0	1	1
1	0	1
1	1	0

inputs are at opposite logic levels. This operation can be stated as follows with reference to inputs A and B and output X:

For an exclusive-OR gate, output X is HIGH when input A is LOW and input B is HIGH, or when input A is HIGH and input B is LOW; X is LOW when A and B are both HIGH or both LOW.

The four possible input combinations and the resulting outputs for an XOR gate are illustrated in Figure 3–43. The HIGH level is the active or asserted output level and occurs only when the inputs are at opposite levels. The operation of an XOR gate is summarized in the truth table shown in Table 3–11.

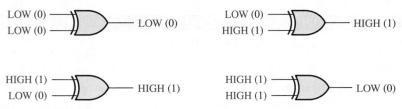

FIGURE 3–43　All possible logic levels for an exclusive-OR gate. Open file F03-43 to verify XOR gate operation.

EXAMPLE 3–20

A certain system contains two identical circuits operating in parallel. As long as both are operating properly, the outputs of both circuits are always the same. If one of the circuits fails, the outputs will be at opposite levels at some time. Devise a way to monitor and detect that a failure has occurred in one of the circuits.

Solution

The outputs of the circuits are connected to the inputs of an XOR gate as shown in Figure 3–44. A failure in either one of the circuits produces differing outputs, which cause the XOR inputs to be at opposite levels. This condition produces a HIGH on the output of the XOR gate, indicating a failure in one of the circuits.

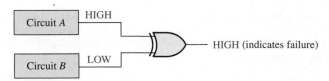

FIGURE 3–44

Related Problem

Will the exclusive-OR gate always detect simultaneous failures in both circuits of Figure 3–44? If not, under what condition?

The Exclusive-NOR Gate

Standard symbols for an **exclusive-NOR** (XNOR) **gate** are shown in Figure 3–45. Like the XOR gate, an XNOR has only two inputs. The bubble on the output of the XNOR symbol indicates that its output is opposite that of the XOR gate. When the two input logic levels are opposite, the output of the exclusive-NOR gate is LOW. The operation can be stated as follows (*A* and *B* are inputs, *X* is the output):

For an exclusive-NOR gate, output *X* is LOW when input *A* is LOW and input *B* is HIGH, or when *A* is HIGH and *B* is LOW; *X* is HIGH when *A* and *B* are both HIGH or both LOW.

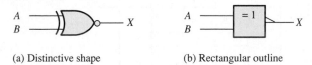

(a) Distinctive shape   (b) Rectangular outline

FIGURE 3–45 Standard logic symbols for the exclusive-NOR gate.

The four possible input combinations and the resulting outputs for an XNOR gate are shown in Figure 3–46. The operation of an XNOR gate is summarized in Table 3–12. Notice that the output is HIGH when the same level is on both inputs.

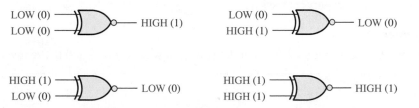

FIGURE 3–46 All possible logic levels for an exclusive-NOR gate. Open file F03-46 to verify XNOR gate operation.

TABLE 3–12

Truth table for an exclusive-NOR gate.

Inputs		Output
A	*B*	*X*
0	0	1
0	1	0
1	0	0
1	1	1

MultiSim

Operation with Waveform Inputs

As we have done with the other gates, let's examine the operation of XOR and XNOR gates with pulse waveform inputs. As before, we apply the truth table operation during each distinct time interval of the pulse waveform inputs, as illustrated in Figure 3–47 for an XOR gate. You can see that the input waveforms *A* and *B* are at opposite levels during time intervals t_2 and t_4. Therefore, the output *X* is HIGH during these two times. Since both inputs are at the same level, either both HIGH or both LOW, during time intervals t_1 and t_3, the output is LOW during those times as shown in the timing diagram.

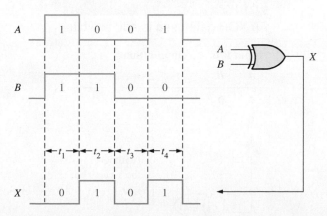

FIGURE 3–47 Example of exclusive-OR gate operation with pulse waveform inputs.

EXAMPLE 3–21

Determine the output waveforms for the XOR gate and for the XNOR gate, given the input waveforms, *A* and *B*, in Figure 3–48.

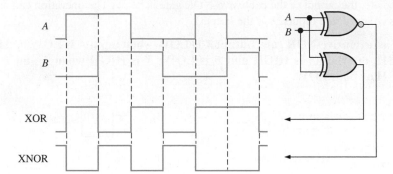

FIGURE 3–48

Solution

The output waveforms are shown in Figure 3–48. Notice that the XOR output is HIGH only when both inputs are at opposite levels. Notice that the XNOR output is HIGH only when both inputs are the same.

Related Problem

Determine the output waveforms if the two input waveforms, *A* and *B*, are inverted.

An Application

An exclusive-OR gate can be used as a two-bit modulo-2 adder. Recall from Chapter 2 that the basic rules for binary addition are as follows: $0 + 0 = 0, 0 + 1 = 1, 1 + 0 = 1$, and $1 + 1 = 10$. An examination of the truth table for an XOR gate shows that its output is the binary sum of the two input bits. In the case where the inputs are both 1s, the output is the sum 0, but you lose the carry of 1. In Chapter 6 you will see how XOR gates are combined to make complete adding circuits. Table 3–13 illustrates an XOR gate used as a modulo-2 adder. It is used in CRC systems to implement the division process that was described in Chapter 2.

TABLE 3–13

An XOR gate used to add two bits.

Input Bits		Output (Sum)
A	*B*	Σ
0	0	0
0	1	1
1	0	1
1	1	0 (without the 1 carry bit)

1. When is the output of an XOR gate HIGH?
2. When is the output of an XNOR gate HIGH?
3. How can you use an XOR gate to detect when two bits are different?

3-7 Programmable Logic

Programmable logic was introduced in Chapter 1. In this section, the basic concept of the programmable AND array, which forms the basis for most programmable logic, is discussed, and the major process technologies are covered. A programmable logic device (PLD) is one that does not initially have a fixed-logic function but that can be programmed to implement just about any logic design. As you have learned, two types of PLD are the SPLD and CPLD. In addition to the PLD, the other major category of programmable logic is the FPGA. Also, basic VHDL programming is introduced.

After completing this section, you should be able to

◆ Describe the concept of a programmable AND array

◆ Discuss various process technologies for programming a PLD

◆ Discuss downloading a design to a programmable logic device

◆ Discuss text entry and graphic entry as two methods for programmable logic design

◆ Explain in-system programming

◆ Write VHDL descriptions of logic gates

The AND Array

Most types of PLDs use some form of **AND array**. Basically, this array consists of AND gates and a matrix of interconnections with a programmable link at each cross point, as shown in Figure 3–49(a). Programmable links allow a connection between a row line and a column line in the interconnection matrix to be opened or left intact. For each input to an AND gate, only one programmable link is left intact in order to connect the desired variable to the gate input. Figure 3–49(b) illustrates an array after it has been programmed.

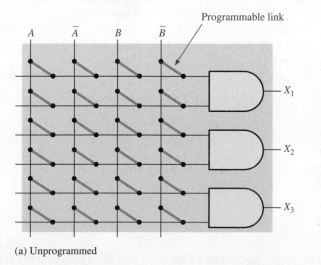

(a) Unprogrammed

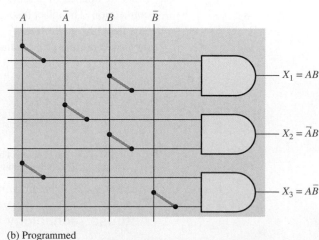

(b) Programmed

FIGURE 3–49 Concept of a programmable AND array.

EXAMPLE 3–22

Show the AND array in Figure 3–49(a) programmed for the following outputs: $X_1 = A\overline{B}$, $X_2 = \overline{A}B$, and $X_3 = \overline{A}\,\overline{B}$

Solution

See Figure 3–50.

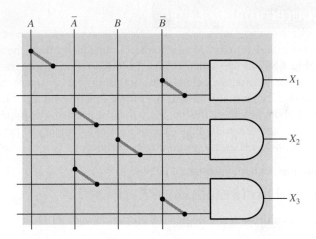

FIGURE 3–50

Related Problem

How many rows, columns, and AND gate inputs are required for three input variables in a 3-AND gate array?

Programmable Link Process Technologies

A process technology is the physical method by which a link is made. Several different process technologies are used for programmable links in PLDs.

Fuse Technology

This was the original programmable link technology. It is still used in some SPLDs. The **fuse** is a metal link that connects a row and a column in the interconnection matrix. Before programming, there is a fused connection at each intersection. To program a device, the selected fuses are opened by passing a current through them sufficient to "blow" the fuse and break the connection. The intact fuses remain and provide a connection between the rows and columns. The fuse link is illustrated in Figure 3–51. Programmable logic devices that use fuse technology are one-time programmable (**OTP**).

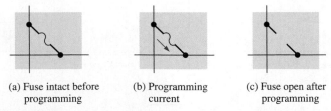

(a) Fuse intact before programming

(b) Programming current

(c) Fuse open after programming

FIGURE 3–51   The programmable fuse link.

Antifuse Technology

An **antifuse** programmable link is the opposite of a fuse link. Instead of breaking the connection, a connection is made during programming. An antifuse starts out as an open circuit whereas the fuse starts out as a short circuit. Before programming, there are no connections between the rows and columns in the interconnection matrix. An antifuse is basically two conductors separated by an insulator. To program a device with antifuse technology, a programmer tool applies a sufficient voltage across selected antifuses to break down the insulation between the two conductive materials, causing the insulator to become a low-resistance link. The antifuse link is illustrated in Figure 3–52. An antifuse device is also a one-time programmable (OTP) device.

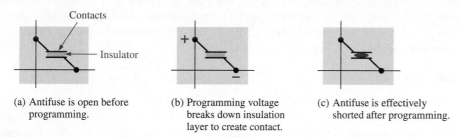

(a) Antifuse is open before programming.

(b) Programming voltage breaks down insulation layer to create contact.

(c) Antifuse is effectively shorted after programming.

FIGURE 3–52 The programmable antifuse link.

EPROM Technology

In certain programmable logic devices, the programmable links are similar to the memory cells in **EPROMs** (electrically programmable read-only memories). This type of PLD is programmed using a special tool known as a device programmer. The device is inserted into the programmer, which is connected to a computer running the programming software. Most EPROM-based PLDs are one-time programmable (OTP). However, those with windowed packages can be erased with UV (ultraviolet) light and reprogrammed using a standard PLD programming fixture. EPROM process technology uses a special type of MOS transistor, known as a floating-gate transistor, as the programmable link. The floating-gate device utilizes a process called Fowler-Nordheim tunneling to place electrons in the floating-gate structure.

In a programmable AND array, the floating-gate transistor acts as a switch to connect the row line to either a HIGH or a LOW, depending on the input variable. For input variables that are not used, the transistor is programmed to be permanently *off* (open). Figure 3–53 shows one AND gate in a simple array. Variable A controls the state of the transistor in the first column, and variable B controls the transistor in the third column. When a transistor is *off*, like an open switch, the input line to the AND gate is at $+V$ (HIGH). When a transistor is *on*, like a closed switch, the input line is connected to ground (LOW). When variable A

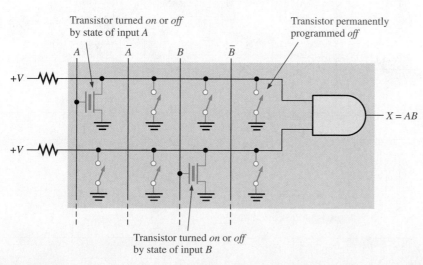

FIGURE 3–53 A simple AND array with EPROM technology. Only one gate in the array is shown for simplicity.

or *B* is 0 (LOW), the transistor is *on,* keeping the input line to the AND gate LOW. When *A* or *B* is 1 (HIGH), the transistor is *off,* keeping the input line to the AND gate HIGH.

EEPROM Technology

Electrically erasable programmable read-only memory technology is similar to EPROM because it also uses a type of floating-gate transistor in E^2CMOS cells. The difference is that **EEPROM** can be erased and reprogrammed electrically without the need for UV light or special fixtures. An E^2CMOS device can be programmed after being installed on a printed circuit board (PCB), and many can be reprogrammed while operating in a system. This is called **in-system programming (ISP)**. Figure 3–53 can also be used as an example to represent an AND array with EEPROM technology.

Flash Technology

Flash technology is based on a single transistor link and is both nonvolatile and reprogrammable. Flash elements are a type of EEPROM but are faster and result in higher density devices than the standard EEPROM link. A detailed discussion of the flash memory element can be found in Chapter 11.

SRAM Technology

Many FPGAs and some CPLDs use a process technology similar to that used in **SRAMs** (static random-access memories). The basic concept of SRAM-based programmable logic arrays is illustrated in Figure 3–54(a). A SRAM-type memory cell is used to turn a transistor *on* or *off* to connect or disconnect rows and columns. For example, when the memory cell contains a 1 (green), the transistor is *on* and connects the associated row and column lines, as shown in part (b). When the memory cell contains a 0 (blue), the transistor is *off* so there is no connection between the lines, as shown in part (c).

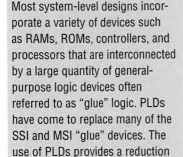

InfoNote

Most system-level designs incorporate a variety of devices such as RAMs, ROMs, controllers, and processors that are interconnected by a large quantity of general-purpose logic devices often referred to as "glue" logic. PLDs have come to replace many of the SSI and MSI "glue" devices. The use of PLDs provides a reduction in package count.

For example, in memory systems, PLDs can be used for memory address decoding and to generate memory write signals as well as other functions.

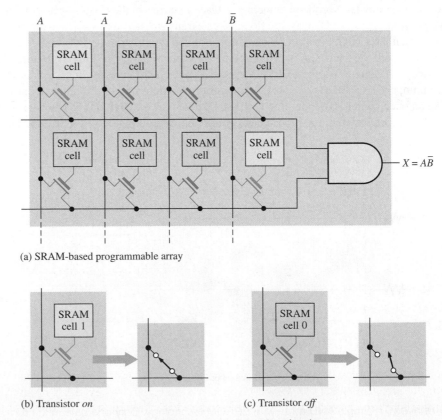

(a) SRAM-based programmable array

(b) Transistor *on*   (c) Transistor *off*

FIGURE 3–54   Concept of an AND array with SRAM technology.

SRAM technology is different from the other process technologies discussed because it is a volatile technology. This means that a SRAM cell does not retain data when power is turned *off*. The programming data must be loaded into a memory; and when power is turned *on*, the data from the memory reprograms the SRAM-based PLD.

The fuse, antifuse, EPROM, EEPROM, and flash process technologies are nonvolatile, so they retain their programming when the power is *off*. A fuse is permanently open, an antifuse is permanently closed, and floating-gate transistors used in EPROM and EEPROM-based arrays can retain their *on* or *off* state indefinitely.

Device Programming

The general concept of programming was introduced in Chapter 1, and you have seen how interconnections can be made in a simple array by opening or closing the programmable links. SPLDs, CPLDs, and FPGAs are programmed in essentially the same way. The devices with OTP (one-time programmable) process technologies (fuse, antifuse, or EPROM) must be programmed with a special hardware fixture called a *programmer*. The programmer is connected to a computer by a standard interface cable. Development software is installed on the computer, and the device is inserted into the programmer socket. Most programmers have adapters that allow different types of packages to be plugged in.

EEPROM, flash, and SRAM-based programmable logic devices are reprogrammable and can be reconfigured multiple times. Although a device programmer can be used for this type of device, it is generally programmed initially on a PLD development board, as shown in Figure 3–55. A logic design can be developed using this approach because any necessary changes during the design process can be readily accomplished by simply reprogramming the PLD. A PLD to which a software logic design can be downloaded is called a **target device**. In addition to the target device, development boards typically provide other circuitry and connectors for interfacing to the computer and other peripheral circuits. Also, test points and display devices for observing the operation of the programmed device are included on the development board.

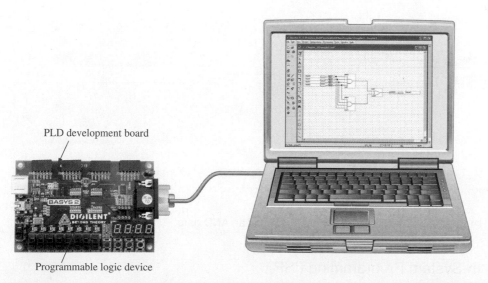

PLD development board

Programmable logic device

FIGURE 3–55 Programming setup for reprogrammable logic devices.   (Photo courtesy of Digilent, Inc.)

Design Entry

As you learned in Chapter 1, design entry is where the logic design is programmed into the development software. The two main ways to enter a design are by text entry or graphic (schematic) entry, and manufacturers of programmable logic provide software packages to support their devices that allow for both methods.

Text entry in most development software, regardless of the manufacturer, supports two or more hardware development languages (HDLs). For example, all software packages support both IEEE standard HDLs, VHDL, and Verilog. Some software packages also support certain proprietary languages such as AHDL.

In **graphic (schematic) entry**, logic symbols such as AND gates and OR gates are placed on the screen and interconnected to form the desired circuit. In this method you use the familiar logic symbols, but the software actually converts each symbol and interconnections to a text file for the computer to use; you do not see this process. A simple example of both a text entry screen and a graphic entry screen for an AND gate is shown in Figure 3–56. As a general rule, graphic entry is used for less-complex logic circuits and text entry, although it can also be used for very simple logic, is used for larger, more complex implementation.

Vhdl1.vhd

```
entity VHDL1 is
    port(A, B: in bit; X: out bit);
end entity VHDL1;
architecture ANDfunction of VHDL1 is
begin
    X <= A and B;
end architecture ANDfunction;
```

(a) VHDL text entry

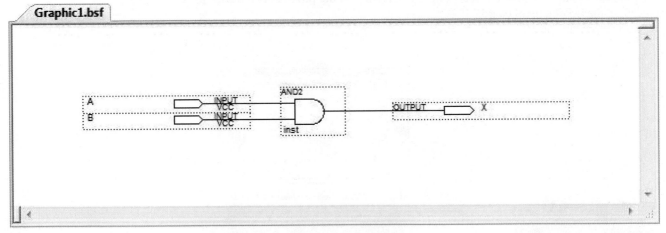

(b) Equivalent graphic (schematic) entry

FIGURE 3–56   Examples of design entry of an AND gate.

In-System Programming (ISP)

Certain CPLDs and FPGAs can be programmed after they have been installed on a system printed circuit board (PCB). After a logic design has been developed and fully tested on a development board, it can then be programmed into a "blank" device that is already soldered onto a system board in which it will be operating. Also, if a design change is required, the device on the system board can be reconfigured to incorporate the design modifications.

In a production situation, programming a device on the system board minimizes handling and eliminates the need for keeping stocks of preprogrammed devices. It also rules out the possibility of wrong parts being placed in a product. Unprogrammed (blank) devices can be kept in the warehouse and programmed on-board as needed. This minimizes the capital a business needs for inventories and enhances the quality of its products.

JTAG

The standard established by the Joint Test Action Group is the commonly used name for IEEE Std. 1149.1. The **JTAG** standard was developed to provide a simple method, called boundary scan, for testing programmable devices for functionality as well as testing circuit boards for bad connections—shorted pins, open pins, bad traces, and the like. Also, JTAG has been used as a convenient way of configuring programmable devices in-system. As the demand for field-upgradable products increases, the use of JTAG as a convenient way of reprogramming CPLDs and FPGAs increases.

JTAG-compliant devices have internal dedicated hardware that interprets instructions and data provided by four dedicated signals. These signals are defined by the JTAG standard to be TDI (Test Data In), TDO (Test Data Out), TMS (Test Mode Select), and TCK (Test Clock). The dedicated JTAG hardware interprets instructions and data on the TDI and TMS signals, and drives data out on the TDO signal. The TCK signal is used to clock the process. A JTAG-compliant PLD is represented in Figure 3–57.

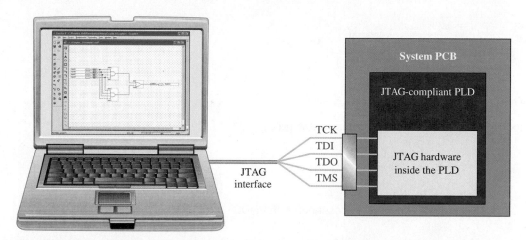

FIGURE 3–57   Simplified illustration of in-system programming via a JTAG interface.

Embedded Processor

Another approach to in-system programming is the use of an embedded microprocessor and memory. The processor is embedded within the system along with the CPLD or FPGA and other circuitry, and it is dedicated to the purpose of in-system configuration of the programmable device.

As you have learned, SRAM-based devices are volatile and lose their programmed data when the power is turned *off*. It is necessary to store the programming data in a PROM (programmable read-only memory), which is nonvolatile. When power is turned *on*, the embedded processor takes control of transferring the stored data from the PROM to the CPLD or FPGA.

Also, an embedded processor is sometimes used for reconfiguration of a programmable device while the system is running. In this case, design changes are done with software, and the new data are then loaded into a PROM without disturbing the operation of the system. The processor controls the transfer of the data to the device "on-the-fly" at an appropriate time.

VHDL Descriptions of Logic Gates

Hardware description languages (HDLs) differ from software programming languages because HDLs include ways of describing logic connections and characteristics. An HDL implements a logic design in hardware (PLD), whereas a software programming language, such as C or BASIC, instructs existing hardware what to do. The two standard HDLs used for programming

PLDs are VHDL and Verilog. Both of these HDLs have their advocates, but VHDL will be used in this textbook. *A VHDL tutorial is available on the website.*

Figure 3–58 shows **VHDL** programs for gates described in this chapter. Two gates are left as Checkup exercises. VHDL has an *entity/architecture* structure. The **entity** defines the logic element and its inputs/outputs or ports; the **architecture** describes the logic operation. Keywords that are part of the VHDL syntax are shown bold for clarity.

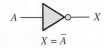

$X = \bar{A}$

entity Inverter **is**
  port (A: **in** bit; X: **out** bit);
end entity Inverter;
architecture NOTfunction **of** Inverter **is**
begin
  X <= **not** A;
end architecture NOTfunction;

(a) Inverter

$X = AB$

entity ANDgate **is**
  port (A, B: **in** bit; X: **out** bit);
end entity ANDgate;
architecture ANDfunction **of** ANDgate **is**
begin
  X <= A **and** B;
end architecture ANDfunction;

(b) AND gate

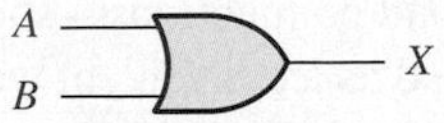

$X = A + B$

entity ORgate **is**
  port (A, B: **in** bit; X: **out** bit);
end entity ORgate;
architecture ORfunction **of** ORgate **is**
begin
  X <= A **or** B;
end architecture ORfunction;

(c) OR gate

$X = \overline{ABC}$

entity NANDgate **is**
  port (A, B, C: **in** bit; X: **out** bit);
end entity NANDgate;
architecture NANDfunction **of** NANDgate **is**
begin
  X <= A **nand** B **nand** C;
end architecture NANDfunction;

(d) NAND gate

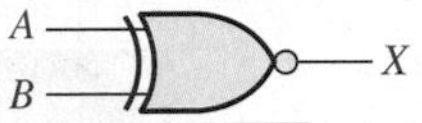

$X = A\bar{B} + \bar{A}B$

entity XNORgate **is**
  port (A, B: **in** bit; X: **out** bit);
end entity XNORgate;
architecture XNORfunction **of** XNORgate **is**
begin
  X <= A **xnor** B;
end architecture XNORfunction;

(e) XNOR gate

FIGURE 3–58 Logic gates described with VHDL.

SECTION 3–7 CHECKUP

1. List six process technologies used for programmable links in programmable logic.
2. What does the term *volatile* mean in relation to PLDs and which process technology is volatile?
3. What are two design entry methods for programming PLDs and FPGAs?
4. Define JTAG.
5. Write a VHDL description of a 3-input NOR gate.
6. Write a VHDL description of an XOR gate.

3–8   Fixed-Function Logic Gates

Fixed-function logic integrated circuits have been around for a long time and are available in a variety of logic functions. Unlike a PLD, a fixed-function IC comes with logic functions that cannot be programmed in and cannot be altered. The fixed-function logic is on a much smaller scale than the amount of logic that can be programmed into a PLD. Although the trend in technology is definitely toward programmable logic, fixed-function logic is used in specialized applications where PLDs are not the optimum choice. Fixed-function logic devices are sometimes called "glue logic" because of their usefulness in tying together larger units of logic such as PLDs in a system.

After completing this section, you should be able to

- List common 74 series gate logic functions
- List the major integrated circuit technologies and name some integrated circuit families
- Obtain data sheet information
- Define *propagation delay time*
- Define *power dissipation*
- Define *unit load* and *fan-out*
- Define *speed-power product*

    All of the various fixed-function logic devices currently available are implemented in two major categories of circuit technology: **CMOS** (complementary metal-oxide semiconductor) and **bipolar** (also known as **TTL**, transistor-transistor logic). A type of bipolar technology that is available in very limited devices is ECL (emitter-coupled logic). BiCMOS is another integrated circuit technology that combines both bipolar and CMOS. CMOS is the most dominant circuit technology.

74 Series Logic Gate Functions

The 74 series is the standard fixed-function logic devices. The device label format includes one or more letters that indentify the type of logic circuit technology family in the IC package and two or more digits that identify the type of logic function. For example, 74HC04 is a fixed-function IC that has six inverters in a package as indicated by 04. The letters, HC, following the prefix 74 identify the circuit technology family as a type of CMOS logic.

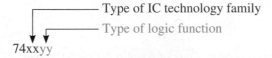

AND Gate

Figure 3–59 shows three configurations of fixed-function AND gates in the 74 series. The 74xx08 is a quad 2-input AND gate device, the 74xx11 is a triple 3-input AND gate device,

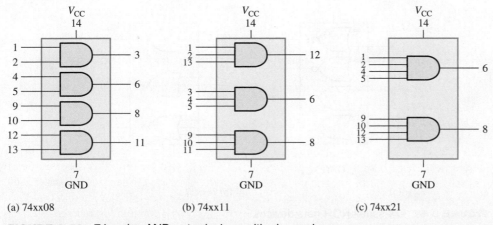

(a) 74xx08

(b) 74xx11

(c) 74xx21

FIGURE 3–59   74 series AND gate devices with pin numbers.

and the 74xx21 is a dual 4-input AND gate device. The label xx can represent any of the integrated circuit technology families such as HC or LS. The numbers on the inputs and outputs are the IC package pin numbers.

NAND Gate

Figure 3–60 shows four configurations of fixed-function NAND gates in the 74 series. The 74xx00 is a quad 2-input NAND gate device, the 74xx10 is a triple 3-input NAND gate device, the 74xx20 is a dual 4-input NAND gate device, and the 74xx30 is a single 8-input NAND gate device.

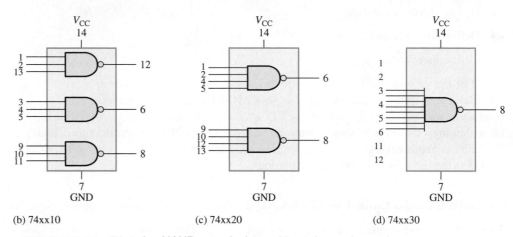

(a) 74xx00   (b) 74xx10   (c) 74xx20   (d) 74xx30

FIGURE 3–60   74 series NAND gate devices with package pin numbers.

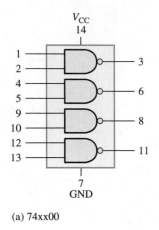

74xx32

FIGURE 3–61   74 series OR gate device.

OR Gate

Figure 3–61 shows a fixed-function OR gate in the 74 series. The 74xx32 is a quad 2-input OR gate device.

NOR Gate

Figure 3–62 shows two configurations of fixed-function NOR gates in the 74 series. The 74xx02 is a quad 2-input NOR gate device, and the 74xx27 is a triple 3-input NOR gate device.

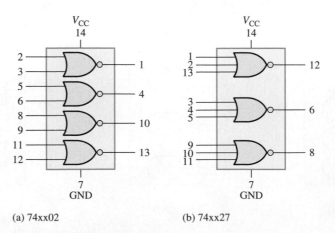

(a) 74xx02   (b) 74xx27

FIGURE 3–62   74 series NOR gate devices.

XOR Gate

Figure 3–63 shows a fixed-function XOR (exclusive-OR) gate in the 74 series. The 74xx86 is a quad 2-input XOR gate.

IC Packages

All of the 74 series CMOS are pin-compatible with the same types of devices in bipolar. This means that a CMOS digital IC such as the 74HC00 (quad 2-input NAND), which contains four 2-input NAND gates in one IC package, has the identical package pin numbers for each input and output as does the corresponding bipolar device. Typical IC gate packages, the dual in-line package (DIP) for plug-in or feedthrough mounting and the small-outline integrated circuit (SOIC) package for surface mounting, are shown in Figure 3–64. In some cases, other types of packages are also available. The SOIC package is significantly smaller than the DIP. Packages with a single gate are known as *little logic*. Most logic gate functions are available and are implemented in a CMOS circuit technology. Typically, the gates have only two inputs and have a different designation than multigate devices. For example, the 74xx1G00 is a single 2-input NAND gate.

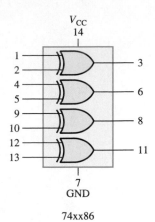

FIGURE 3–63 74 series XOR gate.

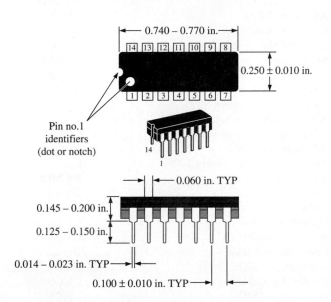

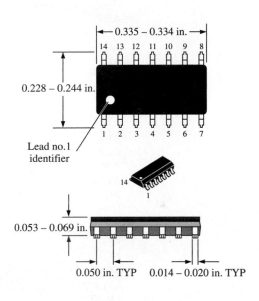

(a) 14-pin dual in-line package (DIP) for feedthrough mounting

(b) 14-pin small outline package (SOIC) for surface mounting

FIGURE 3–64 Typical dual in-line (DIP) and small-outline (SOIC) packages showing pin numbers and basic dimensions.

Handling Precautions for CMOS

CMOS logic is very sensitive to static charge and can be damaged by ESD (electrostatic discharge) if not handled properly as follows:

1. Store and ship in conductive foam.

2. Connect instruments to earth ground.

3. Connect wrist to earth ground through a large series resistor.

4. Do not remove devices from circuit with power on.

5. Do not apply signal voltage when power is off.

74 Series Logic Circuit Families

Although many logic circuit families have become obsolete and some are rapidly on the decline, others are still very active and available. CMOS is the most available and most popular type of logic circuit technology, and the HC (high-speed CMOS) family is the most recommended for new projects. For bipolar, the LS (low-power schottky) family is the most widely used. The HCT, which a variation of the HC family, is compatible with bipolar devices such as LS.

Table 3–14 lists many logic circuit technology families. Because the active status of any given logic family is always in flux, check with a manufacturer, such as Texas Instruments, for information on active/nonactive status and availability for a logic function in a given circuit technology.

TABLE 3–14

74 series logic families based on circuit technology.

Circuit Type	Description	Circuit Technology
ABT	Advanced BiCMOS	BiCMOS
AC	Advanced CMOS	CMOS
ACT	Bipolar compatible AC	CMOS
AHC	Advanced high-speed CMOS	CMOS
AHCT	Bipolar compatible AHC	CMOS
ALB	Advanced low-voltage BiCMOS	BiCMOS
ALS	Advanced low-power Schottky	Bipolar
ALVC	Advanced low-voltage CMOS	CMOS
AUC	Advanced ultra-low-voltage CMOS	CMOS
AUP	Advanced ultra-low-power CMOS	CMOS
AS	Advanced Schottky	Bipolar
AVC	Advanced very-low-power CMOS	CMOS
BCT	Standard BiCMOS	BiCMOS
F	Fast	Bipolar
FCT	Fast CMOS technology	CMOS
HC	High-speed CMOS	CMOS
HCT	Bipolar compatible HC	CMOS
LS	Low-power Schottky	Bipolar
LV-A	Low-voltage CMOS	CMOS
LV-AT	Bipolar compatible LV-A	CMOS
LVC	Low-voltage CMOS	CMOS
LVT	Low-voltage biCMOS	BiCMOS
S	Schottky	Bipolar

The type of integrated circuit technology has nothing to do with the logic function itself. For example, the 74HC00, 74HCT00, and 74LS00 are all quad 2-input NAND gates with identical package pin configurations. The differences among these three logic devices are in the electrical and performance characteristics such as power consumption, dc supply voltage, switching speed, and input/output voltage levels. CMOS and bipolar circuits are implemented with two different types of transistors. Figures 3–65 and 3–66 show partial data sheets for the 74HC00A quad 2-input NAND gate in CMOS and in bipolar technologies, respectively.

Performance Characteristics and Parameters

High-speed logic has a short propagation delay time.

Several things define the performance of a logic circuit. These performance characteristics are the switching speed measured in terms of the propagation delay time, the power

Quad 2-Input NAND Gate High-Performance Silicon–Gate CMOS

The MC54/74HC00A is identical in pinout to the LS00. The device inputs are compatible with Standard CMOS outputs; with pullup resistors, they are compatible with LSTTL outputs.

- Output Drive Capability: 10 LSTTL Loads
- Outputs Directly Interface to CMOS, NMOS and TTL
- Operating Voltage Range: 2 to 6 V
- Low Input Current: 1 μA
- High Noise Immunity Characteristic of CMOS Devices
- In Compliance With the JEDEC Standard No. 7A Requirements
- Chip Complexity: 32 FETs or 8 Equivalent Gates

LOGIC DIAGRAM

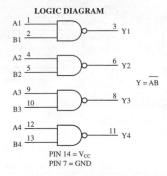

$$Y = \overline{AB}$$

PIN 14 = V_{CC}
PIN 7 = GND

Pinout: 14–Load Packages (Top View)

MC54/74HC00A

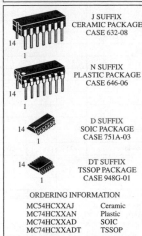

J SUFFIX
CERAMIC PACKAGE
CASE 632-08

N SUFFIX
PLASTIC PACKAGE
CASE 646-06

D SUFFIX
SOIC PACKAGE
CASE 751A-03

DT SUFFIX
TSSOP PACKAGE
CASE 948G-01

ORDERING INFORMATION

MC54HCXXAJ	Ceramic
MC74HCXXAN	Plastic
MC74HCXXAD	SOIC
MC74HCXXADT	TSSOP

FUNCTION TABLE

Inputs		Output
A	B	Y
L	L	H
L	H	H
H	L	H
H	H	L

MAXIMUM RATINGS*

Symbol	Parameter	Value	Unit
V_{CC}	DC Supply Voltage (Referenced to GND)	−0.5 to + 7.0	V
V_{in}	DC Input Voltage (Referenced to GND)	−0.5 to V_{CC} + 0.5	V
V_{out}	DC Output Voltage (Referenced to GND)	−0.5 to V_{CC} + 0.5	V
I_{in}	DC Input Current, per Pin	± 20	mA
I_{out}	DC Output Current, per Pin	± 25	mA
I_{CC}	DC Supply Current, V_{CC} and GND Pins	± 50	mA
P_D	Power Dissipation in Still Air, Plastic or Ceramic DIP†	750	mW
	SOIC Package†	500	
	TSSOP Package†	450	
T_{stg}	Storage Temperature	−65 to + 150	°C
T_L	Lead Temperature, 1 mm from Case for 10 Seconds		°C
	Plastic DIP, SOIC or TSSOP Package	260	
	Ceramic DIP	300	

* Maximum Ratings are those values beyond which damage to the device may occur. Functional operation should be restricted to the Recommended Operating Conditions.
† Derating — Plastic DIP: – 10 mW/°C from 65° to 125° C
Ceramic DIP: – 10 mW/°C from 100° to 125° C
SOIC Package: – 7 mW/°C from 65° to 125° C
TSSOP Package: – 6.1 mW/°C from 65° to 125° C

RECOMMENDED OPERATING CONDITIONS

Symbol	Parameter	in	Max	Unit
V_{CC}	DC Supply Voltage (Referenced to GND)	2.0	6.0	V
V_{in}, V_{out}	DC Input Voltage, Output Voltage (Referenced to GND)	0	V_{CC}	V
T_A	Operating Temperature, All Package Types	−55	+125	°C
t_r, t_f	Input Rise and Fall Time V_{CC} = 2.0 V	0	1000	ns
	V_{CC} = 4.5 V	0	500	
	V_{CC} = 6.0 V	0	400	

DC CHARACTERISTICS (Voltages Referenced to GND)

MC54/74HC00A

Symbol	Parameter	Condition	V_{CC} V	Guaranteed Limit −55 to 25°C	Guaranteed Limit ≤85°C	Guaranteed Limit ≤125°C	Unit
V_{IH}	Minimum High-Level Input Voltage	V_{out} = 0.1V or V_{CC} − 0.1V $|I_{out}| \leq 20\mu A$	2.0	1.50	1.50	1.50	V
			3.0	2.10	2.10	2.10	
			4.5	3.15	3.15	3.15	
			6.0	4.20	4.20	4.20	
V_{IL}	Maximum Low-Level Input Voltage	V_{out} = 0.1V or V_{CC} − 0.1V $|I_{out}| \leq 20\mu A$	2.0	0.50	0.50	0.50	V
			3.0	0.90	0.90	0.90	
			4.5	1.35	1.35	1.35	
			6.0	1.80	1.80	1.80	
V_{OH}	Minimum High-Level Output Voltage	V_{in} = V_{IH} or V_{IL} $|I_{out}| \leq 20\mu A$	2.0	1.9	1.9	1.9	V
			4.5	4.4	4.4	4.4	
			6.0	5.9	5.9	5.9	
		V_{in} = V_{IH} or V_{IL} $|I_{out}| \leq 2.4mA$	3.0	2.48	2.34	2.20	
		$|I_{out}| \leq 4.0mA$	4.5	3.98	3.84	3.70	
		$|I_{out}| \leq 5.2mA$	6.0	5.48	5.34	5.20	
V_{OL}	Maximum Low-Level Output Voltage	V_{in} = V_{IH} or V_{IL} $|I_{out}| \leq 20\mu A$	2.0	0.1	0.1	0.1	V
			4.5	0.1	0.1	0.1	
			6.0	0.1	0.1	0.1	
		V_{in} = V_{IH} or V_{IL} $|I_{out}| \leq 2.4mA$	3.0	0.26	0.33	0.40	
		$|I_{out}| \leq 4.0mA$	4.5	0.26	0.33	0.40	
		$|I_{out}| \leq 5.2mA$	6.0	0.26	0.33	0.40	
I_{in}	Maximum Input Leakage Current	V_{in} = V_{CC} or GND	6.0	±0.1	±1.0	±1.0	μA
I_{CC}	Maximum Quiescent Supply Current (per Package)	V_{in} = V_{CC} or GND I_{out} = 0μA	6.0	1.0	10	40	μA

AC CHARACTERISTICS (C_L = 50 pF, Input t_r = t_f = 6 ns)

Symbol	Parameter	V_{CC} V	Guaranteed Limit −55 to 25°C	Guaranteed Limit ≤85°C	Guaranteed Limit ≤125°C	Unit
t_{PLH}, t_{PHL}	Maximum Propagation Delay, Input A or B to Output Y	2.0	75	95	110	ns
		3.0	30	40	55	
		4.5	15	19	22	
		6.0	13	16	19	
t_{TLH}, t_{THL}	Maximum Output Transition Time, Any Output	2.0	75	95	110	ns
		3.0	27	32	36	
		4.5	15	19	22	
		6.0	13	16	19	
C_{in}	Maximum Input Capacitance		10	10	10	pF

		Typical @ 25°C, V_{CC} = 5.0 V, V_{EE} = 0 V	
C_{PD}	Power Dissipation Capacitance (Per Buffer)	22	pF

FIGURE 3–65   CMOS logic. Partial data sheet for a 54/74HC00A quad 2-input NAND gate. The 54 prefix indicates military grade and the 74 prefix indicates commercial grade.

QUAD 2-INPUT NAND GATE

• ESD > 3500 Volts

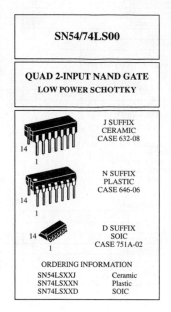

SN54/74LS00

QUAD 2-INPUT NAND GATE
LOW POWER SCHOTTKY

J SUFFIX
CERAMIC
CASE 632-08

N SUFFIX
PLASTIC
CASE 646-06

D SUFFIX
SOIC
CASE 751A-02

ORDERING INFORMATION

SN54LSXXJ	Ceramic
SN74LSXXN	Plastic
SN74LSXXD	SOIC

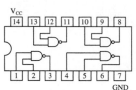

SN54/74LS00

DC CHARACTERISTICS OVER OPERATING TEMPERATURE RANGE (unless otherwise specified)

Symbol	Parameter		Limits			Unit	Test Conditions
			Min	Typ	Max		
V_{IH}	Input HIGH Voltage		2.0			V	Guaranteed Input HIGH Voltage for All Inputs
V_{IL}	Input LOW Voltage	54			0.7	V	Guaranteed Input LOW Voltage for All Inputs
		74			0.8		
V_{IK}	Input Clamp Diode Voltage			−0.65	−1.5	V	V_{CC} = MIN, I_{IN} = −18 mA
V_{OH}	Ouput HIGH Voltage	54	2.5	3.5		V	V_{CC} = MIN, I_{OH} = MAX, V_{IN} = V_{IH} or V_{IL} per Truth Table
		74	2.7	3.5		V	
V_{OL}	Ouput LOW Voltage	54, 74		0.25	0.4	V	I_{OL} = 4.0 mA $\quad$ V_{CC} = V_{CC} MIN, V_{IN} = V_{IL}
		74		0.35	0.5	V	I_{OL} = 8.0 mA $\quad$ or V_{IH} per Truth Table
I_{IH}	Input HIGH Current				20	µA	V_{CC} = MAX, V_{IN} = 2.7 V
					0.1	mA	V_{CC} = MAX, V_{IN} = 7.0 V
I_{IL}	Input LOW Current				−0.4	mA	V_{CC} = MAX, I_{N} = 0.4 V
I_{OS}	Short Circuit Current (Note 1)		−20		−100	mA	V_{CC} = MAX
I_{CC}	Power Supply Current Total, Output HIGH				1.6	mA	V_{CC} = MAX
	Total, Output LOW				4.4		

NOTE 1: Not more than one output should be shorted at a time, nor for more than 1 second.

AC CHARACTERISTICS (T_A = 25°C)

Symbol	Parameter	Limits			Unit	Test Conditions
		Min	Typ	Max		
t_{PLH}	Turn-Off Delay, Input to Output		9.0	15	ns	V_{CC} = 5.0 V
t_{PHL}	Turn-On Delay, Input to Output		10	15	ns	C_L = 15 pF

GUARANTEED OPERATING RANGES

Symbol	Parameter		Min	Typ	Max	Unit
V_{CC}	Supply Voltage	54	4.5	5.0	5.5	V
		74	4.75	5.0	5.25	
T_A	Operating Ambient Temperature Range	54	−55	25	125	°C
		74	0	25	70	
I_{OH}	Output Current — High	54, 74			−0.4	mA
I_{OL}	Output Current — Low	54			4.0	mA
		74			8.0	

FIGURE 3–66   Bipolar logic. Partial data sheet for a 54/74LS00 quad 2-input NAND gate.

dissipation, the fan-out or drive capability, the speed-power product, the dc supply voltage, and the input/output logic levels.

Propagation Delay Time

This parameter is a result of the limitation on switching speed or frequency at which a logic circuit can operate. The terms *low speed* and *high speed,* applied to logic circuits, refer to the propagation delay time. The shorter the propagation delay, the higher the switching speed of the circuit and thus the higher the frequency at which it can operate.

Propagation delay time, t_P, of a logic gate is the time interval between the transition of an input pulse and the occurrence of the resulting transition of the output pulse. There are two different measurements of propagation delay time associated with a logic gate that apply to all the types of basic gates:

• t_{PHL}: The time between a specified reference point on the input pulse and a corresponding reference point on the resulting output pulse, with the output changing from the HIGH level to the LOW level (HL).

• t_{PLH}: The time between a specified reference point on the input pulse and a corresponding reference point on the resulting output pulse, with the output changing from the LOW level to the HIGH level (LH).

For the HCT family CMOS, the propagation delay is 7 ns, for the AC family it is 5 ns, and for the ALVC family it is 3 ns. For standard-family bipolar (TTL) gates, the typical propagation delay is 11 ns and for F family gates it is 3.3 ns. All specified values are dependent on certain operating conditions as stated on a data sheet.

EXAMPLE 3–23

Show the propagation delay times of an inverter.

Solution

An input/output pulse of an inverter is shown in Figure 3–67, and the propagation delay times, t_{PHL} and t_{PLH}, are indicated. In this case, the delays are measured between the 50% points of the corresponding edges of the input and output pulses. The values of t_{PHL} and t_{PLH} are not necessarily equal but in many cases they are the same.

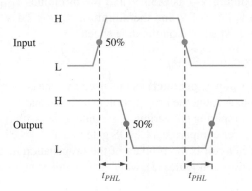

FIGURE 3–67

Related Problem

One type of logic gate has a specified maximum t_{PLH} and t_{PHL} of 10 ns. For another type of gate the value is 4 ns. Which gate can operate at the highest frequency?

DC Supply Voltage (V_{CC})

The typical dc supply voltage for CMOS logic is either 5 V, 3.3 V, 2.5 V, or 1.8 V, depending on the category. An advantage of CMOS is that the supply voltages can vary over a wider range than for bipolar logic. The 5 V CMOS can tolerate supply variations from 2 V to 6 V and still operate properly although propagation delay time and power dissipation are significantly affected. The 3.3 V CMOS can operate with supply voltages from 2 V to 3.6 V. The typical dc supply voltage for bipolar logic is 5.0 V with a minimum of 4.5 V and a maximum of 5.5 V.

Power Dissipation

The **power dissipation**, P_D, of a logic gate is the product of the dc supply voltage and the average supply current. Normally, the supply current when the gate output is LOW is greater than when the gate output is HIGH. The manufacturer's data sheet usually designates the supply current for the LOW output state as I_{CCL} and for the HIGH state as I_{CCH}. The average supply current is determined based on a 50% duty cycle (output LOW half the time and HIGH half the time), so the average power dissipation of a logic gate is

A lower power dissipation means less current from the dc supply.

$$P_D = V_{CC}\left(\frac{I_{CCH} + I_{CCL}}{2}\right) \qquad \text{Equation 3–2}$$

CMOS gates have very low power dissipations compared to the bipolar family. However, the power dissipation of CMOS is dependent on the frequency of operation. At zero frequency the quiescent power is typically in the microwatt/gate range, and at the maximum operating frequency it can be in the low milliwatt range; therefore, power is sometimes specified at a given frequency. The HC family, for example, has a power of 2.75 μW/gate at 0 Hz (quiescent) and 600 μW/gate at 1 MHz.

Power dissipation for bipolar gates is independent of frequency. For example, the ALS family uses 1.4 mW/gate regardless of the frequency and the F family uses 6 mW/gate.

Input and Output Logic Levels

V_{IL} is the LOW level input voltage for a logic gate, and V_{IH} is the HIGH level input voltage. The 5 V CMOS accepts a maximum voltage of 1.5 V as V_{IL} and a minimum voltage of 3.5 V as V_{IH}. Bipolar logic accepts a maximum voltage of 0.8 V as V_{IL} and a minimum voltage of 2 V as V_{IH}.

V_{OL} is the LOW level output voltage and V_{OH} is the HIGH level output voltage. For 5 V CMOS, the maximum V_{OL} is 0.33 V and the minimum V_{OH} is 4.4 V. For bipolar logic, the maximum V_{OL} is 0.4 V and the minimum V_{OH} is 2.4 V. All values depend on operating conditions as specified on the data sheet.

Speed-Power Product (SPP)

This parameter (**speed-power product**) can be used as a measure of the performance of a logic circuit taking into account the propagation delay time and the power dissipation. It is especially useful for comparing the various logic gate series within the CMOS and bipolar technology families or for comparing a CMOS gate to a TTL gate.

The SPP of a logic circuit is the product of the propagation delay time and the power dissipation and is expressed in joules (J), which is the unit of energy. The formula is

$$SPP = t_p P_D \qquad \text{Equation 3–3}$$

EXAMPLE 3–24

A certain gate has a propagation delay of 5 ns and $I_{CCH} = 1$ mA and $I_{CCL} = 2.5$ mA with a dc supply voltage of 5 V. Determine the speed-power product.

Solution

$$P_D = V_{CC}\left(\frac{I_{CCH} + I_{CCL}}{2}\right) = 5\ V\left(\frac{1\ mA + 2.5\ mA}{2}\right) = 5\ V(1.75\ mA) = 8.75\ mW$$

$$SPP = (5\ ns)(8.75\ mW) = \textbf{43.75 pJ}$$

Related Problem

If the propagation delay of a gate is 15 ns and its *SPP* is 150 pJ, what is its average power dissipation?

Fan-Out and Loading

The **fan-out** of a logic gate is the maximum number of inputs of the same series in an IC family that can be connected to a gate's output and still maintain the output voltage levels within specified limits. Fan-out is a significant parameter only for bipolar logic because of the type of circuit technology. Since very high impedances are associated with CMOS circuits, the fan-out is very high but depends on frequency because of capacitive effects.

A higher fan-out means that a gate output can be connected to more gate inputs.

Fan-out is specified in terms of **unit loads**. A unit load for a logic gate equals one input to a like circuit. For example, a unit load for a 74LS00 NAND gate equals *one* input to another logic gate in the 74LS family (not necessarily a NAND gate). Because the current from a LOW input (I_{IL}) of a 74LS00 gate is 0.4 mA and the current that a LOW output (I_{OL}) can accept is 8.0 mA, the number of unit loads that a 74LS00 gate can drive in the LOW state is

$$\text{Unit loads} = \frac{I_{OL}}{I_{IL}} = \frac{8.0\ mA}{0.4\ mA} = 20$$

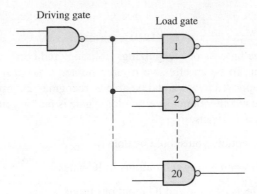

FIGURE 3–68 The LS family NAND gate output fans out to a maximum of 20 LS family gate inputs.

Figure 3–68 shows LS logic gates driving a number of other gates of the same circuit technology, where the number of gates depends on the particular circuit technology. For example, as you have seen, the maximum number of gate inputs (unit loads) that a 74LS family bipolar gate can drive is 20.

Unused gate inputs for bipolar (TTL) and CMOS should be connected to the appropriate logic level (HIGH or LOW). For AND/NAND, it is recommended that unused inputs be connected to V_{CC} (through a 1.0 kΩ resistor with bipolar) and for OR/NOR, unused inputs should be connected to ground.

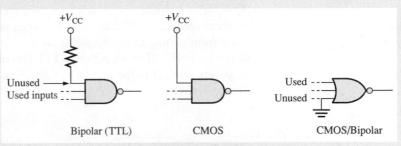

SECTION 3–8 CHECKUP

1. How is fixed-function logic different than PLD logic?

2. List the two types of IC technologies that are the most widely used.

3. Identify the following IC logic designators:

    (a) LS   (b) HC   (c) HCT

4. Which IC technology generally has the lowest power dissipation?

5. What does the term *hex inverter* mean? What does *quad 2-input NAND* mean?

6. A positive pulse is applied to an inverter input. The time from the leading edge of the input to the leading edge of the output is 10 ns. The time from the trailing edge of the input to the trailing edge of the output is 8 ns. What are the values of t_{PLH} and t_{PHL}?

7. A certain gate has a propagation delay time of 6 ns and a power dissipation of 3 mW. Determine the speed-power product?

8. Define I_{CCL} and I_{CCH}.

9. Define V_{IL} and V_{IH}.

10. Define V_{OL} and V_{OH}.

3–9   Troubleshooting

Troubleshooting is the process of recognizing, isolating, and correcting a fault or failure in a circuit or system. To be an effective troubleshooter, you must understand how the circuit or system is supposed to work and be able to recognize incorrect performance. For example, to determine whether or not a certain logic gate is faulty, you must know what the output should be for given inputs.

After completing this section, you should be able to

- ◆ Test for internally open inputs and outputs in IC gates
- ◆ Recognize the effects of a shorted IC input or output
- ◆ Test for external faults on a PCB board
- ◆ Troubleshoot a simple frequency counter using an oscillosope

Internal Failures of IC Logic Gates

Opens and shorts are the most common types of internal gate failures. These can occur on the inputs or on the output of a gate inside the IC package. *Before attempting any troubleshooting, check for proper dc supply voltage and ground.*

Effects of an Internally Open Input

An internal open is the result of an open component on the chip or a break in the tiny wire connecting the IC chip to the package pin. An open input prevents a signal on that input from getting to the output of the gate, as illustrated in Figure 3–69(a) for the case of a 2-input NAND gate. An open TTL (bipolar) input acts effectively as a HIGH level, so pulses applied to the good input get through to the NAND gate output as shown in Figure 3–69(b).

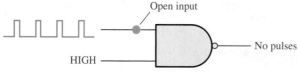

(a) Application of pulses to the open input will produce no pulses on the output.

(b) Application of pulses to the good input will produce output pulses for bipolar NAND and AND gates because an open input typically acts as a HIGH. It is uncertain for CMOS.

FIGURE 3–69   The effect of an open input on a NAND gate.

Conditions for Testing Gates

When testing a NAND gate or an AND gate, always make sure that the inputs that are not being pulsed are HIGH to enable the gate. When checking a NOR gate or an OR gate, always make sure that the inputs that are not being pulsed are LOW. When checking an XOR or XNOR gate, the level of the nonpulsed input does not matter because the pulses on the other input will force the inputs to alternate between the same level and opposite levels.

Troubleshooting an Open Input

Troubleshooting this type of failure is easily accomplished with an oscilloscope and function generator, as demonstrated in Figure 3–70 for the case of a quad 2-input NAND gate package. When measuring digital signals with a scope, always use dc coupling.

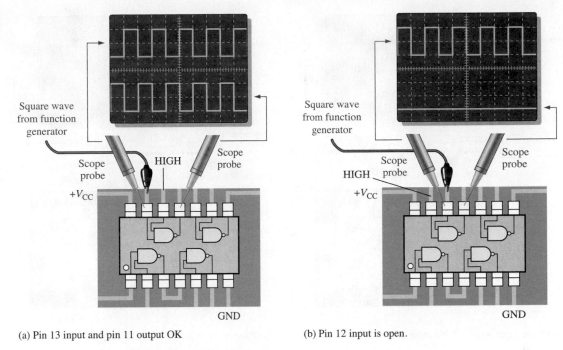

(a) Pin 13 input and pin 11 output OK   (b) Pin 12 input is open.

FIGURE 3–70   Troubleshooting a NAND gate for an open input.

The first step in troubleshooting an IC that is suspected of being faulty is to make sure that the dc supply voltage (V_{CC}) and ground are at the appropriate pins of the IC. Next, apply continuous pulses to one of the inputs to the gate, making sure that the other input is HIGH (in the case of a NAND gate). In Figure 3–70(a), start by applying a pulse waveform to pin 13, which is one of the inputs to the suspected gate. If a pulse waveform is indicated on the output (pin 11 in this case), then the pin 13 input is not open. By the way, this also proves that the output is not open. Next, apply the pulse waveform to the other gate input (pin 12), making sure the other input is HIGH. There is no pulse waveform on the output at pin 11 and the output is LOW, indicating that the pin 12 input is open, as shown in Figure 3–70(b). The input not being pulsed must be HIGH for the case of a NAND gate or AND gate. If this were a NOR gate, the input not being pulsed would have to be LOW.

Effects of an Internally Open Output

An internally open gate output prevents a signal on any of the inputs from getting to the output. Therefore, no matter what the input conditions are, the output is unaffected. The level at the output pin of the IC will depend upon what it is externally connected to. It could be either HIGH, LOW, or floating (not fixed to any reference). In any case, there will be no signal on the output pin.

Troubleshooting an Open Output

Figure 3–71 illustrates troubleshooting an open NOR gate output. In part (a), one of the inputs of the suspected gate (pin 11 in this case) is pulsed, and the output (pin 13) has no pulse waveform. In part (b), the other input (pin 12) is pulsed and again there is no pulse waveform on the output. Under the condition that the input that is not being pulsed is at a LOW level, this test shows that the output is internally open.

Shorted Input or Output

Although not as common as an open, an internal short to the dc supply voltage, ground, another input, or an output can occur. When an input or output is shorted to the supply voltage, it will be stuck in the HIGH state. If an input or output is shorted to ground, it will be

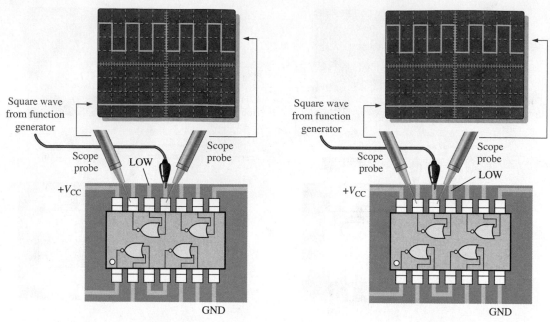

(a) Pulse input on pin 11. No pulse output.   (b) Pulse input on pin 12. No pulse output.

FIGURE 3-71   Troubleshooting a NOR gate for an open output.

stuck in the LOW state (0 V). If two inputs or an input and an output are shorted together, they will always be at the same level.

External Opens and Shorts

Many failures involving digital ICs are due to faults that are external to the IC package. These include bad solder connections, solder splashes, wire clippings, improperly etched printed circuit boards (PCBs), and cracks or breaks in wires or printed circuit interconnections. These open or shorted conditions have the same effect on the logic gate as the internal faults, and troubleshooting is done in basically the same ways. A visual inspection of any circuit that is suspected of being faulty is the first thing a technician should do.

EXAMPLE 3-25

You are checking a 74LS10 triple 3-input NAND gate IC that is one of many ICs located on a PCB. You have checked pins 1 and 2 and they are both HIGH. Now you apply a pulse waveform to pin 13, and place your scope probe first on pin 12 and then on the connecting PCB trace, as indicated in Figure 3–72. Based on your observation of the scope screen, what is the most likely problem?

Solution

The waveform with the probe in position 1 shows that there is pulse activity on the gate output at pin 12, but there are no pulses on the PCB trace as indicated by the probe in position 2. The gate is working properly, but the signal is not getting from pin 12 of the IC to the PCB trace.

Most likely there is a bad solder connection between pin 12 of the IC and the PCB, which is creating an open. You should resolder that point and check it again.

Related Problem

If there are no pulses at either probe position 1 or 2 in Figure 3–72, what fault(s) does this indicate?

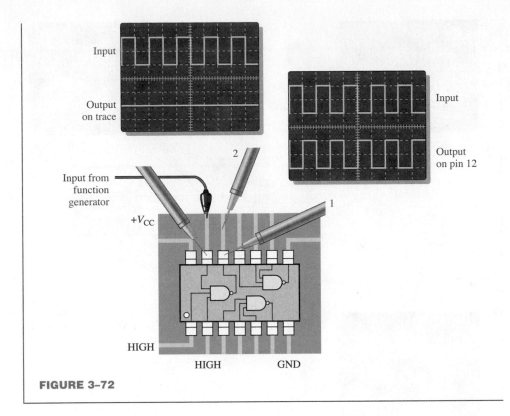

FIGURE 3–72

In most cases, you will be troubleshooting ICs that are mounted on PCBs or proto-type assemblies and interconnected with other ICs. As you progress through this book, you will learn how different types of digital ICs are used together to perform system functions. At this point, however, we are concentrating on individual IC gates. This limitation does not prevent us from looking at the system concept at a very basic and simplified level.

To continue the emphasis on systems, Examples 3–26 and 3–27 deal with troubleshooting the frequency counter that was introduced in Section 3–2.

EXAMPLE 3–26

After trying to operate the frequency counter shown in Figure 3–73, you find that it constantly reads out all 0s on its display, regardless of the input frequency. Determine the cause of this malfunction. The enable pulse has a width of 1 ms.

Figure 3–73(a) gives an example of how the frequency counter should be working with a 12 kHz pulse waveform on the input to the AND gate. Part (b) shows that the display is improperly indicating 0 Hz.

Solution

Three possible causes are

1. A constant active or asserted level on the counter reset input, which keeps the counter at zero.

2. No pulse signal on the input to the counter because of an internal open or short in the counter. This problem would keep the counter from advancing after being reset to zero.

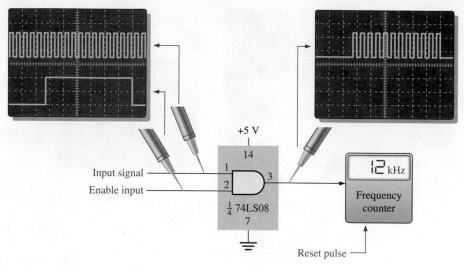

(a) The counter is working properly.

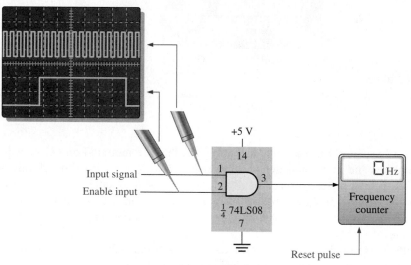

(b) The counter is not measuring a frequency.

FIGURE 3–73

3. No pulse signal on the input to the counter because of an open AND gate output or the absence of input signals, again keeping the counter from advancing from zero.

The first step is to make sure that V_{CC} and ground are connected to all the right places; assume that they are found to be okay. Next, check for pulses on both inputs to the AND gate. The scope indicates that there are proper pulses on both of these inputs. A check of the counter reset shows a LOW level which is known to be the unasserted level and, therefore, this is not the problem. The next check on pin 3 of the 74LS08 shows that there are no pulses on the output of the AND gate, indicating that the gate output is open. Replace the 74LS08 IC and check the operation again.

Related Problem

If pin 2 of the 74LS08 AND gate is open, what indication should you see on the frequency display?

EXAMPLE 3–27

The frequency counter shown in Figure 3–74 appears to measure the frequency of input signals incorrectly. It is found that when a signal with a precisely known frequency is applied to pin 1 of the AND gate, the oscilloscope display indicates a higher frequency. Determine what is wrong. The readings on the screen indicate time per division.

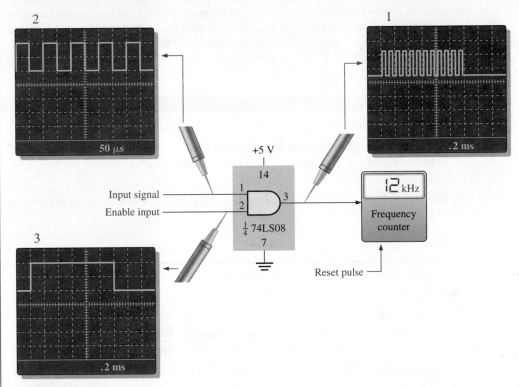

FIGURE 3–74

Solution

Recall from Section 3–2 that the input pulses were allowed to pass through the AND gate for exactly 1 ms. The number of pulses counted in 1 ms is equal to the frequency in hertz. Therefore, the 1 ms interval, which is produced by the enable pulse on pin 2 of the AND gate, is very critical to an accurate frequency measurement. The enable pulses are produced internally by a precision oscillator circuit. The pulse must be exactly 1 ms in width and in this case it occurs every 3 ms to update the count. Just prior to each enable pulse, the counter is reset to zero so that it starts a new count each time.

Since the counter appears to be counting more pulses than it should to produce a frequency readout that is too high, the enable pulse is the primary suspect. Exact time-interval measurements must be made on the oscilloscope.

An input pulse waveform of exactly 10 kHz is applied to pin 1 of the AND gate and the frequency counter incorrectly shows 12 kHz. The first scope measurement, on the output of the AND gate, shows that there are 12 pulses for each enable pulse. In the second scope measurement, the input frequency is verified to be precisely 10 kHz (period = 100 μs). In the third scope measurement, the width of the enable pulse is found to be 1.2 ms rather than 1 ms.

The conclusion is that the enable pulse is out of calibration for some reason.

Related Problem

What would you suspect if the readout were indicating a frequency less than it should be?

Proper grounding is very important when setting up to take measurements or work on a circuit. Properly grounding the oscilloscope protects you from shock and grounding yourself protects your circuits from damage. Grounding the oscilloscope means to connect it to earth ground by plugging the three-prong power cord into a grounded outlet. Grounding yourself means using a wrist-type grounding strap, particularly when you are working with CMOS logic. The wrist strap must have a high-value resistor between the strap and ground for protection against accidental contact with a voltage source.

Also, for accurate measurements, make sure that the ground in the circuit you are testing is the same as the scope ground. This can be done by connecting the ground lead on the scope probe to a known ground point in the circuit, such as the metal chassis or a ground point on the PCB. You can also connect the circuit ground to the GND jack on the front panel of the scope.

SECTION 3–9 CHECKUP

1. What are the most common types of failures in ICs?

2. If two different input waveforms are applied to a 2-input bipolar NAND gate and the output waveform is just like one of the inputs, but inverted, what is the most likely problem?

3. Name two characteristics of pulse waveforms that can be measured on the oscilloscope.

SUMMARY

- The inverter output is the complement of the input.
- The AND gate output is HIGH only when all the inputs are HIGH.
- The OR gate output is HIGH when any of the inputs is HIGH.
- The NAND gate output is LOW only when all the inputs are HIGH.
- The NAND can be viewed as a negative-OR whose output is HIGH when any input is LOW.
- The NOR gate output is LOW when any of the inputs is HIGH.
- The NOR can be viewed as a negative-AND whose output is HIGH only when all the inputs are LOW.
- The exclusive-OR gate output is HIGH when the inputs are not the same.
- The exclusive-NOR gate output is LOW when the inputs are not the same.
- Distinctive shape symbols and truth tables for various logic gates (limited to 2 inputs) are shown in Figure 3–75.

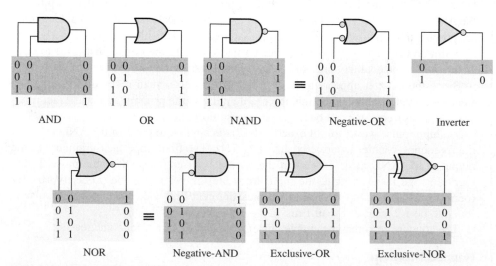

Note: Active states are shown in yellow.

FIGURE 3–75

- Most programmable logic devices (PLDs) are based on some form of AND array.
- Programmable link technologies are fuse, antifuse, EPROM, EEPROM, flash, and SRAM.
- A PLD can be programmed in a hardware fixture called a programmer or mounted on a development printed circuit board.
- PLDs have an associated software development package for programming.
- Two methods of design entry using programming software are text entry (HDL) and graphic (schematic) entry.
- ISP PLDs can be programmed after they are installed in a system, and they can be reprogrammed at any time.
- JTAG stands for Joint Test Action Group and is an interface standard (IEEE Std. 1149.1) used for programming and testing PLDs.
- An embedded processor is used to facilitate in-system programming of PLDs.
- In PLDs, the circuit is programmed in and can be changed by reprogramming.
- The average power dissipation of a logic gate is

$$P_D = V_{CC} \left(\frac{I_{CCH} + I_{CCL}}{2} \right)$$

- The speed-power product of a logic gate is

$$SPP = t_p P_D$$

- As a rule, CMOS has a lower power consumption than bipolar.
- In fixed-function logic, the circuit cannot be altered.

KEY TERMS

Key terms and other bold terms in the chapter are defined in the end-of-book glossary.

AND array   An array of AND gates consisting of a matrix of programmable interconnections.

AND gate   A logic gate that produces a HIGH output only when all of the inputs are HIGH.

Antifuse   A type of PLD nonvolatile programmable link that can be left open or can be shorted once as directed by the program.

Bipolar   A class of integrated logic circuits implemented with bipolar transistors; also known as TTL.

Boolean algebra   The mathematics of logic circuits.

CMOS   Complementary metal-oxide semiconductor; a class of integrated logic circuits that is implemented with a type of field-effect transistor.

Complement   The inverse or opposite of a number. LOW is the complement of HIGH, and 0 is the complement of 1.

EEPROM   A type of nonvolatile PLD reprogrammable link based on electrically erasable programmable read-only memory cells and can be turned on or off repeatedly by programming.

EPROM   A type of PLD nonvolatile programmable link based on electrically programmable read-only memory cells and can be turned either on or off once with programming.

Exclusive-NOR (XNOR) gate   A logic gate that produces a LOW only when the two inputs are at opposite levels.

Exclusive-OR (XOR) gate   A logic gate that produces a HIGH output only when its two inputs are at opposite levels.

Fan-out   The number of equivalent gate inputs of the same family series that a logic gate can drive.

Flash   A type of PLD nonvolatile reprogrammable link technology based on a single transistor cell.

Fuse   A type of PLD nonvolatile programmable link that can be left shorted or can be opened once as directed by the program.

Inverter   A logic circuit that inverts or complements its input.

JTAG   Joint Test Action Group; an interface standard designated IEEE Std. 1149.1.

NAND gate   A logic gate that produces a LOW output only when all the inputs are HIGH.

NOR gate   A logic gate in which the output is LOW when one or more of the inputs are HIGH.

OR gate   A logic gate that produces a HIGH output when one or more inputs are HIGH.

Propagation delay time   The time interval between the occurrence of an input transition and the occurrence of the corresponding output transition in a logic circuit.

SRAM   A type of PLD volatile reprogrammable link based on static random-access memory cells and can be turned on or off repeatedly with programming.

Target device   A PLD mounted on a programming fixture or development board into which a software logic design is to be downloaded.

Truth table   A table showing the inputs and corresponding output(s) of a logic circuit.

Unit load   A measure of fan-out. One gate input represents one unit load to the output of a gate within the same IC family.

VHDL   A standard hardware description language that describes a function with an entity/architecture structure.

TRUE/FALSE QUIZ

Answers are at the end of the chapter.

1. An inverter performs the NOR operation.
2. An AND gate can have only two inputs.
3. If any input to an OR is 1, the output is 1.
4. If all inputs to an AND gate are 1, the output is 0.
5. A NAND gate has an output that is opposite the output of an AND gate.
6. A NOR gate can be considered as an OR gate followed by an inverter.
7. The output of an exclusive-OR is 0 if the inputs are opposite.
8. Two types of fixed-function logic integrated circuits are bipolar and NMOS.
9. Once programmed, PLD logic can be changed.
10. Fan-out is the number of similar gates that a given gate can drive.

SELF-TEST

Answers are at the end of the chapter.

1. When the input to an inverter is HIGH (1), the output is
   (a) HIGH or 1   (b) LOW or 1   (c) HIGH or 0   (d) LOW or 0

2. An inverter performs an operation known as
   (a) complementation   (b) assertion   (c) inversion   (d) both answers (a) and (c)

3. The output of an AND gate with inputs A, B, and C is a 1 (HIGH) when
   (a) $A = 1, B = 1, C = 1$ (b) $A = 1, B = 0, C = 1$ (c) $A = 0, B = 0, C = 0$

4. The output of an OR gate with inputs A, B, and C is a 1 (HIGH) when
   (a) $A = 1, B = 1, C = 1$ (b) $A = 0, B = 0, C = 1$ (c) $A = 0, B = 0, C = 0$
   (d) answers (a), (b), and (c)   (e) only answers (a) and (b)

5. A pulse is applied to each input of a 2-input NAND gate. One pulse goes HIGH at $t = 0$ and goes back LOW at $t = 1$ ms. The other pulse goes HIGH at $t = 0.8$ ms and goes back LOW at $t = 3$ ms. The output pulse can be described as follows:
   (a) It goes LOW at $t = 0$ and back HIGH at $t = 3$ ms.
   (b) It goes LOW at $t = 0.8$ ms and back HIGH at $t = 3$ ms.
   (c) It goes LOW at $t = 0.8$ ms and back HIGH at $t = 1$ ms.
   (d) It goes LOW at $t = 0.8$ ms and back LOW at $t = 1$ ms.

6. A pulse is applied to each input of a 2-input NOR gate. One pulse goes HIGH at $t = 0$ and goes back LOW at $t = 1$ ms. The other pulse goes HIGH at $t = 0.8$ ms and goes back LOW at $t = 3$ ms. The output pulse can be described as follows:
   (a) It goes LOW at $t = 0$ and back HIGH at $t = 3$ ms.
   (b) It goes LOW at $t = 0.8$ ms and back HIGH at $t = 3$ ms.
   (c) It goes LOW at $t = 0.8$ ms and back HIGH at $t = 1$ ms.
   (d) It goes HIGH at $t = 0.8$ ms and back LOW at $t = 1$ ms.

7. A pulse is applied to each input of an exclusive-OR gate. One pulse goes HIGH at $t = 0$ and goes back LOW at $t = 1$ ms. The other pulse goes HIGH at $t = 0.8$ ms and goes back LOW at $t = 3$ ms. The output pulse can be described as follows:
   (a) It goes HIGH at $t = 0$ and back LOW at $t = 3$ ms.
   (b) It goes HIGH at $t = 0$ and back LOW at $t = 0.8$ ms.
   (c) It goes HIGH at $t = 1$ ms and back LOW at $t = 3$ ms.
   (d) both answers (b) and (c)

8. A positive-going pulse is applied to an inverter. The time interval from the leading edge of the input to the leading edge of the output is 7 ns. This parameter is
   (a) speed-power product
   (b) propagation delay, t_{PHL}
   (c) propagation delay, t_{PLH}
   (d) pulse width

9. The purpose of a programmable link in an AND array is to
   (a) connect an input variable to a gate input
   (b) connect a row to a column in the array matrix
   (c) disconnect a row from a column in the array matrix
   (d) do all of the above

10. The term OTP means
    (a) open test point
    (b) one-time programmable
    (c) output test program
    (d) output terminal positive

11. Types of PLD programmable link process technologies are
    (a) antifuse
    (b) flash
    (c) ROM
    (d) both (a) and (b)
    (e) both (a) and (c)

12. A volatile programmable link technology is
    (a) fuse
    (b) EPROM
    (c) SRAM
    (d) EEPROM

13. Two ways to enter a logic design using PLD development software are
    (a) text and numeric
    (b) text and graphic
    (c) graphic and coded
    (d) compile and sort

14. JTAG stands for
    (a) Joint Test Action Group
    (b) Java Top Array Group
    (c) Joint Test Array Group
    (d) Joint Time Analysis Group

15. In-system programming of a PLD typically utilizes
    (a) an embedded clock generator
    (b) an embedded processor
    (c) an embedded PROM
    (d) both (a) and (b)
    (e) both (b) and (c)

16. To measure the period of a pulse waveform, you must use
    (a) a DMM
    (b) a logic probe
    (c) an oscilloscope
    (d) a logic pulser

17. Once you measure the period of a pulse waveform, the frequency is found by
    (a) using another setting
    (b) measuring the duty cycle
    (c) finding the reciprocal of the period
    (d) using another type of instrument

PROBLEMS

Answers to odd-numbered problems are at the end of the book.

Section 3–1 The Inverter

1. The input waveform shown in Figure 3–76 is applied to an inverter. Draw the timing diagram of the output waveform in proper relation to the input.

FIGURE 3–76

2. A combination of inverters is shown in Figure 3–77. If a HIGH is applied to point A, determine the logic levels at points B through F.

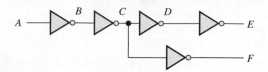

FIGURE 3–77

3. If the waveform in Figure 3–76 is applied to point A in Figure 3–77, determine the waveforms at points B through F.

Section 3–2 The AND Gate

4. Draw the rectangular outline symbol for a 4-input AND gate.

5. Determine the output, X, for a 2-input AND gate with the input waveforms shown in Figure 3–78. Show the proper relationship of output to inputs with a timing diagram.

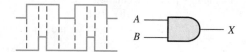

FIGURE 3–78

6. Repeat Problem 5 for the waveforms in Figure 3–79.

FIGURE 3–79

7. The input waveforms applied to a 3-input AND gate are as indicated in Figure 3–80. Show the output waveform in proper relation to the inputs with a timing diagram.

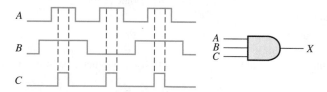

FIGURE 3–80

8. The input waveforms applied to a 4-input AND gate are as indicated in Figure 3–81. Show the output waveform in proper relation to the inputs with a timing diagram.

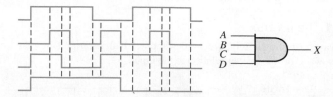

FIGURE 3–81

Section 3–3 The OR Gate

9. Draw the rectangular outline symbol for a 3-input AND gate.

10. Write the expression for a 5-input OR gate with inputs A, B, C, D, E, and output X.

11. Determine the output for a 2-input OR gate when the input waveforms are as in Figure 3–79 and draw a timing diagram.

12. Repeat Problem 7 for a 3-input OR gate.

13. Repeat Problem 8 for a 4-input OR gate.

14. For the five input waveforms in Figure 3–82, determine the output if the five signals are ANDed. Determine the output if the five signals are ORed. Draw the timing diagram for each case.

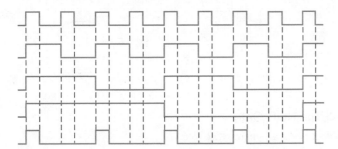

FIGURE 3–82

15. Draw the rectangular outline symbol for a 4-input OR gate.

16. Show the truth table for a 3-input OR gate.

Section 3–4 The NAND Gate

17. For the set of input waveforms in Figure 3–83, determine the output for the gate shown and draw the timing diagram.

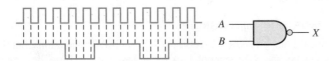

FIGURE 3–83

18. Determine the gate output for the input waveforms in Figure 3–84 and draw the timing diagram.

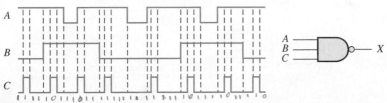

FIGURE 3–84

19. Determine the output waveform in Figure 3–85.

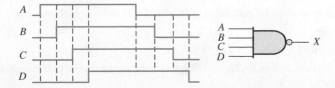

FIGURE 3–85

20. As you have learned, the two logic symbols shown in Figure 3–86 represent equivalent operations. The difference between the two is strictly from a functional viewpoint. For the NAND symbol, look for two HIGHs on the inputs to give a LOW output. For the negative-OR, look for at least one LOW on the inputs to give a HIGH on the output. Using these two functional viewpoints, show that each gate will produce the same output for the given inputs.

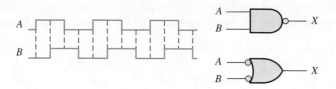

FIGURE 3–86

Section 3–5 The NOR Gate

21. Repeat Problem 17 for a 2-input NOR gate.

22. Determine the output waveform in Figure 3–87 and draw the timing diagram.

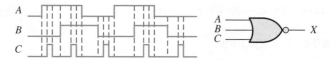

FIGURE 3–87

23. Repeat Problem 19 for a 4-input NOR gate.

24. The NAND and the negative-OR symbols represent equivalent operations, but they are functionally different. For the NOR symbol, look for at least one HIGH on the inputs to give a LOW on the output. For the negative-AND, look for two LOWs on the inputs to give a HIGH output. Using these two functional points of view, show that both gates in Figure 3–88 will produce the same output for the given inputs.

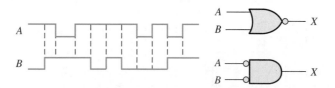

FIGURE 3–88

Section 3–6 The Exclusive-OR and Exclusive-NOR Gates

25. How does an exclusive-OR gate differ from an OR gate in its logical operation?

26. Repeat Problem 17 for an exclusive-OR gate.

27. Repeat Problem 17 for an exclusive-NOR gate.

28. Determine the output of an exclusive-OR gate for the inputs shown in Figure 3–79 and draw a timing diagram.

Section 3–7 Programmable Logic

29. In the simple programmed AND array with programmable links in Figure 3–89, determine the Boolean output expressions.

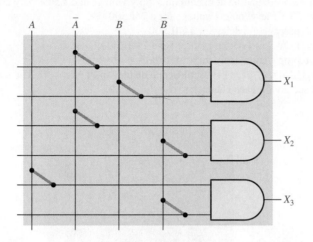

FIGURE 3–89

30. Determine by row and column number which fusible links must be blown in the programmable AND array of Figure 3–90 to implement each of the following product terms: $X_1 = \overline{A}BC, X_2 = A B\overline{C}, X_3 = \overline{A}B\overline{C}$.

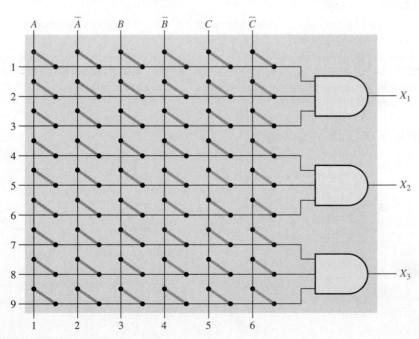

FIGURE 3–90

31. Describe a 4-input AND gate using VHDL.
32. Describe a 5-input NOR gate using VHDL.

Section 3–8 Fixed-Function Logic Gates

33. In the comparison of certain logic devices, it is noted that the power dissipation for one particular type increases as the frequency increases. Is the device bipolar or CMOS?

34. Using the data sheets in Figures 3–65 and 3–66, determine the following:

    (a) 74LS00 power dissipation at maximum supply voltage and a 50% duty cycle

    (b) Minimum HIGH level output voltage for a 74LS00

    (c) Maximum propagation delay for a 74LS00

    (d) Maximum LOW level output voltage for a 74HC00A

    (e) Maximum propagation delay for a 74HC00A

35. Determine t_{PLH} and t_{PHL} from the oscilloscope display in Figure 3–91. The readings indicate volts/div and sec/div for each channel.

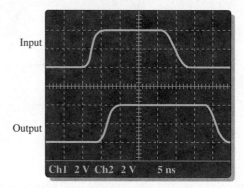

Input

Output

Ch1 2 V Ch2 2 V   5 ns

FIGURE 3–91

36. Gate A has $t_{PLH} = t_{PHL} = 6$ ns. Gate B has $t_{PLH} = t_{PHL} = 10$ ns. Which gate can be operated at a higher frequency?

37. If a logic gate operates on a dc supply voltage of +5 V and draws an average current of 4 mA, what is its power dissipation?

38. The variable I_{CCH} represents the dc supply current from V_{CC} when all outputs of an IC are HIGH. The variable I_{CCL} represents the dc supply current when all outputs are LOW. For a 74LS00 IC, determine the typical power dissipation when all four gate outputs are HIGH. (See data sheet in Figure 3–66.)

Section 3–9 Troubleshooting

39. Examine the conditions indicated in Figure 3–92, and identify the faulty gates.

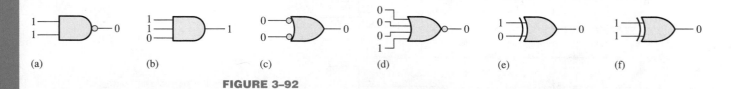

(a)   (b)   (c)   (d)   (e)   (f)

FIGURE 3–92

40. Determine the faulty gates in Figure 3–93 by analyzing the timing diagrams.

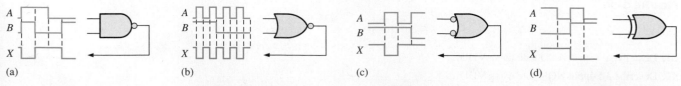

(a)   (b)   (c)   (d)

FIGURE 3–93

41. Using an oscilloscope, you make the observations indicated in Figure 3–94. For each observation determine the most likely gate failure.

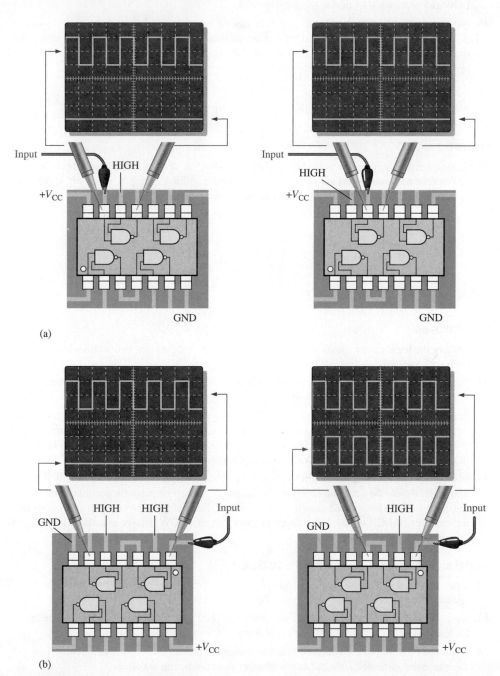

(a)

(b)

FIGURE 3–94

42. The seat belt alarm circuit in Figure 3–17 has malfunctioned. You find that when the ignition switch is turned on and the seat belt is unbuckled, the alarm comes on and will not go off. What is the most likely problem? How do you troubleshoot it?

43. Every time the ignition switch is turned on in the circuit of Figure 3–17, the alarm comes on for thirty seconds, even when the seat belt is buckled. What is the most probable cause of this malfunction?

44. What failure(s) would you suspect if the output of a 3-input NAND gate stays HIGH no matter what the inputs are?

Special Design Problems

45. Modify the frequency counter in Figure 3–16 to operate with an enable pulse that is active-LOW rather than HIGH during the 1 ms interval.

46. Assume that the enable signal in Figure 3–16 has the waveform shown in Figure 3–95. Assume that waveform B is also available. Devise a circuit that will produce an active-HIGH reset pulse to the counter only during the time that the enable signal is LOW.

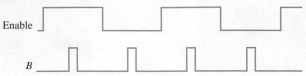

FIGURE 3–95

47. Design a circuit to fit in the beige block of Figure 3–96 that will cause the headlights of an automobile to be turned off automatically 15 s after the ignition switch is turned off, if the light switch is left on. Assume that a LOW is required to turn the lights off.

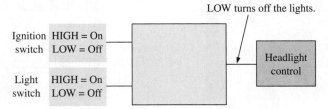

FIGURE 3–96

48. Modify the logic circuit for the intrusion alarm in Figure 3–25 so that two additional rooms, each with two windows and one door, can be protected.

49. Further modify the logic circuit from Problem 48 for a change in the input sensors where Open = LOW and Closed = HIGH.

50. Sensors are used to monitor the pressure and the temperature of a chemical solution stored in a vat. The circuitry for each sensor produces a HIGH voltage when a specified maximum value is exceeded. An alarm requiring a LOW voltage input must be activated when either the pressure or the temperature is excessive. Design a circuit for this application.

51. In a certain automated manufacturing process, electrical components are automatically inserted in a PCB. Before the insertion tool is activated, the PCB must be properly positioned, and the component to be inserted must be in the chamber. Each of these prerequisite conditions is indicated by a HIGH voltage. The insertion tool requires a LOW voltage to activate it. Design a circuit to implement this process.

MultiSim   Multisim Troubleshooting Practice

52. Open file P03-52. For the specified fault, predict the effect on the circuit. Then introduce the fault and verify whether your prediction is correct.

53. Open file P03-53. For the specified fault, predict the effect on the circuit. Then introduce the fault and verify whether your prediction is correct.

54. Open file P03-54. For the observed behavior indicated, predict the fault in the circuit. Then introduce the suspected fault and verify whether your prediction is correct.

55. Open file P03-55. For the observed behavior indicated, predict the fault in the circuit. Then introduce the suspected fault and verify whether your prediction is correct.

ANSWERS

SECTION CHECKUPS

Section 3–1 The Inverter

1. When the inverter input is 1, the output is 0.

2. (a)

(b) A negative-going pulse is on the output (HIGH to LOW and back HIGH).

Section 3–2 The AND Gate

1. An AND gate output is HIGH only when all inputs are HIGH.

2. An AND gate output is LOW when one or more inputs are LOW.

3. Five-input AND: $X = 1$ when $ABCDE = 11111$, and $X = 0$ for all other combinations of $ABCDE$.

Section 3–3 The OR Gate

1. An OR gate output is HIGH when one or more inputs are HIGH.

2. An OR gate output is LOW only when all inputs are LOW.

3. Three-input OR: $X = 0$ when $ABC = 000$, and $X = 1$ for all other combinations of ABC.

Section 3–4 The NAND Gate

1. A NAND gate output is LOW only when all inputs are HIGH.

2. A NAND gate output is HIGH when one or more inputs are LOW.

3. NAND: active-LOW output for all HIGH inputs; negative-OR: active-HIGH output for one or more LOW inputs. They have the same truth tables.

4. $X = \overline{ABC}$

Section 3–5 The NOR Gate

1. A NOR gate output is HIGH only when all inputs are LOW.

2. A NOR gate output is LOW when one or more inputs are HIGH.

3. NOR: active-LOW output for one or more HIGH inputs; negative-AND: active-HIGH output for all LOW inputs. They have the same truth tables.

4. $X = \overline{A + B + C}$

Section 3–6 The Exclusive-OR and Exclusive-NOR Gates

1. An XOR gate output is HIGH when the inputs are at opposite levels.

2. An XNOR gate output is HIGH when the inputs are at the same levels.

3. Apply the bits to the XOR gate inputs; when the output is HIGH, the bits are different.

Section 3–7 Programmable Logic

1. Fuse, antifuse, EPROM, EEPROM, flash, and SRAM

2. Volatile means that all the data are lost when power is off and the PLD must be reprogrammed; SRAM-based

3. Text entry and graphic entry

4. JTAG is Joint Test Action Group; the IEEE Std. 1149.1 for programming and test interfacing.

5. **entity** NORgate **is**
   port (A, B, C: **in** bit; X: **out** bit);
   end entity NORgate;
   architecture NORfunction **of** NORgate **is**
   begin
   　X <= A **nor** B **nor** C;
   end architecture NORfunction;

6. **entity** XORgate **is**
   port (A, B: **in** bit; X: **out** bit);
   end entity XORgate;
   architecture XORfunction **of** XORgate **is**
   begin
   　X <= A **xor** B;
   end architecture XORfunction;

Section 3–8 Fixed-Function Logic Gates

1. Fixed-function logic cannot be changed. PLDs can be programmed for any logic function.

2. CMOS and bipolar (TTL)

3. **(a)** LS—Low-power Schottky

   (b) HC—High-speed CMOS

   (c) HCT—HC CMOS TTL compatible

4. Lowest power—CMOS

5. Six inverters in a package; four 2-input NAND gates in a package

6. $t_{PLH} = 10$ ns; $t_{PHL} = 8$ ns

7. 18 pJ

8. I_{CCL}—dc supply current for LOW output state; I_{CCH}—dc supply current for HIGH output state

9. V_{IL}—LOW input voltage; V_{IH}—HIGH input voltage

10. V_{OL}—LOW output voltage; V_{OH}—HIGH output voltage

Section 3–9 Troubleshooting

1. Opens and shorts are the most common failures.

2. An open input which effectively makes input HIGH

3. Amplitude and period

RELATED PROBLEMS FOR EXAMPLES

3–1 The timing diagram is not affected.

3–2 See Table 3–15.

TABLE 3–15

Inputs	Output	Inputs	Output
ABCD	X	ABCD	X
0000	0	1000	0
0001	0	1001	0
0010	0	1010	0
0011	0	1011	0
0100	0	1100	0
0101	0	1101	0
0110	0	1110	0
0111	0	1111	1

3–3 See Figure 3–97.

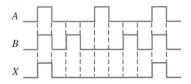

FIGURE 3–97

3–4 The output waveform is the same as input A.

3–5 See Figure 3–98.

3–6 Results are the same as example.

3–7 See Figure 3–99.

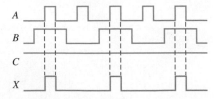

C = HIGH

FIGURE 3–98

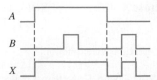

FIGURE 3–99

3–8 See Figure 3–100.
3–9 See Figure 3–101.

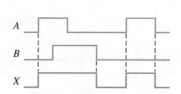

FIGURE 3–100

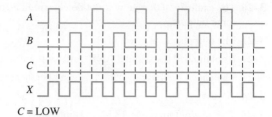

$C = \text{LOW}$

FIGURE 3–101

3–10 See Figure 3–102.
3–11 See Figure 3–103.

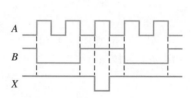

FIGURE 3–102

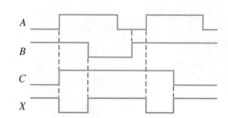

FIGURE 3–103

3–12 Use a 3-input NAND gate.
3–13 Use a 4-input NAND gate operating as a negative-OR gate.
3–14 See Figure 3–104.

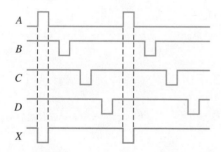

FIGURE 3–104

3–15 See Figure 3–105.
3–16 See Figure 3–106.

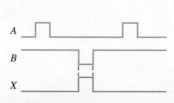

FIGURE 3–105

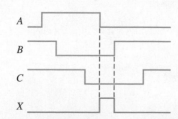

FIGURE 3–106

3–17 Use a 2-input NOR gate.

3–18 A 3-input NAND gate.

3–19 The output is always LOW. The output is a straight line.

3–20 The exclusive-OR gate will not detect simultaneous failures if both circuits produce the same outputs.

3–21 The outputs are unaffected.

3–22 6 columns, 9 rows, and 3 AND gates with three inputs each

3–23 The gate with 4 ns t_{PLH} and t_{PHL} can operate at the highest frequency.

3–24 10 mW

3–25 The gate output or pin 13 input is internally open.

3–26 The display will show an erratic readout because the counter continues until reset.

3–27 The enable pulse is too short or the counter is reset too soon.

TRUE/FALSE QUIZ

1. F   **2.** F   **3.** T   **4.** F   **5.** T

6. T   **7.** F   **8.** F   **9.** T   **10.** T

SELF-TEST

1. (d)   **2.** (d)   **3.** (a)   **4.** (e)   **5.** (c)   **6.** (a)   **7.** (d)   **8.** (b)   **9.** (d)

10. (b)   **11.** (d)   **12.** (c)   **13.** (b)   **14.** (a)   **15.** (d)   **16.** (c)   **17.** (c)

Boolean Algebra and Logic Simplification

CHAPTER OBJECTIVES

■ Apply the basic laws and rules of Boolean algebra
■ Apply DeMorgan's theorems to Boolean expressions
■ Describe gate combinations with Boolean expressions
■ Evaluate Boolean expressions
■ Simplify expressions by using the laws and rules of Boolean algebra
■ Convert any Boolean expression into a sum-of-products (SOP) form
■ Convert any Boolean expression into a product of-sums (POS) form
■ Relate a Boolean expression to a truth table
■ Use a Karnaugh map to simplify Boolean expressions
■ Use a Karnaugh map to simplify truth table functions
■ Utilize "don't care" conditions to simplify logic functions
■ Use the Quine-McCluskey method to simplify Boolean expressions
■ Write a VHDL program for simple logic

■ Apply Boolean algebra and the Karnaugh map method in an application

KEY TERMS

Key terms are in order of appearance in the chapter.

■ Variable
■ Complement
■ Sum term
■ Product term
■ Sum-of-products (SOP)
■ Product-of-sums (POS)
■ Karnaugh map
■ Minimization
■ "Don't care"

VISIT THE WEBSITE

Study aids for this chapter are available at http://www.pearsonhighered.com/careersresources/

INTRODUCTION

In 1854, George Boole published a work titled *An Investigation of the Laws of Thought, on Which Are Founded the Mathematical Theories of Logic and Probabilities.* It was in this publication that a "logical algebra," known today as Boolean algebra, was formulated. Boolean algebra is a convenient and systematic way of expressing and analyzing the operation of logic circuits. Claude Shannon was the first to apply Boole's work to the analysis and design of logic circuits. In 1938, Shannon wrote a thesis at MIT titled *A Symbolic Analysis of Relay and Switching Circuits.*

This chapter covers the laws, rules, and theorems of Boolean algebra and their application to digital circuits. You will learn how to define a given circuit with a Boolean expression and then evaluate its operation. You will also learn how to simplify logic circuits using the methods of Boolean algebra, Karnaugh maps, and the Quine-McCluskey method.

Boolean expressions using the hardware description language VHDL are also covered.

4–1   Boolean Operations and Expressions

Boolean algebra is the mathematics of digital logic. A basic knowledge of Boolean algebra is indispensable to the study and analysis of logic circuits. In the last chapter, Boolean operations and expressions in terms of their relationship to NOT, AND, OR, NAND, and NOR gates were introduced.

After completing this section, you should be able to

- Define *variable*
- Define *literal*
- Identify a sum term
- Evaluate a sum term
- Identify a product term
- Evaluate a product term
- Explain Boolean addition
- Explain Boolean multiplication

InfoNote

In a microprocessor, the arithmetic logic unit (ALU) performs arithmetic and Boolean logic operations on digital data as directed by program instructions. Logical operations are equivalent to the basic gate operations that you are familiar with but deal with a minimum of 8 bits at a time. Examples of Boolean logic instructions are AND, OR, NOT, and XOR, which are called *mnemonics*. An assembly language program uses the mnemonics to specify an operation. Another program called an *assembler* translates the mnemonics into a binary code that can be understood by the microprocessor.

Variable, complement, and *literal* are terms used in Boolean algebra. A **variable** is a symbol (usually an italic uppercase letter or word) used to represent an action, a condition, or data. Any single variable can have only a 1 or a 0 value. The **complement** is the inverse of a variable and is indicated by a bar over the variable (overbar). For example, the complement of the variable A is $\overline{A}$. If $A = 1$, then $\overline{A} = 0$. If $A = 0$, then $\overline{A} = 1$. The complement of the variable A is read as "not A" or "A bar." Sometimes a prime symbol rather than an overbar is used to denote the complement of a variable; for example, B' indicates the complement of B. In this book, only the overbar is used. A **literal** is a variable or the complement of a variable.

Boolean Addition

Recall from Chapter 3 that **Boolean addition** is equivalent to the OR operation. The basic rules are illustrated with their relation to the OR gate in Figure 4–1.

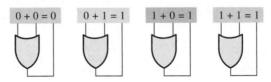

FIGURE 4–1

In Boolean algebra, a **sum term** is a sum of literals. In logic circuits, a sum term is produced by an OR operation with no AND operations involved. Some examples of sum terms are $A + B, A + \overline{B}, A + B + \overline{C}$, and $\overline{A} + B + C + \overline{D}$.

A sum term is equal to 1 when one or more of the literals in the term are 1. A sum term is equal to 0 only if each of the literals is 0.

The OR operation is the Boolean equivalent of addition.

EXAMPLE 4–1

Determine the values of A, B, C, and D that make the sum term $A + \overline{B} + C + \overline{D}$ equal to 0.

Solution

For the sum term to be 0, each of the literals in the term must be 0. Therefore, $A = \mathbf{0}$, $B = \mathbf{1}$ so that $\overline{B} = 0, C = \mathbf{0}$, and $D = \mathbf{1}$ so that $\overline{D} = 0$.

$$A + \overline{B} + C + \overline{D} = 0 + \overline{1} + 0 + \overline{1} = 0 + 0 + 0 + 0 = 0$$

Related Problem*

Determine the values of A and B that make the sum term $\overline{A} + B$ equal to 0.

_____

*Answers are at the end of the chapter.

Boolean Multiplication

Also recall from Chapter 3 that **Boolean multiplication** is equivalent to the AND operation. The basic rules are illustrated with their relation to the AND gate in Figure 4–2.

The AND operation is the Boolean equivalent of multiplication.

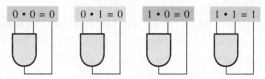

$$0 \cdot 0 = 0 \qquad 0 \cdot 1 = 0 \qquad 1 \cdot 0 = 0 \qquad 1 \cdot 1 = 1$$

FIGURE 4–2

In Boolean algebra, a **product term** is the product of literals. In logic circuits, a product term is produced by an AND operation with no OR operations involved. Some examples of product terms are AB, $A\overline{B}$, ABC, and $A\overline{B}C\overline{D}$.

A product term is equal to 1 only if each of the literals in the term is 1. A product term is equal to 0 when one or more of the literals are 0.

EXAMPLE 4–2

Determine the values of A, B, C, and D that make the product term $A\overline{B}C\overline{D}$ equal to 1.

Solution

For the product term to be 1, each of the literals in the term must be 1. Therefore, $A = \mathbf{1}$, $B = \mathbf{0}$ so that $\overline{B} = 1$, $C = \mathbf{1}$, and $D = \mathbf{0}$ so that $\overline{D} = 1$.

$$A\overline{B}C\overline{D} = 1 \cdot \overline{0} \cdot 1 \cdot \overline{0} = 1 \cdot 1 \cdot 1 \cdot 1 = 1$$

Related Problem

Determine the values of A and B that make the product term $\overline{A}\,\overline{B}$ equal to 1.

SECTION 4–1 CHECKUP

Answers are at the end of the chapter.

1. If $A = 0$, what does $\overline{A}$ equal?
2. Determine the values of A, B, and C that make the sum term $\overline{A} + \overline{B} + C$ equal to 0.
3. Determine the values of A, B, and C that make the product term $A\overline{B}C$ equal to 1.

4–2   Laws and Rules of Boolean Algebra

As in other areas of mathematics, there are certain well-developed rules and laws that must be followed in order to properly apply Boolean algebra. The most important of these are presented in this section.

After completing this section, you should be able to

◆ Apply the commutative laws of addition and multiplication

◆ Apply the associative laws of addition and multiplication

◆ Apply the distributive law

◆ Apply twelve basic rules of Boolean algebra

Laws of Boolean Algebra

The basic laws of Boolean algebra—the **commutative laws** for addition and multiplication, the **associative laws** for addition and multiplication, and the **distributive law**—are the same as in ordinary algebra. Each of the laws is illustrated with two or three variables, but the number of variables is not limited to this.

Commutative Laws

The *commutative law of addition* for two variables is written as

$$A + B = B + A \qquad \text{Equation 4–1}$$

This law states that the order in which the variables are ORed makes no difference. Remember, in Boolean algebra as applied to logic circuits, addition and the OR operation are the same. Figure 4–3 illustrates the commutative law as applied to the OR gate and shows that it doesn't matter to which input each variable is applied. (The symbol ≡ means "equivalent to.")

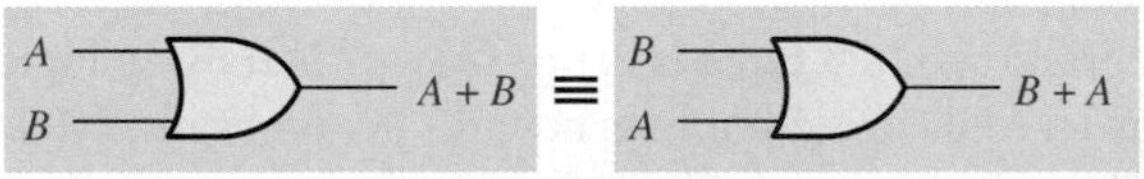

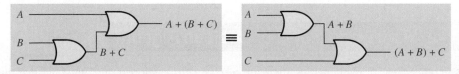

FIGURE 4–3   Application of commutative law of addition.

The *commutative law of multiplication* for two variables is

$$AB = BA \qquad \text{Equation 4–2}$$

This law states that the order in which the variables are ANDed makes no difference. Figure 4–4 illustrates this law as applied to the AND gate. Remember, in Boolean algebra as applied to logic circuits, multiplication and the AND function are the same.

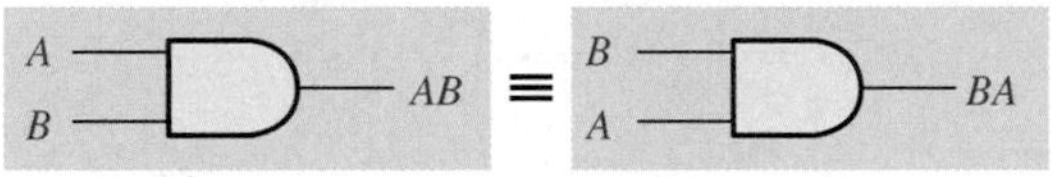

FIGURE 4–4   Application of commutative law of multiplication.

Associative Laws

The *associative law of addition* is written as follows for three variables:

$$A + (B + C) = (A + B) + C \qquad \text{Equation 4–3}$$

This law states that when ORing more than two variables, the result is the same regardless of the grouping of the variables. Figure 4–5 illustrates this law as applied to 2-input OR gates.

FIGURE 4–5   Application of associative law of addition. Open file F04-05 to verify. *A Multisim tutorial is available on the website.*

The *associative law of multiplication* is written as follows for three variables:

$$A(BC) = (AB)C \qquad \text{Equation 4–4}$$

This law states that it makes no difference in what order the variables are grouped when ANDing more than two variables. Figure 4–6 illustrates this law as applied to 2-input AND gates.

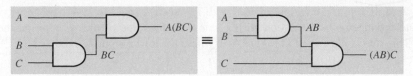

FIGURE 4–6 Application of associative law of multiplication. Open file F04-06 to verify.   **MultiSim**

Distributive Law

The distributive law is written for three variables as follows:

$$A(B + C) = AB + AC \hspace{3cm} \text{Equation 4–5}$$

This law states that ORing two or more variables and then ANDing the result with a single variable is equivalent to ANDing the single variable with each of the two or more variables and then ORing the products. The distributive law also expresses the process of *factoring* in which the common variable A is factored out of the product terms, for example, $AB + AC = A(B + C)$. Figure 4–7 illustrates the distributive law in terms of gate implementation.

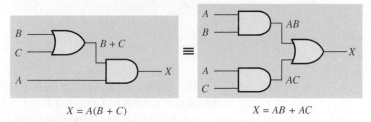

$$X = A(B + C) \hspace{4cm} X = AB + AC$$

FIGURE 4–7 Application of distributive law. Open file F04-07 to verify.   **MultiSim**

Rules of Boolean Algebra

Table 4–1 lists 12 basic rules that are useful in manipulating and simplifying **Boolean expressions**. Rules 1 through 9 will be viewed in terms of their application to logic gates. Rules 10 through 12 will be derived in terms of the simpler rules and the laws previously discussed.

TABLE 4–1

Basic rules of Boolean algebra.

1. $A + 0 = A$	**7.** $A \cdot A = A$ $A \cdot A \cdot A = A$
2. $A + 1 = 1$	**8.** $A \cdot \overline{A} = 0$
3. $A \cdot 0 = 0$	**9.** $\overline{\overline{A}} = A$
4. $A \cdot 1 = A$	**10.** $A + AB = A$
5. $A + A = A$	**11.** $A + \overline{A}B = A + B$
6. $A + \overline{A} = 1$	**12.** $(A + B)(A + C) = A + BC$

$A + A + A = A$

A, B, or C can represent a single variable or a combination of variables.

Rule 1: $A + 0 = A$ A variable ORed with 0 is always equal to the variable. If the input variable A is 1, the output variable X is 1, which is equal to A. If A is 0, the output is 0, which is also equal to A. This rule is illustrated in Figure 4–8, where the lower input is fixed at 0.

$$X = A + 0 = A$$

FIGURE 4–8

Rule 2: $A + 1 = 1$ A variable ORed with 1 is always equal to 1. A 1 on an input to an OR gate produces a 1 on the output, regardless of the value of the variable on the other input. This rule is illustrated in Figure 4–9, where the lower input is fixed at 1.

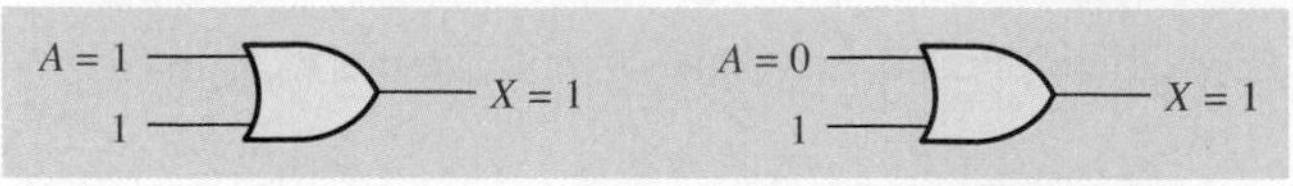

$$X = A + 1 = 1$$

FIGURE 4–9

Rule 3: $A \cdot 0 = 0$ A variable ANDed with 0 is always equal to 0. Any time one input to an AND gate is 0, the output is 0, regardless of the value of the variable on the other input. This rule is illustrated in Figure 4–10, where the lower input is fixed at 0.

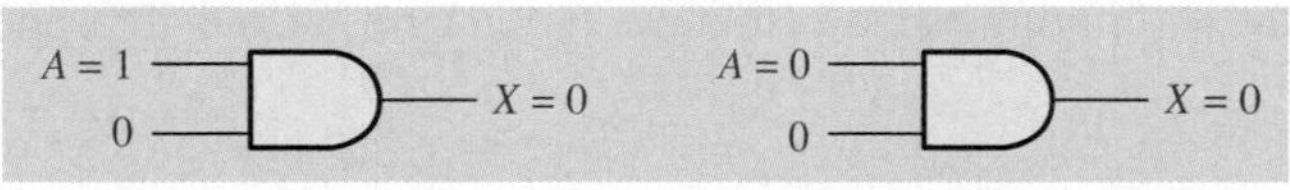

$$X = A \cdot 0 = 0$$

FIGURE 4–10

Rule 4: $A \cdot 1 = A$ A variable ANDed with 1 is always equal to the variable. If A is 0, the output of the AND gate is 0. If A is 1, the output of the AND gate is 1 because both inputs are now 1s. This rule is shown in Figure 4–11, where the lower input is fixed at 1.

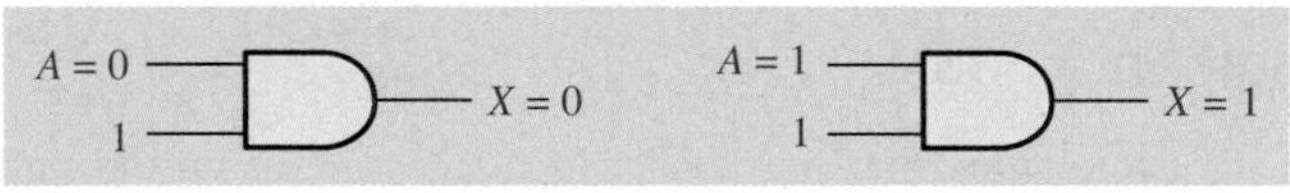

$$X = A \cdot 1 = A$$

FIGURE 4–11

Rule 5: $A + A = A$ A variable ORed with itself is always equal to the variable. If A is 0, then $0 + 0 = 0$; and if A is 1, then $1 + 1 = 1$. This is shown in Figure 4–12, where both inputs are the same variable.

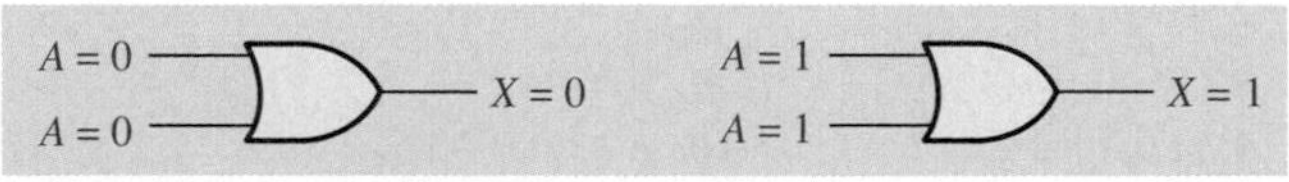

$$X = A + A = A$$

FIGURE 4–12

Rule 6: $A + \bar{A} = 1$ A variable ORed with its complement is always equal to 1. If A is 0, then $0 + \bar{0} = 0 + 1 = 1$. If A is 1, then $1 + \bar{1} = 1 + 0 = 1$. See Figure 4–13, where one input is the complement of the other.

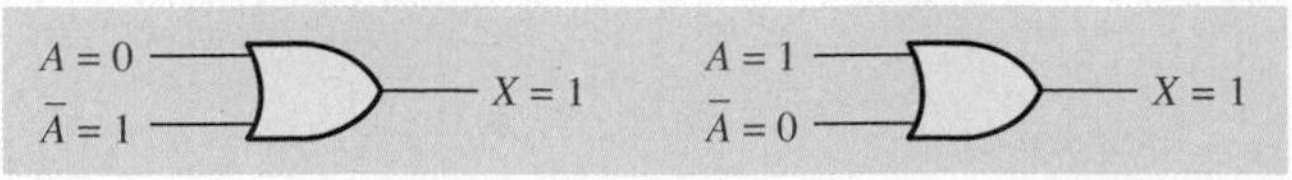

$$X = A + \bar{A} = 1$$

FIGURE 4–13

Rule 7: $A \cdot A = A$ A variable ANDed with itself is always equal to the variable. If $A = 0$, then $0 \cdot 0 = 0$; and if $A = 1$, then $1 \cdot 1 = 1$. Figure 4–14 illustrates this rule.

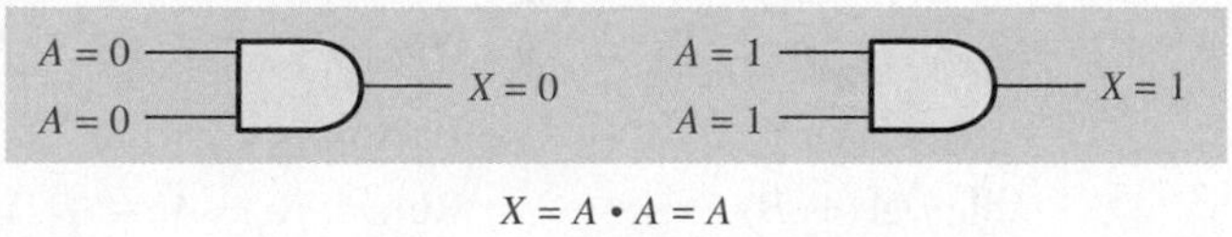

$A = 0$ ──┐
          ├─D─── $X = 0$ $A = 1$ ──┐
$A = 0$ ──┘   ├─D─── $X = 1$
                              $A = 1$ ──┘

$$X = A \cdot A = A$$

FIGURE 4–14

Rule 8: $A \cdot \overline{A} = 0$ A variable ANDed with its complement is always equal to 0. Either A or $\overline{A}$ will always be 0; and when a 0 is applied to the input of an AND gate, the output will be 0 also. Figure 4–15 illustrates this rule.

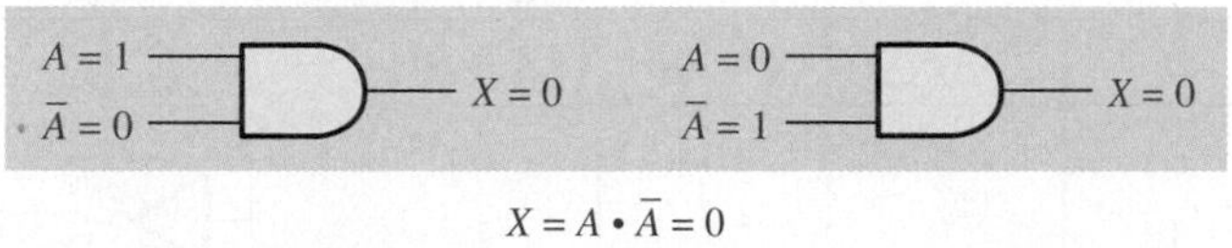

$A = 1$ ──┐
          ├─D─── $X = 0$ $A = 0$ ──┐
$\overline{A} = 0$ ──┘   ├─D─── $X = 0$
                              $\overline{A} = 1$ ──┘

$$X = A \cdot \overline{A} = 0$$

FIGURE 4–15

Rule 9: $\overline{\overline{A}} = A$ The double complement of a variable is always equal to the variable. If you start with the variable A and complement (invert) it once, you get $\overline{A}$. If you then take $\overline{A}$ and complement (invert) it, you get A, which is the original variable. This rule is shown in Figure 4–16 using inverters.

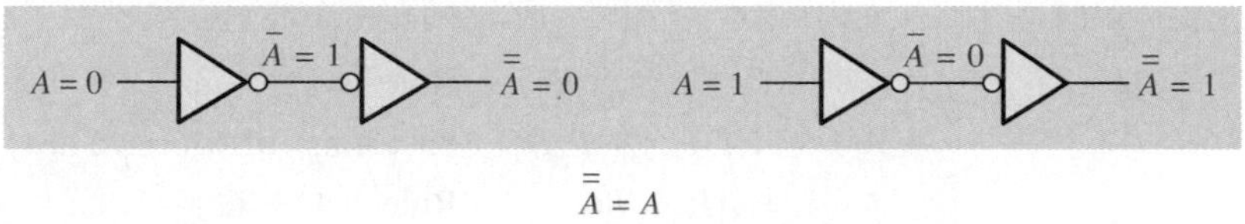

$A = 0$ ──▷○── $\overline{A} = 1$ ──▷○── $\overline{\overline{A}} = 0$ $A = 1$ ──▷○── $\overline{A} = 0$ ──▷○── $\overline{\overline{A}} = 1$

$$\overline{\overline{A}} = A$$

FIGURE 4–16

Rule 10: $A + AB = A$ This rule can be proved by applying the distributive law, rule 2, and rule 4 as follows:

$$A + AB = A \cdot 1 + AB = A(1 + B) \quad \text{Factoring (distributive law)}$$
$$= A \cdot 1 \qquad\qquad\qquad \text{Rule 2: } (1 + B) = 1$$
$$= A \qquad\qquad\qquad\quad \text{Rule 4: } A \cdot 1 = A$$

The proof is shown in Table 4–2, which shows the truth table and the resulting logic circuit simplification.

TABLE 4–2

Rule 10: $A + AB = A$. Open file T04-02 to verify.

A	B	AB	$A + AB$
0	0	0	0
0	1	0	0
1	0	0	1
1	1	1	1

└──────── equal ────────┘

straight connection

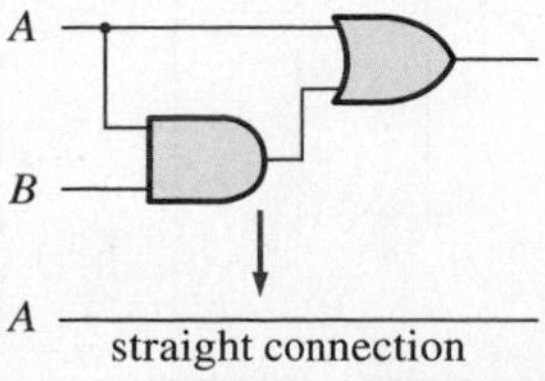

Rule 11: $A + \overline{A}B = A + B$ This rule can be proved as follows:

$$
\begin{aligned}
A + \overline{A}B &= (A + AB) + \overline{A}B && \text{Rule 10: } A = A + AB \\
&= (AA + AB) + \overline{A}B && \text{Rule 7: } A = AA \\
&= AA + AB + A\overline{A} + \overline{A}B && \text{Rule 8: adding } A\overline{A} = 0 \\
&= (A + \overline{A})(A + B) && \text{Factoring} \\
&= 1 \cdot (A + B) && \text{Rule 6: } A + \overline{A} = 1 \\
&= A + B && \text{Rule 4: drop the 1}
\end{aligned}
$$

The proof is shown in Table 4–3, which shows the truth table and the resulting logic circuit simplification.

MultiSim

TABLE 4–3

Rule 11: $A + \overline{A}B = A + B$. Open file T04-03 to verify.

A	B	$\overline{A}B$	$A + \overline{A}B$	$A + B$	
0	0	0	0	0	
0	1	1	1	1	
1	0	0	1	1	
1	1	0	1	1	

Rule 12: $(A + B)(A + C) = A + BC$ This rule can be proved as follows:

$$
\begin{aligned}
(A + B)(A + C) &= AA + AC + AB + BC && \text{Distributive law} \\
&= A + AC + AB + BC && \text{Rule 7: } AA = A \\
&= A(1 + C) + AB + BC && \text{Factoring (distributive law)} \\
&= A \cdot 1 + AB + BC && \text{Rule 2: } 1 + C = 1 \\
&= A(1 + B) + BC && \text{Factoring (distributive law)} \\
&= A \cdot 1 + BC && \text{Rule 2: } 1 + B = 1 \\
&= A + BC && \text{Rule 4: } A \cdot 1 = A
\end{aligned}
$$

The proof is shown in Table 4–4, which shows the truth table and the resulting logic circuit simplification.

MultiSim

TABLE 4–4

Rule 12: $(A + B)(A + C) = A + BC$. Open file T04-04 to verify.

A	B	C	$A + B$	$A + C$	$(A + B)(A + C)$	BC	$A + BC$	
0	0	0	0	0	0	0	0	
0	0	1	0	1	0	0	0	
0	1	0	1	0	0	0	0	
0	1	1	1	1	1	1	1	
1	0	0	1	1	1	0	1	
1	0	1	1	1	1	0	1	
1	1	0	1	1	1	0	1	
1	1	1	1	1	1	1	1	

1. Apply the associative law of addition to the expression $A + (B + C + D)$.

2. Apply the distributive law to the expression $A(B + C + D)$.

4–3 DeMorgan's Theorems

DeMorgan, a mathematician who knew Boole, proposed two theorems that are an important part of Boolean algebra. In practical terms, DeMorgan's theorems provide mathematical verification of the equivalency of the NAND and negative-OR gates and the equivalency of the NOR and negative-AND gates, which were discussed in Chapter 3.

After completing this section, you should be able to

- ◆ State DeMorgan's theorems

- ◆ Relate DeMorgan's theorems to the equivalency of the NAND and negative-OR gates and to the equivalency of the NOR and negative-AND gates

- ◆ Apply DeMorgan's theorems to the simplification of Boolean expressions

DeMorgan's first theorem is stated as follows:

The complement of a product of variables is equal to the sum of the complements of the variables.

Stated another way,

The complement of two or more ANDed variables is equivalent to the OR of the complements of the individual variables.

The formula for expressing this theorem for two variables is

$$\overline{XY} = \overline{X} + \overline{Y} \qquad \text{Equation 4–6}$$

DeMorgan's second theorem is stated as follows:

The complement of a sum of variables is equal to the product of the complements of the variables.

Stated another way,

The complement of two or more ORed variables is equivalent to the AND of the complements of the individual variables.

The formula for expressing this theorem for two variables is

$$\overline{X + Y} = \overline{X}\,\overline{Y} \qquad \text{Equation 4–7}$$

Figure 4–17 shows the gate equivalencies and truth tables for Equations 4–6 and 4–7.

As stated, DeMorgan's theorems also apply to expressions in which there are more than two variables. The following examples illustrate the application of DeMorgan's theorems to 3-variable and 4-variable expressions.

> To apply DeMorgan's theorem, break the bar over the product of variables and change the sign from AND to OR.

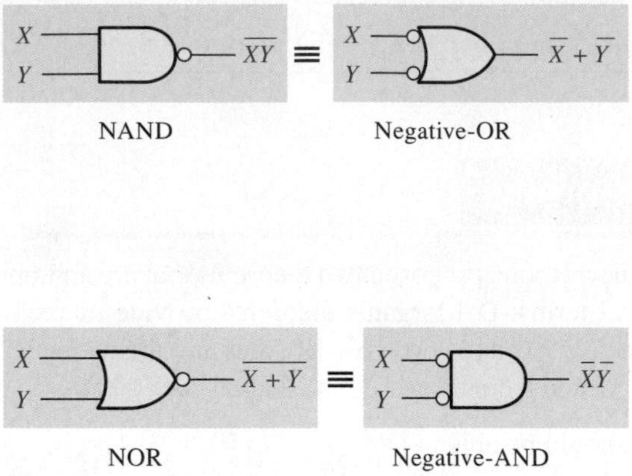

Inputs		Output	
X	Y	$\overline{XY}$	$\overline{X} + \overline{Y}$
0	0	1	1
0	1	1	1
1	0	1	1
1	1	0	0

Inputs		Output	
X	Y	$\overline{X + Y}$	$\overline{X}\,\overline{Y}$
0	0	1	1
0	1	0	0
1	0	0	0
1	1	0	0

FIGURE 4–17 Gate equivalencies and the corresponding truth tables that illustrate DeMorgan's theorems. Notice the equality of the two output columns in each table. This shows that the equivalent gates perform the same logic function.

EXAMPLE 4–3

Apply DeMorgan's theorems to the expressions $\overline{XYZ}$ and $\overline{X + Y + Z}$.

Solution

$$\overline{XYZ} = \overline{X} + \overline{Y} + \overline{Z}$$
$$\overline{X + Y + Z} = \overline{X}\,\overline{Y}\,\overline{Z}$$

Related Problem

Apply DeMorgan's theorem to the expression $\overline{\overline{X} + \overline{Y} + \overline{Z}}$.

EXAMPLE 4–4

Apply DeMorgan's theorems to the expressions $\overline{WXYZ}$ and $\overline{W + X + Y + Z}$.

Solution

$$\overline{WXYZ} = \overline{W} + \overline{X} + \overline{Y} + \overline{Z}$$
$$\overline{W + X + Y + Z} = \overline{W}\,\overline{X}\,\overline{Y}\,\overline{Z}$$

Related Problem

Apply DeMorgan's theorem to the expression $\overline{\overline{W}\,\overline{X}\,\overline{Y}\,\overline{Z}}$.

Each variable in DeMorgan's theorems as stated in Equations 4–6 and 4–7 can also represent a combination of other variables. For example, X can be equal to the term $AB + C$, and Y can be equal to the term $A + BC$. So if you can apply DeMorgan's theorem for two variables as stated by $\overline{XY} = \overline{X} + \overline{Y}$ to the expression $\overline{(AB + C)(A + BC)}$, you get the following result:

$$\overline{(AB + C)(A + BC)} = \overline{(AB + C)} + \overline{(A + BC)}$$

Notice that in the preceding result you have two terms, $\overline{AB + C}$ and $\overline{A + BC}$, to each of which you can again apply DeMorgan's theorem $\overline{X + Y} = \overline{X}\overline{Y}$ individually, as follows:

$$\overline{(AB + C)} + \overline{(A + BC)} = (\overline{AB})\overline{C} + \overline{A}(\overline{BC})$$

Notice that you still have two terms in the expression to which DeMorgan's theorem can again be applied. These terms are $\overline{AB}$ and $\overline{BC}$. A final application of DeMorgan's theorem gives the following result:

$$\overline{(AB)}\overline{C} + \overline{A}\overline{(BC)} = (\overline{A} + \overline{B})\overline{C} + \overline{A}(\overline{B} + \overline{C})$$

Although this result can be simplified further by the use of Boolean rules and laws, DeMorgan's theorems cannot be used any more.

Applying DeMorgan's Theorems

The following procedure illustrates the application of DeMorgan's theorems and Boolean algebra to the specific expression

$$\overline{A + B\overline{C} + D(E + \overline{F})}$$

Step 1: Identify the terms to which you can apply DeMorgan's theorems, and think of each term as a single variable. Let $A + B\overline{C} = X$ and $D(E + \overline{F}) = Y$.

Step 2: Since $\overline{X + Y} = \overline{X}\overline{Y}$,

$$\overline{(A + B\overline{C}) + (D(E + \overline{F}))} = \overline{(A + B\overline{C})}\,\overline{(D(E + \overline{F}))}$$

Step 3: Use rule 9 ($\overline{\overline{A}} = A$) to cancel the double bars over the left term (this is not part of DeMorgan's theorem).

$$\overline{(A + B\overline{C})}\,\overline{(D(E + \overline{F}))} = (A + B\overline{C})\overline{(D(E + \overline{F}))}$$

Step 4: Apply DeMorgan's theorem to the second term.

$$(A + B\overline{C})\overline{(D(E + \overline{F}))} = (A + B\overline{C})(\overline{D} + \overline{(E + \overline{F})})$$

Step 5: Use rule 9 ($\overline{\overline{A}} = A$) to cancel the double bars over the $E + F$ part of the term.

$$(A + B\overline{C})(\overline{D} + \overline{\overline{E + \overline{F}}}) = (A + B\overline{C})(\overline{D} + E + \overline{F})$$

The following three examples will further illustrate how to use DeMorgan's theorems.

EXAMPLE 4–5

Apply DeMorgan's theorems to each of the following expressions:

(a) $\overline{(A + B + C)D}$

(b) $\overline{ABC + DEF}$

(c) $\overline{A\overline{B} + \overline{C}D + EF}$

Solution

(a) Let $A + B + C = X$ and $D = Y$. The expression $\overline{(A + B + C)D}$ is of the form $\overline{XY} = \overline{X} + \overline{Y}$ and can be rewritten as

$$\overline{(A + B + C)D} = \overline{A + B + C} + \overline{D}$$

Next, apply DeMorgan's theorem to the term $\overline{A + B + C}$.

$$\overline{A + B + C} + \overline{D} = \overline{A}\overline{B}\overline{C} + \overline{D}$$

(b) Let $ABC = X$ and $DEF = Y$. The expression $\overline{ABC + DEF}$ is of the form $\overline{X + Y} = \overline{X}\overline{Y}$ and can be rewritten as

$$\overline{ABC + DEF} = \overline{(ABC)}\overline{(DEF)}$$

Next, apply DeMorgan's theorem to each of the terms $\overline{ABC}$ and $\overline{DEF}$.

$$\overline{(ABC)}\overline{(DEF)} = (\overline{A} + \overline{B} + \overline{C})(\overline{D} + \overline{E} + \overline{F})$$

(c) Let $A\overline{B} = X, \overline{C}D = Y$, and $EF = Z$. The expression $\overline{A\overline{B} + \overline{C}D + EF}$ is of the form $\overline{X + Y + Z} = \overline{X}\,\overline{Y}\,\overline{Z}$ and can be rewritten as

$$\overline{A\overline{B} + \overline{C}D + EF} = (\overline{A\overline{B}})(\overline{\overline{C}D})(\overline{EF})$$

Next, apply DeMorgan's theorem to each of the terms $\overline{A\overline{B}}, \overline{\overline{C}D}$, and $\overline{EF}$.

$$(\overline{A\overline{B}})(\overline{\overline{C}D})(\overline{EF}) = (\overline{A} + B)(C + \overline{D})(\overline{E} + \overline{F})$$

Related Problem

Apply DeMorgan's theorems to the expression $\overline{\overline{ABC} + D + E}$.

EXAMPLE 4–6

Apply DeMorgan's theorems to each expression:
(a) $\overline{\overline{(A + B)} + \overline{C}}$
(b) $\overline{\overline{(A + B)} + CD}$
(c) $\overline{(A + B)\overline{C}\overline{D} + E + \overline{F}}$

Solution

(a) $\overline{\overline{(A + B)} + \overline{C}} = \overline{\overline{(A + B)}}\,\overline{\overline{C}} = (A + B)C$
(b) $\overline{\overline{(A + B)} + CD} = \overline{\overline{(A + B)}}\,\overline{CD} = (\overline{A}\,\overline{B})(\overline{C} + \overline{D}) = A\overline{B}(\overline{C} + \overline{D})$
(c) $\overline{(A + B)\overline{C}\overline{D} + E + \overline{F}} = (\overline{(A + B)\overline{C}\overline{D}})(\overline{E + \overline{F}}) = (\overline{A}\,\overline{B} + C + D)\overline{E}F$

Related Problem

Apply DeMorgan's theorems to the expression $\overline{\overline{AB(C + \overline{D})} + E}$.

EXAMPLE 4–7

The Boolean expression for an exclusive-OR gate is $A\overline{B} + \overline{A}B$. With this as a starting point, use DeMorgan's theorems and any other rules or laws that are applicable to develop an expression for the exclusive-NOR gate.

Solution

Start by complementing the exclusive-OR expression and then applying DeMorgan's theorems as follows:

$$\overline{A\overline{B} + \overline{A}B} = (\overline{A\overline{B}})(\overline{\overline{A}B}) = (\overline{A} + \overline{\overline{B}})(\overline{\overline{A}} + \overline{B}) = (\overline{A} + B)(A + \overline{B})$$

Next, apply the distributive law and rule 8 ($A \cdot \overline{A} = 0$).

$$(\overline{A} + B)(A + \overline{B}) = \overline{A}A + \overline{A}\,\overline{B} + AB + B\overline{B} = \overline{A}\,\overline{B} + AB$$

The final expression for the XNOR is $\overline{A}\,\overline{B} + AB$. Note that this expression equals 1 any time both variables are 0s or both variables are 1s.

Related Problem

Starting with the expression for a 4-input NAND gate, use DeMorgan's theorems to develop an expression for a 4-input negative-OR gate.

1. Apply DeMorgan's theorems to the following expressions:

(a) $\overline{ABC} + (\overline{D} + E)$ (b) $\overline{(A + B)C}$ (c) $\overline{A + B + C} + \overline{\overline{DE}}$

4–4   Boolean Analysis of Logic Circuits

Boolean algebra provides a concise way to express the operation of a logic circuit formed by a combination of logic gates so that the output can be determined for various combinations of input values.

After completing this section, you should be able to

◆ Determine the Boolean expression for a combination of gates

◆ Evaluate the logic operation of a circuit from the Boolean expression

◆ Construct a truth table

Boolean Expression for a Logic Circuit

To derive the Boolean expression for a given combinational logic circuit, begin at the left-most inputs and work toward the final output, writing the expression for each gate. For the example circuit in Figure 4–18, the Boolean expression is determined in the following three steps:

A combinational logic circuit can be described by a Boolean equation.

1. The expression for the left-most AND gate with inputs C and D is CD.
2. The output of the left-most AND gate is one of the inputs to the OR gate and B is the other input. Therefore, the expression for the OR gate is $B + CD$.
3. The output of the OR gate is one of the inputs to the right-most AND gate and A is the other input. Therefore, the expression for this AND gate is $A(B + CD)$, which is the final output expression for the entire circuit.

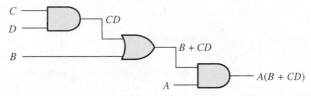

FIGURE 4–18   A combinational logic circuit showing the development of the Boolean expression for the output.

Constructing a Truth Table for a Logic Circuit

Once the Boolean expression for a given logic circuit has been determined, a truth table that shows the output for all possible values of the input variables can be developed. The procedure requires that you evaluate the Boolean expression for all possible combinations of values for the input variables. In the case of the circuit in Figure 4–18, there are four input variables (A, B, C, and D) and therefore sixteen ($2^4 = 16$) combinations of values are possible.

A combinational logic circuit can be described by a truth table.

Evaluating the Expression

To evaluate the expression $A(B + CD)$, first find the values of the variables that make the expression equal to 1, using the rules for Boolean addition and multiplication. In this case, the expression equals 1 only if $A = 1$ and $B + CD = 1$ because

$$A(B + CD) = 1 \cdot 1 = 1$$

Now determine when the $B + CD$ term equals 1. The term $B + CD = 1$ if either $B = 1$ or $CD = 1$ or if both B and CD equal 1 because

$$B + CD = 1 + 0 = 1$$
$$B + CD = 0 + 1 = 1$$
$$B + CD = 1 + 1 = 1$$

The term $CD = 1$ only if $C = 1$ and $D = 1$.

To summarize, the expression $A(B + CD) = 1$ when $A = 1$ and $B = 1$ regardless of the values of C and D or when $A = 1$ and $C = 1$ and $D = 1$ regardless of the value of B. The expression $A(B + CD) = 0$ for all other value combinations of the variables.

Putting the Results in Truth Table Format

The first step is to list the sixteen input variable combinations of 1s and 0s in a binary sequence as shown in Table 4–5. Next, place a 1 in the output column for each combination of input variables that was determined in the evaluation. Finally, place a 0 in the output column for all other combinations of input variables. These results are shown in the truth table in Table 4–5.

TABLE 4–5

Truth table for the logic circuit in Figure 4–18.

Inputs				Output
A	B	C	D	$A(B + CD)$
0	0	0	0	0
0	0	0	1	0
0	0	1	0	0
0	0	1	1	0
0	1	0	0	0
0	1	0	1	0
0	1	1	0	0
0	1	1	1	0
1	0	0	0	0
1	0	0	1	0
1	0	1	0	0
1	0	1	1	1
1	1	0	0	1
1	1	0	1	1
1	1	1	0	1
1	1	1	1	1

EXAMPLE 4–8

Use Multisim to generate the truth table for the logic circuit in Figure 4–18.

Solution

Construct the circuit in Multisim and connect the Multisim Logic Converter to the inputs and output, as shown in Figure 4–19. Click on the ⟶ conversion bar, and the truth table appears in the display as shown.

You can also generate the simplified Boolean expression from the truth table by clicking on  .

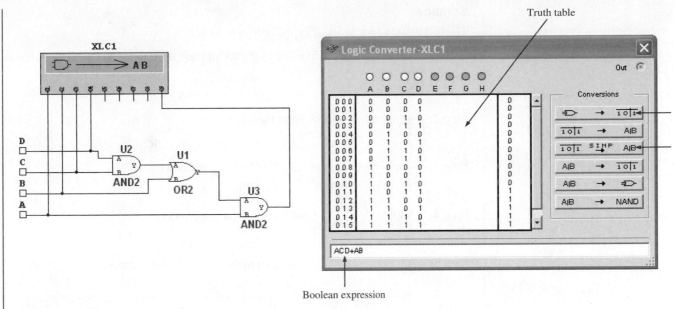

Truth table

Boolean expression

FIGURE 4–19

Related Problem

Open Multisim. Create the setup and do the conversions shown in this example.

MultiSim

4–5   Logic Simplification Using Boolean Algebra

A logic expression can be reduced to its simplest form or changed to a more convenient form to implement the expression most efficiently using Boolean algebra. The approach taken in this section is to use the basic laws, rules, and theorems of Boolean algebra to manipulate and simplify an expression. This method depends on a thorough knowledge of Boolean algebra and considerable practice in its application, not to mention a little ingenuity and cleverness.

After completing this section, you should be able to

◆ Apply the laws, rules, and theorems of Boolean algebra to simplify general expressions

A simplified Boolean expression uses the fewest gates possible to implement a given expression. Examples 4–9 through 4–12 illustrate Boolean simplification.

EXAMPLE 4–9

Using Boolean algebra techniques, simplify this expression:

$$AB + A(B + C) + B(B + C)$$

Solution

The following is not necessarily the only approach.

Step 1: Apply the distributive law to the second and third terms in the expression, as follows:

$$AB + AB + AC + BB + BC$$

Step 2: Apply rule 7 ($BB = B$) to the fourth term.

$$AB + AB + AC + B + BC$$

Step 3: Apply rule 5 ($AB + AB = AB$) to the first two terms.

$$AB + AC + B + BC$$

Step 4: Apply rule 10 ($B + BC = B$) to the last two terms.

$$AB + AC + B$$

Step 5: Apply rule 10 ($AB + B = B$) to the first and third terms.

$$B + AC$$

At this point the expression is simplified as much as possible. Once you gain experience in applying Boolean algebra, you can often combine many individual steps.

Related Problem

Simplify the Boolean expression $A\overline{B} + A(\overline{B + C}) + B(\overline{B + C})$.

Simplification means fewer gates for the same function.

Figure 4–20 shows that the simplification process in Example 4–9 has significantly reduced the number of logic gates required to implement the expression. Part (a) shows that five gates are required to implement the expression in its original form; however, only two gates are needed for the simplified expression, shown in part (b). It is important to realize that these two gate circuits are equivalent. That is, for any combination of levels on the A, B, and C inputs, you get the same output from either circuit.

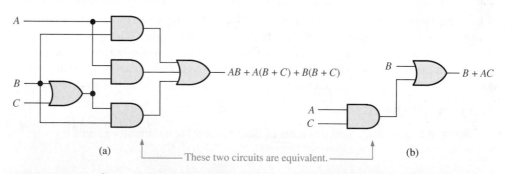

(a)   These two circuits are equivalent.   (b)

 FIGURE 4–20   Gate circuits for Example 4–9. Open file F04-20 to verify equivalency.

EXAMPLE 4–10

Simplify the following Boolean expression:

$$[A\overline{B}(C + BD) + \overline{A}\,\overline{B}]C$$

Note that brackets and parentheses mean the same thing: the term inside is multiplied (ANDed) with the term outside.

Solution

Step 1: Apply the distributive law to the terms within the brackets.

$$(A\overline{B}C + A\overline{B}BD + \overline{A}\,\overline{B})C$$

Step 2: Apply rule 8 ($\overline{B}B = 0$) to the second term within the parentheses.

$$(A\overline{B}C + A \cdot 0 \cdot D + \overline{A}\,\overline{B})C$$

Step 3: Apply rule 3 ($A \cdot 0 \cdot D = 0$) to the second term within the parentheses.

$$(A\overline{B}C + 0 + \overline{A}\,\overline{B})C$$

Step 4: Apply rule 1 (drop the 0) within the parentheses.

$$(A\overline{B}C + \overline{A}\,\overline{B})C$$

Step 5: Apply the distributive law.

$$A\overline{B}CC + \overline{A}\,\overline{B}C$$

Step 6: Apply rule 7 ($CC = C$) to the first term.

$$A\overline{B}C + \overline{A}\,\overline{B}C$$

Step 7: Factor out $\overline{B}C$.

$$\overline{B}C(A + \overline{A})$$

Step 8: Apply rule 6 ($A + \overline{A} = 1$).

$$\overline{B}C \cdot 1$$

Step 9: Apply rule 4 (drop the 1).

$$\overline{B}C$$

Related Problem

Simplify the Boolean expression $[AB(C + \overline{B}D) + \overline{A}B]CD$.

EXAMPLE 4–11

Simplify the following Boolean expression:

$$\overline{A}BC + A\overline{B}\,\overline{C} + \overline{A}\,\overline{B}\,\overline{C} + A\overline{B}C + ABC$$

Solution

Step 1: Factor BC out of the first and last terms.

$$BC(\overline{A} + A) + A\overline{B}\,\overline{C} + \overline{A}\,\overline{B}\,\overline{C} + A\overline{B}C$$

Step 2: Apply rule 6 ($\overline{A} + A = 1$) to the term in parentheses, and factor $A\overline{B}$ from the second and last terms.

$$BC \cdot 1 + A\overline{B}(\overline{C} + C) + \overline{A}\,\overline{B}\,\overline{C}$$

Step 3: Apply rule 4 (drop the 1) to the first term and rule 6 ($\overline{C} + C = 1$) to the term in parentheses.

$$BC + A\overline{B} \cdot 1 + \overline{A}\,\overline{B}\,\overline{C}$$

Step 4: Apply rule 4 (drop the 1) to the second term.

$$BC + A\overline{B} + \overline{A}\,\overline{B}\,\overline{C}$$

Step 5: Factor $\overline{B}$ from the second and third terms.

$$BC + \overline{B}(A + \overline{A}C)$$

Step 6: Apply rule 11 $(A + \overline{A}\,\overline{C} = A + \overline{C})$ to the term in parentheses.

$$BC + \overline{B}(A + \overline{C})$$

Step 7: Use the distributive and commutative laws to get the following expression:

$$BC + A\overline{B} + \overline{B}\,\overline{C}$$

Related Problem

Simplify the Boolean expression $AB\overline{C} + \overline{A}BC + \overline{A}BC + \overline{A}\,\overline{B}\,\overline{C}$.

EXAMPLE 4–12

Simplify the following Boolean expression:

$$\overline{AB + AC} + \overline{A}\,\overline{B}C$$

Solution

Step 1: Apply DeMorgan's theorem to the first term.

$$(\overline{AB})(\overline{AC}) + \overline{A}\,\overline{B}C$$

Step 2: Apply DeMorgan's theorem to each term in parentheses.

$$(\overline{A} + \overline{B})(\overline{A} + \overline{C}) + \overline{A}\,\overline{B}C$$

Step 3: Apply the distributive law to the two terms in parentheses.

$$\overline{A}\,\overline{A} + \overline{A}\,\overline{C} + \overline{A}\,\overline{B} + \overline{B}\,\overline{C} + \overline{A}\,\overline{B}C$$

Step 4: Apply rule 7 $(\overline{A}\,\overline{A} = \overline{A})$ to the first term, and apply rule 10 $[\overline{A}\,\overline{B} + \overline{A}\,\overline{B}C = \overline{A}\,\overline{B}(1 + C) = \overline{A}\,\overline{B}]$ to the third and last terms.

$$\overline{A} + \overline{A}\,\overline{C} + \overline{A}\,\overline{B} + \overline{B}\,\overline{C}$$

Step 5: Apply rule 10 $[\overline{A} + \overline{A}\,\overline{C} = \overline{A}(1 + \overline{C}) = \overline{A}]$ to the first and second terms.

$$\overline{A} + \overline{A}\,\overline{B} + \overline{B}\,\overline{C}$$

Step 6: Apply rule 10 $[\overline{A} + \overline{A}\,\overline{B} = \overline{A}(1 + \overline{B}) = \overline{A}]$ to the first and second terms.

$$\overline{A} + \overline{B}\,\overline{C}$$

Related Problem

Simplify the Boolean expression $\overline{AB} + \overline{AC} + \overline{A}\,\overline{B}\,\overline{C}$.

EXAMPLE 4–13

Use Multisim to perform the logic simplification shown in Figure 4–20.

Solution

Step 1: Connect the Multisim Logic Converter to the circuit as shown in Figure 4–21.

Step 2: Generate the truth table by clicking on ⊸→ 1̄0̄1̄ .

Step 3: Generate the simplified Boolean expression by clicking on 1̄0̄1̄ ⁻ˢᴵᴹᴾ→ A|B .

Step 4: Generate the simplified logic circuit by clicking on A|B → ⊸ .

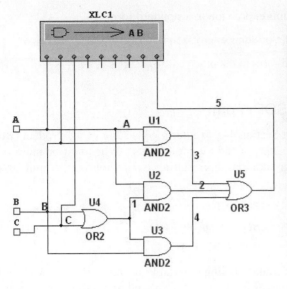

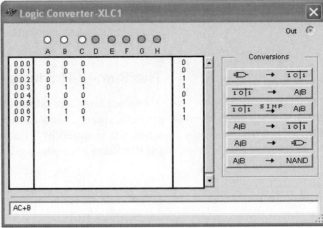

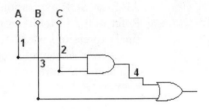

FIGURE 4–21

Related Problem

Open Multisim. Create the setup and perform the logic simplification illustrated in this example.   **MultiSim**

SECTION 4–5 CHECKUP

1. Simplify the following Boolean expressions:

   (a) $A + AB + A\overline{B}C$ **(b)** $(\overline{A} + B)C + ABC$ **(c)** $A\overline{B}C(BD + CDE) + A\overline{C}$

2. Implement each expression in Question 1 as originally stated with the appropriate logic gates. Then implement the simplified expression, and compare the number of gates.

4–6 Standard Forms of Boolean Expressions

All Boolean expressions, regardless of their form, can be converted into either of two standard forms: the sum-of-products form or the product-of-sums form. Standardization makes the evaluation, simplification, and implementation of Boolean expressions much more systematic and easier.

After completing this section, you should be able to

◆ Identify a sum-of-products expression

◆ Determine the domain of a Boolean expression

◆ Convert any sum-of-products expression to a standard form

◆ Evaluate a standard sum-of-products expression in terms of binary values

◆ Identify a product-of-sums expression

- Convert any product-of-sums expression to a standard form
- Evaluate a standard product-of-sums expression in terms of binary values
- Convert from one standard form to the other

The Sum-of-Products (SOP) Form

An SOP expression can be implemented with one OR gate and two or more AND gates.

A product term was defined in Section 4–1 as a term consisting of the product (Boolean multiplication) of literals (variables or their complements). When two or more product terms are summed by Boolean addition, the resulting expression is a **sum-of-products (SOP)**. Some examples are

$$AB + ABC$$
$$ABC + CDE + \overline{B}C\overline{D}$$
$$\overline{A}B + \overline{A}B\overline{C} + AC$$

Also, an SOP expression can contain a single-variable term, as in $A + \overline{A}\,\overline{B}C + BC\overline{D}$. Refer to the simplification examples in the last section, and you will see that each of the final expressions was either a single product term or in SOP form. In an SOP expression, a single overbar cannot extend over more than one variable; however, more than one variable in a term can have an overbar. For example, an SOP expression can have the term $\overline{A}\,\overline{B}\,\overline{C}$ but not $\overline{ABC}$.

Domain of a Boolean Expression

The **domain** of a general Boolean expression is the set of variables contained in the expression in either complemented or uncomplemented form. For example, the domain of the expression $\overline{A}B + A\overline{B}C$ is the set of variables A, B, C and the domain of the expression $AB\overline{C} + C\overline{D}E + \overline{B}C\overline{D}$ is the set of variables A, B, C, D, E.

AND/OR Implementation of an SOP Expression

Implementing an SOP expression simply requires ORing the outputs of two or more AND gates. A product term is produced by an AND operation, and the sum (addition) of two or more product terms is produced by an OR operation. Therefore, an SOP expression can be implemented by AND-OR logic in which the outputs of a number (equal to the number of product terms in the expression) of AND gates connect to the inputs of an OR gate, as shown in Figure 4–22 for the expression $AB + BCD + AC$. The output X of the OR gate equals the SOP expression.

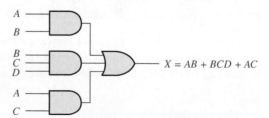

FIGURE 4–22   Implementation of the SOP expression $AB + BCD + AC$.

NAND/NAND Implementation of an SOP Expression

NAND gates can be used to implement an SOP expression. By using only NAND gates, an AND/OR function can be accomplished, as illustrated in Figure 4–23. The first level of NAND gates feed into a NAND gate that acts as a negative-OR gate. The NAND and negative-OR inversions cancel and the result is effectively an AND/OR circuit.

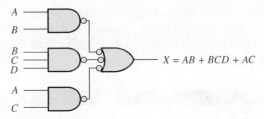

FIGURE 4–23 This NAND/NAND implementation is equivalent to the AND/OR in Figure 4–22.

Conversion of a General Expression to SOP Form

Any logic expression can be changed into SOP form by applying Boolean algebra techniques. For example, the expression $A(B + CD)$ can be converted to SOP form by applying the distributive law:

$$A(B + CD) = AB + ACD$$

EXAMPLE 4–14

Convert each of the following Boolean expressions to SOP form:

(a) $AB + B(CD + EF)$ (b) $(A + B)(B + C + D)$ (c) $\overline{(A + B)} + C$

Solution

(a) $AB + B(CD + EF) = AB + BCD + BEF$

(b) $(A + B)(B + C + D) = AB + AC + AD + BB + BC + BD$

(c) $\overline{(A + B)} + C = \overline{(\overline{A + B})\overline{C}} = (A + B)\overline{C} = A\overline{C} + B\overline{C}$

Related Problem

Convert $\overline{A}B\overline{C} + (A + \overline{B})(B + \overline{C} + A\overline{B})$ to SOP form.

The Standard SOP Form

So far, you have seen SOP expressions in which some of the product terms do not contain all of the variables in the domain of the expression. For example, the expression $\overline{A}B\overline{C} + A\overline{B}D + \overline{A}B\overline{C}D$ has a domain made up of the variables A, B, C, and D. However, notice that the complete set of variables in the domain is not represented in the first two terms of the expression; that is, D or $\overline{D}$ is missing from the first term and C or $\overline{C}$ is missing from the second term.

A *standard SOP expression* is one in which *all* the variables in the domain appear in each product term in the expression. For example, $A\overline{B}CD + \overline{A}\,B\overline{C}D + AB\overline{C}\,\overline{D}$ is a standard SOP expression. Standard SOP expressions are important in constructing truth tables, covered in Section 4–7, and in the Karnaugh map simplification method, which is covered in Section 4–8. Any nonstandard SOP expression (referred to simply as SOP) can be converted to the standard form using Boolean algebra.

Converting Product Terms to Standard SOP

Each product term in an SOP expression that does not contain all the variables in the domain can be expanded to standard form to include all variables in the domain and their complements. As stated in the following steps, a nonstandard SOP expression is converted into standard form using Boolean algebra rule 6 ($A + \overline{A} = 1$) from Table 4–1: A variable added to its complement equals 1.

Step 1: Multiply each nonstandard product term by a term made up of the sum of a missing variable and its complement. This results in two product terms. As you know, you can multiply anything by 1 without changing its value.

Step 2: Repeat Step 1 until all resulting product terms contain all variables in the domain in either complemented or uncomplemented form. In converting a product term to standard form, the number of product terms is doubled for each missing variable, as Example 4–15 shows.

EXAMPLE 4–15

Convert the following Boolean expression into standard SOP form:

$$A\overline{B}C + \overline{A}\overline{B} + AB\overline{C}D$$

Solution

The domain of this SOP expression is A, B, C, D. Take one term at a time. The first term, $A\overline{B}C$, is missing variable D or $\overline{D}$, so multiply the first term by $D + \overline{D}$ as follows:

$$A\overline{B}C = A\overline{B}C(D + \overline{D}) = A\overline{B}CD + A\overline{B}C\overline{D}$$

In this case, two standard product terms are the result.

The second term, $\overline{A}\overline{B}$, is missing variables C or $\overline{C}$ and D or $\overline{D}$, so first multiply the second term by $C + \overline{C}$ as follows:

$$\overline{A}\overline{B} = \overline{A}\overline{B}(C + \overline{C}) = \overline{A}\overline{B}C + \overline{A}\overline{B}\overline{C}$$

The two resulting terms are missing variable D or $\overline{D}$, so multiply both terms by $D + \overline{D}$ as follows:

$$\overline{A}\overline{B} = \overline{A}\overline{B}C + \overline{A}\overline{B}\overline{C} = \overline{A}\overline{B}C(D + \overline{D}) + \overline{A}\overline{B}\overline{C}(D + \overline{D})$$
$$= \overline{A}\overline{B}CD + \overline{A}\overline{B}C\overline{D} + \overline{A}\overline{B}\overline{C}D + \overline{A}\overline{B}\overline{C}\overline{D}$$

In this case, four standard product terms are the result.

The third term, $AB\overline{C}D$, is already in standard form. The complete standard SOP form of the original expression is as follows:

$$A\overline{B}C + \overline{A}\overline{B} + AB\overline{C}D = A\overline{B}CD + A\overline{B}C\overline{D} + \overline{A}\overline{B}CD + \overline{A}\overline{B}C\overline{D} + \overline{A}\overline{B}\overline{C}D + \overline{A}\overline{B}\overline{C}\overline{D} + AB\overline{C}D$$

Related Problem

Convert the expression $W\overline{X}Y + \overline{X}Y\overline{Z} + WX\overline{Y}$ to standard SOP form.

Binary Representation of a Standard Product Term

A standard product term is equal to 1 for only one combination of variable values. For example, the product term $A\overline{B}C\overline{D}$ is equal to 1 when $A = 1$, $B = 0$, $C = 1$, $D = 0$, as shown below, and is 0 for all other combinations of values for the variables.

$$A\overline{B}C\overline{D} = 1 \cdot \overline{0} \cdot 1 \cdot \overline{0} = 1 \cdot 1 \cdot 1 \cdot 1 = 1$$

In this case, the product term has a binary value of 1010 (decimal ten).

Remember, a product term is implemented with an AND gate whose output is 1 only if each of its inputs is 1. Inverters are used to produce the complements of the variables as required.

An SOP expression is equal to 1 only if one or more of the product terms in the expression is equal to 1.

EXAMPLE 4–16

Determine the binary values for which the following standard SOP expression is equal to 1:

$$ABCD + A\overline{B}\overline{C}D + \overline{A}\overline{B}\overline{C}\overline{D}$$

Solution

The term $ABCD$ is equal to 1 when $A = 1$, $B = 1$, $C = 1$, and $D = 1$.

$$ABCD = 1 \cdot 1 \cdot 1 \cdot 1 = 1$$

The term $A\overline{B}\,\overline{C}D$ is equal to 1 when $A = 1$, $B = 0$, $C = 0$, and $D = 1$.

$$A\overline{B}\,\overline{C}D = 1 \cdot \overline{0} \cdot \overline{0} \cdot 1 = 1 \cdot 1 \cdot 1 \cdot 1 = 1$$

The term $\overline{A}\,\overline{B}\,\overline{C}\,\overline{D}$ is equal to 1 when $A = 0$, $B = 0$, $C = 0$, and $D = 0$.

$$\overline{A}\,\overline{B}\,\overline{C}\,\overline{D} = \overline{0} \cdot \overline{0} \cdot \overline{0} \cdot \overline{0} = 1 \cdot 1 \cdot 1 \cdot 1 = 1$$

The SOP expression equals 1 when any or all of the three product terms is 1.

Related Problem

Determine the binary values for which the following SOP expression is equal to 1:

$$\overline{X}YZ + X\overline{Y}Z + XY\overline{Z} + \overline{X}Y\overline{Z} + XYZ$$

Is this a standard SOP expression?

The Product-of-Sums (POS) Form

A sum term was defined in Section 4–1 as a term consisting of the sum (Boolean addition) of literals (variables or their complements). When two or more sum terms are multiplied, the resulting expression is a **product-of-sums (POS)**. Some examples are

$$(\overline{A} + B)(A + \overline{B} + C)$$
$$(\overline{A} + \overline{B} + \overline{C})(C + \overline{D} + E)(\overline{B} + C + D)$$
$$(A + B)(A + \overline{B} + C)(\overline{A} + C)$$

A POS expression can contain a single-variable term, as in $\overline{A}(A + \overline{B} + C)(\overline{B} + \overline{C} + D)$. In a POS expression, a single overbar cannot extend over more than one variable; however, more than one variable in a term can have an overbar. For example, a POS expression can have the term $\overline{A} + \overline{B} + \overline{C}$ but not $\overline{A + B + C}$.

Implementation of a POS Expression

Implementing a POS expression simply requires ANDing the outputs of two or more OR gates. A sum term is produced by an OR operation, and the product of two or more sum terms is produced by an AND operation. Therefore, a POS expression can be implemented by logic in which the outputs of a number (equal to the number of sum terms in the expression) of OR gates connect to the inputs of an AND gate, as Figure 4–24 shows for the expression $(A + B)(B + C + D)(A + C)$. The output X of the AND gate equals the POS expression.

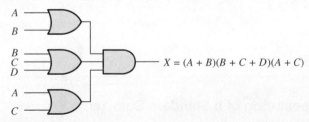

FIGURE 4–24   Implementation of the POS expression $(A + B)(B + C + D)(A + C)$.

The Standard POS Form

So far, you have seen POS expressions in which some of the sum terms do not contain all of the variables in the domain of the expression. For example, the expression

$$(A + \overline{B} + C)(A + B + \overline{D})(A + \overline{B} + \overline{C} + D)$$

has a domain made up of the variables A, B, C, and D. Notice that the complete set of variables in the domain is not represented in the first two terms of the expression; that is, D or $\overline{D}$ is missing from the first term and C or $\overline{C}$ is missing from the second term.

A *standard POS expression* is one in which *all* the variables in the domain appear in each sum term in the expression. For example,

$$(\overline{A} + \overline{B} + \overline{C} + \overline{D})(A + \overline{B} + C + D)(A + B + \overline{C} + D)$$

is a standard POS expression. Any nonstandard POS expression (referred to simply as POS) can be converted to the standard form using Boolean algebra.

Converting a Sum Term to Standard POS

Each sum term in a POS expression that does not contain all the variables in the domain can be expanded to standard form to include all variables in the domain and their complements. As stated in the following steps, a nonstandard POS expression is converted into standard form using Boolean algebra rule 8 ($A \cdot \overline{A} = 0$) from Table 4–1: A variable multiplied by its complement equals 0.

Step 1: Add to each nonstandard product term a term made up of the product of the missing variable and its complement. This results in two sum terms. As you know, you can add 0 to anything without changing its value.

Step 2: Apply rule 12 from Table 4–1: $A + BC = (A + B)(A + C)$

Step 3: Repeat Step 1 until all resulting sum terms contain all variables in the domain in either complemented or uncomplemented form.

EXAMPLE 4–17

Convert the following Boolean expression into standard POS form:

$$(A + \overline{B} + C)(\overline{B} + C + \overline{D})(A + \overline{B} + \overline{C} + D)$$

Solution

The domain of this POS expression is A, B, C, D. Take one term at a time. The first term, $A + \overline{B} + C$, is missing variable D or $\overline{D}$, so add $D\overline{D}$ and apply rule 12 as follows:

$$A + \overline{B} + C = A + \overline{B} + C + D\overline{D} = (A + \overline{B} + C + D)(A + \overline{B} + C + \overline{D})$$

The second term, $\overline{B} + C + \overline{D}$, is missing variable A or $\overline{A}$, so add $A\overline{A}$ and apply rule 12 as follows:

$$\overline{B} + C + \overline{D} = \overline{B} + C + \overline{D} + A\overline{A} = (A + \overline{B} + C + \overline{D})(\overline{A} + \overline{B} + C + \overline{D})$$

The third term, $A + \overline{B} + \overline{C} + D$, is already in standard form. The standard POS form of the original expression is as follows:

$$(A + \overline{B} + C)(\overline{B} + C + \overline{D})(A + \overline{B} + \overline{C} + D) =$$
$$(A + \overline{B} + C + D)(A + \overline{B} + C + \overline{D})(A + \overline{B} + C + \overline{D})(\overline{A} + \overline{B} + C + \overline{D})(A + \overline{B} + \overline{C} + D)$$

Related Problem

Convert the expression $(A + \overline{B})(B + C)$ to standard POS form.

Binary Representation of a Standard Sum Term

A standard sum term is equal to 0 for only one combination of variable values. For example, the sum term $A + \overline{B} + C + \overline{D}$ is 0 when $A = 0$, $B = 1$, $C = 0$, and $D = 1$, as shown below, and is 1 for all other combinations of values for the variables.

$$A + \overline{B} + C + \overline{D} = 0 + \overline{1} + 0 + \overline{1} = 0 + 0 + 0 + 0 = 0$$

In this case, the sum term has a binary value of 0101 (decimal 5). Remember, a sum term is implemented with an OR gate whose output is 0 only if each of its inputs is 0. Inverters are used to produce the complements of the variables as required.

A POS expression is equal to 0 only if one or more of the sum terms in the expression is equal to 0.

EXAMPLE 4–18

Determine the binary values of the variables for which the following standard POS expression is equal to 0:

$$(A + B + C + D)(A + \overline{B} + \overline{C} + D)(\overline{A} + \overline{B} + \overline{C} + \overline{D})$$

Solution

The term $A + B + C + D$ is equal to 0 when $A = 0$, $B = 0$, $C = 0$, and $D = 0$.

$$A + B + C + D = 0 + 0 + 0 + 0 = 0$$

The term $A + \overline{B} + \overline{C} + D$ is equal to 0 when $A = 0$, $B = 1$, $C = 1$, and $D = 0$.

$$A + \overline{B} + \overline{C} + D = 0 + \overline{1} + \overline{1} + 0 = 0 + 0 + 0 + 0 = 0$$

The term $\overline{A} + \overline{B} + \overline{C} + \overline{D}$ is equal to 0 when $A = 1$, $B = 1$, $C = 1$, and $D = 1$.

$$\overline{A} + \overline{B} + \overline{C} + \overline{D} = \overline{1} + \overline{1} + \overline{1} + \overline{1} = 0 + 0 + 0 + 0 = 0$$

The POS expression equals 0 when any of the three sum terms equals 0.

Related Problem

Determine the binary values for which the following POS expression is equal to 0:

$$(X + \overline{Y} + Z)(\overline{X} + Y + Z)(X + Y + \overline{Z})(\overline{X} + \overline{Y} + \overline{Z})(X + \overline{Y} + \overline{Z})$$

Is this a standard POS expression?

Converting Standard SOP to Standard POS

The binary values of the product terms in a given standard SOP expression are not present in the equivalent standard POS expression. Also, the binary values that are not represented in the SOP expression are present in the equivalent POS expression. Therefore, to convert from standard SOP to standard POS, the following steps are taken:

Step 1: Evaluate each product term in the SOP expression. That is, determine the binary numbers that represent the product terms.

Step 2: Determine all of the binary numbers not included in the evaluation in Step 1.

Step 3: Write the equivalent sum term for each binary number from Step 2 and express in POS form.

Using a similar procedure, you can go from POS to SOP.

EXAMPLE 4–19

Convert the following SOP expression to an equivalent POS expression:

$$\overline{A}\,\overline{B}\,\overline{C} + \overline{A}B\overline{C} + \overline{A}BC + A\overline{B}C + ABC$$

Solution

The evaluation is as follows:

$$000 + 010 + 011 + 101 + 111$$

Since there are three variables in the domain of this expression, there are a total of eight (2^3) possible combinations. The SOP expression contains five of these combinations, so the POS must contain the other three which are 001, 100, and 110. Remember, these are the binary values that make the sum term 0. The equivalent POS expression is

$$(A + B + \overline{C})(\overline{A} + B + C)(\overline{A} + \overline{B} + C)$$

Related Problem

Verify that the SOP and POS expressions in this example are equivalent by substituting binary values into each.

1. Identify each of the following expressions as SOP, standard SOP, POS, or standard POS:

   (a) $AB + \overline{A}BD + \overline{A}C\overline{D}$ (b) $(A + \overline{B} + C)(A + B + \overline{C})$

   (c) $\overline{A}BC + AB\overline{C}$ (d) $(A + \overline{C})(A + B)$

2. Convert each SOP expression in Question 1 to standard form.

3. Convert each POS expression in Question 1 to standard form.

4–7   Boolean Expressions and Truth Tables

All standard Boolean expressions can be easily converted into truth table format using binary values for each term in the expression. The truth table is a common way of presenting, in a concise format, the logical operation of a circuit. Also, standard SOP or POS expressions can be determined from a truth table. You will find truth tables in data sheets and other literature related to the operation of digital circuits.

After completing this section, you should be able to

- Convert a standard SOP expression into truth table format

- Convert a standard POS expression into truth table format

- Derive a standard expression from a truth table

- Properly interpret truth table data

Converting SOP Expressions to Truth Table Format

Recall from Section 4–6 that an SOP expression is equal to 1 only if at least one of the product terms is equal to 1. A truth table is simply a list of the possible combinations of input variable values and the corresponding output values (1 or 0). For an expression with a domain of two variables, there are four different combinations of those variables ($2^2 = 4$). For an expression with a domain of three variables, there are eight different combinations of those variables ($2^3 = 8$). For an expression with a domain of four variables, there are sixteen different combinations of those variables ($2^4 = 16$), and so on.

The first step in constructing a truth table is to list all possible combinations of binary values of the variables in the expression. Next, convert the SOP expression to standard form if it is not already. Finally, place a 1 in the output column (X) for each binary value that makes the standard SOP expression a 1 and place a 0 for all the remaining binary values. This procedure is illustrated in Example 4–20.

Develop a truth table for the standard SOP expression $\overline{A}\,\overline{B}C + A\overline{B}\,\overline{C} + ABC$.

Solution

There are three variables in the domain, so there are eight possible combinations of binary values of the variables as listed in the left three columns of Table 4–6. The binary values that make the product terms in the expressions equal to 1 are

TABLE 4–6

	Inputs		Output	
A	B	C	X	Product Term
0	0	0	0	
0	0	1	1	$\overline{A}\,\overline{B}C$
0	1	0	0	
0	1	1	0	
1	0	0	1	$A\overline{B}\,\overline{C}$
1	0	1	0	
1	1	0	0	
1	1	1	1	ABC

$\overline{A}\,\overline{B}C$: 001; $A\overline{B}\,\overline{C}$: 100; and ABC: 111. For each of these binary values, place a 1 in the output column as shown in the table. For each of the remaining binary combinations, place a 0 in the output column.

Related Problem

Create a truth table for the standard SOP expression $\overline{A}\,\overline{B}\,\overline{C} + A\overline{B}C$.

Converting POS Expressions to Truth Table Format

Recall that a POS expression is equal to 0 only if at least one of the sum terms is equal to 0. To construct a truth table from a POS expression, list all the possible combinations of binary values of the variables just as was done for the SOP expression. Next, convert the POS expression to standard form if it is not already. Finally, place a 0 in the output column (X) for each binary value that makes the expression a 0 and place a 1 for all the remaining binary values. This procedure is illustrated in Example 4–21.

EXAMPLE 4–21

Determine the truth table for the following standard POS expression:

$$(A + B + C)(A + \overline{B} + C)(A + \overline{B} + \overline{C})(\overline{A} + B + \overline{C})(\overline{A} + \overline{B} + C)$$

Solution

There are three variables in the domain and the eight possible binary values are listed in the left three columns of Table 4–7. The binary values that make the sum terms in the expression equal to 0 are $A + B + C$: 000; $A + \overline{B} + C$: 010; $A + \overline{B} + \overline{C}$: 011; $\overline{A} + B + \overline{C}$: 101; and $\overline{A} + \overline{B} + C$: 110. For each of these binary values, place a 0 in the output column as shown in the table. For each of the remaining binary combinations, place a 1 in the output column.

TABLE 4–7

	Inputs		Output	
A	B	C	X	Sum Term
0	0	0	0	$(A + B + C)$
0	0	1	1	
0	1	0	0	$(A + \overline{B} + C)$
0	1	1	0	$(A + \overline{B} + \overline{C})$
1	0	0	1	
1	0	1	0	$(\overline{A} + B + \overline{C})$
1	1	0	0	$(\overline{A} + \overline{B} + C)$
1	1	1	1	

Notice that the truth table in this example is the same as the one in Example 4–20. This means that the SOP expression in the previous example and the POS expression in this example are equivalent.

Related Problem

Develop a truth table for the following standard POS expression:

$$(A + \overline{B} + C)(A + B + \overline{C})(\overline{A} + \overline{B} + \overline{C})$$

Determining Standard Expressions from a Truth Table

To determine the standard SOP expression represented by a truth table, list the binary values of the input variables for which the output is 1. Convert each binary value to the corresponding product term by replacing each 1 with the corresponding variable and each 0 with the corresponding variable complement. For example, the binary value 1010 is converted to a product term as follows:

$$1010 \longrightarrow A\overline{B}C\overline{D}$$

If you substitute, you can see that the product term is 1:

$$A\overline{B}C\overline{D} = 1 \cdot \overline{0} \cdot 1 \cdot \overline{0} = 1 \cdot 1 \cdot 1 \cdot 1 = 1$$

To determine the standard POS expression represented by a truth table, list the binary values for which the output is 0. Convert each binary value to the corresponding sum term by replacing each 1 with the corresponding variable complement and each 0 with the corresponding variable. For example, the binary value 1001 is converted to a sum term as follows:

$$1001 \longrightarrow \overline{A} + B + C + \overline{D}$$

If you substitute, you can see that the sum term is 0:

$$\overline{A} + B + C + \overline{D} = \overline{1} + 0 + 0 + \overline{1} = 0 + 0 + 0 + 0 = 0$$

EXAMPLE 4–22

From the truth table in Table 4–8, determine the standard SOP expression and the equivalent standard POS expression.

TABLE 4–8

Inputs			Output
A	B	C	X
0	0	0	0
0	0	1	0
0	1	0	0
0	1	1	1
1	0	0	1
1	0	1	0
1	1	0	1
1	1	1	1

Solution

There are four 1s in the output column and the corresponding binary values are 011, 100, 110, and 111. Convert these binary values to product terms as follows:

$$011 \longrightarrow \overline{A}BC$$
$$100 \longrightarrow A\overline{B}\,\overline{C}$$
$$110 \longrightarrow AB\overline{C}$$
$$111 \longrightarrow ABC$$

The resulting standard SOP expression for the output X is

$$X = \overline{A}BC + A\overline{B}\,\overline{C} + AB\overline{C} + ABC$$

For the POS expression, the output is 0 for binary values 000, 001, 010, and 101. Convert these binary values to sum terms as follows:

$$000 \longrightarrow A + B + C$$
$$001 \longrightarrow A + B + \overline{C}$$
$$010 \longrightarrow A + \overline{B} + C$$
$$101 \longrightarrow \overline{A} + B + \overline{C}$$

The resulting standard POS expression for the output X is

$$X = (A + B + C)(A + B + \overline{C})(A + \overline{B} + C)(\overline{A} + B + \overline{C})$$

Related Problem

By substitution of binary values, show that the SOP and the POS expressions derived in this example are equivalent; that is, for any binary value each SOP and POS term should either both be 1 or both be 0, depending on the binary value.

SECTION 4–7 CHECKUP

1. If a certain Boolean expression has a domain of five variables, how many binary values will be in its truth table?

2. In a certain truth table, the output is a 1 for the binary value 0110. Convert this binary value to the corresponding product term using variables W, X, Y, and Z.

3. In a certain truth table, the output is a 0 for the binary value 1100. Convert this binary value to the corresponding sum term using variables W, X, Y, and Z.

4–8 The Karnaugh Map

A Karnaugh map provides a systematic method for simplifying Boolean expressions and, if properly used, will produce the simplest SOP or POS expression possible, known as the minimum expression. As you have seen, the effectiveness of algebraic simplification depends on your familiarity with all the laws, rules, and theorems of Boolean algebra and on your ability to apply them. The Karnaugh map, on the other hand, provides a "cookbook" method for simplification. Other simplification techniques include the Quine-McCluskey method and the Espresso algorithm.

After completing this section, you should be able to

◆ Construct a Karnaugh map for three or four variables

◆ Determine the binary value of each cell in a Karnaugh map

◆ Determine the standard product term represented by each cell in a Karnaugh map

◆ Explain cell adjacency and identify adjacent cells

A **Karnaugh map** is similar to a truth table because it presents all of the possible values of input variables and the resulting output for each value. Instead of being organized into columns and rows like a truth table, the Karnaugh map is an array of **cells** in which each cell represents a binary value of the input variables. The cells are arranged in a way so that simplification of a given expression is simply a matter of properly grouping the cells. Karnaugh maps can be used for expressions with two, three, four, and five variables, but we will discuss only 3-variable and 4-variable situations to illustrate the principles. *A discussion of 5-variable Karnaugh maps is available on the website.*

The number of cells in a Karnaugh map, as well as the number of rows in a truth table, is equal to the total number of possible input variable combinations. For three variables, the number of cells is $2^3 = 8$. For four variables, the number of cells is $2^4 = 16$.

The 3-Variable Karnaugh Map

The 3-variable Karnaugh map is an array of eight cells, as shown in Figure 4–25(a). In this case, A, B, and C are used for the variables although other letters could be used. Binary values of A and B are along the left side (notice the sequence) and the values of C are across the top. The value of a given cell is the binary values of A and B at the left in the same row combined with the value of C at the top in the same column. For example, the cell in the upper left corner has a binary value of 000 and the cell in the lower right corner has a binary value of 101. Figure 4–25(b) shows the standard product terms that are represented by each cell in the Karnaugh map.

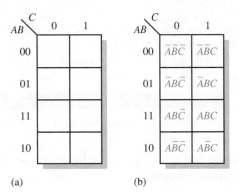

(a)   (b)

FIGURE 4–25 A 3-variable Karnaugh map showing Boolean product terms for each cell.

The 4-Variable Karnaugh Map

The 4-variable Karnaugh map is an array of sixteen cells, as shown in Figure 4–26(a). Binary values of A and B are along the left side and the values of C and D are across the top. The value of a given cell is the binary values of A and B at the left in the same row combined with the binary values of C and D at the top in the same column. For example, the cell in the upper right corner has a binary value of 0010 and the cell in the lower right corner has a binary value of 1010. Figure 4–26(b) shows the standard product terms that are represented by each cell in the 4-variable Karnaugh map.

Cell Adjacency

The cells in a Karnaugh map are arranged so that there is only a single-variable change between adjacent cells. **Adjacency** is defined by a single-variable change. In the 3-variable map the 010 cell is adjacent to the 000 cell, the 011 cell, and the 110 cell. The 010 cell is not adjacent to the 001 cell, the 111 cell, the 100 cell, or the 101 cell.

Physically, each cell is adjacent to the cells that are immediately next to it on any of its four sides. A cell is not adjacent to the cells that diagonally touch any of its corners. Also, the cells in the top row are adjacent to the corresponding cells in the bottom row and

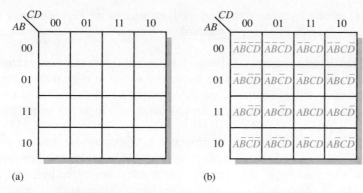

FIGURE 4–26 A 4-variable Karnaugh map.

the cells in the outer left column are adjacent to the corresponding cells in the outer right column. This is called "wrap-around" adjacency because you can think of the map as wrapping around from top to bottom to form a cylinder or from left to right to form a cylinder. Figure 4–27 illustrates the cell adjacencies with a 4-variable map, although the same rules for adjacency apply to Karnaugh maps with any number of cells.

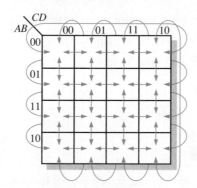

FIGURE 4–27 Adjacent cells on a Karnaugh map are those that differ by only one variable. Arrows point between adjacent cells.

The Quine-McCluskey Method

Minimizing Boolean functions using Karnaugh maps is practical only for up to four or five variables. Also, the Karnaugh map method does not lend itself to be automated in the form of a computer program.

The Quine-McCluskey method is more practical for logic simplification of functions with more than four or five variables. It also has the advantage of being easily implemented with a computer or programmable calculator.

The Quine-McCluskey method is functionally similar to Karnaugh mapping, but the tabular form makes it more efficient for use in computer algorithms, and it also gives a way to check that the minimal form of a Boolean function has been reached. This method is sometimes referred to as the *tabulation method*. An introduction to the Quine-McCluskey method is provided in Section 4–11.

Espresso Algorithm

Although the Quine-McCluskey method is well suited to be implemented in a computer program and can handle more variables than the Karnaugh map method, the result is still far from efficient in terms of processing time and memory usage. Adding a variable to the function will roughly double both of these parameters because the truth table length increases exponentially with the number of variables. Functions with a large number of variables have to be minimized with other methods such as the Espresso logic minimizer, which has become the de facto world standard. *An Espresso algorithm tutorial is available on the website.*

Compared to the other methods, Espresso is essentially more efficient in terms of reducing memory usage and computation time by several orders of magnitude. There is essentially no restrictions to the number of variables, output functions, and product terms of a combinational logic function. In general, tens of variables with tens of output functions can be handled by Espresso.

The Espresso algorithm has been incorporated as a standard logic function minimization step in most logic synthesis tools for programmable logic devices. For implementing a function in multilevel logic, the minimization result is optimized by factorization and mapped onto the available basic logic cells in the target device, such as an FPGA (Field-Programmable Gate Array).

SECTION 4–8 CHECKUP

1. In a 3-variable Karnaugh map, what is the binary value for the cell in each of the following locations:

    (a) upper left corner   (b) lower right corner

    (c) lower left corner   (d) upper right corner

2. What is the standard product term for each cell in Question 1 for variables X, Y, and Z?

3. Repeat Question 1 for a 4-variable map.

4. Repeat Question 2 for a 4-variable map using variables W, X, Y, and Z.

4–9   Karnaugh Map SOP Minimization

As stated in the last section, the Karnaugh map is used for simplifying Boolean expressions to their minimum form. A minimized SOP expression contains the fewest possible terms with the fewest possible variables per term. Generally, a minimum SOP expression can be implemented with fewer logic gates than a standard expression. In this section, Karnaugh maps with up to four variables are covered.

After completing this section, you should be able to

- ◆ Map a standard SOP expression on a Karnaugh map

- ◆ Combine the 1s on the map into maximum groups

- ◆ Determine the minimum product term for each group on the map

- ◆ Combine the minimum product terms to form a minimum SOP expression

- ◆ Convert a truth table into a Karnaugh map for simplification of the represented expression

- ◆ Use "don't care" conditions on a Karnaugh map

Mapping a Standard SOP Expression

For an SOP expression in standard form, a 1 is placed on the Karnaugh map for each product term in the expression. Each 1 is placed in a cell corresponding to the value of a product term. For example, for the product term $A\overline{B}C$, a 1 goes in the 101 cell on a 3-variable map.

When an SOP expression is completely mapped, there will be a number of 1s on the Karnaugh map equal to the number of product terms in the standard SOP expression. The cells that do not have a 1 are the cells for which the expression is 0. Usually, when working with SOP expressions, the 0s are left off the map. The following steps and the illustration in Figure 4–28 show the mapping process.

Step 1: Determine the binary value of each product term in the standard SOP expression. After some practice, you can usually do the evaluation of terms mentally.

Step 2: As each product term is evaluated, place a 1 on the Karnaugh map in the cell having the same value as the product term.

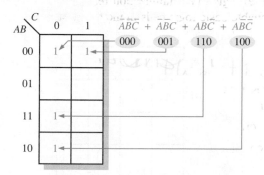

FIGURE 4–28 Example of mapping a standard SOP expression.

EXAMPLE 4–23

Map the following standard SOP expression on a Karnaugh map:

$$\overline{A}\,\overline{B}C + \overline{A}B\overline{C} + AB\overline{C} + ABC$$

Solution

Evaluate the expression as shown below. Place a 1 on the 3-variable Karnaugh map in Figure 4–29 for each standard product term in the expression.

$$\overline{A}\,\overline{B}C + \overline{A}B\overline{C} + AB\overline{C} + ABC$$
$$0\ 0\ 1 \quad\ 0\ 1\ 0 \quad\ 1\ 1\ 0 \quad\ 1\ 1\ 1$$

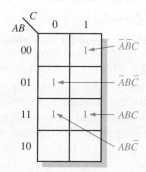

FIGURE 4–29

Related Problem

Map the standard SOP expression $\overline{A}BC + A\overline{B}C + AB\overline{C}$ on a Karnaugh map.

EXAMPLE 4–24

Map the following standard SOP expression on a Karnaugh map:

$$\overline{A}\,\overline{B}CD + \overline{A}B\overline{C}\,\overline{D} + AB\overline{C}D + ABCD + AB\overline{C}\,\overline{D} + \overline{A}\,\overline{B}\,\overline{C}D + A\overline{B}C\overline{D}$$

Solution

Evaluate the expression as shown below. Place a 1 on the 4-variable Karnaugh map in Figure 4–30 for each standard product term in the expression.

$$\overline{A}\,\overline{B}CD + \overline{A}B\overline{C}\,\overline{D} + AB\overline{C}D + ABCD + AB\overline{C}\,\overline{D} + \overline{A}\,\overline{B}\,\overline{C}D + A\overline{B}C\overline{D}$$

     0 0 1 1   0 1 0 0   1 1 0 1   1 1 1 1   1 1 0 0   0 0 0 1   1 0 1 0

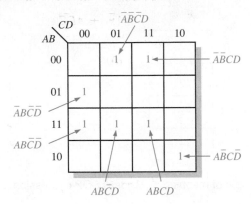

FIGURE 4–30

Related Problem

Map the following standard SOP expression on a Karnaugh map:

$$\overline{A}BC\overline{D} + AB\overline{C}\,\overline{D} + AB\overline{C}\,\overline{D} + ABCD$$

Mapping a Nonstandard SOP Expression

A Boolean expression must first be in standard form before you use a Karnaugh map. If an expression is not in standard form, then it must be converted to standard form by the procedure covered in Section 4–6 or by numerical expansion. Since an expression should be evaluated before mapping anyway, numerical expansion is probably the most efficient approach.

Numerical Expansion of a Nonstandard Product Term

Recall that a nonstandard product term has one or more missing variables. For example, assume that one of the product terms in a certain 3-variable SOP expression is $A\overline{B}$. This term can be expanded numerically to standard form as follows. First, write the binary value of the two variables and attach a 0 for the missing variable $\overline{C}$: 100. Next, write the binary value of the two variables and attach a 1 for the missing variable C: 101. The two resulting binary numbers are the values of the standard SOP terms $A\overline{B}\,\overline{C}$ and $A\overline{B}C$.

As another example, assume that one of the product terms in a 3-variable expression is B (remember that a single variable counts as a product term in an SOP expression). This term can be expanded numerically to standard form as follows. Write the binary value of the variable; then attach all possible values for the missing variables A and C as follows:

$$\begin{array}{c} B \\ 010 \\ 011 \\ 110 \\ 111 \end{array}$$

The four resulting binary numbers are the values of the standard SOP terms $\overline{A}B\overline{C}$, $\overline{A}BC$, $AB\overline{C}$, and ABC.

EXAMPLE 4–25

Map the following SOP expression on a Karnaugh map: $\overline{A} + A\overline{B} + AB\overline{C}$.

Solution

The SOP expression is obviously not in standard form because each product term does not have three variables. The first term is missing two variables, the second term is missing one variable, and the third term is standard. First expand the terms numerically as follows:

$$\overline{A} \quad + A\overline{B} \quad + AB\overline{C}$$

000	100	110
001	101	
010		
011		

Map each of the resulting binary values by placing a 1 in the appropriate cell of the 3-variable Karnaugh map in Figure 4–31.

FIGURE 4–31

Related Problem

Map the SOP expression $BC + \overline{A}\,\overline{C}$ on a Karnaugh map.

EXAMPLE 4–26

Map the following SOP expression on a Karnaugh map:

$$\overline{B}\,\overline{C} + A\overline{B} + AB\overline{C} + A\overline{B}C\overline{D} + \overline{A}\,\overline{B}\,\overline{C}D + ABCD$$

Solution

The SOP expression is obviously not in standard form because each product term does not have four variables. The first and second terms are both missing two variables, the third term is missing one variable, and the rest of the terms are standard. First expand the terms by including all combinations of the missing variables numerically as follows:

$\overline{B}\,\overline{C}$ +	$A\overline{B}$ +	$AB\overline{C}$ +	$A\overline{B}C\overline{D}$ +	$\overline{A}\,\overline{B}\,\overline{C}D$ +	$ABCD$
0000	1000	1100	1010	0001	1011
0001	1001	1101			
1000	1010				
1001	1011				

Map each of the resulting binary values by placing a 1 in the appropriate cell of the 4-variable Karnaugh map in Figure 4–32. Notice that some of the values in the expanded expression are redundant.

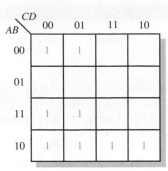

FIGURE 4–32

Related Problem

Map the expression $A + \overline{CD} + AC\overline{D} + \overline{ABCD}$ on a Karnaugh map.

Karnaugh Map Simplification of SOP Expressions

The process that results in an expression containing the fewest possible terms with the fewest possible variables is called **minimization**. After an SOP expression has been mapped, a minimum SOP expression is obtained by grouping the 1s and determining the minimum SOP expression from the map.

Grouping the 1s

You can group 1s on the Karnaugh map according to the following rules by enclosing those adjacent cells containing 1s. The goal is to maximize the size of the groups and to minimize the number of groups.

1. A group must contain either 1, 2, 4, 8, or 16 cells, which are all powers of two. In the case of a 3-variable map, $2^3 = 8$ cells is the maximum group.

2. Each cell in a group must be adjacent to one or more cells in that same group, but all cells in the group do not have to be adjacent to each other.

3. Always include the largest possible number of 1s in a group in accordance with rule 1.

4. Each 1 on the map must be included in at least one group. The 1s already in a group can be included in another group as long as the overlapping groups include noncommon 1s.

EXAMPLE 4–27

Group the 1s in each of the Karnaugh maps in Figure 4–33.

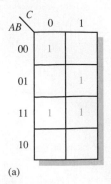

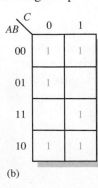

FIGURE 4–33

Solution

The groupings are shown in Figure 4–34. In some cases, there may be more than one way to group the 1s to form maximum groupings.

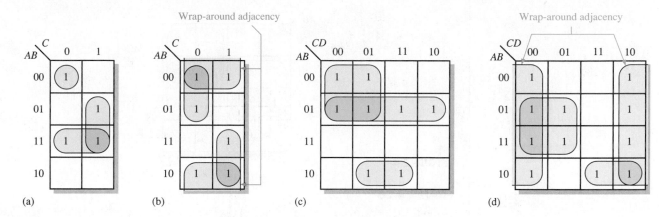

FIGURE 4–34

Related Problem

Determine if there are other ways to group the 1s in Figure 4–34 to obtain a minimum number of maximum groupings.

Determining the Minimum SOP Expression from the Map

When all the 1s representing the standard product terms in an expression are properly mapped and grouped, the process of determining the resulting minimum SOP expression begins. The following rules are applied to find the minimum product terms and the minimum SOP expression:

1. Group the cells that have 1s. Each group of cells containing 1s creates one product term composed of all variables that occur in only one form (either uncomplemented or complemented) within the group. Variables that occur both uncomplemented and complemented within the group are eliminated. These are called *contradictory variables*.

2. Determine the minimum product term for each group.
   (a) For a 3-variable map:
      (1) A 1-cell group yields a 3-variable product term
      (2) A 2-cell group yields a 2-variable product term
      (3) A 4-cell group yields a 1-variable term
      (4) An 8-cell group yields a value of 1 for the expression
   (b) For a 4-variable map:
      (1) A 1-cell group yields a 4-variable product term
      (2) A 2-cell group yields a 3-variable product term
      (3) A 4-cell group yields a 2-variable product term
      (4) An 8-cell group yields a 1-variable term
      (5) A 16-cell group yields a value of 1 for the expression

3. When all the minimum product terms are derived from the Karnaugh map, they are summed to form the minimum SOP expression.

EXAMPLE 4–28

Determine the product terms for the Karnaugh map in Figure 4–35 and write the resulting minimum SOP expression.

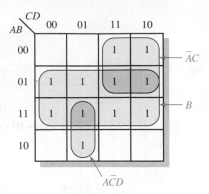

FIGURE 4–35

Solution

Eliminate variables that are in a grouping in both complemented and uncomplemented forms. In Figure 4–35, the product term for the 8-cell group is B because the cells within that group contain both A and $\overline{A}$, C and $\overline{C}$, and D and $\overline{D}$, which are eliminated. The 4-cell group contains B, $\overline{B}$, D, and $\overline{D}$, leaving the variables $\overline{A}$ and C, which form the product term $\overline{A}C$. The 2-cell group contains B and $\overline{B}$, leaving variables A, $\overline{C}$, and D which form the product term $A\overline{C}D$. Notice how overlapping is used to maximize the size of the groups. The resulting minimum SOP expression is the sum of these product terms:

$$B + \overline{A}C + A\overline{C}D$$

Related Problem

For the Karnaugh map in Figure 4–35, add a 1 in the lower right cell (1010) and determine the resulting SOP expression.

EXAMPLE 4–29

Determine the product terms for each of the Karnaugh maps in Figure 4–36 and write the resulting minimum SOP expression.

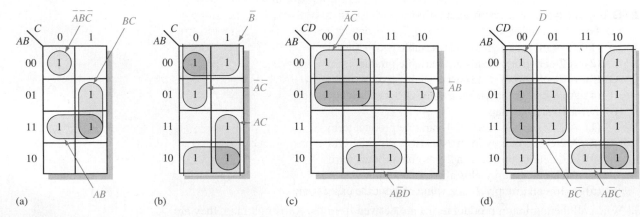

FIGURE 4–36

Solution

The resulting minimum product term for each group is shown in Figure 4–36. The minimum SOP expressions for each of the Karnaugh maps in the figure are

(a) $AB + BC + \overline{A}\,\overline{B}\,\overline{C}$

(b) $\overline{B} + \overline{A}\,\overline{C} + AC$

(c) $\overline{A}B + \overline{A}\,\overline{C} + A\overline{B}D$

(d) $\overline{D} + A\overline{B}C + B\overline{C}$

Related Problem

For the Karnaugh map in Figure 4–36(d), add a 1 in the 0111 cell and determine the resulting SOP expression.

EXAMPLE 4–30

Use a Karnaugh map to minimize the following standard SOP expression:

$$A\overline{B}C + \overline{A}BC + \overline{A}\,\overline{B}C + \overline{A}\,\overline{B}\,\overline{C} + AB\overline{C}$$

Solution

The binary values of the expression are

$$101 + 011 + 001 + 000 + 100$$

Map the standard SOP expression and group the cells as shown in Figure 4–37.

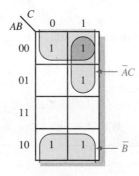

FIGURE 4–37

Notice the "wrap around" 4-cell group that includes the top row and the bottom row of 1s. The remaining 1 is absorbed in an overlapping group of two cells. The group of four 1s produces a single variable term, $\overline{B}$. This is determined by observing that within the group, $\overline{B}$ is the only variable that does not change from cell to cell. The group of two 1s produces a 2-variable term $\overline{A}C$. This is determined by observing that within the group, $\overline{A}$ and C do not change from one cell to the next. The product term for each group is shown. The resulting minimum SOP expression is

$$\overline{B} + \overline{A}C$$

Keep in mind that this minimum expression is equivalent to the original standard expression.

Related Problem

Use a Karnaugh map to simplify the following standard SOP expression:

$$X\overline{Y}Z + XY\overline{Z} + \overline{X}YZ + \overline{X}Y\overline{Z} + X\overline{Y}\,\overline{Z} + XYZ$$

EXAMPLE 4–31

Use a Karnaugh map to minimize the following SOP expression:

$$\overline{B}\,\overline{C}\overline{D} + \overline{A}\,\overline{B}C\overline{D} + AB\overline{C}\overline{D} + \overline{A}\,\overline{B}CD + A\overline{B}CD + \overline{A}BC\overline{D} + \overline{A}BC\overline{D} + ABC\overline{D} + AB\overline{C}D$$

Solution

The first term $\overline{B}\,\overline{C}\,\overline{D}$ must be expanded into $A\overline{B}\,\overline{C}\,\overline{D}$ and $\overline{A}\,\overline{B}\,\overline{C}\,\overline{D}$ to get the standard SOP expression, which is then mapped; the cells are grouped as shown in Figure 4–38.

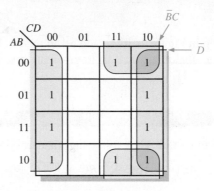

FIGURE 4–38

Notice that both groups exhibit "wrap around" adjacency. The group of eight is formed because the cells in the outer columns are adjacent. The group of four is formed to pick up the remaining two 1s because the top and bottom cells are adjacent. The product term for each group is shown. The resulting minimum SOP expression is

$$\overline{D} + \overline{B}C$$

Keep in mind that this minimum expression is equivalent to the original standard expression.

Related Problem

Use a Karnaugh map to simplify the following SOP expression:

$$\overline{W}\,\overline{X}\,\overline{Y}\overline{Z} + \overline{W}\,\overline{X}YZ + W\overline{X}\,\overline{Y}\overline{Z} + \overline{W}YZ + W\overline{X}\,\overline{Y}\overline{Z}$$

Mapping Directly from a Truth Table

You have seen how to map a Boolean expression; now you will learn how to go directly from a truth table to a Karnaugh map. Recall that a truth table gives the output of a Boolean expression for all possible input variable combinations. An example of a Boolean expression and its truth table representation is shown in Figure 4–39. Notice in the truth table that the output X is 1 for four different input variable combinations. The 1s in the output column of the truth table are mapped directly onto a Karnaugh map into the cells corresponding to the values of the associated input variable combinations, as shown in Figure 4–39. In the figure you can see that the Boolean expression, the truth table, and the Karnaugh map are simply different ways to represent a logic function.

"Don't Care" Conditions

Sometimes a situation arises in which some input variable combinations are not allowed. For example, recall that in the BCD code covered in Chapter 2, there are six invalid combinations: 1010, 1011, 1100, 1101, 1110, and 1111. Since these unallowed states

$$X = \overline{A}\,\overline{B}\,\overline{C} + A\overline{B}\,\overline{C} + AB\overline{C} + ABC$$

Inputs			Output
A	B	C	X
0	0	0	1
0	0	1	0
0	1	0	0
0	1	1	0
1	0	0	1
1	0	1	0
1	1	0	1
1	1	1	1

AB\C	0	1
00	1	
01		
11	1	1
10	1	

FIGURE 4–39 Example of mapping directly from a truth table to a Karnaugh map.

will never occur in an application involving the BCD code, they can be treated as **"don't care"** terms with respect to their effect on the output. That is, for these "don't care" terms either a 1 or a 0 may be assigned to the output; it really does not matter since they will never occur.

The "don't care" terms can be used to advantage on the Karnaugh map. Figure 4–40 shows that for each "don't care" term, an X is placed in the cell. When grouping the 1s, the Xs can be treated as 1s to make a larger grouping or as 0s if they cannot be used to advantage. The larger a group, the simpler the resulting term will be.

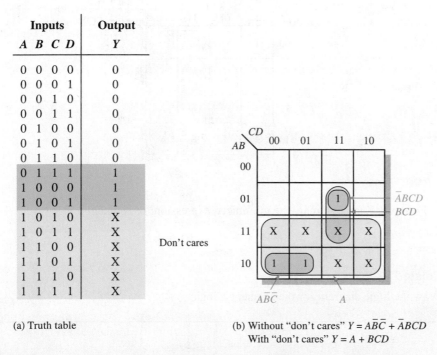

Inputs				Output
A	B	C	D	Y
0	0	0	0	0
0	0	0	1	0
0	0	1	0	0
0	0	1	1	0
0	1	0	0	0
0	1	0	1	0
0	1	1	0	0
0	1	1	1	1
1	0	0	0	1
1	0	0	1	1
1	0	1	0	X
1	0	1	1	X
1	1	0	0	X
1	1	0	1	X
1	1	1	0	X
1	1	1	1	X

Don't cares

(a) Truth table

(b) Without "don't cares" $Y = A\overline{B}\,\overline{C} + \overline{A}BCD$
With "don't cares" $Y = A + BCD$

FIGURE 4–40 Example of the use of "don't care" conditions to simplify an expression.

The truth table in Figure 4–40(a) describes a logic function that has a 1 output only when the BCD code for 7, 8, or 9 is present on the inputs. If the "don't cares" are used as 1s, the resulting expression for the function is $A + BCD$, as indicated in part (b). If the "don't cares" are not used as 1s, the resulting expression is $A\overline{B}\,\overline{C} + \overline{A}BCD$; so you can see the advantage of using "don't care" terms to get the simplest expression.

EXAMPLE 4–32

In a 7-segment display, each of the seven segments is activated for various digits. For example, segment a is activated for the digits 0, 2, 3, 5, 6, 7, 8, and 9, as illustrated in Figure 4–41. Since each digit can be represented by a BCD code, derive an SOP expression for segment a using the variables $ABCD$ and then minimize the expression using a Karnaugh map.

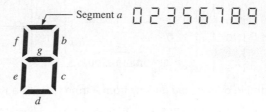

FIGURE 4–41   7-segment display.

Solution

The expression for segment a is

$$a = \overline{A}\,\overline{B}\,\overline{C}\,\overline{D} + \overline{A}\,B\overline{C}\overline{D} + \overline{A}\,BCD + \overline{A}B\overline{C}\overline{D} + \overline{A}BC\overline{D} + \overline{A}BCD + A\overline{B}\,\overline{C}\,\overline{D} + A\overline{B}\,\overline{C}D$$

Each term in the expression represents one of the digits in which segment a is used. The Karnaugh map minimization is shown in Figure 4–42. X's (don't cares) are entered for those states that do not occur in the BCD code.

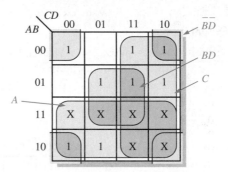

FIGURE 4–42

From the Karnaugh map, the minimized expression for segment a is

$$a = A + C + BD + \overline{B}\,\overline{D}$$

Related Problem

Draw the logic diagram for the segment-a logic.

SECTION 4–9   CHECKUP

1. Lay out Karnaugh maps for three and four variables.
2. Group the 1s and write the simplified SOP expression for the Karnaugh map in Figure 4–29.
3. Write the original standard SOP expressions for each of the Karnaugh maps in Figure 4–36.

4–10 Karnaugh Map POS Minimization

In the last section, you studied the minimization of an SOP expression using a Karnaugh map. In this section, we focus on POS expressions. The approaches are much the same except that with POS expressions, 0s representing the standard sum terms are placed on the Karnaugh map instead of 1s.

After completing this section, you should be able to

- Map a standard POS expression on a Karnaugh map
- Combine the 0s on the map into maximum groups
- Determine the minimum sum term for each group on the map
- Combine the minimum sum terms to form a minimum POS expression
- Use the Karnaugh map to convert between POS and SOP

Mapping a Standard POS Expression

For a POS expression in standard form, a 0 is placed on the Karnaugh map for each sum term in the expression. Each 0 is placed in a cell corresponding to the value of a sum term. For example, for the sum term $A + \bar{B} + C$, a 0 goes in the 010 cell on a 3-variable map.

When a POS expression is completely mapped, there will be a number of 0s on the Karnaugh map equal to the number of sum terms in the standard POS expression. The cells that do not have a 0 are the cells for which the expression is 1. Usually, when working with POS expressions, the 1s are left off. The following steps and the illustration in Figure 4–43 show the mapping process.

Step 1: Determine the binary value of each sum term in the standard POS expression. This is the binary value that makes the term equal to 0.

Step 2: As each sum term is evaluated, place a 0 on the Karnaugh map in the corresponding cell.

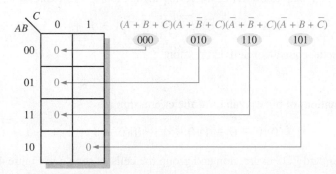

FIGURE 4–43 Example of mapping a standard POS expression.

EXAMPLE 4–33

Map the following standard POS expression on a Karnaugh map:

$$(\bar{A} + \bar{B} + C + D)(\bar{A} + B + \bar{C} + \bar{D})(A + B + \bar{C} + D)(\bar{A} + \bar{B} + \bar{C} + \bar{D})(A + B + \bar{C} + \bar{D})$$

Solution

Evaluate the expression as shown below and place a 0 on the 4-variable Karnaugh map in Figure 4–44 for each standard sum term in the expression.

$$(\bar{A} + \bar{B} + C + D)(\bar{A} + B + \bar{C} + \bar{D})(A + B + \bar{C} + D)(\bar{A} + \bar{B} + \bar{C} + \bar{D})(A + B + \bar{C} + \bar{D})$$

$$\qquad 1100 \qquad\qquad 1011 \qquad\qquad 0010 \qquad\qquad 1111 \qquad\qquad 0011$$

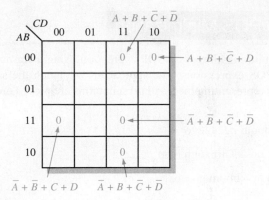

FIGURE 4-44

Related Problem

Map the following standard POS expression on a Karnaugh map:

$$(A + \overline{B} + \overline{C} + D)(A + B + C + \overline{D})(A + B + C + D)(\overline{A} + B + \overline{C} + D)$$

Karnaugh Map Simplification of POS Expressions

The process for minimizing a POS expression is basically the same as for an SOP expression except that you group 0s to produce minimum sum terms instead of grouping 1s to produce minimum product terms. The rules for grouping the 0s are the same as those for grouping the 1s that you learned in Section 4–9.

EXAMPLE 4–34

Use a Karnaugh map to minimize the following standard POS expression:

$$(A + B + C)(A + B + \overline{C})(A + \overline{B} + C)(A + \overline{B} + \overline{C})(\overline{A} + \overline{B} + C)$$

Also, derive the equivalent SOP expression.

Solution

The combinations of binary values of the expression are

$$(0 + 0 + 0)(0 + 0 + 1)(0 + 1 + 0)(0 + 1 + 1)(1 + 1 + 0)$$

Map the standard POS expression and group the cells as shown in Figure 4–45.

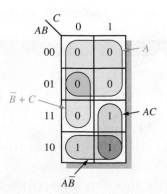

FIGURE 4-45

Notice how the 0 in the 110 cell is included into a 2-cell group by utilizing the 0 in the 4-cell group. The sum term for each blue group is shown in the figure and the resulting minimum POS expression is

$$A(\overline{B} + C)$$

Keep in mind that this minimum POS expression is equivalent to the original standard POS expression.

Grouping the 1s as shown by the gray areas yields an SOP expression that is equivalent to grouping the 0s.

$$AC + A\overline{B} = A(\overline{B} + C)$$

Related Problem

Use a Karnaugh map to simplify the following standard POS expression:

$$(X + \overline{Y} + Z)(X + \overline{Y} + \overline{Z})(\overline{X} + \overline{Y} + Z)(\overline{X} + Y + Z)$$

EXAMPLE 4–35

Use a Karnaugh map to minimize the following POS expression:

$$(B + C + D)(A + B + \overline{C} + D)(\overline{A} + B + C + \overline{D})(A + \overline{B} + C + D)(\overline{A} + \overline{B} + C + D)$$

Solution

The first term must be expanded into $\overline{A} + B + C + D$ and $A + B + C + D$ to get a standard POS expression, which is then mapped; and the cells are grouped as shown in Figure 4–46. The sum term for each group is shown and the resulting minimum POS expression is

$$(C + D)(A + B + D)(\overline{A} + B + C)$$

Keep in mind that this minimum POS expression is equivalent to the original standard POS expression.

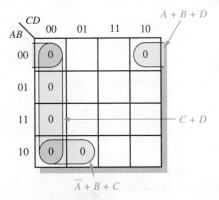

FIGURE 4–46

Related Problem

Use a Karnaugh map to simplify the following POS expression:

$$(W + \overline{X} + Y + \overline{Z})(W + X + Y + Z)(W + \overline{X} + \overline{Y} + Z)(\overline{W} + \overline{X} + Z)$$

Converting Between POS and SOP Using the Karnaugh Map

When a POS expression is mapped, it can easily be converted to the equivalent SOP form directly from the Karnaugh map. Also, given a mapped SOP expression, an equivalent POS expression can be derived directly from the map. This provides a good way to compare both minimum forms of an expression to determine if one of them can be implemented with fewer gates than the other.

For a POS expression, all the cells that do not contain 0s contain 1s, from which the SOP expression is derived. Likewise, for an SOP expression, all the cells that do not contain 1s contain 0s, from which the POS expression is derived. Example 4–36 illustrates this conversion.

EXAMPLE 4–36

Using a Karnaugh map, convert the following standard POS expression into a minimum POS expression, a standard SOP expression, and a minimum SOP expression.

$$(\overline{A} + \overline{B} + C + D)(A + \overline{B} + C + D)(A + B + C + \overline{D})(A + B + \overline{C} + \overline{D})(\overline{A} + B + C + \overline{D})(A + B + \overline{C} + D)$$

Solution

The 0s for the standard POS expression are mapped and grouped to obtain the minimum POS expression in Figure 4–47(a). In Figure 4–47(b), 1s are added to the cells that do not contain 0s. From each cell containing a 1, a standard product term is obtained as indicated. These product terms form the standard SOP expression. In Figure 4–47(c), the 1s are grouped and a minimum SOP expression is obtained.

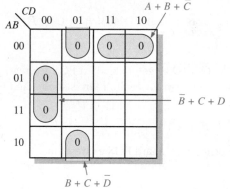

(a) Minimum POS: $(A + B + C)(\overline{B} + \overline{C} + D)(B + C + \overline{D})$

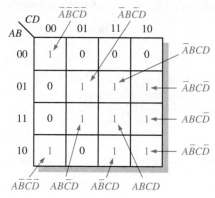

(b) Standard SOP:
$\overline{A}\overline{B}\overline{C}\overline{D} + \overline{A}B\overline{C}\overline{D} + \overline{A}BC\overline{D} + \overline{A}BCD + AB\overline{C}\overline{D} + A\overline{B}C\overline{D} +$
$A\overline{B}\overline{C}\overline{D} + ABC\overline{D} + A\overline{B}CD + ABCD$

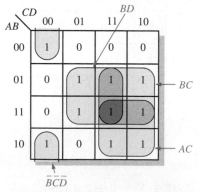

(c) Minimum SOP: $AC + BC + BD + \overline{B}\overline{C}\overline{D}$

FIGURE 4–47

Related Problem

Use a Karnaugh map to convert the following expression to minimum SOP form:

$$(W + \overline{X} + Y + \overline{Z})(\overline{W} + X + \overline{Y} + \overline{Z})(\overline{W} + \overline{X} + \overline{Y} + Z)(\overline{W} + \overline{X} + \overline{Z})$$

1. What is the difference in mapping a POS expression and an SOP expression?

2. What is the standard sum term for a 0 in cell 1011?

3. What is the standard product term for a 1 in cell 0010?

4–11   The Quine-McCluskey Method

For Boolean functions up to four variables, the Karnaugh map method is a powerful minimization method. When there are five variables, the Karnaugh map method is difficult to apply and completely impractical beyond five. The Quine-McCluskey method is a formal tabular method for applying the Boolean distributive law to various terms to find the minimum sum of products by eliminating literals that appear in two terms as complements. (For example, $ABCD + ABC\overline{D} = ABC$). *A Quine-McCluskey method tutorial is available on the website.*

After completing this section, you should be able to

* Describe the Quine-McCluskey method

* Reduce a Boolean expression using the Quine-McCluskey method

Unlike the Karnaugh mapping method, Quine-McCluskey lends itself to the computerized reduction of Boolean expressions, which is its principal use. For simple expressions, with up to four or perhaps even five variables, the Karnaugh map is easier for most people because it is a graphic method.

To apply the Quine-McCluskey method, first write the function in standard **minterm** (SOP) form. To illustrate, we will use the expression

$$X = \overline{A}\,\overline{B}CD + \overline{A}BCD + \overline{A}B\overline{C}\overline{D} + \overline{A}B\overline{C}D + A\overline{B}C\overline{D} + AB\overline{C}\overline{D} + AB\overline{C}D + ABCD$$

and represent it as binary numbers on the truth table shown in Table 4–9. The minterms that appear in the function are listed in the right column.

TABLE 4–9		
ABCD	**X**	**Minterm**
0000	0	
0001	1	m_1
0010	0	
0011	1	m_3
0100	1	m_4
0101	1	m_5
0110	0	
0111	0	
1000	0	
1001	0	
1010	1	m_{10}
1011	0	
1100	1	m_{12}
1101	1	m_{13}
1110	0	
1111	1	m_{15}

The second step in applying the Quine-McCluskey method is to arrange the minterms in the original expression in groups according to the number of 1s in each minterm, as shown in Table 4–10. In this example, there are four groups of minterms. (Note that if m_0 had been in the original expression, there would be five groups.)

TABLE 4–10

Number of 1s	Minterm	ABCD
1	m_1	0001
	m_4	0100
2	m_3	0011
	m_5	0101
	m_{10}	1010
	m_{12}	1100
3	m_{13}	1101
4	m_{15}	1111

Third, compare adjacent groups, looking to see if any minterms are the same in every position *except one*. If they are, place a check mark by those two minterms, as shown in Table 4–11. You should check each minterm against all others in the following group, but it is not necessary to check any groups that are not adjacent. In the column labeled *First Level,* you will have a list of the minterm names and the binary equivalent with an x as the placeholder for the literal that differs. In the example, minterm m_1 in Group 1 (0001) is identical to m_3 in Group 2 (0011) except for the *C* position, so place a check mark by these two minterms and enter 00x1 in the column labeled *First Level*. Minterm m_4 (0100) is identical to m_5 (0101) except for the *D* position, so check these two minterms and enter 010x in the last column. If a given term can be used more than once, it should be. In this case, notice that m_1 can be used again with m_5 in the second row with the x now placed in the *B* position.

TABLE 4–11

Number of 1s in Minterm	Minterm	ABCD	First Level
1	m_1	0001 ✓	(m_1, m_3) 00x1
	m_4	0100 ✓	(m_1, m_5) 0x01
2	m_3	0011 ✓	(m_4, m_5) 010x
	m_5	0101 ✓	(m_4, m_{12}) x100
	m_{10}	1010	(m_5, m_{13}) x101
	m_{12}	1100 ✓	(m_{12}, m_{13}) 110x
3	m_{13}	1101 ✓	(m_{13}, m_{15}) 11x1
4	m_{15}	1111 ✓	

In Table 4–11, minterm m_4 and minterm m_{12} are identical except for the *A* position. Both minterms are checked and x100 is entered in the *First Level* column. Follow this procedure for groups 2 and 3. In these groups, m_5 and m_{13} are combined and so are m_{12} and m_{13} (notice that m_{12} was previously used with m_4 and is used again). For groups 3 and 4, both m_{13} and m_{15} are added to the list in the *First Level* column.

In this example, minterm m_{10} does not have a check mark because no other minterm meets the requirement of being identical except for one position. This term is called an *essential prime implicant,* and it must be included in our final reduced expression.

The terms listed in the *First Level* have been used to form a reduced table (Table 4–12) with one less group than before. The number of 1s remaining in the *First Level* are counted and used to form three new groups.

Terms in the new groups are compared against terms in the adjacent group down. You need to compare these terms only if the x is in the same relative position in adjacent groups; otherwise go on. If the two expressions differ by exactly one position, a check mark is

TABLE 4–12

First Level	Number of 1s in First Level	Second Level
(m_1, m_3) 00x1	1	$(m_4, m_5, m_{12}, m_{13})$ x10x
(m_1, m_5) 0x01		$(m_4, m_5, m_{12}, m_{13})$ x10x
(m_4, m_5) 010x ✓		
(m_4, m_{12}) x100 ✓		
(m_5, m_{13}) x101 ✓	2	
(m_{12}, m_{13}) 110x ✓		
(m_{13}, m_{15}) 11x1	3	

placed next to both terms as before and all of the minterms are listed in the Second Level list. As before, the one position that has changed is entered as an x in the *Second Level*.

For our example, notice that the third term in Group 1 and the second term in Group 2 meet this requirement, differing only with the *A* literal. The fourth term in Group 1 also can be combined with the first term in Group 2, forming a redundant set of minterms. One of these can be crossed off the list and will not be used in the final expression.

With complicated expressions, the process described can be continued. For our example, we can read the *Second Level* expression as $B\overline{C}$. The terms that are unchecked will form other terms in the final reduced expression. The first unchecked term is read as $\overline{A}\,\overline{B}D$. The next one is read as $\overline{A}\,CD$. The last unchecked term is ABD. Recall that m_{10} was an essential prime implicant, so is picked up in the final expression. The reduced expression using the unchecked terms is:

$$X = B\overline{C} + \overline{A}\,\overline{B}D + \overline{A}\,CD + ABD + A\overline{B}C\overline{D}$$

Although this expression is correct, it may not be the minimum possible expression. There is a final check that can eliminate any unnecessary terms. The terms for the expression are written into a prime implicant table, with minterms for each prime implicant checked, as shown in Table 4–13.

TABLE 4–13

Prime Implicants	m_1	m_3	m_4	m_5	m_{10}	m_{12}	m_{13}	m_{15}
$B\overline{C}$ $(m_4, m_5, m_{12}, m_{13})$			✓	✓		✓	✓	
$\overline{A}\,\overline{B}D$ (m_1, m_3)	✓	✓						
$\overline{A}\,CD$ (m_1, m_5)	✓			✓				
ABD (m_{13}, m_{15})							✓	✓
$A\overline{B}C\overline{D}$ (m_{10})					✓			

If a minterm has a single check mark, then the prime implicant is essential and must be included in the final expression. The term ABD must be included because m_{15} is only covered by it. Likewise m_{10} is only covered by $A\overline{B}C\overline{D}$, so it must be in the final expression. Notice that the two minterms in $\overline{A}\,CD$ are covered by the prime implicants in the first two rows, so this term is unnecessary. The final reduced expression is, therefore,

$$X = B\overline{C} + \overline{A}\,\overline{B}D + ABD + A\overline{B}C\overline{D}$$

SECTION 4–11 CHECKUP

1. What is a minterm?

2. What is an essential prime implicant?

4–12   Boolean Expressions with VHDL

The ability to create simple and compact code is important in a VHDL program. By simplifying a Boolean expression for a given logic function, it is easier to write and debug the VHDL code; in addition, the result is a clearer and more concise program. Many VHDL development software packages contain tools that automatically optimize a program when it is compiled and converted to a downloadable file. However, this does not relieve you from creating program code that is clear and concise. You should not only be concerned with the number of lines of code, but you should also be concerned with the complexity of each line of code. In this section, you will see the difference in VHDL code when simplification methods are applied. Also, three levels of abstraction used in the description of a logic function are examined. *A VHDL tutorial is available on the website.*

After completing this section, you should be able to

◆ Write VHDL code to represent a simplified logic expression and compare it to the code for the original expression

◆ Relate the advantages of optimized Boolean expressions as applied to a target device

◆ Understand how a logic function can be described at three levels of abstraction

◆ Relate VHDL approaches to the description of a logic function to the three levels of abstraction

Boolean Algebra in VHDL Programming

The basic rules of Boolean algebra that you have learned in this chapter should be applied to any applicable VHDL code. Eliminating unnecessary gate logic allows you to create compact code that is easier to understand, especially when someone has to go back later and update or modify the program.

In Example 4–37, DeMorgan's theorems are used to simplify a Boolean expression, and VHDL programs for both the original expression and the simplified expression are compared.

EXAMPLE 4–37

First, write a VHDL program for the logic described by the following Boolean expression. Next, apply DeMorgan's theorems and Boolean rules to simplify the expression. Then write a program to reflect the simplified expression.

$$X = \overline{(AC + \overline{B\overline{C}} + D)} + \overline{\overline{BC}}$$

Solution

The VHDL program for the logic represented by the original expression is

```
entity OriginalLogic is
    port (A, B, C, D: in bit; X: out bit);
end entity OriginalLogic;
architecture Expression1 of OriginalLogic is
begin
    X <= not((A and C) or not(B and not C) or D)  or not(not(B and C));
end architecture Expression1;
```

Four inputs and one output are described.

The original logic contains four inputs, 3 AND gates, 2 OR gates, and 3 inverters.

By selectively applying DeMorgan's theorem and the laws of Boolean algebra, you can reduce the Boolean expression to its simplest form.

$$\overline{(AC + \overline{B}\overline{C} + D)} + \overline{\overline{BC}} = (\overline{AC})(\overline{\overline{B}\overline{C}})\overline{D} + \overline{\overline{BC}}$$ Apply DeMorgan

$$= (\overline{AC})(B\overline{C})\overline{D} + BC$$ Cancel double complements

$$= (\overline{A} + \overline{C})B\overline{C}\overline{D} + BC$$ Apply DeMorgan and factor

$$= \overline{A}B\overline{C}\overline{D} + B\overline{C}\overline{D} + BC$$ Distributive law

$$= B\overline{C}\overline{D}(1 + \overline{A}) + BC$$ Factor

$$= B\overline{C}\overline{D} + BC$$ Rule: $1 + A = 1$

The VHDL program for the logic represented by the reduced expression is

entity ReducedLogic **is**   ⟵ 3 inputs and 1 output are described.
    port (B, C, D: **in** bit; X: **out** bit);
end entity ReducedLogic;   The simplified logic contains
architecture Expression2 **of** ReducedLogic **is**   three inputs, 3 AND gates,
begin   ⟵ 1 OR gate, and 2 inverters.
        X <= (B **and not** C **and not** D) **or** (B **and** C);
end architecture Expression2;

As you can see, Boolean simplification is applicable to even simple VHDL programs.

Related Problem

Write the VHDL architecture statement for the expression $X = (\overline{A} + B + C)D$ as stated. Apply any applicable Boolean rules and rewrite the VHDL statement.

Example 4–38 demonstrates a more significant reduction in VHDL code complexity, using a Karnaugh map to reduce an expression.

EXAMPLE 4–38

(a) Write a VHDL program to describe the following SOP expression.

(b) Minimize the expression and show how much the VHDL program is simplified.

$$X = \overline{A}\,\overline{B}\,\overline{C}D + \overline{A}\,\overline{B}\overline{C}D + \overline{A}\,B\overline{C}D + \overline{A}BCD + \overline{A}\overline{B}C\overline{D} + A\overline{B}C\overline{D}$$

$$+ A\overline{B}C\overline{D} + ABC\overline{D} + AB\overline{C}\overline{D} + A\overline{B}\,\overline{C}D + \overline{A}B\overline{C}\overline{D} + AB\overline{C}\overline{D}$$

Solution

(a) The VHDL program for the SOP expression without minimization is large and hard to follow as you can see in the following VHDL code. Code such as this is subject to error. The VHDL program for the original SOP expression is as follows:

entity OriginalSOP **is**
    port (A, B, C, D: **in** bit; X: **out** bit);
end entity OriginalSOP;
architecture Equation1 **of** OriginalSOP **is**
begin
        X <= (**not** A **and not** B **and not** C **and not** D) **or**
            (**not** A **and not** B **and not** C **and** D) **or**
            (**not** A **and** B **and not** C **and not** D) **or**
            (**not** A **and** B **and** C **and not** D) **or**
            (**not** A **and not** B **and** C **and not** D) **or**
            (A **and not** B **and not** C **and not** D) **or**
            (A **and not** B **and** C **and not** D) **or**
            (A **and** B **and** C **and not** D) **or**
            (A **and** B **and not** C **and not** D) **or**

(A **and not** B **and not** C **and** D) **or**
(**not** A **and** B **and not** C **and** D) **or**
(A **and** B **and not** C **and** D);
end architecture Equation1;

(b) Now, use a four-variable Karnaugh map to reduce the original SOP expression to a minimum form. The original SOP expression is mapped in Figure 4–48.

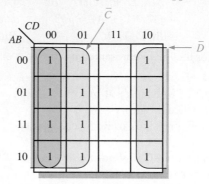

FIGURE 4–48

The original SOP Boolean expression that is plotted on the Karnaugh map in Figure 4–48 contains twelve 4-variable terms as indicated by the twelve 1s on the map. Recall that only the variables that do not change within a group remain in the expression for that group. The simplified expression taken from the map is developed next.

Combining the terms from the Karnaugh map, you get the following simplified expression, which is equivalent to the original SOP expression.

$$X = \overline{C} + \overline{D}$$

Using the simplified expression, the VHDL code can be rewritten with fewer terms, making the code more readable and easier to modify. Also, the logic implemented in a target device by the reduced code consumes much less space in the PLD. The VHDL program for the simplified SOP expression is as follows:

entity SimplifiedSOP **is**
    port (A, B, C, D: **in** bit; X: **out** bit);
end entity SimplifiedSOP;
architecture Equation2 **of** SimplifiedSOP **is**
begin
    X <= **not** C **or not** D
end architecture Equation2;

Related Problem

Write a VHDL architecture statement to describe the logic for the expression

$$X = A(BC + \overline{D})$$

As you have seen, the simplification of Boolean logic is important in the design of any logic function described in VHDL. Target devices have finite capacity and therefore require the creation of compact and efficient program code. Throughout this chapter, you have learned that the simplification of complex Boolean logic can lead to the elimination of unnecessary logic as well as the simplification of VHDL code.

Levels of Abstraction

A given logic function can be described at three different levels. It can be described by a truth table or a state diagram, by a Boolean expression, or by its logic diagram (schematic).

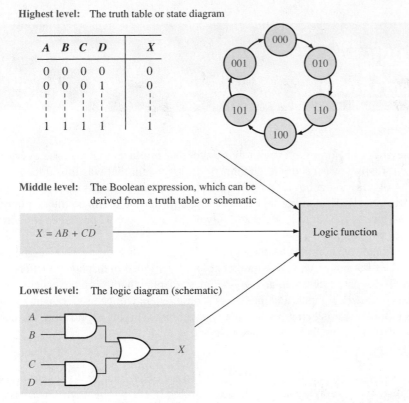

Highest level: The truth table or state diagram

A	B	C	D	X
0	0	0	0	0
0	0	0	1	0
┊	┊	┊	┊	┊
1	1	1	1	1

Middle level: The Boolean expression, which can be derived from a truth table or schematic

$X = AB + CD$

Logic function

Lowest level: The logic diagram (schematic)

FIGURE 4–49 Illustration of the three levels of abstraction for describing a logic function.

The truth table and state diagram are the most abstract ways to describe a logic function. A Boolean expression is the next level of abstraction, and a schematic is the lowest level of abstraction. This concept is illustrated in Figure 4–49 for a simple logic circuit. VHDL provides three approaches for describing functions that correspond to the three levels of abstraction.

- The data flow approach is analogous to describing a logic function with a Boolean expression. The data flow approach specifies each of the logic gates and how the data flows through them. This approach was applied in Examples 4–37 and 4–38.

- The structural approach is analogous to using a logic diagram or schematic to describe a logic function. It specifies the gates and how they are connected, rather than how signals (data) flow through them. The structural approach is used to develop VHDL code for describing logic circuits in Chapter 5.

- The behavioral approach is analogous to describing a logic function using a state diagram or truth table. However, this approach is the most complex; it is usually restricted to logic functions whose operations are time dependent and normally require some type of memory.

SECTION 4–12 CHECKUP

1. What are the advantages of Boolean logic simplification in terms of writing a VHDL program?

2. How does Boolean logic simplification benefit a VHDL program in terms of the target device?

3. Name the three levels of abstraction for a combinational logic function and state the corresponding VHDL approaches for describing a logic function.

Applied Logic

Seven-Segment Display

Seven-segment displays are used in many types of products that you see every day. A 7-segment display was used in the tablet-bottling system that was introduced in Chapter 1. The display in the bottling system is driven by logic circuits that decode a binary coded decimal (BCD) number and activate the appropriate digits on the display. BCD-to-7-segment decoder/drivers are readily available as single IC packages for activating the ten decimal digits.

In addition to the numbers from 0 to 9, the 7-segment display can show certain letters. For the tablet-bottling system, a requirement has been added to display the letters A, b, C, d, and E on a separate common-anode 7-segment display that uses a hexadecimal keypad for both the numerical inputs and the letters. These letters will be used to identify the type of vitamin tablet that is being bottled at any given time. In this application, the decoding logic for displaying the five letters is developed.

The 7-Segment Display

Two types of 7-segment displays are the LED and the LCD. Each of the seven segments in an LED display uses a light-emitting diode to produce a colored light when there is current through it and can be seen in the dark. An LCD or liquid-crystal display operates by polarizing light so that when a segment is not activated by a voltage, it reflects incident light and appears invisible against its background; however, when a segment is activated, it does not reflect light and appears black. LCD displays cannot be seen in the dark.

The seven segments in both LED and LCD displays are arranged as shown in Figure 4–50 and labeled a, b, c, d, e, f, and g as indicated in part (a). Selected segments are activated to create each of the ten decimal digits as well as certain letters of the alphabet, as shown in part (b). The letter b is shown as lowercase because a capital B would be the same as the digit 8. Similarly, for d, a capital letter would appear as a 0.

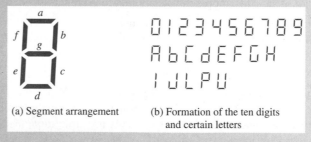

(a) Segment arrangement

(b) Formation of the ten digits and certain letters

FIGURE 4–50   Seven-segment display.

Exercise

1. List the segments used to form the digit 2.
2. List the segments used to form the digit 5.
3. List the segments used to form the letter A.
4. List the segments used to form the letter E.
5. Is there any one segment that is common to all digits?
6. Is there any one segment that is common to all letters?

Display Logic

The segments in a 7-segment display can be used in the formation of various letters as shown in Figure 4–50(b). Each segment must be activated by its own decoding circuit that detects the code for any of the letters in which that segment is used. Because a common-anode display is used, the segments are turned *on* with a LOW (0) logic level and turned *off* with a HIGH (1) logic level. The active segments are shown for each of the letters required for the tablet-bottling system in Table 4–14. Even though the active level is LOW (lighting the LED), the logic expressions are developed exactly the same way as discussed in this chapter, by mapping the desired output (1, 0, or X) for every possible input, grouping the 1s on the map, and reading the SOP expression from the map. In effect, the reduced logic expression is the logic for keeping a given segment OFF. At first, this may sound confusing, but it is simple in practice and it avoids an output current capability issue with bipolar (TTL) logic (discussed in Chapter 15 on the website).

TABLE 4–14

Active segments for each of the five letters used in the system display.

Letter	Segments Activated
A	a, b, c, e, f, g
b	c, d, e, f, g
C	a, d, e, f
d	b, c, d, e, g
E	a, d, e, f, g

A block diagram of a 7-segment logic and display for generating the five letters is shown in Figure 4–51(a), and the truth table is shown in part (b). The logic has four hexadecimal inputs and seven outputs, one for each segment. Because the letter F is not used as an input, we will show it on the truth table with all outputs set to 1 (OFF).

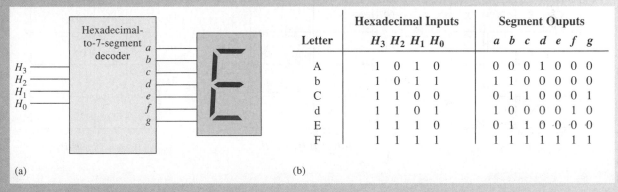

Letter	Hexadecimal Inputs H_3 H_2 H_1 H_0	Segment Ouputs a b c d e f g
A	1 0 1 0	0 0 0 1 0 0 0
b	1 0 1 1	1 1 0 0 0 0 0
C	1 1 0 0	0 1 1 0 0 0 1
d	1 1 0 1	1 0 0 0 0 1 0
E	1 1 1 0	0 1 1 0 0 0 0
F	1 1 1 1	1 1 1 1 1 1 1

(a)   (b)

FIGURE 4–51   Hexadecimal-to-7-segment decoder for letters *A* through *E*, used in the system.

Karnaugh Maps and the Invalid BCD Code Detector

To develop the simplified logic for each segment, the truth table information in Figure 4–51 is mapped onto Karnaugh maps. Recall that the BCD numbers will not be shown on the letter display. For this reason, an entry that represents a BCD number will be entered as an "X" ("don't care") on the K-maps. This makes the logic much simpler but would put some strange outputs on the display unless steps are taken to eliminate that possibility. Because all of the letters are *invalid* BCD characters, the display is activated only when an invalid BCD code is entered into the keypad, thus allowing only letters to be displayed.

Expressions for the Segment Logic

Using the table in 4–51(b), a standard SOP expression can be written for each segment and then minimized using a K-map. The desired outputs from the truth table are entered in the appropriate cells representing the hex inputs. To obtain the minimum SOP expressions for the display logic, the 1s and Xs are grouped.

Segment a   Segment a is used for the letters A, C, and E. For the letter A, the hexadecimal code is 1010 or, in terms of variables, $H_3\overline{H}_2H_1\overline{H}_0$. For the letter C, the hexadecimal code is 1100 or $H_3H_2\overline{H}_1\overline{H}_0$. For the letter E, the code is 1110 or $H_3H_2H_1\overline{H}_0$. The complete standard SOP expression for segment a is

$$a = H_3\overline{H}_2H_1\overline{H}_0 + H_3H_2\overline{H}_1\overline{H}_0 + H_3H_2H_1\overline{H}_0$$

Because a LOW is the active output state for each segment logic circuit, a 0 is entered on the Karnaugh map in each cell that represents the code for the letters in which the segment is *on*. The simplification of the expression for segment a is shown in Figure 4–52(a) after grouping the 1s and Xs.

Segment b   Segment b is used for the letters A and d. The complete standard SOP expression for segment b is

$$b = H_3\overline{H}_2H_1\overline{H}_0 + H_3H_2\overline{H}_1H_0$$

The simplification of the expression for segment b is shown in Figure 4–52(b).

Segment c   Segment c is used for the letters A, b, and d. The complete standard SOP expression for segment c is

$$c = H_3\overline{H}_2H_1\overline{H}_0 + H_3\overline{H}_2H_1H_0 + H_3H_2\overline{H}_1H_0$$

The simplification of the expression for segment c is shown in Figure 4–52(c).

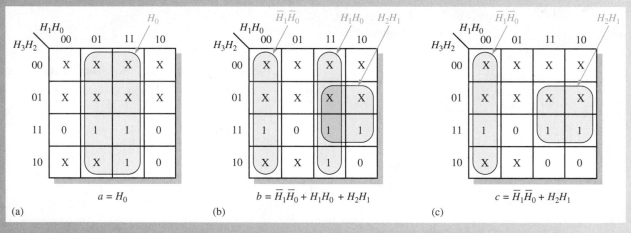

FIGURE 4–52   Minimization of the expressions for segments a, b, and c.

Exercise

7. Develop the minimum expression for segment d.
8. Develop the minimum expression for segment e.
9. Develop the minimum expression for segment f.
10. Develop the minimum expression for segment g.

The Logic Circuits

From the minimum expressions, the logic circuits for each segment can be implemented. For segment a, connect the H_0 input directly (no gate) to the a segment on the display. The segment b and segment c logic are shown in Figure 4–53 using AND or OR gates. Notice that two of the terms (H_2H_1 and $\overline{H}_1\overline{H}_0$) appear in the expressions for both b and c logic so two of the AND gates can be used in both, as indicated.

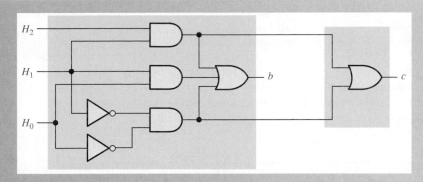

FIGURE 4-53 Segment-*b* and segment-*c* logic circuits.

Exercise

11. Show the logic for segment *d*.
12. Show the logic for segment *e*.
13. Show the logic for segment *f*.
14. Show the logic for segment *g*.

Describing the Decoding Logic with VHDL

The 7-segment decoding logic can be described using VHDL for implementation in a programmable logic device (PLD). The logic expressions for segments *a*, *b*, and *c* of the display are as follows:

$$a = H_0$$
$$b = \overline{H_1}\,\overline{H_0} + H_1 H_0 + H_2 H_1$$
$$c = \overline{H_1}\,\overline{H_0} + H_2 H_1$$

- The VHDL code for segment *a* is

```
entity SEGLOGIC is
    port (H0: in bit; SEGa: out bit);
end entity SEGLOGIC;
architecture LogicFunction of SEGLOGIC is
begin
    SEGa <= H0;
end architecture LogicFunction;
```

- The VHDL code for segment *b* is

```
entity SEGLOGIC is
    port (H0, H1, H2: in bit; SEGb: out bit);
end entity SEGLOGIC;
architecture LogicFunction of SEGLOGIC is
begin
    SEGb <= (not H1 and not H0) or (H1 and H0) or (H2 and H1);
end architecture LogicFunction;
```

- The VHDL code for segment *c* is

```
entity SEGLOGIC is
    port (H0, H1, H2: in bit; SEGc: out bit);
end entity SEGLOGIC;
architecture LogicFunction of SEGLOGIC is
begin
    SEGc <= (not H1 and not H0) or (H2 and H1);
end architecture LogicFunction;
```

Exercise

15. Write the VHDL code for segments *d*, *e*, *f*, and *g*.

Simulation

The decoder simulation using Multisim is shown in Figure 4–54 with the letter E selected. Subcircuits are used for the segment logic to be developed as activities or in the lab. The purpose of simulation is to verify proper operation of the circuit.

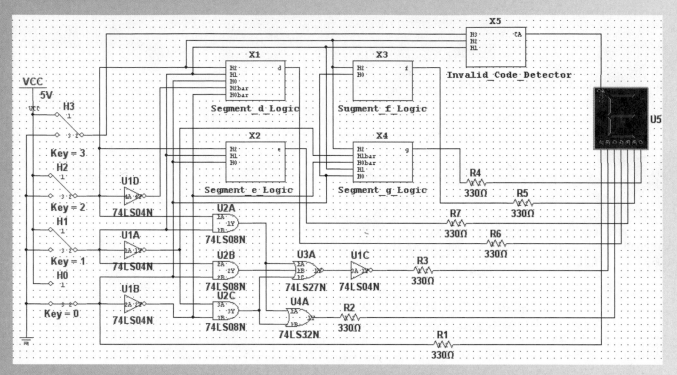

FIGURE 4–54   Multisim circuit screen for decoder and display.

MultiSim

Open file AL04 in the Applied Logic folder on the website. Run the simulation of the decoder and display using your Multisim software. Observe the operation for the specified letters.

Putting Your Knowledge to Work

How would you modify the decoder for a common-cathode 7-segment display?

SUMMARY

- Gate symbols and Boolean expressions for the outputs of an inverter and 2-input gates are shown in Figure 4–55.

FIGURE 4–55

- Commutative laws: $A + B = B + A$
$$AB = BA$$
- Associative laws: $A + (B + C) = (A + B) + C$
$$A(BC) = (AB)C$$
- Distributive law: $A(B + C) = AB + AC$
- Boolean rules:

1. $A + 0 = A$	7. $A \cdot A = A$
2. $A + 1 = 1$	8. $A \cdot \overline{A} = ($
3. $A \cdot 0 = 0$	9. $\overline{\overline{A}} = A$
4. $A \cdot 1 = A$	10. $A + AB = A$
5. $A + A = A$	11. $A + \overline{A}B = A + B$
6. $A + \overline{A} = 1$	12. $(A + B)(A + C) = A + BC$

- DeMorgan's theorems:

  1. The complement of a product is equal to the sum of the complements of the terms in the product.

$$\overline{XY} = \overline{X} + \overline{Y}$$

  2. The complement of a sum is equal to the product of the complements of the terms in the sum.

$$\overline{X + Y} = \overline{X}\,\overline{Y}$$

- Karnaugh maps for 3 variables have 8 cells and for 4 variables have 16 cells.
- Quinn-McCluskey is a method for simplification of Boolean expressions.
- The three levels of abstraction in VHDL are data flow, structural, and behavioral.

KEY TERMS

Key terms and other bold terms in the chapter are defined in the end-of-book glossary.

Complement   The inverse or opposite of a number. In Boolean algebra, the inverse function, expressed with a bar over a variable. The complement of a 1 is 0, and vice versa.

"Don't care"   A combination of input literals that cannot occur and can be used as a 1 or a 0 on a Karnaugh map for simplification.

Karnaugh map   An arrangement of cells representing the combinations of literals in a Boolean expression and used for a systematic simplification of the expression.

Minimization   The process that results in an SOP or POS Boolean expression that contains the fewest possible literals per term.

Product-of-sums (POS)   A form of Boolean expression that is basically the ANDing of ORed terms.

Product term   The Boolean product of two or more literals equivalent to an AND operation.

Sum-of-products (SOP)   A form of Boolean expression that is basically the ORing of ANDed terms.

Sum term   The Boolean sum of two or more literals equivalent to an OR operation.

Variable   A symbol used to represent an action, a condition, or data that can have a value of 1 or 0, usually designated by an italic letter or word.

TRUE/FALSE QUIZ

Answers are at the end of the chapter.

1. *Variable, complement,* and *literal* are all terms used in Boolean algebra.
2. Addition in Boolean algebra is equivalent to the OR function.
3. Multiplication in Boolean algebra is equivalent to the NAND function.
4. The commutative law, associative law, and distributive law are all laws in Boolean algebra.
5. The complement of a 1 is 0.
6. When a Boolean variable is multiplied by its complement, the result is the variable.

7. "The complement of a product of variables is equal to the sum of the complements of each variable" is a statement of DeMorgan's theorem.

8. SOP means series of products.

9. Karnaugh maps can be used to simplify Boolean expressions.

10. A 4-variable Karnaugh map has eight cells.

11. VHDL is a type of hardware definition language.

12. A VHDL program consists of an entity and an architecture.

SELF-TEST

Answers are at the end of the chapter.

1. The complement of a variable is always
   (a) 0   (b) 1
   (c) equal to the variable   (d) the inverse of the variable

2. The Boolean expression $A + \overline{B} + C$ is
   (a) a sum term   (b) a literal term
   (c) a product term   (d) a complemented term

3. The Boolean expression $A\overline{B}C\overline{D}$ is
   (a) a sum term   (b) a product term
   (c) a literal term   (d) always 1

4. The domain of the expression $A\overline{B}CD + A\overline{B} + \overline{C}D + B$ is
   (a) A and D (b) B only
   (c) $A, B, C,$ and D (d) none of these

5. According to the commutative law of addition,
   (a) $AB = BA$ (b) $A = A + A$
   (c) $A + (B + C) = (A + B) + C$ (d) $A + B = B + A$

6. According to the associative law of multiplication,
   (a) $B = BB$ (b) $A(BC) = (AB)C$
   (c) $A + B = B + A$ (d) $B + B(B + 0)$

7. According to the distributive law,
   (a) $A(B + C) = AB + AC$ (b) $A(BC) = ABC$
   (c) $A(A + 1) = A$ (d) $A + AB = A$

8. Which one of the following is *not* a valid rule of Boolean algebra?
   (a) $A + 1 = 1$ (b) $A = \overline{A}$
   (c) $AA = A$ (d) $A + 0 = A$

9. Which of the following rules states that if one input of an AND gate is always 1, the output is equal to the other input?
   (a) $A + 1 = 1$ (b) $A + A = A$
   (c) $A \cdot A = A$ (d) $A \cdot 1 = A$

10. According to DeMorgan's theorems, the following equality(s) is (are) correct:
    (a) $\overline{AB} = \overline{A} + \overline{B}$ (b) $\overline{XYZ} = \overline{X} + \overline{Y} + \overline{Z}$
    (c) $\overline{A + B + C} = \overline{A}\,\overline{B}\,\overline{C}$ (d) all of these

11. The Boolean expression $X = AB + CD$ represents
    (a) two ORs ANDed together   (b) a 4-input AND gate
    (c) two ANDs ORed together   (d) an exclusive-OR

12. An example of a sum-of-products expression is
    (a) $A + B(C + D)$ (b) $\overline{A}B + A\overline{C} + A\overline{B}C$
    (c) $(\overline{A} + B + C)(A + \overline{B} + C)$ (d) both answers (a) and (b)

13. An example of a product-of-sums expression is
    (a) $A(B + C) + A\overline{C}$ (b) $(A + B)(\overline{A} + B + \overline{C})$
    (c) $\overline{A} + \overline{B} + BC$ (d) both answers (a) and (b)

14. An example of a standard SOP expression is
    (a) $\overline{A}B + A\overline{B}C + AB\overline{D}$ (b) $A\overline{B}C + A\overline{C}D$
    (c) $A\overline{B} + \overline{A}B + AB$ (d) $A\overline{B}C\overline{D} + \overline{A}B + \overline{A}$

15. A 3-variable Karnaugh map has
    (a) eight cells   **(b)** three cells
    (c) sixteen cells   **(d)** four cells

16. In a 4-variable Karnaugh map, a 2-variable product term is produced by
    (a) a 2-cell group of 1s   **(b)** an 8-cell group of 1s
    (c) a 4-cell group of 1s   **(d)** a 4-cell group of 0s

17. The Quine-McCluskey method can be used to
    (a) replace the Karnaugh map method   **(b)** simplify expressions with 5 or more variables
    (c) both (a) and (b)   **(d)** none of the above

18. VHDL is a type of
    (a) programmable logic   **(b)** hardware description language
    (c) programmable array   **(d)** logical mathematics

19. In VHDL, a port is
    (a) a type of entity   **(b)** a type of architecture
    (c) an input or output   **(d)** a type of variable

20. Using VDHL, a logic circuit's inputs and outputs are described in the
    (a) architecture   **(b)** component
    (c) entity   **(d)** data flow

PROBLEMS

Answers to odd-numbered problems are at the end of the book.

Section 4–1 Boolean Operations and Expressions

1. Using Boolean notation, write an expression that is a 1 whenever one or more of its variables $(A, B, C,$ and $D)$ are 1s.

2. Write an expression that is a 1 only if all of its variables $(A, B, C, D,$ and $E)$ are 1s.

3. Write an expression that is a 1 when one or more of its variables $(A, B,$ and $C)$ are 0s.

4. Evaluate the following operations:
    (a) $0 + 0 + 1$ **(b)** $1 + 1 + 1$ **(c)** $1 \cdot 0 \cdot 0$
    (d) $1 \cdot 1 \cdot 1$ **(e)** $1 \cdot 0 \cdot 1$ **(f)** $1 \cdot 1 + 0 \cdot 1 \cdot 1$

5. Find the values of the variables that make each product term 1 and each sum term 0.
    (a) AB **(b)** $A\overline{B}C$ **(c)** $A + B$ **(d)** $\overline{A} + B + \overline{C}$
    (e) $\overline{A} + \overline{B} + C$ **(f)** $\overline{A} + B$ **(g)** $A\overline{B}\,\overline{C}$

6. Find the value of X for all possible values of the variables.
    (a) $X = (A + B)C + B$ **(b)** $X = (\overline{A + B})C$ **(c)** $X = A\overline{B}C + AB$
    (d) $X = (A + B)(\overline{A} + B)$ **(e)** $X = (A + BC)(\overline{B} + \overline{C})$

Section 4–2 Laws and Rules of Boolean Algebra

7. Identify the law of Boolean algebra upon which each of the following equalities is based:
    (a) $A\overline{B} + CD + A\overline{C}D + B = B + A\overline{B} + A\overline{C}D + CD$
    (b) $AB\overline{C}D + \overline{AB}C = D\overline{C}BA + \overline{CBA}$
    (c) $AB(CD + E\overline{F} + GH) = ABCD + ABE\overline{F} + ABGH$

8. Identify the Boolean rule(s) on which each of the following equalities is based:
    (a) $\overline{\overline{AB} + \overline{CD}} + \overline{EF} = AB + CD + \overline{EF}$ **(b)** $A\overline{A}B + AB\overline{C} + AB\overline{B} = AB\overline{C}$
    (c) $A(\overline{B}C + BC) + AC = A(BC) + AC$ **(d)** $AB(C + \overline{C}) + AC = AB + AC$
    (e) $A\overline{B} + A\overline{B}C = A\overline{B}$ **(f)** $ABC + \overline{AB} + \overline{ABC}D = ABC + \overline{AB} + D$

Section 4–3 DeMorgan's Theorems

9. Apply DeMorgan's theorems to each expression:
    (a) $\overline{A + \overline{B}}$ **(b)** $\overline{\overline{A}B}$ **(c)** $\overline{A + B + C}$ **(d)** $\overline{ABC}$
    (e) $\overline{A(B + C)}$ **(f)** $\overline{AB + CD}$ **(g)** $\overline{A\overline{B} + C\overline{D}}$ **(h)** $\overline{(A + \overline{B})(\overline{C} + D)}$

10. Apply DeMorgan's theorems to each expression:

    (a) $\overline{\overline{AB}(C + \overline{D})}$ **(b)** $\overline{AB(CD + EF)}$

    (c) $\overline{(A + \overline{B} + C + \overline{D}) + ABC\overline{D}}$ **(d)** $\overline{(\overline{A} + B + C + D)(\overline{AB\,\overline{C}D})}$

    (e) $\overline{\overline{AB}(CD + \overline{EF})(\overline{AB} + CD)}$

11. Apply DeMorgan's theorems to the following:

    (a) $\overline{\overline{(ABC)(EFG)} + \overline{(HIJ)(KLM)}}$ **(b)** $\overline{(A + \overline{BC} + CD) + \overline{BC}}$

    (c) $\overline{\overline{(A + B)(C + D)(E + F)(G + H)}}$

Section 4–4 Boolean Analysis of Logic Circuits

12. Write the Boolean expression for each of the logic gates in Figure 4–56.

 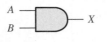 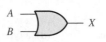 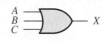

    (a)   (b)   (c)   (d)

FIGURE 4–56

13. Write the Boolean expression for each of the logic circuits in Figure 4–57.

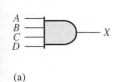 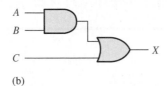 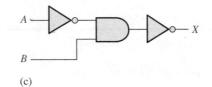 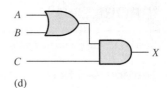

    (a)   (b)   (c)   (d)

FIGURE 4–57

14. Draw the logic circuit represented by each of the following expressions:

    (a) $A + B + C$ **(b)** ABC

    (c) $AB + C$ **(d)** $AB + CD$

15. Draw the logic circuit represented by each expression:

    (a) $A\overline{B} + \overline{A}B$ **(b)** $AB + \overline{A}\,\overline{B} + \overline{A}BC$

    (c) $\overline{A}B(C + \overline{D})$ **(d)** $A + B[C + D(B + \overline{C})]$

16. (a) Draw a logic circuit for the case where the output, ENABLE, is LOW only if the inputs, ASSERT and READY, are both HIGH.

    (b) Draw a logic circuit for the case where the output, HOLD, is LOW only if the input, LOAD, is HIGH and the input, READY, is LOW.

17. Develop the truth table for each of the circuits in Figure 4–58.

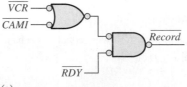 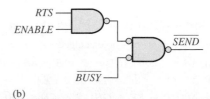

    (a)   (b)

FIGURE 4–58

18. Construct a truth table for each of the following Boolean expressions:

    (a) $A + B$ **(b)** AB **(c)** $AB + BC$

    (d) $(A + B)C$ **(e)** $(A + B)(\overline{B} + C)$

Section 4–5 Logic Simplification Using Boolean Algebra

19. Using Boolean algebra techniques, simplify the following expressions as much as possible:

    (a) $A(A + B)$ **(b)** $A(\overline{A} + AB)$ **(c)** $BC + \overline{B}C$

    (d) $A(A + \overline{A}B)$ **(e)** $A\overline{B}C + \overline{A}BC + \overline{A}\,\overline{B}C$

20. Using Boolean algebra, simplify the following expressions:

   (a) $(A + \overline{B})(A + C)$
   (b) $\overline{A}B + \overline{A}B\overline{C} + \overline{A}BCD + \overline{A}B\overline{C}DE$
   (c) $AB + \overline{A}BC + A$
   (d) $(A + \overline{A})(AB + AB\overline{C})$
   (e) $AB + (\overline{A} + \overline{B})C + AB$

21. Using Boolean algebra, simplify each expression:

   (a) $BD + B(D + E) + \overline{D}(D + F)$
   (b) $\overline{A}\,\overline{B}C + (A + B + \overline{C}) + \overline{A}\,\overline{B}\,\overline{C}D$
   (c) $(B + BC)(B + \overline{B}C)(B + D)$
   (d) $ABCD + AB(\overline{CD}) + (\overline{AB})CD$
   (e) $ABC[AB + \overline{C}(BC + AC)]$

22. Determine which of the logic circuits in Figure 4–59 are equivalent.

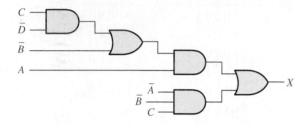

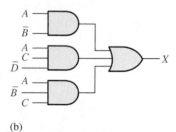

(a)   (b)

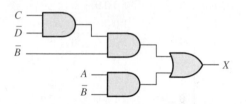

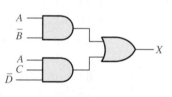

(c)   (d)

FIGURE 4–59

Section 4–6 Standard Forms of Boolean Expressions

23. Convert the following expressions to sum-of-product (SOP) forms:

   (a) $(A + B)(C + \overline{B})$
   (b) $(A + \overline{B}C)C$
   (c) $(A + C)(AB + AC)$

24. Convert the following expressions to sum-of-product (SOP) forms:

   (a) $AB + CD(A\overline{B} + CD)$
   (b) $AB(\overline{B}\,\overline{C} + BD)$
   (c) $A + B[AC + (B + \overline{C})D]$

25. Define the domain of each SOP expression in Problem 23 and convert the expression to standard SOP form.

26. Convert each SOP expression in Problem 24 to standard SOP form.

27. Determine the binary value of each term in the standard SOP expressions from Problem 25.

28. Determine the binary value of each term in the standard SOP expressions from Problem 26.

29. Convert each standard SOP expression in Problem 25 to standard POS form.

30. Convert each standard SOP expression in Problem 26 to standard POS form.

Section 4–7 Boolean Expressions and Truth Tables

31. Develop a truth table for each of the following standard SOP expressions:

   (a) $A\overline{B}C + \overline{A}B\overline{C} + ABC$
   (b) $\overline{X}\,\overline{Y}\overline{Z} + \overline{X}\,\overline{Y}Z + XY\overline{Z} + X\overline{Y}Z + \overline{X}YZ$

32. Develop a truth table for each of the following standard SOP expressions:

   (a) $\overline{A}B\overline{C}D + \overline{A}BC\overline{D} + A\overline{B}\,\overline{C}D + \overline{A}\,\overline{B}C\overline{D}$
   (b) $WXYZ + WX\overline{Y}\overline{Z} + \overline{W}XYZ + W\overline{X}YZ + WX\overline{Y}Z$

33. Develop a truth table for each of the SOP expressions:

   (a) $\overline{A}B + AB\overline{C} + \overline{A}\,\overline{C} + A\overline{B}C$
   (b) $\overline{X} + Y\overline{Z} + WZ + \overline{X}\,\overline{Y}Z$

34. Develop a truth table for each of the standard POS expressions:

(a) $(\overline{A} + \overline{B} + \overline{C})(A + B + C)(A + \overline{B} + C)$

(b) $(\overline{A} + B + \overline{C} + D)(A + \overline{B} + C + \overline{D})(A + \overline{B} + \overline{C} + D)(\overline{A} + B + C + \overline{D})$

35. Develop a truth table for each of the standard POS expressions:

(a) $(A + B)(A + C)(A + B + C)$

(b) $(A + \overline{B})(A + \overline{B} + \overline{C})(B + C + \overline{D})(\overline{A} + B + \overline{C} + D)$

36. For each truth table in Table 4–15, derive a standard SOP and a standard POS expression.

TABLE 4–15

A B C	X		A B C	X		A B C D	X		A B C D	X
						0 0 0 0	1		0 0 0 0	0
						0 0 0 1	1		0 0 0 1	0
						0 0 1 0	0		0 0 1 0	1
						0 0 1 1	1		0 0 1 1	0
						0 1 0 0	0		0 1 0 0	1
						0 1 0 1	1		0 1 0 1	1
						0 1 1 0	1		0 1 1 0	0
0 0 0	0		0 0 0	0		0 1 1 1	0		0 1 1 1	1
0 0 1	1		0 0 1	0		1 0 0 0	0		1 0 0 0	0
0 1 0	0		0 1 0	0		1 0 0 1	1		1 0 0 1	0
0 1 1	0		0 1 1	0		1 0 1 0	0		1 0 1 0	0
1 0 0	1		1 0 0	0		1 0 1 1	0		1 0 1 1	1
1 0 1	1		1 0 1	1		1 1 0 0	1		1 1 0 0	1
1 1 0	0		1 1 0	1		1 1 0 1	0		1 1 0 1	0
1 1 1	1		1 1 1	1		1 1 1 0	0		1 1 1 0	0
						1 1 1 1	0		1 1 1 1	1

   (a)   (b)   (c)   (d)

Section 4–8 The Karnaugh Map

37. Draw a 3-variable Karnaugh map and label each cell according to its binary value.

38. Draw a 4-variable Karnaugh map and label each cell according to its binary value.

39. Write the standard product term for each cell in a 3-variable Karnaugh map.

Section 4–9 Karnaugh Map SOP Minimization

40. Use a Karnaugh map to find the minimum SOP form for each expression:

(a) $\overline{A}\,\overline{B}\,\overline{C} + \overline{A}\,\overline{B}C + A\overline{B}\,\overline{C}$ (b) $AC(\overline{B} + C)$

(c) $\overline{A}(BC + B\overline{C}) + A(BC + B\overline{C})$ (d) $\overline{A}\,\overline{B}\,\overline{C} + A\overline{B}\,\overline{C} + \overline{A}B\overline{C} + AB\overline{C}$

41. Use a Karnaugh map to simplify each expression to a minimum SOP form:

(a) $\overline{A}\,\overline{B}\,\overline{C} + A\overline{B}\,\overline{C} + \overline{A}B\overline{C} + AB\overline{C}$ (b) $AC[\overline{B} + B(B + \overline{C})]$

(c) $DE\overline{F} + \overline{D}E\overline{F} + \overline{D}\,\overline{E}F$

42. Expand each expression to a standard SOP form:

(a) $AB + A\overline{B}C + ABC$ (b) $A + BC$

(c) $\overline{A}B\overline{C}D + AC\overline{D} + B\overline{C}D + \overline{A}BC\overline{D}$ (d) $A\overline{B} + A\overline{B}\,\overline{C}D + CD + B\overline{C}D + ABCD$

43. Minimize each expression in Problem 42 with a Karnaugh map.

44. Use a Karnaugh map to reduce each expression to a minimum SOP form:

(a) $A + B\overline{C} + CD$

(b) $\overline{A}\,\overline{B}\,\overline{C}\,\overline{D} + \overline{A}\,\overline{B}C\overline{D} + ABC\overline{D} + AB\overline{C}\,\overline{D}$

(c) $\overline{A}B(\overline{C}\,\overline{D} + \overline{C}D) + AB(\overline{C}\,\overline{D} + \overline{C}D) + A\overline{B}\,\overline{C}D$

(d) $(\overline{A}\,\overline{B} + A\overline{B})(CD + C\overline{D})$

(e) $\overline{A}\,\overline{B} + A\overline{B} + \overline{C}\,\overline{D} + C\overline{D}$

45. Reduce the function specified in truth Table 4–16 to its minimum SOP form by using a Karnaugh map.

46. Use the Karnaugh map method to implement the minimum SOP expression for the logic function specified in truth Table 4–17.

47. Solve Problem 46 for a situation in which the last six binary combinations are not allowed.

TABLE 4–16

Inputs	Output
A B C	X
0 0 0	1
0 0 1	1
0 1 0	0
0 1 1	1
1 0 0	1
1 0 1	1
1 1 0	0
1 1 1	1

TABLE 4–17

Inputs	Output
A B C D	X
0 0 0 0	0
0 0 0 1	1
0 0 1 0	1
0 0 1 1	0
0 1 0 0	0
0 1 0 1	0
0 1 1 0	1
0 1 1 1	1
1 0 0 0	1
1 0 0 1	0
1 0 1 0	1
1 0 1 1	0
1 1 0 0	1
1 1 0 1	1
1 1 1 0	0
1 1 1 1	1

Section 4–10 Karnaugh Map POS Minimization

48. Use a Karnaugh map to find the minimum POS for each expression:

(a) $(A + B + C)(\overline{A} + \overline{B} + \overline{C})(A + \overline{B} + C)$

(b) $(X + \overline{Y})(\overline{X} + Z)(X + \overline{Y} + \overline{Z})(\overline{X} + \overline{Y} + Z)$

(c) $A(B + \overline{C})(\overline{A} + C)(A + \overline{B} + C)(\overline{A} + B + \overline{C})$

49. Use a Karnaugh map to simplify each expression to minimum POS form:

(a) $(A + \overline{B} + C + \overline{D})(\overline{A} + B + \overline{C} + D)(\overline{A} + \overline{B} + \overline{C} + \overline{D})$

(b) $(X + \overline{Y})(W + \overline{Z})(\overline{X} + \overline{Y} + \overline{Z})(W + X + Y + Z)$

50. For the function specified in Table 4–16, determine the minimum POS expression using a Karnaugh map.

51. Determine the minimum POS expression for the function in Table 4–17.

52. Convert each of the following POS expressions to minimum SOP expressions using a Karnaugh map:

(a) $(A + \overline{B})(A + \overline{C})(\overline{A} + \overline{B} + C)$

(b) $(\overline{A} + B)(\overline{A} + \overline{B} + \overline{C})(B + \overline{C} + D)(A + \overline{B} + C + \overline{D})$

Section 4–11 The Quine-McCluskey Method

53. List the minterms in the expression

$$X = ABC + \overline{A}\,\overline{B}C + AB\overline{C} + A\overline{B}C + \overline{A}BC$$

54. List the minterms in the expression

$$X = \overline{A}\,\overline{B}\,\overline{C}\,\overline{D} + \overline{A}\,\overline{B}\,CD + \overline{A}B\overline{C}D + AB\overline{C}\,\overline{D} + A\overline{B}C\overline{D} + \overline{A}BCD + A\overline{B}\,\overline{C}D$$

55. Create a table for the number of 1s in the minterms for the expression in Problem 54 (similar to Table 4–10).

56. Create a table of first level minterms for the expression in Problem 54 (similar to Table 4–11).

57. Create a table of second level minterms for the expression in Problem 54 (similar to Table 4–12).

58. Create a table of prime implicants for the expression in Problem 54 (similar to Table 4–13).

59. Determine the final reduced expression for the expression in Problem 54.

Section 4–12 Boolean Expressions with VHDL

60. Write a VHDL program for the logic circuit in Figure 4–60.

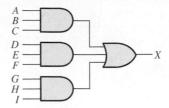

FIGURE 4–60

61. Write a program in VHDL for the expression

$$Y = A\overline{B}C + \overline{A}\,\overline{B}C + A\overline{B}\,\overline{C} + \overline{A}BC$$

Applied Logic

62. If you are required to choose a type of digital display for low light conditions, will you select LED or LCD 7-segment displays? Why?

63. Explain the purpose of the invalid code detector.

64. For segment c, how many fewer gates and inverters does it take to implement the minimum SOP expression than the standard SOP expression?

65. Repeat Problem 64 for the logic for segments d through g.

Special Design Problems

66. The logic for segments b and c in Figure 4–53 produces LOW outputs to activate the segments. If a type of 7-segment display is used that requires a HIGH to activate a segment, modify the logic accordingly.

67. Redesign the logic for segment a in the Applied Logic to include the letter F in the display.

68. Repeat Problem 67 for segments b through g.

69. Design the invalid code detector.

MultiSim Multisim Troubleshooting Practice

70. Open file P04-70. For the specified fault, predict the effect on the circuit. Then introduce the fault and verify whether your prediction is correct.

71. Open file P04-71. For the specified fault, predict the effect on the circuit. Then introduce the fault and verify whether your prediction is correct.

72. Open file P04-72. For the observed behavior indicated, predict the fault in the circuit. Then introduce the suspected fault and verify whether your prediction is correct.

ANSWERS

SECTION CHECKUPS

Section 4–1 Boolean Operations and Expressions

1. $\overline{A} = \overline{0} = 1$

2. $A = 1, B = 1, C = 0; \overline{A} + \overline{B} + C = \overline{1} + \overline{1} + 0 = 0 + 0 + 0 = 0$

3. $A = 1, B = 0, C = 1; A\overline{B}C = 1 \cdot \overline{0} \cdot 1 = 1 \cdot 1 \cdot 1 =$

Section 4–2 Laws and Rules of Boolean Algebra

1. $A + (B + C + D) = (A + B + C) + D$

2. $A(B + C + D) = AB + AC + AD$

Section 4–3 DeMorgan's Theorems

1. (a) $\overline{ABC} + \overline{(D + E)} = \overline{A} + \overline{B} + \overline{C} + D\overline{E}$ (b) $\overline{(A + B)C} = \overline{A}\,\overline{B} + \overline{C}$

   (c) $\overline{A + B + C} + \overline{DE} = \overline{A}\,\overline{B}\,\overline{C} + D + \overline{E}$

Section 4–4 Boolean Analysis of Logic Circuits

1. $(C + D)B + A$

2. Abbreviated truth table: The expression is a 1 when A is 1 or when B and C are 1s or when B and D are 1s. The expression is 0 for all other variable combinations.

Section 4–5 Logic Simplification Using Boolean Algebra

1. (a) $A + AB + A\overline{B}C = A$ (b) $(\overline{A} + B)C + ABC = C(\overline{A} + B)$

   (c) $A\overline{B}C(BD + CDE) + A\overline{C} = A(\overline{C} + \overline{B}DE)$

2. (a) *Original:* 2 AND gates, 1 OR gate, 1 inverter; *Simplified:* No gates (straight connection)

   (b) *Original:* 2 OR gates, 2 AND gates, 1 inverter; *Simplified:* 1 OR gate, 1 AND gate, 1 inverter

   (c) *Original:* 5 AND gates, 2 OR gates, 2 inverters; *Simplified:* 2 AND gates, 1 OR gate, 2 inverters

Section 4–6 Standard Forms of Boolean Expressions

1. (a) SOP   (b) standard POS   (c) standard SOP   (d) POS

2. (a) $AB\overline{C}\,\overline{D} + AB\overline{C}D + ABC\overline{D} + ABCD + \overline{A}B\overline{C}D + \overline{A}BCD + \overline{A}\,\overline{B}C\overline{D} + \overline{A}BC\overline{D}$

   (c) Already standard

3. (b) Already standard

   (d) $(A + \overline{B} + \overline{C})(A + \overline{B} + C)(A + B + \overline{C})(A + B + C)$

Section 4–7 Boolean Expressions and Truth Tables

1. $2^5 = 32$ 2. $0110 \longrightarrow \overline{W}XY\overline{Z}$ 3. $1100 \longrightarrow \overline{W} + \overline{X} + Y + Z$

Section 4–8 The Karnaugh Map

1. (a) upper left cell: 000   (b) lower right cell: 101

   (c) lower left cell: 100   (d) upper right cell: 001

2. (a) upper left cell: $\overline{X}\,\overline{Y}\,\overline{Z}$ (b) lower right cell: $X\overline{Y}Z$

   (c) lower left cell: $X\overline{Y}\,\overline{Z}$ (d) upper right cell: $\overline{X}\,\overline{Y}Z$

3. (a) upper left cell: 0000   (b) lower right cell: 1010

   (c) lower left cell: 1000   (d) upper right cell: 0010

4. (a) upper left cell: $\overline{W}\,\overline{X}\,\overline{Y}\,\overline{Z}$ (b) lower right cell: $W\overline{X}Y\overline{Z}$

   (c) lower left cell: $W\overline{X}\,\overline{Y}\,\overline{Z}$ (d) upper right cell: $\overline{W}\,\overline{X}Y\overline{Z}$

Section 4–9 Karnaugh Map SOP Minimization

1. 8-cell map for 3 variables; 16-cell map for 4 variables

2. $AB + B\overline{C} + \overline{A}BC$

3. (a) $\overline{A}\,\overline{B}\,\overline{C} + \overline{A}BC + ABC + AB\overline{C}$

   (b) $\overline{A}\,\overline{B}\,\overline{C} + \overline{A}\,\overline{B}C + \overline{A}B\overline{C} + \overline{A}BC + A\overline{B}\,\overline{C} + A\overline{B}C$

   (c) $\overline{A}\,\overline{B}\,\overline{C}\,\overline{D} + \overline{A}\,\overline{B}\,\overline{C}D + \overline{A}B\overline{C}\,\overline{D} + \overline{A}BCD + \overline{A}BCD + \overline{A}BC\overline{D} + A\overline{B}\,\overline{C}D + A\overline{B}CD$

   (d) $\overline{A}\,\overline{B}\,\overline{C}\,\overline{D} + \overline{A}\,\overline{B}C\overline{D} + \overline{A}B\overline{C}\,\overline{D} + \overline{A}B\,\overline{C}\,\overline{D} + \overline{A}BC\overline{D} + \overline{A}BCD + A\overline{B}\,\overline{C}D + \overline{A}\,\overline{B}C\overline{D} +$
$\overline{A}BC\overline{D} + ABC\overline{D} + A\overline{B}C\overline{D}$

Section 4–10 Karnaugh Map POS Minimization

1. In mapping a POS expression, 0s are placed in cells whose value makes the standard sum term zero; and in mapping an SOP expression 1s are placed in cells having the same values as the product terms.

2. 0 in the 1011 cell: $\overline{A} + B + \overline{C} + \overline{D}$

3. 1 in the 0010 cell: $\overline{A}\,\overline{B}C\overline{D}$

Section 4–11 The Quine-McCluskey Method

1. A minterm is a product term in which each variable appears once, either complemented or uncomplemented.

2. An essential prime implicant is a product term that cannot be further simplified by combining with other terms.

Section 4–12 Boolean Expressions with VHDL

1. Simplification can make a VHDL program shorter, easier to read, and easier to modify.

2. Code simplification results in less space used in a target device, thus allowing capacity for more complex circuits.

3. Truth table: Behavioral
   Boolean expression: Data flow
   Logic diagram: Structural

RELATED PROBLEMS FOR EXAMPLES

4–1 $\overline{A} + B = 0$ when $A = 1$ and $B = 0$.

4–2 $\overline{A}\,\overline{B} = 1$ when $A = 0$ and $B = 0$.

4–3 XYZ

4–4 $W + X + Y + Z$

4–5 $ABC\overline{D}\,\overline{E}$

4–6 $(A + \overline{B} + \overline{C}D)\overline{E}$

4–7 $\overline{ABCD} = \overline{A} + \overline{B} + \overline{C} + \overline{D}$

4–8 Results should be same as example.

4–9 $A\overline{B}$

4–10 CD

4–11 $AB\overline{C} + \overline{A}C + \overline{A}B$

4–12 $\overline{A} + \overline{B} + \overline{C}$

4–13 Results should be same as example.

4–14 $\overline{A}B\overline{C} + AB + A\overline{C} + A\overline{B} + \overline{B}C$

4–15 $W\overline{X}YZ + W\overline{X}Y\overline{Z} + WX\overline{Y}\overline{Z} + \overline{W}\,\overline{X}Y\overline{Z} + WX\overline{Y}Z + WX\overline{Y}\,\overline{Z}$

4–16 011, 101, 110, 010, 111. Yes

4–17 $(A + \overline{B} + C)(A + \overline{B} + \overline{C})(A + B + C)(\overline{A} + B + C)$

4–18 010, 100, 001, 111, 011. Yes

4–19 SOP and POS expressions are equivalent.

4–20 See Table 4–18.

4–21 See Table 4–19.

TABLE 4–18

A	B	C	X
0	0	0	0
0	0	1	0
0	1	0	1
0	1	1	0
1	0	0	0
1	0	1	1
1	1	0	0
1	1	1	0

TABLE 4–19

A	B	C	X
0	0	0	1
0	0	1	0
0	1	0	0
0	1	1	1
1	0	0	1
1	0	1	1
1	1	0	1
1	1	1	0

4–22 The SOP and POS expressions are equivalent.
4–23 See Figure 4–61.
4–24 See Figure 4–62.

AB \ C	0	1
00		
01		1
11		
10	1	1

FIGURE 4–61

AB \ CD	00	01	11	10
00				
01				1
11	1		1	1
10				

FIGURE 4–62

4–25 See Figure 4–63.
4–26 See Figure 4–64.

AB \ C	0	1
00	1	
01	1	1
11		1
10		

FIGURE 4–63

AB \ CD	00	01	11	10
00		1		
01		1		1
11	1	1	1	1
10	1	1	1	1

FIGURE 4–64

4–27 No other ways
4–28 $X = B + \overline{A}C + A\overline{C}D + C\overline{D}$
4–29 $X = \overline{D} + A\overline{B}C + B\overline{C} + \overline{A}B$
4–30 $Q = X + Y$
4–31 $Q = \overline{X}\overline{Y}\overline{Z} + W\overline{X}Z + \overline{W}YZ$
4–32 See Figure 4–65.
4–33 See Figure 4–66.

FIGURE 4–65

AB \ CD	00	01	11	10
00	0	0		
01				0
11				
10				0

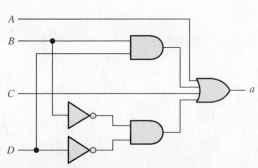

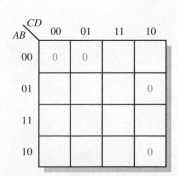

FIGURE 4–66

4–34 $(X + \overline{Y})(X + \overline{Z})(\overline{X} + Y + Z)$

4–35 $(\overline{X} + \overline{Y} + Z)(\overline{W} + \overline{X} + Z)(W + X + Y + Z)(W + \overline{X} + Y + \overline{Z})$

4–36 $\overline{Y}\overline{Z} + \overline{X}\overline{Z} + \overline{W}Y + \overline{X}\overline{Y}Z$

4–37 **architecture** RelProb_1 **of** Example4_37 **is**
    begin
       X $<=$ (**not** A **or** B **or** C) **and** D;
    end architecture RelProb_1;

    architecture RelProb_2 **of** Example4_37 **is**
    begin
       X $<=$ (**not** A **and** D **or** B **and** D **or** C **and** D);
    end architecture RelProb_2;

4–38 **architecture** RelProb **of** Example4_38 **is**
    begin
       X $<=$ **not**(A **and** ((B **and** C) **or not** D))
    end architecture RelProb;

TRUE/FALSE QUIZ

1. T   **2.** T   **3.** F   **4.** T   **5.** T   **6.** F

7. T   **8.** F   **9.** T   **10.** F   **11.** F   **12.** T

SELF-TEST

1. (d)   **2.** (a)   **3.** (b)   **4.** (c)   **5.** (d)   **6.** (b)   **7.** (a)

8. (b)   **9.** (d)   **10.** (d)   **11.** (c)   **12.** (b)   **13.** (b)   **14.** (c)

15. (a)   **16.** (c)   **17.** (c)   **18.** (b)   **19.** (c)   **20.** (c)

Combinational Logic Analysis

CHAPTER OUTLINE

CHAPTER OBJECTIVES

- Analyze basic combinational logic circuits, such as AND-OR, AND-OR-Invert, exclusive-OR, and exclusive-NOR
- Use AND-OR and AND-OR-Invert circuits to implement sum-of-products (SOP) and product-of-sums (POS) expressions
- Write the Boolean output expression for any combinational logic circuit
- Develop a truth table from the output expression for a combinational logic circuit
- Use the Karnaugh map to expand an output expression containing terms with missing variables into a full SOP form
- Design a combinational logic circuit for a given Boolean output expression
- Design a combinational logic circuit for a given truth table
- Simplify a combinational logic circuit to its minimum form
- Use NAND gates to implement any combinational logic function

- Use NOR gates to implement any combinational logic function
- Analyze the operation of logic circuits with pulse inputs
- Write VHDL programs for simple logic circuits
- Troubleshoot faulty logic circuits
- Troubleshoot logic circuits by using signal tracing and waveform analysis
- Apply combinational logic to an application

KEY TERMS

Key terms are in order of appearance in the chapter.

- Universal gate
- Negative-OR
- Negative-AND
- Component
- Signal
- Node
- Signal tracing

VISIT THE WEBSITE

Study aids for this chapter are available at
http://www.pearsonhighered.com/careersresources/

INTRODUCTION

In Chapters 3 and 4, logic gates were discussed on an individual basis and in simple combinations. You were introduced to SOP and POS implementations, which are basic forms of combinational logic. When logic gates are connected together to produce a specified output for certain specified combinations of input variables, with no storage involved, the resulting circuit is in the category of **combinational logic**. In combinational logic, the output level is at all times dependent on the combination of input levels. This chapter expands on the material introduced in earlier chapters with a coverage of the analysis, design, and troubleshooting of various combinational logic circuits. The VHDL structural approach is introduced and applied to combinational logic.

5–1　Basic Combinational Logic Circuits

In Chapter 4, you learned that SOP expressions are implemented with an AND gate for each product term and one OR gate for summing all of the product terms. As you know, this SOP implementation is called AND-OR logic and is the basic form for realizing standard Boolean functions. In this section, the AND-OR and the AND-OR-Invert are examined; the exclusive-OR and exclusive-NOR gates, which are actually a form of AND-OR logic, are also covered.

After completing this section, you should be able to

- ◆ Analyze and apply AND-OR circuits
- ◆ Analyze and apply AND-OR-Invert circuits
- ◆ Analyze and apply exclusive-OR gates
- ◆ Analyze and apply exclusive-NOR gates

AND-OR Logic

AND-OR logic produces an SOP expression.

Figure 5–1(a) shows an AND-OR circuit consisting of two 2-input AND gates and one 2-input OR gate; Figure 5–1(b) is the ANSI standard rectangular outline symbol. The Boolean expressions for the AND gate outputs and the resulting SOP expression for the output X are shown on the diagram. In general, an AND-OR circuit can have any number of AND gates, each with any number of inputs.

The truth table for a 4-input AND-OR logic circuit is shown in Table 5–1. The intermediate AND gate outputs (the AB and CD columns) are also shown in the table.

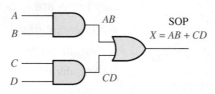

(a) Logic diagram (ANSI standard distinctive shape symbols)

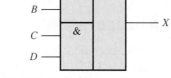

(b) ANSI standard rectangular outline symbol

 FIGURE 5–1　An example of AND-OR logic. Open file F05-01 to verify the operation. *A Multisim tutorial is available on the website.*

TABLE 5–1

Truth table for the AND-OR logic in Figure 5–1.

Inputs				AB	CD	Output
A	B	C	D			X
0	0	0	0	0	0	0
0	0	0	1	0	0	0
0	0	1	0	0	0	0
0	0	1	1	0	1	1
0	1	0	0	0	0	0
0	1	0	1	0	0	0
0	1	1	0	0	0	0
0	1	1	1	0	1	1
1	0	0	0	0	0	0
1	0	0	1	0	0	0
1	0	1	0	0	0	0
1	0	1	1	0	1	1
1	1	0	0	1	0	1
1	1	0	1	1	0	1
1	1	1	0	1	0	1
1	1	1	1	1	1	1

An AND-OR circuit directly implements an SOP expression, assuming the complements (if any) of the variables are available. The operation of the AND-OR circuit in Figure 5–1 is stated as follows:

For a 4-input AND-OR logic circuit, the output X is HIGH (1) if both input A and input B are HIGH (1) or both input C and input D are HIGH (1).

EXAMPLE 5–1

In a certain chemical-processing plant, a liquid chemical is used in a manufacturing process. The chemical is stored in three different tanks. A level sensor in each tank produces a HIGH voltage when the level of chemical in the tank drops below a specified point.

Design a circuit that monitors the chemical level in each tank and indicates when the level in any two of the tanks drops below the specified point.

Solution

The AND-OR circuit in Figure 5–2 has inputs from the sensors on tanks A, B, and C as shown. The AND gate G_1 checks the levels in tanks A and B, gate G_2 checks tanks A and C, and gate G_3 checks tanks B and C. When the chemical level in any two of the tanks gets too low, one of the AND gates will have HIGHs on both of its inputs, causing its output to be HIGH; and so the final output X from the OR gate is HIGH. This HIGH input is then used to activate an indicator such as a lamp or audible alarm, as shown in the figure.

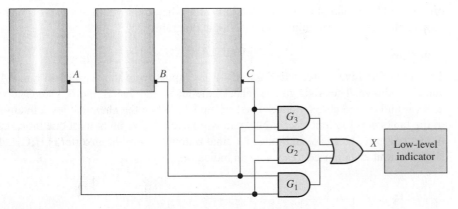

FIGURE 5–2

Related Problem*

Write the Boolean SOP expression for the AND-OR logic in Figure 5–2.

*Answers are at the end of the chapter.

AND-OR-Invert Logic

When the output of an AND-OR circuit is complemented (inverted), it results in an AND-OR-Invert circuit. Recall that AND-OR logic directly implements SOP expressions. POS expressions can be implemented with AND-OR-Invert logic. This is illustrated as follows, starting with a POS expression and developing the corresponding AND-OR-Invert (AOI) expression.

$$X = (\overline{A} + \overline{B})(\overline{C} + \overline{D}) = (\overline{AB})(\overline{CD}) = \overline{\overline{(AB)}\,\overline{(CD)}} = \overline{\overline{\overline{AB}} + \overline{\overline{CD}}} = \overline{AB + CD}$$

The logic diagram in Figure 5–3(a) shows an AND-OR-Invert circuit with four inputs and the development of the POS output expression. The ANSI standard rectangular outline symbol is shown in part (b). In general, an AND-OR-Invert circuit can have any number of AND gates, each with any number of inputs.

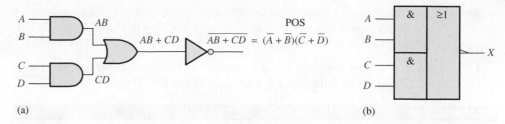

(a)   (b)

MultiSim **FIGURE 5–3**   An AND-OR-Invert circuit produces a POS output. Open file F05-03 to verify the operation.

The operation of the AND-OR-Invert circuit in Figure 5–3 is stated as follows:

For a 4-input AND-OR-Invert logic circuit, the output X is LOW (0) if both input A and input B are HIGH (1) or both input C and input D are HIGH (1).

A truth table can be developed from the AND-OR truth table in Table 5–1 by simply changing all 1s to 0s and all 0s to 1s in the output column.

EXAMPLE 5–2

The sensors in the chemical tanks of Example 5–1 are being replaced by a new model that produces a LOW voltage instead of a HIGH voltage when the level of the chemical in the tank drops below a critical point.

Modify the circuit in Figure 5–2 to operate with the different input levels and still produce a HIGH output to activate the indicator when the level in any two of the tanks drops below the critical point. Show the logic diagram.

Solution

The AND-OR-Invert circuit in Figure 5–4 has inputs from the sensors on tanks A, B, and C as shown. The AND gate G_1 checks the levels in tanks A and B, gate G_2 checks tanks A and C, and gate G_3 checks tanks B and C. When the chemical level in any two of the tanks gets too low, each AND gate will have a LOW on at least one input, causing its output to be LOW and, thus, the final output X from the inverter is HIGH. This HIGH output is then used to activate an indicator.

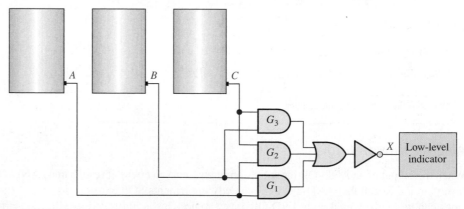

FIGURE 5–4

Related Problem

Write the Boolean expression for the AND-OR-Invert logic in Figure 5–4 and show that the output is HIGH (1) when any two of the inputs A, B, and C are LOW (0).

Exclusive-OR Logic

The exclusive-OR gate was introduced in Chapter 3. Although this circuit is considered a type of logic gate with its own unique symbol, it is actually a combination of two AND gates, one OR gate, and two inverters, as shown in Figure 5–5(a). The two ANSI standard exclusive-OR logic symbols are shown in parts (b) and (c).

The XOR gate is actually a combination of other gates.

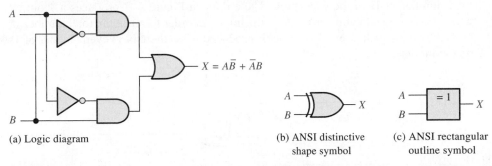

(a) Logic diagram
$X = A\overline{B} + \overline{A}B$

(b) ANSI distinctive shape symbol

(c) ANSI rectangular outline symbol

FIGURE 5–5   Exclusive-OR logic diagram and symbols. Open file F05-05 to verify the operation.

MultiSim

The output expression for the circuit in Figure 5–5 is

$$X = A\overline{B} + \overline{A}B$$

Evaluation of this expression results in the truth table in Table 5–2. Notice that the output is HIGH only when the two inputs are at opposite levels. A special exclusive-OR operator $\oplus$ is often used, so the expression $X = A\overline{B} + \overline{A}B$ can be stated as "X is equal to A exclusive-OR B" and can be written as

$$X = A \oplus B$$

Exclusive-NOR Logic

As you know, the complement of the exclusive-OR function is the exclusive-NOR, which is derived as follows:

$$X = \overline{A\overline{B} + \overline{A}B} = (\overline{A\overline{B}})\,(\overline{\overline{A}B}) = (\overline{A} + B)(A + \overline{B}) = \overline{A}\,\overline{B} + AB$$

Notice that the output X is HIGH only when the two inputs, A and B, are at the same level.

The exclusive-NOR can be implemented by simply inverting the output of an exclusive-OR, as shown in Figure 5–6(a), or by directly implementing the expression $\overline{A}\,\overline{B} + AB$, as shown in part (b).

TABLE 5–2		
Truth table for an exclusive-OR.		
A	*B*	*X*
0	0	0
0	1	1
1	0	1
1	1	0

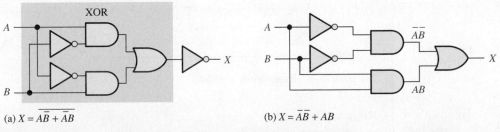

(a) $X = \overline{A\overline{B} + \overline{A}B}$

(b) $X = \overline{A}\,\overline{B} + AB$

FIGURE 5–6   Two equivalent ways of implementing the exclusive-NOR. Open files F05-06 (a) and (b) to verify the operation.

MultiSim

EXAMPLE 5–3

Use exclusive-OR gates to implement an even-parity code generator for an original 4-bit code.

Solution

Recall from Chapter 2 that a parity bit is added to a binary code in order to provide error detection. For even parity, a parity bit is added to the original code to make the total number of 1s in the code even. The circuit in Figure 5–7 produces a 1 output when there is an odd number of 1s on the inputs in order to make the total number of 1s in the output code even. A 0 output is produced when there is an even number of 1s on the inputs.

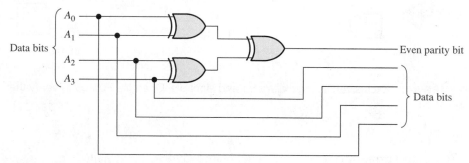

FIGURE 5–7   Even-parity generator.

Related Problem

How would you verify that a correct even-parity bit is generated for each combination of the four data bits?

EXAMPLE 5–4

Use exlusive-OR gates to implement an even-parity checker for the 5-bit code generated by the circuit in Example 5–3.

Solution

The circuit in Figure 5–8 produces a 1 output when there is an error in the five-bit code and a 0 when there is no error.

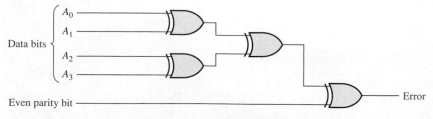

FIGURE 5–8   Even-parity checker.

Related Problem

How would you verify that an error is indicated when the input code is incorrect?

SECTION 5–1 CHECKUP

Answers are at the end of the chapter.

1. Determine the output (1 or 0) of a 4-variable AND-OR-Invert circuit for each of the following input conditions:

   (a) $A = 1, B = 0, C = 1, D = 0$ (b) $A = 1, B = 1, C = 0, D = 1$

   (c) $A = 0, B = 1, C = 1, D = 1$

2. Determine the output (1 or 0) of an exclusive-OR gate for each of the following input conditions:

   (a) $A = 1, B = 0$ (b) $A = 1, B = 1$

   (c) $A = 0, B = 1$ (d) $A = 0, B = 0$

3. Develop the truth table for a certain 3-input logic circuit with the output expression $X = \overline{AB}C + \overline{A}BC + \overline{A}\,\overline{B}\,\overline{C} + AB\overline{C} + ABC.$

4. Draw the logic diagram for an exclusive-NOR circuit.

5–2   Implementing Combinational Logic

In this section, examples are used to illustrate how to implement a logic circuit from a Boolean expression or a truth table. Minimization of a logic circuit using the methods covered in Chapter 4 is also included.

After completing this section, you should be able to

- Implement a logic circuit from a Boolean expression
- Implement a logic circuit from a truth table
- Minimize a logic circuit

> For every Boolean expression there is a logic circuit, and for every logic circuit there is a Boolean expression.

From a Boolean Expression to a Logic Circuit

Let's examine the following Boolean expression:

$$X = AB + CDE$$

A brief inspection shows that this expression is composed of two terms, AB and CDE, with a domain of five variables. The first term is formed by ANDing A with B, and the second term is formed by ANDing C, D, and E. The two terms are then ORed to form the output X. These operations are indicated in the structure of the expression as follows:

$$X = \underset{\uparrow}{AB} + \underset{}{CDE} \quad \text{AND}$$
$$\text{OR}$$

Note that in this particular expression, the AND operations forming the two individual terms, AB and CDE, must be performed *before* the terms can be ORed.

To implement this Boolean expression, a 2-input AND gate is required to form the term AB, and a 3-input AND gate is needed to form the term CDE. A 2-input OR gate is then required to combine the two AND terms. The resulting logic circuit is shown in Figure 5–9.

As another example, let's implement the following expression:

$$X = AB(\overline{CD} + EF)$$

InfoNote

Many control programs require logic operations to be performed by a computer. A driver program is a control program that is used with computer peripherals. For example, a mouse driver requires logic tests to determine if a button has been pressed and further logic operations to determine if it has moved, either horizontally or vertically. Within the heart of a microprocessor is the arithmetic logic unit (ALU), which performs these logic operations as directed by program instructions. All of the logic described in this chapter can also be performed by the ALU, given the proper instructions.

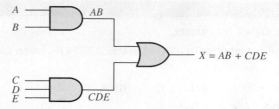

FIGURE 5–9 Logic circuit for $X = AB + CDE$.

A breakdown of this expression shows that the terms AB and $(C\overline{D} + EF)$ are ANDed. The term $C\overline{D} + EF$ is formed by first ANDing C and $\overline{D}$ and ANDing E and F, and then ORing these two terms. This structure is indicated in relation to the expression as follows:

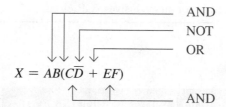

Before you can implement the final expression, you must create the sum term $C\overline{D} + EF$; but before you can get this term; you must create the product terms $C\overline{D}$ and EF; but before you can get the term $C\overline{D}$, you must create $\overline{D}$. So, as you can see, the logic operations must be done in the proper order.

The logic gates required to implement $X = AB(C\overline{D} + EF)$ are as follows:

1. One inverter to form $\overline{D}$

2. Two 2-input AND gates to form $C\overline{D}$ and EF

3. One 2-input OR gate to form $C\overline{D} + EF$

4. One 3-input AND gate to form X

The logic circuit for this expression is shown in Figure 5–10(a). Notice that there is a maximum of four gates and an inverter between an input and output in this circuit (from input D to output). Often the total propagation delay time through a logic circuit is a major consideration. Propagation delays are additive, so the more gates or inverters between input and output, the greater the propagation delay time.

Unless an intermediate term, such as $C\overline{D} + EF$ in Figure 5–10(a), is required as an output for some other purpose, it is usually best to reduce a circuit to its SOP form in order to reduce the overall propagation delay time. The expression is converted to SOP as follows, and the resulting circuit is shown in Figure 5–10(b).

$$AB(C\overline{D} + EF) = ABC\overline{D} + ABEF$$

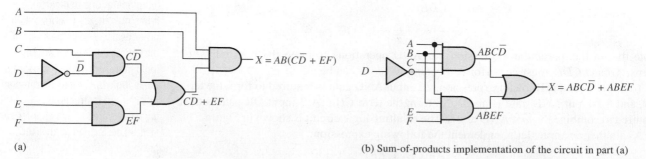

(a)

(b) Sum-of-products implementation of the circuit in part (a)

FIGURE 5–10 Logic circuits for $X = AB(C\overline{D} + EF) = ABC\overline{D} + ABEF$.

TABLE 5–3

Inputs			Output	
A	*B*	*C*	*X*	**Product Term**
0	0	0	0	
0	0	1	0	
0	1	0	0	
0	1	1	1	$\overline{A}BC$
1	0	0	1	$A\overline{B}\,\overline{C}$
1	0	1	0	
1	1	0	0	
1	1	1	0	

From a Truth Table to a Logic Circuit

If you begin with a truth table instead of an expression, you can write the SOP expression from the truth table and then implement the logic circuit. Table 5–3 specifies a logic function.

The Boolean SOP expression obtained from the truth table by ORing the product terms for which $X = 1$ is

$$X = \overline{A}BC + A\overline{B}\,\overline{C}$$

The first term in the expression is formed by ANDing the three variables $\overline{A}$, B, and C. The second term is formed by ANDing the three variables A, $\overline{B}$, and $\overline{C}$.

The logic gates required to implement this expression are as follows: three inverters to form the $\overline{A}$, $\overline{B}$, and $\overline{C}$ variables; two 3-input AND gates to form the terms $\overline{A}BC$ and $A\overline{B}\,\overline{C}$; and one 2-input OR gate to form the final output function, $\overline{A}BC + A\overline{B}\,\overline{C}$.

The implementation of this logic function is illustrated in Figure 5–11.

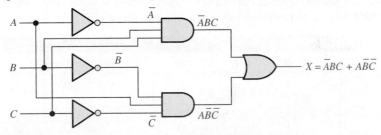

FIGURE 5–11   Logic circuit for $X = \overline{A}BC + A\overline{B}\,\overline{C}$. Open file F05-11 to verify the operation.

MultiSim

EXAMPLE 5–5

Design a logic circuit to implement the operation specified in the truth table of Table 5–4.

TABLE 5–4

Inputs			Output	
A	*B*	*C*	*X*	**Product Term**
0	0	0	0	
0	0	1	0	
0	1	0	0	
0	1	1	1	$\overline{A}BC$
1	0	0	0	
1	0	1	1	$A\overline{B}C$
1	1	0	1	$AB\overline{C}$
1	1	1	0	

Solution

Notice that $X = 1$ for only three of the input conditions. Therefore, the logic expression is

$$X = \overline{A}BC + A\overline{B}C + AB\overline{C}$$

The logic gates required are three inverters, three 3-input AND gates and one 3-input OR gate. The logic circuit is shown in Figure 5–12.

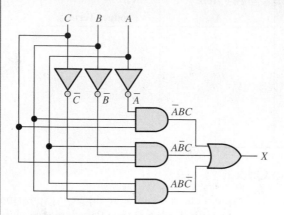

FIGURE 5–12   Open file F05-12 to verify the operation.

MultiSim

Related Problem
Determine if the logic circuit of Figure 5–12 can be simplified.

EXAMPLE 5–6

Develop a logic circuit with four input variables that will only produce a 1 output when exactly three input variables are 1s.

Solution

Out of sixteen possible combinations of four variables, the combinations in which there are exactly three 1s are listed in Table 5–5, along with the corresponding product term for each.

TABLE 5–5

A	B	C	D	Product Term
0	1	1	1	$\overline{A}BCD$
1	0	1	1	$A\overline{B}CD$
1	1	0	1	$AB\overline{C}D$
1	1	1	0	$ABC\overline{D}$

The product terms are ORed to get the following expression:

$$X = \overline{A}BCD + A\overline{B}CD + AB\overline{C}D + ABC\overline{D}$$

This expression is implemented in Figure 5–13 with AND-OR logic.

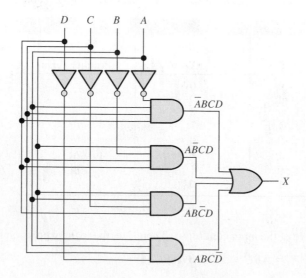

FIGURE 5–13   Open file F05-13 to verify the operation.

MultiSim

Related Problem

Determine if the logic circuit of Figure 5–13 can be simplified.

EXAMPLE 5–7

Reduce the combinational logic circuit in Figure 5–14 to a minimum form.

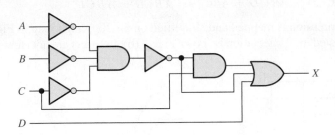

FIGURE 5–14

Open file F05-14 to verify that this circuit is equivalent to the gate in Figure 5–15.

MultiSim

Solution

The expression for the output of the circuit is

$$X = (\overline{\overline{A}\,\overline{B}\,\overline{C}})C + \overline{\overline{A}\,\overline{B}\,\overline{C}} + D$$

Applying DeMorgan's theorem and Boolean algebra,

$$X = (\overline{\overline{A}} + \overline{\overline{B}} + \overline{\overline{C}})C + \overline{\overline{A}} + \overline{\overline{B}} + \overline{\overline{C}} + D$$
$$= AC + BC + CC + A + B + C + D$$
$$= AC + BC + C + A + B + \cancel{C} + D$$
$$= C(A + B + 1) + A + B + D$$
$$X = A + B + C + D$$

The simplified circuit is a 4-input OR gate as shown in Figure 5–15.

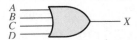

FIGURE 5–15

Related Problem

Verify the minimized expression $A + B + C + D$ using a Karnaugh map.

EXAMPLE 5–8

Minimize the combinational logic circuit in Figure 5–16. Inverters for the complemented variables are not shown.

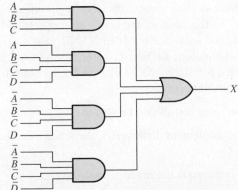

FIGURE 5–16

Solution

The output expression is

$$X = A\overline{B}\,\overline{C} + AB\overline{C}\,\overline{D} + \overline{A}\,\overline{B}CD + \overline{A}\,\overline{B}\,\overline{C}\,\overline{D}$$

Expanding the first term to include the missing variables D and $\overline{D}$,

$$X = A\overline{B}\,\overline{C}(D + \overline{D}) + AB\overline{C}\,\overline{D} + \overline{A}\,\overline{B}CD + \overline{A}\,\overline{B}\,\overline{C}\,\overline{D}$$
$$= A\overline{B}\,\overline{C}D + A\overline{B}\,\overline{C}\,\overline{D} + AB\overline{C}\,\overline{D} + \overline{A}\,\overline{B}CD + \overline{A}\,\overline{B}\,\overline{C}\,\overline{D}$$

This expanded SOP expression is mapped and simplified on the Karnaugh map in Figure 5–17(a). The simplified implementation is shown in part (b). Inverters are not shown.

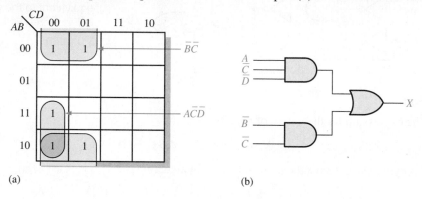

(a)   (b)

FIGURE 5–17

Related Problem

Develop the POS equivalent of the circuit in Figure 5–17(b). See Section 4–10.

SECTION 5–2 CHECKUP

1. Implement the following Boolean expressions as they are stated:

   (a) $X = ABC + AB + AC$ (b) $X = AB(C + DE)$

2. Develop a logic circuit that will produce a 1 on its output only when all three inputs are 1s or when all three inputs are 0s.

3. Reduce the circuits in Question 1 to minimum SOP form.

5–3 The Universal Property of NAND and NOR Gates

Up to this point, you have studied combinational circuits implemented with AND gates, OR gates, and inverters. In this section, the universal property of the NAND gate and the NOR gate is discussed. The universality of the NAND gate means that it can be used as an inverter and that combinations of NAND gates can be used to implement the AND, OR, and NOR operations. Similarly, the NOR gate can be used to implement the inverter (NOT), AND, OR, and NAND operations.

After completing this section, you should be able to

♦ Use NAND gates to implement the inverter, the AND gate, the OR gate, and the NOR gate

♦ Use NOR gates to implement the inverter, the AND gate, the OR gate, and the NAND gate

The NAND Gate as a Universal Logic Element

The NAND gate is a **universal gate** because it can be used to produce the NOT, the AND, the OR, and the NOR functions. An inverter can be made from a NAND gate by connecting all of the inputs together and creating, in effect, a single input, as shown in Figure 5–18(a) for a 2-input gate. An AND function can be generated by the use of NAND gates alone, as shown in Figure 5–18(b). An OR function can be produced with only NAND gates, as illustrated in part (c). Finally, a NOR function is produced as shown in part (d).

Combinations of NAND gates can be used to produce any logic function.

(a) One NAND gate used as an inverter

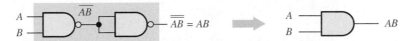

(b) Two NAND gates used as an AND gate

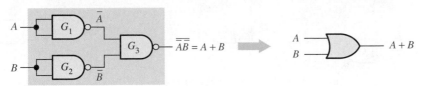

(c) Three NAND gates used as an OR gate

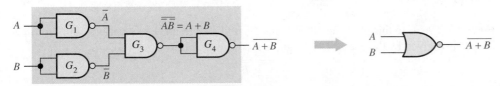

(d) Four NAND gates used as a NOR gate

FIGURE 5–18 Universal application of NAND gates. Open files F05-18(a), (b), (c), and (d) to verify each of the equivalencies.

 MultiSim

In Figure 5–18(b), a NAND gate is used to invert (complement) a NAND output to form the AND function, as indicated in the following equation:

$$X = \overline{\overline{AB}} = AB$$

In Figure 5–18(c), NAND gates G_1 and G_2 are used to invert the two input variables before they are applied to NAND gate G_3. The final OR output is derived as follows by application of DeMorgan's theorem:

$$X = \overline{\overline{A}\,\overline{B}} = A + B$$

In Figure 5–18(d), NAND gate G_4 is used as an inverter connected to the circuit of part (c) to produce the NOR operation $\overline{A + B}$.

The NOR Gate as a Universal Logic Element

Like the NAND gate, the NOR gate can be used to produce the NOT, AND, OR, and NAND functions. A NOT circuit, or inverter, can be made from a NOR gate by connecting all of the inputs together to effectively create a single input, as shown in Figure 5–19(a) with a 2-input example. Also, an OR gate can be produced from NOR gates, as illustrated in Figure 5–19(b). An AND gate can be constructed by the use of NOR gates, as shown in

Combinations of NOR gates can be used to produce any logic function.

(a) One NOR gate used as an inverter

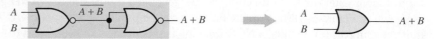

(b) Two NOR gates used as an OR gate

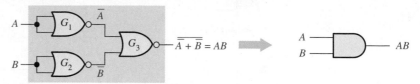

(c) Three NOR gates used as an AND gate

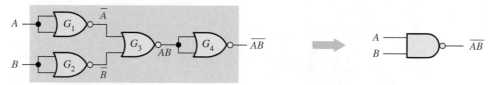

(d) Four NOR gates used as a NAND gate

 FIGURE 5–19 Universal application of NOR gates. Open files F05-19(a), (b), (c), and (d) to verify each of the equivalencies.

Figure 5–19(c). In this case the NOR gates G_1 and G_2 are used as inverters, and the final output is derived by the use of DeMorgan's theorem as follows:

$$X = \overline{\overline{A} + \overline{B}} = AB$$

Figure 5–19(d) shows how NOR gates are used to form a NAND function.

SECTION 5–3 CHECKUP

1. Use NAND gates to implement each expression:
   (a) $X = \overline{A} + B$ (b) $X = A\overline{B}$
2. Use NOR gates to implement each expression:
   (a) $X = \overline{A} + B$ (b) $X = A\overline{B}$

5–4 Combinational Logic Using NAND and NOR Gates

In this section, you will see how NAND and NOR gates can be used to implement a logic function. Recall from Chapter 3 that the NAND gate also exhibits an equivalent operation called the negative-OR and that the NOR gate exhibits an equivalent operation called the negative-AND. You will see how the use of the appropriate symbols to represent the equivalent operations makes "reading" a logic diagram easier.

After completing this section, you should be able to

* Use NAND gates to implement a logic function
* Use NOR gates to implement a logic function
* Use the appropriate dual symbol in a logic diagram

NAND Logic

As you have learned, a NAND gate can function as either a NAND or a negative-OR because, by DeMorgan's theorem,

$$\overline{AB} = \overline{A} + \overline{B}$$

NAND ⎯⎯⎯⎯⎯⎯⎯⎯⎯↑   ↑⎯⎯⎯⎯⎯⎯⎯ negative-OR

Consider the NAND logic in Figure 5–20. The output expression is developed in the following steps:

$$X = \overline{(\overline{AB})(\overline{CD})}$$
$$= \overline{(\overline{A} + \overline{B})(\overline{C} + \overline{D})}$$
$$= \overline{(\overline{A} + \overline{B})} + \overline{(\overline{C} + \overline{D})}$$
$$= \overline{\overline{A}}\,\overline{\overline{B}} + \overline{\overline{C}}\,\overline{\overline{D}}$$
$$= AB + CD$$

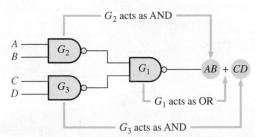

FIGURE 5–20   NAND logic for $X = AB + CD$.

As you can see in Figure 5–20, the output expression, $AB + CD$, is in the form of two AND terms ORed together. This shows that gates G_2 and G_3 act as AND gates and that gate G_1 acts as an OR gate, as illustrated in Figure 5–21(a). This circuit is redrawn in part (b) with NAND symbols for gates G_2 and G_3 and a negative-OR symbol for gate G_1.

Notice in Figure 5–21(b) the bubble-to-bubble connections between the outputs of gates G_2 and G_3 and the inputs of gate G_1. *Since a bubble represents an inversion, two*

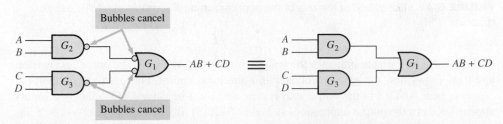

(a) Original NAND logic diagram showing effective
gate operation relative to the output expression

(b) Equivalent NAND/Negative-OR logic diagram   (c) AND-OR equivalent

FIGURE 5–21   Development of the AND-OR equivalent of the circuit in Figure 5–20.

connected bubbles represent a double inversion and therefore cancel each other. This inversion cancellation can be seen in the previous development of the output expression $AB + CD$ and is indicated by the absence of barred terms in the output expression. Thus, the circuit in Figure 5–21(b) is *effectively* an AND-OR circuit, as shown in Figure 5–21(c).

NAND Logic Diagrams Using Dual Symbols

All logic diagrams using NAND gates should be drawn with each gate represented by either a NAND symbol or the equivalent negative-OR symbol to reflect the operation of the gate within the logic circuit. The NAND symbol and the **negative-OR** symbol are called *dual symbols.* When drawing a NAND logic diagram, always use the gate symbols in such a way that every connection between a gate output and a gate input is either bubble-to-bubble or nonbubble-to-nonbubble. In general, a bubble output should not be connected to a nonbubble input or vice versa in a logic diagram.

Figure 5–22 shows an arrangement of gates to illustrate the procedure of using the appropriate dual symbols for a NAND circuit with several gate levels. Although using all NAND symbols as in Figure 5–22(a) is correct, the diagram in part (b) is much easier to "read" and is the preferred method. As shown in Figure 5–22(b), the output gate is represented with a negative-OR symbol. Then the NAND symbol is used for the level of gates right before the output gate and the symbols for successive levels of gates are alternated as you move away from the output.

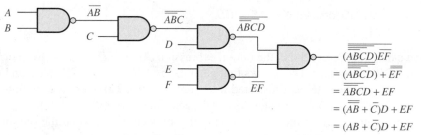

(a) Several Boolean steps are required to arrive at final output expression.

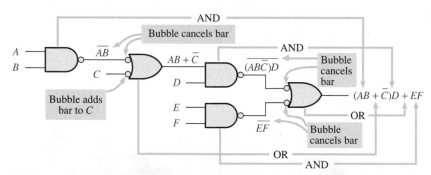

(b) Output expression can be obtained directly from the function of each gate symbol in the diagram.

FIGURE 5–22 Illustration of the use of the appropriate dual symbols in a NAND logic diagram.

The shape of the gate indicates the way its inputs will appear in the output expression and thus shows how the gate functions within the logic circuit. For a NAND symbol, the inputs appear ANDed in the output expression; and for a negative-OR symbol, the inputs appear ORed in the output expression, as Figure 5–22(b) illustrates. The dual-symbol diagram in part (b) makes it easier to determine the output expression directly from the logic diagram because each gate symbol indicates the relationship of its input variables as they appear in the output expression.

EXAMPLE 5–9

Redraw the logic diagram and develop the output expression for the circuit in Figure 5–23 using the appropriate dual symbols.

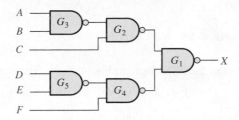

FIGURE 5–23

Solution

Redraw the logic diagram in Figure 5–23 with the use of equivalent negative-OR symbols as shown in Figure 5–24. Writing the expression for X directly from the indicated logic operation of each gate gives $X = (\overline{A} + \overline{B})C + (\overline{D} + \overline{E})F$.

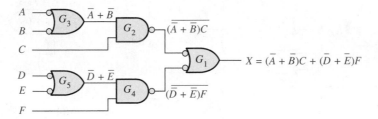

FIGURE 5–24

Related Problem

Derive the output expression from Figure 5–23 and show it is equivalent to the expression in the solution.

EXAMPLE 5–10

Implement each expression with NAND logic using appropriate dual symbols:

(a) $ABC + DE$ (b) $ABC + \overline{D} + \overline{E}$

Solution

See Figure 5–25.

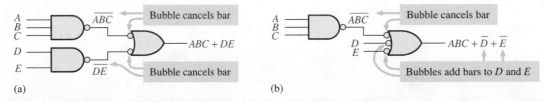

FIGURE 5–25

Related Problem

Convert the NAND circuits in Figure 5–25(a) and (b) to equivalent AND-OR logic.

NOR Logic

A NOR gate can function as either a NOR or a **negative-AND**, as shown by DeMorgan's theorem.

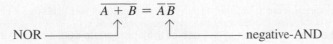

$$\overline{A + B} = \overline{A}\,\overline{B}$$

NOR ⎯⎯⎯⎯⎯⎯⎯↑   ↑⎯⎯⎯⎯⎯⎯ negative-AND

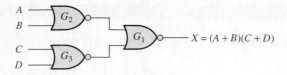

FIGURE 5–26   NOR logic for $X = (A + B)(C + D)$.

Consider the NOR logic in Figure 5–26. The output expression is developed as follows:

$$X = \overline{\overline{A + B} + \overline{C + D}} = (\overline{\overline{A + B}})(\overline{\overline{C + D}}) = (A + B)C + D)$$

As you can see in Figure 5–26, the output expression $(A + B)(C + D)$ consists of two OR terms ANDed together. This shows that gates G_2 and G_3 act as OR gates and gate G_1 acts as an AND gate, as illustrated in Figure 5–27(a). This circuit is redrawn in part (b) with a negative-AND symbol for gate G_1.

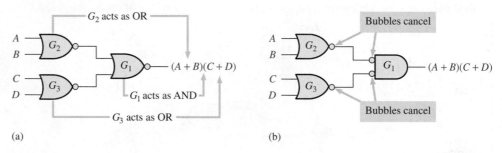

(a)   (b)

FIGURE 5–27

NOR Logic Diagram Using Dual Symbols

As with NAND logic, the purpose for using the dual symbols is to make the logic diagram easier to read and analyze, as illustrated in the NOR logic circuit in Figure 5–28. When the circuit in part (a) is redrawn with dual symbols in part (b), notice that all output-to-input

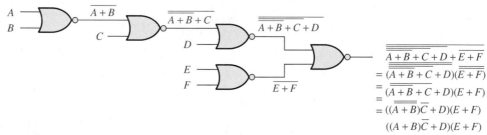

(a) Final output expression is obtained after several Boolean steps.

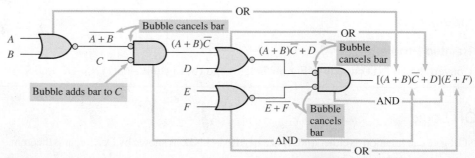

(b) Output expression can be obtained directly from the function of each gate symbol in the diagram.

FIGURE 5–28   Illustration of the use of the appropriate dual symbols in a NOR logic diagram.

connections between gates are bubble-to-bubble or nonbubble-to-nonbubble. Again, you can see that the shape of each gate symbol indicates the type of term (AND or OR) that it produces in the output expression, thus making the output expression easier to determine and the logic diagram easier to analyze.

EXAMPLE 5–11

Using appropriate dual symbols, redraw the logic diagram and develop the output expression for the circuit in Figure 5–29.

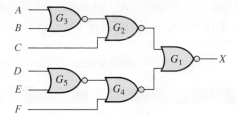

FIGURE 5–29

Solution

Redraw the logic diagram with the equivalent negative-AND symbols as shown in Figure 5–30. Writing the expression for X directly from the indicated operation of each gate,

$$X = (\overline{A}\overline{B} + C)(\overline{D}\overline{E} + F)$$

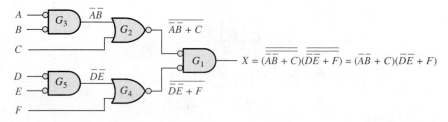

FIGURE 5–30

Related Problem

Prove that the output of the NOR circuit in Figure 5–29 is the same as for the circuit in Figure 5–30.

SECTION 5–4 CHECKUP

1. Implement the expression $X = \overline{(\overline{A} + \overline{B} + \overline{C})DE}$ by using NAND logic.
2. Implement the expression $X = \overline{\overline{A}\overline{B}\overline{C} + (D + E)}$ with NOR logic.

5–5 Pulse Waveform Operation

General combinational logic circuits with pulse waveform inputs are examined in this section. Keep in mind that the operation of each gate is the same for pulse waveform inputs as for constant-level inputs. The output of a logic circuit at any given time depends on the inputs at that particular time, so the relationship of the time-varying inputs is of primary importance.

After completing this section, you should be able to

* Analyze combinational logic circuits with pulse waveform inputs

* Develop a timing diagram for any given combinational logic circuit with specified inputs

The operation of any gate is the same regardless of whether its inputs are pulsed or constant levels. The nature of the inputs (pulsed or constant levels) does not alter the truth table of a circuit. The examples in this section illustrate the analysis of combinational logic circuits with pulse waveform inputs.

The following is a review of the operation of individual gates for use in analyzing combinational circuits with pulse waveform inputs:

1. The output of an AND gate is HIGH only when all inputs are HIGH at the same time.

2. The output of an OR gate is HIGH only when at least one of its inputs is HIGH.

3. The output of a NAND gate is LOW only when all inputs are HIGH at the same time.

4. The output of a NOR gate is LOW only when at least one of its inputs is HIGH.

EXAMPLE 5–12

Determine the final output waveform X for the circuit in Figure 5–31, with input waveforms A, B, and C as shown.

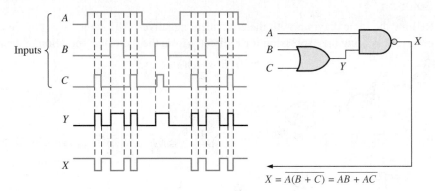

$$X = \overline{A(B + C)} = \overline{AB} + \overline{AC}$$

FIGURE 5–31

Solution

The output expression, $\overline{AB} + \overline{AC}$, indicates that the output X is LOW when both A and B are HIGH or when both A and C are HIGH or when all inputs are HIGH. The output waveform X is shown in the timing diagram of Figure 5–31. The intermediate waveform Y at the output of the OR gate is also shown.

Related Problem

Determine the output waveform if input A is a constant HIGH level.

EXAMPLE 5–13

Draw the timing diagram for the circuit in Figure 5–32 showing the outputs of G_1, G_2, and G_3 with the input waveforms, A, and B, as indicated.

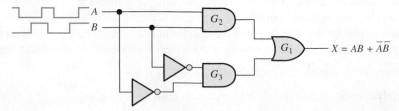

$$X = AB + \overline{A}\overline{B}$$

FIGURE 5–32

Solution

When both inputs are HIGH or when both inputs are LOW, the output X is HIGH as shown in Figure 5–33. Notice that this is an exclusive-NOR circuit. The intermediate outputs of gates G_2 and G_3 are also shown in Figure 5–33.

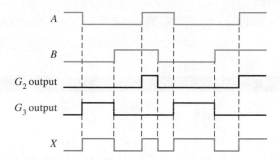

FIGURE 5–33

Related Problem

Determine the output X in Figure 5–32 if input B is inverted.

EXAMPLE 5–14

Determine the output waveform X for the logic circuit in Figure 5–34(a) by first finding the intermediate waveform at each of points Y_1, Y_2, Y_3, and Y_4. The input waveforms are shown in Figure 5–34(b).

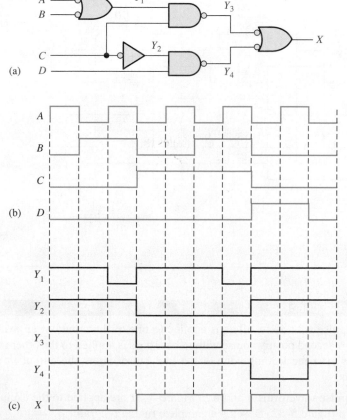

FIGURE 5–34

Solution

All the intermediate waveforms and the final output waveform are shown in the timing diagram of Figure 5–34(c).

Related Problem

Determine the waveforms Y_1, Y_2, Y_3, Y_4 and X if input waveform A is inverted.

EXAMPLE 5–15

Determine the output waveform X for the circuit in Example 5–14, Figure 5–34(a), directly from the output expression.

Solution

The output expression for the circuit is developed in Figure 5–35. The SOP form indicates that the output is HIGH when A is LOW and C is HIGH or when B is LOW and C is HIGH or when C is LOW and D is HIGH.

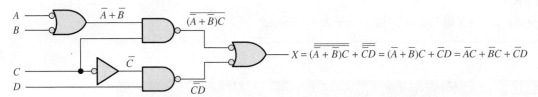

FIGURE 5–35

The result is shown in Figure 5–36 and is the same as the one obtained by the intermediate-waveform method in Example 5–14. The corresponding product terms for each waveform condition that results in a HIGH output are indicated.

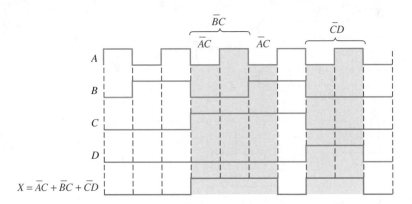

FIGURE 5–36

Related Problem

Repeat this example if all the input waveforms are inverted.

SECTION 5–5 CHECKUP

1. One pulse with $t_W = 50\ \mu s$ is applied to one of the inputs of an exclusive-OR circuit. A second positive pulse with $t_W = 10\ \mu s$ is applied to the other input beginning 15 μs after the leading edge of the first pulse. Show the output in relation to the inputs.

2. The pulse waveforms A and B in Figure 5–31 are applied to the exclusive-NOR circuit in Figure 5–32. Develop a complete timing diagram.

5–6   Combinational Logic with VHDL

The purpose of describing logic using VHDL is so that it can be programmed into a PLD. The data flow approach to writing a VHDL program was described in Chapter 4. In this section, both the data flow approach using Boolean expressions and the structural approach are used to develop VHDL code for describing logic circuits. The VHDL component is introduced and used to illustrate structural descriptions. Some aspects of software development tools are discussed.

After completing this section, you should be able to

◆ Describe a VHDL component and discuss how it is used in a program

◆ Apply the structural approach and the data flow approach to writing VHDL code

◆ Describe two basic software development tools

Structural Approach to VHDL Programming

The structural approach to writing a VHDL description of a logic function can be compared to installing IC devices on a circuit board and interconnecting them with wires. With the structural approach, you describe logic functions and specify how they are connected together. The VHDL **component** is a way to predefine a logic function for repeated use in a program or in other programs. The component can be used to describe anything from a simple logic gate to a complex logic function. The VHDL **signal** can be thought of as a way to specify a "wire" connection between components.

Figure 5–37 provides a simplified comparison of the structural approach to a hardware implementation on a circuit board.

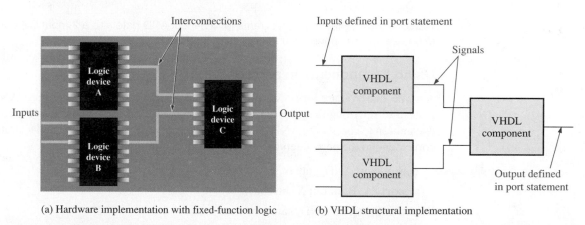

(a) Hardware implementation with fixed-function logic   (b) VHDL structural implementation

FIGURE 5–37   Simplified comparison of the VHDL structural approach to a hardware implementation. The VHDL signals correspond to the interconnections on the circuit board, and the VHDL components correspond to the 74 series IC devices.

VHDL Components

A VHDL component describes predefined logic that can be stored as a package declaration in a VHDL library and called as many times as necessary in a program. You can use components to avoid repeating the same code over and over within a program. For example, you can create a VHDL component for an AND gate and then use it as many times as you wish without having to write a program for an AND gate every time you need one.

VHDL components are stored and are available for use when you write a program. This is similar to having, for example, a storage bin of ICs available when you are constructing a circuit. Every time you need to use one in your circuit, you reach into the storage bin and place it on the circuit board.

The VHDL program for any logic function can become a component and used whenever necessary in a larger program with the use of a component declaration of the following general form. **Component** is a VHDL keyword.

> **component** name_of_component **is**
>
> > **port** (port definitions);
>
> **end component** name_of_component;

For simplicity, let's assume that there are predefined VHDL descriptions of a 2-input AND gate with the entity name AND_gate and a 2-input OR gate with the entity name OR_gate, as shown in Figure 5–38.

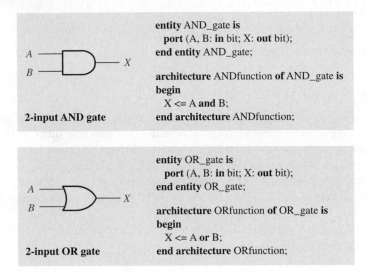

2-input AND gate

```
entity AND_gate is
    port (A, B: in bit; X: out bit);
end entity AND_gate;

architecture ANDfunction of AND_gate is
begin
    X <= A and B;
end architecture ANDfunction;
```

2-input OR gate

```
entity OR_gate is
    port (A, B: in bit; X: out bit);
end entity OR_gate;

architecture ORfunction of OR_gate is
begin
    X <= A or B;
end architecture ORfunction;
```

FIGURE 5–38   Predefined programs for a 2-input AND gate and a 2-input OR gate to be used as components in the structural approach.

Using Components in a Program

Assume that you are writing a program for a logic circuit that has several AND gates. Instead of rewriting the program in Figure 5–38 over and over, you can use a component declaration to specify the AND gate. The port statement in the component declaration must correspond to the port statement in the entity declaration of the AND gate.

> **component** AND_gate **is**
>
> > **port** (A, B: **in** bit; X: **out** bit);
>
> **end component** AND_gate;

To use a component in a program, you must write a component instantiation statement for each instance in which the component is used. You can think of a component instantiation as a request or call for the component to be used in the main program. For example, the simple SOP logic circuit in Figure 5–39 has two AND gates and one OR gate. Therefore, the VHDL program for this circuit will have two components and three component instantiations or calls.

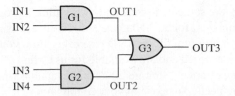

FIGURE 5–39

Signals

In VHDL, signals are analogous to wires that interconnect components on a circuit board. The signals in Figure 5–39 are named OUT1 and OUT2. Signals are the *internal* connections in the logic circuit and are treated differently than the inputs and outputs. Whereas the inputs and outputs are declared in the entity declaration using the port statement, the signals are declared within the architecture using the signal statement. **Signal** is a VHDL keyword.

The Program

The program for the logic in Figure 5–39 begins with an entity declaration as follows:

    entity AND_OR_Logic **is**
        port (IN1, IN2, IN3, IN4: **in** bit; OUT3: **out** bit);
    end entity AND_OR_Logic;

The architecture declaration contains the component declarations for the AND gate and the OR gate, the signal definitions, and the component instantiations.

architecture LogicOperation **of** AND_OR_Logic **is**

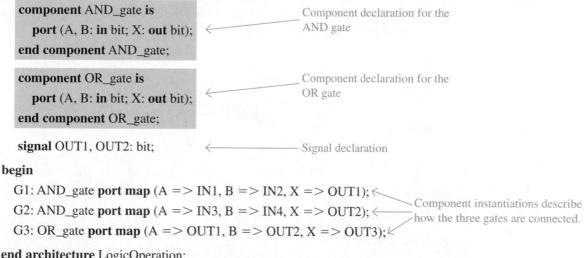

    component AND_gate **is**
        port (A, B: **in** bit; X: **out** bit);   ⟵ Component declaration for the AND gate
    end component AND_gate;

    component OR_gate **is**
        port (A, B: **in** bit; X: **out** bit);   ⟵ Component declaration for the OR gate
    end component OR_gate;

    signal OUT1, OUT2: bit;   ⟵ Signal declaration

begin
    G1: AND_gate **port map** (A => IN1, B => IN2, X => OUT1);
    G2: AND_gate **port map** (A => IN3, B => IN4, X => OUT2);   Component instantiations describe how the three gates are connected.
    G3: OR_gate **port map** (A => OUT1, B => OUT2, X => OUT3);

end architecture LogicOperation;

Component Instantiations

Let's look at the component instantiations. First, notice that the component instantiations appear between the keyword **begin** and the **end architecture** statement. For each instantiation an identifier is defined, such as G1, G2, and G3 in this case. Then the component name is specified. The keyword **port map** essentially makes all the connections for the logic function using the operator =>. For example, the first instantiation,

    G1: AND_gate **port map** (A => IN1, B => IN2, X => OUT1);

can be explained as follows: *Input A of AND gate G1 is connected to input IN1, input B of the gate is connected to input IN2, and the output X of the gate is connected to the signal OUT1.*

The three instantiation statements together completely describe the logic circuit in Figure 5–39, as illustrated in Figure 5–40.

Although the data flow approach using Boolean expressions would have been easier and probably the best way to describe this particular circuit, we have used this simple circuit to explain the concept of the structural approach. Example 5–16 compares the structural and data flow approaches to writing a VHDL program for an SOP logic circuit.

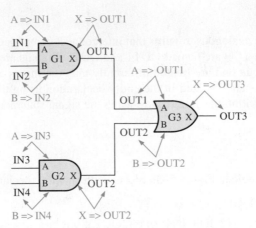

FIGURE 5–40   Illustration of the instantiation statements and port mapping applied to the AND-OR logic. Signals are shown in red.

EXAMPLE 5–16

Write a VHDL program for the SOP logic circuit in Figure 5–41 using the structural approach and compare with the data flow approach. Assume that VHDL components for a 3-input NAND gate and for a 2-input NAND are available. Notice the NAND gate G4 is shown as a negative-OR.

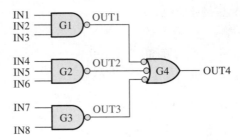

FIGURE 5–41

Solution

The structural approach:

The components and component instantiations are highlighted. Lines preceded by two hyphens are comment lines and are not part of the program.

--Program for the logic circuit in Figure 5–41

entity SOP_Logic **is**
   port (IN1, IN2, IN3, IN4, IN5, IN6, IN7, IN8: **in** bit; OUT4: **out** bit);
end entity SOP_Logic;

architecture LogicOperation **of** SOP_Logic **is**

--component declaration for 3-input NAND gate

component NAND_gate3 **is**
   port (A, B, C: **in** bit X: **out** bit);
end component NAND_gate3;

--component declaration for 2-input NAND gate

component NAND_gate2 **is**
   port (A, B: **in** bit; X: **out** bit);
end component NAND_gate2;

signal OUT1, OUT2, OUT3: bit;

begin

G1: NAND_gate3 **port map**   (A => IN1, B => IN2, C => IN3, X => OUT1);
G2: NAND_gate3 **port map**   (A => IN4, B => IN5, C => IN6, X => OUT2);
G3: NAND_gate2 **port map**   (A => IN7, B => IN8, X => OUT3);
G4: NAND_gate3 **port map**   (A => OUT1, B => OUT2, C => OUT3, X => OUT4);

end architecture LogicOperation;

The data flow approach:

The program for the logic circuit in Figure 5–41 using the data flow approach is written as follows:

entity SOP_Logic **is**
   port (IN1, IN2, IN3, IN4, IN5, IN6, IN7, IN8: **in** bit; OUT4: **out** bit);
end entity SOP_Logic;

architecture LogicOperation **of** SOP_Logic **is**
begin
   OUT4 <= (IN1 **and** IN2 **and** IN3) **or** (IN4 **and** IN5 **and** IN6) **or** (IN7 **and** IN8);
end architecture LogicOperation;

As you can see, the data flow approach results in a much simpler code for this particular logic function. However, in situations where a logic function consists of many blocks of complex logic, the structural approach might have an advantage over the data flow approach.

Related Problem

If another NAND gate is added to the circuit in Figure 5–41 with inputs IN9 and IN10, write a component instantiation to add to the program.

Applying Software Development Tools

A software development package must be used to implement an HDL design in a target device. Once the logic has been described using an HDL and entered via a software tool called a code or text editor, it can be tested using a simulation to verify that it performs properly before actually programming the target device. Using software development tools allows for the design, development, and testing of combinational logic before it is committed to hardware.

Typical software development tools allow you to input VHDL code on a text-based editor specific to the particular development tool that you are using. The VHDL code for a combinational logic circuit has been written using a text-based editor for illustration and appears on the computer screen as shown in Figure 5–42. Many code editors provide enhanced features such as the highlighting of keywords.

After the program has been written into the text editor, it is passed to the compiler. The compiler takes the high-level VHDL code and converts it into a file that can be downloaded to the target device. Once the program has been compiled, you can create a simulation for testing. Simulated input values are inserted into the logic design and allow for verification of the output(s).

You specify the input waveforms on a software tool called a waveform editor, as shown in Figure 5–43. The output waveforms are generated by a simulation of the VHDL code that you entered on the text editor in Figure 5–42. The waveform simulation provides the resulting outputs X and Y for the inputs A, B, C, and D in all sixteen combinations from $0\,0\,0\,0_2$ to $1\,1\,1\,1_2$.

Recall from Chapter 3 that there are several performance characteristics of logic circuits to be considered in the creation of any digital system. Propagation delay, for example, determines the speed or frequency at which a logic circuit can operate. A timing simulation can be used to mimic the propagation delay through the logic design in the target device.

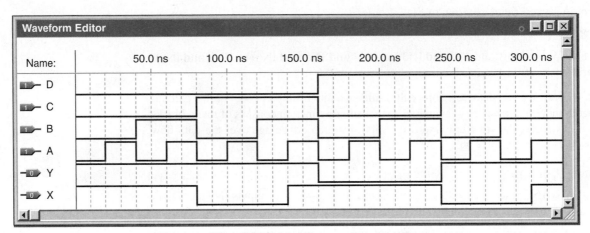

```
Text Editor                                              _ □ ×

File   Edit   View   Project   Assignments   Processing   Tools   Window

    entity Combinational is
      port ( A, B, C, D: in bit; X, Y: out bit );
    end entity Combinational;

    architecture Example of Combinational is
    begin
      X <= ( A and B ) or not C;
      Y <= C or not D;
    end architecture Example;
```

FIGURE 5–42　A VHDL program for a combinational logic circuit after entry on a generic text editor screen that is part of a software development tool.

FIGURE 5–43　A typical waveform editor tool showing the simulated waveforms for the logic circuit described by the VHDL code in Figure 5–42.

SECTION 5–6　CHECKUP

1. What is a VHDL component?

2. State the purpose of a component instantiation in a program architecture.

3. How are interconnections made between components in VHDL?

4. The use of components in a VHDL program represents what approach?

5–7　Troubleshooting

The preceding sections have given you some insight into the operation of combinational logic circuits and the relationships of inputs and outputs. This type of understanding is essential when you troubleshoot digital circuits because you must know what logic levels or waveforms to look for throughout the circuit for a given set of input conditions.

In this section, an oscilloscope is used to troubleshoot a fixed-function logic circuit when a device output is connected to several device inputs. Also, an example of signal tracing and waveform analysis methods is presented using a scope or logic analyzer for locating a fault in a combinational logic circuit.

After completing this section, you should be able to

- ◆ Define a circuit node
- ◆ Use an oscilloscope to find a faulty circuit node
- ◆ Use an oscilloscope to find an open input or output
- ◆ Use an oscilloscope to find a shorted input or output
- ◆ Discuss how to use an oscilloscope or a logic analyzer for signal tracing in a combinational logic circuit

In a combinational logic circuit, the output of a driving device may be connected to two or more load devices as shown in Figure 5–44. The interconnecting paths share a common electrical point known as a **node**.

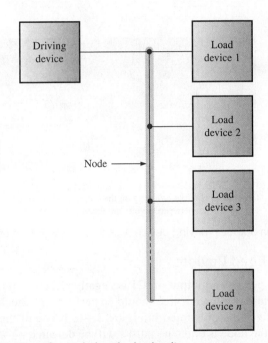

FIGURE 5–44   Illustration of a node in a logic circuit.

The driving device in Figure 5–44 is driving the node, and the other devices represent loads connected to the node. A driving device can drive a number of load device inputs up to its specified fan-out. Several types of failures are possible in this situation. Some of these failure modes are difficult to isolate to a single bad device because all the devices connected to the node are affected. Common types of failures are the following:

1. *Open output in driving device.* This failure will cause a loss of signal to all load devices.

2. *Open input in a load device.* This failure will not affect the operation of any of the other devices connected to the node, but it will result in loss of signal output from the faulty device.

3. *Shorted output in driving device.* This failure can cause the node to be stuck in the LOW state (short to ground) or in the HIGH state (short to V_{CC}).

4. *Shorted input in a load device.* This failure can also cause the node to be stuck in the LOW state (short to ground) or in the HIGH state (short to V_{CC}).

Troubleshooting Common Faults

Open Output in Driving Device

In this situation there is no pulse activity on the node. With circuit power on, an open node will normally result in a "floating" level, as illustrated in Figure 5–45.

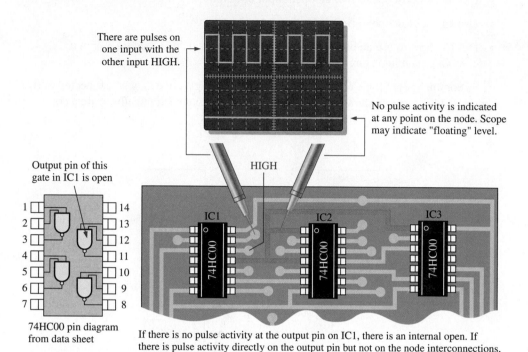

There are pulses on one input with the other input HIGH.

No pulse activity is indicated at any point on the node. Scope may indicate "floating" level.

HIGH

Output pin of this gate in IC1 is open

74HC00 pin diagram from data sheet

If there is no pulse activity at the output pin on IC1, there is an internal open. If there is pulse activity directly on the output pin but not on the node interconnections, the connection between the pin and the board is open.

FIGURE 5–45   Open output in driving device. Assume a HIGH is on one input.

Open Input in a Load Device

If the check for an open driver output in IC1 is negative (there is pulse activity), then a check for an open input in a load device should be performed. Check the output of each device for pulse activity, as illustrated in Figure 5–46. If one of the inputs that is normally connected to the node is open, no pulses will be detected on that device's output.

Output or Input Shorted to Ground

When the output is shorted to ground in the driving device or the input to a load device is shorted to ground, it will cause the node to be stuck LOW, as previously mentioned. A quick check with a scope probe will indicate this, as shown in Figure 5–47. A short to ground in the driving device's output or in any load input will cause this symptom, and further checks must therefore be made to isolate the short to a particular device.

Signal Tracing and Waveform Analysis

Although the methods of isolating an open or a short at a node point are useful from time to time, a more general troubleshooting technique called **signal tracing** is of value in just

When troubleshooting logic circuits, begin with a visual check, looking for obvious problems. In addition to components, visual inspection should include connectors. Edge connectors are frequently used to bring power, ground, and signals to a circuit board. The mating surfaces of the connector need to be clean and have a good mechanical fit. A dirty connector can cause intermittent or complete failure of the circuit. Edge connectors can be cleaned with a common pencil eraser and wiped clean with a Q-tip soaked in alcohol. Also, all connectors should be checked for loose-fitting pins.

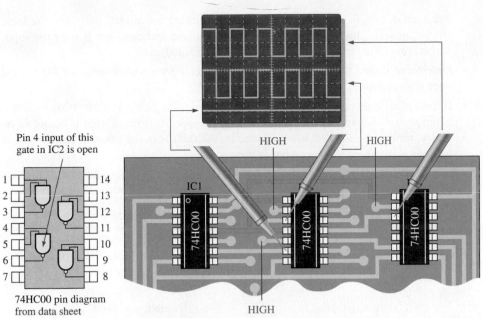

Pin 4 input of this gate in IC2 is open

HIGH   HIGH

IC1

HIGH

74HC00 pin diagram from data sheet

Check the output pin of each device connected to the node with other device inputs HIGH.
No pulse activity on an output indicates an open input or open output.

FIGURE 5–46   Open input in a load device.

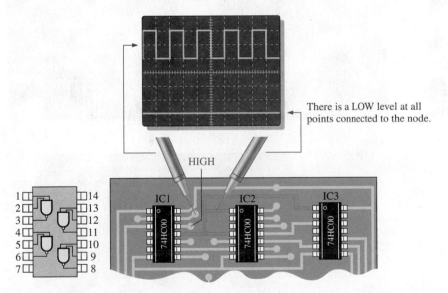

There is a LOW level at all points connected to the node.

HIGH

FIGURE 5–47   Shorted output in the driving device or shorted input in a load.

about every troubleshooting situation. Waveform measurement is accomplished with an oscilloscope or a logic analyzer.

Basically, the signal tracing method requires that you observe the waveforms and their time relationships at all accessible points in the logic circuit. You can begin at the inputs and, from an analysis of the waveform timing diagram for each point, determine where an incorrect waveform first occurs. With this procedure you can usually isolate the fault to a specific device. A procedure beginning at the output and working back toward the inputs can also be used.

The general procedure for signal tracing starting at the inputs is outlined as follows:

- Within a system, define the section of logic that is suspected of being faulty.
- Start at the inputs to the section of logic under examination. We assume, for this discussion, that the input waveforms coming from other sections of the system have been found to be correct.

- For each device, beginning at the input and working toward the output of the logic circuit, observe the output waveform of the device and compare it with the input waveforms by using the oscilloscope or the logic analyzer.

- Determine if the output waveform is correct, using your knowledge of the logical operation of the device.

- If the output is incorrect, the device under test may be faulty. Pull the IC device that is suspected of being faulty, and test it out-of-circuit. If the device is found to be faulty, replace the IC. If it works correctly, the fault is in the external circuitry or in another IC to which the tested one is connected.

- If the output is correct, go to the next device. Continue checking each device until an incorrect waveform is observed.

Figure 5–48 is an example that illustrates the general procedure for a specific logic circuit in the following steps:

Step 1:   Observe the output of gate G_1 (test point 5) relative to the inputs. If it is correct, check the inverter next. If the output is not correct, the gate or its

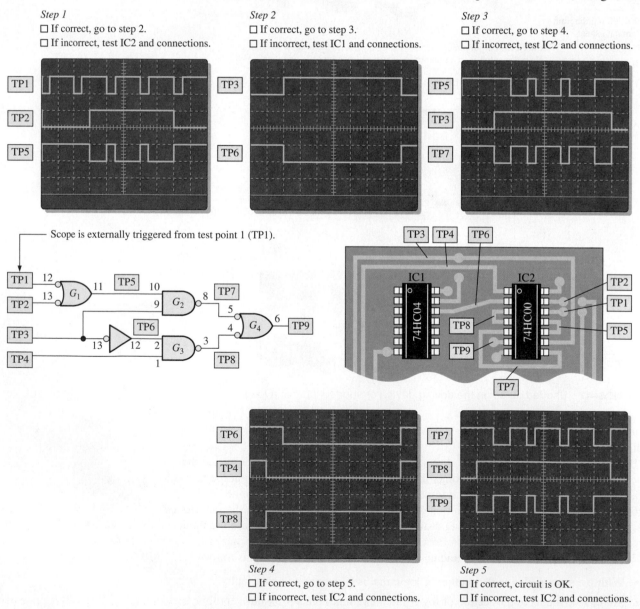

FIGURE 5–48   Example of signal tracing and waveform analysis in a portion of a printed circuit board. TP indicates test point.

connections are bad; or, if the output is LOW, the input to gate G_2 may be shorted.

Step 2: Observe the output of the inverter (TP6) relative to the input. If it is correct, check gate G_2 next. If the output is not correct, the inverter or its connections are bad; or, if the output is LOW, the input to gate G_3 may be shorted.

Step 3: Observe the output of gate G_2 (TP7) relative to the inputs. If it is correct, check gate G_3 next. If the output is not correct, the gate or its connections are bad; or, if the output is LOW, the input to gate G_4 may be shorted.

Step 4: Observe the output of gate G_3 (TP8) relative to the inputs. If it is correct, check gate G_4 next. If the output is not correct, the gate or its connections are bad; or, if the output is LOW, the input to gate G_4 (TP7) may be shorted.

Step 5: Observe the output of gate G_4 (TP9) relative to the inputs. If it is correct, the circuit is okay. If the output is not correct, the gate or its connections are bad.

EXAMPLE 5–17

Determine the fault in the logic circuit of Figure 5–49(a) by using waveform analysis. You have observed the waveforms shown in green in Figure 5–49(b). The red waveforms are correct and are provided for comparison.

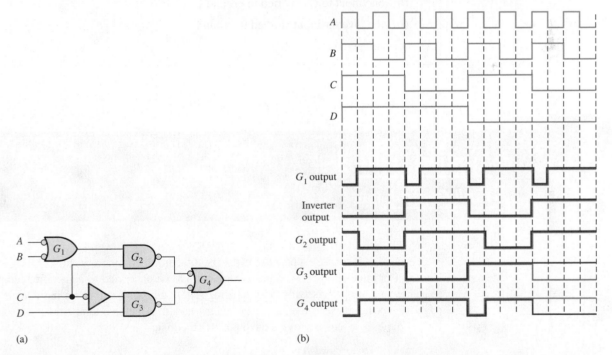

(a)   (b)

FIGURE 5–49

Solution

1. Determine what the correct waveform should be for each gate. The correct waveforms are shown in red, superimposed on the actual measured waveforms, in Figure 5–49(b).

2. Compare waveforms gate by gate until you find a measured waveform that does not match the correct waveform.

In this example, everything tested is correct until gate G_3 is checked. The output of this gate is not correct as the differences in the waveforms indicate. An analysis of the waveforms indicates that if the D input to gate G_3 is open and acting as a HIGH, you will get the output waveform measured (shown in red). Notice that the output of G_4 is also incorrect due to the incorrect input from G_3. Replace the IC containing G_3, and check the circuit's operation again.

Related Problem

For the inputs in Figure 5–49(b), determine the output waveform for the logic circuit (output of G_4) if the inverter has an open output.

As you know, testing and troubleshooting logic circuits often require observing and comparing two digital waveforms simultaneously, such as an input and the output of a device, on an oscilloscope. For digital waveforms, the scope should always be set to DC coupling on each channel input to avoid "shifting" the ground level. You should determine where the 0 V level is on the screen for both channels.

To compare the timing of the waveforms, the scope should be triggered from only one channel (don't use vertical mode or composite triggering). The channel selected for triggering should always be the one that has the lowest frequency waveform, if possible.

SECTION 5–7 CHECKUP

1. List four common internal failures in logic gates.

2. One input of a NOR gate is externally shorted to $+V_{CC}$. How does this condition affect the gate operation?

3. Determine the output of gate G_4 in Figure 5–49(a), with inputs as shown in part (b), for the following faults:

   (a) one input to G_1 shorted to ground

   (b) the inverter input shorted to ground

   (c) an open output in G_3

Applied Logic

Tank Control

A storage tank system for a pancake syrup manufacturing company is shown in Figure 5–50. The control logic allows a volume of corn syrup to be preheated to a specified temperature to achieve the proper viscosity prior to being sent to a mixing vat where ingredients such as sugar, flavoring, preservative, and coloring are added. Level and temperature sensors in the tank and the flow sensor provide the inputs for the logic.

System Operation and Analysis

The tank holds corn syrup for use in a pancake syrup manufacturing process. In preparation for mixing, the temperature of the corn syrup when released from the tank into a mixing vat must be at a specified value for proper viscosity to produce required flow characteristics. This temperature can be selected via a keypad input. The control logic maintains the temperature at this value by turning a heater *on* and *off*. The analog output from the temperature transducer (T_{analog}) is converted to an 8-bit binary code by an analog-to-digital converter and then to an 8-bit BCD code. A temperature controller detects when the temperature falls below the specified value and turns the heater *on*. When the temperature reaches the specified value, the heater is turned *off*.

The level sensors produce a HIGH when the corn syrup is at or above the minimum or at the maximum level. The valve control logic detects when the maximum level (L_{max}) or minimum level (L_{min}) has been reached and when mixture is flowing into the tank (F_{inlet}). Based on these inputs, the control logic opens or closes each valve (V_{inlet} and V_{outlet}). New corn syrup can be

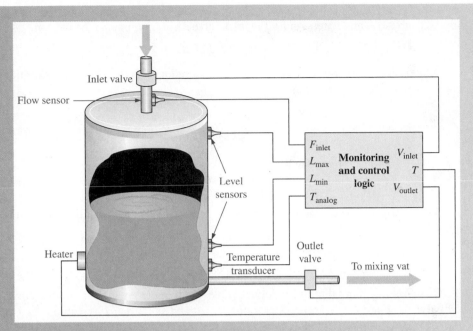

FIGURE 5–50   Tank with level and temperature sensors and controls.

added to the tank via the inlet valve only when the minimum level is reached. Once the inlet valve is opened, the level in the tank must reach the maximum point before the inlet valve is closed. Also, once the outlet valve is opened, the level must reach the minimum point before the outlet valve is closed. New syrup is always cooler than the syrup in the tank. Syrup cannot be released from the tank while it is being filled or its temperature is below the specified value.

Inlet Valve Control   The conditions for which the inlet valve is open, allowing the tank to fill, are

* The solution level is at minimum (L_{min}).
* The tank is filling (F_{inlet}) but the maximum level has not been reached ($\overline{L}_{max}$).

Table 5–6 is the truth table for the inlet valve. A HIGH (1) is the active level for the inlet valve to be open (*on*).

TABLE 5–6

Truth table for inlet valve control.

Inputs			Output	
L_{max}	L_{min}	F_{inlet}	V_{inlet}	**Description**
0	0	0	1	Level below minimum. No inlet flow.
0	0	1	1	Level below minimum. Inlet flow.
0	1	0	0	Level above min and below max. No inlet flow.
0	1	1	1	Level above min and below max. Inlet flow.
1	0	0	X	Invalid
1	0	1	X	Invalid
1	1	0	0	Level at maximum. No inlet flow.
1	1	1	0	Level at maximum. Inlet flow.

Exercise

1. Explain why the two conditions indicated in the truth table are invalid.
2. Under how many input conditions is the inlet valve open?
3. Once the level drops below minimum and the tank starts refilling, when does the inlet valve turn *off*?

From the truth table, an expression for the inlet valve control output can be written.

$$V_{inlet} = \overline{L}_{max}\overline{L}_{min}\overline{F}_{inlet} + \overline{L}_{max}\overline{L}_{min}F_{inlet} + \overline{L}_{max}L_{min}F_{inlet}$$

The SOP expression for the inlet valve logic can be reduced to the following simplified expression using Boolean methods:

$$V_{inlet} = \overline{L}_{min} + \overline{L}_{max}F_{inlet}$$

Exercise

4. Using a K-map, prove that the simplified expression is correct.
5. Using the simplified expression, draw the logic diagram for the inlet valve control.

Outlet Valve Control   The conditions for which the outlet valve is open allowing the tank to drain are

- The syrup level is above minimum and the tank is not filling.
- The temperature of the syrup is at the specified value.

Table 5–7 is the truth table for the outlet valve. A HIGH (1) is the active level for the outlet valve to be open (*on*). (*Note: T* is both an input and an output, T = Temp).

TABLE 5–7

Truth table for outlet valve control.

Inputs				Output	
L_{max}	L_{min}	F_{inlet}	T	V_{outlet}	Description
0	0	0	0	0	Level below minimum. No inlet flow. Temp low.
0	0	0	1	0	Level below minimum. No inlet flow. Temp correct.
0	0	1	0	0	Level below minimum. Inlet flow. Temp low.
0	0	1	1	0	Level below minimum. Inlet flow. Temp correct.
0	1	0	0	0	Level above min and below max. No inlet flow. Temp low.
0	1	0	1	1	Level above min and below max. No inlet flow. Temp correct.
0	1	1	0	0	Level above min and below max. Inlet flow. Temp low.
0	1	1	1	0	Level above min and below max. Inlet flow. Temp correct
1	0	0	0	X	Invalid
1	0	0	1	X	Invalid
1	0	1	0	X	Invalid
1	0	1	1	X	Invalid
1	1	0	0	0	Level at maximum. No inlet flow. Temp low.
1	1	0	1	1	Level at maximum. No inlet flow. Temp correct.
1	1	1	0	0	Level at maximum. Inlet flow. Temp low.
1	1	1	1	0	Level at maximum. Inlet flow. Temp correct.

Exercise

6. Why does the outlet valve control require four inputs and the inlet valve only three?
7. Under how many input conditions is the outlet valve open?
8. Once the level reaches maximum and the tank starts draining, when does the outlet valve turn off?

From the truth table, an expression for the outlet valve control can be written.

$$V_{outlet} = \overline{L}_{max}L_{min}\overline{F}_{inlet}\,T + L_{max}L_{min}\overline{F}_{inlet}T$$

The SOP expression for the outlet valve logic can be reduced to the following simplified expression:

$$V_{outlet} = L_{min}\overline{F}_{inlet}T$$

Exercise

9. Using a K-map, prove that the simplified expression is correct.

10. Using the simplified expression, draw the logic diagram for the outlet valve control.

Temperature Control   The temperature control logic accepts an 8-bit BCD code representing the measured temperature and compares it to the BCD code for the specified temperature. A block diagram is shown in Figure 5–51.

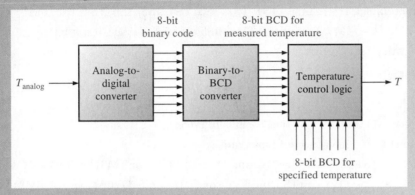

FIGURE 5–51   Block diagram for temperature control circuit.

When the measured temperature and the specified temperature are the same, the two BCD codes are equal and the T output is LOW (0). When the measured temperature falls below the specified value, there is a difference in the BCD codes and the T output is HIGH (1), which turns on the heater. The temperature control logic can be implemented with exclusive-OR gates, as shown in Figure 5–52. Each pair of corresponding bits from the two

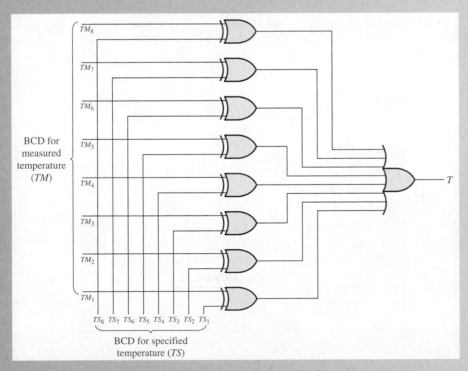

FIGURE 5–52   Logic diagram of the temperature control logic.

BCD codes is applied to an exclusive-OR gate. If the bits are the same, the output of the XOR gate is 0; and if they are different, the output of the XOR gate is 1. When one or more XOR outputs equal 1, the *T* output of the OR gate equals 1, causing the heater to turn on.

VHDL Code for Tank Control Logic

The control logic for the inlet valve, outlet valve, and temperature is described with VHDL using the data flow approach (which is based on the Boolean description of the logic). Exercise 11 requires the structural approach (which is based on the gates and how they are connected) for comparison.

entity TankControl **is**

    port (Finlet, Lmax, Lmin, TS1, TS2, TS3, TS4, TS5, TS6, TS7, TS8, TM1, TM2,

    TM3, TM4, TM5, TM6, TM7, TM8: **in** bit; Vinlet, Voutlet, T: **out** bit);

end entity TankControl;

architecture ValveTempLogic **of** Tank Control **is**

begin

    Vinlet <= **not** Lmin **or** (**not** Lmax **and** Finlet);

    Voutlet <= Lmin **and not** Finlet **and** T;

    T <= (TS1 **xor** TM1) **or** (TS2 **xor** TM2) **or** (TS3 **xor** TM3) **or** (TS4 **xor** TM4)

    or (TS5 **xor** TM5) **or** (TS6 **xor** TM6) **or** (TS7 **xor** TM7) **or** (TS8 **xor** TM8);

    end architecture ValveTempLogic;

Exercise

  11. Write the VHDL code for the tank control logic using the structural approach.

Simulation of the Valve Control Logic

The inlet and outlet valve control logic simulation screen is shown in Figure 5–53. SPDT switches are used to represent the level and flow sensor inputs and the temperature indication. Probes are used to indicate the output states.

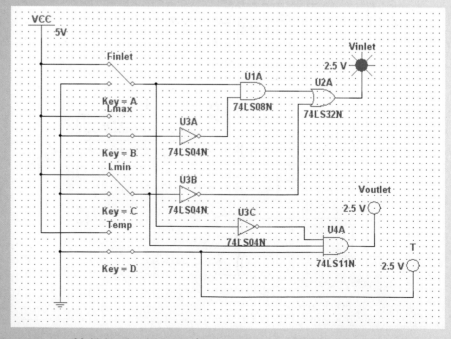

FIGURE 5–53 Multisim circuit screen for the valve control logic.

Open file AL05 in the Applied Logic folder on the website. Run the simulation of the valve-control logic using your Multisim software and observe the operation. Create a new Multisim file, connect the temperature control logic, and run the simulation.

MultiSim

Putting Your Knowledge to Work

If the temperature of the syrup can never be more than 9°C below the specified value, can the temperature control circuit be simplified? If so, how?

SUMMARY

- AND-OR logic produces an output expression in SOP form.
- AND-OR-Invert logic produces a complemented SOP form, which is actually a POS form.
- The operational symbol for exclusive-OR is $\oplus$. An exclusive-OR expression can be stated in two equivalent ways:

$$A\overline{B} + \overline{A}B = A \oplus B$$

- To do an analysis of a logic circuit, start with the logic circuit, and develop the Boolean output expression or the truth table or both.
- Implementation of a logic circuit is the process in which you start with the Boolean output expressions or the truth table and develop a logic circuit that produces the output function.
- All NAND or NOR logic diagrams should be drawn using appropriate dual symbols so that bubble outputs are connected to bubble inputs and nonbubble outputs are connected to nonbubble inputs.
- When two negation indicators (bubbles) are connected, they effectively cancel each other.
- A VHDL component is a predefined logic function stored for use throughout a program or in other programs.
- A component instantiation is used to call for a component in a program.
- A VHDL signal effectively acts as an internal interconnection in a VHDL structural description.

KEY TERMS

Key terms and other bold terms in the chapter are defined in the end-of-book glossary.

Component   A VHDL feature that can be used to predefine a logic function for multiple use throughout a program or programs.

Negative-AND   The dual operation of a NOR gate when the inputs are active-LOW.

Negative-OR   The dual operation of a NAND gate when the inputs are active-LOW.

Node   A common connection point in a circuit in which a gate output is connected to one or more gate inputs.

Signal   A waveform; a type of VHDL object that holds data.

Signal tracing   A troubleshooting technique in which waveforms are observed in a step-by-step manner beginning at the input and working toward the output or vice versa. At each point the observed waveform is compared with the correct signal for that point.

Universal gate   Either a NAND gate or a NOR gate. The term *universal* refers to the property of a gate that permits any logic function to be implemented by that gate or by a combination of that kind.

TRUE/FALSE QUIZ

Answers are at the end of the chapter.

1. AND-OR logic can have only two 2-input AND gates.
2. AOI is an acronym for AND-OR-Invert.
3. If the inputs of an exclusive-OR gate are the same, the output is HIGH (1).
4. If the inputs of an exclusive-NOR gate are different, the output is LOW (0).
5. A parity generator can be implemented using exclusive-OR gates.
6. NAND gates cannot be used to produce the OR function.
7. NOR gates can be used to produce the AND function.
8. Any SOP expression can be implemented using only NAND gates.
9. The dual symbol for a NAND gate is a negative-AND symbol.
10. Negative-OR is equivalent to NAND.

SELF-TEST

Answers are at the end of the chapter.

1. The output expression for an AND-OR circuit having one AND gate with inputs A, B, C, and D and one AND gate with inputs E and F is
   (a) $ABCDEF$
   (b) $A + B + C + D + E + F$
   (c) $(A + B + C + D)(E + F)$
   (d) $ABCD + EF$

2. A logic circuit with an output $X = A\overline{B}C + A\overline{C}$ consists of
   (a) two AND gates and one OR gate
   (b) two AND gates, one OR gate, and two inverters
   (c) two OR gates, one AND gate, and two inverters
   (d) two AND gates, one OR gate, and one inverter

3. To implement the expression $\overline{A}BCD + A\overline{B}CD + ABC\overline{D}$, it takes one OR gate and
   (a) one AND gate
   (b) three AND gates
   (c) three AND gates and four inverters
   (d) three AND gates and three inverters

4. The expression $\overline{A}BCD + ABC\overline{D} + A\overline{B}\,CD$
   (a) cannot be simplified
   (b) can be simplified to $\overline{A}BC + A\overline{B}$
   (c) can be simplified to $ABC\overline{D} + \overline{A}BC$
   (d) None of these answers is correct.

5. The output expression for an AND-OR-Invert circuit having one AND gate with inputs, A, B, C, and D and one AND gate with inputs E and F is
   (a) $ABCD + EF$
   (b) $\overline{A} + \overline{B} + \overline{C} + \overline{D} + \overline{E} + \overline{F}$
   (c) $\overline{(A + B + C + D)(E + F)}$
   (d) $(\overline{A} + \overline{B} + \overline{C} + \overline{D})(\overline{E} + \overline{F})$

6. An exclusive-OR function is expressed as
   (a) $\overline{A}\,\overline{B} + AB$
   (b) $\overline{A}B + A\overline{B}$
   (c) $(\overline{A} + B)(A + \overline{B})$
   (d) $(\overline{A} + \overline{B}) + (A + B)$

7. The AND operation can be produced with
   (a) two NAND gates
   (b) three NAND gates
   (c) one NOR gate
   (d) three NOR gates

8. The OR operation can be produced with
   (a) two NOR gates
   (b) three NAND gates
   (c) four NAND gates
   (d) both answers (a) and (b)

9. When using dual symbols in a logic diagram,
   (a) bubble outputs are connected to bubble inputs
   (b) the NAND symbols produce the AND operations
   (c) the negative-OR symbols produce the OR operations
   (d) All of these answers are true.
   (e) None of these answers is true.

10. All Boolean expressions can be implemented with
    (a) NAND gates only
    (b) NOR gates only
    (c) combinations of NAND and NOR gates
    (d) combinations of AND gates, OR gates, and inverters
    (e) any of these

11. A VHDL component
    (a) can be used once in each program
    (b) is a predefined description of a logic function
    (c) can be used multiple times in a program
    (d) is part of a data flow description
    (e) answers (b) and (c)

12. A VHDL component is called for use in a program by using a
    (a) signal   (b) variable
    (c) component instantiation (d) architecture declaration

PROBLEMS

Answers to odd-numbered problems are at the end of the book.

Section 5–1 Basic Combinational Logic Circuits

1. Draw the ANSI distinctive shape logic diagram for a 3-wide, 4-input AND-OR-Invert circuit.
   Also draw the ANSI standard rectangular outline symbol.

2. Write the output expression for each circuit in Figure 5–54.

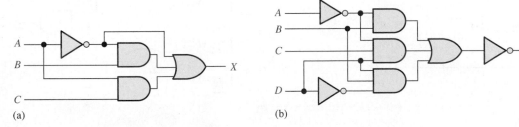

(a)   (b)

FIGURE 5–54

3. Write the output expression for each circuit as it appears in Figure 5–55.

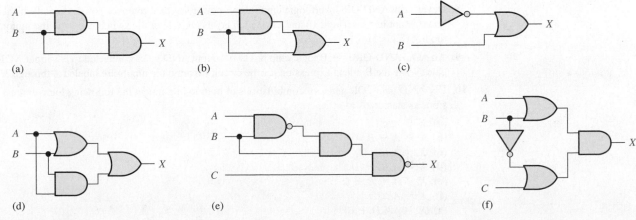

(a)   (b)   (c)

(d)   (e)   (f)

FIGURE 5–55

4. Write the output expression for each circuit as it appears in Figure 5–56 and then change each circuit to an equivalent AND-OR configuration.

5. Develop the truth table for each circuit in Figure 5–55.

6. Develop the truth table for each circuit in Figure 5–56.

7. Show that an exclusive-NOR circuit produces a POS output.

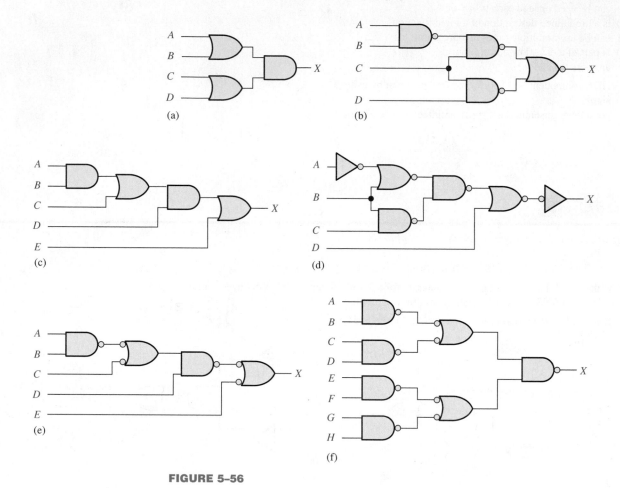

FIGURE 5–56

Section 5–2 Implementing Combinational Logic

8. Develop an AND-OR-Invert logic circuit for a power saw that removes power (logic 0) if the guard is not in place (logic 0) and the switch is *on* (logic 1) or the switch is *on* and the motor is too hot (logic 1).

9. An AOI (AND-OR-Invert) logic chip has two 4-input AND gates connected to a 2-input NOR gate. Write the Boolean expression for the circuit (assume the inputs are labeled *A* through *H*).

10. Use AND gates, OR gates, or combinations of both to implement the following logic expressions as stated:

   (a) $X = AB$
   (b) $X = A + B$
   (c) $X = AB + C$
   (d) $X = ABC + D$
   (e) $X = A + B + C$
   (f) $X = ABCD$
   (g) $X = A(CD + B)$
   (h) $X = AB(C + DEF) + CE(A + B + F)$

11. Use AND gates, OR gates, and inverters as needed to implement the following logic expressions as stated:

   (a) $X = AB + \bar{B}C$
   (b) $X = A(B + \bar{C})$
   (c) $X = A\bar{B} + AB$
   (d) $X = \overline{ABC} + B(EF + \bar{G})$
   (e) $X = A[BC(A + B + C + D)]$
   (f) $X = B(C\bar{D}E + \bar{E}FG)(\overline{AB} + C)$

12. Use NAND gates, NOR gates, or combinations of both to implement the following logic expressions as stated:

   (a) $X = \bar{A}B + CD + (\overline{A + B})(ACD + \bar{B}E)$
   (b) $X = AB\bar{C}\bar{D} + D\bar{E}F + \overline{AF}$
   (c) $X = \bar{A}[B + \bar{C}(D + E)]$

13. Implement a logic circuit for the truth table in Table 5–8.

TABLE 5–8

Inputs			Output
A	B	C	X
0	0	0	1
0	0	1	0
0	1	0	1
0	1	1	0
1	0	0	1
1	0	1	0
1	1	0	1
1	1	1	1

14. Implement a logic circuit for the truth table in Table 5–9.

TABLE 5–9

Inputs				Output
A	B	C	D	X
0	0	0	0	0
0	0	0	1	0
0	0	1	0	1
0	0	1	1	1
0	1	0	0	1
0	1	0	1	0
0	1	1	0	0
0	1	1	1	0
1	0	0	0	1
1	0	0	1	1
1	0	1	0	1
1	0	1	1	1
1	1	0	0	0
1	1	0	1	0
1	1	1	0	0
1	1	1	1	1

15. Simplify the circuit in Figure 5–57 as much as possible, and verify that the simplified circuit is equivalent to the original by showing that the truth tables are identical.

16. Repeat Problem 15 for the circuit in Figure 5–58.

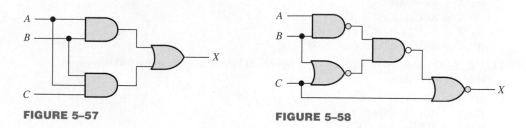

FIGURE 5–57   **FIGURE 5–58**

17. Minimize the gates required to implement the functions in each part of Problem 11 in SOP form.

18. Minimize the gates required to implement the functions in each part of Problem 12 in SOP form.

19. Minimize the gates required to implement the function of the circuit in each part of Figure 5–56 in SOP form.

Section 5–3 The Universal Property of NAND and NOR Gates

20. Implement the logic circuits in Figure 5–54 using only NAND gates.

21. Implement the logic circuit in Figure 5–58 using only NAND gates.

22. Repeat Problem 20 using only NOR gates.

23. Repeat Problem 21 using only NOR gates.

Section 5–4 Combinational Logic Using NAND and NOR Gates

24. Show how the following expressions can be implemented as stated using only NOR gates:

(a) $X = ABC$ (b) $X = \overline{ABC}$ (c) $X = A + B$

(d) $X = A + B + \overline{C}$ (e) $X = \overline{AB} + \overline{CD}$ (f) $X = (A + B)(C + D)$

(g) $X = AB[C(\overline{DE} + \overline{AB}) + \overline{BCE}]$

25. Repeat Problem 24 using only NAND gates.

26. Implement each function in Problem 10 by using only NAND gates.

27. Implement each function in Problem 11 by using only NAND gates.

Section 5–5 Pulse Waveform Operation

28. Given the logic circuit and the input waveforms in Figure 5–59, draw the output waveform.

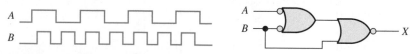

FIGURE 5–59

29. For the logic circuit in Figure 5–60, draw the output waveform in proper relationship to the inputs.

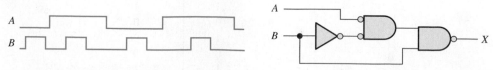

FIGURE 5–60

30. For the input waveforms in Figure 5–61, what logic circuit will generate the output waveform shown?

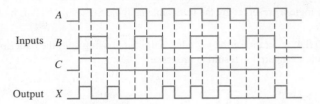

FIGURE 5–61

31. Repeat Problem 30 for the waveforms in Figure 5–62.

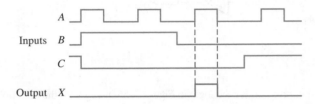

FIGURE 5–62

32. For the circuit in Figure 5–63, draw the waveforms at the numbered points in the proper relationship to each other.

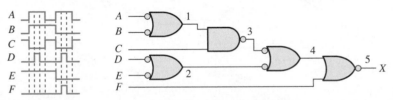

FIGURE 5–63

33. Assuming a propagation delay through each gate of 10 nanoseconds (ns), determine if the *desired* output waveform X in Figure 5–64 (a pulse with a minimum $t_W = 25$ ns positioned as shown) will be generated properly with the given inputs.

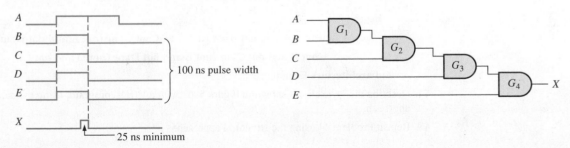

FIGURE 5–64

Section 5–6 Combinational Logic with VHDL

34. Describe a 2-input NAND gate with VHDL.

35. Describe a 3-input AND gate with VHDL.

36. Write a VHDL program using the data flow approach (Boolean expressions) to describe the logic circuit in Figure 5–54(b).

37. Write VHDL programs using the data flow approach (Boolean expressions) for the logic circuits in Figure 5–55(e) and (f).

38. Write a VHDL program using the structural approach for the logic circuit in Figure 5–56(d). Assume component declarations for each type of gate are already available.

39. Repeat Problem 38 for the logic circuit in Figure 5–56(f).

40. Describe the logic represented by the truth table in Table 5–8 using VHDL by first converting it to SOP form.

41. Develop a VHDL program for the logic in Figure 5–65, using both the data flow and the structural approach. Compare the resulting programs.

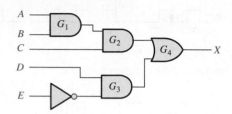

FIGURE 5–65

42. Develop a VHDL program for the logic in Figure 5–66, using both the data flow and the structural approach. Compare the resulting programs.

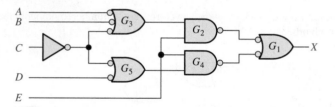

FIGURE 5–66

43. Given the following VHDL program, create the truth table that describes the logic circuit.

> **entity** CombLogic **is**
>
> **port** (A, B, C, D: **in** bit; X: **out** bit);
>
> **end entity** CombLogic;
>
> **architecture** Example **of** CombLogic **is**
>
> **begin**
>
> X <= **not**((**not** A **and not** B) **or** (**not** A **and not** C) **or** (**not** A **and not** D) **or**
>
> (**not** B **and not** C) **or** (**not** B **and not** D) **or** (**not** D **and not** C));
>
> **end architecture** Example;

44. Describe the logic circuit shown in Figure 5–67 with a VHDL program, using the data flow approach.

45. Repeat Problem 44 using the structural approach.

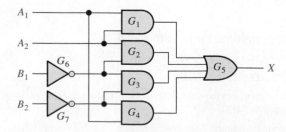

FIGURE 5–67

Section 5–7 Troubleshooting

46. For the logic circuit and the input waveforms in Figure 5–68, the indicated output waveform is observed. Determine if this is the correct output waveform.

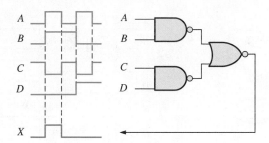

FIGURE 5–68

47. The output waveform in Figure 5–69 is incorrect for the inputs that are applied to the circuit. Assuming that one gate in the circuit has failed, with its output either an apparent constant HIGH or a constant LOW, determine the faulty gate and the type of failure (output open or shorted).

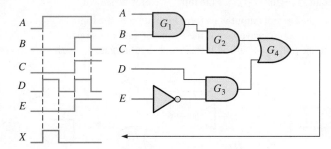

FIGURE 5–69

48. Repeat Problem 47 for the circuit in Figure 5–70, with input and output waveforms as shown.

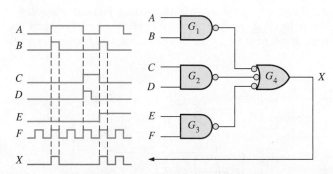

FIGURE 5–70

49. By examining the connections in Figure 5–71, determine the driving gate and load gate(s). Specify by device and pin numbers.

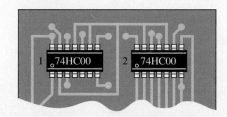

FIGURE 5–71

50. Figure 5–72(a) is a logic circuit under test. Figure 5–72(b) shows the waveforms as observed on a logic analyzer. The output waveform is incorrect for the inputs that are applied to the circuit. Assuming that one gate in the circuit has failed, with its output either an apparent constant HIGH or a constant LOW, determine the faulty gate and the type of failure.

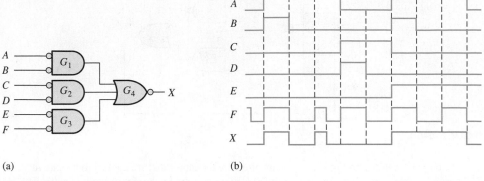

(a)   (b)

FIGURE 5–72

51. The logic circuit in Figure 5–73 has the input waveforms shown.

    (a) Determine the correct output waveform in relation to the inputs.

    (b) Determine the output waveform if the output of gate G_3 is open.

    (c) Determine the output waveform if the upper input to gate G_5 is shorted to ground.

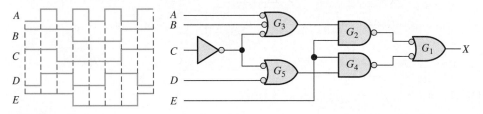

FIGURE 5–73

52. The logic circuit in Figure 5–74 has only one intermediate test point available besides the output, as indicated. For the inputs shown, you observe the indicated waveform at the test point. Is this waveform correct? If not, what are the possible faults that would cause it to appear as it does?

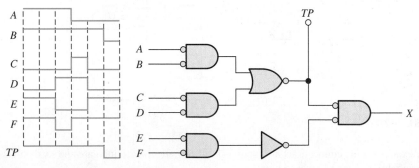

FIGURE 5–74

Applied Logic

53. Describe the function of each of the three sensors in the tank.

54. Implement the inlet valve logic using NOR gates and inverters.

55. Repeat Problem 54 for the outlet valve logic.

56. Implement the temperature control logic using XNOR gates.

57. Design a circuit to enable an additive to be introduced into the syrup through another inlet only when the temperature is at the specified value and the syrup level is at the low-level sensor.

Special Design Problems

58. (a) Design a logic circuit to produce a HIGH output only if the input, represented by a 4-bit binary number, is greater than twelve or less than three. First develop the truth table and then draw the logic diagram.

(b) Describe the logic using VHDL.

59. (a) Develop the logic circuit necessary to meet the following requirements:

A battery-powered lamp in a room is to be operated from two switches, one at the back door and one at the front door. The lamp is to be on if the front switch is on and the back switch is off, or if the front switch is off and the back switch is on. The lamp is to be off if both switches are off or if both switches are on. Let a HIGH output represent the on condition and a LOW output represent the off condition.

(b) Describe the logic using VHDL.

60. (a) Develop the NAND logic for a hexadecimal keypad encoder that will convert each key closure to binary.

(b) Describe the logic using VHDL.

Multisim Troubleshooting Practice

MultiSim

61. Open file P05-61. For the specified fault, predict the effect on the circuit. Then introduce the fault and verify whether your prediction is correct.

62. Open file P05-62. For the specified fault, predict the effect on the circuit. Then introduce the fault and verify whether your prediction is correct.

63. Open file P05-63. For the observed behavior indicated, predict the fault in the circuit. Then introduce the suspected fault and verify whether your prediction is correct.

64. Open file P05-64. For the observed behavior indicated, predict the fault in the circuit. Then introduce the suspected fault and verify whether your prediction is correct.

ANSWERS

SECTION CHECKUPS

Section 5–1 Basic Combinational Logic Circuits

1. (a) $\overline{AB + CD} = \overline{1 \cdot 0 + 1 \cdot 0} = 1$ **(b)** $\overline{AB + CD} = \overline{1 \cdot 1 + 0 \cdot 1} = 0$

(c) $\overline{AB + CD} = \overline{0 \cdot 1 + 1 \cdot 1} = 0$

2. (a) $A\overline{B} + \overline{A}B = 1 \cdot \overline{0} + \overline{1} \cdot 0 = 1$ **(b)** $A\overline{B} + \overline{A}B = 1 \cdot \overline{1} + \overline{1} \cdot 1 = 0$

(c) $A\overline{B} + \overline{A}B = 0 \cdot \overline{1} + \overline{0} \cdot 1 = 1$ **(d)** $A\overline{B} + \overline{A}B = 0 \cdot \overline{0} + \overline{0} \cdot 0 = 0$

3. $X = 1$ when $ABC = 000, 011, 101, 110,$ and 111; $X = 0$ when $ABC = 001, 010,$ and 100

4. $X = AB + \overline{A}\,\overline{B}$; the circuit consists of two AND gates, one OR gate, and two inverters. See Figure 5–6(b) for diagram.

Section 5–2 Implementing Combinational Logic

1. (a) $X = ABC + AB + AC$: three AND gates, one OR gate

(b) $X = AB(C + DE)$: three AND gates, one OR gate

2. $X = ABC + \overline{A}\,\overline{B}\,\overline{C}$; two AND gates, one OR gate, and three inverters

3. (a) $X = AB(C + 1) + AC = AB + AC$

(b) $X = AB(C + DE) = ABC + ABDE$

Section 5–3 The Universal Property of NAND and NOR Gates

1. (a) $X = \overline{A} + B$: a 2-input NAND gate with A and $\overline{B}$ on its inputs.

(b) $X = A\overline{B}$: a 2-input NAND with A and $\overline{B}$ on its inputs, followed by one NAND used as an inverter.

2. (a) $X = \overline{A} + B$: a 2-input NOR with inputs $\overline{A}$ and B, followed by one NOR used as an inverter.

(b) $X = A\overline{B}$: a 2-input NOR with $\overline{A}$ and B on its inputs.

Section 5–4 Combinational Logic Using NAND and NOR Gates

1. $X = \overline{(\overline{A} + \overline{B} + \overline{C})DE}$: a 3-input NAND with inputs, A, B, and C, with its output connected to a second 3-input NAND with two other inputs, D and E

2. $X = \overline{\overline{A}\,\overline{B}\,\overline{C} + (D + E)}$: a 3-input NOR with inputs A, B, and C, with its output connected to a second 3-input NOR with two other inputs, D and E

Section 5–5 Pulse Waveform Operation

1. The exclusive-OR output is a 15 μs pulse followed by a 25 μs pulse, with a separation of 10 μs between the pulses.

2. The output of the exclusive-NOR is HIGH when both inputs are HIGH or when both inputs are LOW.

Section 5–6 Combinational Logic with VHDL

1. A VHDL component is a predefined program describing a specified logic function.

2. A component instantiation is used to call for a specified component in a program architecture.

3. Interconnections between components are made using VHDL signals.

4. Components are used in the structural approach.

Section 5–7 Troubleshooting

1. Common gate failures are input or output open; input or output shorted to ground.

2. Input shorted to V_{CC} causes output to be stuck LOW.

3. (a) G_4 output is HIGH until rising edge of seventh pulse, then it goes LOW.

   (b) G_4 output is the same as input D.

   (c) G_4 output is the inverse of the G_2 output shown in Figure 5–49(b).

RELATED PROBLEMS FOR EXAMPLES

5–1 $X = AB + AC + BC$

5–2 $X = \overline{AB + AC + BC}$

$$\text{If } A = 0 \text{ and } B = 0, X = \overline{0 \cdot 0 + 0 \cdot 1 + 0 \cdot 1} = \overline{0} = 1$$

$$\text{If } A = 0 \text{ and } C = 0, X = \overline{0 \cdot 1 + 0 \cdot 0 + 1 \cdot 0} = \overline{0} = 1$$

$$\text{If } B = 0 \text{ and } C = 0, X = \overline{1 \cdot 0 + 1 \cdot 0 + 0 \cdot 0} = \overline{0} = 1$$

5–3   Determine the even-parity output for all 16 input combinations. Each combination should have an even number of 1s including the parity bit.

5–4   Apply codes with odd number of 1s and verify output is 1.

5–5   Cannot be simplified

5–6   Cannot be simplified

5–7 $X = A + B + C + D$ is valid.

5–8   See Figure 5–75.

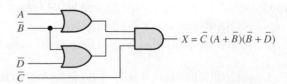

$$X = \overline{C}\,(A + \overline{B})(\overline{B} + \overline{D})$$

FIGURE 5–75

5–9 $X = \overline{(\overline{ABC})(\overline{DEF})} = (\overline{AB})C + (\overline{DE})F = (\overline{A} + \overline{B})C + (\overline{D} + \overline{E})F$

5–10 See Figure 5–76.

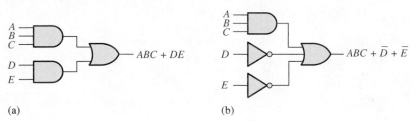

(a)   (b)

FIGURE 5–76

5–11 $X = \overline{\overline{(\overline{A + B} + C)} + \overline{(\overline{D + E} + F)}} = (\overline{A + B} + C)(\overline{D + E} + F) = (\overline{A}\overline{B} + C)(\overline{D}\overline{E} + F)$

5–12 See Figure 5–77.

5–13 See Figure 5–78.

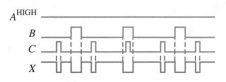

FIGURE 5–77   **FIGURE 5–78**

5–14 See Figure 5–79.

5–15 See Figure 5–80.

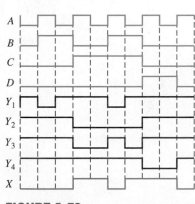

FIGURE 5–79   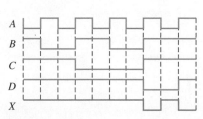 **FIGURE 5–80**

5–16 G5: NAND_gate2 **port map** (A => IN9, B => IN10, X => OUT5);

5–17 See Figure 5–81.

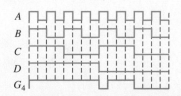

FIGURE 5–81

TRUE/FALSE QUIZ

1. F   **2.** T   **3.** F   **4.** T   **5.** T

6. F   **7.** T   **8.** T   **9.** F   **10.** T

SELF-TEST

1. (d)   **2.** (b)   **3.** (c)   **4.** (a)   **5.** (d)   **6.** (b)   **7.** (a)   **8.** (d)

9. (d)   **10.** (e)   **11.** (e)   **12.** (c)

CHAPTER 6

Functions of Combinational Logic

CHAPTER OUTLINE

CHAPTER OBJECTIVES

- Distinguish between half-adders and full-adders
- Use full-adders to implement multibit parallel binary adders
- Explain the differences between ripple carry and look-ahead carry parallel adders
- Use the magnitude comparator to determine the relationship between two binary numbers and use cascaded comparators to handle the comparison of larger numbers
- Implement a basic binary decoder
- Use BCD-to-7-segment decoders in display systems
- Apply a decimal-to-BCD priority encoder in a simple keyboard application
- Convert from binary to Gray code, and Gray code to binary by using logic devices
- Apply data selectors/multiplexers in multiplexed displays and as a function generator
- Use decoders as demultiplexers
- Explain the meaning of parity
- Use parity generators and checkers to detect bit errors in digital systems
- Describe a simple data communications system
- Write VHDL programs for several logic functions
- Identify glitches, common bugs in digital systems

KEY TERMS

Key terms are in order of appearance in the chapter.

- Half-adder
- Full-adder
- Cascading
- Ripple carry
- Look-ahead carry
- Comparator
- Decoder
- Encoder
- Priority encoder
- Multiplexer (MUX)
- Demultiplexer (DEMUX)
- Parity bit
- Glitch

VISIT THE WEBSITE

Study aids for this chapter are available at
http://www.pearsonhighered.com/careersresources/

INTRODUCTION

In this chapter, several types of combinational logic functions are introduced including adders, comparators, decoders, encoders, code converters, multiplexers (data selectors), demultiplexers, and parity generators/checkers. VHDL implementation of each logic function is provided, and examples of fixed-function IC devices are included. Each device introduced may also be available in other logic families.

6–1   Half and Full Adders

Adders are important in computers and also in other types of digital systems in which numerical data are processed. An understanding of the basic adder operation is fundamental to the study of digital systems. In this section, the half-adder and the full-adder are introduced.

After completing this section, you should be able to

- ◆ Describe the function of a half-adder
- ◆ Draw a half-adder logic diagram
- ◆ Describe the function of the full-adder
- ◆ Draw a full-adder logic diagram using half-adders
- ◆ Implement a full-adder using AND-OR logic

The Half-Adder

A half-adder adds two bits and produces a sum and an output carry.

Recall the basic rules for binary addition as stated in Chapter 2.

$$0 + 0 = 0$$
$$0 + 1 = 1$$
$$1 + 0 = 1$$
$$1 + 1 = 10$$

The operations are performed by a logic circuit called a **half-adder**.

The half-adder accepts two binary digits on its inputs and produces two binary digits on its outputs—a sum bit and a carry bit.

A half-adder is represented by the logic symbol in Figure 6–1.

Input bits $\left\{ \begin{array}{l} A \\ B \end{array} \right.$ $\begin{array}{c} \Sigma \\ C_{out} \end{array}$ $\left. \begin{array}{l} \text{Sum} \\ \text{Carry} \end{array} \right\}$ Outputs

MultiSim   **FIGURE 6–1**   Logic symbol for a half-adder. Open file F06-01 to verify operation. *A Multisim tutorial is available on the website.*

Half-Adder Logic

From the operation of the half-adder as stated in Table 6–1, expressions can be derived for the sum and the output carry as functions of the inputs. Notice that the output carry (C_{out}) is a 1 only when both A and B are 1s; therefore, C_{out} can be expressed as the AND of the input variables.

$$C_{out} = AB \hspace{3cm} \text{Equation 6–1}$$

Now observe that the sum output (Σ) is a 1 only if the input variables, A and B, are not equal. The sum can therefore be expressed as the exclusive-OR of the input variables.

$$\Sigma = A \oplus B \hspace{3cm} \text{Equation 6–2}$$

From Equations 6–1 and 6–2, the logic implementation required for the half-adder function can be developed. The output carry is produced with an AND gate with A and B on the

TABLE 6–1

Half-adder truth table.

A	B	C_{out}	Σ
0	0	0	0
0	1	0	1
1	0	0	1
1	1	1	0

Σ = sum
C_{out} = output carry
A and B = input variables (operands)

inputs, and the sum output is generated with an exclusive-OR gate, as shown in Figure 6–2. Remember that the exclusive-OR can be implemented with AND gates, an OR gate, and inverters.

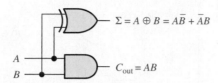

$$\Sigma = A \oplus B = A\overline{B} + \overline{A}B$$

$$C_{out} = AB$$

A
B

FIGURE 6–2   Half-adder logic diagram.

The Full-Adder

The second category of adder is the **full-adder**.

> **The full-adder accepts two input bits and an input carry and generates a sum output and an output carry.**

The basic difference between a full-adder and a half-adder is that the full-adder accepts an input carry. A logic symbol for a full-adder is shown in Figure 6–3, and the truth table in Table 6–2 shows the operation of a full-adder.

A full-adder has an input carry while the half-adder does not.

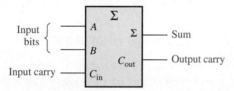

Input bits {
Input carry
A
B
C_{in}
Σ
Σ — Sum
C_{out} — Output carry

FIGURE 6–3   Logic symbol for a full-adder. Open file F06-03 to verify operation.

MultiSim

TABLE 6–2
Full-adder truth table.

A	B	C_{in}	C_{out}	Σ
0	0	0	0	0
0	0	1	0	1
0	1	0	0	1
0	1	1	1	0
1	0	0	0	1
1	0	1	1	0
1	1	0	1	0
1	1	1	1	1

C_{in} = input carry, sometimes designated as CI
C_{out} = output carry, sometimes designated as CO
Σ = sum
A and B = input variables (operands)

Full-Adder Logic

The full-adder must add the two input bits and the input carry. From the half-adder you know that the sum of the input bits A and B is the exclusive-OR of those two variables, $A \oplus B$. For the input carry (C_{in}) to be added to the input bits, it must be exclusive-ORed with $A \oplus B$, yielding the equation for the sum output of the full-adder.

$$\Sigma = (A \oplus B) \oplus C_{in} \qquad\qquad \text{Equation 6–3}$$

This means that to implement the full-adder sum function, two 2-input exclusive-OR gates can be used. The first must generate the term $A \oplus B$, and the second has as its inputs the output of the first XOR gate and the input carry, as illustrated in Figure 6–4(a).

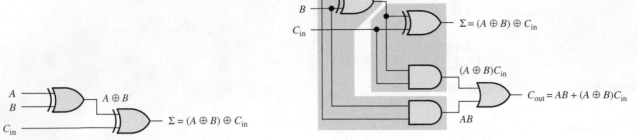

(a) Logic required to form the sum of three bits

(b) Complete logic circuit for a full-adder (each half-adder is enclosed by a shaded area)

MultiSim   **FIGURE 6–4**   Full-adder logic. Open file F06-04 to verify operation.

The output carry is a 1 when both inputs to the first XOR gate are 1s or when both inputs to the second XOR gate are 1s. You can verify this fact by studying Table 6–2. The output carry of the full-adder is therefore produced by input A ANDed with input B and $A \oplus B$ ANDed with C_{in}. These two terms are ORed, as expressed in Equation 6–4. This function is implemented and combined with the sum logic to form a complete full-adder circuit, as shown in Figure 6–4(b).

$$C_{out} = AB + (A \oplus B)C_{in}$$ Equation 6–4

Notice in Figure 6–4(b) there are two half-adders, connected as shown in the block diagram of Figure 6–5(a), with their output carries ORed. The logic symbol shown in Figure 6–5(b) will normally be used to represent the full-adder.

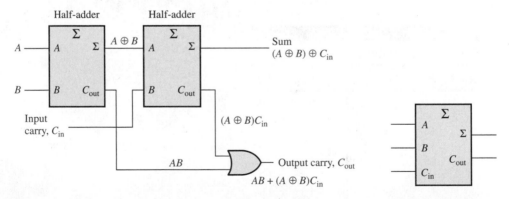

(a) Arrangement of two half-adders to form a full-adder

(b) Full-adder logic symbol

FIGURE 6–5   Full-adder implemented with half-adders.

EXAMPLE 6–1

For each of the three full-adders in Figure 6–6, determine the outputs for the inputs shown.

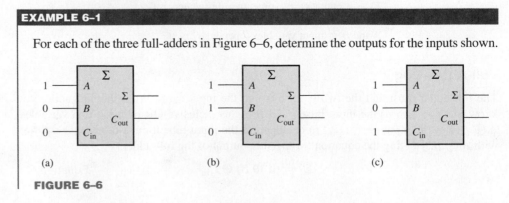

(a)

(b)

(c)

FIGURE 6–6

Solution

(a) The input bits are $A = 1$, $B = 0$, and $C_{in} = 0$.

$$1 + 0 + 0 = 1 \text{ with no carry}$$

Therefore, $\Sigma = \mathbf{1}$ and $C_{out} = \mathbf{0}$.

(b) The input bits are $A = 1$, $B = 1$, and $C_{in} = 0$.

$$1 + 1 + 0 = 0 \text{ with a carry of 1}$$

Therefore, $\Sigma = \mathbf{0}$ and $C_{out} = \mathbf{1}$.

(c) The input bits are $A = 1$, $B = 0$, and $C_{in} = 1$.

$$1 + 0 + 1 = 0 \text{ with a carry of 1}$$

Therefore, $\Sigma = \mathbf{0}$ and $C_{out} = \mathbf{1}$.

Related Problem*

What are the full-adder outputs for $A = 1$, $B = 1$, and $C_{in} = 1$?

*Answers are at the end of the chapter.

SECTION 6–1 CHECKUP

Answers are at the end of the chapter.

1. Determine the sum (Σ) and the output carry (C_{out}) of a half-adder for each set of input bits:

   (a) 01   (b) 00   (c) 10   (d) 11

2. A full-adder has $C_{in} = 1$. What are the sum (Σ) and the output carry (C_{out}) when $A = 1$ and $B = 1$?

6–2 Parallel Binary Adders

Two or more full-adders are connected to form parallel binary adders. In this section, you will learn the basic operation of this type of adder and its associated input and output functions.

After completing this section, you should be able to

+ Use full-adders to implement a parallel binary adder

+ Explain the addition process in a parallel binary adder

+ Use the truth table for a 4-bit parallel adder

+ Apply two 74HC283s for the addition of two 8-bit numbers

+ Expand the 4-bit adder to accommodate 8-bit or 16-bit addition

+ Use VHDL to describe a 4-bit parallel adder

   As you learned in Section 6–1, a single full-adder is capable of adding two 1-bit numbers and an input carry. To add binary numbers with more than one bit, you must use additional full-adders. When one binary number is added to another, each column generates a sum bit and a 1 or 0 carry bit to the next column to the left, as illustrated here with 2-bit numbers.

InfoNote

Addition is performed by processors on two numbers at a time, called *operands*. The *source operand* is a number that is to be added to an existing number called the *destination operand,* which is held in an ALU register, such as the accumulator. The sum of the two numbers is then stored back in the accumulator. Addition is performed on integer numbers or floating-point numbers using ADD or FADD instructions respectively.

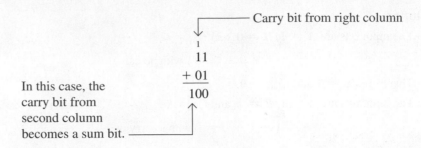

In this case, the carry bit from second column becomes a sum bit.

To add two binary numbers, a full-adder (FA) is required for each bit in the numbers. So for 2-bit numbers, two adders are needed; for 4-bit numbers, four adders are used; and so on. The carry output of each adder is connected to the carry input of the next higher-order adder, as shown in Figure 6–7 for a 2-bit adder. Notice that either a half-adder can be used for the least significant position or the carry input of a full-adder can be made 0 (grounded) because there is no carry input to the least significant bit position.

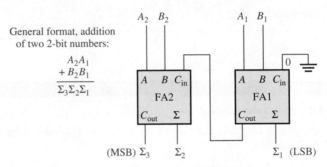

General format, addition of two 2-bit numbers:

$$\begin{array}{r} A_2A_1 \\ + B_2B_1 \\ \hline \Sigma_3\Sigma_2\Sigma_1 \end{array}$$

 FIGURE 6–7   Block diagram of a basic 2-bit parallel adder using two full-adders. Open file F06-07 to verify operation.

In Figure 6–7 the least significant bits (LSB) of the two numbers are represented by A_1 and B_1. The next higher-order bits are represented by A_2 and B_2. The three sum bits are Σ_1, Σ_2, and Σ_3. Notice that the output carry from the left-most full-adder becomes the most significant bit (MSB) in the sum, Σ_3.

EXAMPLE 6–2

Determine the sum generated by the 3-bit parallel adder in Figure 6–8 and show the intermediate carries when the binary numbers 101 and 011 are being added.

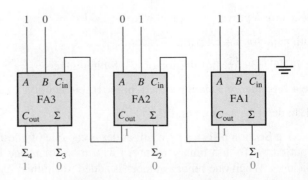

FIGURE 6–8

Solution

The LSBs of the two numbers are added in the right-most full-adder. The sum bits and the intermediate carries are indicated in blue in Figure 6–8.

Related Problem

What are the sum outputs when 111 and 101 are added by the 3-bit parallel adder?

Four-Bit Parallel Adders

A group of four bits is called a **nibble**. A basic 4-bit parallel adder is implemented with four full-adder stages as shown in Figure 6–9. Again, the LSBs (A_1 and B_1) in each number being added go into the right-most full-adder; the higher-order bits are applied as shown to the successively higher-order adders, with the MSBs (A_4 and B_4) in each number being applied to the left-most full-adder. The carry output of each adder is connected to the carry input of the next higher-order adder as indicated. These are called *internal carries*.

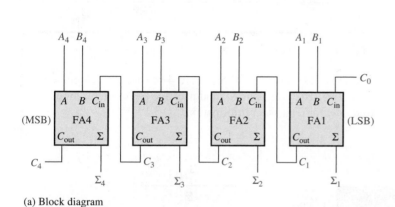

(a) Block diagram   (b) Logic symbol

FIGURE 6–9   A 4-bit parallel adder.

In keeping with most manufacturers' data sheets, the input labeled C_0 is the input carry to the least significant bit adder; C_4, in the case of four bits, is the output carry of the most significant bit adder; and Σ_1 (LSB) through Σ_4 (MSB) are the sum outputs. The logic symbol is shown in Figure 6–9(b).

In terms of the method used to handle carries in a parallel adder, there are two types: the *ripple carry* adder and the *carry look-ahead* adder. These are discussed in Section 6–3.

Truth Table for a 4-Bit Parallel Adder

Table 6–3 is the truth table for a 4-bit adder. On some data sheets, truth tables may be called *function tables* or *functional truth tables*. The subscript n represents the adder bits and can be 1, 2, 3, or 4 for the 4-bit adder. C_{n-1} is the carry from the previous adder. Carries C_1, C_2, and C_3 are generated internally. C_0 is an external carry input and C_4 is an output. Example 6–3 illustrates how to use Table 6–3.

TABLE 6–3

Truth table for each stage of a 4-bit parallel adder.

C_{n-1}	A_n	B_n	Σ_n	C_n
0	0	0	0	0
0	0	1	1	0
0	1	0	1	0
0	1	1	0	1
1	0	0	1	0
1	0	1	0	1
1	1	0	0	1
1	1	1	1	1

EXAMPLE 6–3

Use the 4-bit parallel adder truth table (Table 6–3) to find the sum and output carry for the addition of the following two 4-bit numbers if the input carry (C_{n-1}) is 0:

$$A_4A_3A_2A_1 = 1100 \quad \text{and} \quad B_4B_3B_2B_1 = 1100$$

Solution

For $n = 1$: $A_1 = 0$, $B_1 = 0$, and $C_{n-1} = 0$. From the 1st row of the table,

$$\Sigma_1 = \mathbf{0} \quad \text{and} \quad C_1 = 0$$

For $n = 2$: $A_2 = 0$, $B_2 = 0$, and $C_{n-1} = 0$. From the 1st row of the table,

$$\Sigma_2 = \mathbf{0} \quad \text{and} \quad C_2 = 0$$

For $n = 3$: $A_3 = 1$, $B_3 = 1$, and $C_{n-1} = 0$. From the 4th row of the table,

$$\Sigma_3 = \mathbf{0} \quad \text{and} \quad C_3 = 1$$

For $n = 4$: $A_4 = 1$, $B_4 = 1$, and $C_{n-1} = 1$. From the last row of the table,

$$\Sigma_4 = \mathbf{1} \quad \text{and} \quad C_4 = \mathbf{1}$$

C_4 becomes the output carry; the sum of 1100 and 1100 is 11000.

Related Problem

Use the truth table (Table 6–3) to find the result of adding the binary numbers 1011 and 1010.

IMPLEMENTATION: 4-BIT PARALLEL ADDER

Fixed-Function Device   The 74HC283 and the 74LS283 are 4-bit parallel adders with identical package pin configurations. The logic symbol and package pin configuration are shown in Figure 6–10. Go to *ti.com* for data sheet information.

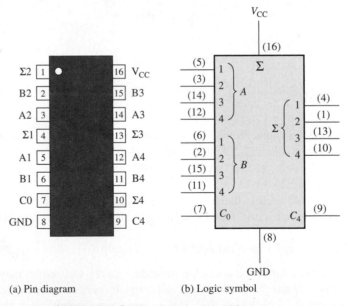

(a) Pin diagram   (b) Logic symbol

FIGURE 6–10   The 74HC283/74LS283 4-bit parallel adder.

Programmable Logic Device (PLD)   A 4-bit adder can be described using VHDL and implemented in a PLD. First, the data flow approach is used to describe the full adder, which is shown in Figure 6–4(b), for use as a component. (Blue text comments are not part of the program.)

entity FullAdder **is**
   port (A, B, CIN: **in** bit; SUM, COUT: **out** bit);   Inputs and outputs declared
end entity FullAdder;

```
architecture LogicOperation of FullAdder is
begin
  SUM <= (A xor B) xor CIN;
  COUT <= ((A xor B) and CIN) or (A and B);
end architecture LogicOperation;
```

⎫
⎬ Boolean expressions for
⎭ the outputs

Next, the FullAdder program code is used as a component in a VHDL structural approach to the 4-bit full-adder in Figure 6–9(a).

A1-A4: Inputs
B1-B4: Inputs
C0: Carry input
S1-S4: Sum outputs
C4: Carry output

```
entity 4BitFullAdder is
  port (A1, A2, A3, A4, B1, B2, B3, B4, C0: in bit; S1, S2, S3, S4, C4: out bit);
end entity 4BitFullAdder;

architecture LogicOperation of 4BitFullAdder is
  component FullAdder is
    port (A, B, CIN: in bit; SUM, COUT: out bit);
  end component FullAdder;

  signal C1, C2, C3: bit;
begin
```

⎫
⎬ Full-adder component
⎭ declaration

Instantiations for each of
the four full adders

```
  FA1: FullAdder port map (A => A1, B => B1, CIN => C0, SUM => S1, COUT => C1);
  FA2: FullAdder port map (A => A2, B => B2, CIN => C1, SUM => S2, COUT => C2);
  FA3: FullAdder port map (A => A3, B => B3, CIN => C2, SUM => S3, COUT => C3);
  FA4: FullAdder port map (A => A4, B => B4, CIN => C3, SUM => S4, COUT => C4);

end architecture LogicOperation;
```

Adder Expansion

The 4-bit parallel adder can be expanded to handle the addition of two 8-bit numbers by using two 4-bit adders. The carry input of the low-order adder (C_0) is connected to ground because there is no carry into the least significant bit position, and the carry output of the low-order adder is connected to the carry input of the high-order adder, as shown in Figure 6–11. This process is known as **cascading**. Notice that, in this case, the output carry is designated C_8 because it is generated from the eighth bit position. The low-order adder is

Adders can be expanded to handle more bits by cascading.

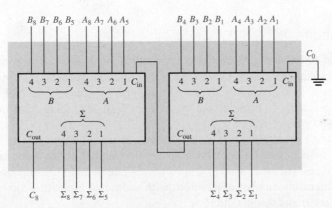

FIGURE 6–11   Cascading of two 4-bit adders to form an 8-bit adder.

the one that adds the lower or less significant four bits in the numbers, and the high-order adder is the one that adds the higher or more significant four bits in the 8-bit numbers. Similarly, four 4-bit adders can be cascaded to handle two 16-bit numbers.

EXAMPLE 6–4

Show how two 74HC283 adders can be connected to form an 8-bit parallel adder. Show output bits for the following 8-bit input numbers:

$$A_8A_7A_6A_5A_4A_3A_2A_1 = 10111001 \quad \text{and} \quad B_8B_7B_6B_5B_4B_3B_2B_1 = 10011110$$

Solution

Two 74HC283 4-bit parallel adders are used to implement the 8-bit adder. The only connection between the two 74HC283s is the carry output (pin 9) of the low-order adder to the carry input (pin 7) of the high-order adder, as shown in Figure 6–12. Pin 7 of the low-order adder is grounded (no carry input).

The sum of the two 8-bit numbers is

$$\Sigma_9\Sigma_8\Sigma_7\Sigma_6\Sigma_5\Sigma_4\Sigma_3\Sigma_2\Sigma_1 = 101010111$$

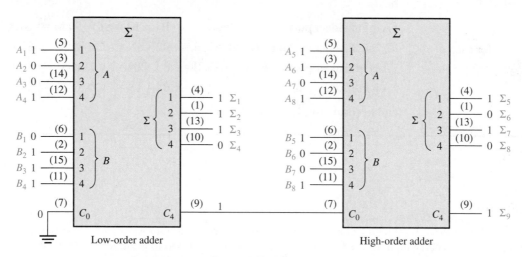

FIGURE 6–12   Two 74HC283 adders connected as an 8-bit parallel adder (pin numbers are in parentheses).

Related Problem

Use 74HC283 adders to implement a 12-bit parallel adder.

An Application

An example of full-adder and parallel adder application is a simple voting system that can be used to simultaneously provide the number of "yes" votes and the number of "no" votes. This type of system can be used where a group of people are assembled and there is a need for immediately determining opinions (for or against), making decisions, or voting on certain issues or other matters.

In its simplest form, the system includes a switch for "yes" or "no" selection at each position in the assembly and a digital display for the number of yes votes and one for the number of no votes. The basic system is shown in Figure 6–13 for a 6-position setup, but it can be expanded to any number of positions with additional 6-position modules and additional parallel adder and display circuits.

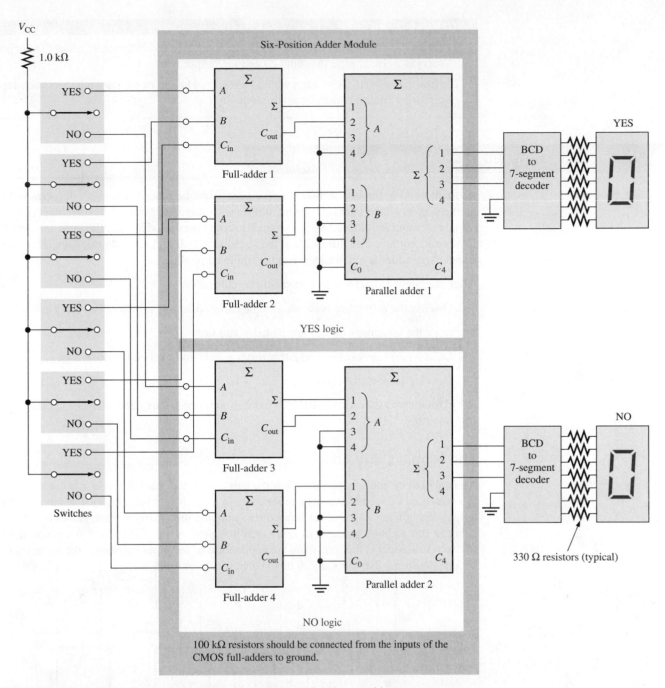

FIGURE 6–13   A voting system using full-adders and parallel binary adders.

In Figure 6–13 each full-adder can produce the sum of up to three votes. The sum and output carry of each full-adder then goes to the two lower-order inputs of a parallel binary adder. The two higher-order inputs of the parallel adder are connected to ground (0) because there is never a case where the binary input exceeds 0011 (decimal 3). For this basic 6-position system, the outputs of the parallel adder go to a BCD-to-7-segment decoder that drives the 7-segment display. As mentioned, additional circuits must be included when the system is expanded.

The resistors from the inputs of each full-adder to ground assure that each input is LOW when the switch is in the neutral position (CMOS logic is used). When a switch is moved to the "yes" or to the "no" position, a HIGH level (V_{CC}) is applied to the associated full-adder input.

6–3   Ripple Carry and Look-Ahead Carry Adders

As mentioned in the last section, parallel adders can be placed into two categories based on the way in which internal carries from stage to stage are handled. Those categories are ripple carry and look-ahead carry. Externally, both types of adders are the same in terms of inputs and outputs. The difference is the speed at which they can add numbers. The look-ahead carry adder is much faster than the ripple carry adder.

After completing this section, you should be able to

◆ Discuss the difference between a ripple carry adder and a look-ahead carry adder

◆ State the advantage of look-ahead carry addition

◆ Define *carry generation* and *carry propagation* and explain the difference

◆ Develop look-ahead carry logic

◆ Explain why cascaded 74HC283s exhibit both ripple carry and look-ahead carry properties

The Ripple Carry Adder

A **ripple carry** adder is one in which the carry output of each full-adder is connected to the carry input of the next higher-order stage (a stage is one full-adder). The sum and the output carry of any stage cannot be produced until the input carry occurs; this causes a time delay in the addition process, as illustrated in Figure 6–14. The carry propagation delay for each full-adder is the time from the application of the input carry until the output carry occurs, assuming that the A and B inputs are already present.

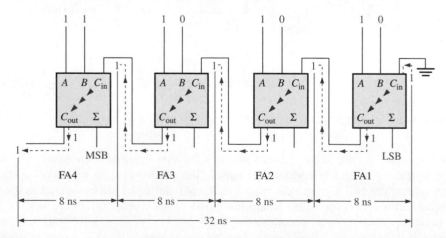

FIGURE 6–14   A 4-bit parallel ripple carry adder showing "worst-case" carry propagation delays.

Full-adder 1 (FA1) cannot produce a potential output carry until an input carry is applied. Full-adder 2 (FA2) cannot produce a potential output carry until FA1 produces an output carry. Full-adder 3 (FA3) cannot produce a potential output carry until an output carry is produced by FA1 followed by an output carry from FA2, and so on. As you can see in Figure 6–14, the input carry to the least significant stage has to ripple through all the adders before a final sum is produced. The cumulative delay through all the adder stages is a "worst-case" addition time. The total delay can vary, depending on the carry bit produced by each full-adder. If two numbers are added such that no carries (0) occur between stages, the addition time is simply the propagation time through a single full-adder from the application of the data bits on the inputs to the occurrence of a sum output; however, worst-case addition time must always be assumed.

The Look-Ahead Carry Adder

The speed with which an addition can be performed is limited by the time required for the carries to propagate, or ripple, through all the stages of a parallel adder. One method of speeding up the addition process by eliminating this ripple carry delay is called **look-ahead carry** addition. The look-ahead carry adder anticipates the output carry of each stage, and based on the inputs, produces the output carry by either carry generation or carry propagation.

Carry generation occurs when an output carry is produced (generated) internally by the full-adder. A carry is generated only when both input bits are 1s. The generated carry, C_g, is expressed as the AND function of the two input bits, A and B.

$$C_g = AB \qquad \text{Equation 6–5}$$

Carry propagation occurs when the input carry is rippled to become the output carry. An input carry may be propagated by the full-adder when either or both of the input bits are 1s. The propagated carry, C_p, is expressed as the OR function of the input bits.

$$C_p = A + B \qquad \text{Equation 6–6}$$

The conditions for carry generation and carry propagation are illustrated in Figure 6–15. The three arrowheads symbolize ripple (propagation).

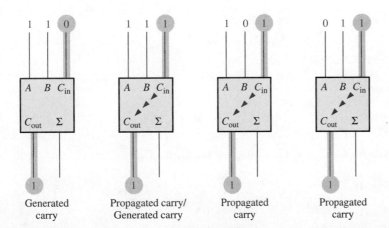

FIGURE 6–15   Illustration of conditions for carry generation and carry propagation.

The output carry of a full-adder can be expressed in terms of both the generated carry (C_g) and the propagated carry (C_p). The output carry (C_{out}) is a 1 if the generated carry is a 1 OR if the propagated carry is a 1 AND the input carry (C_{in}) is a 1. In other words, we get an output carry of 1 if it is generated by the full-adder $(A = 1 \text{ AND } B = 1)$ or if the adder propagates the input carry $(A = 1 \text{ OR } B = 1)$ AND $C_{in} = 1$. This relationship is expressed as

$$C_{out} = C_g + C_p C_{in} \qquad \text{Equation 6–7}$$

Now let's see how this concept can be applied to a parallel adder, whose individual stages are shown in Figure 6–16 for a 4-bit example. For each full-adder, the output carry is dependent on the generated carry (C_g), the propagated carry (C_p), and its input carry (C_{in}). The C_g and C_p functions for each stage are *immediately* available as soon as the input bits A and B and the input carry to the LSB adder are applied because they are dependent only on these bits. The input carry to each stage is the output carry of the previous stage.

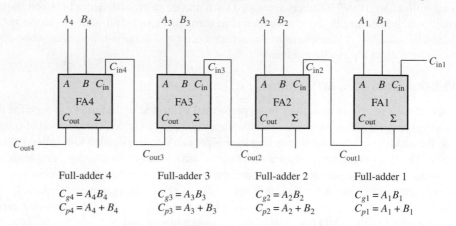

FIGURE 6–16 Carry generation and carry propagation in terms of the input bits to a 4-bit adder.

Based on this analysis, we can now develop expressions for the output carry, C_{out}, of each full-adder stage for the 4-bit example.

Full-adder 1:

$$C_{out1} = C_{g1} + C_{p1}C_{in1}$$

Full-adder 2:

$$C_{in2} = C_{out1}$$
$$C_{out2} = C_{g2} + C_{p2}C_{in2} = C_{g2} + C_{p2}C_{out1} = C_{g2} + C_{p2}(C_{g1} + C_{p1}C_{in1})$$
$$= C_{g2} + C_{p2}C_{g1} + C_{p2}C_{p1}C_{in1}$$

Full-adder 3:

$$C_{in3} = C_{out2}$$
$$C_{out3} = C_{g3} + C_{p3}C_{in3} = C_{g3} + C_{p3}C_{out2} = C_{g3} + C_{p3}(C_{g2} + C_{p2}C_{g1} + C_{p2}C_{p1}C_{in1})$$
$$= C_{g3} + C_{p3}C_{g2} + C_{p3}C_{p2}C_{g1} + C_{p3}C_{p2}C_{p1}C_{in1}$$

Full-adder 4:

$$C_{in4} = C_{out3}$$
$$C_{out4} = C_{g4} + C_{p4}C_{in4} = C_{g4} + C_{p4}C_{out3}$$
$$= C_{g4} + C_{p4}(C_{g3} + C_{p3}C_{g2} + C_{p3}C_{p2}C_{g1} + C_{p3}C_{p2}C_{p1}C_{in1})$$
$$= C_{g4} + C_{p4}C_{g3} + C_{p4}C_{p3}C_{g2} + C_{p4}C_{p3}C_{p2}C_{g1} + C_{p4}C_{p3}C_{p2}C_{p1}C_{in1}$$

Notice that in each of these expressions, the output carry for each full-adder stage is dependent only on the initial input carry (C_{in1}), the C_g and C_p functions of that stage, and the C_g and C_p functions of the preceding stages. Since each of the C_g and C_p functions can be expressed in terms of the A and B inputs to the full-adders, all the output carries are immediately available (except for gate delays), and you do not have to wait for a carry to ripple through all the stages before a final result is achieved. Thus, the look-ahead carry technique speeds up the addition process.

The C_{out} equations are implemented with logic gates and connected to the full-adders to create a 4-bit look-ahead carry adder, as shown in Figure 6–17.

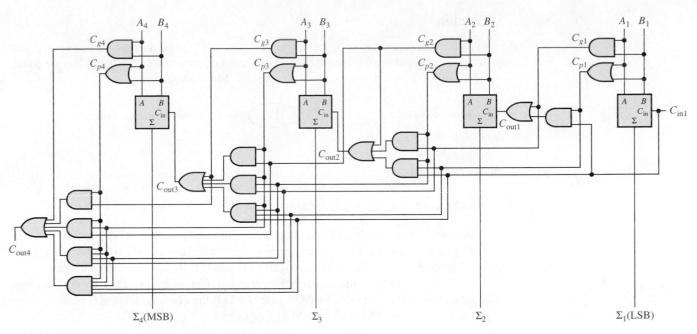

FIGURE 6–17 Logic diagram for a 4-stage look-ahead carry adder.

Combination Look-Ahead and Ripple Carry Adders

As with most fixed-function IC adders, the 74HC283 4-bit adder that was introduced in Section 6–2 is a look-ahead carry adder. When these adders are cascaded to expand their capability to handle binary numbers with more than four bits, the output carry of one adder is connected to the input carry of the next. This creates a ripple carry condition between the 4-bit adders so that when two or more 74HC283s are cascaded, the resulting adder is actually a combination look-ahead and ripple carry adder. The look-ahead carry operation is internal to each MSI adder and the ripple carry feature comes into play when there is a carry out of one of the adders to the next one.

SECTION 6–3 CHECKUP

1. The input bits to a full-adder are $A = 1$ and $B = 0$. Determine C_g and C_p.
2. Determine the output carry of a full-adder when $C_{in} = 1$, $C_g = 0$, and $C_p = 1$.

6–4 Comparators

The basic function of a **comparator** is to compare the magnitudes of two binary quantities to determine the relationship of those quantities. In its simplest form, a comparator circuit determines whether two numbers are equal.

After completing this section, you should be able to

- Use the exclusive-NOR gate as a basic comparator
- Analyze the internal logic of a magnitude comparator that has both equality and inequality outputs
- Apply the 74HC85 comparator to compare the magnitudes of two 4-bit numbers
- Cascade 74HC85s to expand a comparator to eight or more bits
- Use VHDL to describe a 4-bit magnitude comparator

Equality

As you learned in Chapter 3, the exclusive-NOR gate can be used as a basic comparator because its output is a 0 if the two input bits are not equal and a 1 if the input bits are equal. Figure 6–18 shows the exclusive-NOR gate as a 2-bit comparator.

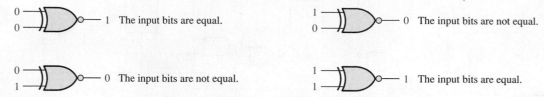

FIGURE 6–18   Basic comparator operation.

In order to compare binary numbers containing two bits each, an additional exclusive-NOR gate is necessary. The two least significant bits (LSBs) of the two numbers are compared by gate G_1, and the two most significant bits (MSBs) are compared by gate G_2, as shown in Figure 6–19. If the two numbers are equal, their corresponding bits are the same, and the output of each exclusive-NOR gate is a 1. If the corresponding sets of bits are not equal, a 0 occurs on that exclusive-NOR gate output.

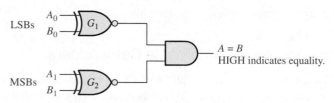

General format: Binary number $A \rightarrow A_1 A_0$
Binary number $B \rightarrow B_1 B_0$

MultiSim   **FIGURE 6–19**   Logic diagram for equality comparison of two 2-bit numbers. Open file F06-19 to verify operation.

In order to produce a single output indicating an equality or inequality of two numbers, an AND gate can be combined with XNOR gates, as shown in Figure 6–19. The output of each exclusive-NOR gate is applied to the AND gate input. When the two input bits for each exclusive-NOR are equal, the corresponding bits of the numbers are equal, producing a 1 on both inputs to the AND gate and thus a 1 on the output. When the two numbers are not equal, one or both sets of corresponding bits are unequal, and a 0 appears on at least one input to the AND gate to produce a 0 on its output. Thus, the output of the AND gate indicates equality (1) or inequality (0) of the two numbers. Example 6–5 illustrates this operation for two specific cases.

A comparator determines if two binary numbers are equal or unequal.

EXAMPLE 6–5

Apply each of the following sets of binary numbers to the comparator inputs in Figure 6–20, and determine the output by following the logic levels through the circuit.

(a) 10 and 10   **(b)** 11 and 10

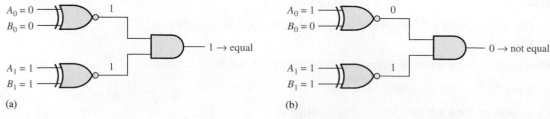

(a)   (b)

FIGURE 6–20

Solution

(a) The output is **1** for inputs 10 and 10, as shown in Figure 6–20(a).

(b) The output is **0** for inputs 11 and 10, as shown in Figure 6–20(b).

Related Problem

Repeat the process for binary inputs of 01 and 10.

As you know from Chapter 3, the basic comparator can be expanded to any number of bits. The AND gate sets the condition that all corresponding bits of the two numbers must be equal if the two numbers themselves are equal.

Inequality

In addition to the equality output, fixed-function comparators can provide additional outputs that indicate which of the two binary numbers being compared is the larger. That is, there is an output that indicates when number A is greater than number B ($A > B$) and an output that indicates when number A is less than number B ($A < B$), as shown in the logic symbol for a 4-bit comparator in Figure 6–21.

To determine an inequality of binary numbers A and B, you first examine the highest-order bit in each number. The following conditions are possible:

1. If $A_3 = 1$ and $B_3 = 0$, number A is greater than number B.

2. If $A_3 = 0$ and $B_3 = 1$, number A is less than number B.

3. If $A_3 = B_3$, then you must examine the next lower bit position for an inequality.

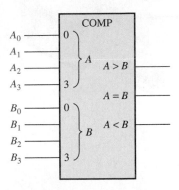

FIGURE 6–21   Logic symbol for a 4-bit comparator with inequality indication.

These three operations are valid for each bit position in the numbers. The general procedure used in a comparator is to check for an inequality in a bit position, starting with the highest-order bits (MSBs). When such an inequality is found, the relationship of the two numbers is established, and any other inequalities in lower-order bit positions must be ignored because it is possible for an opposite indication to occur; *the highest-order indication must take precedence.*

InfoNote

In a computer, the *cache* is a very fast intermediate memory between the central processing unit (CPU) and the slower main memory. The CPU requests data by sending out its *address* (unique location) in memory. Part of this address is called a *tag*. The *tag address comparator* compares the tag from the CPU with the tag from the cache directory. If the two agree, the addressed data is already in the cache and is retrieved very quickly. If the tags disagree, the data must be retrieved from the main memory at a much slower rate.

EXAMPLE 6–6

Determine the $A = B$, $A > B$, and $A < B$ outputs for the input numbers shown on the comparator in Figure 6–22.

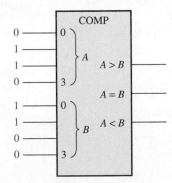

FIGURE 6–22

Solution

The number on the A inputs is 0110 and the number on the B inputs is 0011. The $A > B$ **output is HIGH and the other outputs are LOW.**

Related Problem

What are the comparator outputs when $A_3A_2A_1A_0 = 1001$ and $B_3B_2B_1B_0 = 1010$?

IMPLEMENTATION: 4-BIT MAGNITUDE COMPARATOR

Fixed-Function Device The 74HC85/74LS85 pin diagram and logic symbol are shown in Figure 6–23. Notice that this device has all the inputs and outputs of the generalized comparator previously discussed and, in addition, has three cascading inputs: $A < B, A = B, A > B$. These inputs allow several comparators to be cascaded for comparison of any number of bits greater than four. To expand the comparator, the $A < B$, $A = B$, and $A > B$ outputs of the lower-order comparator are connected to the corresponding cascading inputs of the next higher-order comparator. The lowest-order comparator must have a HIGH on the $A = B$ input and LOWs on the $A < B$ and $A > B$ inputs.

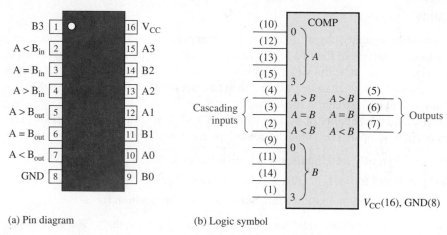

(a) Pin diagram   (b) Logic symbol

FIGURE 6–23   The 74HC85/74LS85 4-bit magnitude comparator.

Programmable Logic Device (PLD) A 4-bit magnitude comparator can be described using VHDL and implemented in a PLD. The following VHDL program uses the data flow approach to implement a simplified comparator ($A = B$ output only) in Figure 6–24. (The blue comments are not part of the program.)

entity 4BitComparator **is**

  port (A0, A1, A2, A3, B0, B1, B2, B3: **in** bit; AequalB: **out** bit);

end entity 4BitComparator;   Inputs and outputs declared

architecture LogicOperation **of** 4BitComparator **is**

begin

AequalB <= (A0 **xnor** B0) **and** (A1 **xnor** B1) **and**   ⎫ Output in terms of a
(A2 **xnor** B2) **and** (A3 **xnor** B);   ⎬ Boolean expression
                                                              ⎭

end architecture LogicOperation;

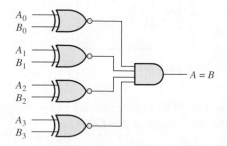

FIGURE 6–24

EXAMPLE 6–7

Use 74HC85 comparators to compare the magnitudes of two 8-bit numbers. Show the comparators with proper interconnections.

Solution

Two 74HC85s are required to compare two 8-bit numbers. They are connected as shown in Figure 6–25 in a cascaded arrangement.

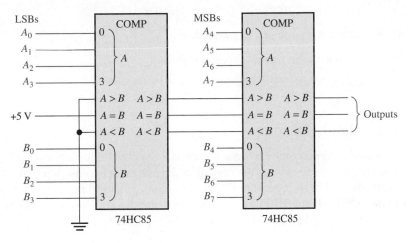

FIGURE 6–25 An 8-bit magnitude comparator using two 74HC85s.

Related Problem

Expand the circuit in Figure 6–25 to a 16-bit comparator.

Most CMOS devices contain protection circuitry to guard against damage from high static voltages or electric fields. However, precautions must be taken to avoid applications of any voltages higher than maximum rated voltages. For proper operation, input and output voltages should be between ground and V_{CC}. Also, remember that unused inputs must always be connected to an appropriate logic level (ground or V_{CC}). Unused outputs may be left open.

SECTION 6–4 CHECKUP

1. The binary numbers $A = 1011$ and $B = 1010$ are applied to the inputs of a 74HC85. Determine the outputs.

2. The binary numbers $A = 11001011$ and $B = 11010100$ are applied to the 8-bit comparator in Figure 6–25. Determine the states of the outputs on each comparator.

6–5   Decoders

A **decoder** is a digital circuit that detects the presence of a specified combination of bits (code) on its inputs and indicates the presence of that code by a specified output level. In its general form, a decoder has n input lines to handle n bits and from one to 2^n output lines to indicate the presence of one or more n-bit combinations. In this section, three fixed-function IC decoders are introduced. The basic principles can be extended to other types of decoders.

After completing this section, you should be able to

◆ Define *decoder*

◆ Design a logic circuit to decode any combination of bits

◆ Describe the 74HC154 binary-to-decimal decoder

◆ Expand decoders to accommodate larger numbers of bits in a code

◆ Describe the 74HC42 BCD-to-decimal decoder

◆ Describe the 74HC47 BCD-to-7-segment decoder

◆ Discuss zero suppression in 7-segment displays

◆ Use VHDL to describe various types of decoders

◆ Apply decoders to specific applications

The Basic Binary Decoder

Suppose you need to determine when a binary 1001 occurs on the inputs of a digital circuit. An AND gate can be used as the basic decoding element because it produces a HIGH output only when all of its inputs are HIGH. Therefore, you must make sure that all of the inputs to the AND gate are HIGH when the binary number 1001 occurs; this can be done by inverting the two middle bits (the 0s), as shown in Figure 6–26.

InfoNote

An *instruction* tells the processor what operation to perform. Instructions are in machine code (1s and 0s) and, in order for the processor to carry out an instruction, the instruction must be decoded. Instruction decoding is one of the steps in instruction *pipelining*, which are as follows: Instruction is read from the memory (instruction fetch), instruction is decoded, operand(s) is (are) read from memory (operand fetch), instruction is executed, and result is written back to memory. Basically, pipelining allows the next instruction to begin processing before the current one is completed.

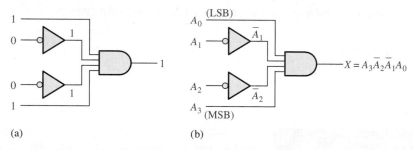

FIGURE 6–26   Decoding logic for the binary code 1001 with an active-HIGH output.

The logic equation for the decoder of Figure 6–26(a) is developed as illustrated in Figure 6–26(b). You should verify that the output is 0 except when $A_0 = 1$, $A_1 = 0$, $A_2 = 0$, and $A_3 = 1$ are applied to the inputs. A_0 is the LSB and A_3 is the MSB. *In the representation of a binary number or other weighted code in this book, the LSB is the right-most bit in a horizontal arrangement and the topmost bit in a vertical arrangement, unless specified otherwise.*

If a NAND gate is used in place of the AND gate in Figure 6–26, a LOW output will indicate the presence of the proper binary code, which is 1001 in this case.

EXAMPLE 6–8

Determine the logic required to decode the binary number 1011 by producing a HIGH level on the output.

Solution

The decoding function can be formed by complementing only the variables that appear as 0 in the desired binary number, as follows:

$$X = A_3\overline{A}_2A_1A_0 \quad (1011)$$

This function can be implemented by connecting the true (uncomplemented) variables A_0, A_1, and A_3 directly to the inputs of an AND gate, and inverting the variable A_2 before applying it to the AND gate input. The decoding logic is shown in Figure 6–27.

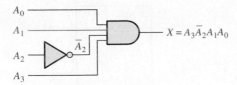

FIGURE 6–27 Decoding logic for producing a HIGH output when 1011 is on the inputs.

Related Problem

Develop the logic required to detect the binary code 10010 and produce an active-LOW output.

The 4-Bit Decoder

In order to decode all possible combinations of four bits, sixteen decoding gates are required ($2^4 = 16$). This type of decoder is commonly called either a *4-line-to-16-line decoder* because there are four inputs and sixteen outputs or a *1-of-16 decoder* because for any given code on the inputs, one of the sixteen outputs is activated. A list of the sixteen binary codes and their corresponding decoding functions is given in Table 6–4.

TABLE 6–4

Decoding functions and truth table for a 4-line-to-16-line (1-of-16) decoder with active-LOW outputs.

Decimal Digit	Binary Inputs				Decoding Function	Outputs															
	A_3	A_2	A_1	A_0		0	1	2	3	4	5	6	7	8	9	10	11	12	13	14	15
0	0	0	0	0	$\overline{A}_3\overline{A}_2\overline{A}_1\overline{A}_0$	0	1	1	1	1	1	1	1	1	1	1	1	1	1	1	1
1	0	0	0	1	$\overline{A}_3\overline{A}_2\overline{A}_1 A_0$	1	0	1	1	1	1	1	1	1	1	1	1	1	1	1	1
2	0	0	1	0	$\overline{A}_3\overline{A}_2 A_1\overline{A}_0$	1	1	0	1	1	1	1	1	1	1	1	1	1	1	1	1
3	0	0	1	1	$\overline{A}_3\overline{A}_2 A_1 A_0$	1	1	1	0	1	1	1	1	1	1	1	1	1	1	1	1
4	0	1	0	0	$\overline{A}_3 A_2\overline{A}_1\overline{A}_0$	1	1	1	1	0	1	1	1	1	1	1	1	1	1	1	1
5	0	1	0	1	$\overline{A}_3 A_2\overline{A}_1 A_0$	1	1	1	1	1	0	1	1	1	1	1	1	1	1	1	1
6	0	1	1	0	$\overline{A}_3 A_2 A_1\overline{A}_0$	1	1	1	1	1	1	0	1	1	1	1	1	1	1	1	1
7	0	1	1	1	$\overline{A}_3 A_2 A_1 A_0$	1	1	1	1	1	1	1	0	1	1	1	1	1	1	1	1
8	1	0	0	0	$A_3\overline{A}_2\overline{A}_1\overline{A}_0$	1	1	1	1	1	1	1	1	0	1	1	1	1	1	1	1
9	1	0	0	1	$A_3\overline{A}_2\overline{A}_1 A_0$	1	1	1	1	1	1	1	1	1	0	1	1	1	1	1	1
10	1	0	1	0	$A_3\overline{A}_2 A_1\overline{A}_0$	1	1	1	1	1	1	1	1	1	1	0	1	1	1	1	1
11	1	0	1	1	$A_3\overline{A}_2 A_1 A_0$	1	1	1	1	1	1	1	1	1	1	1	0	1	1	1	1
12	1	1	0	0	$A_3 A_2\overline{A}_1\overline{A}_0$	1	1	1	1	1	1	1	1	1	1	1	1	0	1	1	1
13	1	1	0	1	$A_3 A_2\overline{A}_1 A_0$	1	1	1	1	1	1	1	1	1	1	1	1	1	0	1	1
14	1	1	1	0	$A_3 A_2 A_1\overline{A}_0$	1	1	1	1	1	1	1	1	1	1	1	1	1	1	0	1
15	1	1	1	1	$A_3 A_2 A_1 A_0$	1	1	1	1	1	1	1	1	1	1	1	1	1	1	1	0

If an active-LOW output is required for each decoded number, the entire decoder can be implemented with NAND gates and inverters. In order to decode each of the sixteen binary codes, sixteen NAND gates are required (AND gates can be used to produce active-HIGH outputs).

A logic symbol for a 4-line-to-16-line (1-of-16) decoder with active-LOW outputs is shown in Figure 6–28. The BIN/DEC label indicates that a binary input makes the corresponding decimal output active. The input labels 8, 4, 2, and 1 represent the binary weights of the input bits ($2^3 2^2 2^1 2^0$).

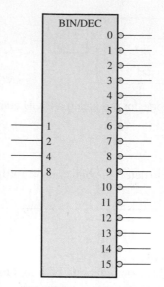

FIGURE 6–28   Logic symbol for a 4-line-to-16-line (1-of-16) decoder. Open file F06-28 to verify operation.

IMPLEMENTATION: 1-OF-16 DECODER

Fixed-Function Device   The 74HC154 is a good example of a fixed-function IC decoder. The logic symbol is shown in Figure 6–29. There is an enable function (*EN*) provided on this device, which is implemented with a NOR gate used as a negative-AND. A LOW level on each chip select input, $\overline{CS}_1$ and $\overline{CS}_2$, is required in order to make the enable gate output (*EN*) HIGH. The enable gate output is connected to an input of *each* NAND gate in the decoder, so it must be HIGH for the NAND gates to be enabled. If the enable gate is not activated by a LOW on both inputs, then all sixteen decoder outputs (*OUT*) will be HIGH regardless of the states of the four input variables, A_0, A_1, A_2, and A_3.

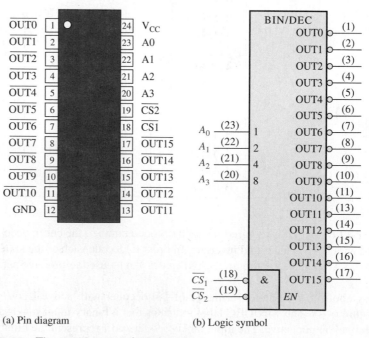

(a) Pin diagram   (b) Logic symbol

FIGURE 6–29   The 74HC154 1-of-16 decoder.

Programmable Logic Device (PLD)   The 1-of-16 decoder can be described using VHDL and implemented as hardware in a PLD. The decoder consists of sixteen 5-input NAND gates for decoding, a 2-input negative-AND for the enable function, and four inverters. The following VHDL program code uses the data flow approach. (Blue text comments are not part of the program.)

entity 1of16Decoder **is**

    port (A0, A1, A2, A3, CS1, CS2: **in** bit; OUT0, OUT1, OUT2, OUT3, OUT4, OUT5, OUT6, OUT7, OUT8, OUT9, OUT10, OUT11, OUT12, OUT13, OUT14, OUT15: **out** bit);   } Inputs and outputs declared

end entity 1of16Decoder;

architecture LogicOperation **of** 1of16Decoder **is**

signal EN: bit;

begin

    OUT0 <= **not**(**not** A0 **and not** A1 **and not** A2 **and not** A3 **and** EN);

    OUT1 <= **not**(A0 **and not** A1 **and not** A2 **and not** A3 **and** EN);

    OUT2 <= **not**(**not** A0 **and** A1 **and not** A2 **and not** A3 **and** EN);

    OUT3 <= **not**(A0 **and** A1 **and not** A2 **and not** A3 **and** EN);

    OUT4 <= **not**(**not** A0 **and not** A1 **and** A2 **and not** A3 **and** EN);

    OUT5 <= **not**(A0 **and not** A1 **and** A2 **and not** A3 **and** EN);

    OUT6 <= **not**(**not** A0 **and** A1 **and** A2 **and not** A3 **and** EN);

    OUT7 <= **not**(A0 **and** A1 **and** A2 **and not** A3 **and** EN);

    OUT8 <= **not**(**not** A0 **and not** A1 **and not** A2 **and** A3 **and** EN);   } Boolean expressions for the sixteen outputs

    OUT9 <= **not**(A0 **and not** A1 **and not** A2 **and** A3 **and** EN);

    OUT10 <= **not**(**not** A0 **and** A1 **and not** A2 **and** A3 **and** EN);

    OUT11 <= **not**(A0 **and** A1 **and not** A2 **and** A3 **and** EN);

    OUT12 <= **not**(**not** A0 **and not** A1 **and** A2 **and** A3 **and** EN);

    OUT13 <= **not**(A0 **and not** A1 **and** A2 **and** A3 **and** EN);

    OUT14 <= **not**(**not** A0 **and** A1 **and** A2 **and** A3 **and** EN);

    OUT15 <= **not**(A0 **and** A1 **and** A2 **and** A3 **and** EN);

    EN <= **not** CS1 **and not** CS2;

end architecture LogicOperation;

EXAMPLE 6–9

A certain application requires that a 5-bit number be decoded. Use 74HC154 decoders to implement the logic. The binary number is represented by the format $A_4A_3A_2A_1A_0$.

Solution

Since the 74HC154 can handle only four bits, two decoders must be used to form a 5-bit expansion. The fifth bit, A_4, is connected to the chip select inputs, $\overline{CS}_1$ and $\overline{CS}_2$, of one decoder, and $\overline{A}_4$ is connected to the $\overline{CS}_1$ and $\overline{CS}_2$ inputs of the other decoder, as shown in Figure 6–30. When the decimal number is 15 or less, $A_4 = 0$, the low-order decoder is enabled, and the high-order decoder is disabled. When the decimal number is greater than 15, $A_4 = 1$ so $\overline{A}_4 = 0$, the high-order decoder is enabled, and the low-order decoder is disabled.

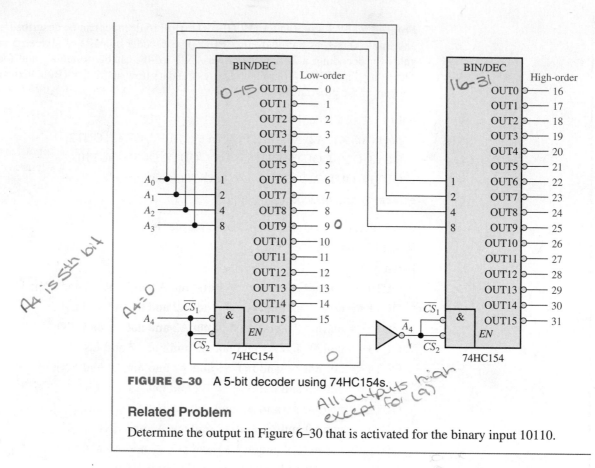

FIGURE 6–30 A 5-bit decoder using 74HC154s.

Related Problem

Determine the output in Figure 6–30 that is activated for the binary input 10110.

The BCD-to-Decimal Decoder

The BCD-to-decimal decoder converts each BCD code (8421 code) into one of ten possible decimal digit indications. It is frequently referred as a *4-line-to-10-line decoder* or a *1-of-10 decoder*.

The method of implementation is the same as for the 1-of-16 decoder previously discussed, except that only ten decoding gates are required because the BCD code represents only the ten decimal digits 0 through 9. A list of the ten BCD codes and their corresponding decoding functions is given in Table 6–5. Each of these decoding functions is implemented with NAND gates to provide active-LOW outputs. If an active-HIGH output is required, AND gates are used for decoding. The logic is identical to that of the first ten decoding gates in the 1-of-16 decoder (see Table 6–4).

TABLE 6–5

BCD decoding functions.

Decimal Digit	BCD Code				Decoding Function
	A_3	A_2	A_1	A_0	
0	0	0	0	0	$\overline{A_3}\,\overline{A_2}\,\overline{A_1}\,\overline{A_0}$
1	0	0	0	1	$\overline{A_3}\,\overline{A_2}\,\overline{A_1}\,A_0$
2	0	0	1	0	$\overline{A_3}\,\overline{A_2}\,A_1\,\overline{A_0}$
3	0	0	1	1	$\overline{A_3}\,\overline{A_2}\,A_1\,A_0$
4	0	1	0	0	$\overline{A_3}\,A_2\,\overline{A_1}\,\overline{A_0}$
5	0	1	0	1	$\overline{A_3}\,A_2\,\overline{A_1}\,A_0$
6	0	1	1	0	$\overline{A_3}\,A_2\,A_1\,\overline{A_0}$
7	0	1	1	1	$\overline{A_3}\,A_2\,A_1\,A_0$
8	1	0	0	0	$A_3\,\overline{A_2}\,\overline{A_1}\,\overline{A_0}$
9	1	0	0	1	$A_3\,\overline{A_2}\,\overline{A_1}\,A_0$

IMPLEMENTATION: BCD-TO-DECIMAL DECODER

Fixed-Function Device   The 74HC42 is a fixed-function IC decoder with four BCD inputs and ten active-LOW decimal outputs. The logic symbol is shown in Figure 6–31.

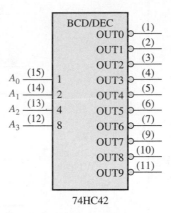

FIGURE 6–31 The 74HC42 BCD-to-decimal decoder.

Programmable Logic Device (PLD)   The logic of the BCD-to-decimal decoder is similar to the 1-of-16 decoder except simpler. In this case, there are ten gates and four inverters instead of sixteen gates and four inverters. This decoder does not have an enable function. Using the data flow approach, the VHDL program code for the 1-of-16 decoder can be simplified to implement the BCD-to-decimal decoder.

entity BCDdecoder **is**

     port (A0, A1, A2, A3: **in** bit; OUT0, OUT1, OUT2, OUT3,   ⎫   Inputs and outputs

     OUT4, OUT5, OUT6, OUT7, OUT8, OUT9: **out** bit);   ⎬   declared

end entity BCDdecoder;

architecture LogicOperation **of** BCDdecoder **is**

begin

     OUT0 <= **not**(**not** A0 **and not** A1 **and not** A2 **and not** A3);

     OUT1 <= **not**(A0 **and not** A1 **and not** A2 **and not** A3);

     OUT2 <= **not**(**not** A0 **and** A1 **and not** A2 **and not** A3);

     OUT3 <= **not**(A0 **and** A1 **and not** A2 **and not** A3);

     OUT4 <= **not**(**not** A0 **and not** A1 **and** A2 **and not** A3);   Boolean expressions

     OUT5 <= **not**(A0 **and not** A1 **and** A2 **and not** A3);   for the ten outputs

     OUT6 <= **not**(**not** A0 **and** A1 **and** A2 **and not** A3);

     OUT7 <= **not**(A0 **and** A1 **and** A2 **and not** A3);

     OUT8 <= **not**(**not** A0 **and not** A1 **and not** A2 **and** A3);

     OUT9 <= **not**(A0 **and not** A1 **and not** A2 **and** A3);

end architecture LogicOperation;

EXAMPLE 6–10

If the input waveforms in Figure 6–32(a) are applied to the inputs of the 74HC42, show the output waveforms.

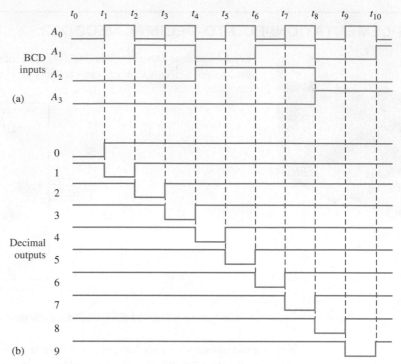

FIGURE 6–32

Solution

The output waveforms are shown in Figure 6–32(b). As you can see, the inputs are sequenced through the BCD for digits 0 through 9. The output waveforms in the timing diagram indicate that sequence on the decimal-value outputs.

Related Problem

Construct a timing diagram showing input and output waveforms for the case where the BCD inputs sequence through the decimal numbers as follows: 0, 2, 4, 6, 8, 1, 3, 5, and 9.

The BCD-to-7-Segment Decoder

The BCD-to-7-segment decoder accepts the BCD code on its inputs and provides outputs to drive 7-segment display devices to produce a decimal readout. The logic diagram for a basic 7-segment decoder is shown in Figure 6–33.

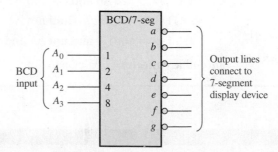

 FIGURE 6–33   Logic symbol for a BCD-to-7-segment decoder/driver with active-LOW outputs. Open file F06-33 to verify operation.

IMPLEMENTATION: BCD-TO-7-SEGMENT DECODER/DRIVER

Fixed-Function Device The 74HC47 is an example of an IC device that decodes a BCD input and drives a 7-segment display. In addition to its decoding and segment drive capability, the 74HC47 has several additional features as indicated by the $\overline{LT}$, $\overline{RBI}$, $\overline{BI}/\overline{RBO}$ functions in the logic symbol of Figure 6–34. As indicated by the bubbles on the logic symbol, all of the outputs (*a* through *g*) are active-LOW as are the $\overline{LT}$ (lamp test), $\overline{RBI}$ (ripple blanking input), and $\overline{BI}/\overline{RBO}$ (blanking input/ripple blanking output) functions. The outputs can drive a common-anode 7-segment display directly. Recall that 7-segment displays were discussed in Chapter 4. In addition to decoding a BCD input and producing the appropriate 7-segment outputs, the 74HC47 has lamp test and zero suppression capability.

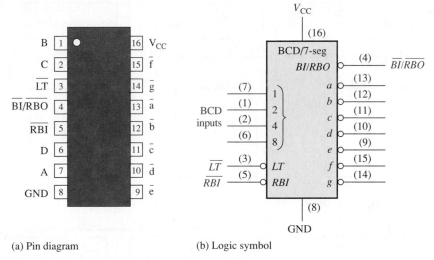

(a) Pin diagram   (b) Logic symbol

FIGURE 6–34 The 74HC47 BCD-to-7-segment decoder/driver.

Lamp Test When a LOW is applied to the $\overline{LT}$ input and the $\overline{BI}/\overline{RBO}$ is HIGH, all of the seven segments in the display are turned on. Lamp test is used to verify that no segments are burned out.

Zero Suppression **Zero suppression** is a feature used for multidigit displays to blank out unnecessary zeros. For example, in a 6-digit display the number 6.4 may be displayed as 006.400 if the zeros are not blanked out. Blanking the zeros at the front of a number is called *leading zero suppression* and blanking the zeros at the back of the number is called *trailing zero suppression.* Keep in mind that only nonessential zeros are blanked. With zero suppression, the number 030.080 will be displayed as 30.08 (the essential zeros remain).

Zero suppression in the 74HC47 is accomplished using the $\overline{RBI}$ and $\overline{BI}/\overline{RBO}$ functions. $\overline{RBI}$ is the ripple blanking input and $\overline{RBO}$ is the ripple blanking output on the 74HC47; these are used for zero suppression. $\overline{BI}$ is the blanking input that shares the same pin with $\overline{RBO}$; in other words, the $\overline{BI}/\overline{RBO}$ pin can be used as an input or an output. When used as a $\overline{BI}$ (blanking input), all segment outputs are HIGH (nonactive) when $\overline{BI}$ is LOW, which overrides all other inputs. The $\overline{BI}$ function is not part of the zero suppression capability of the device.

All of the segment outputs of the decoder are nonactive (HIGH) if a zero code (0000) is on its BCD inputs and if its $\overline{RBI}$ is LOW. This causes the display to be blank and produces a LOW $\overline{RBO}$.

Programmable Logic Device (PLD) The VHDL program code is the same as for the 74HC42 BCD-to-decimal decoder, except the 74HC47 has fewer outputs.

Zero Suppression for a 4-Digit Display

Zero suppression results in leading or trailing zeros in a number not showing on a display.

The logic diagram in Figure 6–35(a) illustrates leading zero suppression for a whole number. The highest-order digit position (left-most) is always blanked if a zero code is on its BCD inputs because the $\overline{RBI}$ of the most-significant decoder is made LOW by connecting it to ground. The $\overline{RBO}$ of each decoder is connected to the $\overline{RBI}$ of the next lowest-order decoder so that all zeros to the left of the first nonzero digit are blanked. For example, in part (a) of the figure the two highest-order digits are zeros and therefore are blanked. The remaining two digits, 3 and 0 are displayed.

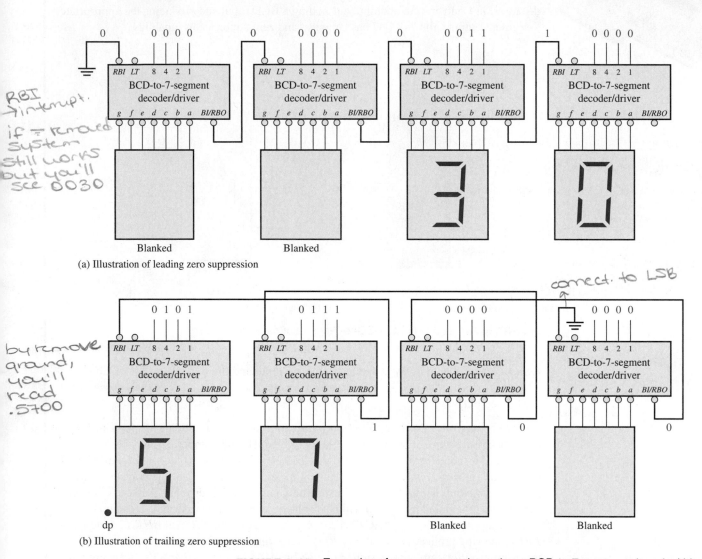

(a) Illustration of leading zero suppression

(b) Illustration of trailing zero suppression

FIGURE 6–35 Examples of zero suppression using a BCD-to-7-segment decoder/driver.

The logic diagram in Figure 6–35(b) illustrates trailing zero suppression for a fractional number. The lowest-order digit (right-most) is always blanked if a zero code is on its BCD inputs because the $\overline{RBI}$ is connected to ground. The $\overline{RBO}$ of each decoder is connected to the $\overline{RBI}$ of the next highest-order decoder so that all zeros to the right of the first nonzero digit are blanked. In part (b) of the figure, the two lowest-order digits are zeros and therefore are blanked. The remaining two digits, 5 and 7 are displayed. To combine both leading and trailing zero suppression in one display and to have decimal point capability, additional logic is required.

SECTION 6–5 CHECKUP

1. A 3-line-to-8-line decoder can be used for octal-to-decimal decoding. When a binary 101 is on the inputs, which output line is activated?

2. How many 74HC154 1-of-16 decoders are necessary to decode a 6-bit binary number?

3. Would you select a decoder/driver with active-HIGH or active-LOW outputs to drive a common-cathode 7-segment LED display?

6–6   Encoders

An **encoder** is a combinational logic circuit that essentially performs a "reverse" decoder function. An encoder accepts an active level on one of its inputs representing a digit, such as a decimal or octal digit, and converts it to a coded output, such as BCD or binary. Encoders can also be devised to encode various symbols and alphabetic characters. The process of converting from familiar symbols or numbers to a coded format is called *encoding*.

After completing this section, you should be able to

* Determine the logic for a decimal-to-BCD encoder

* Explain the purpose of the priority feature in encoders

* Describe the 74HC147 decimal-to-BCD priority encoder

* Use VHDL to describe a decimal-to-BCD encoder

* Apply the encoder to a specific application

The Decimal-to-BCD Encoder

This type of encoder has ten inputs—one for each decimal digit—and four outputs corresponding to the BCD code, as shown in Figure 6–36. This is a basic 10-line-to-4-line encoder.

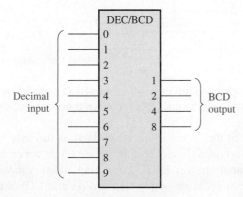

FIGURE 6–36   Logic symbol for a decimal-to-BCD encoder.

The BCD (8421) code is listed in Table 6–6. From this table you can determine the relationship between each BCD bit and the decimal digits in order to analyze the logic. For instance, the most significant bit of the BCD code, A_3, is always a 1 for decimal digit 8 or 9. An OR expression for bit A_3 in terms of the decimal digits can therefore be written as

$$A_3 = 8 + 9$$

TABLE 6–6

Decimal Digit	BCD Code			
	A_3	A_2	A_1	A_0
0	0	0	0	0
1	0	0	0	1
2	0	0	1	0
3	0	0	1	1
4	0	1	0	0
5	0	1	0	1
6	0	1	1	0
7	0	1	1	1
8	1	0	0	0
9	1	0	0	1

Bit A_2 is always a 1 for decimal digit 4, 5, 6 or 7 and can be expressed as an OR function as follows:

$$A_2 = 4 + 5 + 6 + 7$$

Bit A_1 is always a 1 for decimal digit 2, 3, 6, or 7 and can be expressed as

$$A_1 = 2 + 3 + 6 + 7$$

Finally, A_0 is always a 1 for decimal digit 1, 3, 5, 7, or 9. The expression for A_0 is

$$A_0 = 1 + 3 + 5 + 7 + 9$$

Now let's implement the logic circuitry required for encoding each decimal digit to a BCD code by using the logic expressions just developed. It is simply a matter of ORing the appropriate decimal digit input lines to form each BCD output. The basic encoder logic resulting from these expressions is shown in Figure 6–37.

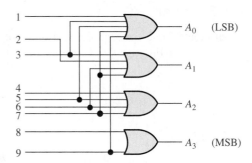

FIGURE 6–37 Basic logic diagram of a decimal-to-BCD encoder. A 0-digit input is not needed because the BCD outputs are all LOW when there are no HIGH inputs.

The basic operation of the circuit in Figure 6–37 is as follows: When a HIGH appears on *one* of the decimal digit input lines, the appropriate levels occur on the four BCD output lines. For instance, if input line 9 is HIGH (assuming all other input lines are LOW), this condition will produce a HIGH on outputs A_0 and A_3 and LOWs on outputs A_1 and A_2, which is the BCD code (1001) for decimal 9.

The Decimal-to-BCD Priority Encoder

This type of encoder performs the same basic encoding function as previously discussed. A **priority encoder** also offers additional flexibility in that it can be used in applications that require priority detection. The priority function means that the encoder will produce a BCD output corresponding to the *highest-order decimal digit* input that is active and will ignore any other lower-order active inputs. For instance, if the 6 and the 3 inputs are both active, the BCD output is 0110 (which represents decimal 6).

IMPLEMENTATION: DECIMAL-TO-BCD ENCODER

Fixed-Function Device The 74HC147 is a priority encoder with active-LOW inputs (0) for decimal digits 1 through 9 and active-LOW BCD outputs as indicated in the logic symbol in Figure 6–38. A BCD zero output is represented when none of the inputs is active. The device pin numbers are in parentheses.

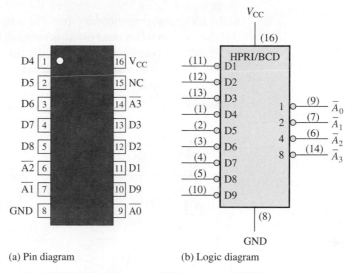

(a) Pin diagram   (b) Logic diagram

FIGURE 6–38 The 74HC147 decimal-to-BCD encoder (HPRI means highest value input has priority.

Programmable Logic Device (PLD) The logic of the decimal-to-BCD encoder shown in Figure 6–38 can be described in VHDL for implementation in a PLD. The data flow approach is used in this case.

entity DecBCDencoder **is**

   port (D1, D2, D3, D4, D5, D6, D7, D8, D9: $\left.\vphantom{\begin{array}{c}a\\b\end{array}}\right\}$ Inputs and outputs declared
   in bit; A0, A1, A2, A3: **out** bit);

end entity DecBCDencoder;

architecture LogicFunction **of** DecBCDencoder **is**

begin

   A0 <= (D1 **or** D3 **or** D5 **or** D7 **or** D9); $\left.\vphantom{\begin{array}{c}a\\b\\c\\d\end{array}}\right\}$ Boolean expressions for the
   A1 <= (D2 **or** D3 **or** D6 **or** D7);
   A2 <= (D4 **or** D5 **or** D6 **or** D7); four BCD outputs
   A3 <= (D8 **or** D9);

end architecture LogicFunction;

EXAMPLE 6–11

If LOW levels appear on pins, 1, 4, and 13 of the 74HC147 shown in Figure 6–38, indicate the state of the four outputs. All other inputs are HIGH.

Solution

Pin 4 is the highest-order decimal digit input having a LOW level and represents decimal 7. Therefore, the output levels indicate the BCD code for decimal 7 where $\overline{A}_0$ is the LSB and $\overline{A}_3$ is the MSB. Output $\overline{A}_0$ is LOW, $\overline{A}_1$ is LOW, $\overline{A}_2$ is LOW, and $\overline{A}_3$ is HIGH.

Related Problem

What are the outputs of the 74HC147 if all its inputs are LOW? If all its inputs are HIGH?

An Application

The ten decimal digits on a numeric keypad must be encoded for processing by the logic circuitry. In this example, when one of the keys is pressed, the decimal digit is encoded to the corresponding BCD code. Figure 6–39 shows a simple keyboard encoder arrangement using a priority encoder. The keys are represented by ten push-button switches, each with a **pull-up resistor** to $+V$. The pull-up resistor ensures that the line is HIGH when a key is not depressed. When a key is depressed, the line is connected to ground, and a LOW is applied to the corresponding encoder input. The zero key is not connected because the BCD output represents zero when none of the other keys is depressed.

The BCD complement output of the encoder goes into a storage device, and each successive BCD code is stored until the entire number has been entered. Methods of storing BCD numbers and binary data are covered in Chapter 11.

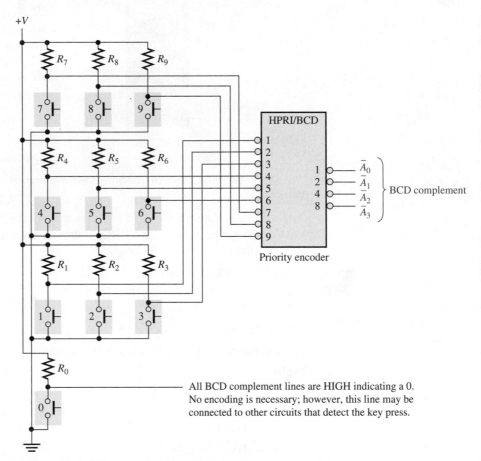

All BCD complement lines are HIGH indicating a 0. No encoding is necessary; however, this line may be connected to other circuits that detect the key press.

FIGURE 6–39   A simplified keyboard encoder.

SECTION 6–6 CHECKUP

1. Suppose the HIGH levels are applied to the 2 input and the 9 input of the circuit in Figure 6–37.

   (a) What are the states of the output lines?

   (b) Does this represent a valid BCD code?

   (c) What is the restriction on the encoder logic in Figure 6–37?

2. (a) What is the $\overline{A}_3 \overline{A}_2 \overline{A}_1 \overline{A}_0$ output when LOWs are applied to pins 1 and 5 of the 74HC147 in Figure 6–38?

   (b) What does this output represent?

6–7 Code Converters

In this section, we will examine some methods of using combinational logic circuits to convert from one code to another.

After completing this section, you should be able to

- ◆ Explain the process for converting BCD to binary
- ◆ Use exclusive-OR gates for conversions between binary and Gray codes

BCD-to-Binary Conversion

One method of BCD-to-binary code conversion uses adder circuits. The basic conversion process is as follows:

1. The value, or weight, of each bit in the BCD number is represented by a binary number.
2. All of the binary representations of the weights of bits that are 1s in the BCD number are added.
3. The result of this addition is the binary equivalent of the BCD number.

A more concise statement of this operation is

The binary numbers representing the weights of the BCD bits are summed to produce the total binary number.

Let's examine an 8-bit BCD code (one that represents a 2-digit decimal number) to understand the relationship between BCD and binary. For instance, you already know that the decimal number 87 can be expressed in BCD as

$$\underbrace{1000}_{8} \quad \underbrace{0111}_{7}$$

The left-most 4-bit group represents 80, and the right-most 4-bit group represents 7. That is, the left-most group has a weight of 10, and the right-most group has a weight of 1. Within each group, the binary weight of each bit is as follows:

	Tens Digit				Units Digit			
Weight:	80	40	20	10	8	4	2	1
Bit designation:	B_3	B_2	B_1	B_0	A_3	A_2	A_1	A_0

The binary equivalent of each BCD bit is a binary number representing the weight of that bit within the total BCD number. This representation is given in Table 6–7.

TABLE 6–7

Binary representations of BCD bit weights.

BCD Bit	BCD Weight	(MSB) 64	32	Binary Representation 16	8	4	2	(LSB) 1
A_0	1	0	0	0	0	0	0	1
A_1	2	0	0	0	0	0	1	0
A_2	4	0	0	0	0	1	0	0
A_3	8	0	0	0	1	0	0	0
B_0	10	0	0	0	1	0	1	0
B_1	20	0	0	1	0	1	0	0
B_2	40	0	1	0	1	0	0	0
B_3	80	1	0	1	0	0	0	0

If the binary representations for the weights of all the 1s in the BCD number are added, the result is the binary number that corresponds to the BCD number. Example 6–12 illustrates this.

EXAMPLE 6–12

Convert the BCD numbers 00100111 (decimal 27) and 10011000 (decimal 98) to binary.

Solution

Write the binary representations of the weights of all 1s appearing in the numbers, and then add them together.

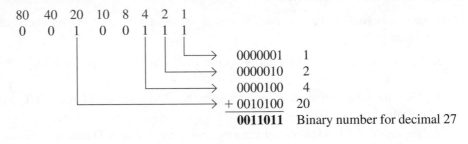

$$
\begin{array}{cccccccc}
80 & 40 & 20 & 10 & 8 & 4 & 2 & 1 \\
0 & 0 & 1 & 0 & 0 & 1 & 1 & 1
\end{array}
$$

0000001	1
0000010	2
0000100	4
+ 0010100	20
0011011	Binary number for decimal 27

$$
\begin{array}{cccccccc}
80 & 40 & 20 & 10 & 8 & 4 & 2 & 1 \\
1 & 0 & 0 & 1 & 1 & 0 & 0 & 0
\end{array}
$$

0001000	8
0001010	10
+ 1010000	80
1100010	Binary number for decimal 98

Related Problem

Show the process of converting 01000001 in BCD to binary.

MultiSim Open file EX06-12 and run the simulation to observe the operation of a BCD-to-binary logic circuit.

Binary-to-Gray and Gray-to-Binary Conversion

The basic process for Gray-binary conversions was covered in Chapter 2. Exclusive-OR gates can be used for these conversions. Programmable logic devices (PLDs) can also be programmed for these code conversions. Figure 6–40 shows a 4-bit binary-to-Gray code converter, and Figure 6–41 illustrates a 4-bit Gray-to-binary converter.

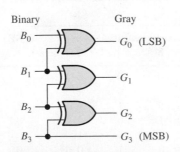

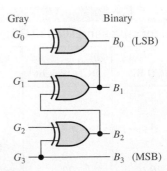

MultiSim **FIGURE 6–40** Four-bit binary-to-Gray conversion logic. Open file F06-40 to verify operation.

MultiSim **FIGURE 6–41** Four-bit Gray-to-binary conversion logic. Open file F06-41 to verify operation.

EXAMPLE 6–13

(a) Convert the binary number 0101 to Gray code with exclusive-OR gates.

(b) Convert the Gray code 1011 to binary with exclusive-OR gates.

Solution

(a) 0101_2 is 0111 Gray. See Figure 6–42(a).

(b) 1011 Gray is 1101_2. See Figure 6–42(b).

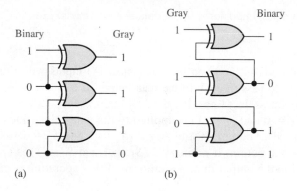

(a)   (b)

FIGURE 6–42

Related Problem

How many exclusive-OR gates are required to convert 8-bit binary to Gray?

SECTION 6–7 CHECKUP

1. Convert the BCD number 10000101 to binary.

2. Draw the logic diagram for converting an 8-bit binary number to Gray code.

6–8 Multiplexers (Data Selectors) "MUX"

A **multiplexer (MUX)** is a device that allows digital information from several sources to be routed onto a single line for transmission over that line to a common destination. The basic multiplexer has several data-input lines and a single output line. It also has data-select inputs, which permit digital data on any one of the inputs to be switched to the output line. Multiplexers are also known as data selectors.

After completing this section, you should be able to

◆ Explain the basic operation of a multiplexer

◆ Describe the 74HC153 and the 74HC151 multiplexers

◆ Expand a multiplexer to handle more data inputs

◆ Use the multiplexer as a logic function generator

◆ Use VHDL to describe 4-input and 8-input multiplexers

A logic symbol for a 4-input multiplexer (MUX) is shown in Figure 6–43. Notice that there are two data-select lines because with two select bits, any one of the four data-input lines can be selected.

In a multiplexer, data are switched from several lines to one line.

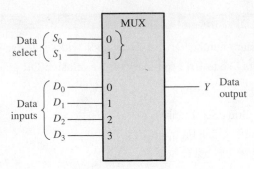

FIGURE 6–43 Logic symbol for a 1-of-4 data selector/multiplexer.

In Figure 6–43, a 2-bit code on the data-select (S) inputs will allow the data on the selected data input to pass through to the data output. If a binary 0 ($S_1 = 0$ and $S_0 = 0$) is applied to the data-select lines, the data on input D_0 appear on the data-output line. If a binary 1 ($S_1 = 0$ and $S_0 = 1$) is applied to the data-select lines, the data on input D_1 appear on the data output. If a binary 2 ($S_1 = 1$ and $S_0 = 0$) is applied, the data on D_2 appear on the output. If a binary 3 ($S_1 = 1$ and $S_0 = 1$) is applied, the data on D_3 are switched to the output line. A summary of this operation is given in Table 6–8.

TABLE 6–8

Data selection for a 1-of-4-multiplexer.

Data-Select Inputs		Input Selected
S_1	S_0	
0	0	D_0
0	1	D_1
1	0	D_2
1	1	D_3

InfoNote

A *bus* is a multiple conductor pathway along which electrical signals are sent from one part of a computer to another. In computer networks, a *shared bus* is one that is connected to all the microprocessors in the system in order to exchange data. A shared bus may contain memory and input/output devices that can be accessed by all the microprocessors in the system. Access to the shared bus is controlled by a *bus arbiter* (a multiplexer of sorts) that allows only one microprocessor at a time to use the system's shared bus.

Now let's look at the logic circuitry required to perform this multiplexing operation. The data output is equal to the state of the *selected* data input. You can therefore, derive a logic expression for the output in terms of the data input and the select inputs.

The data output is equal to D_0 only if $S_1 = 0$ and $S_0 = 0$: $Y = D_0 \overline{S}_1 \overline{S}_0$.
The data output is equal to D_1 only if $S_1 = 0$ and $S_0 = 1$: $Y = D_1 \overline{S}_1 S_0$.
The data output is equal to D_2 only if $S_1 = 1$ and $S_0 = 0$: $Y = D_2 S_1 \overline{S}_0$.
The data output is equal to D_3 only if $S_1 = 1$ and $S_0 = 1$: $Y = D_3 S_1 S_0$.

When these terms are ORed, the total expression for the data output is

$$Y = D_0 \overline{S}_1 \overline{S}_0 + D_1 \overline{S}_1 S_0 + D_2 S_1 \overline{S}_0 + D_3 S_1 S_0$$

The implementation of this equation requires four 3-input AND gates, a 4-input OR gate, and two inverters to generate the complements of S_1 and S_0, as shown in Figure 6–44. Because data can be selected from any one of the input lines, this circuit is also referred to as a **data selector**.

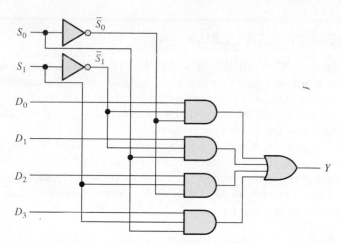

FIGURE 6–44 Logic diagram for a 4-input multiplexer. Open file F06-44 to verify operation.

MultiSim

EXAMPLE 6–14

The data-input and data-select waveforms in Figure 6–45(a) are applied to the multiplexer in Figure 6–44. Determine the output waveform in relation to the inputs.

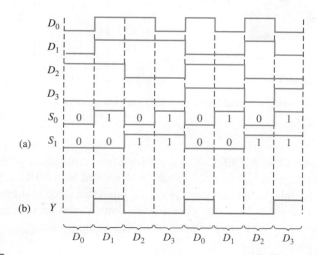

FIGURE 6–45

Solution

The binary state of the data-select inputs during each interval determines which data input is selected. Notice that the data-select inputs go through a repetitive binary sequence 00, 01, 10, 11, 00, 01, 10, 11, and so on. The resulting output waveform is shown in Figure 6–45(b).

Related Problem

Construct a timing diagram showing all inputs and the output if the S_0 and S_1 waveforms in Figure 6–45 are interchanged.

IMPLEMENTATION: DATA SELECTOR/MULTIPLEXER

Fixed-Function Device   The 74HC153 is a dual four-input data selector/multiplexer. The pin diagram is shown in Figure 6–46(a). The inputs to one of the multiplexers are 1I0 through 1I3 and the inputs to the second multiplexer are 2I0 through 2I3. The data select inputs are S0 and S1 and the active-LOW enable inputs are 1E and 2E. Each of the multiplexers has an active-LOW enable input.

The ANSI/IEEE logic symbol with dependency notation is shown in Figure 6–46(b). The two multiplexers are indicated by the partitioned outline, and the inputs common to both multiplexers are inputs to the notched block (common control block) at the top. The G_3^0 dependency notation indicates an AND relationship between the two select inputs (A and B) and the inputs to each multiplexer block.

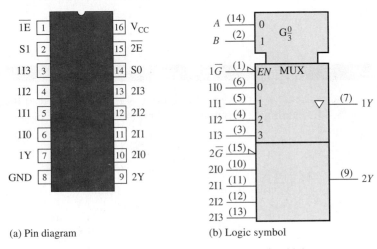

(a) Pin diagram   (b) Logic symbol

FIGURE 6–46   The 74HC153 dual four-input data selector/multiplexer.

Programmable Logic Device (PLD)   The logic for a four-input multiplexer like the one shown in the logic diagram of Figure 6–44 can be described with VHDL. The data flow approach is used for this particular circuit. Keep in mind that once you have written the VHDL program for a given logic, the code is then downloaded into a PLD device and becomes actual hardware just as fixed-function devices are hardware.

entity FourInputMultiplexer **is**

   port (S0, S1, D0, D1, D2, D3; **in** bit; Y: **out** bit);   Inputs and outputs declared

end entity FourInputMultiplexer;

architecture LogicFunction **of** FourInputMultiplexer **is**

begin

   Y <= (D0 **and not** S0 **and not** S1) **or** (Dl **and** S0 **and not** S1) } Boolean expression
   or (D2 **and not** S0 **and** S1) **or** (D3 **and** S0 **and** S1);   for the output

end architecture LogicFunction;

IMPLEMENTATION: EIGHT-INPUT DATA SELECTOR/MULTIPLEXER

Fixed-Function Device   The 74HC151 has eight data inputs (D_0–D_7) and, therefore, three data-select or address input lines (S_0–S_2). Three bits are required to select any one of the eight data inputs ($2^3 = 8$). A LOW on the $\overline{Enable}$ input allows the selected input data to pass through to the output. Notice that the data output and its complement are both available. The pin diagram is shown in Figure 6–47(a), and the ANSI/IEEE logic symbol is shown in part (b). In this case there is no need for a common control block on the logic symbol because there is only one multiplexer to be controlled, not two as in the 74HC153. The G_7^0 label within the logic symbol indicates the AND relationship between the data-select inputs and each of the data inputs 0 through 7.

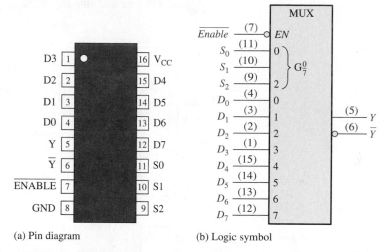

(a) Pin diagram   (b) Logic symbol

FIGURE 6–47   The 74HC151 eight-input data selector/multiplexer.

Programmable Logic Device (PLD)   The logic for the eight-input multiplexer is implemented by first writing the VHDL code. For the 74HC151, eight 5-input AND gates, one 8-input OR gate, and four inverters are required.

entity EightInputMUX **is**

   port (S0, S1, S2, D0, D1, D2, D3, D4, D5, D6, D7, EN: **in** bit; Y: **inout** bit; YI: **out** bit);   } Inputs and outputs declared

end entity EightInputMUX;

architecture LogicOperation **of** EightInputMUX **is**

Internal signals (outputs of AND gates) declared

   signal AND0, AND1, AND2, AND3, AND4, AND5, AND6, AND7: bit;

   begin

   AND0 <= **not** S0 **and not** S1 **and not** S2 **and** D0 **and not** EN;

   AND1 <= S0 **and not** S1 **and not** S2 **and** D1 **and not** EN;

   AND2 <= **not** S0 **and** S1 **and not** S2 **and** D2 **and not** EN;

   AND3 <= S0 **and** S1 **and not** S2 **and** D3 **and not** EN;

   AND4 <= **not** S0 **and not** S1 **and** S2 **and** D4 **and not** EN;

   AND5 <= S0 **and not** S1 **and** S2 **and** D5 **and not** EN;

   AND6 <= **not** S0 **and** S1 **and** S2 **and** D6 **and not** EN;

   AND7 <= S0 **and** S1 **and** S2 **and** D7 **and not** EN;

Boolean expressions for fixed outputs

   Y <= AND0 **or** AND1 **or** AND2 **or** AND3 **or** AND4 **or** AND5 **or** AND6 **or** AND7;

   YI <= **not** Y;

end architecture LogicOperation;

Boolean expressions for internal AND gate outputs

EXAMPLE 6–15

Use 74HC151s and any other logic necessary to multiplex 16 data lines onto a single data-output line.

Solution

An expansion of two 74HC151s is shown in Figure 6–48. Four bits are required to select one of 16 data inputs ($2^4 = 16$). In this application the $\overline{Enable}$ input is used as the most significant data-select bit. When the MSB in the data-select code is LOW, the left 74HC151 is enabled, and one of the data inputs (D_0 through D_7) is selected by the other three data-select bits. When the data-select MSB is HIGH, the right 74HC151 is enabled, and one of the data inputs (D_8 through D_{15}) is selected. The selected input data are then passed through to the negative-OR gate and onto the single output line.

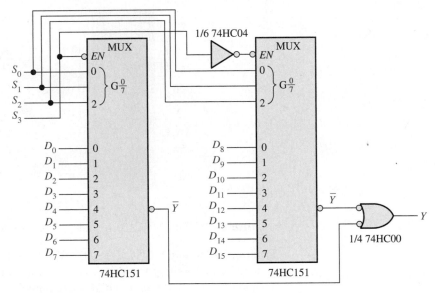

FIGURE 6–48   A 16-input multiplexer.

Related Problem

Determine the codes on the select inputs required to select each of the following data inputs: D_0, D_4, D_8, and D_{13}.

Applications

A 7-Segment Display Multiplexer

Figure 6–49 shows a simplified method of multiplexing BCD numbers to a 7-segment display. In this example, 2-digit numbers are displayed on the 7-segment readout by the use of a single BCD-to-7-segment decoder. This basic method of display multiplexing can be extended to displays with any number of digits. The 74HC157 is a quad 2-input multiplexer.

The basic operation is as follows. Two BCD digits ($A_3A_2A_1A_0$ and $B_3B_2B_1B_0$) are applied to the multiplexer inputs. A square wave is applied to the data-select line, and when it is LOW, the A bits ($A_3A_2A_1A_0$) are passed through to the inputs of the 74HC47 BCD-to-7-segment decoder. The LOW on the data-select also puts a LOW on the A_1 input of the 74HC139 2-line-to-4-line decoder, thus activating its 0 output and enabling the A-digit display by effectively connecting its common terminal to ground. The A digit is now *on* and the B digit is *off*.

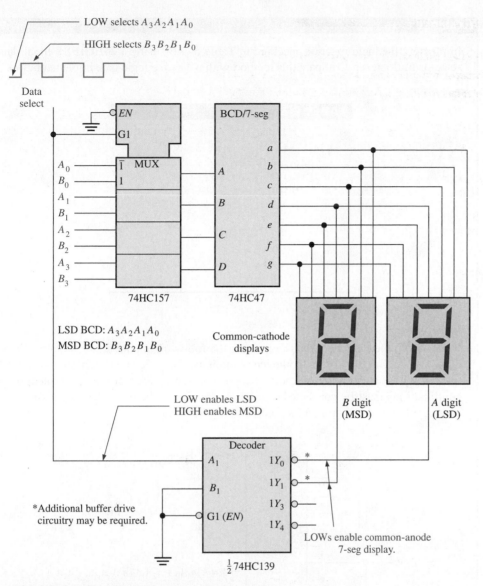

FIGURE 6–49 Simplified 7-segment display multiplexing logic.

When the data-select line goes HIGH, the B bits ($B_3B_2B_1B_0$) are passed through to the inputs of the BCD-to-7-segment decoder. Also, the 74HC139 decoder's 1 output is activated, thus enabling the B-digit display. The B digit is now *on* and the A digit is *off*. The cycle repeats at the frequency of the data-select square wave. This frequency must be high enough to prevent visual flicker as the digit displays are multiplexed.

A Logic Function Generator

A useful application of the data selector/multiplexer is in the generation of combinational logic functions in sum-of-products form. When used in this way, the device can replace discrete gates, can often greatly reduce the number of ICs, and can make design changes much easier.

To illustrate, a 74HC151 8-input data selector/multiplexer can be used to implement any specified 3-variable logic function if the variables are connected to the data-select inputs and each data input is set to the logic level required in the truth table for that function. For example, if the function is a 1 when the variable combination is $\overline{A_2}A_1\overline{A_0}$, the 2 input (selected by 010) is connected to a HIGH. This HIGH is passed through to the output when this particular combination of variables occurs on the data-select lines. Example 6–16 will help clarify this application.

EXAMPLE 6–16

Implement the logic function specified in Table 6–9 by using a 74HC151 8-input data selector/multiplexer. Compare this method with a discrete logic gate implementation.

TABLE 6–9

Inputs			Output
A_2	A_1	A_0	Y
0	0	0	0
0	0	1	1
0	1	0	0
0	1	1	1
1	0	0	0
1	0	1	1
1	1	0	1
1	1	1	0

Solution

Notice from the truth table that Y is a 1 for the following input variable combinations: 001, 011, 101, and 110. For all other combinations, Y is 0. For this function to be implemented with the data selector, the data input selected by each of the above-mentioned combinations must be connected to a HIGH (5 V). All the other data inputs must be connected to a LOW (ground), as shown in Figure 6–50.

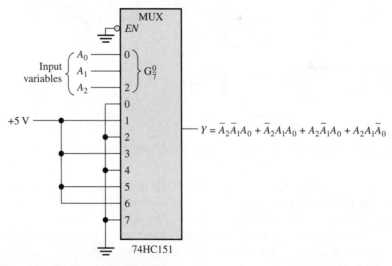

$$Y = \bar{A}_2\bar{A}_1A_0 + \bar{A}_2A_1A_0 + A_2\bar{A}_1A_0 + A_2A_1\bar{A}_0$$

FIGURE 6–50   Data selector/multiplexer connected as a 3-variable logic function generator.

The implementation of this function with logic gates would require four 3-input AND gates, one 4-input OR gate, and three inverters unless the expression can be simplified.

Related Problem

Use the 74HC151 to implement the following expression:

$$Y = \bar{A}_2\bar{A}_1\bar{A}_0 + A_2\bar{A}_1\bar{A}_0 + \bar{A}_2A_1\bar{A}_0$$

Example 6–16 illustrated how the 8-input data selector can be used as a logic function generator for three variables. Actually, this device can be also used as a 4-variable logic function generator by the utilization of one of the bits (A_0) in conjunction with the data inputs.

A 4-variable truth table has sixteen combinations of input variables. When an 8-bit data selector is used, each input is selected twice: the first time when A_0 is 0 and the second time when A_0 is 1. With this in mind, the following rules can be applied (Y is the output, and A_0 is the least significant bit):

1. If $Y = 0$ both times a given data input is selected by a certain combination of the input variables, $A_3A_2A_1$, connect that data input to ground (0).

2. If $Y = 1$ both times a given data input is selected by a certain combination of the input variables, $A_3A_2A_1$, connect the data input to $+V$ (1).

3. If Y is different the two times a given data input is selected by a certain combination of the input variables, $A_3A_2A_1$, and if $Y = A_0$, connect that data input to A_0.

4. If Y is different the two times a given data input is selected by a certain combination of the input variables, $A_3A_2A_1$, and if $Y = \overline{A_0}$, connect that data input to $\overline{A_0}$.

EXAMPLE 6–17

Implement the logic function in Table 6–10 by using a 74HC151 8-input data selector/multiplexer. Compare this method with a discrete logic gate implementation.

TABLE 6–10

Decimal Digit	Inputs A_3	A_2	A_1	A_0	Output Y
0	0	0	0	0	0
1	0	0	0	1	1
2	0	0	1	0	1
3	0	0	1	1	0
4	0	1	0	0	0
5	0	1	0	1	1
6	0	1	1	0	1
7	0	1	1	1	1
8	1	0	0	0	1
9	1	0	0	1	0
10	1	0	1	0	1
11	1	0	1	1	0
12	1	1	0	0	1
13	1	1	0	1	1
14	1	1	1	0	0
15	1	1	1	1	1

Solution

The data-select inputs are $A_3A_2A_1$. In the first row of the table, $A_3A_2A_1 = 000$ and $Y = A_0$. In the second row, where $A_3A_2A_1$ again is 000, $Y = A_0$. Thus, A_0 is connected to the 0 input. In the third row of the table, $A_3A_2A_1 = 001$ and $Y = \overline{A_0}$. Also, in the fourth row, when $A_3A_2A_1$ again is 001, $Y = \overline{A_0}$. Thus, A_0 is inverted and connected to the 1 input. This analysis is continued until each input is properly connected according to the specified rules. The implementation is shown in Figure 6–51.

If implemented with logic gates, the function would require as many as ten 4-input AND gates, one 10-input OR gate, and four inverters, although possible simplification would reduce this requirement.

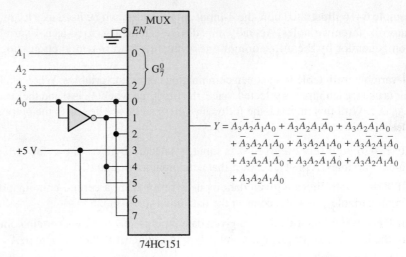

FIGURE 6–51   Data selector/multiplexer connected as a 4-variable logic function generator.

Related Problem

In Table 6–10, if $Y = 0$ when the inputs are all zeros and is alternately a 1 and a 0 for the remaining rows in the table, use a 74HC151 to implement the resulting logic function.

SECTION 6–8 CHECKUP

1. In Figure 6–44, $D_0 = 1$, $D_1 = 0$, $D_2 = 1$, $D_3 = 0$, $S_0 = 1$, and $S_1 = 0$. What is the output?

2. Identify each device.

   (a) 74HC153   (b) 74HC151

3. A 74HC151 has alternating LOW and HIGH levels on its data inputs beginning with $D_0 = 0$. The data-select lines are sequenced through a binary count (000, 001, 010, and so on) at a frequency of 1 kHz. The enable input is LOW. Describe the data output waveform.

4. Briefly describe the purpose of each of the following devices in Figure 6–49:

   (a) 74HC157   (b) 74HC47   (c) 74HC139

6–9   Demultiplexers

A **demultiplexer (DEMUX)** basically reverses the multiplexing function. It takes digital information from one line and distributes it to a given number of output lines. For this reason, the demultiplexer is also known as a data distributor. As you will learn, decoders can also be used as demultiplexers.

After completing this section, you should be able to

 ◆ Explain the basic operation of a demultiplexer

 ◆ Describe how a 4-line-to-16-line decoder can be used as a demultiplexer

 ◆ Develop the timing diagram for a demultiplexer with specified data and data selection inputs

Figure 6–52 shows a 1-line-to-4-line demultiplexer (DEMUX) circuit. The data-input line goes to all of the AND gates. The two data-select lines enable only one gate at a time, and the data appearing on the data-input line will pass through the selected gate to the associated data-output line.

In a demultiplexer, data are switched from one line to several lines.

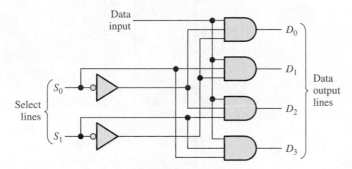

FIGURE 6–52   A 1-line-to-4-line demultiplexer.

EXAMPLE 6–18

The serial data-input waveform (Data in) and data-select inputs (S_0 and S_1) are shown in Figure 6–53. Determine the data-output waveforms on D_0 through D_3 for the demultiplexer in Figure 6–52.

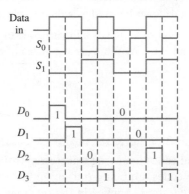

FIGURE 6–53

Solution

Notice that the select lines go through a binary sequence so that each successive input bit is routed to D_0, D_1, D_2, and D_3 in sequence, as shown by the output waveforms in Figure 6–53.

Related Problem

Develop the timing diagram for the demultiplexer if the S_0 and S_1 waveforms are both inverted.

4-Line-to-16-Line Decoder as a Demultiplexer

We have already discussed a 4-line-to-16-line decoder (Section 6–5). This device and other decoders can also be used in demultiplexing applications. The logic symbol for this device when used as a demultiplexer is shown in Figure 6–54. In demultiplexer applications, the input lines are used as the data-select lines. One of the chip select inputs is used as the data-input line, with the other chip select input held LOW to enable the internal negative-AND gate at the bottom of the diagram.

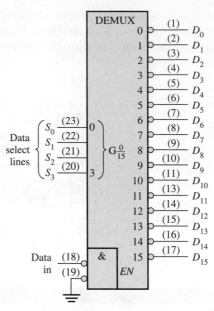

FIGURE 6–54 The decoder used as a demultiplexer.

1. Generally, how can a decoder be used as a demultiplexer?

2. The demultiplexer in Figure 6–54 has a binary code of 1010 on the data-select lines, and the data-input line is LOW. What are the states of the output lines?

6–10 Parity Generators/Checkers

Errors can occur as digital codes are being transferred from one point to another within a digital system or while codes are being transmitted from one system to another. The errors take the form of undesired changes in the bits that make up the coded information; that is, a 1 can change to a 0, or a 0 to a 1, because of component malfunctions or electrical noise. In most digital systems, the probability that even a single bit error will occur is very small, and the likelihood that more than one will occur is even smaller. Nevertheless, when an error occurs undetected, it can cause serious problems in a digital system.

After completing this section, you should be able to

♦ Explain the concept of parity

♦ Implement a basic parity circuit with exclusive-OR gates

♦ Describe the operation of basic parity generating and checking logic

♦ Discuss the 74HC280 9-bit parity generator/checker

♦ Use VHDL to describe a 9-bit parity generator/checker

♦ Discuss how error detection can be implemented in a data transmission system

The parity method of error detection in which a **parity bit** is attached to a group of information bits in order to make the total number of 1s either even or odd (depending on the system) was covered in Chapter 2. In addition to parity bits, several specific codes also provide inherent error detection.

Basic Parity Logic

In order to check for or to generate the proper parity in a given code, a basic principle can be used:

> **The sum (disregarding carries) of an even number of 1s is always 0, and the sum of an odd number of 1s is always 1.**

Therefore, to determine if a given code has **even parity** or **odd parity,** all the bits in that code are summed. As you know, the modulo-2 sum of two bits can be generated by an exclusive-OR gate, as shown in Figure 6–55(a); the modulo-2 sum of four bits can be formed by three exclusive-OR gates connected as shown in Figure 6–55(b); and so on. When the number of 1s on the inputs is even, the output X is 0 (LOW). When the number of 1s is odd, the output X is 1 (HIGH).

A parity bit indicates if the number of 1s in a code is even or odd for the purpose of error detection.

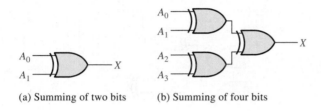

(a) Summing of two bits   (b) Summing of four bits

FIGURE 6–55

IMPLEMENTATION: 9-BIT PARITY GENERATOR/CHECKER

Fixed-Function Device   The logic symbol and function table for a 74HC280 are shown in Figure 6–56. This particular device can be used to check for odd or even parity on a 9-bit code (eight data bits and one parity bit), or it can be used to generate a parity bit for a binary code with up to nine bits. The inputs are A through I; when there is an even number of 1s on the inputs, the Σ Even output is HIGH and the Σ Odd output is LOW.

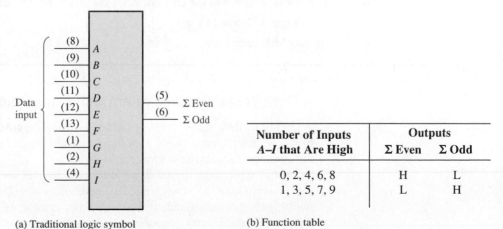

Number of Inputs *A–I* that Are High	Outputs	
	Σ **Even**	Σ **Odd**
0, 2, 4, 6, 8	H	L
1, 3, 5, 7, 9	L	H

(a) Traditional logic symbol   (b) Function table

FIGURE 6–56   The 74HC280 9-bit parity generator/checker.

Parity Checker   When this device is used as an even parity checker, the number of input bits should always be even; and when a parity error occurs, the Σ Even output goes LOW and the Σ Odd output goes HIGH. When it is used as an odd parity checker, the number of input bits should always be odd; and when a parity error occurs, the Σ Odd output goes LOW and the Σ Even output goes HIGH.

Parity Generator   If this device is used as an even parity generator, the parity bit is taken at the Σ Odd output because this output is a 0 if there is an even number of input bits and it is a 1 if there is an odd number. When used as an odd parity generator, the parity bit is taken at the Σ Even output because it is a 0 when the number of inputs bits is odd.

Programmable Logic Device (PLD)   The 9-bit parity generator/checker can be described using VHDL and implemented in a PLD. We will expand the 4-bit logic circuit in Figure 6–55(b) as shown in Figure 6–57. The data flow approach is used.

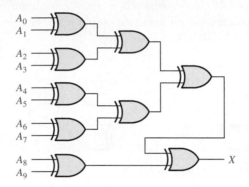

FIGURE 6–57

entity ParityCheck **is**

port (A0, A1, A2, A3, A4, A5, A6, A7, A8, A9: **in** bit;

X: **out** bit);   } Inputs and output declared

end entity ParityCheck;

architecture LogicOperation **of** ParityCheck **is**

begin

X $<=$ ((A0 **xor** A1) **xor** (A2 **xor** A3)) **xor** ((A4 **xor** A5) **xor**   } Output defined by

(A6 **xor** A7)) **xor** (A8 **xor** A9);   Boolean expression

end architecture LogicOperation;

A Data Transmission System with Error Detection

A simplified data transmission system is shown in Figure 6–58 to illustrate an application of parity generators/checkers, as well as multiplexers and demultiplexers, and to illustrate the need for data storage in some applications.

In this application, digital data from seven sources are multiplexed onto a single line for transmission to a distant point. The seven data bits (D_0 through D_6) are applied to the multiplexer data inputs and, at the same time, to the even parity generator inputs. The Σ Odd output of the parity generator is used as the even parity bit. This bit is 0 if the number of 1s on the inputs A through I is even and is a 1 if the number of 1s on A through I is odd. This bit is D_7 of the transmitted code.

The data-select inputs are repeatedly cycled through a binary sequence, and each data bit, beginning with D_0, is serially passed through and onto the transmission line ($\overline{Y}$). In this example, the transmission line consists of four conductors: one carries the serial data and three carry the timing signals (data selects). There are more sophisticated ways of sending the timing information, but we are using this direct method to illustrate a basic principle.

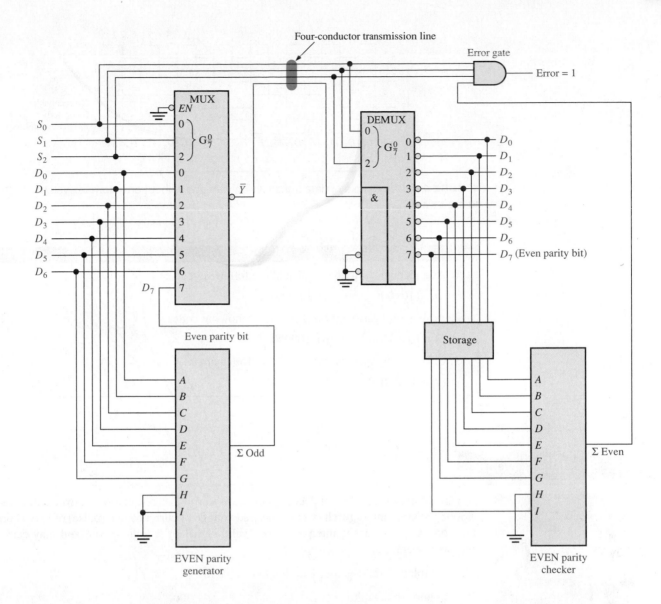

FIGURE 6–58   Simplified data transmission system with error detection.

At the demultiplexer end of the system, the data-select signals and the serial data stream are applied to the demultiplexer. The data bits are distributed by the demultiplexer onto the output lines in the order in which they occurred on the multiplexer inputs. That is, D_0 comes out on the D_0 output, D_1 comes out on the D_1 output, and so on. The parity bit comes out on the D_7 output. These eight bits are temporarily stored and applied to the even parity checker. Not all of the bits are present on the parity checker inputs until the parity bit D_7 comes out and is stored. At this time, the error gate is enabled by the data-select code 111. If the parity is correct, a 0 appears on the Σ Even output, keeping the Error output at 0. If the parity is incorrect, all 1s appear on the error gate inputs, and a 1 on the Error output results.

This particular application has demonstrated the need for data storage. Storage devices will be introduced in Chapter 7 and covered in Chapter 11.

The timing diagram in Figure 6–59 illustrates a specific case in which two 8-bit words are transmitted, one with correct parity and one with an error.

InfoNote

Microprocessors perform internal parity checks as well as parity checks of the external data and address buses. In a read operation, the external system can transfer the parity information together with the data bytes. The microprocessor checks whether the resulting parity is even and sends out the corresponding signal. When it sends out an address code, the microprocessor does not perform an address parity check, but it does generate an even parity bit for the address.

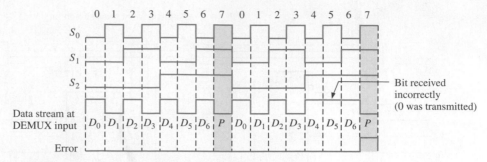

FIGURE 6–59 Example of data transmission with and without error for the system in Figure 6–58.

SECTION 6–10 CHECKUP

1. Add an even parity bit to each of the following codes:
   (a) 110100   (b) 01100011
2. Add an odd parity bit to each of the following codes:
   (a) 1010101   (b) 1000001
3. Check each of the even parity codes for an error.
   (a) 100010101   (b) 1110111001

6–11   Troubleshooting

In this section, the problem of decoder glitches is introduced and examined from a troubleshooting standpoint. A **glitch** is any undesired voltage or current spike (pulse) of very short duration. A glitch can be interpreted as a valid signal by a logic circuit and may cause improper operation.

After completing this section, you should be able to

♦ Explain what a glitch is

♦ Determine the cause of glitches in a decoder application

♦ Use the method of output strobing to eliminate glitches

The 74HC138 is used as a 3-line-to-8-line decoder (binary-to-octal) in Figure 6–60 to illustrate how glitches occur and how to identify their cause. The $A_2A_1A_0$ inputs of the decoder are sequenced through a binary count, and the resulting waveforms of the inputs and outputs can be displayed on the screen of a logic analyzer, as shown in Figure 6–60. A_2 transitions are delayed from A_1 transitions and A_1 transitions are delayed from A_0 transitions. This commonly occurs when waveforms are generated by a binary counter, as you will learn in Chapter 9.

The output waveforms are correct except for the glitches that occur on some of the output signals. A logic analyzer or an oscilloscope can be used to display glitches, which are normally very difficult to see. Generally, the logic analyzer is preferred, especially for low repetition rates (less than 10 kHz) and/or irregular occurrence because most logic analyzers have a *glitch capture* capability. Oscilloscopes can be used to observe glitches with reasonable success, particularly if the glitches occur at a regular high repetition rate (greater than 10 kHz).

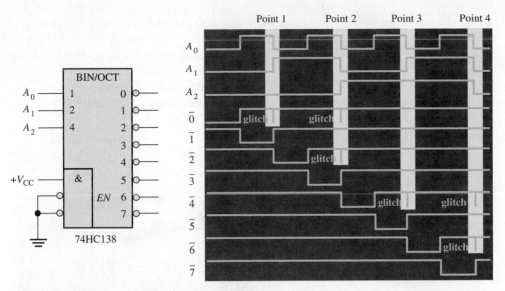

FIGURE 6–60 Decoder waveforms with output glitches.

The points of interest indicated by the highlighted areas on the input waveforms in Figure 6–60 are displayed as shown in Figure 6–61. At point 1 there is a transitional state of 000 due to delay differences in the waveforms. This causes the first glitch on the $\overline{0}$ output of the decoder. At point 2 there are two transitional states, 010 and 000. These cause the glitch on the $\overline{2}$ output of the decoder and the second glitch on

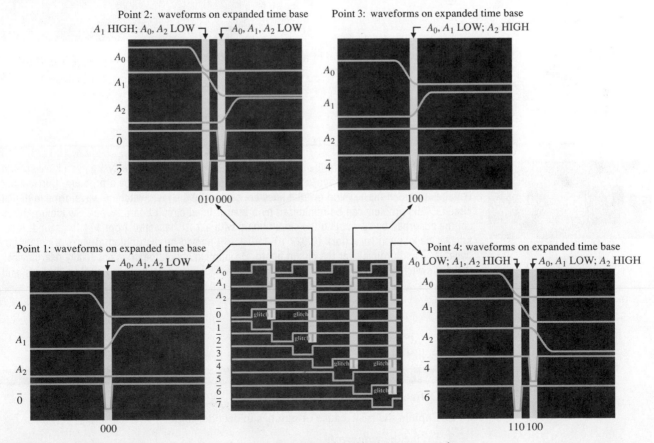

FIGURE 6–61 Decoder waveform displays showing how transitional input states produce glitches in the output waveforms.

the $\bar{0}$ output, respectively. At point 3 the transitional state is 100, which causes the first glitch on the $\bar{4}$ output of the decoder. At point 4 the two transitional states, 110 and 100, result in the glitch on the $\bar{6}$ output and the second glitch on the $\bar{4}$ output, respectively.

One way to eliminate the glitch problem is a method called **strobing**, in which the decoder is enabled by a strobe pulse only during the times when the waveforms are not in transition. This method is illustrated in Figure 6–62.

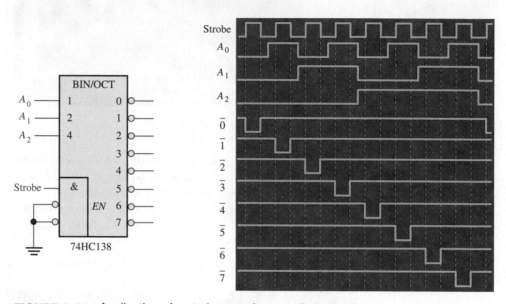

FIGURE 6–62 Application of a strobe waveform to eliminate glitches on decoder outputs.

In addition to glitches that are the result of differences in propagation delays, as you have seen in the case of a decoder, other types of unwanted noise spikes can also be a problem. Current and voltage spikes on the V_{CC} and ground lines are caused by the fast switching waveforms in digital circuits. This problem can be minimized by proper printed circuit board layout. Switching spikes can be absorbed by decoupling the circuit board with a 1 μF capacitor from V_{CC} to ground. Also, smaller decoupling capacitors (0.022 μF to 0.1 μF) should be distributed at various points between V_{CC} and ground over the circuit board. Decoupling should be done especially near devices that are switching at higher rates or driving more loads such as oscillators, counters, buffers, and bus drivers.

SECTION 6–11 CHECKUP

1. Define the term *glitch*.

2. Explain the basic cause of glitches in decoder logic.

3. Define the term *strobe*.

Applied Logic

Traffic Signal Controller: Part 1

The control logic is developed for a traffic signal at the intersection of a busy main street and a lightly used side street. The system requirements are established, and a general block diagram is developed. Also, a state diagram is introduced to define the sequence of operation. The combinational logic unit of the controller is developed in this chapter, and the remaining units are developed in Chapter 7.

Timing Requirements

The control logic establishes the sequencing of the lights for a traffic signal at the intersection of a busy main street and an occasionally used side street. The following are the timing requirements:

- The green light for the main street will stay on for a minimum of 25 s or as long as there is no vehicle on the side street.
- The green light for the side street will stay on until there is no vehicle on the side street up to a maximum of 25 s.
- The yellow caution light will stay on for 4 s between changes from green to red on both the main street and the side street.

The State Diagram

From the timing requirements, a state diagram can be developed to describe the complete operation. A state diagram graphically shows the sequence of states, the conditions for each state, and the requirements for transitions from one state to the next.

Defining the Variables   The variables that determine how the system sequences through the various states are defined as follows:

- V_s A vehicle is present on the side street.
- T_L The 25 s timer (long timer) is *on*.
- T_S The 4 s timer (short timer) is *on*.

A complemented variable indicates the opposite condition.

State Descriptions   A state diagram is shown in Figure 6–63. Each of the four states is assigned a 2-bit Gray code as indicated. A looping arrow means that the system remains in a state, and an arrow between states means that the system transitions to the next state. The Boolean expression or variable associated with each of the arrows in the state diagram indicate the condition under which the system remains in a state or transitions to the next state.

First State   The Gray code is 00. In this state, the light is green on the main street and red on the side street for 25 s when the long timer is *on* or there is no vehicle on the side street. This condition is expressed as $T_L + \overline{V}_s$. The system transitions to the next state when the long timer goes *off* and there is a vehicle on the side street. This condition is expressed as $\overline{T}_L V_s$.

Second State   The Gray code is 01. In this state, the light is yellow on the main street and red on the side street. The system remains in this state for 4 s when the short timer is *on*. This condition is expressed as T_S. The system transitions to the next state when the short timer goes *off*. This condition is expressed as $\overline{T}_S$.

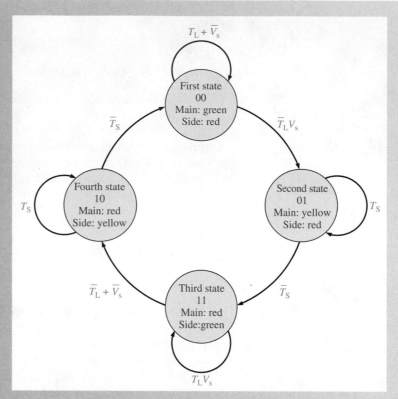

FIGURE 6–63 State diagram for the traffic signal control.

Third State The Gray code is 11. In this state, the light is red on the main street and green on the side street for 25 s when the long timer is *on* as long as there is a vehicle on the side street. This condition is expressed as $T_L V_s$. The system transitions to the next state when the long timer goes *off* or when there is no vehicle on the side street. This condition is expressed as $\overline{T}_L + \overline{V}_s$.

Fourth State The Gray code is 10. In this state, the light is red on the main street and yellow on the side street. The system remains in this state for 4 s when the short timer is *on*. This condition is expressed as T_S. The system transitions back to the first state when the short timer goes *off*. This condition is expressed as $\overline{T}_S$.

Exercise

1. How long can the system remain in the first state?
2. How long can the system remain in the fourth state?
3. Write the expression for the condition that produces a transition from the first state to the second state.
4. Write the expression for the condition that keeps the system in the second state.

Block Diagram

The traffic signal controller consists of three units: combinational logic, sequential logic, and timing circuits, as shown in Figure 6–64. The combinational logic unit provides outputs to turn the signal lights on and off. It also provides trigger outputs to start the long and short timers. The input sequence to this logic represents the four states described by the state diagram. The timing circuits unit provides the 25 s and the 4 s timing outputs. A frequency divider in the timing circuits unit divides the system clock down to a 1 Hz clock for use in producing the 25 s and 4 s signals. The sequential logic unit produces the sequence of 2-bit Gray codes representing the four states.

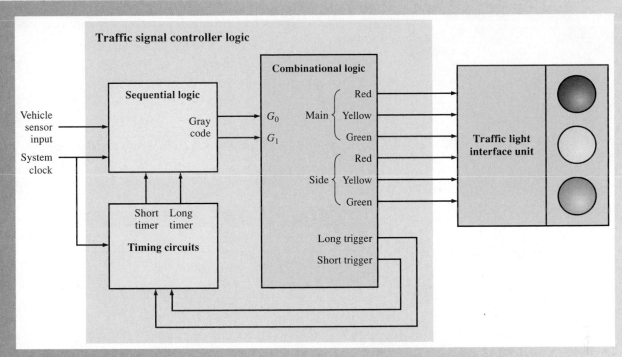

FIGURE 6-64   Block diagram of the traffic signal controller.

The Combinational Logic

The combinational logic consists of a state decoder, light output logic, and trigger logic, as shown in Figure 6–65.

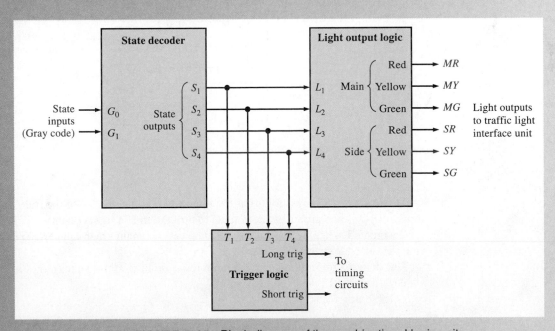

FIGURE 6-65   Block diagram of the combinational logic unit.

State Decoder   This logic decodes the 2-bit Gray code from the sequential logic to determine which of the four states the system is in. The inputs to the state decoder are the two Gray code bits G_1 and G_0. There are four state outputs S_1, S_2, S_3, and S_4. For each of the four input codes, one and only one of the outputs is activated. The Boolean expressions for the state outputs in terms of the inputs are

$$S_1 = \overline{G}_1 \overline{G}_0$$
$$S_2 = \overline{G}_1 G_0$$
$$S_3 = G_1 G_0$$
$$S_4 = G_1 \overline{G}_0$$

The truth table for the state decoder logic is shown in Table 6–11, and the logic diagram is shown in Figure 6–66.

TABLE 6–11

Truth table for the state decoder.

State Inputs (Gray Code)		State Outputs			
G_1	G_0	S_1	S_2	S_3	S_4
0	0	1	0	0	0
0	1	0	1	0	0
1	1	0	0	1	0
1	0	0	0	0	1

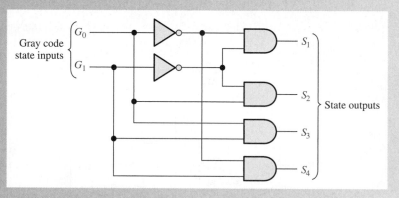

FIGURE 6–66   State decoder logic.

Light Output Logic   This logic has the four state outputs (S_1–S_4) of the state decoder as its inputs (L_1–L_4) and produces six outputs to turn the traffic lights on and off. These outputs are designated *MR, MY, MG* (main red, main yellow, main green) and *SR, SY, SG* (side red, side yellow, side green).

The state diagram shows that the main red is *on* in the third state (L_3) or in the fourth state (L_4), so the Boolean expression is

$$MR = L_3 + L_4$$

The main yellow is *on* in the second state (L_2), so the expression is

$$MY = L_2$$

The main green is *on* in the first state (L_1), so the expression is

$$MG = L_1$$

Similarly, the state diagram is used to obtain the following expressions for the side street:

$$SR = L_1 + L_2$$
$$SY = L_4$$
$$SG = L_3$$

The logic circuit is shown in Figure 6–67.

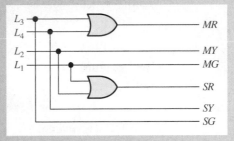

FIGURE 6–67   Light output logic.

Exercise

5. Show the logic diagram for the light output logic using specific IC devices with pin numbers.

6. Develop a truth table for the light output logic.

Trigger Logic   The trigger logic produces two outputs, the long trigger output and the short trigger output. The long trigger output initiates the 25 s timer on a LOW-to-HIGH transition at the beginning of the first or third states. The short trigger output initiates the 4 s timer on a LOW-to-HIGH transition at the beginning of the second or fourth states. The Boolean expressions for this logic are

$$LongTrig = T_1 + T_3$$
$$ShortTrig = T_2 + T_4$$

Equivalently,

$$LongTrig = T_1 + T_3$$
$$ShortTrig = \overline{T_1 + T_3}$$

The logic circuit is shown in Figure 6–68.

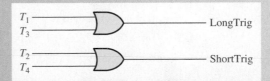

FIGURE 6–68   Trigger logic.

Exercise

7. Show the logic diagram for the trigger logic using specific IC devices with pin numbers.

8. Develop a truth table for the trigger logic.

9. Show the complete combinational logic by combining the state decoder, light output logic, and trigger logic. Include specific IC devices and pin numbers.

VHDL Descriptions

The VHDL program for the combinational logic unit of the traffic signal controller can be written using the data flow approach to describe each of the three functional blocks of the combinational logic unit. These functional blocks are the state decoder, the light output logic, and the trigger logic, as shown in Figure 6–65.

◆ The VHDL program code for the state decoder is as follows:

entity StateDecoder **is**
    port (G0, G1: **in** bit; S1, S2, S3, S4: **out** bit);   G0, G1: Gray code inputs
    S1–S4: State outputs
end entity StateDecoder;

architecture LogicOperation **of** StateDecoder **is**
begin
    S1 <= **not** G0 **and not** G1;
    S2 <= G0 **and not** G1;   } Boolean expressions for
    S3 <= G0 **and** G1;   state decoder outputs
    S4 <= **not** G0 **and** G1;

end architecture LogicOperation;

◆ The VHDL program code for the light output logic is as follows:

entity LightOutputLogic **is**
    port (L1, L2, L3, L4: **in** bit; MR, MY, MG, SR, SY, SG: **out** bit);   Inputs and outputs declared
end entity LightOutputLogic;

architecture LogicOperation **of** LightOutputLogic **is**
begin
    MR <= L3 **or** L4;
    MY <= L2;
    MG <= L1;   } Boolean expressions for
    SR <= L1 **or** L2;   light output logic outputs
    SY <= L4;
    SG <= L3;

end architecture LogicOperation;

◆ The VHDL program code for the trigger logic is as follows:

entity TriggerLogic **is**
    port (T1, T2, T3, T4: **in** bit; LongTrig, ShortTrig: **out** bit);   Inputs and outputs declared
end entity TriggerLogic;

architecture LogicOperation **of** TriggerLogic **is**
begin
    LongTrig <= T1 **or** T3;   } Boolean expressions for
    ShortTrig <= T2 **or** T4; } trigger logic outputs
end architecture LogicOperation;

Development of the traffic signal controller will continue in the Applied Logic in Chapter 7.

Simulation

Open Multisim file AL06 in the Applied Logic folder on the website. Run the simulation for the combinational logic unit of the traffic signal controller and observe the operation for each of the four states in the light sequence.

Putting Your Knowledge to Work

There is a requirement for a pedestrian push button that would activate the yellow caution light for 4 s and the red light for 15 s on both the main street and the side street. (a) Modify the state diagram for this additional feature. (b) Develop the additional logic required.

SUMMARY

- Half-adder and full-adder operations are summarized in truth Tables 6–12 and 6–13.

TABLE 6–12

Inputs		Carry Out	Sum
A	B	C_{out}	Σ
0	0	0	0
0	1	0	1
1	0	0	1
1	1	1	0

TABLE 6–13

Inputs		Carry In	Carry Out	Sum
A	B	C_{in}	C_{out}	Σ
0	0	0	0	0
0	0	1	0	1
0	1	0	0	1
0	1	1	1	0
1	0	0	0	1
1	0	1	1	0
1	1	0	1	0
1	1	1	1	1

- Combination logic functions include comparators, decoders, encoders, code converters, multiplexers, demultiplexers, and parity generators/checkers.
- Software versions of standard logic functions from the 74XX series are available for use in a programmable logic design.

KEY TERMS

Key terms and other bold terms in the chapter are defined in the end-of-book glossary.

Cascading   Connecting two or more similar devices in a manner that expands the capability of one device.

Comparator   A digital circuit that compares the magnitudes of two quantities and produces an output indicating the relationship of the quantities.

Decoder   A digital circuit that converts coded information into a familiar or noncoded form.

Demultiplexer (DEMUX)   A circuit that switches digital data from one input line to several output lines in a specified time sequence.

Encoder   A digital circuit that converts information to a coded form.

Full-adder   A digital circuit that adds two bits and an input carry to produce a sum and an output carry.

Glitch   A voltage or current spike of short duration, usually unintentionally produced and unwanted.

Half-adder   A digital circuit that adds two bits and produces a sum and an output carry. It cannot handle input carries.

Look-ahead carry   A method of binary addition whereby carries from preceding adder stages are anticipated, thus eliminating carry propagation delays.

Multiplexer (MUX)   A circuit that switches digital data from several input lines onto a single output line in a specified time sequence.

Parity bit   A bit attached to each group of information bits to make the total number of 1s odd or even for every group of bits.

Priority encoder   An encoder in which only the highest value input digit is encoded and any other active input is ignored.

Ripple carry   A method of binary addition in which the output carry from each adder becomes the input carry of the next higher-order adder.

TRUE/FALSE QUIZ

Answers are at the end of the chapter.

1. A half-adder adds two binary bits.

2. A half-adder has a sum output only.

3. A full-adder adds three bits and produces two outputs.

4. Two 4-bit numbers can be added using two full-adders.

5. When the two input bits are both 1 and the carry input bit is a 1, the sum output of a full adder is 0.

6. A comparator determines when two binary numbers are equal.

7. A decoder detects the presence of a specified combination of input bits.

8. The 4-line-to-10-line decoder and the 1-of-10 decoder are two different types.

9. An encoder essentially performs a reverse decoder function.

10. A multiplexer is a logic circuit that allows digital information from a single source to be routed onto several lines.

SELF-TEST

Answers are at the end of the chapter.

1. A half-adder is characterized by
   (a) two inputs and two outputs   (b) three inputs and two outputs
   (c) two inputs and three outputs   (d) two inputs and one output

2. A full-adder is characterized by
   (a) two inputs and two outputs   (b) three inputs and two outputs
   (c) two inputs and three outputs   (d) two inputs and one output

3. The inputs to a full-adder are $A = 1$, $B = 1$, $C_{in} = 0$. The outputs are
   (a) $\Sigma = 1, C_{out} = 1$ (b) $\Sigma = 1, C_{out} = 0$
   (c) $\Sigma = 0, C_{out} = 1$ (d) $\Sigma = 0, C_{out} = 0$

4. A 4-bit parallel adder can add
   (a) two 4-bit binary numbers   (b) two 2-bit binary numbers
   (c) four bits at a time   (d) four bits in sequence

5. To expand a 4-bit parallel adder to an 8-bit parallel adder, you must
   (a) use four 4-bit adders with no interconnections
   (b) use two 4-bit adders and connect the sum outputs of one to the bit inputs of the other
   (c) use eight 4-bit adders with no interconnections
   (d) use two 4-bit adders with the carry output of one connected to the carry input of the other

6. If a 74HC85 magnitude comparator has $A = 1011$ and $B = 1001$ on its inputs, the outputs are
   (a) $A > B = 0, A < B = 1, A = B = 0$ (b) $A > B = 1, A < B = 0, A = B = 0$
   (c) $A > B = 1, A < B = 1, A = B = 0$ (d) $A > B = 0, A < B = 0, A = B = 1$

7. If a 1-of-16 decoder with active-LOW outputs exhibits a LOW on the decimal 12 output, what are the inputs?
   - (a) $A_3A_2A_1A_0 = 1010$
   - (b) $A_3A_2A_1A_0 = 1110$
   - (c) $A_3A_2A_1A_0 = 1100$
   - (d) $A_3A_2A_1A_0 = 0100$

8. A BCD-to-7 segment decoder has 0100 on its inputs. The active outputs are
   - (a) a, c, f, g
   - (b) b, c, f, g
   - (c) b, c, e, f
   - (d) b, d, e, g

9. If an octal-to-binary priority encoder has its 0, 2, 5, and 6 inputs at the active level, the active-HIGH binary output is
   - (a) 110
   - (b) 010
   - (c) 101
   - (d) 000

10. In general, a multiplexer has
    - (a) one data input, several data outputs, and selection inputs
    - (b) one data input, one data output, and one selection input
    - (c) several data inputs, several data outputs, and selection inputs
    - (d) several data inputs, one data output, and selection inputs

11. Data selectors are basically the same as
    - (a) decoders
    - (b) demultiplexers
    - (c) multiplexers
    - (d) encoders

12. Which of the following codes exhibit even parity?
    - (a) 10011000
    - (b) 01111000
    - (c) 11111111
    - (d) 11010101
    - (e) all
    - (f) both answers (b) and (c)

PROBLEMS

Answers to odd-numbered problems are at the end of the book.

Section 6–1 Half and Full Adders

1. For the full-adder of Figure 6–4, determine the logic state (1 or 0) at each gate output for the following inputs:
   - (a) $A = 1, B = 1, C_{in} = 1$
   - (b) $A = 0, B = 1, C_{in} = 1$
   - (c) $A = 0, B = 1, C_{in} = 0$

2. What are the full-adder inputs that will produce each of the following outputs:
   - (a) $\Sigma = 0, C_{out} = 0$
   - (b) $\Sigma = 1, C_{out} = 0$
   - (c) $\Sigma = 1, C_{out} = 1$
   - (d) $\Sigma = 0, C_{out} = 1$

3. Determine the outputs of a full-adder for each of the following inputs:
   - (a) $A = 1, B = 0, C_{in} = 0$
   - (b) $A = 0, B = 0, C_{in} = 1$
   - (c) $A = 0, B = 1, C_{in} = 1$
   - (d) $A = 1, B = 1, C_{in} = 1$

Section 6–2 Parallel Binary Adders

4. For the parallel adder in Figure 6–69, determine the complete sum by analysis of the logical operation of the circuit. Verify your result by longhand addition of the two input numbers.

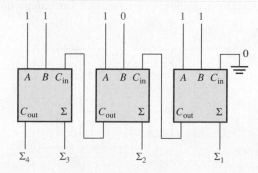

FIGURE 6–69

5. Repeat Problem 4 for the circuit and input conditions in Figure 6–70.

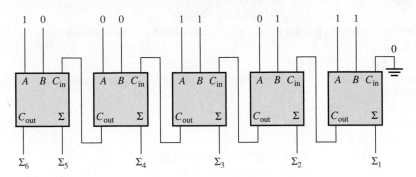

FIGURE 6–70

6. The circuit shown in Figure 6–71 is a 4-bit circuit that can add or subtract numbers in a form used in computers (positive numbers in true form; negative numbers in complement form). (a) Explain what happens when the $\overline{Add}/Subt.$ input is HIGH. (b) What happens when $\overline{Add}/Subt.$ is LOW?

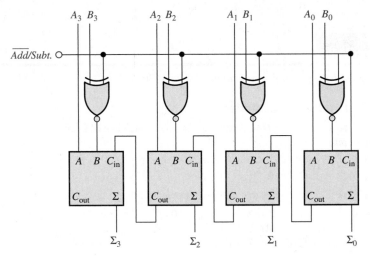

FIGURE 6–71

7. For the circuit in Figure 6–71, assume the inputs are $\overline{Add}/Subt. = 1$, $A = 1001$ and $B = 1100$. What is the output?

8. The input waveforms in Figure 6–72 are applied to a 2-bit adder. Determine the waveforms for the sum and the output carry in relation to the inputs by constructing a timing diagram.

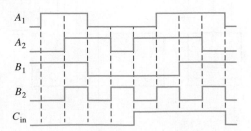

FIGURE 6–72

9. The following sequences of bits (right-most bit first) appear on the inputs to a 4-bit parallel adder. Determine the resulting sequence of bits on each sum output.

A_1	1001
A_2	1110
A_3	0000
A_4	1011
B_1	1111
B_2	1100
B_3	1010
B_4	0010

10. In the process of checking a 74HC283 4-bit parallel adder, the following logic levels are observed on its pins: 1-HIGH, 2-HIGH, 3-HIGH, 4-HIGH, 5-LOW, 6-LOW, 7-LOW, 9-HIGH, 10-LOW, 11-HIGH, 12-LOW, 13-HIGH, 14-HIGH, and 15-HIGH. Determine if the IC is functioning properly.

Section 6–3 Ripple Carry and Look-Ahead Carry Adders

11. Each of the eight full-adders in an 8-bit parallel ripple carry adder exhibits the following propagation delays:

A to Σ and C_{out}:	40 ns
B to Σ and C_{out}:	40 ns
C_{in} to Σ:	35 ns
C_{in} to C_{out}:	25 ns

Determine the maximum total time for the addition of two 8-bit numbers.

12. Show the additional logic circuitry necessary to make the 4-bit look-ahead carry adder in Figure 6–17 into a 5-bit adder.

Section 6–4 Comparators

13. The waveforms in Figure 6–73 are applied to the comparator as shown. Determine the output $(A = B)$ waveform.

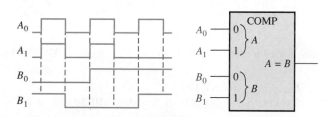

FIGURE 6–73

14. For the 4-bit comparator in Figure 6–74, plot each output waveform for the inputs shown. The outputs are active-HIGH.

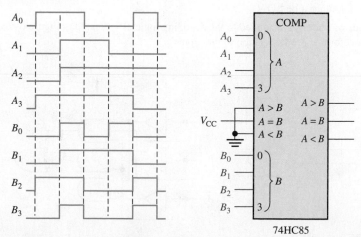

FIGURE 6–74

15. For each set of binary numbers, determine the output states for the comparator of Figure 6–21.

    (a) $A_3A_2A_1A_0 = 1100$ (b) $A_3A_2A_1A_0 = 1000$ (c) $A_3A_2A_1A_0 = 0100$
       $B_3B_2B_1B_0 = 1001$ $B_3B_2B_1B_0 = 1011$ $B_3B_2B_1B_0 = 0100$

Section 6–5 Decoders

16. When a HIGH is on the output of each of the decoding gates in Figure 6–75, what is the binary code appearing on the inputs? The MSB is A_3.

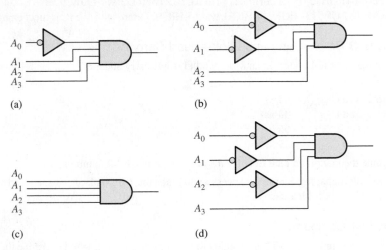

FIGURE 6–75

17. Show the decoding logic for each of the following codes if an active-HIGH (1) output is required:

    (a) 1101   (b) 1000   (c) 11011   (d) 11100
    (e) 101010   (f) 111110   (g) 000101   (h) 1110110

18. Solve Problem 17, given that an active-LOW (0) output is required.

19. You wish to detect only the presence of the codes 1010, 1100, 0001, and 1011. An active-HIGH output is required to indicate their presence. Develop the minimum decoding logic with a single output that will indicate when any one of these codes is on the inputs. For any other code, the output must be LOW.

20. If the input waveforms are applied to the decoding logic as indicated in Figure 6–76, sketch the output waveform in proper relation to the inputs.

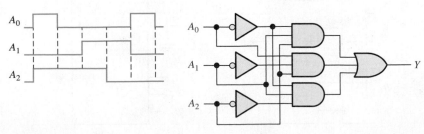

FIGURE 6–76

21. BCD numbers are applied sequentially to the BCD-to-decimal decoder in Figure 6–77. Draw a timing diagram, showing each output in the proper relationship with the others and with the inputs.

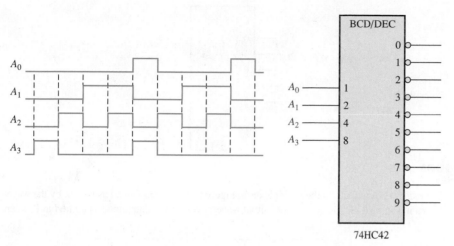

FIGURE 6–77

22. A 7-segment decoder/driver drives the display in Figure 6–78. If the waveforms are applied as indicated, determine the sequence of digits that appears on the display.

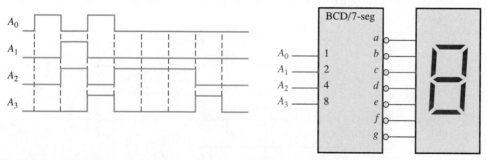

FIGURE 6–78

Section 6–6 Encoders

23. For the decimal-to-BCD encoder logic of Figure 6–37, assume that the 9 input and the 3 input are both HIGH. What is the output code? Is it a valid BCD (8421) code?

24. A 74HC147 encoder has LOW levels on pins 2, 5, and 12. What BCD code appears on the outputs if all the other inputs are HIGH?

Section 6–7 Code Converters

25. Convert the following decimal numbers to BCD and then to binary.

   (a) 2   **(b)** 8   **(c)** 13   **(d)** 26   **(e)** 33

26. Show the logic required to convert a 10-bit binary number to Gray code, and use that logic to convert the following binary numbers to Gray code:

   (a) 1010101010   **(b)** 1111100000   **(c)** 0000001110   **(d)** 1111111111

27. Show the logic required to convert a 10-bit Gray code to binary, and use that logic to convert the following Gray code words to binary:

   (a) 1010000000   **(b)** 0011001100   **(c)** 1111000111   **(d)** 0000000001

Section 6–8 Multiplexers (Data Selectors)

28. For the multiplexer in Figure 6–79, determine the output for the following input states: $D_0 = 0$, $D_1 = 1$, $D_2 = 1$, $D_3 = 0$, $S_0 = 1$, $S_1 = 0$.

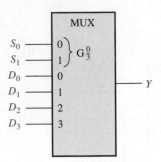

FIGURE 6–79

29. If the data-select inputs to the multiplexer in Figure 6–79 are sequenced as shown by the waveforms in Figure 6–80, determine the output waveform with the data inputs specified in Problem 28.

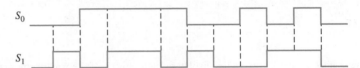

FIGURE 6–80

30. The waveforms in Figure 6–81 are observed on the inputs of a 74HC151 8-input multiplexer. Sketch the Y output waveform.

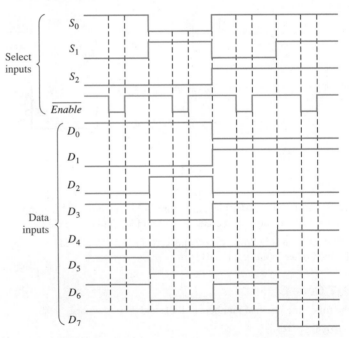

FIGURE 6–81

Section 6–9 Demultiplexers

31. Develop the total timing diagram (inputs and outputs) for a 74HC154 used in a demultiplexing application in which the inputs are as follows: The data-select inputs are repetitively sequenced through a straight binary count beginning with 0000, and the data input is a serial data stream carrying BCD data representing the decimal number 2468. The least significant digit (8) is first in the sequence, with its LSB first, and it should appear in the first 4-bit positions of the output.

Section 6–10 Parity Generators/Checkers

32. The waveforms in Figure 6–82 are applied to the 4-bit parity logic. Determine the output waveform in proper relation to the inputs. For how many bit times does even parity occur, and how is it indicated? The timing diagram includes eight bit times.

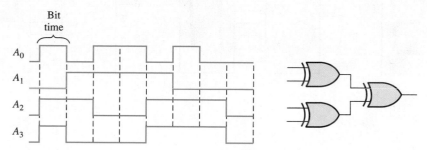

FIGURE 6–82

33. Determine the Σ Even and the Σ Odd outputs of a 74HC280 9-bit parity generator/checker for the inputs in Figure 6–83. Refer to the function table in Figure 6–56.

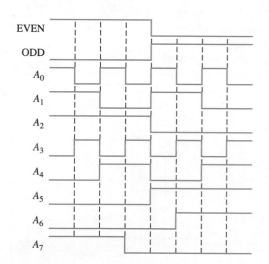

FIGURE 6–83

Section 6–11 Troubleshooting

34. The full-adder in Figure 6–84 is tested under all input conditions with the input waveforms shown. From your observation of the Σ and C_{out} waveforms, is it operating properly, and if not, what is the most likely fault?

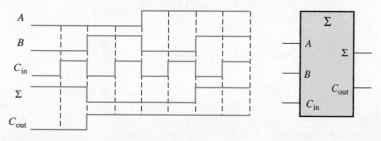

FIGURE 6–84

35. List the possible faults for each decoder/display in Figure 6–85.

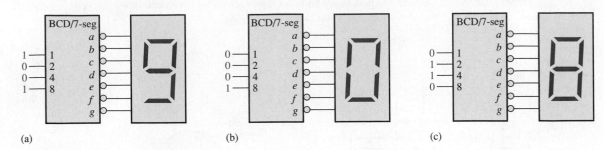

(a)   (b)   (c)

FIGURE 6–85

36. Develop a systematic test procedure to check out the complete operation of the keyboard encoder in Figure 6–39.

37. You are testing a BCD-to-binary converter consisting of 4-bit adders as shown in Figure 6–86. First verify that the circuit converts BCD to binary. The test procedure calls for applying BCD numbers in sequential order beginning with 0_{10} and checking for the correct binary output. What symptom or symptoms will appear on the binary outputs in the event of each of the following faults? For what BCD number is each fault *first* detected?

(a) The A_1 input is open (top adder).

(b) The C_{out} is open (top adder).

(c) The Σ_4 output is shorted to ground (top adder).

(d) The 32 output is shorted to ground (bottom adder).

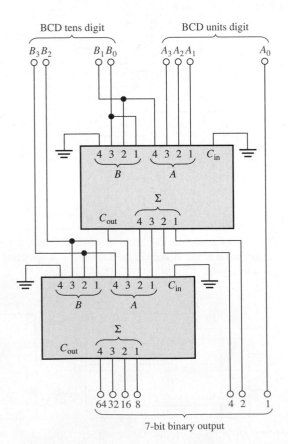

FIGURE 6–86

38. For the 7-segment display multiplexing system in Figure 6–49, determine the most likely cause or causes for each of the following symptoms:

   (a) The B-digit (MSD) display does not turn on at all.
   (b) Neither 7-segment display turns on.
   (c) The f-segment of both displays appears to be on all the time.
   (d) There is a visible flicker on the displays.

39. Develop a systematic procedure to fully test the 74HC151 data selector IC.

40. During the testing of the data transmission system in Figure 6–58, a code is applied to the D_0 through D_6 inputs that contains an odd number of 1s. A single bit error is deliberately introduced on the serial data transmission line between the MUX and the DEMUX, but the system does not indicate an error (error output = 0). After some investigation, you check the inputs to the even parity checker and find that D_0 through D_6 contain an even number of 1s, as you would expect. Also, you find that the D_7 parity bit is a 1. What are the possible reasons for the system not indicating the error?

41. In general, describe how you would fully test the data transmission system in Figure 6–58, and specify a method for the introduction of parity errors.

Applied Logic

42. Use a 74HC00 (quad NAND gates) and any other devices that may be required to produce active-HIGH outputs for the given inputs of the state decoder.

43. Implement the light output logic with the 74HC00 if active-LOW outputs are required.

Special Design Problems

44. Modify the design of the 7-segment display multiplexing system in Figure 6–49 to accommodate two additional digits.

45. Using Table 6–2, write the SOP expressions for the Σ and C_{out} of a full-adder. Use a Karnaugh map to minimize the expressions and then implement them with inverters and AND-OR logic. Show how you can replace the AND-OR logic with 74HC151 data selectors.

46. Implement the logic function specified in Table 6–14 by using a 74HC151 data selector.

TABLE 6–14				
Inputs				**Output**
A_3	A_2	A_1	A_0	Y
0	0	0	0	0
0	0	0	1	0
0	0	1	0	1
0	0	1	1	1
0	1	0	0	0
0	1	0	1	0
0	1	1	0	1
0	1	1	1	1
1	0	0	0	1
1	0	0	1	0
1	0	1	0	1
1	0	1	1	1
1	1	0	0	0
1	1	0	1	1
1	1	1	0	0
1	1	1	1	1

47. Using two of the 6-position adder modules from Figure 6–13, design a 12-position voting system.

48. The adder block in the tablet-bottling system in Figure 6–87 performs the addition of the 8-bit binary number from the counter and the 16-bit binary number from Register B. The result from

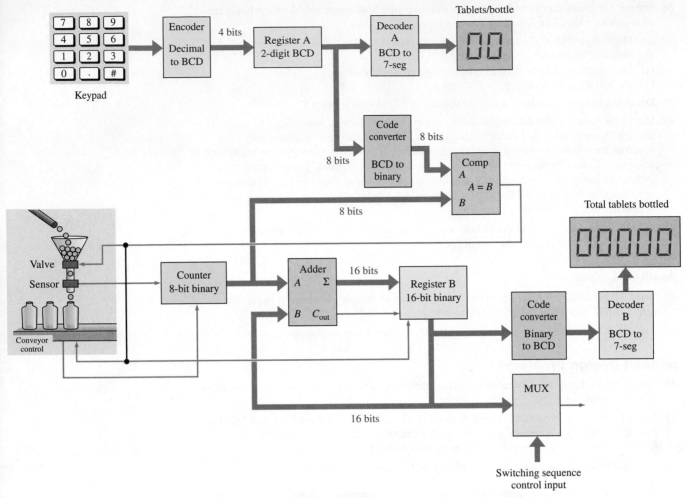

FIGURE 6–87

the adder goes back into Register B. Use 74HC283s to implement this function and draw a complete logic diagram including pin numbers. This is similar to the system in Section 1–4.

49. Use 74HC85s to implement the comparator block in the tablet-bottling system in Figure 6–87 and draw a complete logic diagram including pin numbers. The comparator compares the 8-bit binary number (actually only seven bits are required) from the BCD-to-binary converter with the 8-bit binary number from the counter.

50. Two BCD-to-7-segment decoders are used in the tablet-bottling system in Figure 6–87. One is required to drive the 2-digit *tablets/bottle* display and the other to drive the 5-digit *total tablets bottled* display. Use 74HC47s to implement each decoder and draw a complete logic diagram including pin numbers.

51. The encoder shown in the system block diagram of Figure 6–87 encodes each decimal key closure and converts it to BCD. Use a 74HC147 to implement this function and draw a complete logic diagram including pin numbers.

52. The system in Figure 6–87 requires two code converters. The BCD-to-binary converter changes the 2-digit BCD number in Register A to an 8-bit binary code (actually only 7 bits are required because the MSB is always 0). Use appropriate fixed-function IC code converters to implement the BCD-to-binary converter function and draw a complete logic diagram including pin numbers.

MultiSim   Multisim Troubleshooting Practice

53. Open file P06-53. For the specified fault, predict the effect on the circuit. Then introduce the fault and verify whether your prediction is correct.

54. Open file P06-54. For the specified fault, predict the effect on the circuit. Then introduce the fault and verify whether your prediction is correct.

55. Open file P06-55. For the observed behavior indicated, predict the fault in the circuit. Then introduce the suspected fault and verify whether your prediction is correct.

56. Open file P06-56. For the observed behavior indicated, predict the fault in the circuit. Then introduce the suspected fault and verify whether your prediction is correct.

ANSWERS

SECTION CHECKUPS

Section 6–1 Half and Full Adders

1. (a) $\Sigma = 1, C_{out} = 0$

(b) $\Sigma = 0, C_{out} = 0$

(c) $\Sigma = 1, C_{out} = 0$

(d) $\Sigma = 0, C_{out} = 1$

2. $\Sigma = 1, C_{out} = 1$

Section 6–2 Parallel Binary Adders

1. $C_{out}\Sigma_4\Sigma_3\Sigma_2\Sigma_1 = 11001$

2. Three 74HC283s are required to add two 10-bit numbers.

Section 6–3 Ripple Carry and Look-Ahead Carry Adders

1. $C_g = 0, C_p = 1$

2. $C_{out} = 1$

Section 6–4 Comparators

1. $A > B = 1, A < B = 0, A = B = 0$ when $A = 1011$ and $B = 1010$

2. Right comparator: $A < B = 1; A = B = 0; A > B = 0$
Left comparator: $A < B = 0; A = B = 0; A > B = 1$

Section 6–5 Decoders

1. Output 5 is active when 101 is on the inputs.

2. Four 74HC154s are used to decode a 6-bit binary number.

3. Active-HIGH output drives a common-cathode LED display.

Section 6–6 Encoders

1. (a) $A_0 = 1, A_1 = 1, A_2 = 0, A_3 = 1$

(b) No, this is not a valid BCD code.

(c) Only one input can be active for a valid output.

2. (a) $\overline{A}_3 = 0, \overline{A}_2 = 1, \overline{A}_1 = 1, \overline{A}_0 = 1$

(b) The output is 0111, which is the complement of 1000 (8).

Section 6–7 Code Converters

1. 10000101 (BCD) $= 1010101_2$

2. An 8-bit binary-to-Gray converter consists of seven exclusive-OR gates in an arrangement like that in Figure 6–40 but with inputs B_0–B_7.

Section 6–8 Multiplexers (Data Selectors)

1. The output is 0.

2. (a) 74HC153: Dual 4-input data selector/multiplexer

(b) 74HC151: 8-input data selector/multiplexer

3. The data output alternates between LOW and HIGH as the data-select inputs sequence through the binary states.

4. (a) The 74HC157 multiplexes the two BCD codes to the 7-segment decoder.

    (b) The 74HC47 decodes the BCD to energize the display.

    (c) The 74HC139 enables the 7-segment displays alternately.

Section 6–9 Demultiplexers

1. A decoder can be used as a multiplexer by using the input lines for data selection and an Enable line for data input.

2. The outputs are all HIGH except D_{10}, which is LOW.

Section 6–10 Parity Generators/Checkers

1. (a) Even parity: $\underline{1}110100$ **(b)** Even parity: $\underline{0}01100011$

2. (a) Odd parity: $\underline{1}1010101$ **(b)** Odd parity: $\underline{1}1000001$

3. (a) Code is correct, four 1s.   **(b)** Code is in error, seven 1s

Section 6–11 Troubleshooting

1. A glitch is a very short-duration voltage spike (usually unwanted).

2. Glitches are caused by transition states.

3. Strobe is the enabling of a device for a specified period of time when the device is not in transition.

RELATED PROBLEMS FOR EXAMPLES

6–1 $\Sigma = 1, C_{out} = 1$

6–2 $\Sigma_1 = 0, \Sigma_2 = 0, \Sigma_3 = 1, \Sigma_4 = 1$

6–3 $1011 + 1010 = 10101$

6–4   See Figure 6–88.

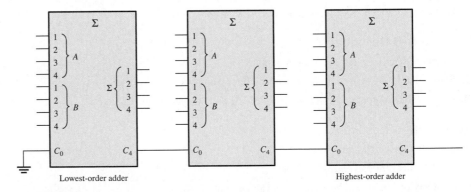

FIGURE 6–88

6–5   See Figure 6–89.

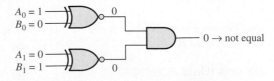

FIGURE 6–89

6–6 $A > B = 0, A = B = 0, A < B = 1$

6–7   See Figure 6–90.

6–8   See Figure 6–91.

6–9   Output 22

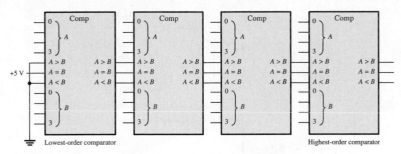

FIGURE 6–90

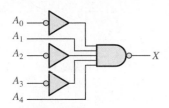

FIGURE 6–91

6–10   See Figure 6–92.

6–11   All inputs LOW: $\overline{A}_0 = 0, \overline{A}_1 = 1, \overline{A}_2 = 1, \overline{A}_3 = 0$

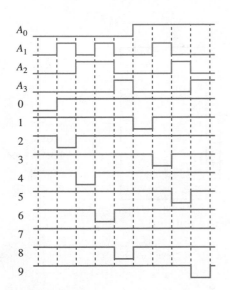

FIGURE 6–92

All inputs HIGH: All outputs HIGH.

6–12   BCD 01000001

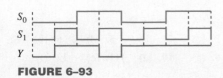

$$00000001 \quad 1$$
$$00101000 \quad 40$$
Binary   00101001   41

6–13   Seven exclusive-OR gates

6–14   See Figure 6–93.

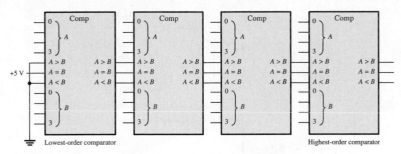

FIGURE 6–93

6–15 D_0: $S_3 = 0, S_2 = 0, S_1 = 0, S_0 = 0$
D_4: $S_3 = 0, S_2 = 1, S_1 = 0, S_0 = 0$
D_8: $S_3 = 1, S_2 = 0, S_1 = 0, S_0 = 0$
D_{13}: $S_3 = 1, S_2 = 1, S_1 = 0, S_0 = 1$

6–16 See Figure 6–94.

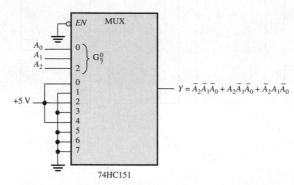

FIGURE 6–94

6–17 See Figure 6–95.

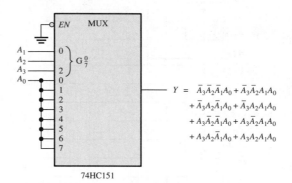

FIGURE 6–95

6–18 See Figure 6–96.

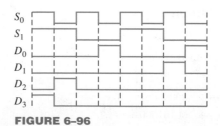

FIGURE 6–96

TRUE/FALSE QUIZ

1. T	**2.** F	**3.** T	**4.** F	**5.** F
6. T	**7.** T	**8.** F	**9.** T	**10.** F

SELF-TEST

1. (a)	**2.** (b)	**3.** (c)	**4.** (a)	**5.** (d)	**6.** (b)
7. (c)	**8.** (b)	**9.** (a)	**10.** (d)	**11.** (c)	**12.** (f)

Latches, Flip-Flops, and Timers

CHAPTER OBJECTIVES

- Use logic gates to construct basic latches
- Explain the difference between an S-R latch and a D latch
- Recognize the difference between a latch and a flip-flop
- Explain how D and J-K flip-flops differ
- Understand the significance of propagation delays, set-up time, hold time, maximum operating frequency, minimum clock pulse widths, and power dissipation in the application of flip-flops
- Apply flip-flops in basic applications
- Explain how retriggerable and nonretriggerable one-shots differ
- Connect a 555 timer to operate as either an astable multivibrator or a one-shot
- Describe latches, flip-flops, and timers using VHDL
- Troubleshoot basic flip-flop circuits

KEY TERMS

Key terms are in order of appearance in the chapter.

- Latch
- Bistable
- SET
- RESET
- Clock
- Edge-triggered flip-flop
- D flip-flop
- Synchronous
- J-K flip-flop
- Toggle
- Preset
- Clear
- Propagation delay time
- Set-up time
- Hold time
- Power dissipation
- One-shot
- Monostable
- Timer
- Astable

VISIT THE WEBSITE

Study aids for this chapter are available at
http://www.pearsonhighered.com/careersresources/

INTRODUCTION

This chapter begins a study of the fundamentals of sequential logic. Bistable, monostable, and astable logic devices called *multivibrators* are covered. Two categories of bistable devices are the latch and the flip-flop. Bistable devices have two stable states, called SET and RESET; they can retain either of these states indefinitely, making them useful as storage devices. The basic difference between latches and flip-flops is the way in which they are changed from one state to the other. The flip-flop is a basic building block for counters, registers, and other sequential control logic and is used in certain types of memories. The monostable multivibrator, commonly known as the one-shot, has only one stable state. A one-shot produces a single controlled-width pulse when activated or triggered. The astable multivibrator has no stable state and is used primarily as an oscillator, which is a self-sustained waveform generator. Pulse oscillators are used as the sources for timing waveforms in digital systems.

7–1   Latches

The **latch** is a type of temporary storage device that has two stable states (bistable) and is normally placed in a category separate from that of flip-flops. Latches are similar to flip-flops because they are bistable devices that can reside in either of two states using a feedback arrangement, in which the outputs are connected back to the opposite inputs. The main difference between latches and flip-flops is in the method used for changing their state.

After completing this section, you should be able to

- ◆ Explain the operation of a basic S-R latch
- ◆ Explain the operation of a gated S-R latch
- ◆ Explain the operation of a gated D latch
- ◆ Implement an S-R or D latch with logic gates
- ◆ Describe the 74HC279A and 74HC75 quad latches

InfoNote

Latches are sometimes used for multiplexing data onto a bus. For example, data being input to a computer from an external source have to share the data bus with data from other sources. When the data bus becomes unavailable to the external source, the existing data must be temporarily stored, and latches placed between the external source and the data bus may be used to do this.

The S-R (SET-RESET) Latch

A latch is a type of **bistable** logic device or **multivibrator.** An active-HIGH input S-R (SET-RESET) latch is formed with two cross-coupled NOR gates, as shown in Figure 7–1(a); an active-LOW input $\overline{S}$-$\overline{R}$ latch is formed with two cross-coupled NAND gates, as shown in Figure 7–1(b). Notice that the output of each gate is connected to an input of the opposite gate. This produces the regenerative **feedback** that is characteristic of all latches and flip-flops.

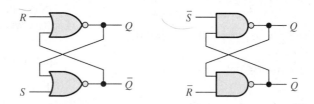

(a) Active-HIGH input S-R latch   (b) Active-LOW input $\overline{S}$-$\overline{R}$ latch

MultiSim

FIGURE 7–1   Two versions of SET-RESET (S-R) latches. Open files F07-01(a) and (b) and verify the operation of both latches. *A Multisim tutorial is available on the website.*

To explain the operation of the latch, we will use the NAND gate $\overline{S}$-$\overline{R}$ latch in Figure 7–1(b). This latch is redrawn in Figure 7–2 with the negative-OR equivalent symbols used for the NAND gates. This is done because LOWs on the $\overline{S}$ and $\overline{R}$ lines are the activating inputs.

The latch in Figure 7–2 has two inputs, $\overline{S}$ and $\overline{R}$, and two outputs, Q and $\overline{Q}$. Let's start by assuming that both inputs and the Q output are HIGH, which is the normal latched state. Since the Q output is connected back to an input of gate G_2, and the $\overline{R}$ input is HIGH, the output of G_2 must be LOW. This LOW output is coupled back to an input of gate G_1, ensuring that its output is HIGH.

When the Q output is HIGH, the latch is in the **SET** state. It will remain in this state indefinitely until a LOW is temporarily applied to the $\overline{R}$ input. With a LOW on the $\overline{R}$ input and a HIGH on $\overline{S}$, the output of gate G_2 is forced HIGH. This HIGH on the $\overline{Q}$ output is coupled back to an input of G_1, and since the $\overline{S}$ input is HIGH, the output of G_1 goes LOW. This LOW on the Q output is then coupled back to an input of G_2, ensuring that the $\overline{Q}$ output remains HIGH even when the LOW on the $\overline{R}$ input is removed. When the Q output is LOW, the latch is in the **RESET** state. Now the latch remains indefinitely in the RESET state until a momentary LOW is applied to the $\overline{S}$ input.

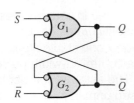

FIGURE 7–2   Negative-OR equivalent of the NAND gate $\overline{S}$-$\overline{R}$ latch in Figure 7–1(b).

A latch can reside in either of its two states, SET or RESET.

In normal operation, the outputs of a latch are always complements of each other.

When Q is HIGH, $\overline{Q}$ is LOW, and when Q is LOW, $\overline{Q}$ is HIGH.

An invalid condition in the operation of an active-LOW input $\overline{S}\text{-}\overline{R}$ latch occurs when LOWs are applied to both $\overline{S}$ and $\overline{R}$ at the same time. As long as the LOW levels are simultaneously held on the inputs, both the Q and $\overline{Q}$ outputs are forced HIGH, thus violating the basic complementary operation of the outputs. Also, if the LOWs are released simultaneously, both outputs will attempt to go LOW. Since there is always some small difference in the propagation delay time of the gates, one of the gates will dominate in its transition to the LOW output state. This, in turn, forces the output of the slower gate to remain HIGH. In this situation, you cannot reliably predict the next state of the latch.

Figure 7–3 illustrates the active-LOW input $\overline{S}\text{-}\overline{R}$ latch operation for each of the four possible combinations of levels on the inputs. (The first three combinations are valid, but the last is not.) Table 7–1 summarizes the logic operation in truth table form. Operation of the active-HIGH input NOR gate latch in Figure 7–1(a) is similar but requires the use of opposite logic levels.

SET means that the Q output is HIGH.

RESET means that the Q output is LOW.

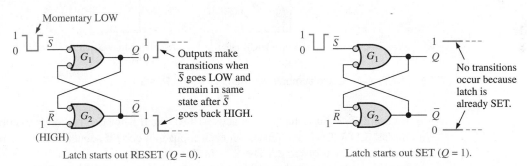

(a) Two possibilities for the SET operation

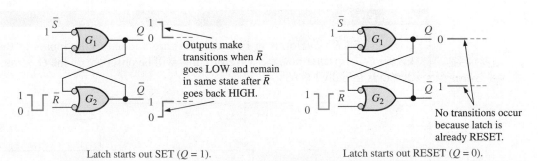

(b) Two possibilities for the RESET operation

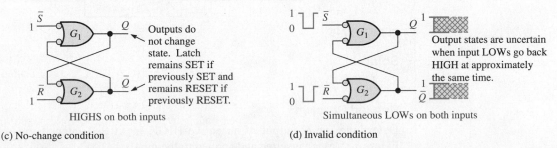

(c) No-change condition   (d) Invalid condition

FIGURE 7–3   The three modes of basic $\overline{S}\text{-}\overline{R}$ latch operation (SET, RESET, no-change) and the invalid condition.

TABLE 7–1

Truth table for an active-LOW input $\overline{S}$-$\overline{R}$ latch.

Inputs		Outputs		
$\overline{S}$	$\overline{R}$	Q	$\overline{Q}$	Comments
1	1	NC	NC	No change. Latch remains in present state.
0	1	1	0	Latch SET.
1	0	0	1	Latch RESET.
0	0	1	1	Invalid condition

Logic symbols for both the active-HIGH input and the active-LOW input latches are shown in Figure 7–4.

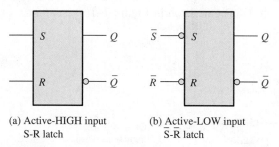

(a) Active-HIGH input
S-R latch

(b) Active-LOW input
$\overline{S}$-$\overline{R}$ latch

FIGURE 7–4   Logic symbols for the S-R and $\overline{S}$-$\overline{R}$ latch.

Example 7–1 illustrates how an active-LOW input $\overline{S}$-$\overline{R}$ latch responds to conditions on its inputs. LOW levels are pulsed on each input in a certain sequence and the resulting Q output waveform is observed. The $\overline{S} = 0, \overline{R} = 0$ condition is avoided because it results in an invalid mode of operation and is a major drawback of any SET-RESET type of latch.

EXAMPLE 7–1

If the $\overline{S}$ and $\overline{R}$ waveforms in Figure 7–5(a) are applied to the inputs of the latch in Figure 7–4(b), determine the waveform that will be observed on the Q output. Assume that Q is initially LOW.

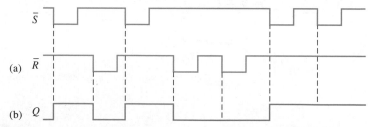

(a) $\overline{S}$

$\overline{R}$

(b) Q

FIGURE 7–5

Solution

See Figure 7–5(b).

Related Problem*

Determine the Q output of an active-HIGH input S-R latch if the waveforms in Figure 7–5(a) are inverted and applied to the inputs.

*Answers are at the end of the chapter.

An Application

The Latch as a Contact-Bounce Eliminator

A good example of an application of an $\overline{S}\text{-}\overline{R}$ latch is in the elimination of mechanical switch contact "bounce." When the pole of a switch strikes the contact upon switch closure, it physically vibrates or bounces several times before finally making a solid contact. Although these bounces are very short in duration, they produce voltage spikes that are often not acceptable in a digital system. This situation is illustrated in Figure 7–6(a).

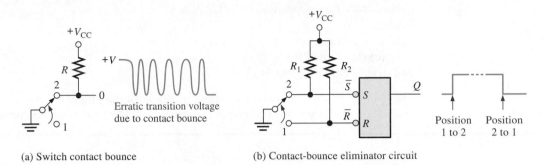

(a) Switch contact bounce   (b) Contact-bounce eliminator circuit

FIGURE 7–6   The $\overline{S}\text{-}\overline{R}$ latch used to eliminate switch contact bounce.

An $\overline{S}\text{-}\overline{R}$ latch can be used to eliminate the effects of switch bounce as shown in Figure 7–6(b). The switch is normally in position 1, keeping the $\overline{R}$ input LOW and the latch RESET. When the switch is thrown to position 2, $\overline{R}$ goes HIGH because of the pull-up resistor to V_{CC}, and $\overline{S}$ goes LOW on the first contact. Although $\overline{S}$ remains LOW for only a very short time before the switch bounces, this is sufficient to set the latch. Any further voltage spikes on the $\overline{S}$ input due to switch bounce do not affect the latch, and it remains SET. Notice that the Q output of the latch provides a clean transition from LOW to HIGH, thus eliminating the voltage spikes caused by contact bounce. Similarly, a clean transition from HIGH to LOW is made when the switch is thrown back to position 1.

IMPLEMENTATION: $\overline{S}\text{-}\overline{R}$ LATCH

Fixed-Function Device   The 74HC279A is a quad $\overline{S}\text{-}\overline{R}$ latch represented by the logic diagram of Figure 7–7(a) and the pin diagram in part (b). Notice that two of the latches each have two $\overline{S}$ inputs.

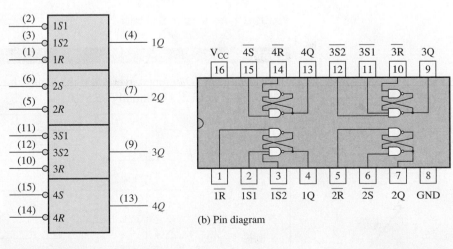

(a) Logic diagram   (b) Pin diagram

FIGURE 7–7   The 74HC279A quad $\overline{S}\text{-}\overline{R}$ latch.

Programmable Logic Device (PLD) An $\overline{\text{S}}\text{-}\overline{\text{R}}$ latch can be described using VHDL and implemented as hardware in a PLD. VHDL statements and keywords not used in previous chapters are introduced in this chapter. These are **library, use, std_logic, all,** and **inout**. The data flow approach is used in this program to describe a single $\overline{\text{S}}\text{-}\overline{\text{R}}$ latch. (The blue comments are not part of the program.)

entity SRLatch **is**
    port (SNot, RNot: **in std_logic**; Q, QNot: **inout std_logic**);
end entity SRLatch;
 	
architecture LogicOperation **of** SRLatch **is**
begin
    Q <= QNot **nand** SNot; ⎱ Boolean expressions
    QNot <= Q **nand** RNot; ⎰ define the outputs
end architecture LogicOperation;
	SNot: SET complement
	RNot: RESET complement
	Q: Latch output
	QNot: Latch output complement

The two inputs SNot and RNot are defined as **std_logic** from the IEEE library. The **inout** keyword allows the Q and QNot outputs of the latch to be used also as inputs for cross-coupling.

The Gated S-R Latch

A gated latch requires an enable input, *EN* (*G* is also used to designate an enable input). The logic diagram and logic symbol for a gated S-R latch are shown in Figure 7–8. The *S* and *R* inputs control the state to which the latch will go when a HIGH level is applied to the *EN* input. The latch will not change until *EN* is HIGH; but as long as it remains HIGH, the output is controlled by the state of the *S* and *R* inputs. The gated latch is a *level-sensitive* device. In this circuit, the invalid state occurs when both *S* and *R* are simultaneously HIGH and *EN* is also HIGH.

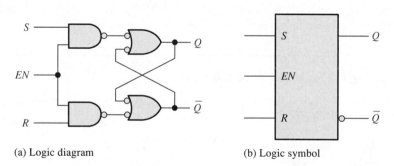

(a) Logic diagram   (b) Logic symbol

FIGURE 7–8   A gated S-R latch.

EXAMPLE 7–2

Determine the *Q* output waveform if the inputs shown in Figure 7–9(a) are applied to a gated S-R latch that is initially RESET.

(a)

(b)

FIGURE 7–9

Solution

The Q waveform is shown in Figure 7–9(b). When S is HIGH and R is LOW, a HIGH on the EN input sets the latch. When S is LOW and R is HIGH, a HIGH on the EN input resets the latch. When both S and R are LOW, the Q output does not change from its present state.

Related Problem

Determine the Q output of a gated S-R latch if the S and R inputs in Figure 7–9(a) are inverted.

The Gated D Latch

Another type of gated latch is called the D latch. It differs from the S-R latch because it has only one input in addition to EN. This input is called the D (data) input. Figure 7–10 contains a logic diagram and logic symbol of a D latch. When the D input is HIGH and the EN input is HIGH, the latch will set. When the D input is LOW and EN is HIGH, the latch will reset. Stated another way, the output Q follows the input D when EN is HIGH.

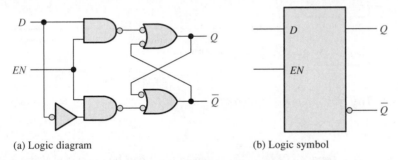

(a) Logic diagram   (b) Logic symbol

FIGURE 7–10   A gated D latch. Open file F07-10 and verify the operation.

MultiSim

EXAMPLE 7–3

Determine the Q output waveform if the inputs shown in Figure 7–11(a) are applied to a gated D latch, which is initially RESET.

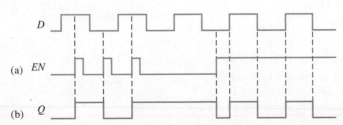

FIGURE 7–11

Solution

The Q waveform is shown in Figure 7–11(b). When D is HIGH and EN is HIGH, Q goes HIGH. When D is LOW and EN is HIGH, Q goes LOW. When EN is LOW, the state of the latch is not affected by the D input.

Related Problem

Determine the Q output of the gated D latch if the D input in Figure 7–11(a) is inverted.

IMPLEMENTATION: GATED D LATCH

Fixed-Function Device   An example of a gated D latch is the 74HC75 represented by the logic symbol in Figure 7–12(a). The device has four latches. Notice that each active-HIGH *EN* input is shared by two latches and is designated as a control input (*C*). The truth table for each latch is shown in Figure 7–12(b). The X in the truth table represents a "don't care" condition. In this case, when the *EN* input is LOW, it does not matter what the *D* input is because the outputs are unaffected and remain in their prior states.

Inputs		Outputs		
D	**EN**	**Q**	**$\overline{Q}$**	**Comments**
0	1	0	1	RESET
1	1	1	0	SET
X	0	Q_0	$\overline{Q}_0$	No change

Note: Q_0 is the prior output level before the indicated input conditions were established.

(a) Logic symbol   (b) Truth table (each latch)

FIGURE 7–12   The 74HC75 quad D latch.

Programmable Logic Device (PLD)   The gated D latch can be described using VHDL and implemented as hardware in a PLD. The data flow approach is used in this program to describe a single D latch.

```
library ieee;
use ieee.std_logic_1164.all;

entity DLatch1 is
    port (D, EN: in std_logic; Q, QNot: inout std_logic);
end entity DLatch1;

architecture LogicOperation of DLatch1 is
begin
    Q <= QNot nand (D nand EN);
    QNot <= Q nand (not D nand EN);
end architecture LogicOperation;
```

D: Data input
EN: Enable
Q: Latch output
QNot: Latch output complement

} Boolean expressions define the outputs

SECTION 7–1 CHECKUP

Answers are at the end of the chapter.

1. List three types of latches.

2. Develop the truth table for the active-HIGH input S-R latch in Figure 7–1(a).

3. What is the *Q* output of a D latch when *EN* = 1 and *D* = 1?

7–2   Flip-Flops

Flip-flops are synchronous bistable devices, also known as *bistable multivibrators*. In this case, the term *synchronous* means that the output changes state only at a specified point (leading or trailing edge) on the triggering input called the **clock** (CLK), which is designated as a control input, C; that is, changes in the output occur in synchronization with the clock. Flip-flops are edge-triggered or edge-sensitive whereas gated latches are level-sensitive.

After completing this section, you should be able to

- Define *clock*

- Define *edge-triggered flip-flop*

- Explain the difference between a flip-flop and a latch

- Identify an edge-triggered flip-flop by its logic symbol

- Discuss the difference between a positive and a negative edge-triggered flip-flop

- Discuss and compare the operation of D and J-K edge-triggered flip-flops and explain the differences in their truth tables

- Discuss the asynchronous inputs of a flip-flop

An **edge-triggered flip-flop** changes state either at the positive edge (rising edge) or at the negative edge (falling edge) of the clock pulse and is sensitive to its inputs only at this transition of the clock. Two types of edge-triggered flip-flops are covered in this section: D and J-K. The logic symbols for these flip-flops are shown in Figure 7–13. Notice that each type can be either positive edge-triggered (no bubble at C input) or negative edge-triggered (bubble at C input). The key to identifying an edge-triggered flip-flop by its logic symbol is the small triangle inside the block at the clock (C) input. This triangle is called the *dynamic input indicator.*

The dynamic input indicator ▷ means the flip-flop changes state only on the edge of a clock pulse.

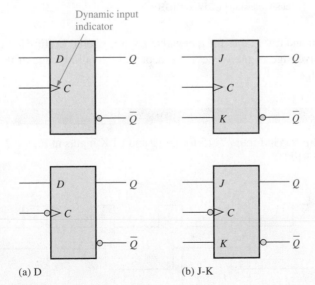

Dynamic input indicator

(a) D   (b) J-K

FIGURE 7–13   Edge-triggered flip-flop logic symbols (top: positive edge-triggered; bottom: negative edge-triggered).

The D Flip-Flop

The D input of the **D flip-flop** is a **synchronous** input because data on the input are transferred to the flip-flop's output only on the triggering edge of the clock pulse. When D is HIGH, the Q output goes HIGH on the triggering edge of the clock pulse, and the flip-flop

D flip-flop but D as variable.

is SET. When D is LOW, the Q output goes LOW on the triggering edge of the clock pulse, and the flip-flop is RESET.

This basic operation of a positive edge-triggered D flip-flop is illustrated in Figure 7–14, and Table 7–2 is the truth table for this type of flip-flop. Remember, *the flip-flop cannot change state except on the triggering edge of a clock pulse.* The D input can be changed at any time when the clock input is LOW or HIGH (except for a very short interval around the triggering transition of the clock) without affecting the output. Just remember, Q follows D at the triggering edge of the clock.

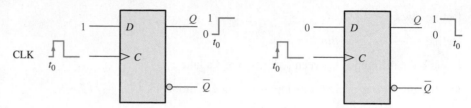

(a) $D = 1$ flip-flop SETS on positive clock edge. (If already SET, it remains SET.)

(b) $D = 0$ flip-flop RESETS on positive clock edge. (If already RESET, it remains RESET.)

FIGURE 7–14 Operation of a positive edge-triggered D flip-flop.

TABLE 7–2

Truth table for a positive edge-triggered D flip-flop.

Inputs		Outputs		
D	CLK	Q	$\overline{Q}$	Comments
0	↑	0	1	RESET
1	↑	1	0	SET

↑ = clock transition LOW to HIGH

The operation and truth table for a negative edge-triggered D flip-flop are the same as those for a positive edge-triggered device except that the falling edge of the clock pulse is the triggering edge.

EXAMPLE 7–4

Determine the Q and $\overline{Q}$ output waveforms of the flip-flop in Figure 7–15 for the D and CLK inputs in Figure 7–16(a). Assume that the positive edge-triggered flip-flop is initially RESET.

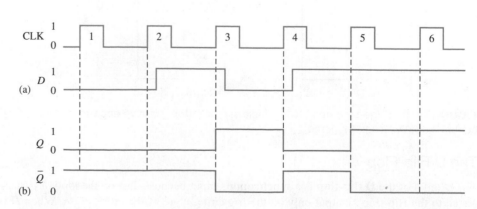

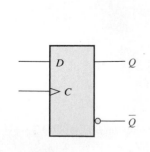

FIGURE 7–15

FIGURE 7–16

Solution

1. At clock pulse 1, D is LOW, so Q remains LOW (RESET).
2. At clock pulse 2, D is LOW, so Q remains LOW (RESET).
3. At clock pulse 3, D is HIGH, so Q goes HIGH (SET).
4. At clock pulse 4, D is LOW, so Q goes LOW (RESET).
5. At clock pulse 5, D is HIGH, so Q goes HIGH (SET).
6. At clock pulse 6, D is HIGH, so Q remains HIGH (SET).

Once Q is determined, $\overline{Q}$ is easily found since it is simply the complement of Q. The resulting waveforms for Q and $\overline{Q}$ are shown in Figure 7–16(b) for the input waveforms in part (a).

Related Problem

Determine Q and $\overline{Q}$ for the D input in Figure 7–16(a) if the flip-flop is a negative edge-triggered device.

The J-K Flip-Flop

The J and K inputs of the **J-K flip-flop** are synchronous inputs because data on these inputs are transferred to the flip-flop's output only on the triggering edge of the clock pulse. When J is HIGH and K is LOW, the Q output goes HIGH on the triggering edge of the clock pulse, and the flip-flop is SET. When J is LOW and K is HIGH, the Q output goes LOW on the triggering edge of the clock pulse, and the flip-flop is RESET. When both J and K are LOW, the output does not change from its prior state. When J and K are both HIGH, the flip-flop changes state. This called the **toggle** mode.

This basic operation of a positive edge-triggered flip-flop is illustrated in Figure 7–17, and Table 7–3 is the truth table for this type of flip-flop. Remember, *the flip-flop cannot change state except on the triggering edge of a clock pulse.* The J and K inputs can be changed at any time when the clock input is LOW or HIGH (except for a very short interval around the triggering transition of the clock) without affecting the output.

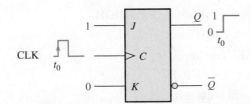

(a) $J = 1$, $K = 0$ flip-flop SETS on positive clock edge. (If already SET, it remains SET.)

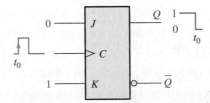

(b) $J = 0$, $K = 1$ flip-flop RESETS on positive clock edge. (If already RESET, it remains RESET.)

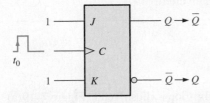

(c) $J = 1$, $K = 1$ flip-flop changes state (toggle).

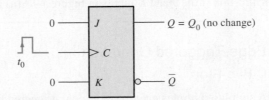

(d) $J = 0$, $K = 0$ flip-flop does not change. (If SET, it remains SET; if RESET, it remains RESET.)

FIGURE 7–17 Operation of a positive edge-triggered J-K flip-flop.

TABLE 7–3

Truth table for a positive edge-triggered J-K flip-flop.

Inputs			Outputs		
J	K	**CLK**	Q	$\overline{Q}$	**Comments**
0	0	↑	Q_0	$\overline{Q_0}$	No change
0	1	↑	0	1	RESET
1	0	↑	1	0	SET
1	1	↑	$\overline{Q_0}$	Q_0	Toggle

↑ = clock transition LOW to HIGH
Q_0 = output level prior to clock transition

EXAMPLE 7–5

The waveforms in Figure 7–18(a) are applied to the J, K, and clock inputs as indicated. Determine the Q output, assuming that the flip-flop is initially RESET.

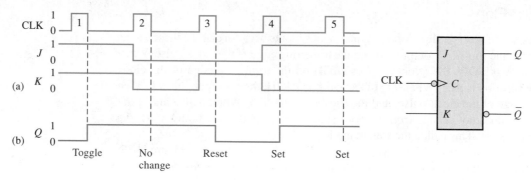

FIGURE 7–18

Solution

Since this is a negative edge-triggered flip-flop, as indicated by the "bubble" at the clock input, the Q output will change only on the negative-going edge of the clock pulse.

1. At the first clock pulse, both J and K are HIGH; and because this is a toggle condition, Q goes HIGH.
2. At clock pulse 2, a no-change condition exists on the inputs, keeping Q at a HIGH level.
3. When clock pulse 3 occurs, J is LOW and K is HIGH, resulting in a RESET condition; Q goes LOW.
4. At clock pulse 4, J is HIGH and K is LOW, resulting in a SET condition; Q goes HIGH.
5. A SET condition still exists on J and K when clock pulse 5 occurs, so Q will remain HIGH.

The resulting Q waveform is indicated in Figure 7–18(b).

Related Problem

Determine the Q output of the J-K flip-flop if the J and K inputs in Figure 7–18(a) are inverted.

Edge-Triggered Operation

D Flip-Flop

A simplified implementation of an edge-triggered D flip-flop is illustrated in Figure 7–19(a) and is used to demonstrate the concept of edge-triggering. Notice that the basic D flip-flop differs from the gated D latch only in that it has a pulse transition detector.

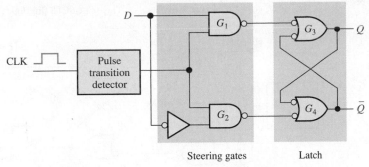

(a) A simplified logic diagram for a positive edge-triggered D flip-flop

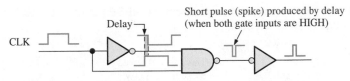

(b) A type of pulse transition detector

FIGURE 7–19   Edge triggering.

One basic type of pulse transition detector is shown in Figure 7–19(b). As you can see, there is a small delay through the inverter on one input to the NAND gate so that the inverted clock pulse arrives at the gate input a few nanoseconds after the true clock pulse. This circuit produces a very short-duration spike on the positive-going transition of the clock pulse. In a negative edge-triggered flip-flop the clock pulse is inverted first, thus producing a narrow spike on the negative-going edge.

The circuit in Figure 7–19(a) is partitioned into two sections, one labeled Steering gates and the other labeled Latch. The steering gates direct, or steer, the clock spike either to the input to gate G_3 or to the input to gate G_4, depending on the state of the D input. To understand the operation of this flip-flop, begin with the assumptions that it is in the RESET state $(Q = 0)$ and that the D and CLK inputs are LOW. For this condition, the outputs of gate G_1 and gate G_2 are both HIGH. The LOW on the Q output is coupled back into one input of gate G_4, making the $\overline{Q}$ output HIGH. Because $\overline{Q}$ is HIGH, both inputs to gate G_3 are HIGH (remember, the output of gate G_1 is HIGH), holding the Q output LOW. If a pulse is applied to the CLK input, the outputs of gates G_1 and G_2 remain HIGH because they are disabled by the LOW on the D input; therefore, there is no change in the state of the flip-flop—it remains in the RESET state.

Let's now make D HIGH and apply a clock pulse. Because the D input to gate G_1 is now HIGH, the output of gate G_1 goes LOW for a very short time (spike) when CLK goes HIGH, causing the Q output to go HIGH. Both inputs to gate G_4 are now HIGH (remember, gate G_2 output is HIGH because D is HIGH), forcing the $\overline{Q}$ output LOW. This LOW on $\overline{Q}$ is coupled back into one input of gate G_3, ensuring that the Q output will remain HIGH. The flip-flop is now in the SET state. Figure 7–20 illustrates the logic level transitions that take place within the flip-flop for this condition.

Next, let's make D LOW and apply a clock pulse. The positive-going edge of the clock produces a negative-going spike on the output of gate G_2, causing the $\overline{Q}$ output to go HIGH. Because of this HIGH on $\overline{Q}$, both inputs to gate G_3 are now HIGH (remember, the output of gate G_1 is HIGH because of the LOW on D), forcing the Q output to go LOW. This LOW on Q is coupled back into one input of gate G_4, ensuring that $\overline{Q}$ will remain HIGH. The flip-flop is now in the RESET state. Figure 7–21 illustrates the logic level transitions that occur within the flip-flop for this condition.

The Q output of a D flip-flop assumes the state of the *D* input on the triggering edge of the clock.

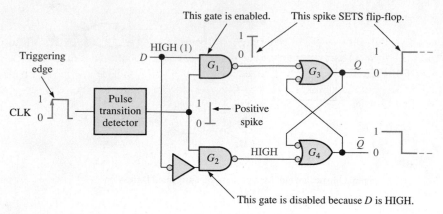

FIGURE 7–20 Flip-flop making a transition from the RESET state to the SET state on the positive-going edge of the clock pulse.

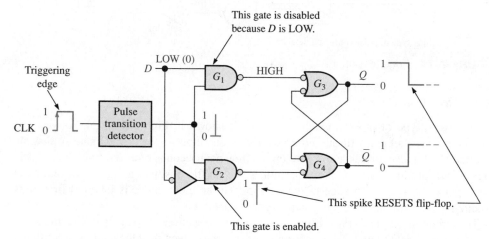

FIGURE 7–21 Flip-flop making a transition from the SET state to the RESET state on the positive-going edge of the clock pulse.

EXAMPLE 7–6

Given the waveforms in Figure 7–22(a) for the D input and the clock, determine the Q output waveform if the flip-flop starts out RESET.

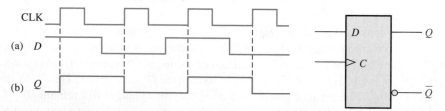

FIGURE 7–22

Solution

The Q output goes to the state of the D input at the time of the positive-going clock edge. The resulting output is shown in Figure 7–22(b).

Related Problem

Determine the Q output for the D flip-flop if the D input in Figure 7–22(a) is inverted.

J-K Flip-Flop

Figure 7–23 shows the basic internal logic for a positive edge-triggered J-K flip-flop. The Q output is connected back to the input of gate G_2, and the $\overline{Q}$ output is connected back to the input of gate G_1. The two control inputs are labeled J and K in honor of Jack Kilby, who invented the integrated circuit. A J-K flip-flop can also be of the negative edge-triggered type, in which case the clock input is inverted.

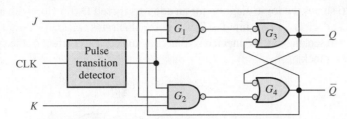

FIGURE 7–23   A simplified logic diagram for a positive edge-triggered J-K flip-flop.

Let's assume that the flip-flop in Figure 7–24 is RESET and that the J input is HIGH and the K input is LOW rather than as shown. When a clock pulse occurs, a leading-edge spike indicated by ① is passed through gate G_1 because $\overline{Q}$ is HIGH and J is HIGH. This will cause the latch portion of the flip-flop to change to the SET state. The flip-flop is now SET.

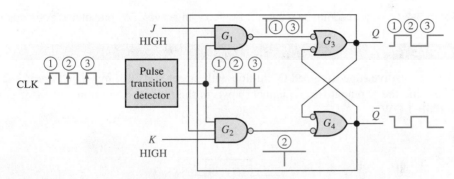

FIGURE 7–24   Transitions illustrating flip-flop operation.

If you make J LOW and K HIGH, the next clock spike indicated by ② will pass through gate G_2 because Q is HIGH and K is HIGH. This will cause the latch portion of the flip-flop to change to the RESET state.

If you apply a LOW to both the J and K inputs, the flip-flop will stay in its present state when a clock pulse occurs. A LOW on both J and K results in a *no-change* condition.

When both the J and K inputs are HIGH and the flip-flop is RESET, the HIGH on the $\overline{Q}$ enables gate G_1; so the clock spike indicated by ③ passes through to set the flip-flop. Now there is a HIGH on Q, which allows the next clock spike to pass through gate G_2 and reset the flip-flop.

As you can see, on each successive clock spike, the flip-flop toggles to the opposite state. Figure 7–24 illustrates the transitions when the flip-flop is in the toggle mode. A J-K flip-flop connected for toggle operation is sometimes called a *T flip-flop*.

In the toggle mode, a J-K flip-flop changes state on every clock pulse.

Asynchronous Preset and Clear Inputs

For the flip-flops just discussed, the D and J-K inputs are called *synchronous inputs* because data on these inputs are transferred to the flip-flop's output only on the triggering edge of the clock pulse; that is, the data are transferred synchronously with the clock.

An active preset input makes the Q output HIGH (SET).

An active clear input makes the Q output LOW (RESET).

Most integrated circuit flip-flops also have **asynchronous** inputs. These are inputs that affect the state of the flip-flop *independent of the clock*. They are normally labeled **preset** (*PRE*) and **clear** (*CLR*), or *direct set* (S_D) and *direct reset* (R_D) by some manufacturers. An active level on the preset input will set the flip-flop, and an active level on the clear input will reset it. A logic symbol for a D flip-flop with preset and clear inputs is shown in Figure 7–25. These inputs are active-LOW, as indicated by the bubbles. These preset and clear inputs must both be kept HIGH for synchronous operation. In normal operation, preset and clear would not be LOW at the same time.

Figure 7–26 shows the logic diagram for an edge-triggered D flip-flop with active-LOW preset ($\overline{PRE}$) and clear ($\overline{CLR}$) inputs. This figure illustrates basically how these inputs work. As you can see, they are connected so that they override the effect of the synchronous input, D and the clock.

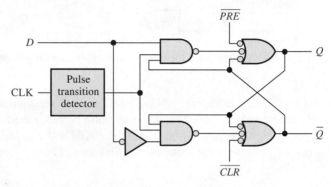

FIGURE 7–25 Logic symbol for a D flip-flop with active-LOW preset and clear inputs.

FIGURE 7–26 Logic diagram for a basic D flip-flop with active-LOW preset and clear inputs.

EXAMPLE 7–7

For the positive edge-triggered D flip-flop with preset and clear inputs in Figure 7–27, determine the Q output for the inputs shown in the timing diagram in part (a) if Q is initially LOW.

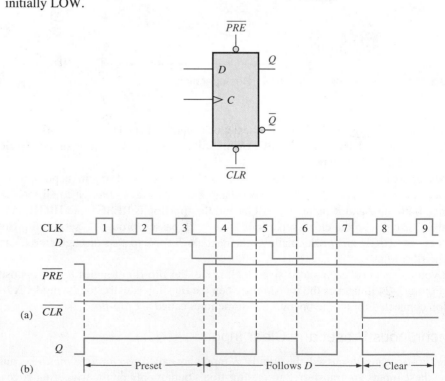

(a)

(b)

MultiSim

FIGURE 7–27 Open file F07-27 to verify the operation.

Solution

1. During clock pulses 1, 2, and 3, the preset ($\overline{PRE}$) is LOW, keeping the flip-flop SET regardless of the synchronous D input.

2. For clock pulses 4, 5, 6, and 7, the output follows the input on the clock pulse because both $\overline{PRE}$ and $\overline{CLR}$ are HIGH.

3. For clock pulses 8 and 9, the clear ($\overline{CLR}$) input is LOW, keeping the flip-flop RESET regardless of the synchronous inputs.

The resulting Q output is shown in Figure 7–27(b).

Related Problem

If you interchange the $\overline{PRE}$ and $\overline{CLR}$ waveforms in Figure 7–27(a), what will the Q output look like?

Let's look at two specific edge-triggered flip-flops. They are representative of the various types of flip-flops available in fixed-function IC form and, like most other devices, are available in CMOS and in bipolar (TTL) logic families.

Also, you will learn how VHDL is used to describe the types of flip-flops.

IMPLEMENTATION: D FLIP-FLOP

Fixed-Function Device   The 74HC74 dual D flip-flop contains two identical D flip-flops that are independent of each other except for sharing V_{CC} and ground. The flip-flops are positive edge-triggered and have active-LOW asynchronous preset and clear inputs. The logic symbols for the individual flip-flops within the package are shown in Figure 7–28(a), and an ANSI/IEEE standard single block symbol that represents the entire device is shown in part (b). The pin numbers are shown in parentheses.

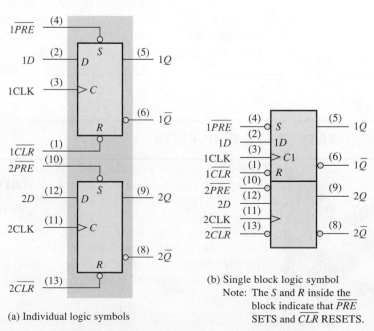

(a) Individual logic symbols

(b) Single block logic symbol
Note: The S and R inside the block indicate that $\overline{PRE}$ SETS and $\overline{CLR}$ RESETS.

FIGURE 7–28   The 74HC74 dual positive edge-triggered D flip-flop.

Programmable Logic Device (PLD)   The positive edge-triggered D flip-flop can be described using VHDL and implemented as hardware in a PLD. In this program, the behavioral approach will be used for the first time because it lends itself to describing sequential operations. A new VHDL statement, **wait until rising_edge**, is introduced. This statement allows the program to wait for the rising edge of a clock pulse to process the *D* input to create the desired results. Also the **if then else** statement is introduced. The keyword **process** is a block of code placed between the **begin** and **end** statements of the architecture to allow statements to be sequentially processed. The program code for a single D flip-flop is as follows:

```
library ieee;
use ieee.std_logic_1164.all;

entity dffl is                                              D: Flip-flop input
    port (D, Clock, Pre, Clr: in std_logic; Q: inout std_logic);    Clock: System clock
end entity dffl;                                           Pre: Preset input
                                                           Clr: Clear input
                                                           Q: Flip-flop output

architecture LogicOperation of dffl is
begin
process
    begin
        wait until rising_edge (Clock);
            if Clr = '1' then          ⎫  Check for Preset and Clear conditions
                if Pre = '1' then       ⎬
                    if D = '1' then     ⎭
                        Q <= '1';        Q input follows D input when Clr and Pre inputs
                    else                 are HIGH.
                        Q <= '0';
                    end if;
                else
                    Q <= '1';   Q is set HIGH when Pre input is LOW.
                end if;
            else
                Q <= '0';   Q is set LOW when Clr input is LOW.
            end if;
    end process;
end architecture LogicOperation;
```

IMPLEMENTATION: J-K FLIP-FLOP

Fixed-Function Device   The 74HC112 dual J-K flip-flop has two identical flip-flops that are negative edge-triggered and have active-LOW asynchronous preset and clear inputs. The logic symbols are shown in Figure 7–29.

Programmable Logic Device (PLD)   The negative edge-triggered J-K flip-flop can be described using VHDL and implemented as hardware in a PLD. In this program, the behavioral approach will be used. A new VHDL statement, **if falling edge then,** is introduced. This statement allows the program to wait for the falling edge of a clock pulse

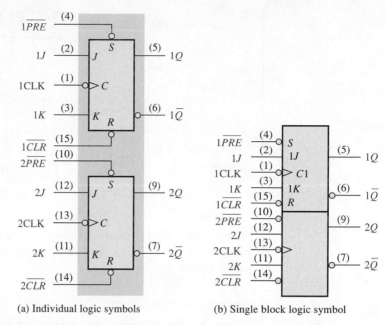

FIGURE 7–29   The 74HC112 dual negative edge-triggered J-K flip-flop.

to process the *J* and *K* inputs to create the desired results. The following program code describes a single J-K flip-flop with no preset or clear inputs.

```
library ieee;
use ieee.std_logic_1164.all;
entity JKFlipFlop is
    port (J, K, Clock: in std_logic; Q, QNot: inout std_logic);    }  Inputs and outputs
end entity JKFlipFlop;                                                  declared


architecture LogicOperation of JKFlipFlop is
signal J1, K1: std_logic;

begin
process (J, K, Clock, J1, K1, Q, QNot)
    begin
        if falling_edge(Clock) and Clock = '0' then
            J1 <= not (J and not Clock and QNot);    }  Identifies with Boolean expressions
            K1 <= not (K and not Clock and Q);           the inputs (J1 and K1) to the latch
        end if;                                          portion of the flip-flop
            Q <= J1 nand QNot;    }  Defines the outputs in terms of J1 and
            QNot <= K1 nand Q;        K1 with Boolean expressions
end process;
end architecture LogicOperation;
```

EXAMPLE 7–8

The $1J$, $1K$, 1CLK, $1\overline{PRE}$, and $1\overline{CLR}$ waveforms in Figure 7–30(a) are applied to one of the negative edge-triggered flip-flops in a 74HC112 package. Determine the $1Q$ output waveform.

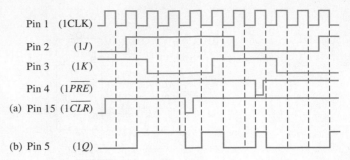

FIGURE 7–30

Solution

The resulting $1Q$ waveform is shown in Figure 7–30(b). Notice that each time a LOW is applied to the $1\overline{PRE}$ or $1\overline{CLR}$, the flip-flop is set or reset regardless of the states of the other inputs.

Related Problem

Determine the $1Q$ output waveform if the waveforms for $1\overline{PRE}$ and $1\overline{CLR}$ are interchanged.

SECTION 7–2 CHECKUP

1. Describe the main difference between a gated D latch and an edge-triggered D flip-flop.

2. How does a J-K flip-flop differ from a D flip-flop in its basic operation?

3. Assume that the flip-flop in Figure 7–22 is negative edge-triggered. Describe the output waveform for the same CLK and D waveforms.

7–3   Flip-Flop Operating Characteristics

The performance, operating requirements, and limitations of flip-flops are specified by several operating characteristics or parameters found on the data sheet for the device. Generally, the specifications are applicable to all CMOS and bipolar (TTL) flip-flops.

After completing this section, you should be able to

- Define *propagation delay time*
- Explain the various propagation delay time specifications
- Define *set-up time* and discuss how it limits flip-flop operation
- Define *hold time* and discuss how it limits flip-flop operation
- Discuss the significance of maximum clock frequency
- Discuss the various pulse width specifications
- Define *power dissipation* and calculate its value for a specific device
- Compare various series of flip-flops in terms of their operating parameters

Propagation Delay Times

A **propagation delay time** is the interval of time required after an input signal has been applied for the resulting output change to occur. Four categories of propagation delay times are important in the operation of a flip-flop:

1. Propagation delay t_{PLH} as measured from the triggering edge of the clock pulse to the LOW-to-HIGH transition of the output. This delay is illustrated in Figure 7–31(a).

2. Propagation delay t_{PHL} as measured from the triggering edge of the clock pulse to the HIGH-to-LOW transition of the output. This delay is illustrated in Figure 7–31(b).

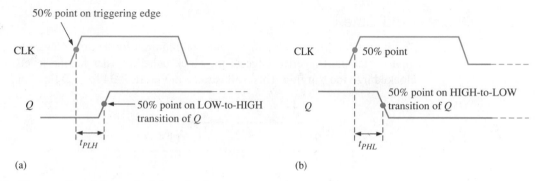

(a)　　　　　　　　　　(b)

FIGURE 7–31　Propagation delays, clock to output.

3. Propagation delay t_{PLH} as measured from the leading edge of the preset input to the LOW-to-HIGH transition of the output. This delay is illustrated in Figure 7–32(a) for an active-LOW preset input.

4. Propagation delay t_{PHL} as measured from the leading edge of the clear input to the HIGH-to-LOW transition of the output. This delay is illustrated in Figure 7–32(b) for an active-LOW clear input.

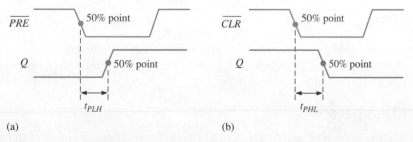

(a)　　　　　　　　　　(b)

FIGURE 7–32　Propagation delays, preset input to output and clear input to output.

Set-up Time

The **set-up time** (t_s) is the minimum interval required for the logic levels to be maintained constantly on the inputs (J and K, or D) prior to the triggering edge of the clock pulse in order for the levels to be reliably clocked into the flip-flop. This interval is illustrated in Figure 7–33 for a D flip-flop.

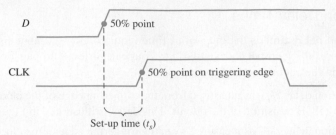

FIGURE 7–33 Set-up time (t_s). The logic level must be present on the D input for a time equal to or greater than t_s before the triggering edge of the clock pulse for reliable data entry.

Hold Time

The **hold time** (t_h) is the minimum interval required for the logic levels to remain on the inputs after the triggering edge of the clock pulse in order for the levels to be reliably clocked into the flip-flop. This is illustrated in Figure 7–34 for a D flip-flop.

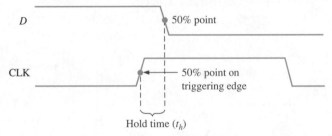

FIGURE 7–34 Hold time (t_h). The logic level must remain on the D input for a time equal to or greater than t_h after the triggering edge of the clock pulse for reliable data entry.

Maximum Clock Frequency

The maximum clock frequency (f_{max}) is the highest rate at which a flip-flop can be reliably triggered. At clock frequencies above the maximum, the flip-flop would be unable to respond quickly enough, and its operation would be impaired.

Pulse Widths

Minimum pulse widths (t_W) for reliable operation are usually specified by the manufacturer for the clock, preset, and clear inputs. Typically, the clock is specified by its minimum HIGH time and its minimum LOW time.

Power Dissipation

The **power dissipation** of any digital circuit is the total power consumption of the device. For example, if the flip-flop operates on a +5 V dc source and draws 5 mA of current, the power dissipation is

$$P = V_{CC} \times I_{CC} = 5\,V \times 5\,mA = 25\,mW$$

The power dissipation is very important in most applications in which the capacity of the dc supply is a concern. As an example, let's assume that you have a digital system that requires a total of ten flip-flops, and each flip-flop dissipates 25 mW of power. The total power requirement is

$$P_T = 10 \times 25\,mW = 250\,mW = 0.25\,W$$

An advantage of CMOS is that it can operate over a wider range of dc supply voltages (typically 2 V to 6 V) than bipolar and, therefore, less expensive power supplies that do not have precise regulation can be used. Also, batteries can be used as secondary or primary sources for CMOS circuits. In addition, lower voltages mean that the IC dissipates less power. The drawback is that the performance of CMOS is degraded with lower supply voltages. For example, the guaranteed maximum clock frequency of a CMOS flip-flop is much less at $V_{CC} = 2$ V than at $V_{CC} = 6$ V.

This tells you the output capacity required of the dc supply. If the flip-flops operate on +5 V dc, then the amount of current that the supply must provide is

$$I = \frac{250 \text{ mW}}{5 \text{ V}} = 50 \text{ mA}$$

You must use a +5 V dc supply that is capable of providing at least 50 mA of current.

Comparison of Specific Flip-Flops

Table 7–4 provides a comparison, in terms of the operating parameters discussed in this section, of four CMOS and bipolar (TTL) flip-flops of the same type but with different IC families (HC, AHC, LS, and F).

TABLE 7–4

Comparison of operating parameters for four IC families of flip-flops of the same type at 25°C.

Parameter	CMOS		Bipolar (TTL)	
	74HC74A	74AHC74	74LS74A	74F74
t_{PHL} (CLK to Q)	17 ns	4.6 ns	40 ns	6.8 ns
t_{PLH} (CLK to Q)	17 ns	4.6 ns	25 ns	8.0 ns
$t_{PHL}(\overline{CLR}$ to Q)	18 ns	4.8 ns	40 ns	9.0 ns
$t_{PLH}(\overline{PRE}$ to Q)	18 ns	4.8 ns	25 ns	6.1 ns
t_s (set-up time)	14 ns	5.0 ns	20 ns	2.0 ns
t_h (hold time)	3.0 ns	0.5 ns	5 ns	1.0 ns
t_W (CLK HIGH)	10 ns	5.0 ns	25 ns	4.0 ns
t_W (CLK LOW)	10 ns	5.0 ns	25 ns	5.0 ns
$t_W(\overline{CLR/PRE})$	10 ns	5.0 ns	25 ns	4.0 ns
f_{max}	35 MHz	170 MHz	25 MHz	100 MHz
Power, quiescent	0.012 mW	1.1 mW		
Power, 50% duty cycle			44 mW	88 mW

SECTION 7–3 CHECKUP

1. Define the following:

   (a) set-up time   (b) hold time

2. Which specific flip-flop in Table 7–4 can be operated at the highest frequency?

7–4 Flip-Flop Applications

In this section, three general applications of flip-flops are discussed to give you an idea of how they can be used. In Chapters 8 and 9, flip-flop applications in registers and counters are covered in detail.

After completing this section, you should be able to

+ Discuss the application of flip-flops in data storage
+ Describe how flip-flops are used for frequency division
+ Explain how flip-flops are used in basic counter applications

Parallel Data Storage

A common requirement in digital systems is to store several bits of data from parallel lines simultaneously in a group of flip-flops. This operation is illustrated in Figure 7–35(a) using four flip-flops. Each of the four parallel data lines is connected to the D input of a flip-flop. The clock inputs of the flip-flops are connected together, so that each flip-flop is triggered by the same clock pulse. In this example, positive edge-triggered flip-flops are used, so the data on the D inputs are stored simultaneously by the flip-flops on the positive edge of the clock, as indicated in the timing diagram in Figure 7–35(b). Also, the asynchronous reset (R) inputs are connected to a common $\overline{CLR}$ line, which initially resets all the flip-flops.

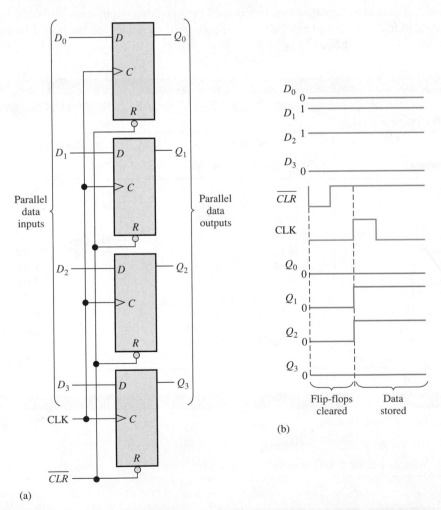

(a)

(b)

FIGURE 7–35   Example of flip-flops used in a basic register for parallel data storage.

This group of four flip-flops is an example of a basic register used for data storage. In digital systems, data are normally stored in groups of bits (usually eight or multiples thereof) that represent numbers, codes, or other information. Registers are covered in Chapter 8.

Frequency Division

Another application of a flip-flop is dividing (reducing) the frequency of a periodic wave-form. When a pulse waveform is applied to the clock input of a D or J-K flip-flop that is connected to toggle ($D = \overline{Q}$ or $J = K = 1$), the Q output is a square wave with one-half the frequency of the clock input. Thus, a single flip-flop can be applied as a divide-by-2 device, as is illustrated in Figure 7–36 for both a D and a J-K flip-flop. As you can see in part (c), the flip-flop changes state on each triggering clock edge (positive edge-triggered in this case). This results in an output that changes at half the frequency of the clock waveform.

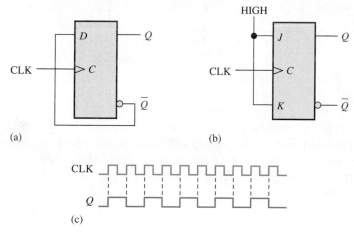

FIGURE 7–36   The D flip-flop and J-K flip-flop as a divide-by-2 device. Q is one-half the frequency of CLK. Open file F07-36 and verify the operation.

MultiSim

Further division of a clock frequency can be achieved by using the output of one flip-flop as the clock input to a second flip-flop, as shown in Figure 7–37. The frequency of the Q_A output is divided by 2 by flip-flop B. The Q_B output is, therefore, one-fourth the frequency of the original clock input. Propagation delay times are not shown on the timing diagrams.

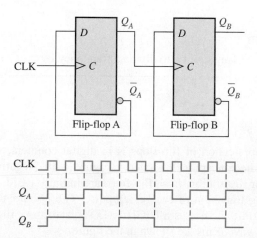

FIGURE 7–37   Example of two D flip-flops used to divide the clock frequency by 4. Q_A is one-half and Q_B is one-fourth the frequency of CLK. Open file F07-37 and verify the operation.

By connecting flip-flops in this way, a frequency division of 2^n is achieved, where n is the number of flip-flops. For example, three flip-flops divide the clock frequency by $2^3 = 8$; four flip-flops divide the clock frequency by $2^4 = 16$; and so on.

EXAMPLE 7–9

Develop the f_{out} waveform for the circuit in Figure 7–38 when an 8 kHz square wave input is applied to the clock input of flip-flop A.

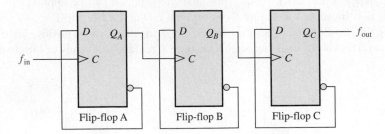

FIGURE 7–38

Solution

The three flip-flops are connected to divide the input frequency by eight ($2^3 = 8$) and the Q_C (f_{out}) waveform is shown in Figure 7–39. Since these are positive edge-triggered flip-flops, the outputs change on the positive-going clock edge. There is one output pulse for every eight input pulses, so the output frequency is 1 kHz. Waveforms of Q_A and Q_B are also shown.

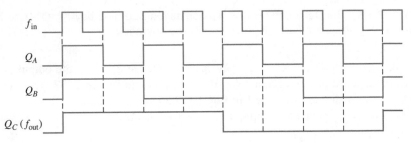

FIGURE 7–39

Related Problem

How many flip-flops are required to divide a frequency by thirty-two?

Counting

Another important application of flip-flops is in digital counters, which are covered in detail in Chapter 9. The concept is illustrated in Figure 7–40. Negative edge-triggered J-K flip-flops are used for illustration. Both flip-flops are initially RESET. Flip-flop A toggles on the negative-going transition of each clock pulse. The Q output of flip-flop A clocks flip-flop B, so each time Q_A makes a HIGH-to-LOW transition, flip-flop B toggles. The resulting Q_A and Q_B waveforms are shown in the figure.

Observe the sequence of Q_A and Q_B in Figure 7–40. Prior to clock pulse 1, $Q_A = 0$ and $Q_B = 0$; after clock pulse 1, $Q_A = 1$ and $Q_B = 0$; after clock pulse 2, $Q_A = 0$ and $Q_B = 1$; and after clock pulse 3, $Q_A = 1$ and $Q_B = 1$. If we take Q_A as the least significant bit, a 2-bit sequence is produced as the flip-flops are clocked. This binary sequence repeats every four clock pulses, as shown in the timing diagram of Figure 7–40. Thus, the flip-flops are counting in sequence from 0 to 3 (00, 01, 10, 11) and then recycling back to 0 to begin the sequence again.

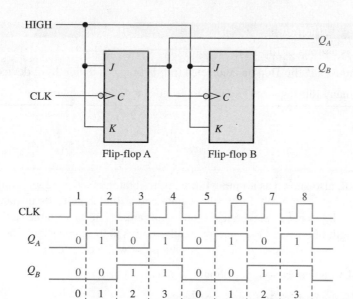

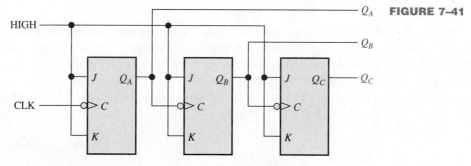

FIGURE 7–40 J-K flip-flops used to generate a binary count sequence (00, 01, 10, 11). Two repetitions are shown.

EXAMPLE 7–10

Determine the output waveforms in relation to the clock for Q_A, Q_B, and Q_C in the circuit of Figure 7–41 and show the binary sequence represented by these waveforms.

Q_A **FIGURE 7–41**

Solution

The output timing diagram is shown in Figure 7–42. Notice that the outputs change on the negative-going edge of the clock pulses. The outputs go through the binary sequence 000, 001, 010, 011, 100, 101, 110, and 111 as indicated.

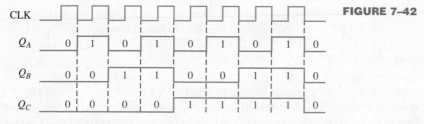

FIGURE 7–42

Related Problem

How many flip-flops are required to produce a binary sequence representing decimal numbers 0 through 15?

1. What is a group of flip-flops used for data storage called?

2. How must a D flip-flop be connected to function as a divide-by-2 device?

3. How many flip-flops are required to produce a divide-by-64 device?

7–5   One-Shots

The **one-shot,** also known as a **monostable** multivibrator, is a device with only one stable state. A one-shot is normally in its stable state and will change to its unstable state only when triggered. Once it is triggered, the one-shot remains in its unstable state for a predetermined length of time and then automatically returns to its stable state. The time that the device stays in its unstable state determines the pulse width of its output.

After completing this section, you should be able to

- Describe the basic operation of a one-shot

- Explain how a nonretriggerable one-shot works

- Explain how a retriggerable one-shot works

- Set up the 74121 and the 74LS122 one-shots to obtain a specified output pulse width

- Recognize a Schmitt trigger symbol and explain basically what it means

- Describe the basic elements of a 555 timer

- Set up a 555 timer as a one-shot

A one-shot produces a single pulse each time it is triggered.

Figure 7–43 shows a basic one-shot (monostable multivibrator) that is composed of a logic gate and an inverter. When a pulse is applied to the **trigger** input, the output of gate G_1 goes LOW. This HIGH-to-LOW transition is coupled through the capacitor to the input of inverter G_2. The apparent LOW on G_2 makes its output go HIGH. This HIGH is connected back into G_1, keeping its output LOW. Up to this point the trigger pulse has caused the output of the one-shot, Q, to go HIGH.

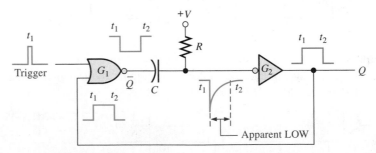

FIGURE 7–43   A simple one-shot circuit.

The capacitor immediately begins to charge through R toward the high voltage level. The rate at which it charges is determined by the RC time constant. When the capacitor charges to a certain level, which appears as a HIGH to G_2, the output goes back LOW.

To summarize, the output of inverter G_2 goes HIGH in response to the trigger input. It remains HIGH for a time set by the RC time constant. At the end of this time, it goes LOW. A single narrow trigger pulse produces a single output pulse whose time duration is controlled by the RC time constant. This operation is illustrated in Figure 7–43.

A typical one-shot logic symbol is shown in Figure 7–44(a), and the same symbol with an external R and C is shown in Figure 7–44(b). The two basic types of IC one-shots are nonretriggerable and retriggerable.

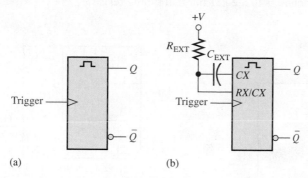

FIGURE 7–44 Basic one-shot logic symbols. *CX* and *RX* stand for external components.

A nonretriggerable one-shot will not respond to any additional trigger pulses from the time it is triggered into its unstable state until it returns to its stable state. In other words, it will ignore any trigger pulses occurring before it times out. The time that the one-shot remains in its unstable state is the pulse width of the output.

Figure 7–45 shows the nonretriggerable one-shot being triggered at intervals greater than its pulse width and at intervals less than the pulse width. Notice that in the second case, the additional pulses are ignored.

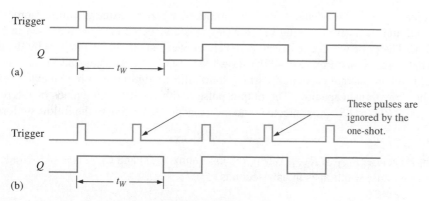

FIGURE 7–45 Nonretriggerable one-shot action.

A retriggerable one-shot can be triggered before it times out. The result of retriggering is an extension of the pulse width as illustrated in Figure 7–46.

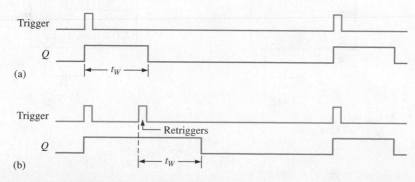

FIGURE 7–46 Retriggerable one-shot action.

Nonretriggerable One-Shot

The 74121 is an example of a nonretriggerable IC one-shot. It has provisions for external R and C, as shown in Figure 7–47. The inputs labeled A_1, A_2, and B are gated trigger inputs. The R_{INT} input connects to a 2 kΩ internal timing resistor.

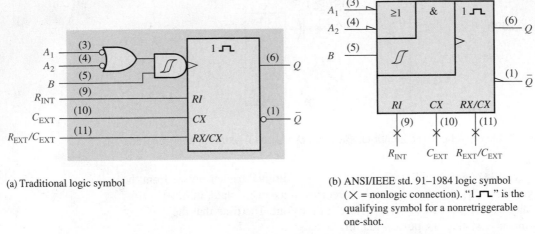

(a) Traditional logic symbol

(b) ANSI/IEEE std. 91–1984 logic symbol ($\times$ = nonlogic connection). "1$\sqcap$" is the qualifying symbol for a nonretriggerable one-shot.

FIGURE 7–47 Logic symbols for the 74121 nonretriggerable one-shot.

Setting the Pulse Width

A typical pulse width of about 30 ns is produced when no external timing components are used and the internal timing resistor (R_{INT}) is connected to V_{CC}, as shown in Figure 7–48(a). The pulse width can be set anywhere between about 30 ns and 28 s by the use of external components. Figure 7–48(b) shows the configuration using the internal resistor (2 kΩ) and an external capacitor. Part (c) shows the configuration using an external resistor and an external capacitor. The output pulse width is set by the values of the resistor ($R_{INT} = 2$ kΩ, and R_{EXT} is selected) and the capacitor according to the following formula:

$$t_W = 0.7RC_{EXT}$$

Equation 7–1

where R is either R_{INT} or R_{EXT}. When R is in kilohms (kΩ) and C_{EXT} is in picofarads (pF), the output pulse width t_W is in nanoseconds (ns).

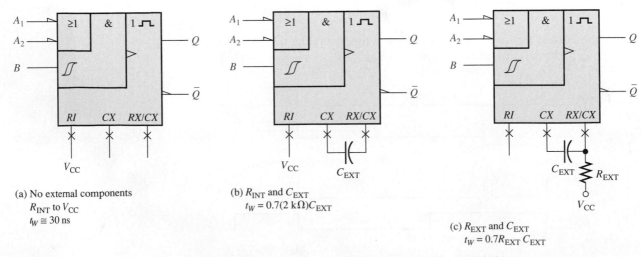

(a) No external components
R_{INT} to V_{CC}
$t_W \cong 30$ ns

(b) R_{INT} and C_{EXT}
$t_W = 0.7(2$ k$\Omega)C_{EXT}$

(c) R_{EXT} and C_{EXT}
$t_W = 0.7R_{EXT}C_{EXT}$

FIGURE 7–48 Three ways to set the pulse width of a 74121.

The Schmitt-Trigger Symbol

The symbol $\int\!\!\int$ indicates a Schmitt-trigger input. This type of input uses a special threshold circuit that produces **hysteresis**, a characteristic that prevents erratic switching between states when a slow-changing trigger voltage hovers around the critical input level. This allows reliable triggering to occur even when the input is changing as slowly as 1 volt/second.

Retriggerable One-Shot

The 74LS122 is an example of a retriggerable IC one-shot with a clear input. It also has provisions for external R and C, as shown in Figure 7–49. The inputs labeled A_1, A_2, B_1, and B_2 are the gated trigger inputs.

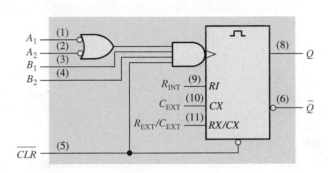

(a) Traditional logic symbol

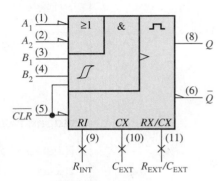

(b) ANSI/IEEE std. 91–1984 logic symbol ($\times$ = nonlogic connection). $\sqcap$ is the qualifying symbol for a retriggerable one-shot.

FIGURE 7–49   Logic symbol for the 74LS122 retriggerable one-shot.

A minimum pulse width of approximately 45 ns is obtained with no external components. Wider pulse widths are achieved by using external components. A general formula for calculating the values of these components for a specified pulse width (t_W) is

$$t_W = 0.32RC_{\text{EXT}}\left(1 + \frac{0.7}{R}\right)$$

Equation 7–2

where 0.32 is a constant determined by the particular type of one-shot, R is in $k\Omega$ and is either the internal or the external resistor, C_{EXT} is in pF, and t_W is in ns. The internal resistance is 10 $k\Omega$ and can be used instead of an external resistor. (Notice the difference between this formula and that for the 74121, shown in Equation 7–1.)

EXAMPLE 7–11

A certain application requires a one-shot with a pulse width of approximately 100 ms. Using a 74121, show the connections and the component values.

Solution

Arbitrarily select $R_{\text{EXT}} = \mathbf{39\ k\Omega}$ and calculate the necessary capacitance.

$$t_W = 0.7R_{\text{EXT}}C_{\text{EXT}}$$

$$C_{\text{EXT}} = \frac{t_W}{0.7R_{\text{EXT}}}$$

where C_{EXT} is in pF, R_{EXT} is in $k\Omega$, and t_W is in ns. Since 100 ms = 1×10^8 ns,

$$C_{\text{EXT}} = \frac{1 \times 10^8\ \text{ns}}{0.7(39\ k\Omega)} = 3.66 \times 10^{-6}\ \text{pF} = \mathbf{3.66\ \mu F}$$

A standard 3.3 μF capacitor will give an output pulse width of 91 ms. The proper connections are shown in Figure 7–50. To achieve a pulse width closer to 100 ms, other combinations of values for R_{EXT} and C_{EXT} can be tried. For example, $R_{EXT} = 68$ kΩ and $C_{EXT} = 2.2$ μF gives a pulse width of 105 ms.

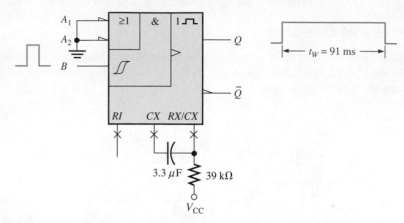

FIGURE 7–50

Related Problem

Use an external capacitor in conjunction with R_{INT} to produce an output pulse width of 10 μs from the 74121.

EXAMPLE 7–12

Determine the values of R_{EXT} and C_{EXT} that will produce a pulse width of 1 μs when connected to a 74LS122.

Solution

Assume a value of $C_{EXT} = $ **560 pF** and then solve for R_{EXT}. The pulse width must be expressed in ns and C_{EXT} in pF. R_{EXT} will be in kΩ.

$$t_w = 0.32 R_{EXT} C_{EXT} \left(1 + \frac{0.7}{R_{EXT}} \right) = 0.32 R_{EXT} C_{EXT} + 0.7 \left(\frac{0.32 R_{EXT} C_{EXT}}{R_{EXT}} \right)$$

$$= 0.32 R_{EXT} C_{EXT} + (0.7)(0.32) C_{EXT}$$

$$R_{EXT} = \frac{t_W - (0.7)(0.32) C_{EXT}}{0.32 C_{EXT}} = \frac{t_W}{0.32 C_{EXT}} - 0.7$$

$$= \frac{1000 \text{ ns}}{(0.32)560 \text{ pF}} - 0.7 = \textbf{4.88 k}\boldsymbol{\Omega}$$

Use a standard value of **4.7 kΩ.**

Related Problem

Show the connections and component values for a 74LS122 one-shot with an output pulse width of 5 μs. Assume $C_{EXT} = 560$ pF.

An Application

One practical one-shot application is a sequential timer that can be used to illuminate a series of lights. This type of circuit can be used, for example, in a lane change directional indicator for highway construction projects or in sequential turn signals on automobiles.

Figure 7–51 shows three 74LS122 one-shots connected as a sequential timer. This particular circuit produces a sequence of three 1 s pulses. The first one-shot is triggered by a switch closure or a low-frequency pulse input, producing a 1 s output pulse. When the first one-shot (OS 1) times out and the 1 s pulse goes LOW, the second one-shot (OS 2) is triggered, also producing a 1 s output pulse. When this second pulse goes LOW, the third one-shot (OS 3) is triggered and the third 1 s pulse is produced. The output timing is illustrated in the figure. Variations of this basic arrangement can be used to produce a variety of timed outputs.

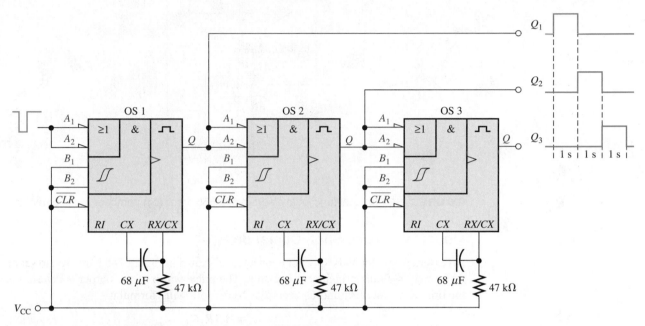

FIGURE 7–51   A sequential timing circuit using three 74LS122 one-shots.

The 555 Timer as a One-Shot

The 555 **timer** is a versatile and widely used IC device because it can be configured in two different modes as either a monostable multivibrator (one-shot) or as an astable multivibrator (pulse oscillator). The astable multivibrator is discussed in Section 7–6.

The 555 Timer Operation

A functional diagram showing the internal components of a 555 timer is shown in Figure 7–52. The comparators are devices whose outputs are HIGH when the voltage on the positive (+) input is greater than the voltage on the negative (−) input and LOW when the − input voltage is greater than the + input voltage. The voltage divider consisting of three 5 kΩ resistors provides a trigger level of $\frac{1}{3} V_{CC}$ and a threshold level of $\frac{2}{3} V_{CC}$. The control voltage input (pin 5) can be used to externally adjust the trigger and threshold levels to other values if necessary. When the normally HIGH trigger input momentarily goes below $\frac{1}{3} V_{CC}$, the output of comparator B switches from LOW to HIGH and sets the S-R latch, causing the output (pin 3) to go HIGH and turning the discharge transistor Q_1 off. The output will stay HIGH until the normally LOW threshold input goes above $\frac{2}{3} V_{CC}$ and causes the output of comparator A to switch from LOW to HIGH. This resets the latch, causing the output to go back LOW and turning the discharge transistor on. The external reset input can be used to reset the latch independent of the threshold circuit. The trigger and threshold inputs (pins 2 and 6) are controlled by external components connected to produce either monostable or astable action.

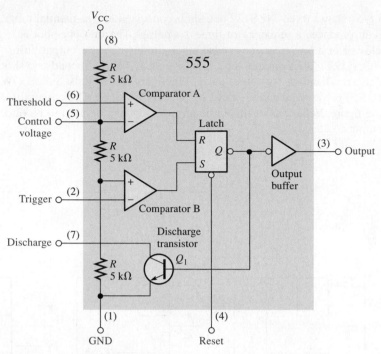

FIGURE 7–52   Internal functional diagram of a 555 timer (pin numbers are in parentheses).

Monostable (One-Shot) Operation

An external resistor and capacitor connected as shown in Figure 7–53 are used to set up the 555 timer as a nonretriggerable one-shot. The pulse width of the output is determined by the time constant of R_1 and C_1 according to the following formula:

$$t_W = 1.1R_1C_1$$ Equation 7–3

The control voltage input is not used and is connected to a decoupling capacitor C_2 to prevent noise from affecting the trigger and threshold levels.

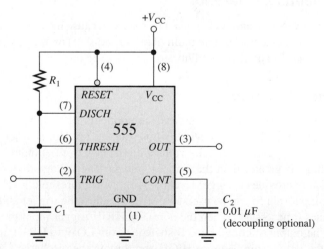

FIGURE 7–53   The 555 timer connected as a one-shot.

Before a trigger pulse is applied, the output is LOW and the discharge transistor Q_1 is *on*, keeping C_1 discharged as shown in Figure 7–54(a). When a negative-going trigger pulse is applied at t_0, the output goes HIGH and the discharge transistor turns *off*, allowing capacitor C_1 to begin charging through R_1 as shown in part (b). When C_1 charges to $\frac{1}{3} V_{CC}$, the output goes back LOW at t_1 and Q_1 turns *on* immediately, discharging C_1 as shown in part (c). As you can see, the charging rate of C_1 determines how long the output is HIGH.

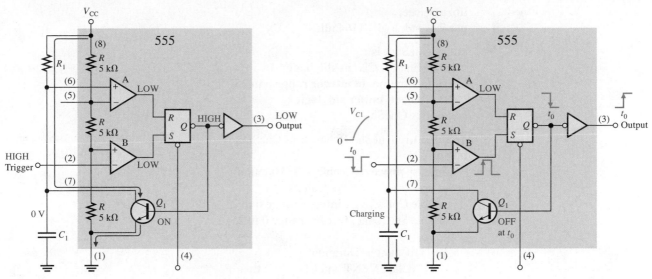

(a) Prior to triggering. (The current path is indicated by the red arrow.)   (b) When triggered

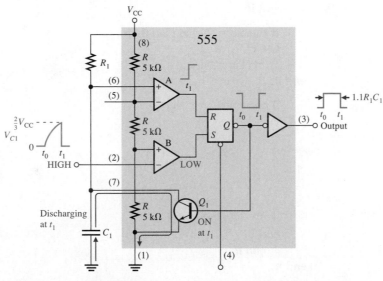

(c) At end of charging interval

FIGURE 7–54   One-shot operation of the 555 timer.

EXAMPLE 7–13

What is the output pulse width for a 555 monostable circuit with $R_1 = 2.2 \text{ k}\Omega$ and $C_1 = 0.01 \text{ }\mu\text{F}$?

Solution

From Equation 7–3 the pulse width is

$$t_W = 1.1R_1C_1 = 1.1(2.2 \text{ k}\Omega)(0.01 \text{ }\mu\text{F}) = \textbf{24.2 }\mu\textbf{s}$$

Related Problem

For $C_1 = 0.01 \text{ }\mu\text{F}$, determine the value of R_1 for a pulse width of 1 ms.

One-Shot with VHDL

An example of a VHDL program code for a one-shot is as follows:

```vhdl
library ieee;
use ieee.std_logic_1164.all;

entity OneShot is
    port (Enable, Clk: in std_logic;
          Duration: in integer range 0 to 25;
          QOut: buffer std_logic);
end entity OneShot;

architecture OneShotBehavior of OneShot is
begin
    Counter: process (Enable, Clk, Duration)
    variable Flag       : boolean := true;
    variable Cnt         : integer range 0 to 25;
    variable SetCount : integer range 0 to 25;
    begin
        SetCount := Duration;
        if (Clk'EVENT and Clk = '1') then
        if Enable = '0' then
                Flag := true;
            end if;

            if Enable = '1' and Flag then
            Cnt := 1;
                Flag :=False;
        end if;

            if cnt = SetCount then
            Qout <= '0';
                Cnt := 0;
                    Flag := false;
        else
                if Cnt > 0 then
                Cnt := Cnt + 1;
                    Qout <= '1';
            end if;
                end if;
            end if;
        end process;
end architecture OneShotBehavior;
```

In normal operation, a one-shot produces only a single pulse, which can be difficult to measure on an oscilloscope because the pulse does not occur regularly. To obtain a stable display for test purposes, it is useful to trigger the one-shot from a pulse generator that is set to a longer period than the expected pulse width and trigger the oscilloscope from the same pulse. For very long pulses, either store the waveform using a digital storage oscilloscope or shorten the time constant by some known factor. For example, replace a 1000 μF capacitor with a 1 μF capacitor to shorten the time by a factor of 1000. A faster pulse is easier to see and measure with an oscilloscope.

1. Describe the difference between a nonretriggerable and a retriggerable one-shot.

2. How is the output pulse width set in most IC one-shots?

3. What is the pulse width of a 555 timer one-shot when $C = 1\ \mu F$ and $R = 10\ k\Omega$?

7–6 The Astable Multivibrator

An **astable** multivibrator is a device that has no stable states; it changes back and forth (oscillates) between two unstable states without any external triggering. The resulting output is typically a square wave that is used as a clock signal in many types of sequential logic circuits. Astable multivibrators are also known as pulse **oscillators**.

After completing this section, you should be able to

◆ Describe the operation of a simple astable multivibrator using a Schmitt trigger circuit.

◆ Set up a 555 timer as an astable multivibrator.

Figure 7–55(a) shows a simple form of astable multivibrator using an inverter with hysteresis (Schmitt trigger) and an *RC* circuit connected in a feedback arrangement. When power is first applied, the capacitor has no charge; so the input to the Schmitt trigger inverter is LOW and the output is HIGH. The capacitor charges through *R* until the inverter input voltage reaches the upper trigger point (UTP), as shown in Figure 7–55(b). At this point, the inverter output goes LOW, causing the capacitor to discharge back through *R*, shown in part (b). When the inverter input voltage decreases to the lower trigger point (LTP), its output goes HIGH and the capacitor charges again. This charging/discharging cycle continues to repeat as long as power is applied to the circuit, and the resulting output is a pulse waveform, as indicated.

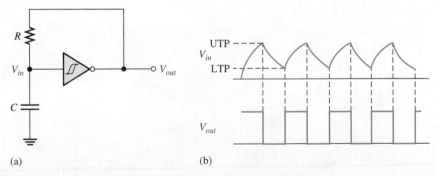

(a)   (b)

FIGURE 7–55   Basic astable multivibrator using a Schmitt trigger.

The 555 Timer as an Astable Multivibrator

A 555 timer connected to operate as an astable multivibrator is shown in Figure 7–56. Notice that the threshold input (*THRESH*) is now connected to the trigger input (*TRIG*). The external components R_1, R_2, and C_1 form the timing network that sets the frequency of oscillation. The 0.01 μF capacitor, C_2, connected to the control (*CONT*) input is strictly for decoupling and has no effect on the operation; in some cases it can be left off.

> **InfoNote**
>
> Most systems require a timing source to provide accurate clock waveforms. The timing section controls all system timing and is responsible for the proper operation of the system hardware. The timing section usually consists of a crystal-controlled oscillator and counters for frequency division. Using a high-frequency oscillator divided down to a lower frequency provides for greater accuracy and frequency stability.

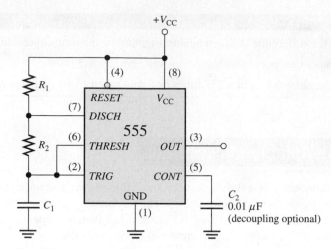

FIGURE 7–56   The 555 timer connected as an astable multivibrator (oscillator).

Initially, when the power is turned on, the capacitor (C_1) is uncharged and thus the trigger voltage (pin 2) is at 0 V. This causes the output of comparator B to be HIGH and the output of comparator A to be LOW, forcing the output of the latch, and thus the base of Q_1, LOW and keeping the transistor off. Now, C_1 begins charging through R_1 and R_2, as indicated in Figure 7–57. When the capacitor voltage reaches $\frac{1}{3}$ V_{CC}, comparator B switches to its LOW output state; and when the capacitor voltage reaches $\frac{2}{3}$ V_{CC}, comparator A switches to its HIGH output state. This resets the latch, causing the base of Q_1 to go HIGH and turning on the transistor. This sequence creates a discharge path for the capacitor through R_2 and the transistor, as indicated. The capacitor now begins to discharge, causing comparator A to go LOW. At the point where the capacitor discharges down to $\frac{1}{3}$ V_{CC}, comparator B switches HIGH; this sets the latch, making the base of Q_1 LOW and turning off the transistor. Another charging cycle begins, and the entire process repeats. The

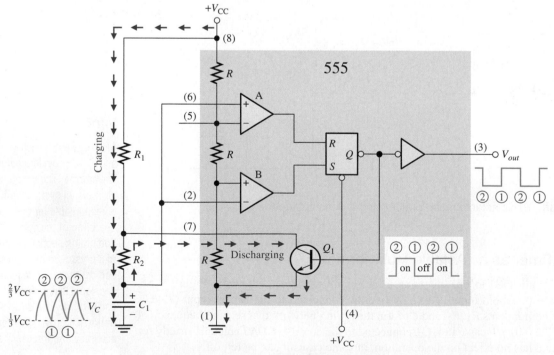

FIGURE 7–57   Operation of the 555 timer in the astable mode.

result is a rectangular wave output whose duty cycle depends on the values of R_1 and R_2. The frequency of oscillation is given by the following formula, or it can be found using the graph in Figure 7–58.

$$f = \frac{1.44}{(R_1 + 2R_2)C_1}$$

Equation 7–4

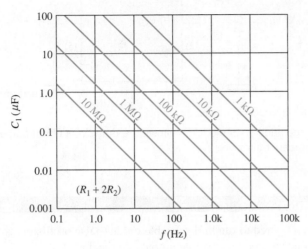

FIGURE 7–58 Frequency of oscillation as a function of C_1 and $R_1 + 2R_2$. The sloped lines are values of $R_1 + 2R_2$.

By selecting R_1 and R_2, the duty cycle of the output can be adjusted. Since C_1 charges through $R_1 + R_2$ and discharges only through R_2, duty cycles approaching a minimum of 50 percent can be achieved if $R_2 \gg R_1$ so that the charging and discharging times are approximately equal.

An expression for the duty cycle is developed as follows. The time that the output is HIGH (t_H) is how long it takes C_1 to charge from $\frac{1}{3} V_{CC}$ to $\frac{2}{3} V_{CC}$. It is expressed as

$$t_H = 0.7(R_1 + R_2)C_1$$

Equation 7–5

The time that the output is LOW (t_L) is how long it takes C_1 to discharge from $\frac{1}{3} V_{CC}$ to $\frac{2}{3} V_{CC}$. It is expressed as

$$t_L = 0.7R_2C_1$$

Equation 7–6

The period, T, of the output waveform is the sum of t_H and t_L. This is the reciprocal of f in Equation 7–4.

$$T = t_H + t_L = 0.7(R_1 + 2R_2)C_1$$

Finally, the duty cycle is

$$\text{Duty cycle} = \frac{t_H}{T} = \frac{t_H}{t_H + t_L}$$

$$\textbf{Duty cycle} = \left(\frac{R_1 + R_2}{R_1 + 2R_2}\right)100\%$$

Equation 7–7

To achieve duty cycles of less than 50 percent, the circuit in Figure 7–56 can be modified so that C_1 charges through only R_1 and discharges through R_2. This is achieved with a diode, D_1, placed as shown in Figure 7–59. The duty cycle can be made less than 50 percent by making R_1 less than R_2. Under this condition, the expression for the duty cycle is

$$\textbf{Duty cycle} = \left(\frac{R_1}{R_1 + R_2}\right)100\%$$

Equation 7–8

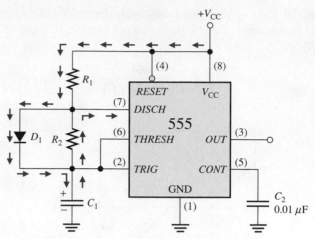

FIGURE 7–59   The addition of diode D_1 allows the duty cycle of the output to be adjusted to less than 50 percent by making $R_1 < R_2$.

EXAMPLE 7–14

A 555 timer configured to run in the astable mode (pulse oscillator) is shown in Figure 7–60. Determine the frequency of the output and the duty cycle.

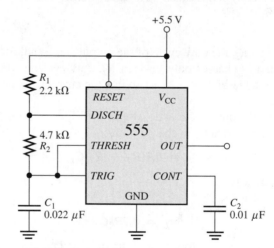

MultiSim

FIGURE 7–60   Open file F07-60 to verify operation.

Solution

Use Equations 7–4 and 7–7.

$$f = \frac{1.44}{(R_1 + 2R_2)C_1} = \frac{1.44}{(2.2 \text{ k}\Omega + 9.4 \text{ k}\Omega)0.022 \text{ }\mu\text{F}} = \textbf{5.64 kHz}$$

$$\text{Duty cycle} = \left(\frac{R_1 + R_2}{R_1 + 2R_2}\right)100\% = \left(\frac{2.2 \text{ k}\Omega + 4.7 \text{ k}\Omega}{2.2 \text{ k}\Omega + 9.4 \text{ k}\Omega}\right)100\% = \textbf{59.5\%}$$

Related Problem

Determine the duty cycle in Figure 7–60 if a diode is connected across R_2 as indicated in Figure 7–59.

SECTION 7–6 CHECKUP

1. Explain the difference in operation between an astable multivibrator and a monostable multivibrator.

2. For a certain astable multivibrator, $t_H = 15$ ms and $T = 20$ ms. What is the duty cycle of the output?

7–7 Troubleshooting

It is standard practice to test a new circuit design to be sure that it is operating as specified. New fixed-function designs are "breadboarded" and tested before the design is finalized. The term *breadboard* refers to a method of temporarily hooking up a circuit so that its operation can be verified and any design flaws worked out before a prototype unit is built.

After completing this section, you should be able to

- ◆ Describe how the timing of a circuit can produce erroneous glitches

- ◆ Approach the troubleshooting of a new design with greater insight and awareness of potential problems

The circuit shown in Figure 7–61(a) generates two clock waveforms (CLK A and CLK B) that have an alternating occurrence of pulses. Each waveform is to be one-half the frequency of the original clock (CLK), as shown in the ideal timing diagram in part (b).

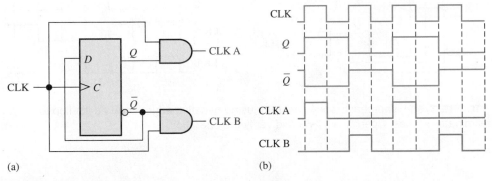

(a)   (b)

FIGURE 7–61 Two-phase clock generator with ideal waveforms. Open file F07-61 and verify the operation.

MultiSim

When the circuit is tested with an oscilloscope or logic analyzer, the CLK A and CLK B waveforms appear on the display screen as shown in Figure 7–62(a). Since glitches occur on both waveforms, something is wrong with the circuit either in its basic design or in the way it is connected. Further investigation reveals that the glitches are caused by a **race** condition between the CLK signal and the Q and $\overline{Q}$ signals at the inputs of the AND gates. As displayed in Figure 7–62(b), the propagation delays between CLK and Q and $\overline{Q}$ create a short-duration coincidence of HIGH levels at the leading edges of alternate clock pulses. Thus, there is a basic design flaw.

The problem can be corrected by using a negative edge-triggered flip-flop in place of the positive edge-triggered device, as shown in Figure 7–63(a). Although the propagation delays between CLK and Q and $\overline{Q}$ still exist, they are initiated on the trailing edges of the clock (CLK), thus eliminating the glitches, as shown in the timing diagram of Figure 7–63(b).

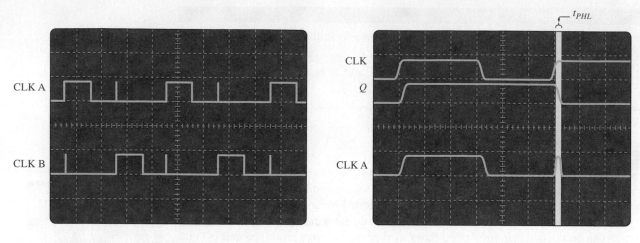

(a) Oscilloscope display of CLK A and CLK B waveforms with glitches indicated by the "spikes".

(b) Oscilloscope display showing propagation delay that creates glitch on CLK A waveform

FIGURE 7–62   Oscilloscope displays for the circuit in Figure 7–61.

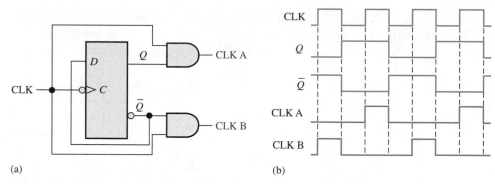

(a)

(b)

FIGURE 7–63   Two-phase clock generator using negative edge-triggered flip-flop to eliminate glitches. Open file F07-63 and verify the operation.

Glitches that occur in digital systems are very fast (extremely short in duration) and can be difficult to see on an oscilloscope, particularly at lower sweep rates. A logic analyzer, however, can show a glitch easily. To look for glitches using a logic analyzer, select "latch" mode or (if available) transitional sampling. In the latch mode, the analyzer looks for a voltage level change. When a change occurs, even if it is of extremely short duration (a few nanoseconds), the information is "latched" into the analyzer's memory as another sampled data point. When the data are displayed, the glitch will show as an obvious change in the sampled data, making it easy to identify.

SECTION 7–7 CHECKUP

1. Can a negative edge-triggered J-K flip-flop be used in the circuit of Figure 7–63?
2. What device can be used to provide the clock for the circuit in Figure 7–63?

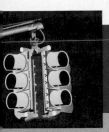

Applied Logic

Traffic Signal Controller: Part 2

The combinational logic unit of the traffic signal controller was completed in Chapter 6. Now, the timing circuits and sequential logic are developed. Recall that the timing circuits produce a 25 s time interval for the red and green lights and a 4 s interval for the yellow caution light. These outputs will be used by the sequential logic. The block diagram of the complete traffic signal controller is shown in Figure 7–64.

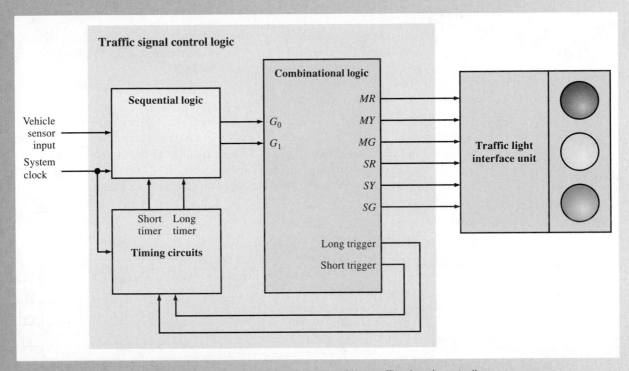

FIGURE 7–64 Block diagram of the traffic signal controller.

Timing Circuits

The timing circuits unit of the traffic signal controller consists of a 25 s timer and a 4 s timer and a clock generator. One way to implement this unit is with two 555 timers configured as one-shots and one 555 timer configured as an astable multivibrator (oscillator), as discussed earlier in this chapter. Component values are calculated based on the formulas given.

Another way to implement the timing circuits is shown in Figure 7–65. An external 24 MHz system clock (arbitrary value) is divided down to an accurate 1 Hz clock by the frequency divider. The 1 Hz clock is then used to establish the 25 s and the 4 s intervals by counting the 1 Hz pulses. This approach lends itself better to a VHDL description.

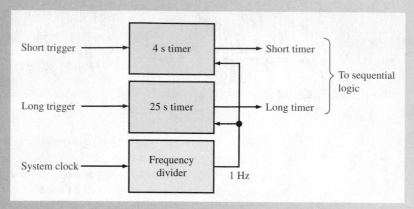

FIGURE 7–65   Block diagram of the timing circuits unit.

Exercise

1. Determine the values for the resistor and capacitor in a 25 s 555 timer.
2. Determine the values for the resistor and capacitor in a 4 s 555 timer.
3. What is the purpose of the frequency divider?

Controller Programming with VHDL

A programming model for the traffic signal controller is shown in Figure 7–66, where all the input and output labels are given. Notice that the Timing circuits block is split into two parts; the Frequency divider and the Timer circuits; and the Combinational logic block is divided into the State decoder and two logic sections (Light output logic and Trigger logic). This model will be used to develop the VHDL program codes.

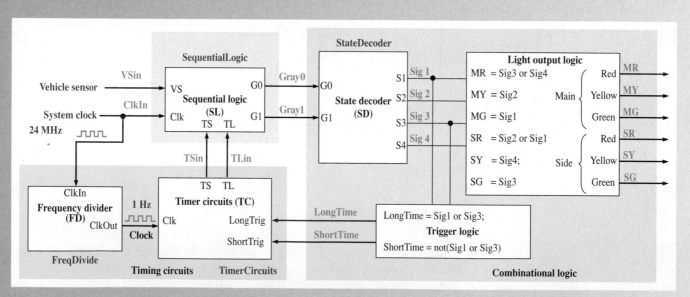

FIGURE 7–66   Programming model for the traffic signal controller.

Frequency Divider   The purpose of the frequency divider is to produce a 1 Hz clock for the timer circuits. The input ClkIn in this application is a 24.00 MHz oscillator that drives the program code. SetCount is used to initialize the count for a 1 Hz interval. The program

FreqDivide counts up from zero to the value assigned to SetCount (one-half the oscillator speed) and inverts the output identifier ClkOut.

The integer value Cnt is set to zero prior to operation. The clock pulses are counted and compared to the value assigned to SetCount. When the number of pulses counted reaches the value in SetCount, the output ClkOut is checked to see if it is currently set to a 1 or 0. If ClkOut is currently 0, ClkOut is assigned a 1; otherwise, ClkIn is set to 1. Cnt is assigned a value of 0 and the process repeats. Toggling the output ClkOut each time the value of SetCount is reached creates a 1 Hz clock output with a 50% duty cycle.

The VHDL program code for the frequency divider is as follows:

```vhdl
library ieee;                          Clkln: 24.00 MHz clock driver
use ieee.std_logic_1164.all;           ClkOut: Output at 1 Hz

entity FreqDivide is
port(Clkln, in std_logic;                 Cnt: Counts up to value in SetCount
     ClkOut: buffer std_logic);       SetCount: Holds 1/2 timer interval value
end entity FreqDivide;

architecture FreqDivide Behavior of FreqDivide is
begin
   FreqDivide: process(Clkln)
   variable Cnt: integer := 0;
   variable SetCount: integer;

   begin
      SetCount := 12000000; -- 1/2 duty cycle
      if (ClkIn'EVENT and ClkIn = '1') then
         if (Cnt = SetCount) then
            if ClkOut = '0' then
               ClkOut <= '1'; --Output high 50%
            else
               ClkOut <= '0'; --Output Low 50%
            end if;
            Cnt := 0;
         else
            Cnt := Cnt + 1;
         end if;
      end if;
   end process;
end architecture FreqDivideBehavior;
```

SetCount is assigned a value equal to half the system clock to produce a 1 Hz output. In this case, a 24 MHz system clock is used.

The if statement causes program to wait for a clock event and clock = 1 to start operation.

Check that the terminal value in SetCount has been reached at which time ClkOut is toggled and Cnt is reset to 0.

If terminal value has not been reached, Cnt is incremented.

Timer Circuits   The program TimerCircuits uses two one-shot instances consisting of a 25 s timer (TLong) and a 4 s timer (TShort). The 25 s and the 4 s timers are triggered by long trigger (LongTrig) and short trigger (ShortTrig). In the VHDL program, countdown timers driven by a 1 Hz clock input (Clk) replicate the one-shot components TLong and TShort. The values stored in SetCountLong and SetCountShort are assigned to the Duration inputs of one-shot components TLong and TShort, setting the 25-second and 4-second timeouts. When Enable is set LOW, the one-shot timer is initiated and output QOut is set HIGH. When the one-shot timers time out, QOut is set LOW. The output of one-shot component TLong is sent to TimerCircuits identifier TL. The output of one-shot component TShort is sent to TimerCircuits identifier TS.

The VHDL program code for the timing circuits is as follows:

```
library ieee;
use ieee.std_logic_1164.all;

entity TimerCircuits is
    port(LongTrig, ShortTrig, Clk: in std_logic;
        TS, TL: buffer std_logic);
end entity TimerCircuits;

architecture TimerBehavior of TimerCircuits is
component OneShot is
    port(Enable, Clk: in std_logic;
        Duration :in integer range 0 to 25;
        QOut    :buffer std_logic);
end component OneShot;

signal SetCountLong, SetCountShort: integer range 0 to 25;
begin
    SetCountLong <= 25;
    SetCountShort <= 4;
    TLong:OneShot port map(Enable=>LongTrig, Clk=>Clk, Duration=>SetCountLong, QOut=>TL);
    TShort:OneShot port map(Enable=>ShortTrig, Clk=>Clk, Duration=>SetCountShort, QOut=>TS);
end architecture TimerBehavior;
```

LongTrig: Long timeout timer enable input
ShortTrig: Short timeout timer enable input
Clk: 1 Hz Clock input
TS: Short timer timeout signal
TL: Long timer timeout signal

Component declaration for OneShot.

SetCountLong: Holds long timer duration
SetCountShort: Holds short timer duration

Long and short count times are hard-coded to 25 and 4 based on a 1 Hz clock.

← Instantiation TLong
← Instantiation TShort

Sequential Logic

The sequential logic unit controls the sequencing of the traffic lights, based on inputs from the timing circuits and the side street vehicle sensor. The sequential logic produces a 2-bit Gray code sequence for each of the four states that were described in Chapter 6.

The Counter   The sequential logic consists of a 2-bit Gray code counter and the associated input logic, as shown in Figure 7–67. The counter produces the four-state sequence on outputs G_0 and G_1. Transitions from one state to the next are determined by the short timer (T_S), the long timer (T_L), and vehicle sensor (V_s) inputs.

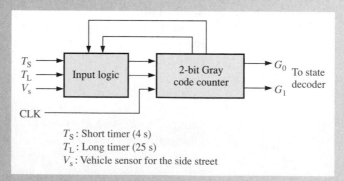

T_S : Short timer (4 s)
T_L : Long timer (25 s)
V_s : Vehicle sensor for the side street

FIGURE 7–67   Block diagram of the sequential logic.

The diagram in Figure 7–68 shows how two D flip-flops can be used to implement the Gray code counter. Outputs from the input logic provide the *D* inputs to the flip-flops so they sequence through the proper states.

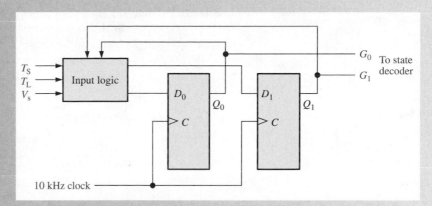

FIGURE 7–68 Sequential logic diagram with two D flip-flops used to implement the 2-bit Gray code counter.

The D flip-flop transition table is shown in Table 7–5. A next-state table developed from the state diagram in Chapter 6 Applied Logic is shown in Table 7–6. The subject of counter design is covered further in Chapter 8.

TABLE 7–5

D flip-flop transition table. Q_N is the output before clock pulse. Q_{N+1} is output after clock pulse.

Output Transitions		Flip-Flop Input
Q_N	Q_{N+1}	D
0 $\longrightarrow$	0	0
0 $\longrightarrow$	1	1
1 $\longrightarrow$	0	0
1 $\longrightarrow$	1	1

TABLE 7–6

Next-state table for the counter.

Present State		Next State		Input Conditions	FF Inputs	
Q_1	Q_0	Q_1	Q_0		D_1	D_0
0	0	0	0	$T_L + \overline{V}_s$	0	0
0	0	0	1	$\overline{T}_L V_s$	0	1
0	1	0	1	T_S	0	1
0	1	1	1	$\overline{T}_S$	1	1
1	1	1	1	$T_L V_s$	1	1
1	1	1	0	$\overline{T}_L + \overline{V}_s$	1	0
1	0	1	0	T_S	1	0
1	0	0	0	$\overline{T}_S$	0	0

The Input Logic   Using Tables 7–5 and 7–6, the conditions required for each flip-flop to go to the 1 state can be determined. For example, G_0 goes from 0 to 1 when the present state is 00 and the condition on input D_0 is $\overline{T}_L V_s$, as indicated on the second row of Table 7–6. D_0 must be a 1 to make G_0 go to a 1 or to remain a 1 on the next clock pulse. A Boolean expression describing the conditions that make D_0 a 1 is derived from Table 7–6 as follows:

$$D_0 = \overline{G}_1 \overline{G}_0 \overline{T}_L V_s + \overline{G}_1 G_0 T_S + \overline{G}_1 G_0 \overline{T}_S + G_1 G_0 T_L V_s$$

In the two middle terms, the T_S and the $\overline{T}_S$ variables cancel, leaving the expression

$$D_0 = \overline{G}_1 \overline{G}_0 \overline{T}_L V_s + \overline{G}_1 G_0 + G_1 G_0 T_L V_s$$

Also, from Table 7–6, an expression for D_1 can be developed as follows:

$$D_1 = \overline{G}_1 G_0 \overline{T}_S + G_1 G_0 T_L V_s + G_1 G_0 \overline{T}_L + G_1 G_0 \overline{V}_s + G_1 \overline{G}_0 T_S$$

Based on the minimized expression for D_0 and D_1, the complete sequential logic diagram is shown in Figure 7–69.

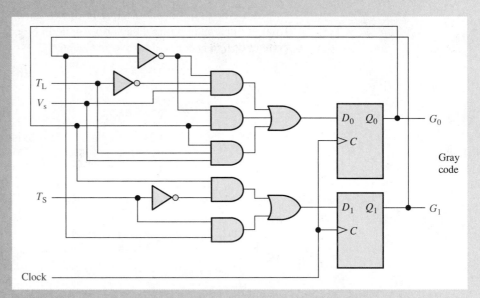

FIGURE 7–69   Complete diagram for the sequential logic.

Exercise

4. State the Boolean law and rule that permits the cancellation of T_S and $\overline{T}_S$ in the expression for D_0.
5. Use the Karnaugh map to reduce the D_0 expression further to a minimum form.
6. Use Boolean laws, rules, and/or the Karnaugh map to reduce the D_1 expression to a minimum form.
7. Do your minimized expressions for D_0 and D_1 agree with the logic shown in Figure 7–69?

The Sequential Logic with VHDL

The program SequentialLogic describes the Gray code logic needed to drive the traffic signal controller based on input from the timing circuits and the side street vehicle sensor. The sequential logic code produces a 2-bit Gray code sequence for each of the four sequence states. The component definition dff is used to instantiate two D flip-flop instances DFF0 and DFF1. DFF0 and DFF1 produce the two-bit Gray code. The Gray code output sequences the traffic signal controller through each of four states. Internal variables D0 and D1 store the results of the D0 and D1 Boolean expressions developed in this chapter. The stored results in D0 and D1 are assigned to D flip-flops DFF0 and DFF1 along with the system clock to drive outputs G0 and G1 from the D flip-flop Q outputs.

The VHDL program code for the sequential logic is as follows:

```vhdl
library ieee;
use ieee.std_logic_1164.all;

entity SequentialLogic is
port(VS, TL, TS, Clk: in std_logic; G0, G1: inout std_logic);
end entity SequentialLogic;

architecture SequenceBehavior of SequentialLogic is

component dff is
port (D, Clk: in std_logic; Q: out std_logic);
end component dff;

signal D0, D1: std_logic;
begin
D1 <= (G0 and not TS) or (G1 and TS);
D0 <= (not G1 and not TL and VS) or (not G1 and G0)
      or (G0 and TL and VS);

DFF0: dff port map(D=> D0, Clk => Clk, Q => G0);
DFF1: dff port map(D=> D1, Clk => Clk, Q => G1);

end architecture SequenceBehavior;
```

VS: Vehicle sensor input
TL: Long timer input
TS: Short timer input
Clk: System clock
G0: Gray code output bit 0
G1: Gray code output bit 1
D0: Logic for DFlipFlop DFF0
D1: Logic for DFlipFlop DFF1

Component declaration for D flip-flop (dff)

Logic definitions for D flip-flop inputs D0 and D1 derived from Boolean expressions developed in this chapter.

Component instantiations

The Complete Traffic Signal Controller

The program TrafficLights completes the traffic signal controller. Components FreqDivide, TimerCircuits, SequentialLogic, and StateDecoder are used to compose the completed system. Signal CLKin from the TrafficLights program source code is the clock input to the FreqDivide component. The frequency divided output ClkOut is stored as local variable Clock and is the divided clock input to the TimerCircuits and SequentialLogic components. TimerCircuits is controlled by local variables LongTime and ShortTime, which are controlled by the outputs Sig1 and Sig3 from component StateDecoder. StateDecoder also provides outputs Sig1 through Sig4 to control the traffic lights MG, SG, MY, SY, MR, and SR. TimerCircuit timeout signals TS and TL are stored in variables TLin (timer long in) and TSin (timer short in).

Signals TSin and TLin from TimerCircuits are used along with vehicle sensor VSin as inputs to the SequentialLogic component. The outputs from SequentialLogic G0 and G1 are stored in variables Gray0 and Gray1 as inputs to component StateDecoder. Component StateDecoder returns signals S1 through S4 which are in turn passed to variables Sig1 through Sig4. The light output logic and trigger logic developed in Chapter 6 are not used as components in this program, but are stated as logic expressions. The values stored in variables Sig1 through Sig4 provide the logic for outputs MG, SG, MY, SY, MR, SR; and local timer triggers LongTime and ShortTime are sent to TimerCircuits.

The VHDL program code for the traffic signal controller is as follows:

VSin	: Vehicle sensor input
CLKin	: System Clock
MR	: Main red light output
SR	: Side red light output
MY	: Main yellow light output
SY	: Side yellow light output
MG	: Main green light output
SG	: Side green light output

```vhdl
library ieee;
use ieee.std_logic_1164.all;

entity TrafficLights is
port(VSin, ClkIn: in std_logic; MR, SR, MY, SY, MG, SG: out std_logic);
end entity TrafficLights;

architecture TrafficLightsBehavior of TrafficLights is

component StateDecoder is
port(G0, G1: in std_logic; S1, S2, S3, S4: out std_logic);       } Component declaration for StateDecoder
end component StateDecoder;

component SequentialLogic is
port(VS, TL, TS, Clk: in std_logic; G0, G1: inout std_logic);    } Component declaration for SequentialLogic
end component SequentialLogic;

component TimerCircuits is
port(LongTrig, ShortTrig, Clk: In std_logic; TS, TL: buffer std_logic);  } Component declaration for TimerCircuits
end component TimerCircuits;

component FreqDivide is
port(Clkin: in std_logic; ClkOut: buffer std_logic);             } Component declaration for FreqDivider
end component FreqDivide;

signal Sig1, Sig2, Sig3, Sig4, Gray0, Gray1: std_logic;
signal LongTime, ShortTime, TLin, TSin, Clock: std_logic;
```

Sig1-4	: Return values from StateDecoder
Gray0-1	: SequentialLogic Gray code return
LongTime	: Trigger input to TimerCircuits
ShortTime	: Trigger input to TimerCircuits
TLin	: Store TimerCircuits long timeout
TSin	: Store TimerCircuits Short timeout
Clock	: Divided clock from FreqDivide

```vhdl
begin
MR <= Sig3 or Sig4;
SR <= Sig2 or Sig1;
MY <= Sig2;                  } Logic definitions for the
SY <= Sig4;                    light output logic
MG <= Sig1;
SG <= Sig3;

LongTime <= Sig1 or Sig3;    } Logic definitions for the trigger logic
ShortTime <= not(Sig1 or Sig3);

SD: StateDecoder     port map (G0 => Gray0, G1 => Gray1, S1 => Sig1, S2 => Sig2, S3 => Sig3, S4 => Sig4);
SL: SequentialLogic  port map (VS => VSin, TL => TLin, TS => TSin, Clk => Fout, G0 => Gray0, G1 => Gray1);
TC: TimerCircuits    port map (LongTrig=>LongTime, ShortTrig=>ShortTime, Clk=>Clock, TS=>TSin, TL=>TLin);
FD: FreqDivide       port map (ClkIn => CLKin, ClkOut =>-Clock);

end architecture TrafficLightsBehavior;
```

Component instantiations

Simulation

MultiSim

> Open file AL07 in the Applied Logic folder on the website. Run the traffic signal controller simulation using your Multisim software and observe the operation. Lights will appear randomly when first turned on. Simulation times may vary.

Putting Your Knowledge to Work

Add your modification for the pedestrian input developed in Chapter 6 and run a simulation.

SUMMARY

- Latches are bistable devices whose state normally depends on asynchronous inputs.
- Edge-triggered flip-flops are bistable devices with synchronous inputs whose state depends on the inputs only at the triggering transition of a clock pulse. Changes in the outputs occur at the triggering transition of the clock.
- Monostable multivibrators (one-shots) have one stable state. When the one-shot is triggered, the output goes to its unstable state for a time determined by an RC circuit.
- Astable multivibrators have no stable states and are used as oscillators to generate timing waveforms in digital systems.

KEY TERMS

Key terms and other bold terms in the chapter are defined in the end-of-book glossary.

Astable   Having no stable state. An astable multivibrator oscillates between two quasi-stable states.

Bistable   Having two stable states. Flip-flops and latches are bistable multivibrators.

Clear   An asynchronous input used to reset a flip-flop (make the Q output 0).

Clock   The triggering input of a flip-flop.

D flip-flop   A type of bistable multivibrator in which the output assumes the state of the D input on the triggering edge of a clock pulse.

Edge-triggered flip-flop   A type of flip-flop in which the data are entered and appear on the output on the same clock edge.

Hold time   The time interval required for the control levels to remain on the inputs to a flip-flop after the triggering edge of the clock in order to reliably activate the device.

J-K flip-flop   A type of flip-flop that can operate in the SET, RESET, no-change, and toggle modes.

Latch   A bistable digital circuit used for storing a bit.

Monostable   Having only one stable state. A monostable multivibrator, commonly called a *one-shot,* produces a single pulse in response to a triggering input.

One-shot   A monostable multivibrator.

Power dissipation   The amount of power required by a circuit.

Preset   An asynchronous input used to set a flip-flop (make the Q output 1).

Propagation delay time   The interval of time required after an input signal has been applied for the resulting output change to occur.

RESET   The state of a flip-flop or latch when the output is 0; the action of producing a RESET state.

SET   The state of a flip-flop or latch when the output is 1; the action of producing a SET state.

Set-up time   The time interval required for the control levels to be on the inputs to a digital circuit, such as a flip-flop, prior to the triggering edge of a clock pulse.

Synchronous   Having a fixed time relationship.

Timer   A circuit that can be used as a one-shot or as an oscillator.

Toggle   The action of a flip-flop when it changes state on each clock pulse.

TRUE/FALSE QUIZ

Answers are at the end of the chapter.

1. A latch has two stable states.
2. A latch is considered to be in the SET state when the Q output is LOW.

3. A gated D latch must be enabled in order to change state.

4. Flip-flops and latches are both bistable devices.

5. An edge-triggered D flip-flop changes state whenever the D input changes.

6. A clock input is necessary for an edge-triggered flip-flop.

7. When both the J and K inputs are HIGH, an edge-triggered J-K flip-flop changes state on each clock pulse.

8. A one-shot is also known as an astable multivibrator.

9. When triggered, a one-shot produces a single pulse.

10. The 555 timer can be used as a one-shot or as a pulse oscillator.

SELF-TEST

Answers are at the end of the chapter.

1. If an S-R latch has a 1 on the S input and a 0 on the R input and then the S input goes to 0, the latch will be
   (a) set   (b) reset   (c) invalid   (d) clear

2. The invalid state of an S-R latch occurs when
   (a) $S = 1, R = 0$ (b) $S = 0, R = 1$
   (c) $S = 1, R = 1$ (d) $S = 0, R = 0$

3. For a gated D latch, the Q output always equals the D input
   (a) before the enable pulse
   (b) during the enable pulse
   (c) immediately after the enable pulse
   (d) answers (b) and (c)

4. Like the latch, the flip-flop belongs to a category of logic circuits known as
   (a) monostable multivibrators
   (b) bistable multivibrators
   (c) astable multivibrators
   (d) one-shots

5. The purpose of the clock input to a flip-flop is to
   (a) clear the device
   (b) set the device
   (c) always cause the output to change states
   (d) cause the output to assume a state dependent on the controlling (J-K or D) inputs.

6. For an edge-triggered D flip-flop,
   (a) a change in the state of the flip-flop can occur only at a clock pulse edge
   (b) the state that the flip-flop goes to depends on the D input
   (c) the output follows the input at each clock pulse
   (d) all of these answers

7. A feature that distinguishes the J-K flip-flop from the D flip-flop is the
   (a) toggle condition   (b) preset input
   (c) type of clock   (d) clear input

8. A flip-flop is in the toggle condition when
   (a) $J = 1, K = 0$ (b) $J = 1, K = 1$
   (c) $J = 0, K = 0$ (d) $J = 0, K = 1$

9. A J-K flip-flop with $J = 1$ and $K = 1$ has a 10 kHz clock input. The Q output is
   (a) constantly HIGH   (b) constantly LOW
   (c) a 10 kHz square wave   (d) a 5 kHz square wave

10. A one-shot is a type of
    (a) monostable multivibrator   (b) astable multivibrator
    (c) timer   (d) answers (a) and (c)
    (e) answers (b) and (c)

11. The output pulse width of a nonretriggerable one-shot depends on

    (a) the trigger intervals   **(b)** the supply voltage

    (c) a resistor and capacitor   **(d)** the threshold voltage

12. An astable multivibrator

    (a) requires a periodic trigger input   **(b)** has no stable state

    (c) is an oscillator   **(d)** produces a periodic pulse output

    (e) answers (a), (b), (c), and (d)   **(f)** answers (b), (c), and (d) only

PROBLEMS

Answers to odd-numbered problems are at the end of the book.

Section 7–1 Latches

1. If the waveforms in Figure 7–70 are applied to an active-LOW input $\overline{S}$-$\overline{R}$ latch, draw the resulting Q output waveform in relation to the inputs. Assume that Q starts LOW.

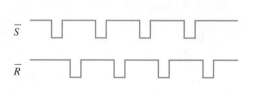
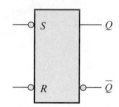

FIGURE 7–70

2. Solve Problem 1 for the input waveforms in Figure 7–71 applied to an active-HIGH S-R latch.

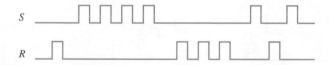

FIGURE 7–71

3. Solve Problem 1 for the input waveforms in Figure 7–72.

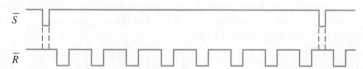

FIGURE 7–72

4. For a gated S-R latch, determine the Q and $\overline{Q}$ outputs for the inputs in Figure 7–73. Show them in proper relation to the enable input. Assume that Q starts LOW.

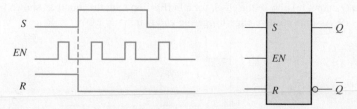

FIGURE 7–73

5. Determine the output of a gated D latch for the inputs in Figure 7–74.

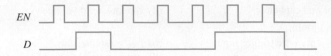

FIGURE 7–74

6. Determine the output of a gated D latch for the inputs in Figure 7–75.

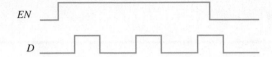

FIGURE 7–75

7. For a gated D latch, the waveforms shown in Figure 7–76 are observed on its inputs. Draw the timing diagram showing the output waveform you would expect to see at Q if the latch is initially RESET.

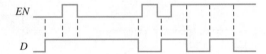

FIGURE 7–76

Section 7–2 Flip-Flops

8. Two edge-triggered J-K flip-flops are shown in Figure 7–77. If the inputs are as shown, draw the Q output of each flip-flop relative to the clock, and explain the difference between the two. The flip-flops are initially RESET.

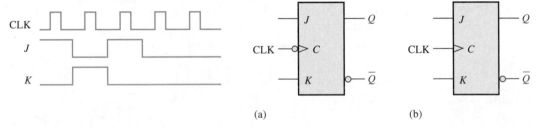

(a)   (b)

FIGURE 7–77

9. The Q output of an edge-triggered D flip-flop is shown in relation to the clock signal in Figure 7–78. Determine the input waveform on the D input that is required to produce this output if the flip-flop is a positive edge-triggered type.

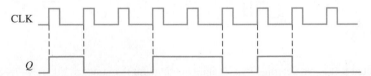

FIGURE 7–78

10. Draw the Q output relative to the clock for a D flip-flop with the inputs as shown in Figure 7–79. Assume positive edge-triggering and Q initially LOW.

FIGURE 7–79

11. Solve Problem 10 for the inputs in Figure 7–80.

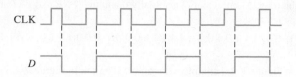

FIGURE 7–80

12. For a positive edge-triggered D flip-flop with the input as shown in Figure 7–81, determine the Q output relative to the clock. Assume that Q starts LOW.

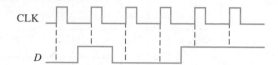

FIGURE 7–81

13. Solve Problem 12 for the input in Figure 7–82.

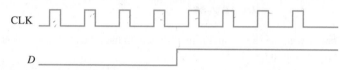

FIGURE 7–82

14. Determine the Q waveform relative to the clock if the signals shown in Figure 7–83 are applied to the inputs of the J-K flip-flop. Assume that Q is initially LOW.

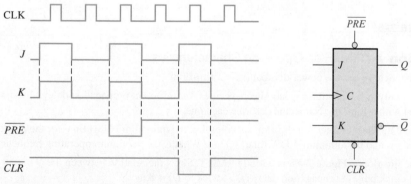

FIGURE 7–83

15. For a negative edge-triggered J-K flip-flop with the inputs in Figure 7–84, develop the Q output waveform relative to the clock. Assume that Q is initially LOW.

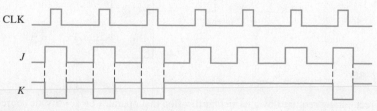

FIGURE 7–84

16. The following serial data are applied to the flip-flop through the AND gates as indicated in Figure 7–85. Determine the resulting serial data that appear on the Q output. There is one clock pulse for each bit time. Assume that Q is initially 0 and that $\overline{PRE}$ and $\overline{CLR}$ are HIGH. Right-most bits are applied first.

J_1: 1 0 1 0 0 1 1; J_2: 0 1 1 1 0 1 0; J_3: 1 1 1 1 0 0 0; K_1: 0 0 0 1 1 1 0; K_2: 1 1 0 1 1 0 0; K_3: 1 0 1 0 1 0 1

17. For the circuit in Figure 7–85, complete the timing diagram in Figure 7–86 by showing the Q output (which is initially LOW). Assume $\overline{PRE}$ and $\overline{CLR}$ remain HIGH.

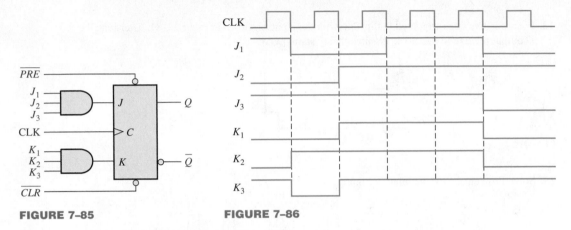

FIGURE 7–85   **FIGURE 7–86**

18. Solve Problem 17 with the same J and K inputs but with the $\overline{PRE}$ and $\overline{CLR}$ inputs as shown in Figure 7–87 in relation to the clock.

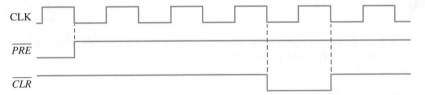

FIGURE 7–87

Section 7–3 Flip-Flop Operating Characteristics

19. What determines the power dissipation of a flip-flop?

20. Typically, a manufacturer's data sheet specifies four different propagation delay times associated with a flip-flop. Name and describe each one.

21. The data sheet of a certain flip-flop specifies that the minimum HIGH time for the clock pulse is 30 ns and the minimum LOW time is 37 ns. What is the maximum operating frequency?

22. The flip-flop in Figure 7–88 is initially RESET. Show the relation between the Q output and the clock pulse if propagation delay t_{PLH} (clock to Q) is 8 ns.

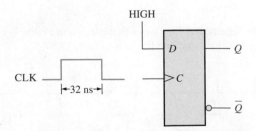

FIGURE 7–88

23. The direct current required by a particular flip-flop that operates on a +5 V dc source is found to be 10 mA. A certain digital device uses 15 of these flip-flops. Determine the current capacity required for the +5 V dc supply and the total power dissipation of the system.

24. For the circuit in Figure 7–89, determine the maximum frequency of the clock signal for reliable operation if the set-up time for each flip-flop is 2 ns and the propagation delays (t_{PLH} and t_{PHL}) from clock to output are 5 ns for each flip-flop.

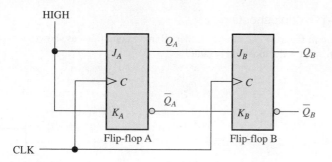

HIGH

FIGURE 7–89

Section 7–4 Flip-Flop Applications

25. A D flip-flop is connected as shown in Figure 7–90. Determine the Q output in relation to the clock. What specific function does this device perform?

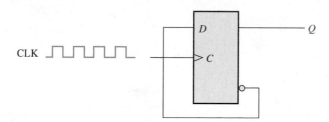

FIGURE 7–90

26. For the circuit in Figure 7–89, develop a timing diagram for eight clock pulses, showing the Q_A and Q_B outputs in relation to the clock.

Section 7–5 One-Shots

27. Determine the pulse width of a 74121 one-shot if the external resistor is 3.3 kΩ and the external capacitor is 2000 pF.

28. An output pulse of 5 μs duration is to be generated by a 74LS122 one-shot. Using a capacitor of 10,000 pF, determine the value of external resistance required.

29. Create a one-shot, using a 555 timer that will produce a 0.25 s output pulse.

Section 7–6 The Astable Multivibrator

30. A 555 timer is configured to run as an astable multivibrator as shown in Figure 7–91. Determine its frequency.

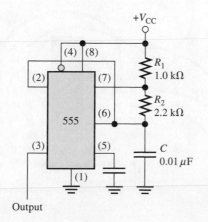

FIGURE 7–91

31. Determine the values of the external resistors for a 555 timer used as an astable multivibrator with an output frequency of 20 kHz, if the external capacitor C is 0.002 μF and the duty cycle is to be approximately 75%.

Section 7–7 Troubleshooting

32. The flip-flop in Figure 7–92 is tested under all input conditions as shown. Is it operating properly? If not, what is the most likely fault?

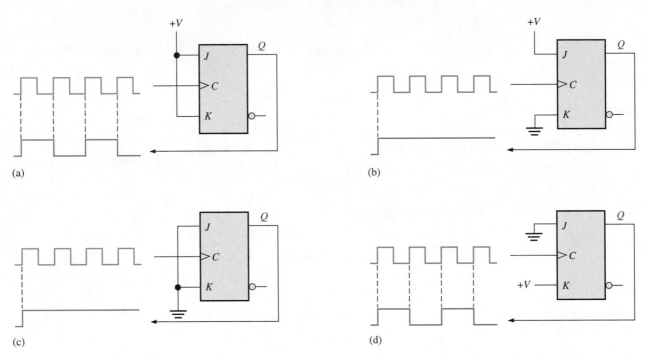

(a)   (b)

(c)   (d)

FIGURE 7–92

33. A 74HC00 quad NAND gate IC is used to construct a gated S-R latch on a protoboard in the lab as shown in Figure 7–93. The schematic in part (a) is used to connect the circuit in part (b). When you try to operate the latch, you find that the Q output stays HIGH no matter what the inputs are. Determine the problem.

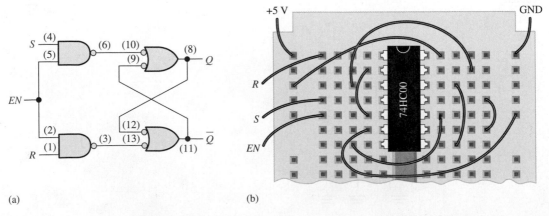

(a)   (b)

FIGURE 7–93

34. Determine if the flip-flop in Figure 7–94 is operating properly, and if not, identify the most probable fault.

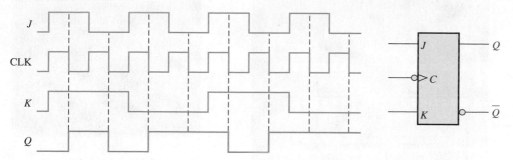

FIGURE 7–94

35. The parallel data storage circuit in Figure 7–35 does not operate properly. To check it out, you first make sure that V_{CC} and ground are connected, and then you apply LOW levels to all the D inputs and pulse the clock line. You check the Q outputs and find them all to be LOW; so far, so good. Next you apply HIGHs to all the D inputs and again pulse the clock line. When you check the Q outputs, they are still all LOW. What is the problem, and what procedure will you use to isolate the fault to a single device?

36. The flip-flop circuit in Figure 7–95(a) is used to generate a binary count sequence. The gates form a decoder that is supposed to produce a HIGH when a binary zero or a binary three state occurs (00 or 11). When you check the Q_A and Q_B outputs, you get the display shown in part (b), which reveals glitches on the decoder output (X) in addition to the correct pulses. What is causing these glitches, and how can you eliminate them?

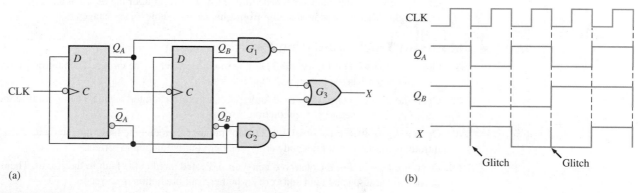

(a)   (b)

FIGURE 7–95

37. Determine the Q_A, Q_B and X outputs over six clock pulses in Figure 7–95(a) for each of the following faults in the bipolar (TTL) circuits. Start with both Q_A and Q_B LOW.
  (a) D input open   **(b)** Q_B output open
  (c) clock input to flip-flop B shorted   **(d)** gate G_2 output open

38. Two 74121 one-shots are connected on a circuit board as shown in Figure 7–96. After observing the oscilloscope display, do you conclude that the circuit is operating properly? If not, what is the most likely problem?

Applied Logic

39. Using 555 timers, redesign the timing circuits portion of the traffic signal controller for an approximate 6 s caution light and 40 s red and green lights.

40. Repeat Problem 39 using 74121 one-shots.

41. Repeat Problem 39 using 74122 one-shots.

42. Implement the input logic in the sequential circuit unit of the traffic signal controller using only NAND gates.

43. Specify how you would change the time interval for the green light from 25 s to 60 s.

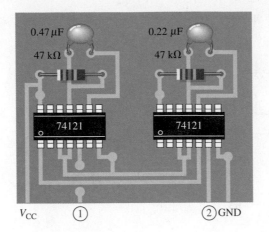

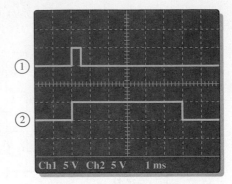

FIGURE 7–96

Special Design Problems

44. Design a basic counting circuit that produces a binary sequence from zero through seven by using negative edge-triggered J-K flip-flops.

45. In the shipping department of a softball factory, the balls roll down a conveyor and through a chute single file into boxes for shipment. Each ball passing through the chute activates a switch circuit that produces an electrical pulse. The capacity of each box is 32 balls. Design a logic circuit to indicate when a box is full so that an empty box can be moved into position.

46. List the design changes that would be necessary in the traffic signal controller to add a 15 s left turn arrow for the main street. The turn arrow will occur after the red light and prior to the green light. Modify the state diagram from Chapter 6 to show these changes.

MultiSim

Multisim Troubleshooting Practice

47. Open file P07-47. For the specified fault, predict the effect on the circuit. Then introduce the fault and verify whether your prediction is correct.

48. Open file P07-48. For the specified fault, predict the effect on the circuit. Then introduce the fault and verify whether your prediction is correct.

49. Open file P07-49. For the observed behavior indicated, predict the fault in the circuit. Then introduce the suspected fault and verify whether your prediction is correct.

50. Open file P07-50. For the observed behavior indicated, predict the fault in the circuit. Then introduce the suspected fault and verify whether your prediction is correct.

51. Open file P07-51. For the observed behavior indicated, predict the fault in the circuit. Then introduce the suspected fault and verify whether your prediction is correct.

ANSWERS

SECTION CHECKUPS

Section 7–1 Latches

1. Three types of latches are S-R, gated S-R, and gated D.
2. $SR = 00$, NC; $SR = 01$, $Q = 0$; $SR = 10$, $Q = 1$; $SR = 11$, invalid
3. $Q = 1$

Section 7–2 Flip-Flops

1. The output of a gated D latch can change any time the gate enable (EN) input is active. The output of an edge-triggered D flip-flop can change only on the triggering edge of a clock pulse.
2. The output of a J-K flip-flop is determined by the state of its two inputs whereas the output of a D flip-flop follows the input.
3. Output Q goes HIGH on the trailing edge of the first clock pulse, LOW on the trailing edge of the second pulse, HIGH on the trailing edge of the third pulse, and LOW on the trailing edge of the fourth pulse.

Section 7–3 Flip-Flop Operating Characteristics

1. **(a)** Set-up time is the time required for input data to be present before the triggering edge of the clock pulse.

   (b) Hold time is the time required for data to remain on the inputs after the triggering edge of the clock pulse.

2. The 74AHC74 can be operated at the highest frequency, according to Table 7–4.

Section 7–4 Flip-Flop Applications

1. A group of data storage flip-flops is a register.

2. For divide-by-2 operation, the flip-flop must toggle ($D = \overline{Q}$).

3. Six flip-flops are used in a divide-by-64 device.

Section 7–5 One-Shots

1. A nonretriggerable one-shot times out before it can respond to another trigger input. A retriggerable one-shot responds to each trigger input.

2. Pulse width is set with external R and C components.

3. 11 ms.

Section 7–6 The Astable Multivibrator

1. An astable multivibrator has no stable state. A monostable multivibrator has one stable state.

2. Duty cycle $= (15 \text{ ms}/20 \text{ ms})100\% = 75\%$

Section 7–7 Troubleshooting

1. Yes, a negative edge-triggered J-K flip-flop can be used.

2. An astable multivibrator using a 555 timer can be used to provide the clock.

RELATED PROBLEMS FOR EXAMPLES

7–1 The Q output is the same as shown in Figure 7–5(b).

7–2 See Figure 7–97.

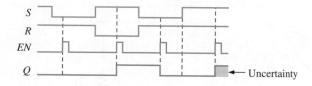

FIGURE 7–97

7–3 See Figure 7–98.

7–4 See Figure 7–99.

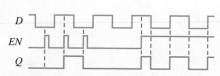

FIGURE 7–98

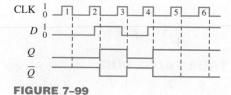

FIGURE 7–99

7–5 See Figure 7–100.

7–6 See Figure 7–101.

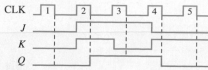

FIGURE 7–100

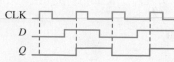

FIGURE 7–101

7–7   See Figure 7–102.

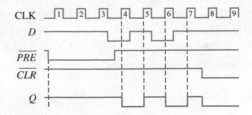

FIGURE 7–102

7–8   See Figure 7–103.

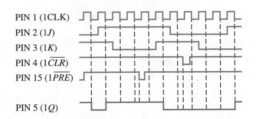

FIGURE 7–103

7–9 $2^5 = 32$. Five flip-flops are required.

7–10   Sixteen states require four flip-flops ($2^4 = 16$).

7–11 $C_{EXT} = 7143$ pF connected from CX to RX/CX of the 74121 with no external resistor.

7–12 $C_{EXT} = 560$ pF, $R_{EXT} = 27$ kΩ. See Figure 7–104.

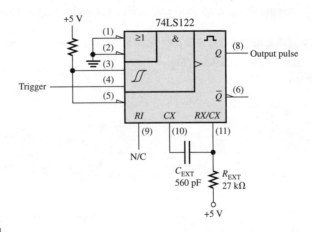

FIGURE 7–104

7–13 $R_1 = 91$ kΩ

7–14   Duty cycle $\cong 32\%$

TRUE/FALSE QUIZ

1. T   **2.** F   **3.** T   **4.** T   **5.** F   **6.** T   **7.** T   **8.** F   **9.** T   **10.** T

SELF-TEST

1. (a)   **2.** (c)   **3.** (d)   **4.** (b)   **5.** (d)   **6.** (d)

7. (a)   **8.** (b)   **9.** (d)   **10.** (d)   **11.** (c)   **12.** (f)

Shift Registers

CHAPTER OBJECTIVES

■ Identify the basic forms of data movement in shift registers

■ Explain how serial in/serial out, serial in/parallel out, parallel in/serial out, and parallel in/parallel out shift registers operate

■ Describe how a bidirectional shift register operates

■ Determine the sequence of a Johnson counter

■ Set up a ring counter to produce a specified sequence

■ Construct a ring counter from a shift register

■ Use a shift register as a time-delay device

■ Use a shift register to implement a serial-to-parallel data converter

■ Implement a basic shift-register-controlled keyboard encoder

■ Interpret ANSI/IEEE Standard 91-1984 shift register symbols with dependency notation

■ Use shift registers in a system application

KEY TERMS

Key terms are in order of appearance in the chapter.

■ Register
■ Stage
■ Load
■ Bidirectional

VISIT THE WEBSITE

Study aids for this chapter are available at http://www.pearsonhighered.com/careersresources/

INTRODUCTION

Shift registers are a type of sequential logic circuit used primarily for the storage of digital data and typically do not possess a characteristic internal sequence of states. There are exceptions, however, and these are covered in Section 8–4.

In this chapter, the basic types of shift registers are studied and several applications are presented. Also, a troubleshooting method is introduced.

8–1   Shift Register Operations

Shift registers consist of arrangements of flip-flops and are important in applications involving the storage and transfer of data in a digital system. A register has no specified sequence of states, except in certain very specialized applications. A register, in general, is used solely for storing and shifting data (1s and 0s) entered into it from an external source and typically possesses no characteristic internal sequence of states.

After completing this section, you should be able to

* ◆ Explain how a flip-flop stores a data bit
* ◆ Define the storage capacity of a shift register
* ◆ Describe the shift capability of a register

A register can consist of one or more flip-flops used to store and shift data.

A **register** is a digital circuit with two basic functions: data storage and data movement. The storage capability of a register makes it an important type of memory device. Figure 8–1 illustrates the concept of storing a 1 or a 0 in a D flip-flop. A 1 is applied to the data input as shown, and a clock pulse is applied that stores the 1 by *setting* the flip-flop. When the 1 on the input is removed, the flip-flop remains in the SET state, thereby storing the 1. A similar procedure applies to the storage of a 0 by *resetting* the flip-flop, as also illustrated in Figure 8–1.

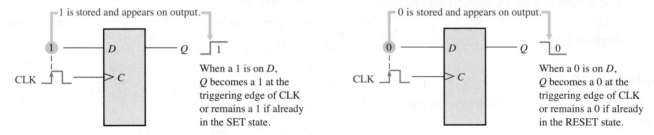

When a 1 is on D, Q becomes a 1 at the triggering edge of CLK or remains a 1 if already in the SET state.

When a 0 is on D, Q becomes a 0 at the triggering edge of CLK or remains a 0 if already in the RESET state.

FIGURE 8–1   The flip-flop as a storage element.

The *storage capacity* of a register is the total number of bits (1s and 0s) of digital data it can retain. Each **stage** (flip-flop) in a shift register represents one bit of storage capacity; therefore, the number of stages in a register determines its storage capacity.

The *shift capability* of a register permits the movement of data from stage to stage within the register or into or out of the register upon application of clock pulses. Figure 8–2

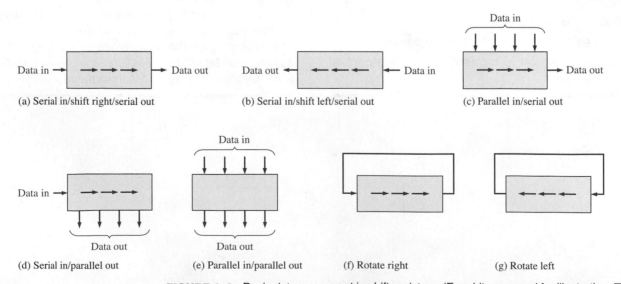

(a) Serial in/shift right/serial out

(b) Serial in/shift left/serial out

(c) Parallel in/serial out

(d) Serial in/parallel out

(e) Parallel in/parallel out

(f) Rotate right

(g) Rotate left

FIGURE 8–2   Basic data movement in shift registers. (Four bits are used for illustration. The bits move in the direction of the arrows.)

illustrates the types of data movement in shift registers. The block represents any arbitrary 4-bit register, and the arrows indicate the direction of data movement.

    1. What determines the storage capacity of a shift register?

    2. What two principal functions are performed by a shift register?

8–2 Types of Shift Register Data I/Os

In this section, four types of shift registers based on data input and output (inputs/outputs) are discussed: serial in/serial out, serial in/parallel out, parallel in/serial out, and parallel in/parallel out.

After completing this section, you should be able to

◆ Describe the operation of four types of shift registers

◆ Explain how data bits are entered into a shift register

◆ Describe how data bits are shifted through a register

◆ Explain how data bits are taken out of a shift register

◆ Develop and analyze timing diagrams for shift registers

Serial In/Serial Out Shift Registers

The serial in/serial out shift register accepts data serially—that is, one bit at a time on a single line. It produces the stored information on its output also in serial form. Let's first look at the serial entry of data into a typical shift register. Figure 8–3 shows a 4-bit device implemented with D flip-flops. With four stages, this register can store up to four bits of data.

InfoNote

Frequently, it is necessary to *clear* an internal register in a processor. For example, a register may be cleared prior to an arithmetic or other operation. One way that registers in a processor are cleared is using software to subtract the contents of the register from itself. The result, of course, will always be zero. For example, a processor instruction that performs this operation is SUB AL,AL. with this instruction, the register named AL is cleared.

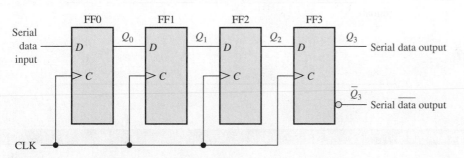

FIGURE 8–3   Serial in/serial out shift register.

Table 8–1 shows the entry of the four bits 1010 into the register in Figure 8–3, beginning with the least significant bit. The register is initially clear. The 0 is put onto the data input line, making $D = 0$ for FF0. When the first clock pulse is applied, FF0 is reset, thus storing the 0.

TABLE 8-1

Shifting a 4-bit code into the shift register in Figure 8–3. Data bits are indicated by a beige screen.

CLK	FF0 (Q_0)	FF1 (Q_1)	FF2 (Q_2)	FF3 (Q_3)
Initial	0	0	0	0
1	0	0	0	0
2	1	0	0	0
3	0	1	0	0
4	1	0	1	0

Next the second bit, which is a 1, is applied to the data input, making $D = 1$ for FF0 and $D = 0$ for FF1 because the D input of FF1 is connected to the Q_0 output. When the second clock pulse occurs, the 1 on the data input is shifted into FF0, causing FF0 to set; and the 0 that was in FF0 is shifted into FF1.

The third bit, a 0, is now put onto the data-input line, and a clock pulse is applied. The 0 is entered into FF0, the 1 stored in FF0 is shifted into FF1, and the 0 stored in FF1 is shifted into FF2.

The last bit, a 1, is now applied to the data input, and a clock pulse is applied. This time the 1 is entered into FF0, the 0 stored in FF0 is shifted into FF1, the 1 stored in FF1 is shifted into FF2, and the 0 stored in FF2 is shifted into FF3. This completes the serial entry of the four bits into the shift register, where they can be stored for any length of time as long as the flip-flops have dc power.

If you want to get the data out of the register, the bits must be shifted out serially to the Q_3 output, as Table 8–2 illustrates. After CLK4 in the data-entry operation just described, the LSB, 0, appears on the Q_3 output. When clock pulse CLK5 is applied, the second bit appears on the Q_3 output. Clock pulse CLK6 shifts the third bit to the output, and CLK7 shifts the fourth bit to the output. While the original four bits are being shifted out, more bits can be shifted in. All zeros are shown being shifted in, after CLK8.

For serial data, one bit at a time is transferred.

TABLE 8-2

Shifting a 4-bit code out of the shift register in Figure 8–3. Data bits are indicated by a beige screen.

CLK	FF0 (Q_0)	FF1 (Q_1)	FF2 (Q_2)	FF3 (Q_3)
Initial	1	0	1	0
5	0	1	0	1
6	0	0	1	0
7	0	0	0	1
8	0	0	0	0

EXAMPLE 8-1

Show the states of the 5-bit register in Figure 8–4(a) for the specified data input and clock waveforms. Assume that the register is initially cleared (all 0s).

Solution

The first data bit (1) is entered into the register on the first clock pulse and then shifted from left to right as the remaining bits are entered and shifted. The register contains $Q_4Q_3Q_2Q_1Q_0 = 11010$ after five clock pulses. See Figure 8–4(b).

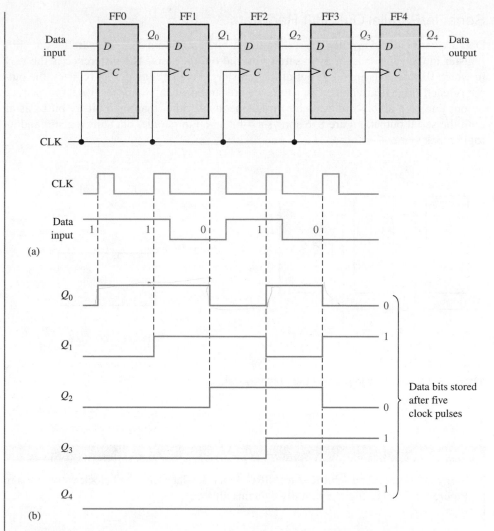

(a)

(b)

FIGURE 8–4 Open file F08-04 to verify operation. *A Multisim tutorial is available on the website.*

MultiSim

Related Problem*

Show the states of the register if the data input is inverted. The register is initially cleared.

———————————

*Answers are at the end of the chapter.

A traditional logic block symbol for an 8-bit serial in/serial out shift register is shown in Figure 8–5. The "SRG 8" designation indicates a shift register (SRG) with an 8-bit capacity.

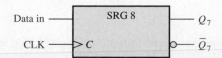

FIGURE 8–5 Logic symbol for an 8-bit serial in/serial out shift register.

Serial In/Parallel Out Shift Registers

Data bits are entered serially (least-significant bit first) into a serial in/parallel out shift register in the same manner as in serial in/serial out registers. The difference is the way in which the data bits are taken out of the register; in the parallel output register, the output of each stage is available. Once the data are stored, each bit appears on its respective output line, and all bits are available simultaneously, rather than on a bit-by-bit basis as with the serial output. Figure 8–6 shows a 4-bit serial in/parallel out shift register and its logic block symbol.

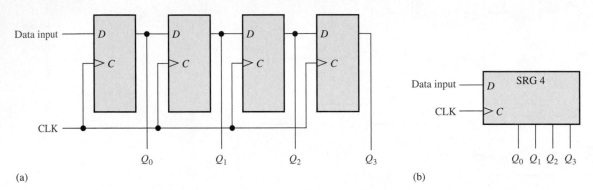

(a)   (b)

FIGURE 8–6   A serial in/parallel out shift register.

EXAMPLE 8–2

Show the states of the 4-bit register (SRG 4) for the data input and clock waveforms in Figure 8–7(a). The register initially contains all 1s.

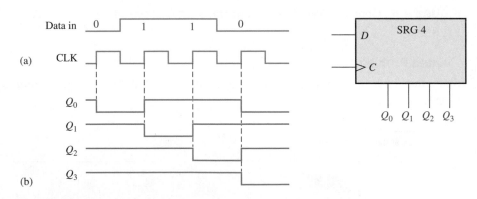

FIGURE 8–7

Solution

The register contains 0110 after four clock pulses. See Figure 8–7(b).

Related Problem

If the data input remains 0 after the fourth clock pulse, what is the state of the register after three additional clock pulses?

IMPLEMENTATION: 8-BIT SERIAL IN/PARALLEL OUT SHIFT REGISTER

Fixed-Function Device   The 74HC164 is an example of a fixed-function IC shift register having serial in/parallel out operation. The logic block symbol is shown in Figure 8–8. This device has two gated serial inputs, A and B, and an asynchronous clear ($\overline{CLR}$) input that is active-LOW. The parallel outputs are Q_0 through Q_7.

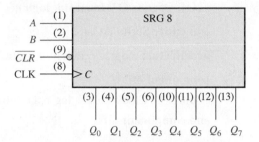

FIGURE 8–8   The 74HC164 8-bit serial in/parallel out shift register.

A sample timing diagram for the 74HC164 is shown in Figure 8–9. Notice that the serial input data on input A are shifted into and through the register after input B goes HIGH.

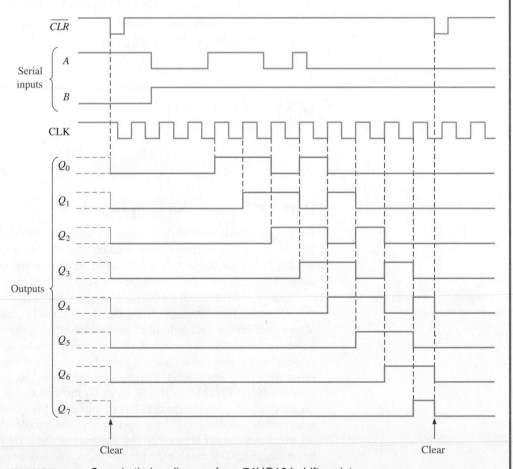

FIGURE 8–9   Sample timing diagram for a 74HC164 shift register.

Programmable Logic Device (PLD)   The 8-bit serial in/parallel out shift register can be described using VHDL and implemented as hardware in a PLD. The program code is as follows. (Blue comments are not part of the program.)

library ieee;

use ieee.std_logic_1164.**all**;

entity SerInParOutShift **is**

    port (D0, Clock, Clr: **in std_logic**; Q0, Q1, Q2, Q3,   D0: Data input
    Q4, Q5, Q6, Q7: **inout std_logic**);   Clock: System clock

end entity SerInParOutShift;   Clr: Clear
                                          Q0–Q7: Register outputs

architecture LogicOperation **of** SerInParOutShift **is**

component dff1 **is**   D flip-flop with preset and

    port (D, Clock: **in std_logic**; Q: **inout std_logic**);   clear inputs was described in
                                             Chapter 7 and is used as a
end component dff1;   component.

begin

    FF0: dff1 **port map**(D=>D0 **and** Clr, Clock=>Clock, Q=>Q0);
    FF1: dff1 **port map**(D=>Q0 **and** Clr, Clock=>Clock, Q=>Q1);
    FF2: dff1 **port map**(D=>Q1 **and** Clr, Clock=>Clock, Q=>Q2);   Instantiations
    FF3: dff1 **port map**(D=>Q2 **and** Clr, Clock=>Clock, Q=>Q3);   describe how
                                                         the flip-flops
    FF4: dff1 **port map**(D=>Q3 **and** Clr, Clock=>Clock, Q=>Q4);   are connected
    FF5: dff1 **port map**(D=>Q4 **and** Clr, Clock=>Clock, Q=>Q5);   to form the
    FF6: dff1 **port map**(D=>Q5 **and** Clr, Clock=>Clock, Q=>Q6);   register.
    FF7: dff1 **port map**(D=>Q6 **and** Clr, Clock=>Clock, Q=>Q7);

end architecture LogicOperation;

Parallel In/Serial Out Shift Registers

For a register with parallel data inputs, the bits are entered simultaneously into their respective stages on parallel lines rather than on a bit-by-bit basis on one line as with serial data inputs. The serial output is the same as in serial in/serial out shift registers, once the data are completely stored in the register.

Figure 8–10 illustrates a 4-bit parallel in/serial out shift register and a typical logic symbol. There are four data-input lines, D_0, D_1, D_2, and D_3, and a $SHIFT/\overline{LOAD}$ input, which allows four bits of data to **load** in parallel into the register. When $SHIFT/\overline{LOAD}$ is LOW, gates G_1 through G_4 are enabled, allowing each data bit to be applied to the D input of its respective flip-flop. When a clock pulse is applied, the flip-flops with $D = 1$ will set and those with $D = 0$ will reset, thereby storing all four bits simultaneously.

When $SHIFT/\overline{LOAD}$ is HIGH, gates G_1 through G_4 are disabled and gates G_5 through G_7 are enabled, allowing the data bits to shift right from one stage to the next. The OR gates allow either the normal shifting operation or the parallel data-entry operation, depending on which AND gates are enabled by the level on the $SHIFT/\overline{LOAD}$ input. Notice that FF0 has a single AND to disable the parallel input, D_0. It does not require an AND/OR arrangement because there is no serial data in.

For parallel data, multiple bits are transferred at one time.

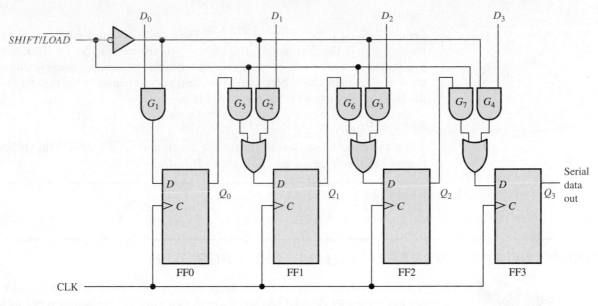

(a) Logic diagram

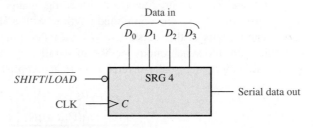

(b) Logic symbol

FIGURE 8–10   A 4-bit parallel in/serial out shift register. Open file F08-10 to verify operation.

MultiSim

EXAMPLE 8–3

Show the data-output waveform for a 4-bit register with the parallel input data and the clock and *SHIFT/LOAD* waveforms given in Figure 8–11(a). Refer to Figure 8–10(a) for the logic diagram.

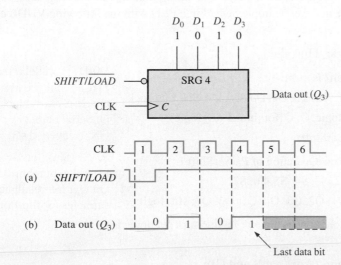

FIGURE 8–11

Solution

On clock pulse 1, the parallel data ($D_0D_1D_2D_3 = 1010$) are loaded into the register, making Q_3 a 0. On clock pulse 2 the 1 from Q_2 is shifted onto Q_3; on clock pulse 3 the 0 is shifted onto Q_3; on clock pulse 4 the last data bit (1) is shifted onto Q_3; and on clock pulse 5, all data bits have been shifted out, and only 1s remain in the register (assuming the D_0 input remains a 1). See Figure 8–11(b).

Related Problem

Show the data-output waveform for the clock and $SHIFT/\overline{LOAD}$ inputs shown in Figure 8–11(a) if the parallel data are $D_0D_1D_2D_3 = 0101$.

IMPLEMENTATION: 8-BIT PARALLEL LOAD SHIFT REGISTER

Fixed-Function Device The 74HC165 is an example of a fixed-function IC shift register that has a parallel in/serial out operation (it can also be operated as serial in/serial out). Figure 8–12 shows a typical logic block symbol. A LOW on the $SHIFT/\overline{LOAD}$ input ($SH/\overline{LD}$) enables asynchronous parallel loading. Data can be entered serially on the *SER* input. Also, the clock can be inhibited anytime with a HIGH on the *CLK INH* input. The serial data outputs of the register are Q_7 and its complement $\overline{Q}_7$. This implementation is different from the synchronous method of parallel loading previously discussed, demonstrating that there are usually several ways to accomplish the same function.

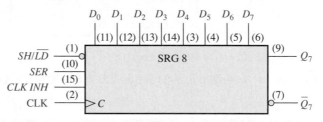

FIGURE 8–12 The 74HC165 8-bit parallel load shift register.

Figure 8–13 is a timing diagram showing an example of the operation of a 74HC165 shift register.

Programmable Logic Device (PLD) The 8-bit parallel load shift register is a parallel in/serial out device and can be implemented in a PLD with the following VHDL code:

```
library ieee:
use ieee.std_logic_1164.all;

entity ParSerShift is
    port (D0, D1, D2, D3, D4, D5, D6, D7, SHLD, Clock:
        in std_logic; Q, QNot: inout std_logic);
end entity ParSerShift;

architecture LogicOperation of ParSerShift is
    signal S1, S2, S3, S4, S5, S6, S7,
        Q0, Q1, Q2, Q3, Q4, Q5, Q6, Q7: std_logic;
function ShiftLoad (A,B,C: in std_logic)return std_logic is
begin
    return ((A and B) or (not B and C));
end function ShiftLoad;
```

Function ShiftLoad provides the AND-OR function shown in Figure 8–10 to allow the parallel load of data or data shift from one flip-flop stage to the next.

D0-D7: Parallel input
SHLD: Shift Load input
Clock: System clock
Q: Serial output
QNot: Inverted serial output

S1-S7: Shift load signals from function ShiftLoad
Q0-Q7: Intermediate variables for flip-flop stages

```
component dff1 is
port (D, Clock: in std_logic;                    D flip-flop component used as
      Q: inout std_logic);                       storage for shift register
end component dff1;

begin

    SL1:S1 <=ShiftLoad(Q0, SHLD, D1);
    SL2:S2 <=ShiftLoad(Q1, SHLD, D2);
    SL3:S3 <=ShiftLoad(Q2, SHLD, D3);
    SL4:S4 <=ShiftLoad(Q3, SHLD, D4);
    SL5:S5 <=ShiftLoad(Q4, SHLD, D5);
    SL6:S6 <=ShiftLoad(Q5, SHLD, D6);
    SL7:S7 <=ShiftLoad(Q6, SHLD, D7);
    FF0: dff1 port map(D=>D0 and not SHLD, Clock=>Clock, Q=>Q0);
    FF1: dff1 port map(D=>S1, Clock=>Clock, Q=>Q1);
    FF2: dff1 port map(D=>S2, Clock=>Clock, Q=>Q2);
    FF3: dff1 port map(D=>S3, Clock=>Clock, Q=>Q3);
    FF4: dff1 port map(D=>S4, Clock=>Clock, Q=>Q4);
    FF5: dff1 port map(D=>S5, Clock=>Clock, Q=>Q5);
    FF6: dff1 port map(D=>S6, Clock=>Clock, Q=>Q6);
    FF7: dff1 port map(D=>S7, Clock=>Clock, Q=>Q);
    QNot <=not Q;

end architecture LogicOperation;
```

ShiftLoad instances SL1–SL7 allow eight bits of data to load into flip-flop stages FF0–FF7 or to shift through the register providing the parallel load serial out function.

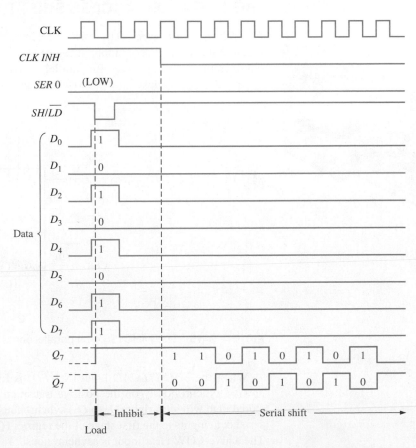

FIGURE 8–13 Sample timing diagram for a 74HC165 shift register.

Parallel In/Parallel Out Shift Registers

Parallel entry and parallel output of data have been discussed. The parallel in/parallel out register employs both methods. Immediately following the simultaneous entry of all data bits, the bits appear on the parallel outputs. Figure 8–14 shows a parallel in/parallel out shift register.

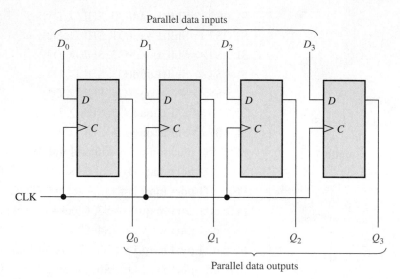

FIGURE 8–14   A parallel in/parallel out register.

IMPLEMENTATION: 4-BIT PARALLEL-ACCESS SHIFT REGISTER

Fixed-Function Device   The 74HC195 can be used for parallel in/parallel out operation. Because it also has a serial input, it can be used for serial in/serial out and serial in/parallel out operations. It can be used for parallel in/serial out operation by using Q_3 as the output. A typical logic block symbol is shown in Figure 8–15.

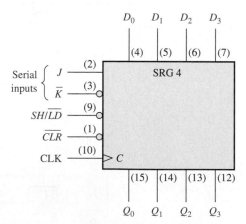

FIGURE 8–15   The 74HC195 4-bit parallel access shift register.

When the *SHIFT/LOAD* input ($SH/\overline{LD}$) is LOW, the data on the parallel inputs are entered synchronously on the positive transition of the clock. When ($SH/\overline{LD}$) is HIGH, stored data will shift right (Q_0 to Q_3) synchronously with the clock. Inputs J and $\overline{K}$ are the serial data inputs to the first stage of the register (Q_0); Q_3 can be used for serial output data. The active-LOW clear input is asynchronous.

The timing diagram in Figure 8–16 illustrates the operation of this register.

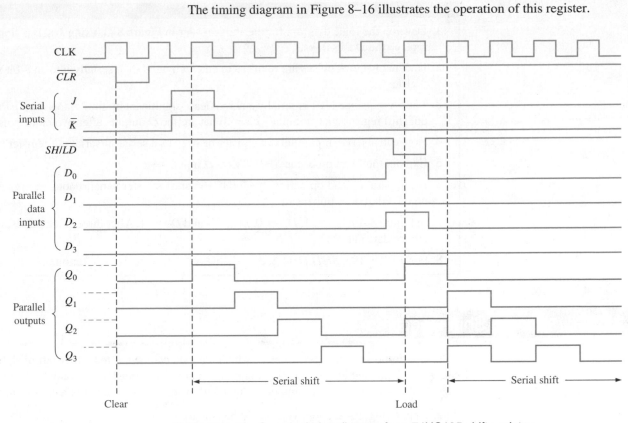

FIGURE 8–16 Sample timing diagram for a 74HC195 shift register.

Programmable Logic Device (PLD) The VHDL code for a 4-bit parallel in/parallel out shift register is as follows:

```
library ieee;
use ieee.std logic_1164.all;

entity ParInParOut is
    port (D0, D1, D2, D3, Clock: in std_logic;
            Q0, Q1, Q2, Q3: inout std_logic);
end entity ParInParOut;

architecture LogicOperation of ParInParOut is
    component dff1 is
        port (D, Clock: in std_logic;
                Q: inout std_logic);
    end component dff1;
begin
    FF0: dff1 port map (D=>D0, Clock=>Clock, Q=>Q0);
    FF1: dff1 port map (D=>D1, Clock=>Clock, Q=>Q1);
    FF2: dff1 port map (D=>D2, Clock=>Clock, Q=>Q2);
    FF3: dff1 port map (D=>D3, Clock=>Clock, Q=>Q3);
end architecture LogicOperation;
```

SECTION 8–2 CHECKUP

1. Develop the logic diagram for the shift register in Figure 8–3, using J-K flip-flops to replace the D flip-flops.

2. How many clock pulses are required to enter a byte of data serially into an 8-bit shift register?

3. The bit sequence 1101 is serially entered (least-significant bit first) into a 4-bit parallel out shift register that is initially clear. What are the Q outputs after two clock pulses?

4. How can a serial in/parallel out register be used as a serial in/serial out register?

5. Explain the function of the $SHIFT/\overline{LOAD}$ input.

6. Is the parallel load operation in a 74HC165 shift register synchronous or asynchronous? What does this mean?

7. In Figure 8–14, $D_0 = 1, D_1 = 0, D_2 = 0,$ and $D_3 = 1$. After three clock pulses, what are the data outputs?

8. For a 74HC195, $SH/\overline{LD} = 1, J = 1,$ and $\overline{K} = 1$. What is Q_0 after one clock pulse?

8–3 Bidirectional Shift Registers

A **bidirectional** shift register is one in which the data can be shifted either left or right. It can be implemented by using gating logic that enables the transfer of a data bit from one stage to the next stage to the right or to the left, depending on the level of a control line.

After completing this section, you should be able to

◆ Explain the operation of a bidirectional shift register

◆ Discuss the 74HC194 4-bit bidirectional universal shift register

◆ Develop and analyze timing diagrams for bidirectional shift registers

A 4-bit bidirectional shift register is shown in Figure 8–17. A HIGH on the $RIGHT/\overline{LEFT}$ control input allows data bits inside the register to be shifted to the right, and a LOW

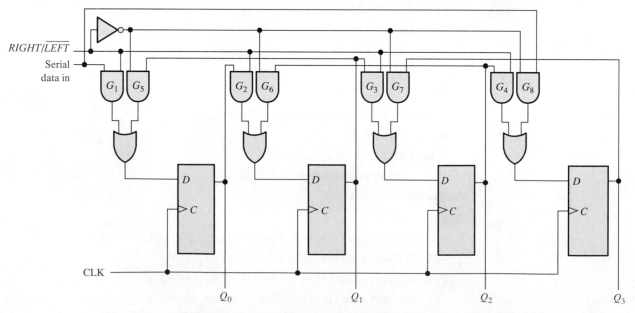

MultiSim   **FIGURE 8–17**   Four-bit bidirectional shift register. Open file F08-17 to verify the operation.

enables data bits inside the register to be shifted to the left. An examination of the gating logic will make the operation apparent. When the $RIGHT/\overline{LEFT}$ control input is HIGH, gates G_1 through G_4 are enabled, and the state of the Q output of each flip-flop is passed through to the D input of the *following* flip-flop. When a clock pulse occurs, the data bits are shifted one place to the *right*. When the $RIGHT/\overline{LEFT}$ control input is LOW, gates G_5 through G_8 are enabled, and the Q output of each flip-flop is passed through to the D input of the *preceding* flip-flop. When a clock pulse occurs, the data bits are then shifted one place to the *left*.

EXAMPLE 8–4

Determine the state of the shift register of Figure 8–17 after each clock pulse for the given $RIGHT/\overline{LEFT}$ control input waveform in Figure 8–18(a). Assume that $Q_0 = 1$, $Q_1 = 1$, $Q_2 = 0$, and $Q_3 = 1$ and that the serial data-input line is LOW.

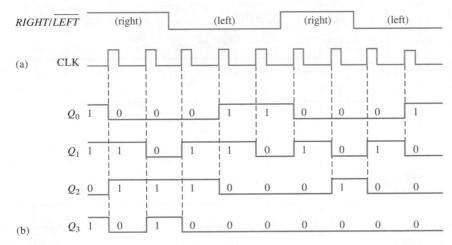

FIGURE 8–18

Solution

See Figure 8–18(b).

Related Problem

Invert the $RIGHT/\overline{LEFT}$ waveform, and determine the state of the shift register in Figure 8–17 after each clock pulse.

IMPLEMENTATION: 4-BIT BIDIRECTIONAL UNIVERSAL SHIFT REGISTER

Fixed-Function Device   The 74HC194 is an example of a universal bidirectional shift register in integrated circuit form. A **universal shift register** has both serial and parallel input and output capability. A logic block symbol is shown in Figure 8–19, and a sample timing diagram is shown in Figure 8–20.

Parallel loading, which is synchronous with a positive transition of the clock, is accomplished by applying the four bits of data to the parallel inputs and a HIGH to the S_0 and S_1 inputs. Shift right is accomplished synchronously with the positive edge of the clock when S_0 is HIGH and S_1 is LOW. Serial data in this mode are entered at the shift-right serial input (*SR SER*). When S_0 is LOW and S_1 is HIGH, data bits shift left synchronously with the clock, and new data are entered at the shift-left serial input (*SL SER*). Input *SR SER* goes into the Q_0 stage, and *SL SER* goes into the Q_3 stage.

FIGURE 8–19 The 74HC194 4-bit bidirectional universal shift register.

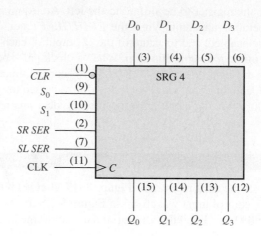

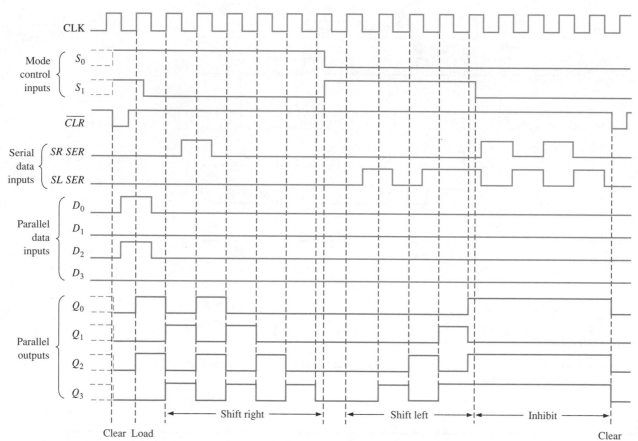

FIGURE 8–20 Sample timing diagram for a 74HC194 shift register.

Programmable Logic Device (PLD) The following code describes a 4-bit bidirectional shift register with a serial input:

library ieee;
use ieee.std_logic_1164.**all**;

entity FourBitBiDirSftReg **is**
port (R_L, DataIn, Clock: **in std_logic**;
  Q0, Q1, Q2, Q3: **buffer std_logic**);
end entity FourBitBiDirSftReg;

R_L: Right/left
DataIn: Serial input data
Clock: System clock
Q0-Q3: Register outputs

architecture LogicOperation **of** FourBitBiDirSftReg **is**

component dff1 is

    port(D,Clock: **in std_logic**; Q: **out std_logic**); } D flip-flop component declaration

end component dff1;

signal D0, D1, D2, D3: **std_logic**;   Internal flip-flop inputs

begin

    DO <= (DataIn **and** R_L) **or** (**not** R_L **and** Q1); ⎤

    D1 <= (Q0 **and** R_L) **or** (**not** R_L **and** Q2); ⎬ Describes the internal signals

    D2 <= (Q1 **and** R_L) **or** (**not** R_L **and** Q3); ⎥ with Boolean equations

    D3 <= (Q2 **and** R_L) **or** (**not** R_L **and** DataIn); ⎦

    FF0: dff1 **port map**(D => D0, Clock => Clock, Q => Q0); ⎤

    FF1: dff1 **port map**(D => D1, Clock => Clock, Q => Q1); ⎬ Describes how the

    FF2: dff1 **port map**(D => D2, Clock => Clock, Q => Q2); ⎥ flip-flops are connected

    FF3: dff1 **port map**(D => D3, Clock => Clock, Q => Q3); ⎦

end architecture LogicOperation;

SECTION 8–3 CHECKUP

1. Assume that the 4-bit bidirectional shift register in Figure 8–17 has the following contents: $Q_0 = 1, Q_1 = 1, Q_2 = 0$, and $Q_3 = 0$. There is a 1 on the serial data-input line. If $RIGHT/\overline{LEFT}$ is HIGH for three clock pulses and LOW for two more clock pulses, what are the contents after the fifth clock pulse?

8–4   Shift Register Counters

A shift register counter is basically a shift register with the serial output connected back to the serial input to produce special sequences. These devices are often classified as counters because they exhibit a specified sequence of states. Two of the most common types of shift register counters, the Johnson counter and the ring counter, are introduced in this section.

After completing this section, you should be able to

- Discuss how a shift register counter differs from a basic shift register
- Explain the operation of a Johnson counter
- Specify a Johnson sequence for any number of bits
- Explain the operation of a ring counter and determine the sequence of any specific ring counter

The Johnson Counter

In a **Johnson counter** the complement of the output of the last flip-flop is connected back to the D input of the first flip-flop (it can be implemented with other types of flip-flops as well). If the counter starts at 0, this feedback arrangement produces a characteristic sequence of states, as shown in Table 8–3 for a 4-bit device and in Table 8–4 for a 5-bit device. Notice that the 4-bit sequence has a total of eight states, or bit patterns, and that the 5-bit sequence has a total of ten states. In general, a Johnson counter will produce a modulus of $2n$, where n is the number of stages in the counter.

TABLE 8–3

Four-bit Johnson sequence.

Clock Pulse	Q_0	Q_1	Q_2	Q_3
0	0	0	0	0
1	1	0	0	0
2	1	1	0	0
3	1	1	1	0
4	1	1	1	1
5	0	1	1	1
6	0	0	1	1
7	0	0	0	1

TABLE 8–4

Five-bit Johnson sequence.

Clock Pulse	Q_0	Q_1	Q_2	Q_3	Q_4
0	0	0	0	0	0
1	1	0	0	0	0
2	1	1	0	0	0
3	1	1	1	0	0
4	1	1	1	1	0
5	1	1	1	1	1
6	0	1	1	1	1
7	0	0	1	1	1
8	0	0	0	1	1
9	0	0	0	0	1

The implementations of the 4-stage and 5-stage Johnson counters are shown in Figure 8–21. The implementation of a Johnson counter is very straightforward and is the same regardless of the number of stages. The Q output of each stage is connected to the D input of the next

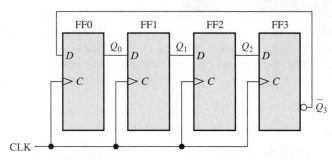

(a) Four-bit Johnson counter

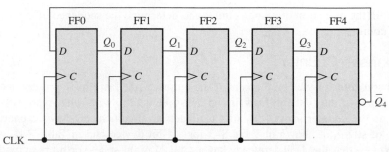

(b) Five-bit Johnson counter

FIGURE 8–21   Four-bit and 5-bit Johnson counters.

stage (assuming that D flip-flops are used). The single exception is that the $\overline{Q}$ output of the last stage is connected back to the D input of the first stage. As the sequences in Table 8–3 and 8–4 show, if the counter starts at 0, it will "fill up" with 1s from left to right, and then it will "fill up" with 0s again.

Diagrams of the timing operations of the 4-bit and 5-bit counters are shown in Figures 8–22 and 8–23, respectively.

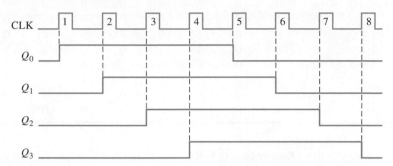

FIGURE 8–22   Timing sequence for a 4-bit Johnson counter.

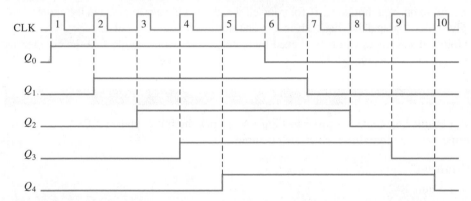

FIGURE 8–23   Timing sequence for a 5-bit Johnson counter.

The Ring Counter

A **ring counter** utilizes one flip-flop for each state in its sequence. It has the advantage that decoding gates are not required. In the case of a 10-bit ring counter, there is a unique output for each decimal digit.

A logic diagram for a 10-bit ring counter is shown in Figure 8–24. The sequence for this ring counter is given in Table 8–5. Initially, a 1 is preset into the first flip-flop, and the rest of the flip-flops are cleared. Notice that the interstage connections are the same as those for a

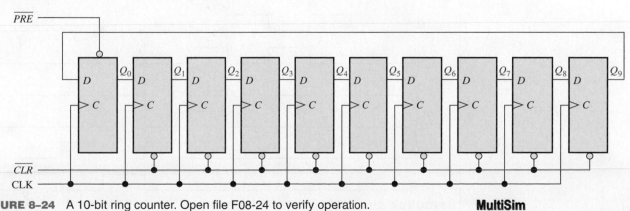

FIGURE 8–24   A 10-bit ring counter. Open file F08-24 to verify operation.   **MultiSim**

TABLE 8–5

Ten-bit ring counter sequence.

Clock Pulse	Q_0	Q_1	Q_2	Q_3	Q_4	Q_5	Q_6	Q_7	Q_8	Q_9
0	1	0	0	0	0	0	0	0	0	0
1	0	1	0	0	0	0	0	0	0	0
2	0	0	1	0	0	0	0	0	0	0
3	0	0	0	1	0	0	0	0	0	0
4	0	0	0	0	1	0	0	0	0	0
5	0	0	0	0	0	1	0	0	0	0
6	0	0	0	0	0	0	1	0	0	0
7	0	0	0	0	0	0	0	1	0	0
8	0	0	0	0	0	0	0	0	1	0
9	0	0	0	0	0	0	0	0	0	1

Johnson counter, except that Q rather than $\overline{Q}$ is fed back from the last stage. The ten outputs of the counter indicate directly the decimal count of the clock pulse. For instance, a 1 on Q_0 represents a zero, a 1 on Q_1 represents a one, a 1 on Q_2 represents a two, a 1 on Q_3 represents a three, and so on. You should verify for yourself that the 1 is always retained in the counter and simply shifted "around the ring," advancing one stage for each clock pulse.

Modified sequences can be achieved by having more than a single 1 in the counter, as illustrated in Example 8–5.

EXAMPLE 8–5

If a 10-bit ring counter similar to Figure 8–24 has the initial state 1010000000, determine the waveform for each of the Q outputs.

Solution

See Figure 8–25.

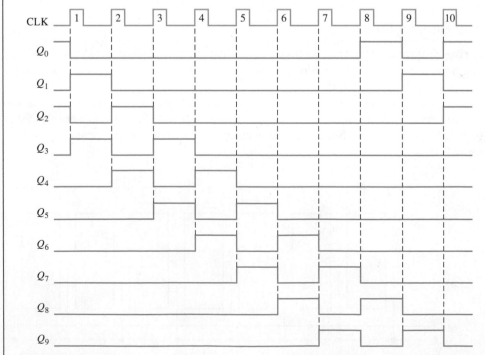

FIGURE 8–25

Related Problem

If a 10-bit ring counter has an initial state 0101001111, determine the waveform for each Q output.

---

1. How many states are there in an 8-bit Johnson counter sequence?

2. Write the sequence of states for a 3-bit Johnson counter starting with 000.

8–5 Shift Register Applications

Shift registers are found in many types of applications, a few of which are presented in this section.

After completing this section, you should be able to

◆ Use a shift register to generate a time delay

◆ Implement a specified ring counter sequence using a 74HC195 shift register

◆ Discuss how shift registers are used for serial-to-parallel conversion of data

◆ Define *UART*

◆ Explain the operation of a keyboard encoder and how registers are used in this application

Time Delay

A serial in/serial out shift register can be used to provide a time delay from input to output that is a function of both the number of stages (n) in the register and the clock frequency.

When a data pulse is applied to the serial input as shown in Figure 8–26, it enters the first stage on the triggering edge of the clock pulse. It is then shifted from stage to stage on each successive clock pulse until it appears on the serial output n clock periods later. This time-delay operation is illustrated in Figure 8–26, in which an 8-bit serial in/serial out shift register is used with a clock frequency of 1 MHz to achieve a time delay (t_d) of 8 μs (8 × 1 μs). This time can be adjusted up or down by changing the clock frequency. The time delay can also be increased by cascading shift registers and decreased by taking the outputs from successively lower stages in the register if the outputs are available, as illustrated in Example 8–6.

InfoNote

Microprocessors have special instructions that can emulate a serial shift register. The accumulator register can shift data to the left or right. A right shift is equivalent to a divide-by-2 operation and a left shift is equivalent to a multiply-by-2 operation. Data in the accumulator can be shifted left or right with the rotate instructions; ROR is the rotate right instruction, and ROL is the rotate left instruction. Two other instructions treat the carry flag bit as an additional bit for the rotate operation. These are the RCR for rotate carry right and RCL for rotate carry left.

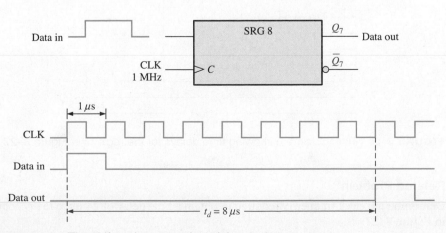

FIGURE 8–26 The shift register as a time-delay device.

EXAMPLE 8–6

Determine the amount of time delay between the serial input and each output in Figure 8–27. Show a timing diagram to illustrate.

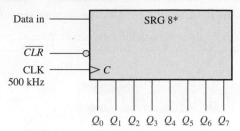

* Data shifts from Q_0 toward Q_7.

FIGURE 8–27

Solution

The clock period is 2 μs. Thus, the time delay can be increased or decreased in 2 μs increments from a minimum of 2 μs to a maximum of 16 μs, as illustrated in Figure 8–28.

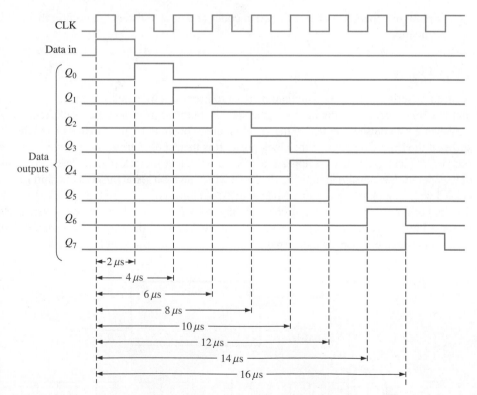

FIGURE 8–28   Timing diagram showing time delays for the register in Figure 8–27.

Related Problem

Determine the clock frequency required to obtain a time delay of 24 μs to the Q_7 output in Figure 8–27.

IMPLEMENTATION: A RING COUNTER

Fixed-Function Device   If the output is connected back to the serial input, a shift register can be used as a ring counter. Figure 8–29 illustrates this application with a 74HC195 4-bit shift register.

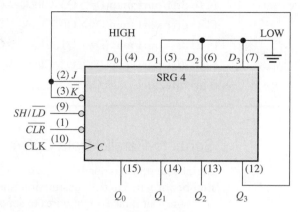

FIGURE 8–29   74HC195 connected as a ring counter.

Initially, a bit pattern of 1000 (or any other pattern) can be synchronously preset into the counter by applying the bit pattern to the parallel data inputs, taking the $SH/\overline{LD}$ input LOW, and applying a clock pulse. After this initialization, the 1 continues to circulate through the ring counter, as the timing diagram in Figure 8–30 shows.

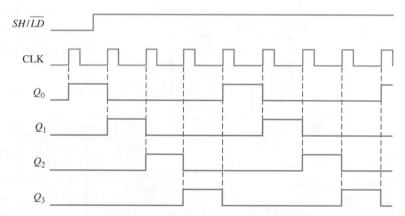

FIGURE 8–30   Timing diagram showing two complete cycles of the ring counter in Figure 8–29 when it is initially preset to 1000.

Programmable Logic Device (PLD)   The VHDL code for a 4-bit ring counter using D flip-flops is as follows:

library ieee;

use ieee.std_logic_1164.**all**;

entity RingCtr **is**
   port (I, Clr, Clock: **in std_logic**;
      Q0, Q1, Q2, Q3: **inout std_logic**);
end entity RingCtr;

architecture LogicOperation **of** RingCtr **is**

I: Serial input bit to clock data into the shift register
Clr: Ring counter clear input
Clock: System clock
Q0-Q3: Ring counter output stages

FF0-FF3 flip-flop instantiations show how flip-flops are connected and represent one flip-flop for each state in the ring counter sequence. FF0 Pre input acts as a serial input when I is high. FF1-FF3 Clr input clears flip-flop stages when Clr is low.

```
component dff1 is
    port (D, Clock, Pre, Clr: in std_logic;
        Q: inout std_logic);
end component dff1;
begin
    FF0: dff1 port map(D=> Q3, Clock=>Clock, Q=>Q0, Pre=> not I, Clr=>'1');
    FF1: dff1 port map(D=> Q0, Clock=>Clock, Q=>Q1, Pre=>'1', Clr=>not Clr);
    FF2: dff1 port map(D=> Q1, Clock=>Clock, Q=>Q2, Pre=>'1', Clr=>not Clr);
    FF3: dff1 port map(D=> Q2, Clock=>Clock, Q=>Q3, Pre=>'1', Clr=>not Clr);
end architecture LogicOperation;
```

D flip-flop component used as storage for shift register

Serial-to-Parallel Data Converter

Serial data transmission from one digital system to another is commonly used to reduce the number of wires in the transmission line. For example, eight bits can be sent serially over one wire, but it takes eight wires to send the same data in parallel.

Serial data transmission is widely used by peripherals to pass data back and forth to a computer. For example, USB (universal serial bus) is used to connect keyboards printers, scanners, and more to the computer. All computers process data in parallel form, thus the requirement for serial-to-parallel conversion. A simplified serial-to-parallel data converter, in which two types of shift registers are used, is shown in Figure 8–31.

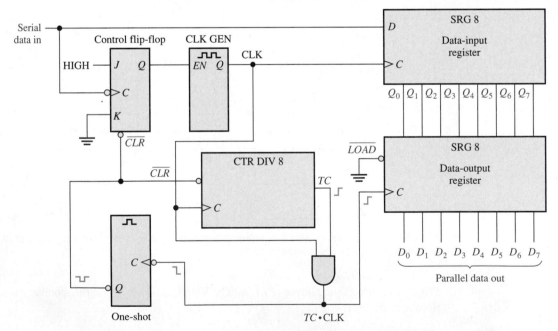

FIGURE 8–31 Simplified logic diagram of a serial-to-parallel converter.

To illustrate the operation of this serial-to-parallel converter, the serial data format shown in Figure 8–32 is used. It consists of eleven bits. The first bit (start bit) is always 0 and always begins with a HIGH-to-LOW transition. The next eight bits (D_7 through D_0) are the data bits (one of the bits can be parity), and the last one or two bits (stop bits) are always 1s. When no data are being sent, there is a continuous HIGH on the serial data line.

FIGURE 8–32 Serial data format.

The HIGH-to-LOW transition of the start bit sets the control flip-flop, which enables the clock generator. After a fixed delay time, the clock generator begins producing a pulse waveform, which is applied to the data-input register and to the divide-by-8 counter. The clock has a frequency precisely equal to that of the incoming serial data, and the first clock pulse after the start bit occurs during the first data bit.

The timing diagram in Figure 8–33 illustrates the following basic operation: The eight data bits (D_7 through D_0) are serially shifted into the data-input register. Shortly after the

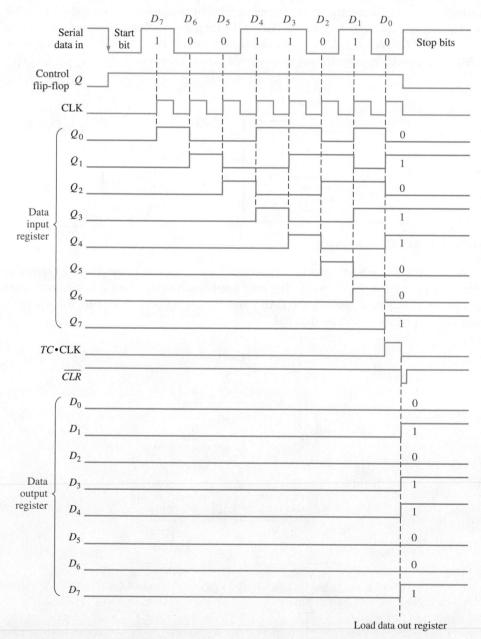

FIGURE 8–33 Timing diagram illustrating the operation of the serial-to-parallel data converter in Figure 8–31.

eighth clock pulse, the terminal count (*TC*) goes from LOW to HIGH, indicating the counter is at the last state. This rising edge is ANDed with the clock pulse, which is still HIGH, producing a rising edge at $TC \cdot CLK$. This parallel loads the eight data bits from the data-input shift register to the data-output register. A short time later, the clock pulse goes LOW and this HIGH-to-LOW transition triggers the one-shot, which produces a short-duration pulse to clear the counter and reset the control flip-flop and thus disable the clock generator. The system is now ready for the next group of eleven bits, and it waits for the next HIGH-to-LOW transition at the beginning of the start bit.

By reversing the process just stated, parallel-to-serial data conversion can be accomplished. Since the serial data format must be produced, start and stop bits must be added to the sequence.

Universal Asynchronous Receiver Transmitter (UART)

As mentioned, computers and microprocessor-based systems often send and receive data in a parallel format. Frequently, these systems must communicate with external devices that send and/or receive serial data. An interfacing device used to accomplish these conversions is the UART (Universal Asynchronous Receiver Transmitter). Figure 8–34 illustrates the UART in a general microprocessor-based system application.

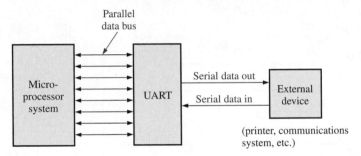

FIGURE 8–34 UART interface.

A UART includes both serial-to-parallel and parallel-to-serial conversion, as shown in the block diagram in Figure 8–35. The data bus is basically a set of parallel conductors along which data move between the UART and the microprocessor system. Buffers interface the data registers with the data bus.

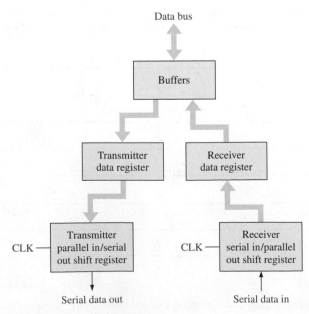

FIGURE 8–35 Basic UART block diagram.

The UART receives data in serial format, converts the data to parallel format, and places them on the data bus. The UART also accepts parallel data from the data bus, converts the data to serial format, and transmits them to an external device.

Keyboard Encoder

The keyboard encoder is a good example of the application of a shift register used as a ring counter in conjunction with other devices. Recall that a simplified computer keyboard encoder without data storage was presented in Chapter 6.

Figure 8–36 shows a simplified keyboard encoder for encoding a key closure in a 64-key matrix organized in eight rows and eight columns. Two parallel in/parallel out 4-bit shift

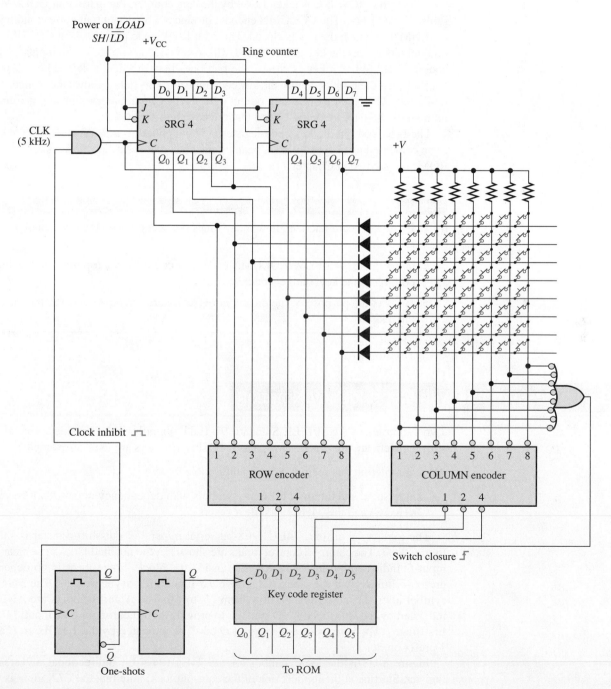

FIGURE 8–36 Simplified keyboard encoding circuit.

registers are connected as an 8-bit ring counter with a fixed bit pattern of seven 1s and one 0 preset into it when the power is turned on. Two priority encoders (introduced in Chapter 6) are used as eight-line-to-three-line encoders (9 input HIGH, 8 output unused) to encode the ROW and COLUMN lines of the keyboard matrix. A parallel in/parallel out register (key code) stores the ROW/COLUMN code from the priority encoders.

The basic operation of the keyboard encoder in Figure 8–36 is as follows: The ring counter "scans" the rows for a key closure as the clock signal shifts the 0 around the counter at a 5 kHz rate. The 0 (LOW) is sequentially applied to each ROW line, while all other ROW lines are HIGH. All the ROW lines are connected to the ROW encoder inputs, so the 3-bit output of the ROW encoder at any time is the binary representation of the ROW line that is LOW. When there is a key closure, one COLUMN line is connected to one ROW line. When the ROW line is taken LOW by the ring counter, that particular COLUMN line is also pulled LOW. The COLUMN encoder produces a binary output corresponding to the COLUMN in which the key is closed. The 3-bit ROW code plus the 3-bit COLUMN code uniquely identifies the key that is closed. This 6-bit code is applied to the inputs of the key code register. When a key is closed, the two one-shots produce a delayed clock pulse to parallel-load the 6-bit code into the key code register. This delay allows the contact bounce to die out. Also, the first one-shot output inhibits the ring counter to prevent it from scanning while the data are being loaded into the key code register.

The 6-bit code in the key code register is now applied to a ROM (read-only memory) to be converted to an appropriate alphanumeric code that identifies the keyboard character. ROMs are studied in Chapter 11.

SECTION 8–5 CHECKUP

1. In the keyboard encoder, how many times per second does the ring counter scan the keyboard?

2. What is the 6-bit ROW/COLUMN code (key code) for the top row and the left-most column in the keyboard encoder?

3. What is the purpose of the diodes in the keyboard encoder? What is the purpose of the resistors?

8–6   Logic Symbols with Dependency Notation

Two examples of ANSI/IEEE Standard 91-1984 symbols with dependency notation for shift registers are presented. Two specific IC shift registers are used as examples.

After completing this section, you should be able to

- Understand and interpret the logic symbols with dependency notation for the 74HC164 and the 74HC194 shift registers

The logic symbol for a 74HC164 8-bit serial in/parallel out shift register is shown in Figure 8–37. The common control inputs are shown on the notched block. The clear ($\overline{CLR}$) input is indicated by an R (for RESET) inside the block. Since there is no dependency prefix to link R with the clock (C1), the clear function is asynchronous. The right arrow symbol after C1 indicates data flow from Q_0 to Q_7. The A and B inputs are ANDed, as indicated by the embedded AND symbol, to provide the synchronous data input, $1D$, to the first stage (Q_0). Note the dependency of D on C, as indicated by the 1 suffix on C and the 1 prefix on D.

Figure 8–38 is the logic symbol for the 74HC194 4-bit bidirectional universal shift register. Starting at the top left side of the control block, note that the $\overline{CLR}$ input is active-LOW and is asynchronous (no prefix link with C). Inputs S_0 and S_1 are mode inputs that

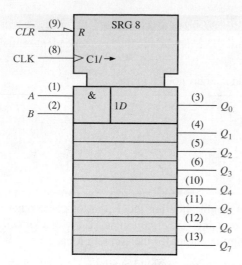

FIGURE 8–37 Logic symbol for the 74HC164.

determine the *shift-right, shift-left,* and *parallel load* modes of operation, as indicated by the $\frac{0}{3}$ dependency designation following the *M*. The $\frac{0}{3}$ represents the binary states of 0, 1, 2, and 3 on the S_0 and S_1 inputs. When one of these digits is used as a prefix for another input, a dependency is established. The $1 \rightarrow /2 \leftarrow$ symbol on the clock input indicates the following: $1 \rightarrow$ indicates that a right shift (Q_0 toward Q_3) occurs when the mode inputs (S_0, S_1) are in the binary 1 state ($S_0 = 1$, $S_1 = 0$), $2 \leftarrow$ indicates that a left shift (Q_3 toward Q_0) occurs when the mode inputs are in the binary 2 state ($S_0 = 0$, $S_1 = 1$). The shift-right serial input (*SR SER*) is both mode-dependent and clock-dependent, as indicated by 1, 4*D*. The parallel inputs (D_0, D_1, D_2, and D_3) are all mode-dependent (prefix 3 indicates parallel load mode) and clock-dependent, as indicated by 3, 4*D*. The shift-left serial input (*SL SER*) is both mode-dependent and clock-dependent, as indicated by 2, 4*D*.

The four modes for the 74HC194 are summarized as follows:

Do nothing:	$S_0 = 0, S_1 = 0$	(mode 0)
Shift right:	$S_0 = 1, S_1 = 0$	(mode 1, as in 1, 4*D*)
Shift left:	$S_0 = 0, S_1 = 1$	(mode 2, as in 2, 4*D*)
Parallel load:	$S_0 = 1, S_1 = 1$	(mode 3, as in 3, 4*D*)

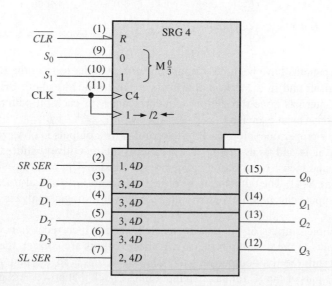

FIGURE 8–38 Logic symbol for the 74HC194.

8-7   Troubleshooting

A traditional method of troubleshooting sequential logic and other more complex systems uses a procedure of "exercising" the circuit under test with a known input waveform (stimulus) and then observing the output for the correct bit pattern.

After completing this section, you should be able to

- Explain the procedure of "exercising" as a troubleshooting technique
- Discuss exercising of a serial-to-parallel converter

The serial-to-parallel data converter in Figure 8–31 is used to illustrate the "exercising" procedure. The main objective in exercising the circuit is to force all elements (flip-flops and gates) into all of their states to be certain that nothing is stuck in a given state as a result of a fault. The input test pattern, in this case, must be designed to force each flip-flop in the registers into both states, to clock the counter through all of its eight states, and to take the control flip-flop, the clock generator, the one-shot, and the AND gate through their paces.

The input test pattern that accomplishes this objective for the serial-to-parallel data converter is based on the serial data format in Figure 8–32. It consists of the pattern 10101010 in one serial group of data bits followed by 01010101 in the next group, as shown in Figure 8–39. These patterns are generated on a repetitive basis by a special test-pattern generator. The basic test setup is shown in Figure 8–40.

FIGURE 8–39   Sample test pattern.

After both patterns have been run through the circuit under test, all the flip-flops in the data-input register and in the data-output register have resided in both SET and RESET states, the counter has gone through its sequence (once for each bit pattern), and all the other devices have been exercised.

To check for proper operation, each of the parallel data outputs is observed for an alternating pattern of 1s and 0s as the input test patterns are repetitively shifted into the data-input register and then loaded into the data-output register. The proper timing diagram is shown in Figure 8–41. The outputs can be observed in pairs with a dual-trace oscilloscope, or all eight outputs can be observed simultaneously with a logic analyzer configured for timing analysis.

If one or more outputs of the data-output register are incorrect, then you must back up to the outputs of the data-input register. If these outputs are correct, then the problem is associated with the data-output register. Check the inputs to the data-output register directly on the pins of the IC for an open input line. Check that power and ground are correct (look for the absence of noise on the ground line). Verify that the load line is a solid LOW and that there are clock pulses on the clock input of the correct amplitude. Make

Circuit under Test

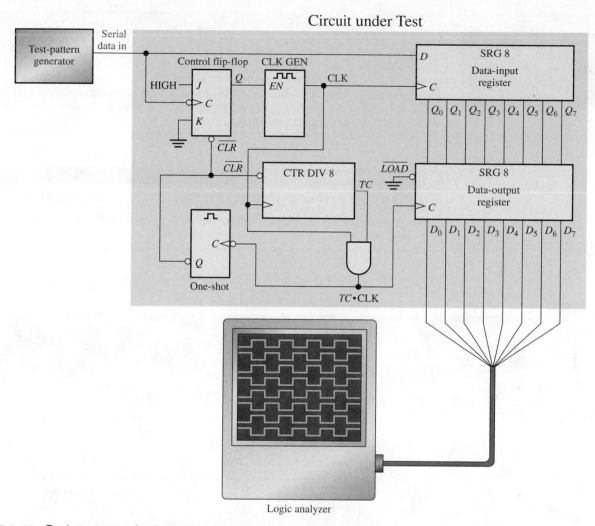

FIGURE 8-40   Basic test setup for the serial-to-parallel data converter of Figure 8–31.

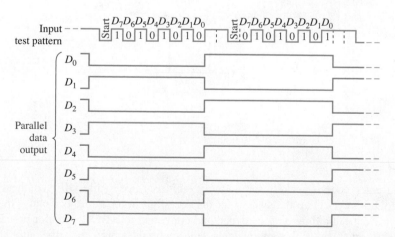

FIGURE 8-41   Proper outputs for the circuit under test in Figure 8–40. The input test pattern is shown.

sure that the connection to the logic analyzer did not short two output lines together. If all of these checks pass inspection, then it is likely that the output register is defective. If the data-input register outputs are also incorrect, the fault could be associated with the input register itself or with any of the other logic, and additional investigation is necessary to isolate the problem.

When measuring digital signals with an oscilloscope, you should always use dc coupling, rather than ac coupling. The reason that ac coupling is not best for viewing digital signals is that the 0 V level of the signal will appear at the *average* level of the signal, not at true ground or 0 V level. It is much easier to find a "floating" ground or incorrect logic level with dc coupling. If you suspect an open ground in a digital circuit, increase the sensitivity of the scope to the maximum possible. A good ground will never appear to have noise under this condition, but an open will likely show some noise, which appears as a random fluctuation in the 0 V level.

SECTION 8–7 CHECKUP

1. What is the purpose of providing a test input to a sequential logic circuit?
2. Generally, when an output waveform is found to be incorrect, what is the next step to be taken?

Applied Logic

Security System

A security system that provides coded access to a secured area is developed. Once a 4-digit security code is stored in the system, access is achieved by entering the correct code on a keypad. A block diagram for the security system is shown in Figure 8–42. The system consists of the security code logic, the code-selection logic, and the keypad. The keypad is a standard numeric keypad.

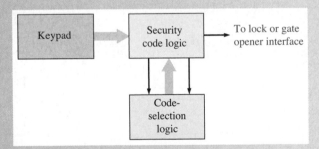

FIGURE 8–42 Block diagram of the security system.

Basic Operation

A 4-digit entry code is set into the memory with user-accessible DIP switches. Initially pressing the # key sets up the system for the first digit in the code. For entry, the code is entered one digit at a time on the keypad and converted to a BCD code for processing by the security code logic. If the entered code agrees with the stored code, the output activates the access mechanism and allows the door or gate, depending on the type of area that is secured, to be opened.

Exercise

1. Write the BCD code sequence produced by the code generator if the 4-digit access number 4739 is entered on the keypad.

The Security Code Logic

The security code logic compares the code entered on the keypad with the predetermined code from the code-selection logic. A logic diagram of the security code logic is shown in Figure 8–43.

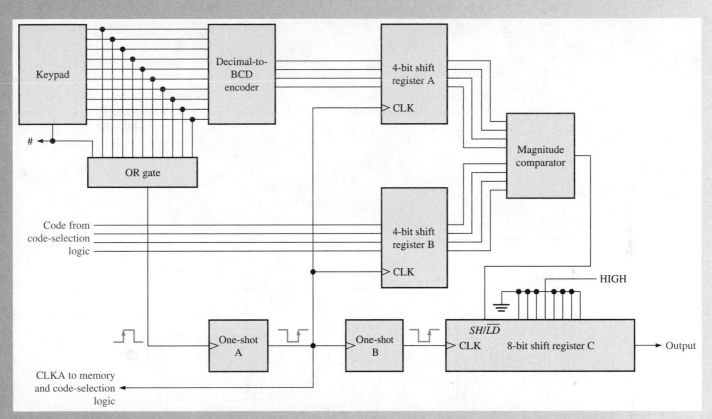

FIGURE 8–43   Block diagram of the security code logic with keypad.

In order to gain entry, first the # key on the keypad is pressed to trigger the one-shots, thus initializing the 8-bit register C with a preset pattern (00010000). Next the four digits of the code are entered in proper sequence on the keypad. As each digit is entered, it is converted to BCD by the decimal-to-BCD encoder, and a clock pulse is produced by one-shot A that shifts the 4-bit code into register A. The one-shot is triggered by a transition on the output of the OR gate when a key is pressed. At the same time, the corresponding digit from the code generator is shifted into register B. Also, one-shot B is triggered after one-shot A to provide a delayed clock pulse for register C to serially shift the preloaded pattern (00010000). The left-most three 0s are simply "fillers" and serve no purpose in the operation of the system. The outputs of registers A and B are applied to the comparator; if the codes are the same, the output of the comparator goes HIGH, placing shift register C in the SHIFT mode.

Each time an entered digit agrees with the preset digit, the 1 in shift register C is shifted right one position. On the fourth code agreement, the 1 appears on the output of the shift register and activates the mechanism to unlock the door or open the gate. If the code digits do not agree, the output of the comparator goes LOW, placing shift register C in the LOAD mode so the shift register is reinitialized to the preset pattern (00010000).

Exercise

2. What is the state of shift register C after two correct code digits are entered?
3. Explain the purpose of the OR gate.
4. If the digit 4 is entered on the keypad, what appears on the outputs of register A?

The Code-Selection Logic

A logic diagram of the code-selection logic is shown in Figure 8–44. This part of the system includes a set of DIP switches into which a 4-digit entry code is set. Initially pressing the # key sets up the system for the first digit in the code by causing a preset pattern to be loaded into the 4-bit shift register (0001). The four bits in the first code digit are selected by a HIGH on the Q_0 output of the shift register, enabling the four AND gates labeled A1–A4. As each digit of the code is entered on the keypad, the clock from the security code logic shifts the 1 in the shift register to sequentially enable each set of four AND gates. As a result, the BCD digits in the security code appear sequentially on the outputs.

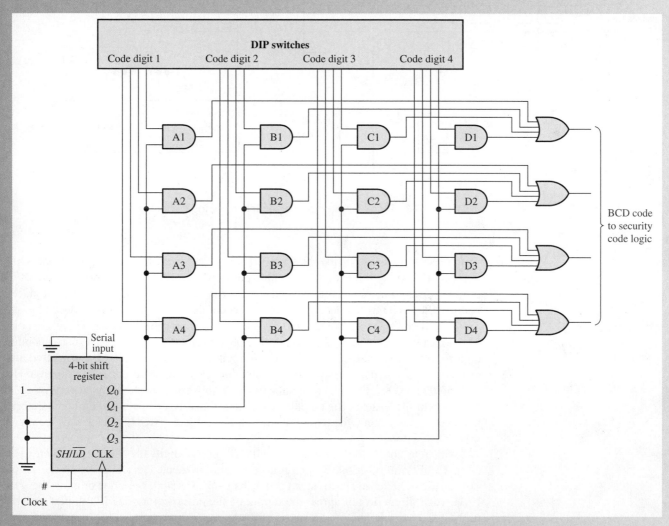

FIGURE 8–44   Logic diagram of the code-selection logic.

Security System with VHDL

The security system can be described using VHDL for implementation in a PLD. The three blocks of the system (keypad, security code logic, and code-selection logic) are combined in the program code to describe the complete system.

The security system block diagram is shown in Figure 8–45 as a programming model. Six program components perform the logical operations of the security system. Each component corresponds to a block or blocks in the figure. The security system program SecuritySystem contains the code that defines how the components interact.

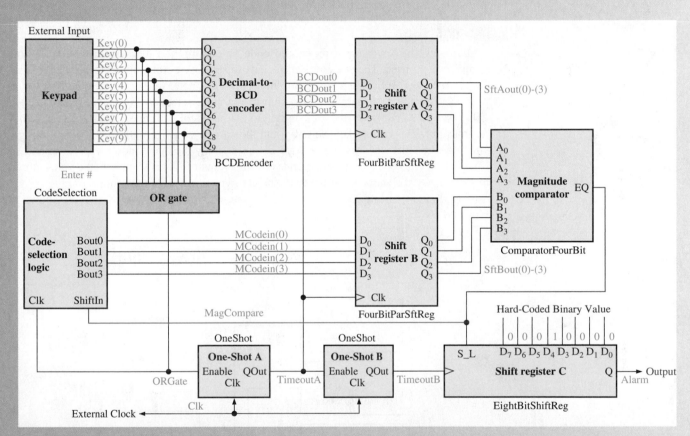

FIGURE 8–45   Security system block diagram as a programming model.

The security system includes a ten-bit input vector Key—one input bit for each decimal digit—and an input Enter, representing a typical numeric keypad. Once a key is pressed, the data stored in input array Key are sent to the decimal-to-BCD encoder (BCDEncoder). Its 4-bit output is then sent to the inputs of the 4-bit parallel in/parallel out shift register A (FourBitParSftReg). An external system clock applied to input Clk drives the overall security system. The Alarm output signal is set HIGH upon a successful arming operation.

Pressing the Enter key sends an initial HIGH clock signal to the code-selection logic block (CodeSelection), which loads an initial binary value of 1000 to shift register B. At this time, a binary 0000 is stored in shift register A, and the output of the magnitude comparator (ComparatorFourBit) is set LOW. The code-selection logic is now ready to present the first stored code value that is to be compared to the value of the first numeric keypad entry. At this time a LOW on the 8-bit parallel in/serial out shift register C (EightBitShiftReg) S_L input loads an initial value of 00010000.

When a numeric key is pressed, the output of the OR gate (ORGate) clocks the first stored value to the inputs of shift register B, and the output of the decimal-to-BCD encoder is sent to the inputs of shift register A. If the values in shift registers A and B match, the output of the magnitude comparator is set HIGH; and the code-selection logic is ready to clock in the next stored code value.

At the conclusion of four successful comparisons of stored code values against four correct keypad entries, the value 00010000 initially in shift register C will shift four places to the right, setting the Alarm output to a HIGH. An incorrect keypad entry will not match the stored code value in shift register B and the magnitude comparator will output a LOW. With the comparator output LOW, the code-selection logic will reset to the first stored code value; and the value 00010000 is reloaded into shift register C, starting the process over again.

To clock the keypad and the stored code values through the system, two one-shots (OneShot) are used. The one-shots allow data to stabilize before any action is taken. One-shot A receives an Enable signal from the keypad OR gate, which initiates the first timed process. The OR gate output is also sent to the code-selection logic, and the first code value from the code-selection logic is sent to the inputs of shift register A. When one-shot A times out, the selected keypad entry and the current code from the code-selection logic are stored in shift registers A and B for comparison by the magnitude comparator, and an Enable is sent to one-shot B. If the codes in shift registers A and B match, the value stored in shift register C shifts one place to the right after one-shot B times out.

The six components used in the security system program SecuritySystem are shown in Figure 8–46.

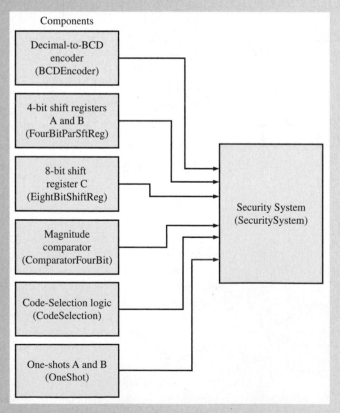

FIGURE 8–46 Security system components.

The VHDL program code for the security system is as follows:

```
library ieee;
use ieee.std_logic_1164.all;
entity SecuritySystem is
port (key: in std_logic_vector(0 to 9); Enter: in std_logic;
    Clk: in std_logic; Alarm: out std_logic);
end entity SecuritySystem;

architecture SecuritySystemBehavior of SecuritySystem is
```

Key : 10 - Key input
Enter : # - Key input
Clk : System clock
Alarm : Alarm output

```
component BCDEncoder is
port(D: in std_logic_vector(0 to 9);
    Q: out std_logic_vector(0 to 3));
end component BCDEncoder;
```
Component declaration for BCDEncoder

```
component FourBitParSftReg is
port(D: in std_logic_vector(0 to 3);
    Clk: in std_logic;
    Q: out std_logic_vector(0 to 3));
end component FourBitParSftReg;
```
Component declaration for FourBitParSftReg

```
component ComparatorFourBit is
port(A, B: in std_logic_vector(0 to 3);
    EQ: out std_logic);
end component ComparatorFourBit;
```
Component declaration for ComparatorFourBit

```
component OneShot is
port(Enable, Clk: in std_logic;
    QOut: buffer std_logic);
end component OneShot;
```
Component declaration for OneShot

```
component EightBitShiftReg is
port(S_L, Clk: in std_logic;
    D: in std_logic_vector(0 to 7);
    Q: buffer std_logic);
end component EightBitShiftReg;
```
Component declaration for EightBitShiftReg

```
component CodeSelection is
port(ShiftIn, Clk: in std_logic;
    Bout: out std_logic_vector(1 to 4));
end component CodeSelection;
```
Component declaration for CodeSelection

```
signal BCDout: std_logic_vector(0 to 3);
signal SftAout: std_logic_vector(0 to 3);
signal SftBout: std_logic_vector(0 to 3);
signal MCodein: std_logic_vector(0 to 3);
signal ORgate: std_logic;
signal MagCompare: std_logic;
signal TimeoutA, TimeoutB: std_logic;
```
BDCout: BCD encoder return
SftAout: Shift Register A return
SftBout: Shift Register B return
MCodein: Security Code value
ORgate: OR output from 10-keypad
MagCompare: Key entry to code compare
TimeoutA/B: One-shot timer variables

begin

ORgate <= (Key(0) **or** Key(1) **or** Key(2) **or** Key(3) **or** Key(4)   Logic definition for ORGate

        or key(5) **or** Key(6) **or** Key(7) **or** Key(8) **or** Key(9));

BCD: BCDEncoder

port map(D(0)=>Key(0),D(1)=>Key(1),D(2)=>Key(2),D(3)=>Key(3),

     D(4)=>Key(4),D(5)=>Key(5),D(6)=>Key(6),D(7)=>Key(7),D(8)=>Key(8),D(9)=>Key(9),

     Q(0)=>BCDout(0),Q(1)=>BCDout(1),Q(2)=>BCDout(2),Q(3)=>BCDout(3));

ShiftRegisterA: FourBitParSftReg

port map(D(0)=>BCDout(0),D(1)=>BCDout(1),D(2)=>BCDout(2),D(3)=>BCDout(3),

     Clk=>not TimeoutA,Q(0)=>SftAout(0),Q(1)=>SftAout(1),Q(2)=>SftAout(2),Q(3)=>SftAout(3));

ShiftRegisterB: FourBitParSftReg

port map(D(0)=>MCodein(0),D(1)=>MCodein(1),D(2)=>MCodein(2),D(3)=>MCodein(3),   Component

     Clk=>not TimeoutA,Q(0)=>SftBout(0),Q(1)=>SftBout(1),Q(2)=>SftBout(2),Q(3)=>SftBout(3));   instantiations

Magnitude Comparator: ComparatorFourBit **port map**(A=>SftAout,B=>SftBout,EQ=>MagCompare);

OSA:OneShot **port map**(Enable=>Enter or ORgate,Clk=>Clk,QOut=>TimeoutA);

OSB:OneShot **port map**(Enable=>not TimeoutA,Clk=>Clk,QOut=>TimeoutB);

ShiftRegisterC:EightBitShiftReg

port map(S_L=>MagCompare,Clk=> TimeoutB,D(0)=>'0',D(1)=>'0',

     D(2)=>'0',D(3)=>'1',D(4)=>'0',D(5)=>'0',D(6)=>'0',D(7)=>'0',Q=>Alarm);

CodeSelectionA: CodeSelection

port map(ShiftIn=>MagCompare,Clk=>Enter or ORGate,Bout=>MCodein);

end architecture SecuritySystemBehavior;

Simulation

Open File AL08 in the Applied Logic folder on the website. Run the security code logic simulation using your Multisim software and observe the operation. A DIP switch is used to simulate the 10-digit keypad and switch J1 simulates the # key. Switches J2–J5 are used for test purposes to enter the code that is produced by the code selection logic in the complete system. Probe lights are used only for test purposes to indicate the states of registers A and B, the output of the comparator, and the output of register C.

Putting Your Knowledge to Work

Explain how the security code logic can be modified to accommodate a 5-digit code.

SUMMARY

- The basic types of data movement in shift registers are
    1. Serial in/shift right/serial out
    2. Serial in/shift left/serial out
    3. Parallel in/serial out
    4. Serial in/parallel out

5. Parallel in/parallel out

6. Rotate right

7. Rotate left

- Shift register counters are shift registers with feedback that exhibit special sequences. Examples are the Johnson counter and the ring counter.
- The Johnson counter has $2n$ states in its sequence, where n is the number of stages.
- The ring counter has n states in its sequence.

KEY TERMS

Key terms and other bold terms in the chapter are defined in the end-of-book glossary.

Bidirectional   Having two directions. In a bidirectional shift register, the stored data can be shifted right or left.

Load   To enter data into a shift register.

Register   One or more flip-flops used to store and shift data.

Stage   One storage element in a register.

TRUE/FALSE QUIZ

Answers are at the end of the chapter.

1. Shift registers consist of an arrangement of flip-flops.
2. Two functions of a shift register are data storage and data movement.
3. In a serial shift register, several data bits are entered at the same time.
4. All shift registers are defined by specified sequences.
5. A shift register can have both parallel and serial outputs.
6. A shift register with four stages can store a maximum count of fifteen.
7. The Johnson counter is a special type of shift register.
8. The modulus of an 8-bit Johnson counter is eight.
9. A ring counter uses one flip-flop for each state in its sequence.
10. A shift register can be used as a time delay device.

SELF-TEST

Answers are at the end of the chapter.

1. A stage in a shift register consists of
   - (a) a latch
   - (b) a flip-flop
   - (c) a byte of storage
   - (d) four bits of storage

2. To serially shift a byte of data into a shift register, there must be
   - (a) one clock pulse
   - (b) one load pulse
   - (c) eight clock pulses
   - (d) one clock pulse for each 1 in the data

3. To parallel load a byte of data into a shift register with a synchronous load, there must be
   - (a) one clock pulse
   - (b) one clock pulse for each 1 in the data
   - (c) eight clock pulses
   - (d) one clock pulse for each 0 in the data

4. The group of bits 10110101 is serially shifted (right-most bit first) into an 8-bit parallel output shift register with an initial state of 11100100. After two clock pulses, the register contains
   - (a) 01011110
   - (b) 10110101
   - (c) 01111001
   - (d) 00101101

5. With a 100 kHz clock frequency, eight bits can be serially entered into a shift register in
   (a) 80 μs   (b) 8 μs
   (c) 80 ms   (d) 10 μs

6. With a 1 MHz clock frequency, eight bits can be parallel entered into a shift register
   (a) in 8 μs
   (b) in the propagation delay time of eight flip-flops
   (c) in 1 μs
   (d) in the propagation delay time of one flip-flop

7. A modulus-10 Johnson counter requires
   (a) ten flip-flops   (b) four flip-flops
   (c) five flip-flops   (d) twelve flip-flops

8. A modulus-10 ring counter requires a minimum of
   (a) ten flip-flops   (b) five flip-flops
   (c) four flip-flops   (d) twelve flip-flops

9. When an 8-bit serial in/serial out shift register is used for a 24 μs time delay, the clock frequency must be
   (a) 41.67 kHz   (b) 333 kHz
   (c) 125 kHz   (d) 8 MHz

10. The purpose of the ring counter in the keyboard encoding circuit of Figure 8–36 is
    (a) to sequentially apply a HIGH to each row for detection of key closure
    (b) to provide trigger pulses for the key code register
    (c) to sequentially apply a LOW to each row for detection of key closure
    (d) to sequentially reverse bias the diodes in each row

PROBLEMS

Answers to odd-numbered problems are at the end of the book.

Section 8–1 Shift Register Operations

1. Why are shift registers considered basic memory devices?

2. What is the storage capacity of a register that can retain two bytes of data?

3. Name two functions of a shift register.

Section 8–2 Types of Shift Register Data I/Os

4. The sequence 1011 is applied to the input of a 4-bit serial shift register that is initially cleared. What is the state of the shift register after three clock pulses?

5. For the data input and clock in Figure 8–47, determine the states of each flip-flop in the shift register of Figure 8–3 and show the *Q* waveforms. Assume that the register contains all 1s initially.

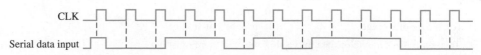

FIGURE 8–47

6. Solve Problem 5 for the waveforms in Figure 8–48.

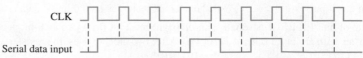

FIGURE 8–48

7. What is the state of the register in Figure 8–49 after each clock pulse if it starts in the 101001111000 state?

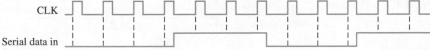
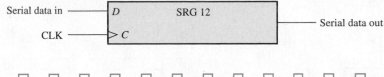

FIGURE 8–49

8. For the serial in/serial out shift register, determine the data-output waveform for the data-input and clock waveforms in Figure 8–50. Assume that the register is initially cleared.

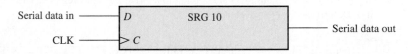
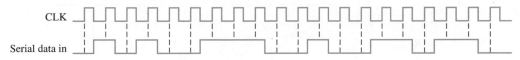

FIGURE 8–50

9. Solve Problem 8 for the waveforms in Figure 8–51.

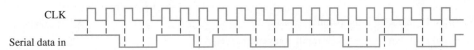

FIGURE 8–51

10. A leading-edge clocked serial in/serial out shift register has a data-output waveform as shown in Figure 8–52. What binary number is stored in the 8-bit register if the first data bit out (leftmost) is the LSB?

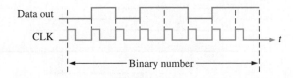

FIGURE 8–52

11. Show a complete timing diagram including the parallel outputs for the shift register in Figure 8–6. Use the waveforms in Figure 8–50 with the register initially clear.

12. Solve Problem 11 for the input waveforms in Figure 8–51.

13. Develop the Q_0 through Q_7 outputs for a 74HC164 shift register with the input waveforms shown in Figure 8–53.

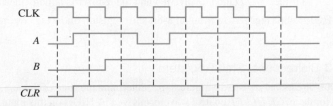

FIGURE 8–53

14. The shift register in Figure 8–54(a) has $SHIFT/\overline{LOAD}$ and CLK inputs as shown in part (b). The serial data input (*SER*) is a 0. The parallel data inputs are $D_0 = 1$, $D_1 = 0$, $D_2 = 1$, and $D_3 = 0$ as shown. Develop the data-output waveform in relation to the inputs.

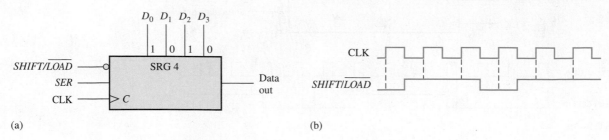

(a)   (b)

FIGURE 8–54

15. The waveforms in Figure 8–55 are applied to a 74HC165 shift register. The parallel inputs are all 0. Determine the Q_7 waveform.

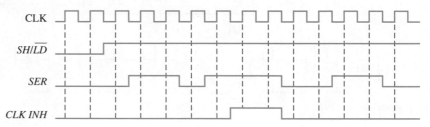

FIGURE 8–55

16. Solve Problem 15 if the parallel inputs are all 1.

17. Solve Problem 15 if the *SER* input is inverted.

18. Determine all the Q output waveforms for a 74HC195 4-bit shift register when the inputs are as shown in Figure 8–56.

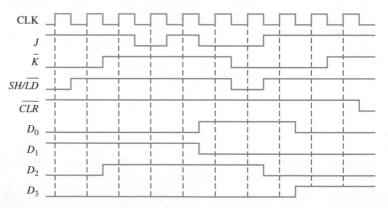

FIGURE 8–56

19. Solve Problem 18 if the $SH/\overline{LD}$ input is inverted and the register is initially clear.

20. Use two 74HC195 shift registers to form an 8-bit shift register. Show the required connections.

Section 8–3 Bidirectional Shift Registers

21. For the 8-bit bidirectional register in Figure 8–57, determine the state of the register after each clock pulse for the $RIGHT/\overline{LEFT}$ control waveform given. A HIGH on this input enables a shift to the right, and a LOW enables a shift to the left. Assume that the register is initially storing the decimal number seventy-six in binary, with the right-most position being the LSB. There is a LOW on the data-input line.

FIGURE 8–57

22. Solve Problem 21 for the waveforms in Figure 8–58.

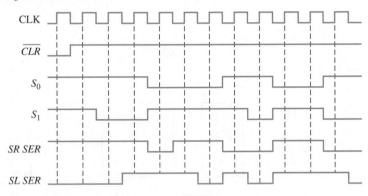

FIGURE 8–58

23. Use two 74HC194 4-bit bidirectional shift registers to create an 8-bit bidirectional shift register. Show the connections.

24. Determine the Q outputs of a 74HC194 with the inputs shown in Figure 8–59. Inputs D_0, D_1, D_2, and D_3 are all HIGH.

FIGURE 8–59

Section 8–4 Shift Register Counters

25. How many flip-flops are required to implement each of the following in a Johnson counter configuration:
    (a) modulus-6
    (b) modulus-10
    (c) modulus-14
    (d) modulus-16

26. Draw the logic diagram for a modulus-18 Johnson counter. Show the timing diagram and write the sequence in tabular form.

27. For the ring counter in Figure 8–60, show the waveforms for each flip-flop output with respect to the clock. Assume that FF0 is initially SET and that the rest are RESET. Show at least ten clock pulses.

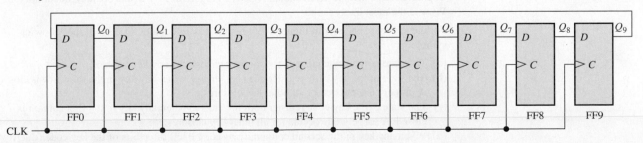

FIGURE 8–60

28. The waveform pattern in Figure 8–61 is required. Devise a ring counter, and indicate how it can be preset to produce this waveform on its Q_9 output. At CLK16 the pattern begins to repeat.

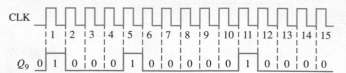

FIGURE 8–61

Section 8–5 Shift Register Applications

29. Use 74HC195 4-bit shift registers to implement a 16-bit ring counter. Show the connections.

30. What is the purpose of the power-on $\overline{LOAD}$ input in Figure 8–36?

31. What happens when two keys are pressed simultaneously in Figure 8–36?

Section 8–7 Troubleshooting

32. Based on the waveforms in Figure 8–62(a), determine the most likely problem with the register in part (b) of the figure.

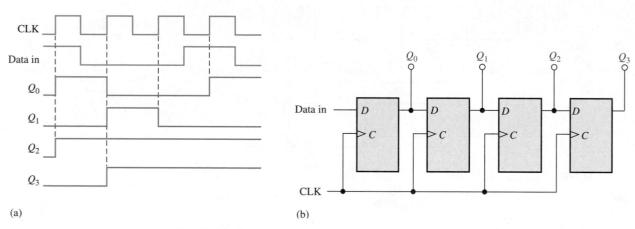

(a)   (b)

FIGURE 8–62

33. Refer to the parallel in/serial out shift register in Figure 8–10. The register is in the state where $Q_0Q_1Q_2Q_3 = 1001$, and $D_0D_1D_2D_3 = 1010$ is loaded in. When the $SHIFT/\overline{LOAD}$ input is taken HIGH, the data shown in Figure 8–63 are shifted out. Is this operation correct? If not, what is the most likely problem?

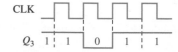

FIGURE 8–63

34. You have found that the bidirectional register in Figure 8–17 will shift data right but not left. What is the most likely fault?

35. For the keyboard encoder in Figure 8–36, list the possible faults for each of the following symptoms:

(a) The state of the key code register does not change for any key closure.

(b) The state of the key code register does not change when any key in the third row is closed. A proper code occurs for all other key closures.

(c) The state of the key code register does not change when any key in the first column is closed. A proper code occurs for all other key closures.

(d) When any key in the second column is closed, the left three bits of the key code ($Q_0Q_1Q_2$) are correct, but the right three bits are all 1s.

36. Develop a test procedure for exercising the keyboard encoder in Figure 8–36. Specify the procedure on a step-by-step basis, indicating the output code from the key code register that should be observed at each step in the test.

37. What symptoms are observed for the following failures in the serial-to-parallel converter in Figure 8–31:

   (a) AND gate output stuck in HIGH state

   (b) clock generator output stuck in LOW state

   (c) third stage of data-input register stuck in SET state

   (d) terminal count output of counter stuck in HIGH state

Applied Logic

38. What is the major purpose of the security code logic?

39. Assume the entry code is 1939. Determine the states of shift register A and shift register C after the second correct digit has been entered in Figure 8–43.

40. Assume the entry code is 7646 and the digits 7645 are entered. Determine the states of shift register A and shift register C after each of the digits is entered.

Special Design Problems

41. Specify the devices that can be used to implement the serial-to-parallel data converter in Figure 8–31. Develop the complete logic diagram, showing any modifications necessary to accommodate the specific devices used.

42. Modify the serial-to-parallel converter in Figure 8–31 to provide 16-bit conversion.

43. Design an 8-bit parallel-to-serial data converter that produces the data format in Figure 8–32. Show a logic diagram and specify the devices.

44. Design a power-on $\overline{LOAD}$ circuit for the keyboard encoder in Figure 8–36. This circuit must generate a short-duration LOW pulse when the power switch is turned on.

45. Implement the test-pattern generator used in Figure 8–40 to troubleshoot the serial-to-parallel converter.

46. Review the tablet-bottling system that was introduced in Chapter 1. Utilizing the knowledge gained in this chapter, implement registers A and B in that system using specific fixed-function IC devices.

Multisim Troubleshooting Practice

MultiSim

47. Open file P08-47. For the specified fault, predict the effect on the circuit. Then introduce the fault and verify whether your prediction is correct.

48. Open file P08-48. For the specified fault, predict the effect on the circuit. Then introduce the fault and verify whether your prediction is correct.

49. Open file P08-49. For the specified fault, predict the effect on the circuit. Then introduce the fault and verify whether your prediction is correct.

50. Open file P08-50. For the observed behavior indicated, predict the fault in the circuit. Then introduce the suspected fault and verify whether your prediction is correct.

51. Open file P08-51. For the observed behavior indicated, predict the fault in the circuit. Then introduce the suspected fault and verify whether your prediction is correct.

ANSWERS

SECTION CHECKUPS

Section 8–1 Shift Register Operations

   1. The number of stages.

   2. Storage and data movement are two functions of a shift register.

Section 8–2 Types of Shift Register Data I/Os

1. FF0: data input to J_0, $\overline{\text{data input}}$ to K_0; FF1: Q_0 to J_1, $\overline{Q}_0$ to K_1; FF2: Q_1 to J_2, $\overline{Q}_1$ to K_2; FF3: Q_2 to J_3, $\overline{Q}_2$ to K_3

2. Eight clock pulses

3. 0100 after 2 clock pulses

4. Take the serial output from the right-most flip-flop for serial out operation.

5. When $SHIFT/\overline{LOAD}$ is HIGH, the data are shifted right one bit per clock pulse. When $SHIFT/\overline{LOAD}$ is LOW, the data on the parallel inputs are loaded into the register.

6. The parallel load operation is asynchronous, so it is not dependent on the clock.

7. The data outputs are 1001.

8. $Q_0 = 1$ after one clock pulse

Section 8–3 Bidirectional Shift Registers

1. 1111 after the fifth clock pulse

Section 8–4 Shift Register Counters

1. Sixteen states are in an 8-bit Johnson counter sequence.

2. For a 3-bit Johnson counter: 000, 100, 110, 111, 011, 001, 000

Section 8–5 Shift Register Applications

1. 625 scans/second

2. $Q_5Q_4Q_3Q_2Q_1Q_0 = 011011$

3. The diodes provide unidirectional paths for pulling the ROWs LOW and preventing HIGHs on the ROW lines from being connected to the switch matrix. The resistors pull the COLUMN lines HIGH.

Section 8–6 Logic Symbols with Dependency Notation

1. No inputs are dependent on the mode inputs being in the 0 state.

2. Yes, the parallel load is synchronous with the clock as indicated by the 4D label.

Section 8–7 Troubleshooting

1. A test input is used to sequence the circuit through all of its states.

2. Check the input to that portion of the circuit. If the signal on that input is correct, the fault is isolated to the circuitry between the good input and the bad output.

RELATED PROBLEMS FOR EXAMPLES

8–1 See Figure 8–64.

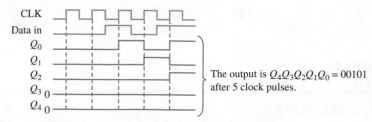

FIGURE 8–64

8–2 The state of the register after three additional clock pulses is 0000.

8–3 See Figure 8–65.

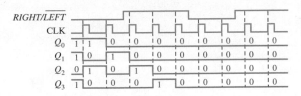

FIGURE 8–65

8–4 See Figure 8–66.

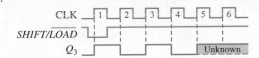

FIGURE 8–66

8–5 See Figure 8–67.

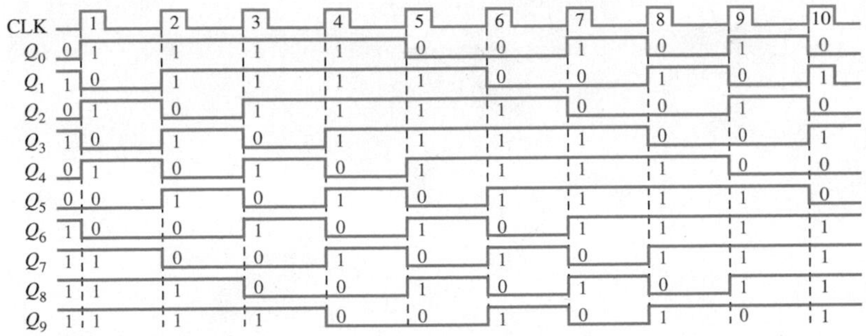

FIGURE 8–67

8–6 $f = 1/3\ \mu s = 333$ kHz

TRUE/FALSE QUIZ

1. T   **2.** T   **3.** F   **4.** F   **5.** T   **6.** T   **7.** T   **8.** F   **9.** T   **10.** T

SELF-TEST

1. (b)   **2.** (c)   **3.** (a)   **4.** (c)   **5.** (a)   **6.** (d)   **7.** (c)   **8.** (a)   **9.** (b)   **10.** (c)

Counters

CHAPTER OUTLINE

CHAPTER OBJECTIVES

- Discuss the types of state machines
- Describe the difference between an asynchronous and a synchronous counter
- Analyze counter timing diagrams
- Analyze counter circuits
- Explain how propagation delays affect the operation of a counter
- Determine the modulus of a counter
- Modify the modulus of a counter
- Recognize the difference between a 4-bit binary counter and a decade counter
- Use an up/down counter to generate forward and reverse binary sequences
- Determine the sequence of a counter
- Use IC counters in various applications
- Design a counter that will have any specified sequence of states
- Use cascaded counters to achieve a higher modulus
- Use logic gates to decode any given state of a counter
- Eliminate glitches in counter decoding
- Explain how a digital clock operates
- Interpret counter logic symbols that use dependency notation
- Troubleshoot counters for various types of faults

KEY TERMS

Key terms are in order of appearance in the chapter.

- State machine
- Asynchronous
- Recycle
- Modulus
- Decade
- Synchronous
- Terminal count
- State diagram
- Cascade

VISIT THE WEBSITE

Study aids for this chapter are available at http://www.pearsonhighered.com/careersresources

INTRODUCTION

As you learned in Chapter 7, flip-flops can be connected together to perform counting operations. Such a group of flip-flops is a counter, which is a type of finite state machine. The number of flip-flops used and the way in which they are connected determine the number of states (called the modulus) and also the specific sequence of states that the counter goes through during each complete cycle.

Counters are classified into two broad categories according to the way they are clocked: asynchronous and synchronous. In asynchronous counters, commonly called *ripple counters,* the first flip-flop is clocked by the external clock pulse and then each successive flip-flop is clocked by the output of the preceding flip-flop. In synchronous counters, the clock input is connected to all of the flip-flops so that they are clocked simultaneously. Within each of these two categories, counters are classified primarily by the type of sequence, the number of states, or the number of flip-flops in the counter. VHDL codes for various types of counters are presented.

9–1   Finite State Machines

A **state machine** is a sequential circuit having a limited (finite) number of states occuring in a prescribed order. A counter is an example of a state machine; the number of states is called the *modulus*. Two basic types of state machines are the Moore and the Mealy. The **Moore state machine** is one where the outputs depend only on the internal present state. The **Mealy state machine** is one where the outputs depend on both the internal present state and on the inputs. Both types have a timing input (clock) that is not considered a controlling input. A design approach to counters is presented in this section.

After completing this section, you should be able to

- Describe a Moore state machine

- Describe a Mealy state machine

- Discuss examples of Moore and Mealy state machines

General Models of Finite State Machines

A Moore state machine consists of combinational logic that determines the sequence and memory (flip-flops), as shown in Figure 9–1(a). A Mealy state machine is shown in part (b).

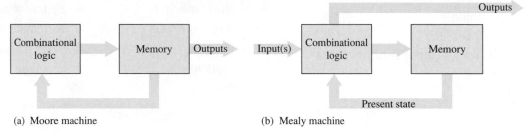

(a) Moore machine   (b) Mealy machine

FIGURE 9–1   Two types of sequential logic.

In the Moore machine, the combinational logic is a gate array with outputs that determine the next state of the flip-flops in the memory. There may or may not be inputs to the combinational logic. There may also be output combinational logic, such as a decoder. If there is an input(s), it does not affect the outputs because they always correspond to and are dependent only on the present state of the memory. For the Mealy machine, the present state affects the outputs, just as in the Moore machine; but in addition, the inputs also affect the outputs. The outputs come directly from the combinational logic and not the memory.

Example of a Moore Machine

Figure 9–2(a) shows a Moore machine (modulus-26 binary counter with states 0 through 25) that is used to control the number of tablets (25) that go into each bottle in an assembly line. When the binary number in the memory (flip-flops) reaches binary twenty-five (11001), the counter recycles to 0 and the tablet flow and clock are cut off until the next bottle is in place. The combinational logic for the state transitions sets the modulus of the counter so that it sequences from binary state 0 to binary state 25, where 0 is the reset or rest state and the output combinational logic decodes binary state 25. There is no input in this case, other than the clock, so the next state is determined only by the present state, which makes this a Moore machine. One tablet is bottled for each clock pulse. Once a bottle is in place, the first tablet is inserted at binary state 1, the second at binary state 2, and the twenty-fifth tablet when the binary state is 25. Count 25 is decoded and used to stop the flow of tablets and the clock. The counter stays in the 0 state until the next bottle is in position (indicated by a 1). Then the clock resumes, the count goes to 1, and the cycle repeats, as illustrated by the state diagram in Figure 9–2(b).

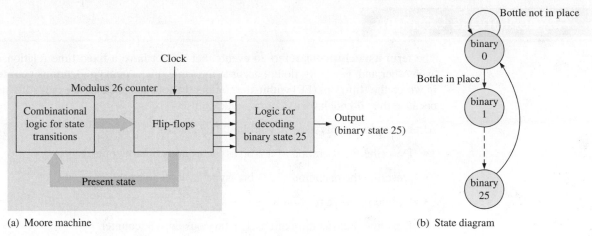

(a) Moore machine   (b) State diagram

FIGURE 9–2 A fixed-modulus binary counter as an example of a Moore state machine. The dashed line in the state diagram means the states between binary 1 and 25 are not shown for simplicity.

Example of a Mealy Machine

Let's assume that the tablet-bottling system uses three different sizes of bottles: a 25-tablet bottle, a 50-tablet bottle, and a 100-tablet bottle. This operation requires a state machine with three different terminal counts: 25, 50, and 100. One approach is illustrated in Figure 9–3(a). The combinational logic sets the modulus of the counter depending on the modulus-select inputs. The output of the counter depends on both the present state and the modulus-select inputs, making this a Mealy machine. The state diagram is shown in part (b).

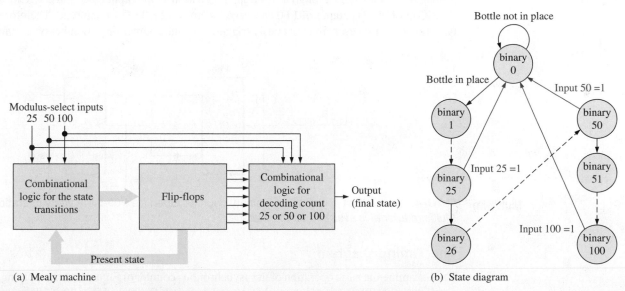

(a) Mealy machine   (b) State diagram

FIGURE 9–3 A variable-modulus binary counter as an example of a Mealy state machine. The red arrows in the state diagram represent the recycle paths that depend on the input number. The black dashed lines mean the interim states are not shown for simplicity.

SECTION 9–1 CHECKUP

Answers are at the end of the chapter.

  1. What characterizes a finite state machine?

  2. Name the types of finite state machines.

  3. Explain the difference between the two types of state machines.

9–2   Asynchronous Counters

The term **asynchronous** refers to events that do not have a fixed time relationship with each other and, generally, do not occur at the same time. An **asynchronous counter** is one in which the flip-flops (FF) within the counter do not change states at exactly the same time because they do not have a common clock pulse.

After completing this section, you should be able to

*   Describe the operation of a 2-bit asynchronous binary counter
*   Describe the operation of a 3-bit asynchronous binary counter
*   Define *ripple* in relation to counters
*   Describe the operation of an asynchronous decade counter
*   Develop counter timing diagrams
*   Discuss the implementation of a 4-bit asynchronous binary counter

A 2-Bit Asynchronous Binary Counter

> The clock input of an asynchronous counter is always connected only to the LSB flip-flop.

Figure 9–4 shows a 2-bit counter connected for asynchronous operation. Notice that the clock (CLK) is applied to the clock input (C) of *only* the first flip-flop, FF0, which is always the least significant bit (LSB). The second flip-flop, FF1, is triggered by the $\overline{Q}_0$ output of FF0. FF0 changes state at the positive-going edge of each clock pulse, but FF1 changes only when triggered by a positive-going transition of the $\overline{Q}_0$ output of FF0. Because of the inherent propagation delay time through a flip-flop, a transition of the input clock pulse (CLK) and a transition of the $\overline{Q}_0$ output of FF0 can never occur at exactly the same time. Therefore, the two flip-flops are never simultaneously triggered, so the counter operation is asynchronous.

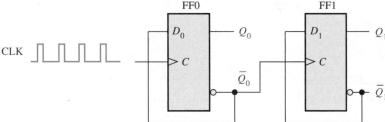

 FIGURE 9–4 A 2-bit asynchronous binary counter. Open file F09-04 to verify operation. *A Multisim tutorial is available on the website.*

The Timing Diagram

Let's examine the basic operation of the asynchronous counter of Figure 9–4 by applying four clock pulses to FF0 and observing the Q output of each flip-flop. Figure 9–5 illustrates the changes in the state of the flip-flop outputs in response to the clock pulses. Both flip-flops are connected for toggle operation ($D = \overline{Q}$) and are assumed to be initially RESET (Q LOW).

> Asynchronous counters are also known as ripple counters.

The positive-going edge of CLK1 (clock pulse 1) causes the Q_0 output of FF0 to go HIGH, as shown in Figure 9–5. At the same time the $\overline{Q}_0$ output goes LOW, but it has no effect on FF1 because a positive-going transition must occur to trigger the flip-flop. After the leading edge of CLK1, $Q_0 = 1$ and $Q_1 = 0$. The positive-going edge of CLK2 causes Q_0 to go LOW. Output $\overline{Q}_0$ goes HIGH and triggers FF1, causing Q_1 to go HIGH. After the leading edge of CLK2, $Q_0 = 0$ and $Q_1 = 1$. The positive-going edge of CLK3 causes Q_0 to go HIGH again. Output $\overline{Q}_0$ goes LOW and has no effect on FF1. Thus, after the leading edge of CLK3, $Q_0 = 1$ and $Q_1 = 1$. The positive-going edge of CLK4 causes Q_0 to go LOW, while $\overline{Q}_0$ goes HIGH and triggers FF1, causing Q_1 to go LOW. After the leading

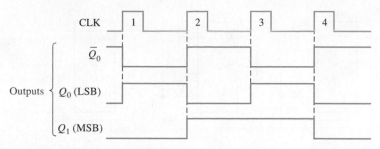

FIGURE 9-5   Timing diagram for the counter of Figure 9–4. As in previous chapters, output waveforms are shown in green.

edge of CLK4, $Q_0 = 0$ and $Q_1 = 0$. The counter has now recycled to its original state (both flip-flops are RESET).

In the timing diagram, the waveforms of the Q_0 and Q_1 outputs are shown relative to the clock pulses as illustrated in Figure 9–5. For simplicity, the transitions of Q_0, Q_1, and the clock pulses are shown as simultaneous even though this is an asynchronous counter. There is, of course, some small delay between the CLK and the Q_0 transition and between the $\overline{Q_0}$ transition and the Q_1 transition.

Note in Figure 9–5 that the 2-bit counter exhibits four different states, as you would expect with two flip-flops ($2^2 = 4$). Also, notice that if Q_0 represents the least significant bit (LSB) and Q_1 represents the most significant bit (MSB), the sequence of counter states represents a sequence of binary numbers as listed in Table 9–1.

In digital logic, Q_0 is always the LSB unless otherwise specified.

TABLE 9-1

Binary state sequence for the counter in Figure 9–4.

Clock Pulse	Q_1	Q_0
Initially	0	0
1	0	1
2	1	0
3	1	1
4 (recycles)	0	0

Since it goes through a binary sequence, the counter in Figure 9–4 is a binary counter. It actually counts the number of clock pulses up to three, and on the fourth pulse it recycles to its original state ($Q_0 = 0$, $Q_1 = 0$). The term **recycle** is commonly applied to counter operation; it refers to the transition of the counter from its final state back to its original state.

A 3-Bit Asynchronous Binary Counter

The state sequence for a 3-bit binary counter is listed in Table 9–2, and a 3-bit asynchronous binary counter is shown in Figure 9–6(a). The basic operation is the same as that of the 2-bit

TABLE 9-2

State sequence for a 3-bit binary counter.

Clock Pulse	Q_2	Q_1	Q_0
Initially	0	0	0
1	0	0	1
2	0	1	0
3	0	1	1
4	1	0	0
5	1	0	1
6	1	1	0
7	1	1	1
8 (recycles)	0	0	0

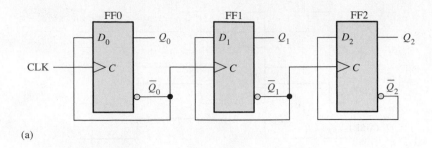

(a)

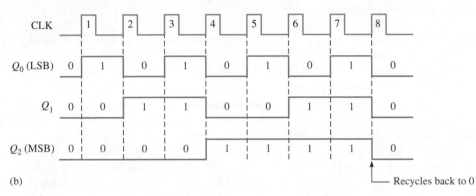

(b)

                                                                 — Recycles back to 0

MultiSim   **FIGURE 9–6**   Three-bit asynchronous binary counter and its timing diagram for one cycle. Open file F09-06 to verify operation.

counter except that the 3-bit counter has eight states, due to its three flip-flops. A timing diagram is shown in Figure 9–6(b) for eight clock pulses. Notice that the counter progresses through a binary count of zero through seven and then recycles to the zero state. This counter can be easily expanded for higher count, by connecting additional toggle flip-flops.

Propagation Delay

Asynchronous counters are commonly referred to as **ripple counters** for the following reason: The effect of the input clock pulse is first "felt" by FF0. This effect cannot get to FF1 immediately because of the propagation delay through FF0. Then there is the propagation delay through FF1 before FF2 can be triggered. Thus, the effect of an input clock pulse "ripples" through the counter, taking some time, due to propagation delays, to reach the last flip-flop.

    To illustrate, notice that all three flip-flops in the counter of Figure 9–6 change state on the leading edge of CLK4. This ripple clocking effect is shown in Figure 9–7 for the first four clock pulses, with the propagation delays indicated. The LOW-to-HIGH transition of

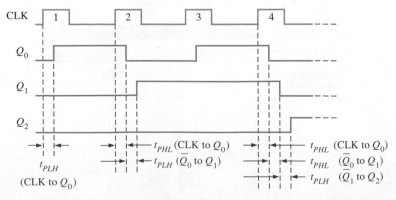

FIGURE 9–7   Propagation delays in a 3-bit asynchronous (ripple-clocked) binary counter.

Q_0 occurs one delay time (t_{PLH}) after the positive-going transition of the clock pulse. The LOW-to-HIGH transition of Q_1 occurs one delay time (t_{PLH}) after the positive-going transition of $\overline{Q}_0$. The LOW-to-HIGH transition of Q_2 occurs one delay time (t_{PLH}) after the positive-going transition of $\overline{Q}_1$. As you can see, FF2 is not triggered until two delay times after the positive-going edge of the clock pulse, CLK4. Thus, it takes three propagation delay times for the effect of the clock pulse, CLK4, to ripple through the counter and change Q_2 from LOW to HIGH.

This cumulative delay of an asynchronous counter is a major disadvantage in many applications because it limits the rate at which the counter can be clocked and creates decoding problems. The maximum cumulative delay in a counter must be less than the period of the clock waveform.

EXAMPLE 9–1

A 4-bit asynchronous binary counter is shown in Figure 9–8(a). Each D flip-flop is negative edge-triggered and has a propagation delay for 10 nanoseconds (ns). Develop a timing diagram showing the Q output of each flip-flop, and determine the total propagation delay time from the triggering edge of a clock pulse until a corresponding change can occur in the state of Q_3. Also determine the maximum clock frequency at which the counter can be operated.

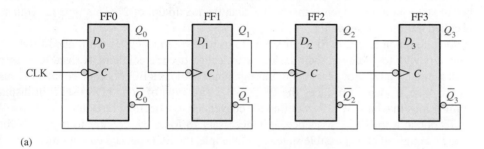

(a)

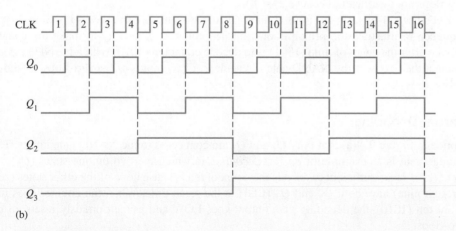

(b)

FIGURE 9–8 Four-bit asynchronous binary counter and its timing diagram. Open file F09-08 and verify the operation.

MultiSim

Solution

The timing diagram with delays omitted is as shown in Figure 9–8(b). For the total delay time, the effect of CLK8 or CLK16 must propagate through four flip-flops before Q_3 changes, so

$$t_{p(tot)} = 4 \times 10\,\text{ns} = \mathbf{40\ ns}$$

The maximum clock frequency is

$$f_{max} = \frac{1}{t_{p(tot)}} = \frac{1}{40 \text{ ns}} = \textbf{25 MHz}$$

The counter should be operated below this frequency to avoid problems due to the propagation delay.

Related Problem*

Show the timing diagram if all of the flip-flops in Figure 9–8(a) are positive edge-triggered.

---

*Answers are at the end of the chapter.

Asynchronous Decade Counters

A counter can have 2^n states, where n is the number of flip-flops.

The **modulus** of a counter is the number of unique states through which the counter will sequence. The maximum possible number of states (maximum modulus) of a counter is 2^n, where n is the number of flip-flops in the counter. Counters can be designed to have a number of states in their sequence that is less than the maximum of 2^n. This type of sequence is called a *truncated sequence*.

One common modulus for counters with truncated sequences is ten (called MOD10). Counters with ten states in their sequence are called **decade** counters. A **decade counter** with a count sequence of zero (0000) through nine (1001) is a BCD decade counter because its ten-state sequence produces the BCD code. This type of counter is useful in display applications in which BCD is required for conversion to a decimal readout.

To obtain a truncated sequence, it is necessary to force the counter to recycle before going through all of its possible states. For example, the BCD decade counter must recycle back to the 0000 state after the 1001 state. A decade counter requires four flip-flops (three flip-flops are insufficient because $2^3 = 8$).

Let's use a 4-bit asynchronous counter such as the one in Example 9–1 and modify its sequence to illustrate the principle of truncated counters. One way to make the counter recycle after the count of nine (1001) is to decode count ten (1010) with a NAND gate and connect the output of the NAND gate to the clear ($\overline{CLR}$) inputs of the flip-flops, as shown in Figure 9–9(a).

Partial Decoding

Notice in Figure 9–9(a) that only Q_1 and Q_3 are connected to the NAND gate inputs. This arrangement is an example of *partial decoding,* in which the two unique states ($Q_1 = 1$ and $Q_3 = 1$) are sufficient to decode the count of ten because none of the other states (zero through nine) have both Q_1 and Q_3 HIGH at the same time. When the counter goes into count ten (1010), the decoding gate output goes LOW and asynchronously resets all the flip-flops.

The resulting timing diagram is shown in Figure 9–9(b). Notice that there is a glitch on the Q_1 waveform. The reason for this glitch is that Q_1 must first go HIGH before the count of ten can be decoded. Not until several nanoseconds after the counter goes to the count of ten does the output of the decoding gate go LOW (both inputs are HIGH). Thus, the counter is in the 1010 state for a short time before it is reset to 0000, thus producing the glitch on Q_1 and the resulting glitch on the $\overline{CLR}$ line that resets the counter.

Other truncated sequences can be implemented in a similar way, as Example 9–2 shows.

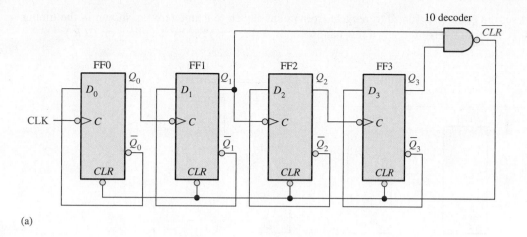

(a)

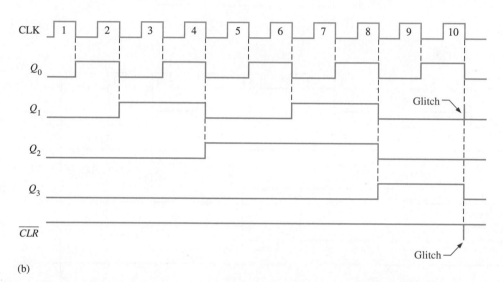

(b)

FIGURE 9–9   An asynchronously clocked decade counter with asynchronous recycling.

EXAMPLE 9–2

Show how an asynchronous counter with J-K flip-flops can be implemented having a modulus of twelve with a straight binary sequence from 0000 through 1011.

Solution

Since three flip-flops can produce a maximum of eight states, four flip-flops are required to produce any modulus greater than eight but less than or equal to sixteen.

When the counter gets to its last state, 1011, it must recycle back to 0000 rather than going to its normal next state of 1100, as illustrated in the following sequence chart:

Q_3	Q_2	Q_1	Q_0	
0	0	0	0	←
.	.	.	.	
.	.	.	.	Recycles
.	.	.	.	
1	0	1	1	
1	1	0	0	← Normal next state

Observe that Q_0 and Q_1 both go to 0 anyway, but Q_2 and Q_3 must be forced to 0 on the twelfth clock pulse. Figure 9–10(a) shows the modulus-12 counter. The NAND gate partially decodes count twelve (1100) and resets flip-flop 2 and flip-flop 3.

Thus, on the twelfth clock pulse, the counter is forced to recycle from count eleven to count zero, as shown in the timing diagram of Figure 9–10(b). (It is in count twelve for only a few nanoseconds before it is reset by the glitch on $\overline{CLR}$.)

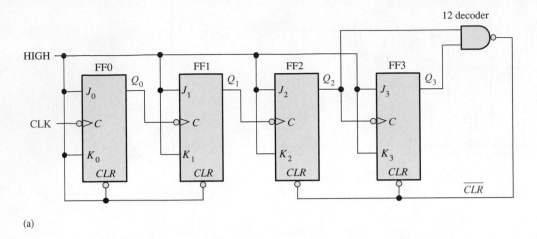

(a)

(b)

FIGURE 9–10   Asynchronously clocked modulus-12 counter with asynchronous recycling.

Related Problem

How can the counter in Figure 9–10(a) be modified to make it a modulus-13 counter?

IMPLEMENTATION: 4-BIT ASYNCHRONOUS BINARY COUNTER

Fixed-Function Device   The 74HC93 is an example of a specific integrated circuit asynchronous counter. This device actually consists of a single flip-flop (CLK A) and a 3-bit asynchronous counter (CLK B). This arrangement is for flexibility. It can be used as a divide-by-2 device if only the single flip-flop is used, or it can be used as a modulus-8 counter if only the 3-bit counter portion is used. This device also provides gated reset inputs, *RO(1)* and *RO(2)*. When both of these inputs are HIGH, the counter is reset to the 0000 state $\overline{CLR}$.

Additionally, the 74HC93 can be used as a 4-bit modulus-16 counter (counts 0 through 15) by connecting the Q_0 output to the CLK B input as shown by the logic symbol in Figure 9–11(a). It can also be configured as a decade counter (counts 0 through 9) with asynchronous recycling by using the gated reset inputs for partial decoding of count ten, as shown by the logic symbol in Figure 9–11(b).

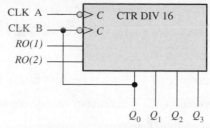 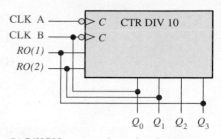

(a) 74HC93 connected as a modulus-16 counter   (b) 74HC93 connected as a decade counter

FIGURE 9–11 Two configurations of the 74HC93 asynchronous counter. (The qualifying label, CTR DIV *n*, indicates a counter with *n* states.)

Programmable Logic Device (PLD)   The VHDL code for a generic 4-bit asynchronous binary counter using J-K flip flops with preset (PRN) and clear (CLRN) inputs is as follows:

```
library ieee;
use ieee.std_logic_1164.all;

entity AsyncFourBitBinCntr is
    port (Clock, Clr: in std_logic; Q0, Q1, Q2, Q3: inout std_logic);    Inputs and outputs declared
end entity AsyncFourBitBinCntr;

architecture LogicOperation of AsyncFourBitBinCntr is
component jkff is
    port (J, K, Clk, PRN, CLRN: in std_logic; Q: out std_logic);    J-K flip-flop component
end component jkff;                                                    declaration

begin
    FF0: jkff port map(J=>'1', K=>'1', Clk=>Clock,  CLRN=>Clr, PRN=>'1', Q=>Q0);
    FF1: jkff port map(J=>'1', K=>'1', Clk=>not Q0, CLRN=>Clr, PRN=>'1', Q=>Q1);    Instantiations define
    FF2: jkff port map(J=>'1', K=>'1', Clk=>not Q1, CLRN=>Clr, PRN=>'1', Q=>Q2);    how each flip-flop is
    FF3: jkff port map(J=>'1', K=>'1', Clk=>not Q2, CLRN=>Clr, PRN=>'1', Q=>Q3);    connected.
end architecture LogicOperation;
```

SECTION 9–2 CHECKUP

1. What does the term *asynchronous* mean in relation to counters?
2. How many states does a modulus-14 counter have? What is the minimum number of flip-flops required?

9–3 Synchronous Counters

The term **synchronous** refers to events that have a fixed time relationship with each other. A **synchronous counter** is one in which all the flip-flops in the counter are clocked at the same time by a common clock pulse. J-K flip-flops are used to illustrate most synchronous counters. D flip-flops can also be used but generally require more logic because of having no direct toggle or no-change states.

After completing this section, you should be able to

- ◆ Describe the operation of a 2-bit synchronous binary counter
- ◆ Describe the operation of a 3-bit synchronous binary counter
- ◆ Describe the operation of a 4-bit synchronous binary counter
- ◆ Describe the operation of a synchronous decade counter
- ◆ Develop counter timing diagrams

A 2-Bit Synchronous Binary Counter

Figure 9–12 shows a 2-bit synchronous binary counter. Notice that an arrangement different from that for the asynchronous counter must be used for the J_1 and K_1 inputs of FF1 in order to achieve a binary sequence. A D flip-flop implementation is shown in part (b).

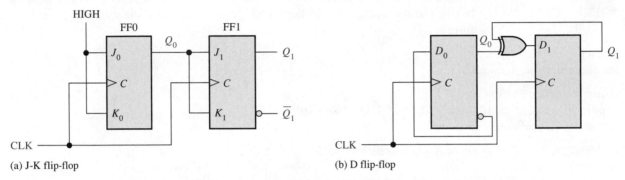

FIGURE 9–12   2-bit synchronous binary counters.

The clock input goes to each flip-flop in a synchronous counter.

The operation of a J-K flip-flop synchronous counter is as follows: First, assume that the counter is initially in the binary 0 state; that is, both flip-flops are RESET. When the positive edge of the first clock pulse is applied, FF0 will toggle and Q_0 will therefore go HIGH. What happens to FF1 at the positive-going edge of CLK1? To find out, let's look at the input conditions of FF1. Inputs J_1 and K_1 are both LOW because Q_0, to which they are connected, has not yet gone HIGH. Remember, there is a propagation delay from the triggering edge of the clock pulse until the Q output actually makes a transition. So, $J = 0$ and $K = 0$ when the leading edge of the first clock pulse is applied. This is a no-change condition, and therefore FF1 does not change state. A timing detail of this portion of the counter operation is shown in Figure 9–13(a).

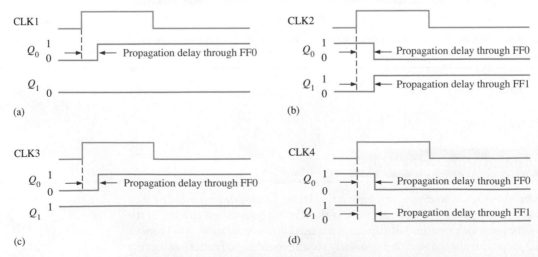

FIGURE 9–13   Timing details for the 2-bit synchronous counter operation (the propagation delays of both flip-flops are assumed to be equal).

After CLK1, $Q_0 = 1$ and $Q_1 = 0$ (which is the binary 1 state). When the leading edge of CLK2 occurs, FF0 will toggle and Q_0 will go LOW. Since FF1 has a HIGH ($Q_0 = 1$) on its J_1 and K_1 inputs at the triggering edge of this clock pulse, the flip-flop toggles and Q_1 goes HIGH. Thus, after CLK2, $Q_0 = 0$ and $Q_1 = 1$ (which is a binary 2 state). The timing detail for this condition is shown in Figure 9–13(b).

When the leading edge of CLK3 occurs, FF0 again toggles to the SET state ($Q_0 = 1$), and FF1 remains SET ($Q_1 = 1$) because its J_1 and K_1 inputs are both LOW ($Q_0 = 0$). After this triggering edge, $Q_0 = 1$ and $Q_1 = 1$ (which is a binary 3 state). The timing detail is shown in Figure 9–13(c).

Finally, at the leading edge of CLK4, Q_0 and Q_1 go LOW because they both have a toggle condition on their J and K inputs. The timing detail is shown in Figure 9–13(d). The counter has now recycled to its original state, binary 0. Examination of the D flip-flop counter in Figure 9–12(b) will show the timing diagram is the same as for the J-K flip-flop counter.

The complete timing diagram for the counters in Figure 9–12 is shown in Figure 9–14. Notice that all the waveform transitions appear coincident; that is, the propagation delays are not indicated. Although the delays are an important factor in the synchronous counter operation, in an overall timing diagram they are normally omitted for simplicity. Major waveform relationships resulting from the normal operation of a circuit can be conveyed completely without showing small delay and timing differences. However, in high-speed digital circuits, these small delays are an important consideration in design and troubleshooting.

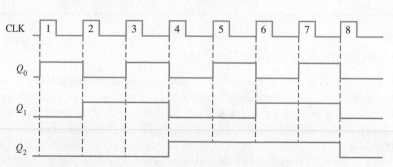

FIGURE 9–14 Timing diagram for the counters of Figure 9–12.

A 3-Bit Synchronous Binary Counter

A 3-bit synchronous binary counter is shown in Figure 9–15, and its timing diagram is shown in Figure 9–16. You can understand this counter operation by examining its sequence of states as shown in Table 9–3.

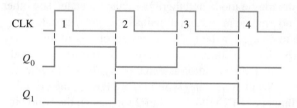

FIGURE 9–15 A 3-bit synchronous binary counter. Open file F09-15 to verify the operation.

MultiSim

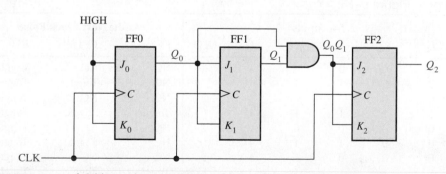

FIGURE 9–16 Timing diagram for the counter of Figure 9–15.

InfoNote

The TSC or *time stamp counter* in some microprocessors is used for performance monitoring, which enables a number of parameters important to the overall performance of a system to be determined exactly. By reading the TSC before and after the execution of a procedure, the precise time required for the procedure can be determined based on the processor cycle time. In this way, the TSC forms the basis for all time evaluations in connection with optimizing system operation. For example, it can be accurately determined which of two or more programming sequences is more efficient. This is a very useful tool for compiler developers and system programmers in producing the most effective code.

TABLE 9–3

State sequence for a 3-bit binary counter.

Clock Pulse	Q_2	Q_1	Q_0
Initially	0	0	0
1	0	0	1
2	0	1	0
3	0	1	1
4	1	0	0
5	1	0	1
6	1	1	0
7	1	1	1
8 (recycles)	0	0	0

First, let's look at Q_0. Notice that Q_0 changes on each clock pulse as the counter progresses from its original state to its final state and then back to its original state. To produce this operation, FF0 must be held in the toggle mode by constant HIGHs on its J_0 and K_0 inputs. Notice that Q_1 goes to the opposite state following each time Q_0 is a 1. This change occurs at CLK2, CLK4, CLK6, and CLK8. The CLK8 pulse causes the counter to recycle. To produce this operation, Q_0 is connected to the J_1 and K_1 inputs of FF1. When Q_0 is a 1 and a clock pulse occurs, FF1 is in the toggle mode and therefore changes state. The other times, when Q_0 is a 0, FF1 is in the no-change mode and remains in its present state.

Next, let's see how FF2 is made to change at the proper times according to the binary sequence. Notice that both times Q_2 changes state, it is preceded by the unique condition in which both Q_0 and Q_1 are HIGH. This condition is detected by the AND gate and applied to the J_2 and K_2 inputs of FF2. Whenever both Q_0 and Q_1 are HIGH, the output of the AND gate makes the J_2 and K_2 inputs of FF2 HIGH, and FF2 toggles on the following clock pulse. At all other times, the J_2 and K_2 inputs of FF2 are held LOW by the AND gate output, and FF2 does not change state.

The analysis of the counter in Figure 9–15 is summarized in Table 9–4.

TABLE 9–4

Summary of the analysis of the counter in Figure 9–15.

Clock Pulse	Outputs			J-K Inputs						At the Next Clock Pulse		
	Q_2	Q_1	Q_0	J_2	K_2	J_1	K_1	J_0	K_0	FF2	FF1	FF0
Initially	0	0	0	0	0	0	0	1	1	NC*	NC	Toggle
1	0	0	1	0	0	1	1	1	1	NC	Toggle	Toggle
2	0	1	0	0	0	0	0	1	1	NC	NC	Toggle
3	0	1	1	1	1	1	1	1	1	Toggle	Toggle	Toggle
4	1	0	0	0	0	0	0	1	1	NC	NC	Toggle
5	1	0	1	0	0	1	1	1	1	NC	Toggle	Toggle
6	1	1	0	0	0	0	0	1	1	NC	NC	Toggle
7	1	1	1	1	1	1	1	1	1	Toggle	Toggle	Toggle
										Counter recycles back to 000.		

*NC indicates *No Change*.

A 4-Bit Synchronous Binary Counter

Figure 9–17(a) shows a 4-bit synchronous binary counter, and Figure 9–17(b) shows its timing diagram. This particular counter is implemented with negative edge-triggered flip-flops. The reasoning behind the J and K input control for the first three flip-flops is the same as previously discussed for the 3-bit counter. The fourth stage, FF3, changes only twice in the sequence. Notice that both of these transitions occur following the times that Q_0, Q_1, and Q_2 are all HIGH. This condition is decoded by AND gate G_2 so that when a

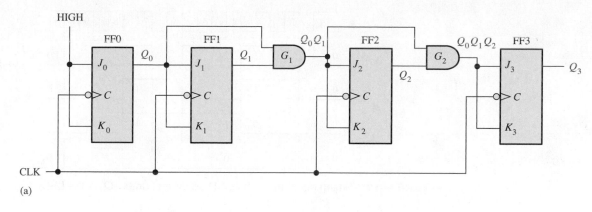

(a)

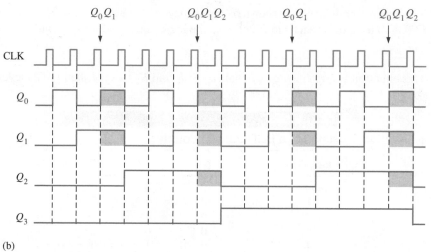

(b)

FIGURE 9–17   A 4-bit synchronous binary counter and timing diagram. Times where the AND gate outputs are HIGH are indicated by the shaded areas.

clock pulse occurs, FF3 will change state. For all other times the J_3 and K_3 inputs of FF3 are LOW, and it is in a no-change condition.

A 4-Bit Synchronous Decade Counter

As you know, a BCD decade counter exhibits a truncated binary sequence and goes from 0000 through the 1001 state. Rather than going from the 1001 state to the 1010 state, it recycles to the 0000 state. A synchronous BCD decade counter is shown in Figure 9–18. The timing diagram for the decade counter is shown in Figure 9–19.

A decade counter has ten states.

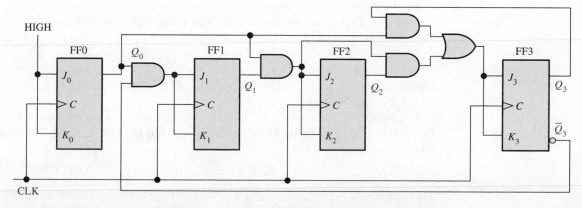

FIGURE 9–18   A synchronous BCD decade counter. Open file F09-18 to verify operation.

MultiSim

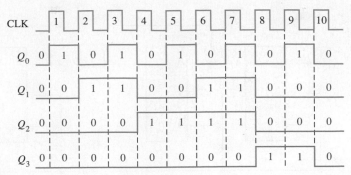

FIGURE 9–19 Timing diagram for the BCD decade counter (Q_0 is the LSB).

The counter operation is shown by the sequence of states in Table 9–5. First, notice that FF0 (Q_0) toggles on each clock pulse, so the logic equation for its J_0 and K_0 inputs is

$$J_0 = K_0 = 1$$

This equation is implemented by connecting J_0 and K_0 to a constant HIGH level.

TABLE 9–5

States of a BCD decade counter.

Clock Pulse	Q_3	Q_2	Q_1	Q_0
Initially	0	0	0	0
1	0	0	0	1
2	0	0	1	0
3	0	0	1	1
4	0	1	0	0
5	0	1	0	1
6	0	1	1	0
7	0	1	1	1
8	1	0	0	0
9	1	0	0	1
10 (recycles)	0	0	0	0

Next, notice in Table 9–5 that FF1 (Q_1) changes on the next clock pulse each time $Q_0 = 1$ and $Q_3 = 0$, so the logic equation for the J_1 and K_1 inputs is

$$J_1 = K_1 = Q_0\overline{Q}_3$$

This equation is implemented by ANDing Q_0 and $\overline{Q}_3$ and connecting the gate output to the J_1 and K_1 inputs of FF1.

Flip-flop 2 (Q_2) changes on the next clock pulse each time both $Q_0 = 1$ and $Q_1 = 1$. This requires an input logic equation as follows:

$$J_2 = K_2 = Q_0Q_1$$

This equation is implemented by ANDing Q_0 and Q_1 and connecting the gate output to the J_2 and K_2 inputs of FF2.

Finally, FF3 (Q_3) changes to the opposite state on the next clock pulse each time $Q_0 = 1$, $Q_1 = 1$, and $Q_2 = 1$ (state 7), or when $Q_0 = 1$ and $Q_3 = 1$ (state 9). The equation for this is as follows:

$$J_3 = K_3 = Q_0Q_1Q_2 + Q_0Q_3$$

This function is implemented with the AND/OR logic connected to the J_3 and K_3 inputs of FF3 as shown in the logic diagram in Figure 9–18. Notice that the differences between this decade counter and the modulus-16 binary counter in Figure 9–17(a) are the $Q_0\overline{Q}_3$ AND gate, the Q_0Q_3 AND gate, and the OR gate; this arrangement detects the occurrence of the 1001 state and causes the counter to recycle properly on the next clock pulse.

IMPLEMENTATION: 4-BIT SYNCHRONOUS BINARY COUNTER

Fixed-Function Device   The 74HC163 is an example of an integrated circuit 4-bit synchronous binary counter. A logic symbol is shown in Figure 9–20 with pin numbers in parentheses. This counter has several features in addition to the basic functions previously discussed for the general synchronous binary counter.

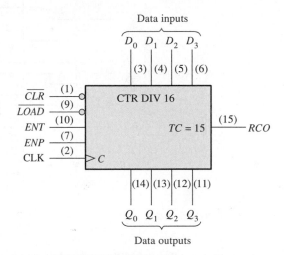

FIGURE 9–20   The 74HC163 4-bit synchronous binary counter. (The qualifying label CTR DIV 16 indicates a counter with sixteen states.)

First, the counter can be synchronously preset to any 4-bit binary number by applying the proper levels to the parallel data inputs. When a LOW is applied to the $\overline{LOAD}$ input, the counter will assume the state of the data inputs on the next clock pulse. Thus, the counter sequence can be started with any 4-bit binary number.

Also, there is an active-LOW clear input ($\overline{CLR}$), which synchronously resets all four flip-flops in the counter. There are two enable inputs, *ENP* and *ENT*. These inputs must both be HIGH for the counter to sequence through its binary states. When at least one input is LOW, the counter is disabled. The ripple clock output (RCO) goes HIGH when the counter reaches the last state in its sequence of fifteen, called the **terminal count** ($TC = 15$). This output, in conjunction with the enable inputs, allows these counters to be cascaded for higher count sequences.

Figure 9–21 shows a timing diagram of this counter being preset to twelve (1100) and then counting up to its terminal count, fifteen (1111). Input D_0 is the least significant input bit, and Q_0 is the least significant output bit.

Let's examine this timing diagram in detail. This will aid you in interpreting timing diagrams in this chapter or on manufacturers' data sheets. To begin, the LOW level pulse on the $\overline{CLR}$ input causes all the outputs (Q_0, Q_1, Q_2, and Q_3) to go LOW.

Next, the LOW level pulse on the $\overline{LOAD}$ input synchronously enters the data on the data inputs (D_0, D_1, D_2, and D_3) into the counter. These data appear on the Q outputs at the time of the first positive-going clock edge after $\overline{LOAD}$ goes LOW. This is the preset operation. In this particular example, Q_0 is LOW, Q_1 is LOW, Q_2 is HIGH, and Q_3 is HIGH. This, of course, is a binary 12 (Q_0 is the LSB).

The counter now advances through states 13, 14, and 15 on the next three positive-going clock edges. It then recycles to 0, 1, 2 on the following clock pulses. Notice that

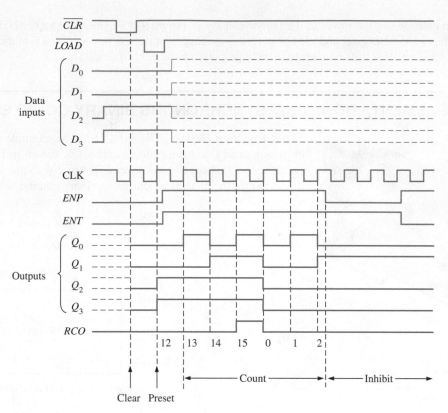

FIGURE 9–21   Timing example for a 74HC163.

both *ENP* and *ENT* inputs are HIGH during the state sequence. When *ENP* goes LOW, the counter is inhibited and remains in the binary 2 state.

Programmable Logic Device (PLD)   The VHDL code for a 4-bit synchronous decade counter using J-K flip flops is as follows:

library ieee;
use ieee.std_logic_1164.**all**;

entity FourBitSynchDecadeCounter **is**
    port (Clk: **in std_logic**; Q0, Q1, Q2, Q3: **inout std_logic**);   Input and outputs
end entity FourBitSynchDecadeCounter;   declared

architecture LogicOperation **of** FourBitSynchDecadeCounter **is**

component jkff **is**
    port (J, K, Clk: **in std_logic**; Q: **out std_logic**);   Component declaration for
end component jkff;   the J-K flip-flop

signal J1, J2, J3: **std_logic**;
begin
J1 <= Q0 **and not** Q3;   Boolean expressions for *J* input
J2 <= Q1 **and** Q0;   of each flip-flop (*J* = *K*)
J3 <= (Q2 **and** J2) **or** (Q0 **and** Q3);

FF0: jkff **port map** (J => '1', K => '1', Clk => Clk, Q => Q0);
FF1: jkff **port map** (J => J1, K => J1, Clk => Clk, Q => Q1);   Instantiations define
FF2: jkff **port map** (J => J2, K => J2, Clk => Clk, Q => Q2);   connections for each
FF3: jkff **port map** (J => J3, K => J3, Clk => Clk, Q => Q3);   flip-flop.
end architecture LogicOperation;

9–4 Up/Down Synchronous Counters

An **up/down counter** is one that is capable of progressing in either direction through a certain sequence. An up/down counter, sometimes called a bidirectional counter, can have any specified sequence of states. A 3-bit binary counter that advances upward through its sequence (0, 1, 2, 3, 4, 5, 6, 7) and then can be reversed so that it goes through the sequence in the opposite direction (7, 6, 5, 4, 3, 2, 1, 0) is an illustration of up/down sequential operation.

After completing this section, you should be able to

- Explain the basic operation of an up/down counter
- Discuss the 74HC190 up/down decade counter

In general, most up/down counters can be reversed at any point in their sequence. For instance, the 3-bit binary counter can be made to go through the following sequence:

$$\overbrace{\quad\quad}^{\text{UP}}\quad\quad\overbrace{\quad\quad}^{\text{UP}}$$

$$0, 1, 2, 3, 4, 5, \underbrace{4, 3, 2,}_{\text{DOWN}} \underbrace{3, 4, 5, 6, 7,}_{\text{DOWN}} 6, 5, \text{ etc.}$$

Table 9–6 shows the complete up/down sequence for a 3-bit binary counter. The arrows indicate the state-to-state movement of the counter for both its UP and its DOWN modes of operation. An examination of Q_0 for both the up and down sequences shows that FF0 toggles on each clock pulse. Thus, the J_0 and K_0 inputs of FF0 are

$$J_0 = K_0 = 1$$

TABLE 9–6

Up/Down sequence for a 3-bit binary counter.

Clock Pulse	Up	Q_2	Q_1	Q_0	Down
0	(	0	0	0	)
1	(	0	0	1	)
2	(	0	1	0	)
3	(	0	1	1	)
4	(	1	0	0	)
5	(	1	0	1	)
6	(	1	1	0	)
7	(	1	1	1	)

For the up sequence, Q_1 changes state on the next clock pulse when $Q_0 = 1$. For the down sequence, Q_1 changes on the next clock pulse when $Q_0 = 0$. Thus, the J_1 and K_1 inputs of FF1 must equal 1 under the conditions expressed by the following equation:

$$J_1 = K_1 = (Q_0 \cdot \text{UP}) + (\overline{Q}_0 \cdot \text{DOWN})$$

For the up sequence, Q_2 changes state on the next clock pulse when $Q_0 = Q_1 = 1$. For the down sequence, Q_2 changes on the next clock pulse when $Q_0 = Q_1 = 0$. Thus, the J_2 and K_2 inputs of FF2 must equal 1 under the conditions expressed by the following equation:

$$J_2 = K_2 = (Q_0 \cdot Q_1 \cdot UP) + (\overline{Q}_0 \cdot \overline{Q}_1 \cdot DOWN)$$

Each of the conditions for the J and K inputs of each flip-flop produces a toggle at the appropriate point in the counter sequence.

Figure 9–22 shows a basic implementation of a 3-bit up/down binary counter using the logic equations just developed for the J and K inputs of each flip-flop. Notice that the $UP/\overline{DOWN}$ control input is HIGH for UP and LOW for DOWN.

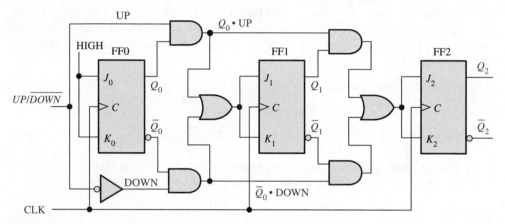

 FIGURE 9–22   A basic 3-bit up/down synchronous counter. Open file F09-22 to verify operation.

EXAMPLE 9–3

Show the timing diagram and determine the sequence of a 4-bit synchronous binary up/down counter if the clock and $UP/\overline{DOWN}$ control inputs have waveforms as shown in Figure 9–23(a). The counter starts in the all-0s state and is positive edge-triggered.

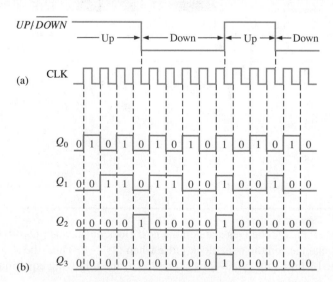

FIGURE 9–23

Solution

The timing diagram showing the Q outputs is shown in Figure 9–23(b). From these waveforms, the counter sequence is as shown in Table 9–7.

TABLE 9–7

Q_3	Q_2	Q_1	Q_0	
0	0	0	0	
0	0	0	1	
0	0	1	0	UP
0	0	1	1	
0	1	0	0	
0	0	1	1	
0	0	1	0	
0	0	0	1	DOWN
0	0	0	0	
1	1	1	1	
0	0	0	0	
0	0	0	1	UP
0	0	1	0	
0	0	0	1	DOWN
0	0	0	0	

Related Problem

Show the timing diagram if the $UP/\overline{DOWN}$ control waveform in Figure 9–23(a) is inverted.

IMPLEMENTATION: UP/DOWN DECADE COUNTER

Fixed-Function Device   Figure 9–24 shows a logic diagram for the 74HC190, an example of an integrated circuit up/down synchronous decade counter. The direction of the count is determined by the level of the up/down input ($D/\overline{U}$). When this input is HIGH, the counter counts down; when it is LOW, the counter counts up. Also, this device can be preset to any desired BCD digit as determined by the states of the data inputs when the $\overline{LOAD}$ input is LOW.

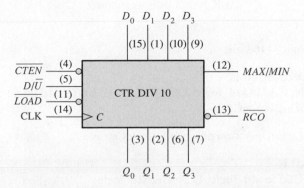

FIGURE 9–24   The 74HC190 up/down synchronous decade counter.

The *MAX/MIN* output produces a HIGH pulse when the terminal count nine (1001) is reached in the UP mode or when the terminal count zero (0000) is reached in the DOWN mode. The *MAX/MIN* output, the ripple clock output ($\overline{RCO}$), and the count enable input ($\overline{CTEN}$) are used when cascading counters. (Cascaded counters are discussed in Section 9–6.)

Figure 9–25 is a timing diagram that shows the 74HC190 counter preset to seven (0111) and then going through a count-up sequence followed by a count-down sequence. The *MAX/MIN* output is HIGH when the counter is in either the all-0s state (*MIN*) or the 1001 state (*MAX*).

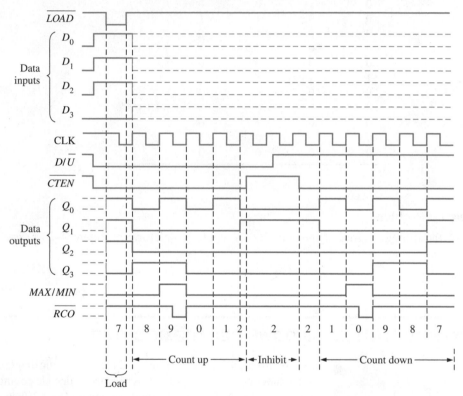

FIGURE 9–25   Timing example for a 74HC190.

Programmable Logic Device (PLD)   A VHDL code for an up/down decade counter using J-K flip-flops is as follows:

```
library ieee;
use ieee.std_logic_1164.all;

entity UpDnDecadeCntr is                                    UPDN:   Counter direction
    port (UPDN, Clk: in std_logic; Q0, Q1, Q2, Q3: buffer std_logic);   Clk:      System clock
end entity UpDnDecadeCntr;                                  Q0-Q3:  Counter output

architecture LogicOperation of UpDnDecadeCntr is

component jkff is
    port (J, K, Clk: in std_logic; Q: buffer std_logic);  } J-K flip flop component
end component jkff;
```

```
function UpDown(A, B, C, D: in std_logic)
    return std_logic is
begin
    return((A and B) or (C and D));
end function UpDown;

signal J1Up, J1Dn, J1, J2, J3: std_logic;

begin
    J1Up <= UPDN and Q0; J1Dn <= not UPDN and not Q0;
    UpDn1: J1 <= UpDown(UPDN, Q0, not UPDN, not Q0);
    UpDn2: J2 <= UpDown(J1Up, Q1, J1Dn, not Q1);
    UpDn3: J3 <= UpDown(J1Up and Q1, Q2, J1Dn and not Q1, not Q2);

    FF0: jkff port map (J =>'1', K =>'1', Clk => Clk, Q => Q0);
    FF1: jkff port map (J => J1, K => J1, Clk => Clk, Q => Q1);
    FF2: jkff port map (J => J2, K => J2, Clk => Clk, Q => Q2);
    FF3: jkff port map (J => J3, K => J3, Clk => Clk, Q => Q3);
end architecture LogicOperation;
```

Function UpDown is a helper function performing the common logic between stages performed by the two AND gates applied to the OR gate supplying the J K inputs of the next stage. See Figure 9–22.

J1Up: Initial Up logic for FF1.
J1Dn: Initial Down logic for FF1.
J1-J3: Variable for combined UpDown applied to FF1-FF3.

Identifiers J1, J2, and J3 complete the up/down logic applied to the J and K inputs of flip-flop stages FF0-FF1. Using a function to perform operations common to multiple tasks simplifies the overall code design and implementation.

Flip-flop stages FF0-FF3 complete the Up/Down counter.

SECTION 9–4 CHECKUP

1. A 4-bit up/down binary counter is in the DOWN mode and in the 1010 state. On the next clock pulse, to what state does the counter go?

2. What is the terminal count of a 4-bit binary counter in the UP mode? In the DOWN mode? What is the next state after the terminal count in the DOWN mode?

9–5 Design of Synchronous Counters

In this section, you will learn the six steps to design a counter (state machine). As you learned in Section 9–1, sequential circuits can be classified into two types: (1) those in which the output or outputs depend only on the present internal state (Moore state machines) and (2) those in which the output or outputs depend on both the present state and the input or inputs (Mealy state machines). This section is recommended for those who want an introduction to counter design or to state machine design in general. It is not a prerequisite for any other material.

After completing this section, you should be able to

- Develop a state diagram for a given sequence

- Develop a next-state table for a specified counter sequence

- Create a flip-flop transition table

- Use the Karnaugh map method to derive the logic requirements for a synchronous counter

- Implement a counter to produce a specified sequence of states

Step 1: State Diagram

The first step in the design of a state machine (counter) is to create a state diagram. A **state diagram** shows the progression of states through which the counter advances when it is clocked. As an example, Figure 9–26 is a state diagram for a basic 3-bit Gray code counter. This particular circuit has no inputs other than the clock and no outputs other than the outputs taken off each flip-flop in the counter. You may wish to review the coverage of the Gray code in Chapter 2 at this time.

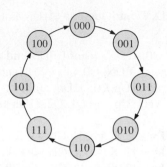

FIGURE 9–26 State diagram for a 3-bit Gray code counter.

Step 2: Next-State Table

Once the sequential circuit is defined by a state diagram, the second step is to derive a next-state table, which lists each state of the counter (present state) along with the corresponding next state. *The next state is the state that the counter goes to from its present state upon application of a clock pulse.* The next-state table is derived from the state diagram and is shown in Table 9–8 for the 3-bit Gray code counter. Q_0 is the least significant bit.

TABLE 9–8

Next-state table for 3-bit Gray code counter.

Present State			Next State		
Q_2	Q_1	Q_0	Q_2	Q_1	Q_0
0	0	0	0	0	1
0	0	1	0	1	1
0	1	1	0	1	0
0	1	0	1	1	0
1	1	0	1	1	1
1	1	1	1	0	1
1	0	1	1	0	0
1	0	0	0	0	0

TABLE 9–9

Transition table for a J-K flip-flop.

Output Transitions			Flip-Flop Inputs	
Q_N		Q_{N+1}	J	K
0	$\longrightarrow$	0	0	X
0	$\longrightarrow$	1	1	X
1	$\longrightarrow$	0	X	1
1	$\longrightarrow$	1	X	0

Q_N: present state
Q_{N+1}: next state
X: "don't care"

Step 3: Flip-Flop Transition Table

Table 9–9 is a transition table for the J-K flip-flop. All possible output transitions are listed by showing the Q output of the flip-flop going from present states to next states. Q_N is the present state of the flip-flop (before a clock pulse) and Q_{N+1} is the next state (after a clock pulse). For each output transition, the J and K inputs that will cause the transition to occur are listed. An X indicates a "don't care" (the input can be either a 1 or a 0).

To design the counter, the transition table is applied to each of the flip-flops in the counter, based on the next-state table (Table 9–8). For example, for the present state 000,

Q_0 goes from a present state of 0 to a next state of 1. To make this happen, J_0 must be a 1 and you don't care what K_0 is ($J_0 = 1$, $K_0 = X$), as you can see in the transition table (Table 9–9). Next, Q_1 is 0 in the present state and remains a 0 in the next state. For this transition, $J_1 = 0$ and $K_1 = X$. Finally, Q_2 is 0 in the present state and remains a 0 in the next state. Therefore, $J_2 = 0$ and $K_2 = X$. This analysis is repeated for each present state in Table 9–8.

Step 4: Karnaugh Maps

Karnaugh maps can be used to determine the logic required for the J and K inputs of each flip-flop in the counter. There is a Karnaugh map for the J input and a Karnaugh map for the K input of each flip-flop. In this design procedure, each cell in a Karnaugh map represents one of the present states in the counter sequence listed in Table 9–8.

From the J and K states in the transition table (Table 9–9) a 1, 0, or X is entered into each present-state cell on the maps depending on the transition of the Q output for a particular flip-flop. To illustrate this procedure, two sample entries are shown for the J_0 and the K_0 inputs to the least significant flip-flop (Q_0) in Figure 9–27.

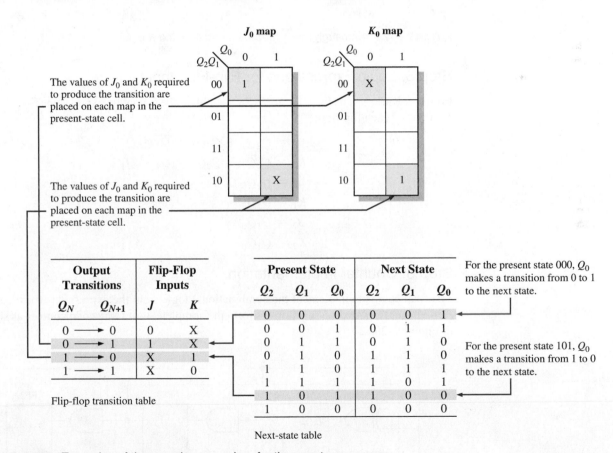

FIGURE 9–27 Examples of the mapping procedure for the counter sequence represented in Table 9–8 and Table 9–9.

The completed Karnaugh maps for all three flip-flops in the counter are shown in Figure 9–28. The cells are grouped as indicated and the corresponding Boolean expressions for each group are derived.

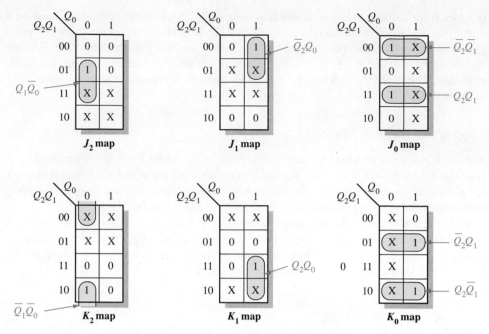

FIGURE 9–28   Karnaugh maps for present-state J and K inputs.

Step 5: Logic Expressions for Flip-Flop Inputs

From the Karnaugh maps of Figure 9–28 you obtain the following expressions for the J and K inputs of each flip-flop:

$$J_0 = Q_2Q_1 + \overline{Q}_2\overline{Q}_1 = \overline{Q_2 \oplus Q_1}$$
$$K_0 = Q_2\overline{Q}_1 + \overline{Q}_2Q_1 = Q_2 \oplus Q_1$$
$$J_1 = \overline{Q}_2Q_0$$
$$K_1 = Q_2Q_0$$
$$J_2 = Q_1\overline{Q}_0$$
$$K_2 = \overline{Q}_1\overline{Q}_0$$

Step 6: Counter Implementation

The final step is to implement the combinational logic from the expressions for the J and K inputs and connect the flip-flops to form the complete 3-bit Gray code counter as shown in Figure 9–29.

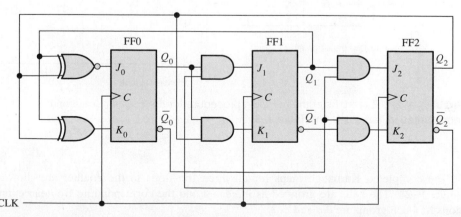

FIGURE 9–29   Three-bit Gray code counter. Open file F09-29 to verify operation.

A summary of steps used in the design of the 3-bit Gray code counter follows. In general, these steps can be applied to any state machine.

1. Specify the counter sequence and draw a state diagram.

2. Derive a next-state table from the state diagram.

3. Develop a transition table showing the flip-flop inputs required for each transition. The transition table is always the same for a given type of flip-flop.

4. Transfer the J and K states from the transition table to Karnaugh maps. There is a Karnaugh map for each input of each flip-flop.

5. Group the Karnaugh map cells to generate and derive the logic expression for each flip-flop input.

6. Implement the expressions with combinational logic, and combine with the flip-flops to create the counter.

This procedure is now applied to the design of other synchronous counters in Examples 9–4 and 9–5.

EXAMPLE 9–4

Design a counter with the irregular binary count sequence shown in the state diagram of Figure 9–30. Use D flip-flops.

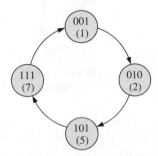

FIGURE 9–30

Solution

Step 1: The state diagram is as shown. Although there are only four states, a 3-bit counter is required to implement this sequence because the maximum binary count is seven. Since the required sequence does not include all the possible binary states, the invalid states (0, 3, 4, and 6) can be treated as "don't cares" in the design. However, if the counter should erroneously get into an invalid state, you must make sure that it goes back to a valid state.

Step 2: The next-state table is developed from the state diagram and is given in Table 9–10.

TABLE 9–10

Next-state table.

Present State			Next State		
Q_2	Q_1	Q_0	Q_2	Q_1	Q_0
0	0	1	0	1	0
0	1	0	1	0	1
1	0	1	1	1	1
1	1	1	0	0	1

Step 3: The transition table for the D flip-flop is shown in Table 9–11.

TABLE 9–11
Transition table for a D flip-flop.

Output Transitions			Flip-Flop Input
Q_N		Q_{N+1}	D
0	$\longrightarrow$	0	0
0	$\longrightarrow$	1	1
1	$\longrightarrow$	0	0
1	$\longrightarrow$	1	1

Step 4: The D inputs are plotted on the present-state Karnaugh maps in Figure 9–31. Also "don't cares" can be placed in the cells corresponding to the invalid states of 000, 011, 100, and 110, as indicated by the red Xs.

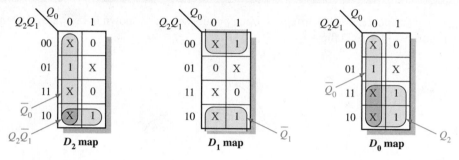

FIGURE 9–31

Step 5: Group the 1s, taking advantage of as many of the "don't care" states as possible for maximum simplification, as shown in Figure 9–31. The expression for each D input taken from the maps is as follows:

$$D_0 = \overline{Q}_0 + Q_2$$
$$D_1 = \overline{Q}_1$$
$$D_2 = \overline{Q}_0 + Q_2\overline{Q}_1$$

Step 6: The implementation of the counter is shown in Figure 9–32.

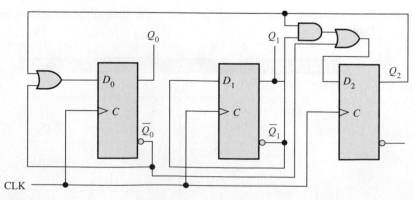

FIGURE 9–32

An analysis shows that if the counter, by accident, gets into one of the invalid states (0, 3, 4, 6), it will always return to a valid state according to the following sequences: $0 \rightarrow 3 \rightarrow 4 \rightarrow 7$, and $6 \rightarrow 1$.

Related Problem

Verify the analysis that proves the counter will always return (eventually) to a valid state from an invalid state.

EXAMPLE 9–5

Develop a synchronous 3-bit up/down counter with a Gray code sequence using J-K flip-flops. The counter should count up when an UP/$\overline{\text{DOWN}}$ control input is 1 and count down when the control input is 0.

Solution

Step 1: The state diagram is shown in Figure 9–33. The 1 or 0 beside each arrow indicates the state of the UP/$\overline{\text{DOWN}}$ control input, Y.

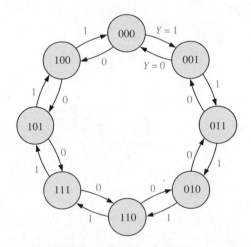

FIGURE 9–33 State diagram for a 3-bit up/down Gray code counter.

Step 2: The next-state table is derived from the state diagram and is shown in Table 9–12. Notice that for each present state there are two possible next states, depending on the UP/$\overline{\text{DOWN}}$ control variable, Y.

TABLE 9–12

Next-state table for 3-bit up/down Gray code counter.

			Next State					
Present State			$Y = 0$ (DOWN)			$Y = 1$ (UP)		
Q_2	Q_1	Q_0	Q_2	Q_1	Q_0	Q_2	Q_1	Q_0
0	0	0	1	0	0	0	0	1
0	0	1	0	0	0	0	1	1
0	1	1	0	0	1	0	1	0
0	1	0	0	1	1	1	1	0
1	1	0	0	1	0	1	1	1
1	1	1	1	1	0	1	0	1
1	0	1	1	1	1	1	0	0
1	0	0	1	0	1	0	0	0

$Y = $ UP/$\overline{\text{DOWN}}$ control input.

Step 3: The transition table for the J-K flip-flops is repeated in Table 9–13.

TABLE 9–13				
Transition table for a J-K flip-flop.				
Output Transitions			**Flip-Flop Inputs**	
Q_N		Q_{N+1}	J	K
0	$\longrightarrow$	0	0	X
0	$\longrightarrow$	1	1	X
1	$\longrightarrow$	0	X	1
1	$\longrightarrow$	1	X	0

Step 4: The Karnaugh maps for the J and K inputs of the flip-flops are shown in Figure 9–34. The UP/$\overline{\text{DOWN}}$ control input, Y, is considered one of the state variables along with Q_0, Q_1, and Q_2. Using the next-state table, the information in the "Flip-Flop Inputs" column of Table 9–13 is transferred onto the maps as indicated for each present state of the counter.

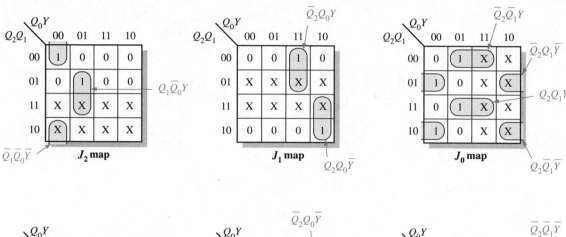

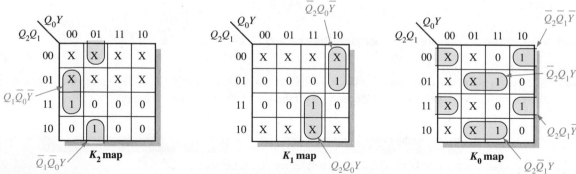

FIGURE 9–34 J and K maps for Table 9–12. The UP/$\overline{\text{DOWN}}$ control input, Y, is treated as a fourth variable.

Step 5: The 1s are combined in the largest possible groupings, with "don't cares" (Xs) used where possible. The groups are factored, and the expressions for the J and K inputs are as follows:

$$J_0 = Q_2Q_1Y + Q_2\overline{Q}_1\overline{Y} + \overline{Q}_2\overline{Q}_1Y + \overline{Q}_2Q_1\overline{Y} \qquad K_0 = \overline{Q}_2\overline{Q}_1\overline{Y} + \overline{Q}_2Q_1Y + Q_2\overline{Q}_1Y + Q_2Q_1\overline{Y}$$
$$J_1 = \overline{Q}_2Q_0Y + Q_2Q_0\overline{Y} \qquad\qquad\qquad K_1 = \overline{Q}_2Q_0\overline{Y} + Q_2Q_0Y$$
$$J_2 = Q_1\overline{Q}_0Y + \overline{Q}_1\overline{Q}_0\overline{Y} \qquad\qquad\qquad K_2 = Q_1\overline{Q}_0\overline{Y} + \overline{Q}_1\overline{Q}_0Y$$

Step 6: The J and K equations are implemented with combinational logic. This step is the Related Problem.

Related Problem

Specify the number of flip-flops, gates, and inverters that are required to implement the logic described in Step 5.

1. A flip-flop is presently in the RESET state and must go to the SET state on the next clock pulse. What must J and K be?

2. A flip-flop is presently in the SET state and must remain SET on the next clock pulse. What must J and K be?

3. A binary counter is in the $Q_3\overline{Q}_2Q_1\overline{Q}_0 = 1010$ state.

    (a) What is its next state?

    (b) What condition must exist on each flip-flop input to ensure that it goes to the proper next state on the clock pulse?

9–6 Cascaded Counters

Counters can be connected in cascade to achieve higher-modulus operation. In essence, **cascading** means that the last-stage output of one counter drives the input of the next counter.

After completing this section, you should be able to

- ◆ Determine the overall modulus of cascaded counters
- ◆ Analyze the timing diagram of a cascaded counter configuration
- ◆ Use cascaded counters as a frequency divider
- ◆ Use cascaded counters to achieve specified truncated sequences

Asynchronous Cascading

An example of two asynchronous counters connected in cascade is shown in Figure 9–35 for a 2-bit and a 3-bit ripple counter. The timing diagram is shown in Figure 9–36. Notice

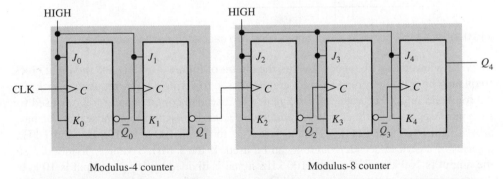

FIGURE 9–35 Two cascaded asynchronous counters (all J and K inputs are HIGH).

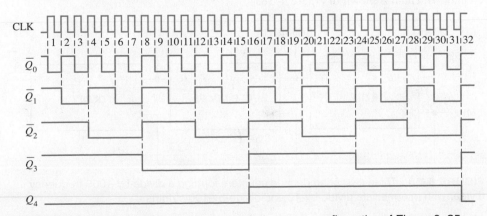

FIGURE 9–36 Timing diagram for the cascaded counter configuration of Figure 9–35.

The overall modulus of cascaded counters is equal to the product of the individual moduli.

that the final output of the modulus-8 counter, Q_4, occurs once for every 32 input clock pulses. The overall modulus of the two cascaded counters is $4 \times 8 = 32$; that is, they act as a divide-by-32 counter.

Synchronous Cascading

When operating synchronous counters in a cascaded configuration, it is necessary to use the count enable and the terminal count functions to achieve higher-modulus operation. On some devices the count enable is labeled simply *CTEN* (or some other designation such as *G*), and terminal count (*TC*) is analogous to ripple clock output (*RCO*) on some IC counters.

Figure 9–37 shows two decade counters connected in cascade. The terminal count (*TC*) output of counter 1 is connected to the count enable (*CTEN*) input of counter 2. Counter 2 is inhibited by the LOW on its *CTEN* input until counter 1 reaches its last, or terminal, state and its terminal count output goes HIGH. This HIGH now enables counter 2, so that when the first clock pulse after counter 1 reaches its terminal count (CLK10), counter 2 goes from its initial state to its second state. Upon completion of the entire second cycle of counter 1 (when counter 1 reaches terminal count the second time), counter 2 is again enabled and advances to its next state. This sequence continues. Since these are decade counters, counter 1 must go through ten complete cycles before counter 2 completes its first cycle. In other words, for every ten cycles of counter 1, counter 2 goes through one cycle. Thus, counter 2 will complete one cycle after one hundred clock pulses. The overall modulus of these two cascaded counters is $10 \times 10 = 100$.

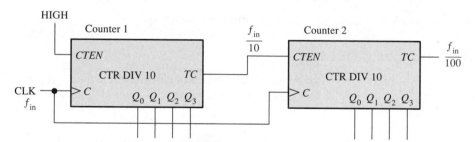

FIGURE 9–37 A modulus-100 counter using two cascaded decade counters.

When viewed as a frequency divider, the circuit of Figure 9–37 divides the input clock frequency by 100. Cascaded counters are often used to divide a high-frequency clock signal to obtain highly accurate pulse frequencies. Cascaded counter configurations used for such purposes are sometimes called *countdown chains*. For example, suppose that you have a basic clock frequency of 1 MHz and you wish to obtain 100 kHz, 10 kHz, and 1 kHz; a series of cascaded decade counters can be used. If the 1 MHz signal is divided by 10, the output is 100 kHz. Then if the 100 kHz signal is divided by 10, the output is 10 kHz. Another division by 10 produces the 1 kHz frequency. The general implementation of this countdown chain is shown in Figure 9–38.

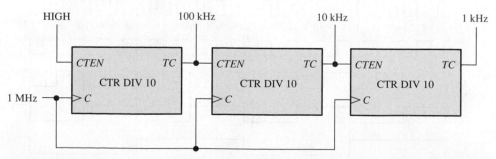

FIGURE 9–38 Three cascaded decade counters forming a divide-by-1000 frequency divider with intermediate divide-by-10 and divide-by-100 outputs.

EXAMPLE 9–6

Determine the overall modulus of the two cascaded counter configurations in Figure 9–39.

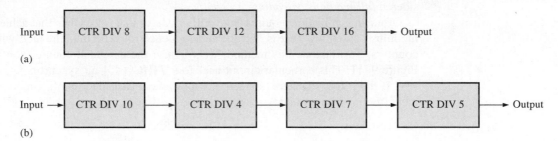

(a)

(b)

FIGURE 9–39

Solution

In Figure 9–39(a), the overall modulus for the 3-counter configuration is

$$8 \times 12 \times 16 = \mathbf{1536}$$

In Figure 9–39(b), the overall modulus for the 4-counter configuration is

$$10 \times 4 \times 7 \times 5 = \mathbf{1400}$$

Related Problem

How many cascaded decade counters are required to divide a clock frequency by 100,000?

EXAMPLE 9–7

Use 74HC190 up/down decade counters connected in the UP mode to obtain a 10 kHz waveform from a 1 MHz clock. Show the logic diagram.

Solution

To obtain 10 kHz from a 1 MHz clock requires a division factor of 100. Two 74HC190 counters must be cascaded as shown in Figure 9–40. The left counter produces a terminal count (*MAX/MIN*) pulse for every 10 clock pulses. The right counter produces a terminal count (*MAX/MIN*) pulse for every 100 clock pulses.

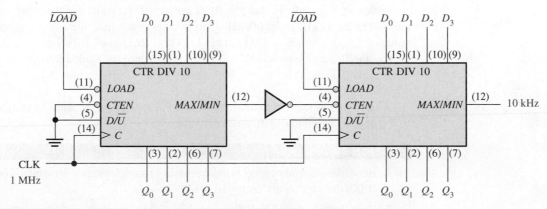

FIGURE 9–40   A divide-by-100 counter using two 74HC190 up/down decade counters connected for the up sequence.

Related Problem

Determine the frequency of the waveform at the Q_0 output of the second counter (the one on the right) in Figure 9–40.

Cascaded Counters with Truncated Sequences

The preceding discussion has shown how to achieve an overall modulus (divide-by-factor) that is the product of the individual moduli of all the cascaded counters. This can be considered *full-modulus cascading*.

Often an application requires an overall modulus that is less than that achieved by full-modulus cascading. That is, a truncated sequence must be implemented with cascaded counters. To illustrate this method, we will use the cascaded counter configuration in Figure 9–41. This particular circuit uses four 74HC161 4-bit synchronous binary counters. If these four counters (sixteen bits total) were cascaded in a full-modulus arrangement, the modulus would be

$$2^{16} = 65,536$$

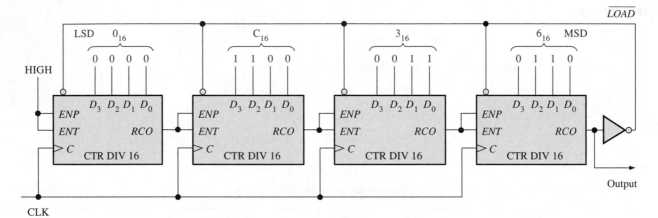

FIGURE 9–41   A divide-by-40,000 counter using 74HC161 4-bit binary counters. Note that each of the parallel data inputs is shown in binary order (the right-most bit D_0 is the LSB in each counter).

Let's assume that a certain application requires a divide-by-40,000 counter (modulus 40,000). The difference between 65,536 and 40,000 is 25,536, which is the number of states that must be *deleted* from the full-modulus sequence. The technique used in the circuit of Figure 9–41 is to preset the cascaded counter to 25,536 (63C0 in hexadecimal) each time it recycles, so that it will count from 25,536 up to 65,535 on each full cycle. Therefore, each full cycle of the counter consists of 40,000 states.

Notice in Figure 9–41 that the *RCO* output of the right-most counter is inverted and applied to the $\overline{LOAD}$ input of each 4-bit counter. Each time the count reaches its terminal value of 65,535, which is 11111111111111111_2, *RCO* goes HIGH and causes the number on the parallel data inputs ($63C0_{16}$) to be synchronously loaded into the counter with the clock pulse. Thus, there is one *RCO* pulse from the right-most 4-bit counter for every 40,000 clock pulses.

With this technique any modulus can be achieved by synchronous loading of the counter to the appropriate initial state on each cycle.

SECTION 9–6 CHECKUP

1. How many decade counters are necessary to implement a divide-by-1000 (modulus-1000) counter? A divide-by-10,000?

2. Show with general block diagrams how to achieve each of the following, using a flip-flop, a decade counter, and a 4-bit binary counter, or any combination of these:

    (a) Divide-by-20 counter   **(b)** Divide-by-32 counter

    (c) Divide-by-160 counter   **(d)** Divide-by-320 counter

9–7   Counter Decoding

In many applications, it is necessary that some or all of the counter states be decoded. The decoding of a counter involves using decoders or logic gates to determine when the counter is in a certain binary state in its sequence. For instance, the terminal count function previously discussed is a single decoded state (the last state) in the counter sequence.

After completing this section, you should be able to

- Implement the decoding logic for any given state in a counter sequence
- Explain why glitches occur in counter decoding logic
- Use the method of strobing to eliminate decoding glitches

Suppose that you wish to decode binary state 6 (110) of a 3-bit binary counter. When $Q_2 = 1$, $Q_1 = 1$, and $Q_0 = 0$, a HIGH appears on the output of the decoding gate, indicating that the counter is at state 6. This can be done as shown in Figure 9–42. This is called *active-HIGH decoding*. Replacing the AND gate with a NAND gate provides active-LOW decoding.

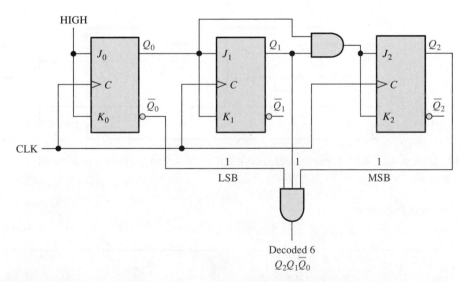

Decoded 6
$Q_2 Q_1 \overline{Q}_0$

FIGURE 9–42   Decoding of state 6 (110). Open file F09-42 to verify operation.   **MultiSim**

EXAMPLE 9–8

Implement the decoding of binary state 2 and binary state 7 of a 3-bit synchronous counter. Show the entire counter timing diagram and the output waveforms of the decoding gates. Binary 2 = $\overline{Q}_2 Q_1 \overline{Q}_0$ and binary 7 = $Q_2 Q_1 Q_0$.

Solution

See Figure 9–43. The 3-bit counter was originally discussed in Section 9–3 (Figure 9–15).

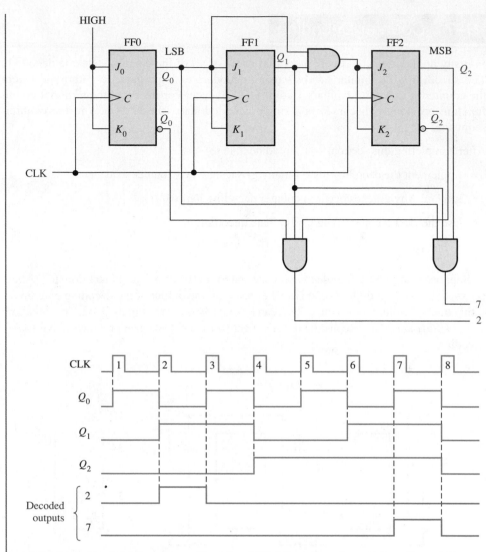

FIGURE 9–43 A 3-bit counter with active-HIGH decoding of count 2 and count 7. Open file F09-43 to verify operation.

Related Problem

Show the logic for decoding state 5 in the 3-bit counter.

Decoding Glitches

A glitch is an unwanted spike of voltage.

The problem of glitches produced by the decoding process was discussed in Chapter 6. As you have learned, the propagation delays due to the ripple effect in asynchronous counters create transitional states in which the counter outputs are changing at slightly different times. These transitional states produce undesired voltage spikes of short duration (glitches) on the outputs of a decoder connected to the counter. The glitch problem can also occur to some degree with synchronous counters because the propagation delays from the clock to the Q outputs of each flip-flop in a counter can vary slightly.

Figure 9–44 shows a basic asynchronous BCD decade counter connected to a BCD-to-decimal decoder. To see what happens in this case, let's look at a timing diagram in which the propagation delays are taken into account, as shown in Figure 9–45. Notice that these delays cause false states of short duration. The value of the false binary state at each critical transition is indicated on the diagram. The resulting glitches can be seen on the decoder outputs.

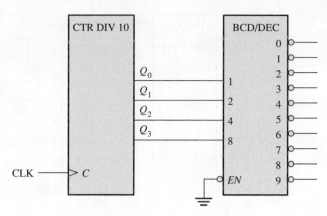

FIGURE 9–44 A basic decade (BCD) counter and decoder.

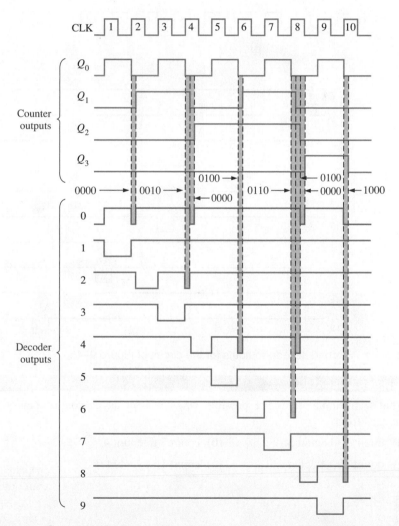

FIGURE 9–45 Outputs with glitches from the decoder in Figure 9–44. Glitch widths are exaggerated for illustration and are usually only a few nanoseconds wide.

One way to eliminate the glitches is to enable the decoded outputs at a time after the glitches have had time to disappear. This method is known as *strobing* and can be accomplished in the case of an active-HIGH clock by using the LOW level of the clock to enable the decoder, as shown in Figure 9–46. The resulting improved timing diagram is shown in Figure 9–47.

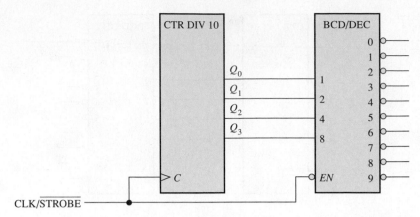

FIGURE 9–46   The basic decade counter and decoder with strobing to eliminate glitches.

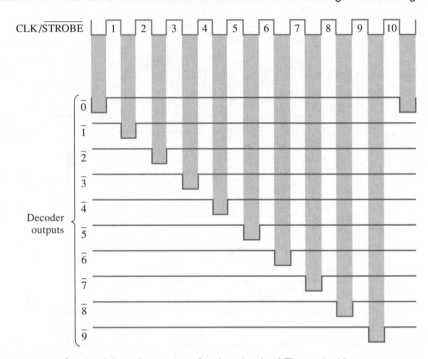

FIGURE 9–47   Strobed decoder outputs for the circuit of Figure 9–46.

SECTION 9–7 CHECKUP

1. What transitional states are possible when a 4-bit asynchronous binary counter changes from

   (a) count 2 to count 3   (b) count 3 to count 4

   (c) count 10_{10} to count 11_{10} (d) count 15 to count 0

9–8   Counter Applications

The digital counter is a useful and versatile device that is found in many applications. In this section, some representative counter applications are presented.

After completing this section, you should be able to

- Describe how counters are used in a basic digital clock system

- Explain how a divide-by-60 counter is implemented and how it is used in a digital clock

◆ Explain how the hours counter is implemented

◆ Discuss the application of a counter in an automobile parking control system

◆ Describe how a counter is used in the process of parallel-to-serial data conversion

A Digital Clock

A common example of a counter application is in timekeeping systems. Figure 9–48 is a simplified logic diagram of a digital clock that displays seconds, minutes, and hours. First, a 60 Hz sinusoidal ac voltage is converted to a 60 Hz pulse waveform and divided down to a 1 Hz pulse waveform by a divide-by-60 counter formed by a divide-by-10 counter followed by a divide-by-6 counter. Both the *seconds* and *minutes* counts are also produced by divide-by-60 counters, the details of which are shown in Figure 9–49. These counters count from 0 to 59 and then recycle to 0; synchronous decade counters are used in this particular implementation. Notice that the divide-by-6 portion is formed with a decade counter with a truncated sequence achieved by using the decoder count 6 to asynchronously clear the counter. The terminal count, 59, is also decoded to enable the next counter in the chain.

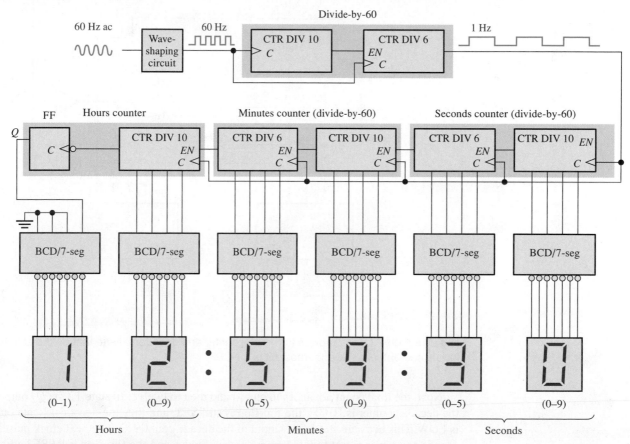

FIGURE 9–48 Simplified logic diagram for a 12-hour digital clock. Logic details using specific devices are shown in Figures 9–49 and 9–50.

The *hours* counter is implemented with a decade counter and a flip-flop as shown in Figure 9–50. Consider that initially both the decade counter and the flip-flop are RESET, and the decode-12 gate and decode-9 gate outputs are HIGH. The decade counter advances through all of its states from zero to nine, and on the clock pulse that recycles it from nine back to zero, the flip-flop goes to the SET state ($J = 1$, $K = 0$). This illuminates a 1 on the tens-of-hours display. The total count is now ten (the decade counter is in the zero state and the flip-flop is SET).

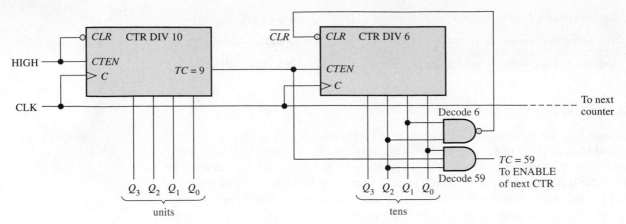

FIGURE 9–49 Logic diagram of typical divide-by-60 counter using synchronous decade counters. Note that the outputs are in binary order (the right-most bit is the LSB).

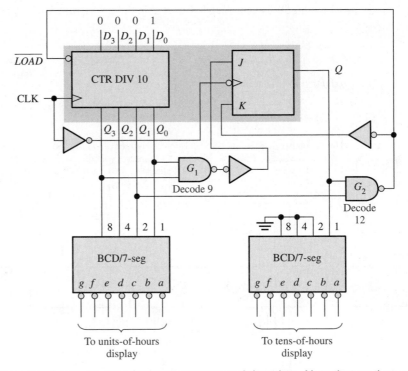

FIGURE 9–50 Logic diagram for hours counter and decoders. Note that on the counter inputs and outputs, the right-most bit is the LSB.

Next, the total count advances to eleven and then to twelve. In state 12 the Q_2 output of the decade counter is HIGH, the flip-flop is still SET, and thus the decode-12 gate output is LOW. This activates the $\overline{LOAD}$ input of the decade counter. On the next clock pulse, the decade counter is preset to 0001 from the data inputs, and the flip-flop is RESET ($J = 0$, $K = 1$). As you can see, this logic always causes the counter to recycle from twelve back to one rather than back to zero.

Automobile Parking Control

This counter example illustrates the use of an up/down counter to solve an everyday problem. The problem is to devise a means of monitoring available spaces in a one-hundred-space parking garage and provide for an indication of a full condition by illuminating a display sign and lowering a gate bar at the entrance.

A system that solves this problem consists of optoelectronic sensors at the entrance and exit of the garage, an up/down counter and associated circuitry, and an interface circuit that uses the counter output to turn the FULL sign on or off as required and lower or raise the gate bar at the entrance. A general block diagram of this system is shown in Figure 9–51.

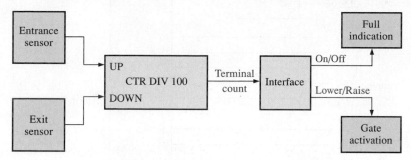

FIGURE 9–51   Functional block diagram for parking garage control.

A logic diagram of the up/down counter is shown in Figure 9–52. It consists of two cascaded up/down decade counters. The operation is described in the following paragraphs.

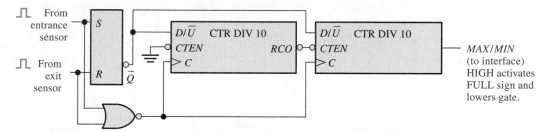

FIGURE 9–52   Logic diagram for modulus-100 up/down counter for automobile parking control.

The counter is initially preset to 0 using the parallel data inputs, which are not shown. Each automobile entering the garage breaks a light beam, activating a sensor that produces an electrical pulse. This positive pulse sets the S-R latch on its leading edge. The LOW on the $\overline{Q}$ output of the latch puts the counter in the UP mode. Also, the sensor pulse goes through the NOR gate and clocks the counter on the LOW-to-HIGH transition of its trailing edge. Each time an automobile enters the garage, the counter is advanced by one **(incremented)**. When the one-hundredth automobile enters, the counter goes to its last state (100_{10}). The *MAX/MIN* output goes HIGH and activates the interface circuit (no detail), which lights the FULL sign and lowers the gate bar to prevent further entry.

> Incrementing a counter increases its count by one.

When an automobile exits, an optoelectronic sensor produces a positive pulse, which resets the S-R latch and puts the counter in the DOWN mode. The trailing edge of the clock decreases the count by one **(decremented)**. If the garage is full and an automobile leaves, the *MAX/MIN* output of the counter goes LOW, turning off the FULL sign and raising the gate.

> Decrementing a counter decreases its count by one.

Parallel-to-Serial Data Conversion (Multiplexing)

A simplified example of data transmission using multiplexing and demultiplexing techniques was introduced in Chapter 6. Essentially, the parallel data bits on the multiplexer inputs are converted to serial data bits on the single transmission line. A group of bits appearing simultaneously on parallel lines is called *parallel data*. A group of bits appearing on a single line in a time sequence is called *serial data*.

Parallel-to-serial conversion is normally accomplished by the use of a counter to provide a binary sequence for the data-select inputs of a data selector/multiplexer, as illustrated in Figure 9–53. The Q outputs of the modulus-8 counter are connected to the data-select inputs of an 8-bit multiplexer.

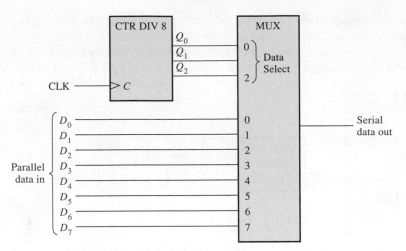

FIGURE 9–53   Parallel-to-serial data conversion logic.

Figure 9–54 is a timing diagram illustrating the operation of this circuit. The first byte (eight-bit group) of parallel data is applied to the multiplexer inputs. As the counter goes through a binary sequence from zero to seven, each bit, beginning with D_0, is sequentially

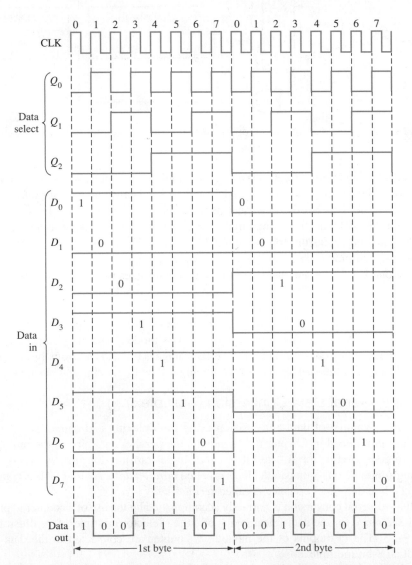

FIGURE 9–54   Example of parallel-to-serial conversion timing for the circuit in Figure 9–53.

selected and passed through the multiplexer to the output line. After eight clock pulses the data byte has been converted to a serial format and sent out on the transmission line. When the counter recycles back to 0, the next byte is applied to the data inputs and is sequentially converted to serial form as the counter cycles through its eight states. This process continues repeatedly as each parallel byte is converted to a serial byte.

SECTION 9–8 CHECKUP

1. Explain the purpose of each NAND gate in Figure 9–50.

2. Identify the two recycle conditions for the hours counter in Figure 9–48, and explain the reason for each.

9–9 Logic Symbols with Dependency Notation

Up to this point, the logic symbols with dependency notation specified in ANSI/IEEE Standard 91-1984 have been introduced on a limited basis. In many cases, the symbols do not deviate greatly from the traditional symbols. A significant departure does occur, however, for some devices, including counters and other more complex devices. Although we will continue to use primarily the more traditional symbols throughout this book, a brief coverage of logic symbols with dependency notation is provided. A specific IC counter is used as an example.

After completing this section, you should be able to

- Interpret logic symbols that include dependency notation
- Identify the common block and the individual elements of a counter symbol
- Interpret the qualifying symbol
- Discuss control dependency
- Discuss mode dependency
- Discuss AND dependency

Dependency notation is fundamental to the ANSI/IEEE standard. Dependency notation is used in conjunction with the logic symbols to specify the relationships of inputs and outputs so that the logical operation of a given device can be determined entirely from its logic symbol without a prior knowledge of the details of its internal structure and without a detailed logic diagram for reference. This coverage of a specific logic symbol with dependency notation is intended to aid in the interpretation of other such symbols that you may encounter in the future.

The 74HC163 4-bit synchronous binary counter is used for illustration. For comparison, Figure 9–55 shows a traditional block symbol and the ANSI/IEEE symbol with dependency notation. Basic descriptions of the symbol and the dependency notation follow.

Common Control Block

The upper block with notched corners in Figure 9–55(b) has inputs and an output that are considered common to all elements in the device and not unique to any one of the elements.

Individual Elements

The lower block in Figure 9–55(b), which is partitioned into four abutted sections, represents the four storage elements (D flip-flops) in the counter, with inputs D_0, D_1, D_2, and D_3 and outputs Q_0, Q_1, Q_2, and Q_3.

Qualifying Symbol

The label "CTR DIV 16" in Figure 9–55(b) identifies the device as a counter (CTR) with sixteen states (DIV 16).

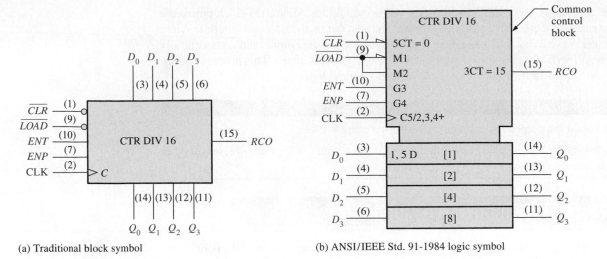

FIGURE 9–55   The 74HC163 4-bit synchronous counter.

Control Dependency (C)

As shown in Figure 9–55(b), the letter C denotes control dependency. Control inputs usually enable or disable the data inputs (D, J, K, S, and R) of a storage element. The C input is usually the clock input. In this case the digit 5 following C (C5/2,3,4+) indicates that the inputs labeled with a 5 prefix are dependent on the clock (synchronous with the clock). For example, 5CT = 0 on the $\overline{CLR}$ input indicates that the clear function is dependent on the clock; that is, it is a synchronous clear. When the $\overline{CLR}$ input is LOW (0), the counter is reset to zero ($CT = 0$) on the triggering edge of the clock pulse. Also, the 5 D label at the input of storage element [1] indicates that the data storage is dependent on (synchronous with) the clock. All labels in the [1] storage element apply to the [2], [4], and [8] elements below it since they are not labeled differently.

Mode Dependency (M)

As shown in Figure 9–55(b), the letter M denotes mode dependency. This label is used to indicate how the functions of various inputs or outputs depend on the mode in which the device is operating. In this case the device has two modes of operation. When the $\overline{LOAD}$ input is LOW (0), as indicated by the triangle input, the counter is in a preset mode (M1) in which the input data (D_0, D_1, D_2, and D_3) are synchronously loaded into the four flip-flops. The digit 1 following M in M1 and the 1 in the label 1, 5 D show a dependency relationship and indicate that input data are stored only when the device is in the preset mode (M1), in which $\overline{LOAD} = 0$. When the $\overline{LOAD}$ input is HIGH (1), the counter advances through its normal binary sequence, as indicated by M2 and the 2 in C5/2,3,4+.

AND Dependency (G)

As shown in Figure 9–55(b), the letter G denotes AND dependency, indicating that an input designated with G followed by a digit is ANDed with any other input or output having the same digit as a prefix in its label. In this particular example, the G3 at the ENT input and the 3CT = 15 at the RCO output are related, as indicated by the 3, and that relationship is an AND dependency, indicated by the G. This tells us that ENT must be HIGH (no triangle on the input) *and* the count must be fifteen ($CT = 15$) for the RCO output to be HIGH.

   Also, the digits 2, 3, and 4 in the label C5/2,3,4+ indicate that the counter advances through its states when $\overline{LOAD} = 1$, as indicated by the mode dependency label M2, and when $ENT = 1$ and $ENP = 1$, as indicated by the AND dependency labels G3 and G4. The + indicates that the counter advances by one count when these conditions exist.

1. In dependency notation, what do the letters *C*, *M*, and *G* stand for?
2. By what letter is data storage denoted?

9–10 Troubleshooting

The troubleshooting of counters can be simple or quite involved, depending on the type of counter and the type of fault. This section will give you some insight into how to approach the troubleshooting of sequential circuits.

After completing this section, you should be able to

◆ Detect a faulty counter

◆ Isolate faults in maximum-modulus cascaded counters

◆ Isolate faults in cascaded counters with truncated sequences

◆ Determine faults in counters implemented with individual flip-flops

Counters

The symptom for a faulty counter is usually that it does not advance its count. If this is the case, then check power and ground on the chip. Look at these lines with a scope to make sure there is no noise present (a noisy ground may actually be open). Check that there are clock pulses and that they have the correct amplitude and rise time and that there is not extraneous noise on the line. (Sometimes clock pulses can be loaded down by other ICs, making it appear that the counter is faulty when it is not). If power, ground, and the clock pulses are okay, check all inputs (including enable, load, and clear inputs), to see that they are connected correctly and that the logic is correct. An open input can cause a counter to work correctly some of the time—inputs should never be left open, even if they are not used. (An unused input should be connected to an inactive level). If the counter is stuck in a state and the clock is present, determine what input should be present to advance the counter. This may point to a faulty input (including clear or load inputs), which can be caused by logic elsewhere in the circuit. If inputs are all checked okay, an output may be pulled LOW or HIGH by an external short or open (or another faulty IC), keeping the output from advancing.

Cascaded Counters with Maximum Modulus

A failure in one of the counters in a chain of cascaded counters can affect all the counters that follow it. For example, if a count enable input opens, it effectively acts as a HIGH (for TTL logic), and the counter is always enabled. This type of failure in one of the counters will cause that counter to run at the full clock rate and will also cause all the succeeding counters to run at higher than normal rates. This is illustrated in Figure 9–56 for a divide-by-1000 cascaded counter arrangement where an open enable (*CTEN*) input acts as a TTL HIGH and continuously enables the second counter. Other faults that can affect "downstream" counter stages are open or shorted clock inputs or terminal count outputs. In some of these situations, pulse activity can be observed, but it may be at the wrong frequency. Exact frequency or frequency ratio measurements must be made.

Cascaded Counters with Truncated Sequences

The count sequence of a cascaded counter with a truncated sequence, such as that in Figure 9–57, can be affected by other types of faults in addition to those mentioned for maximum-modulus cascaded counters. For example, a failure in one of the parallel data inputs, the $\overline{LOAD}$ input, or the inverter can alter the preset count and thus change the modulus of the counter.

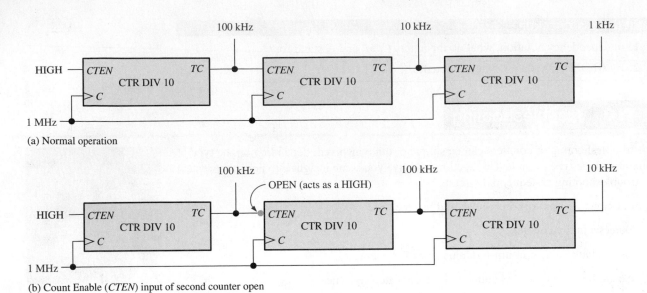

(a) Normal operation

(b) Count Enable (*CTEN*) input of second counter open

FIGURE 9–56   Example of a failure that affects following counters in a cascaded arrangement.

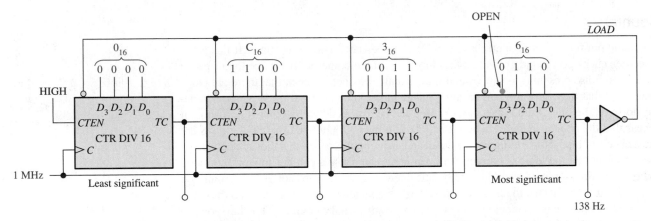

FIGURE 9–57   Example of a failure in a cascaded counter with a truncated sequence.

For example, suppose the D_3 input of the most significant counter in Figure 9–57 is open and acts as a HIGH. Instead of 6_{16} (0110) being preset into the counter, E_{16} (1110) is preset in. So, instead of beginning with $63C0_{16}$ ($25,536_{10}$) each time the counter recycles, the sequence will begin with $E3C0_{16}$ ($58,304_{10}$). This changes the modulus of the counter from 40,000 to $65,536 - 58,304 = 7232$.

To check this counter, apply a known clock frequency, for example 1 MHz, and measure the output frequency at the final terminal count output. If the counter is operating properly, the output frequency is

$$f_{\text{out}} = \frac{f_{\text{in}}}{\text{modulus}} = \frac{1\text{ MHz}}{40,000} = 25\text{ Hz}$$

In this case, the specific failure described in the preceding paragraph will cause the output frequency to be

$$f_{\text{out}} = \frac{f_{\text{in}}}{\text{modulus}} = \frac{1\text{ MHz}}{7232} \cong 138\text{ Hz}$$

EXAMPLE 9–9

Frequency measurements are made on the truncated counter in Figure 9–58 as indicated. Determine if the counter is working properly, and if not, isolate the fault.

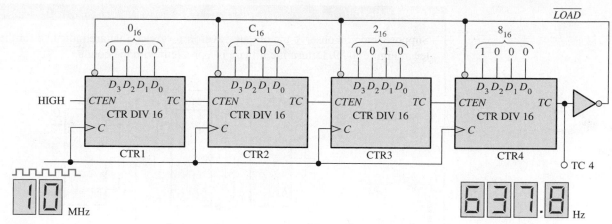

FIGURE 9–58

Solution

Check to see if the frequency measured at TC 4 is correct. If it is, the counter is working properly.

$$\text{truncated modulus} = \text{full modulus} - \text{preset count}$$
$$= 16^4 - 82C0_{16}$$
$$= 65{,}536 - 33{,}472 = 32{,}064$$

The correct frequency at TC 4 is

$$f_4 = \frac{10\text{ MHz}}{32{,}064} \cong 312\text{ Hz}$$

There is a problem. The measured frequency of 637.8 Hz does not agree with the correct calculated frequency of 312 Hz. To find the faulty counter, determine the actual truncated modulus as follows:

$$\text{modulus} = \frac{f_{\text{in}}}{f_{\text{out}}} = \frac{10\text{ MHz}}{637.8\text{ Hz}} = 15{,}679$$

Because the truncated modulus should be 32,064, most likely the counter is being preset to the wrong count when it recycles. The actual preset count is determined as follows:

$$\text{truncated modulus} = \text{full modulus} - \text{preset count}$$
$$\text{preset count} = \text{full modulus} - \text{truncated modulus}$$
$$= 65{,}536 - 15{,}679$$
$$= 49{,}857$$
$$= C2C0_{16}$$

This shows that the counter is being preset to $C2C0_{16}$ instead of $82C0_{16}$ each time it recycles.

Counters 1, 2, and 3 are being preset properly but counter 4 is not. Since $C_{16} = 1100_2$, the D_2 input to counter 4 is HIGH when it should be LOW. This is most likely caused by an **open input.** Check for an external open caused by a bad solder connection, a broken conductor, or a bent pin on the IC. If none can be found, replace the IC and the counter should work properly.

Related Problem

Determine what the output frequency at TC 4 would be if the D_3 input of counter 3 were open.

Counters Implemented with Individual Flip-Flops

Counters implemented with individual flip-flop and gate ICs are sometimes more difficult to troubleshoot because there are many more inputs and outputs with external connections than there are in an IC counter. The sequence of a counter can be altered by a single open or short on an input or output, as Example 9–10 illustrates.

EXAMPLE 9–10

Suppose that you observe the output waveforms (green) that are indicated for the counter in Figure 9–59. Determine if there is a problem with the counter.

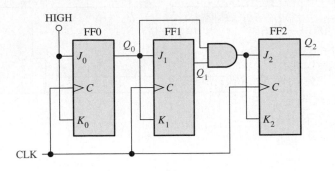

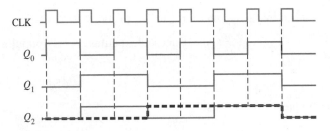

FIGURE 9–59

Solution

The Q_2 waveform is incorrect. The correct waveform is shown as a red dashed line. You can see that the Q_2 waveform looks exactly like the Q_1 waveform, so whatever is causing FF1 to toggle appears to also be controlling FF2.

Checking the J and K inputs to FF2, you find a waveform that looks like Q_0. This result indicates that Q_0 is somehow getting through the AND gate. The only way this can happen is if the Q_1 input to the AND gate is always HIGH. However, you have seen that Q_1 has a correct waveform. This observation leads to the conclusion that the lower input to the AND gate must be internally open and acting as a HIGH. Replace the AND gate and retest the circuit.

Related Problem

Describe the Q_2 output of the counter in Figure 9–59 if the Q_1 output of FF1 is open.

To observe the time relationship between two digital signals with a dual-trace analog oscilloscope, the proper way to trigger the scope is with the slower of the two signals. The reason for this is that the slower signal has fewer possible trigger points than the faster signal and there will be no ambiguity for starting the sweep. Vertical mode triggering uses a composite of both channels and should never be used for determining absolute time information. Since clock signals are usually the fastest signal in a digital system, they should not be used for triggering.

SECTION 9–10 CHECKUP

1. What failures can cause the counter in Figure 9–56 to have no pulse activity on any of the *TC* outputs?

2. What happens if the inverter in Figure 9–58 develops an open output?

Applied Logic

Elevator Controller: Part 1

This Applied Logic describes the operation and implementation of a service elevator controller for a seven-story building. The controller consists of logic that controls the elevator operation, a counter that determines the floor at which the elevator is located at any given time, and a floor number display. For simplicity, there is only one floor call and one floor request for each elevator cycle. A cycle occurs when the elevator is called to a given floor to pick up a passenger and the passenger is delivered to a requested floor. The elevator sequence for one cycle is shown in Figure 9–60.

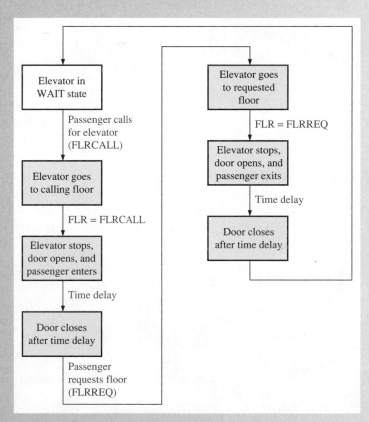

FIGURE 9–60 One cycle of the elevator operation.

The five states in the elevator control sequence are WAIT, DOWN, UP, STOP/OPEN, and CLOSE. In the WAIT state, the elevator is waiting on the last floor serviced for an external call button (FLRCALL) on any floor to be pressed. When there is a call for the elevator from any floor, the appropriate command (UP or DOWN) is issued. When the elevator arrives and stops at the calling floor, the door opens; the person enters and presses a number to request a destination floor. If the number of the requested floor is less than the number of the current floor, the elevator goes into the DOWN mode. If the number of the requested floor is greater than the number of the current floor, the elevator goes into the UP mode. The elevator goes to the STOP/OPEN mode at the requested floor to allow exit. After the door is open for a specified time, it closes and then goes back to the WAIT state until another floor call is received.

FIGURE 9–61 Elevator controller state diagram.

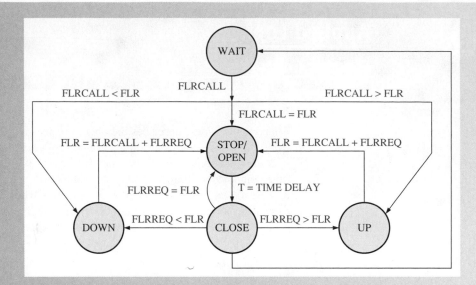

The following states are shown in the state diagram of Figure 9–61:

WAIT   The system always begins in the WAIT state on the floor last serviced. When a floor call (FLRCALL) signal is received, the control logic determines if the number of the calling floor is greater than the current floor (FLRCALL > FLR), less than the current floor (FLRCALL < FLR), or equal to the current floor (FLRCALL = FLR) and puts the system in the UP mode, DOWN mode, or OPEN mode, respectively.

DOWN   In this state, the elevator moves down toward the calling floor.

UP   In this mode, the elevator moves up toward the calling floor.

STOP/OPEN   This state occurs when the calling floor has been reached. When the number of the floor where the elevator is equals the number of the calling or requested floor, a signal is issued to stop the elevator and open the door.

CLOSE   After a preset time (T) to allow entry or exit, the door closes.

The signals used by the elevator controller are defined as follows:

FLR   Number of floor represented by a 3-bit binary code.

Floor sensor pulse   A pulse issued at each floor to clock the floor counter to the next state.

FLRCALL   Number of floor where a call for elevator service originates, represented by a 3-bit binary code.

Call pulse   A pulse issued in conjunction with FLRCALL to clock the 3-bit code into a register.

FLRREQ   Number of floor to which the passenger desires to go, represented by a 3-bit binary code.

Request pulse   A pulse issued in conjunction with FLRREQ to clock the 3-bit code into a register.

UP   A signal issued to the elevator motor control to cause the elevator to move from a lower floor to a higher floor.

DOWN   A signal issued to the elevator motor control to cause the elevator to move from a higher floor to a lower floor.

STOP   A signal issued to the elevator motor control to cause the elevator to stop.

OPEN   A signal issued to door motor control to cause the door to open.

CLOSE   A signal issued to the door motor control to cause the door to close.

Elevator Controller Block Diagram

Figure 9–62 shows the elevator controller block diagram, which consists of controller logic, a floor counter, and a floor number display. Assume that the elevator is on the first floor in

FIGURE 9–62 Elevator
controller block diagram.

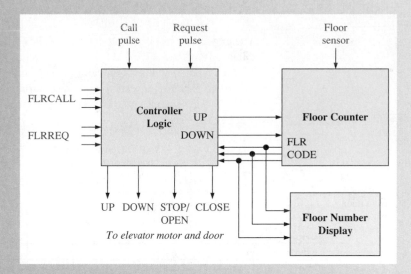

the WAIT state. The floor counter contains 001, which is the first floor code. Suppose the
FLRCALL (101) comes in from the call button on the fifth floor. Since FLRCALL > FLR
(101 > 001), the controller issues an UP command to the elevator motor. As the elevator moves
up, the floor counter receives a floor sensor pulse as it reaches each floor which advances its state
(001, 010, 011, 100, 101). When the fifth floor is reached and FLR = FLRCALL, the controller
logic stops the elevator and opens its door. The process is repeated for a FLRREQ input.

The floor counter sequentially tracks the number of the floor and always contains the
number of the current floor. It can count up or down and can reverse its state at any point
under the direction of the state controller and the floor sensor input. A 3-bit counter is required since there are eight floors ($2^3 = 8$) including the basement, as shown in the floor
counter state diagram in Figure 9–63.

FIGURE 9–63 Floor counter
state diagram.

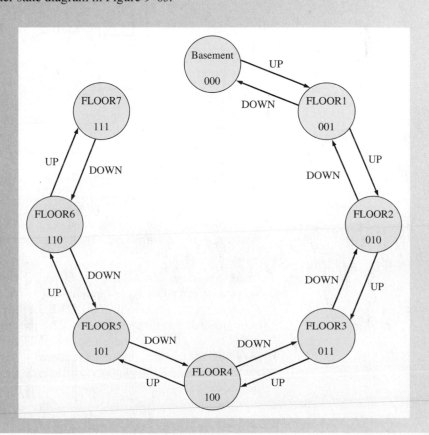

Operation of Elevator Controller

The elevator controller logic diagram is shown in Figure 9–64. Elevator action is initiated by either a floor call (FLRCALL) or a floor request (FLRREQ). Keep in mind that FLRCALL is when a person calls the elevator to come to a particular floor. FLRREQ is when a passenger in the elevator requests to go to a specified floor. This simplified operation is based on a CALL/REQ sequence; that is, a call followed by a request followed by a call.

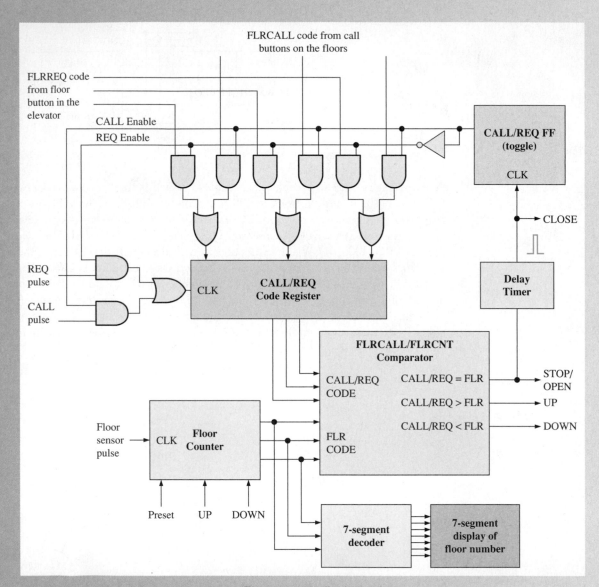

FIGURE 9–64 Elevator controller logic diagram.

As you know, FLRCALL and FLRREQ are 3-bit codes representing specific floors. When a person presses a call button on a given floor, the specific 3-bit code for that floor is placed on the inputs to the CALL/REQ code register and a CALL pulse is generated to enter the code into the register. The same process occurs when a request button is pressed inside the elevator. The code is input to the CALL/REQ code register, and a REQ pulse is generated to store the code in the register.

The elevator does not know the difference between a call and a request. The comparator determines if the destination floor number is greater than, less than, or equal to the current floor where the elevator is located. As a result of this comparison, either an UP command, a DOWN command, or an OPEN command is issued to the elevator motor control. As the elevator moves toward the desired floor, the floor counter is either incremented at each floor as it goes up or decremented at each floor as it goes down. Once the elevator reaches the desired floor, a STOP/OPEN command is issued to the elevator motor control and to the door control. After a preset time, the delay timer issues a CLOSE signal to the elevator door control. As mentioned, this elevator design is limited to one floor call and one floor request per cycle.

Initialization   The initial one-time setup requires that the elevator be placed at the basement level and the floor counter be preset to 000. After this, the counter will automatically move through the sequence of states determined by the elevator position.

Exercise

1. Explain the purpose of the floor counter.
2. Describe what happens during the WAIT mode.
3. How does the system know when the desired floor has been reached?
4. Discuss the limitations of the elevator design in Figure 9–64.

Implementation

The elevator controller can be implemented using fixed-function logic devices, a PLD programmed with a VHDL (or Verilog) code, or a programmed microcontroller or microprocessor. In the Chapter 10 Applied Logic, the VHDL program code for the elevator controller is presented. You will see how to program a PLD step by step.

Putting Your Knowledge to Work

What changes are required in the logic diagram of Figure 9–64 to upgrade the elevator controller for a ten-story building?

SUMMARY

- Asynchronous and synchronous counters differ only in the way in which they are clocked. The first stage of an asynchronous counter is driven by a clock pulse. Each succeeding stage is clocked by the output of the previous stage. In a synchronous counter, all stages are clocked by the same clock pulse. Synchronous counters can run at faster clock rates than asynchronous counters.
- The maximum modulus of a counter is the maximum number of possible states and is a function of the number of stages (flip-flops). Thus,

$$\text{Maximum modulus} = 2^n$$

where n is the number of stages in the counter. The modulus of a counter is the *actual* number of states in its sequence and can be equal to or less than the maximum modulus.
- The overall modulus of cascaded counters is equal to the product of the moduli of the individual counters.

KEY TERMS

Key terms and other bold terms in the chapter are defined in the end-of-book glossary.

Asynchronous   Not occurring at the same time.

Cascade   To connect "end-to-end" as when several counters are connected from the terminal count output of one counter to the enable input of the next counter.

Decade   Characterized by ten states or values.

Modulus   The number of unique states through which a counter will sequence.

Recycle   To undergo transition (as in a counter) from the final or terminal state back to the initial state.

State diagram   A graphic depiction of a sequence of states or values.

State machine   A logic system or circuit exhibiting a sequence of states conditioned by internal logic and external inputs; any sequential circuit exhibiting a specified sequence of states. Two types of state machine are Moore and Mealy.

Synchronous   Occurring at the same time.

Terminal count   The final state in a counter's sequence.

TRUE/FALSE QUIZ

Answers are at the end of the chapter.

1. In an asynchronous counter, all flip-flops change state at the same time.

2. In a synchronous counter, all flip-flops are clocked simultaneously.

3. An asynchronous counter is also known as a ripple counter.

4. A decade counter has sixteen states.

5. A counter with four stages has a maximum modulus of sixteen.

6. To achieve a maximum modulus of 32, sixteen stages are required.

7. If the present state is 1000, the next state of a 4-bit up/down counter in the DOWN mode is 0111.

8. Two cascaded decade counters divide the clock frequency by 20.

9. A counter with a truncated sequence has less than its maximum number of states.

10. To achieve a modulus of 100, ten decade counters are required.

SELF-TEST

Answers are at the end of the chapter.

1. The output of a Moore machine depends only on its
   - (a) inputs
   - (b) next state
   - (c) present state
   - (d) number of states

2. The output of a Mealy machine depends on its
   - (a) inputs
   - (b) next state
   - (c) present state
   - (d) answers (a) and (c)

3. Asynchronous counters are known as
   - (a) ripple counters
   - (b) multiple clock counters
   - (c) decade counters
   - (d) modulus counters

4. An asynchronous counter differs from a synchronous counter in
   - (a) the number of states in its sequence
   - (b) the method of clocking
   - (c) the type of flip-flops used
   - (d) the value of the modulus

5. The modulus of a counter is
   - (a) the number of flip-flops
   - (b) the actual number of states in its sequence
   - (c) the number of times it recycles in a second
   - (d) the maximum possible number of states

6. A 3-bit binary counter has a maximum modulus of
   - (a) 3   (b) 6   (c) 8   (d) 16

7. A 4-bit binary counter has a maximum modulus of
   - (a) 16   (b) 32   (c) 8   (d) 4

8. A modulus-12 counter must have
   - (a) 12 flip-flops
   - (b) 3 flip-flops
   - (c) 4 flip-flops
   - (d) synchronous clocking

9. Which one of the following is an example of a counter with a truncated modulus?
   - (a) Modulus 8
   - (b) Modulus 14
   - (c) Modulus 16
   - (d) Modulus 32

10. A 4-bit ripple counter consists of flip-flops that each have a propagation delay from clock to Q output of 12 ns. For the counter to recycle from 1111 to 0000, it takes a total of
    - (a) 12 ns
    - (b) 24 ns
    - (c) 48 ns
    - (d) 36 ns

11. A BCD counter is an example of
    - (a) a full-modulus counter
    - (b) a decade counter
    - (c) a truncated-modulus counter
    - (d) answers (b) and (c)

12. Which of the following is an invalid state in an 8421 BCD counter?
    - (a) 1100
    - (b) 0010
    - (c) 0101
    - (d) 1000

13. Three cascaded modulus-10 counters have an overall modulus of
    - (a) 30
    - (b) 100
    - (c) 1000
    - (d) 10,000

14. A 10 MHz clock frequency is applied to a cascaded counter consisting of a modulus-5 counter, a modulus-8 counter, and two modulus-10 counters. The lowest output frequency possible is
    - (a) 10 kHz
    - (b) 2.5 kHz
    - (c) 5 kHz
    - (d) 25 kHz

15. A 4-bit binary up/down counter is in the binary state of zero. The next state in the DOWN mode is
    - (a) 0001
    - (b) 1111
    - (c) 1000
    - (d) 1110

16. The terminal count of a modulus-13 binary counter is
    - (a) 0000
    - (b) 1111
    - (c) 1101
    - (d) 1100

PROBLEMS

Answers to odd-numbered problems are at the end of the book.

Section 9–1 Finite State Machines

1. Represent a decade counter with the terminal state decoded as a state machine. Identify the type and show the block diagram and the state diagram.

2. Identify the type of state machine for the traffic signal controller in Chapter 6. State the reason why it is the type you specified.

Section 9–2 Asynchronous Counters

3. For the ripple counter shown in Figure 9–65, show the complete timing diagram for eight clock pulses, showing the clock, Q_0, and Q_1 waveforms.

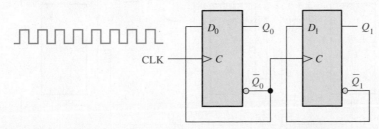

FIGURE 9–65

4. For the ripple counter in Figure 9–66, show the complete timing diagram for sixteen clock pulses. Show the clock, Q_0, Q_1, and Q_2 waveforms.

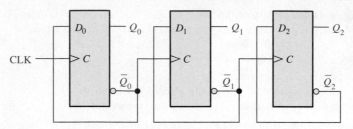

FIGURE 9–66

5. In the counter of Problem 4, assume that each flip-flop has a propagation delay from the triggering edge of the clock to a change in the Q output of 8 ns. Determine the worst-case (longest) delay time from a clock pulse to the arrival of the counter in a given state. Specify the state or states for which this worst-case delay occurs.

6. Show how to connect a 74HC93 4-bit asynchronous counter for each of the following moduli:

   (a) 9   **(b)** 11   **(c)** 13   **(d)** 14   **(e)** 15

Section 9–3 Synchronous Counters

7. If the counter of Problem 5 were synchronous rather than asynchronous, what would be the longest delay time?

8. Show the complete timing diagram for the 5-stage synchronous binary counter in Figure 9–67. Verify that the waveforms of the Q outputs represent the proper binary number after each clock pulse.

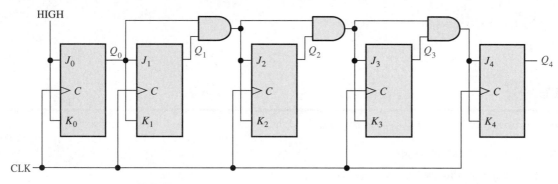

FIGURE 9–67

9. By analyzing the J and K inputs to each flip-flop prior to each clock pulse, prove that the decade counter in Figure 9–68 progresses through a BCD sequence. Explain how these conditions in each case cause the counter to go to the next proper state.

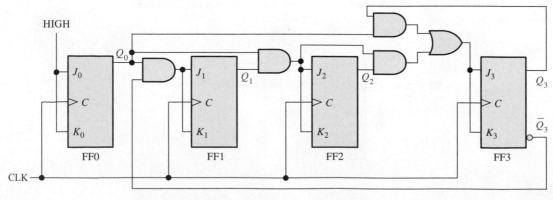

FIGURE 9–68

10. The waveforms in Figure 9–69 are applied to the count enable, clear, and clock inputs as indicated. Show the counter output waveforms in proper relation to these inputs. The clear input is asynchronous.

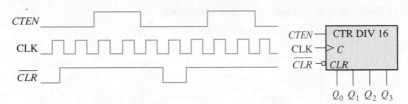

FIGURE 9–69

11. A BCD decade counter is shown in Figure 9–70. The waveforms are applied to the clock and clear inputs as indicated. Determine the waveforms for each of the counter outputs (Q_0, Q_1, Q_2, and Q_3). The clear is synchronous, and the counter is initially in the binary 1000 state.

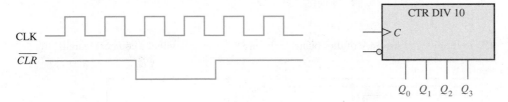

FIGURE 9–70

12. The waveforms in Figure 9–71 are applied to a 74HC163 binary counter. Determine the Q outputs and the RCO. The inputs are $D_0 = 1$, $D_1 = 1$, $D_2 = 0$, and $D_3 = 1$.

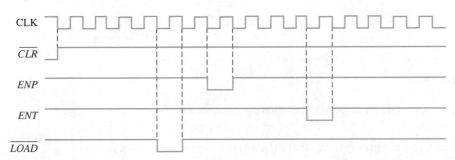

FIGURE 9–71

13. The waveforms in Figure 9–71 are applied to a 74HC161 counter. Determine the Q outputs and the RCO. The inputs are $D_0 = 1$, $D_1 = 0$, $D_2 = 0$, and $D_3 = 1$.

Section 9–4 Up/Down Synchronous Counters

14. Show a complete timing diagram for a 3-bit up/down counter that goes through the following sequence. Indicate when the counter is in the UP mode and when it is in the DOWN mode. Assume positive edge-triggering.

$$0, 1, 2, 3, 2, 1, 2, 3, 4, 5, 6, 5, 4, 3, 2, 1, 0$$

15. Develop the Q output waveforms for a 74HC190 up/down counter with the input waveforms shown in Figure 9–72. A binary 0 is on the data inputs. Start with a count of 0000.

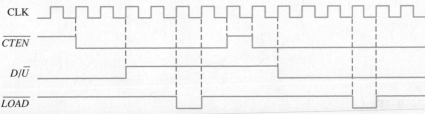

FIGURE 9–72

16. Repeat Problem 15 if the $D/\overline{U}$ input signal is inverted with the other inputs the same.

17. Repeat Problem 15 if the $\overline{CTEN}$ is inverted with the other inputs the same.

Section 9–5 Design of Synchronous Counters

18. Determine the sequence of the counter in Figure 9–73.

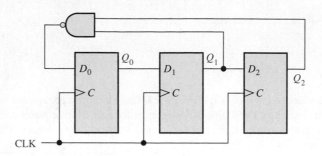

FIGURE 9–73

19. Determine the sequence of the counter in Figure 9–74. Begin with the counter cleared.

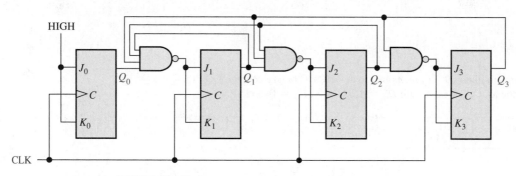

FIGURE 9–74

20. Design a counter to produce the following sequence. Use J-K flip-flops.

$$00, 10, 01, 11, 00, \ldots$$

21. Design a counter to produce the following binary sequence. Use J-K flip-flops.

$$1, 4, 3, 5, 7, 6, 2, 1, \ldots$$

22. Design a counter to produce the following binary sequence. Use J-K flip-flops.

$$0, 9, 1, 8, 2, 7, 3, 6, 4, 5, 0, \ldots$$

23. Design a binary counter with the sequence shown in the state diagram of Figure 9–75.

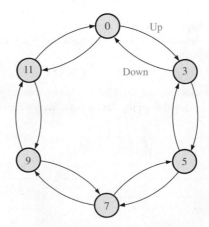

FIGURE 9–75

Section 9–6 Cascaded Counters

24. For each of the cascaded counter configurations in Figure 9–76, determine the frequency of the waveform at each point indicated by a circled number, and determine the overall modulus.

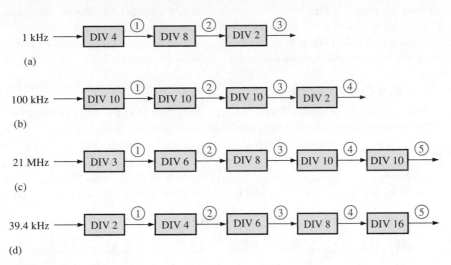

(a)

(b)

(c)

(d)

FIGURE 9–76

25. Expand the counter in Figure 9–38 to create a divide-by-10,000 counter and a divide-by-100,000 counter.

26. With general block diagrams, show how to obtain the following frequencies from a 10 MHz clock by using single flip-flops, modulus-5 counters, and decade counters:

(a) 5 MHz   (b) 2.5 MHz   (c) 2 MHz   (d) 1 MHz   (e) 500 kHz
(f) 250 kHz   (g) 62.5 kHz   (h) 40 kHz   (i) 10 kHz   (j) 1 kHz

Section 9–7 Counter Decoding

27. Given a BCD decade counter with only the Q outputs available, show what decoding logic is required to decode each of the following states and how it should be connected to the counter. A HIGH output indication is required for each decoded state. The MSB is to the left.

(a) 0001   (b) 0011   (c) 0101   (d) 0111   (e) 1000

28. For the 4-bit binary counter connected to the decoder in Figure 9–77, determine each of the decoder output waveforms in relation to the clock pulses.

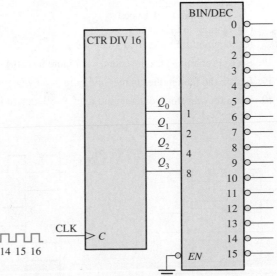

FIGURE 9–77

29. If the counter in Figure 9–77 is asynchronous, determine where the decoding glitches occur on the decoder output waveforms.

30. Modify the circuit in Figure 9–77 to eliminate decoding glitches.

31. Analyze the counter in Figure 9–42 for the occurrence of glitches on the decode gate output. If glitches occur, suggest a way to eliminate them.

32. Analyze the counter in Figure 9–43 for the occurrence of glitches on the outputs of the decoding gates. If glitches occur, make a design change that will eliminate them.

Section 9–8 Counter Applications

33. Assume that the digital clock of Figure 9–48 is initially reset to 12 o'clock. Determine the binary state of each counter after sixty-two 60 Hz pulses have occurred.

34. What is the output frequency of each counter in the digital clock circuit of Figure 9–48?

35. For the automobile parking control system in Figure 9–51, a pattern of entrance and exit sensor pulses during a given 24-hour period are shown in Figure 9–78. If there were 53 cars already in the garage at the beginning of the period, what is the state of the counter at the end of the 24 hours?

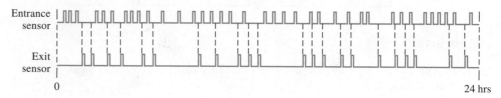

FIGURE 9–78

36. The binary number for decimal 57 appears on the parallel data inputs of the parallel-to-serial converter in Figure 9–53 (D_0 is the LSB). The counter initially contains all zeros and a 10 kHz clock is applied. Develop the timing diagram showing the clock, the counter outputs, and the serial data output.

Section 9–10 Troubleshooting

37. For the counter in Figure 9–4, show the timing diagram for the Q_0 and Q_1 waveforms for each of the following faults (assume Q_0 and Q_1 are initially LOW):

    (a) clock input to FF0 shorted to ground

    (b) Q_0 output open

    (c) clock input to FF1 open

    (d) D input to FF0 open

    (e) D input to FF1 shorted to ground

38. Solve Problem 37 for the counter in Figure 9–12(b).

39. Isolate the fault in the counter in Figure 9–6 by analyzing the waveforms in Figure 9–79.

40. From the waveform diagram in Figure 9–80, determine the most likely fault in the counter of Figure 9–15.

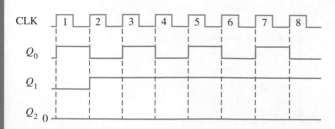

FIGURE 9–79

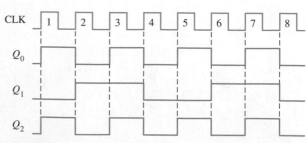

FIGURE 9–80

41. Solve Problem 40 if the Q_2 output has the waveform observed in Figure 9–81. Outputs Q_0 and Q_1 are the same as in Figure 9–80.

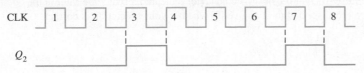

FIGURE 9–81

42. You apply a 5 MHz clock to the cascaded counter in Figure 9–41 and measure a frequency of 76.2939 Hz at the last *RCO* output. Is this correct, and if not, what is the most likely problem?

43. Develop a table for use in testing the counter in Figure 9–41 that will show the frequency at the final *RCO* output for all possible open failures of the parallel data inputs (D_0, D_1, D_2, and D_3) taken one at a time. Use 10 MHz as the test frequency for the clock.

44. The tens-of-hours 7-segment display in the digital clock system of Figure 9–48 continuously displays a 1. All the other digits work properly. What could be the problem?

45. What would be the visual indication of an open Q_1 output in the tens portion of the minutes counter in Figure 9–48? Also see Figure 9–49.

46. One day (perhaps a Monday) complaints begin flooding in from patrons of a parking garage that uses the control system depicted in Figures 9–51 and 9–52. The patrons say that they enter the garage because the gate is up and the FULL sign is off but that, once in, they can find no empty space. As the technician in charge of this facility, what do you think the problem is, and how will you troubleshoot and repair the system as quickly as possible?

Applied Logic

47. Propose a general design for generation of the 3-bit FLRCALL code and the Call pulse by the pressing of a single button.

48. Propose a general design for generation of the 3-bit FLRREQ code and the Request pulse by the pressing of one of seven buttons.

49. What changes are required to the logic diagram in Figure 9–64 to modify the elevator controller for a four-story building?

Special Design Problems

50. Design a modulus-1000 counter by using decade counters.

51. Modify the design of the counter in Figure 9–41 to achieve a modulus of 30,000.

52. Repeat Problem 51 for a modulus of 50,000.

53. Modify the digital clock in Figures 9–48, 9–49, and 9–50 so that it can be preset to any desired time.

54. Design an alarm circuit for the digital clock that can detect a predetermined time (hours and minutes only) and produce a signal to activate an audio alarm.

55. Modify the design of the circuit in Figure 9–52 for a 1000-space parking garage and a 3000-space parking garage.

56. Implement the parallel-to-serial data conversion logic in Figure 9–53 with specific fixed-function devices.

57. In Problem 19 it was found that the counter locks up and alternates between two states. It turns out that this operation is the result of a design flaw. Redesign the counter so that when it goes into the second of the lock-up states, it will recycle to the all-0s state on the next clock pulse.

Multisim Troubleshooting Practice

MultiSim

58. Open file P09-58. For the specified fault, predict the effect on the circuit. Then introduce the fault and verify whether your prediction is correct.

59. Open file P09-59. For the specified fault, predict the effect on the circuit. Then introduce the fault and verify whether your prediction is correct.

60. Open file P09-60. For the specified fault, predict the effect on the circuit. Then introduce the fault and verify whether your prediction is correct.

61. Open file P09-61. For the observed behavior indicated, predict the fault in the circuit. Then introduce the suspected fault and verify whether your prediction is correct.

62. Open file P09-62. For the observed behavior indicated, predict the fault in the circuit. Then introduce the suspected fault and verify whether your prediction is correct.

ANSWERS

SECTION CHECKUPS

Section 9–1 Checkup

1. A finite state machine is a sequential circuit having a finite number of states that occur in a specified order.

2. Moore state machine and Mealy state machine

3. The Moore state machine has an output(s) that is dependent on the present internal state only. The Mealy state machine has an output(s) that is dependent on both the present internal state and the value of the inputs.

Section 9–2 Asynchronous Counters

1. Asynchronous means that each flip-flop after the first one is enabled by the output of the preceding flip-flop.

2. A modulus-14 counter has fourteen states requiring four flip-flops.

Section 9–3 Synchronous Counters

1. All flip-flops in a synchronous counter are clocked simultaneously.

2. The counter can be preset (initialized) to any given state.

3. Counter is enabled when *ENP* and *ENT* are both HIGH; *RCO* goes HIGH when final state in sequence is reached.

Section 9–4 Up/Down Synchronous Counters

1. The counter goes to 1001.

2. UP: 1111: DOWN: 0000; the next state is 1111.

Section 9–5 Design of Synchronous Counters

1. $J = 1, K = X$ ("don't care")

2. $J = X$ ("don't care"), $K = 0$

3. **(a)** The next state is 1011.

(b) Q_3 (MSB): no-change or SET; Q_2: no-change or RESET; Q_1: no change or SET; Q_0 (LSB): SET or toggle

Section 9–6 Cascaded Counters

1. Three decade counters produce ÷ 1000; 4 decade counters produce ÷ 10,000.

2. **(a)** ÷ 20: flip-flop and DIV 10

(b) ÷ 32: flip-flop and DIV 16

(c) ÷ 160: DIV 16 and DIV 10

(d) ÷ 320: DIV 16 and DIV 10 and flip-flop

Section 9–7 Counter Decoding

1. **(a)** No transitional states because there is a single bit change

(b) 0000, 0001, 0010, 0101, 0110, 0111

(c) No transitional states because there is a single bit change

(d) 0001, 0010, 0011, 0100, 0101, 0110, 0111, 1000, 1001, 1010, 1011, 1100, 1101, 1110

Section 9–8 Counter Applications

1. Gate G_1 resets flip-flop on first clock pulse after count 9. Gate G_2 decodes count 12 to preset counter to 0001.

2. The hours decade counter advances through each state from zero to nine, and as it recycles from nine back to zero, the flip-flop is toggled to the SET state. This produces a ten (10) on the display. When the hours decade counter is in state 12, the decode NAND gate causes the counter to recycle to state 1 on the next clock pulse. The flip-flop resets. This results in a one (01) on the display.

Section 9–9 Logic Symbols with Dependency Notation

1. *C:* control, usually clock; *M:* mode; *G:* AND

2. *D* indicates data storage.

Section 9–10 Troubleshooting

1. No pulses on *TC* outputs: *CTEN* of first counter shorted to ground or to a LOW; clock input of first counter open; clock line shorted to ground or to a LOW; *TC* output of first counter shorted to ground or to a LOW.

2. With inverter output open, the counter does not recycle at the preset count but acts as a full-modulus counter.

RELATED PROBLEMS FOR EXAMPLES

9–1   See Figure 9–82.

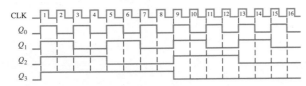

FIGURE 9–82

9–2   Connect Q_0 to the NAND gate as a third input (Q_2 and Q_3 are two of the inputs). Connect the $\overline{CLR}$ line to the $\overline{CLR}$ input of FF0 as well as FF2 and FF3.

9–3   See Figure 9–83.

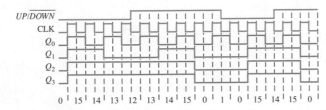

FIGURE 9–83

9–4   See Table 9–14.

TABLE 9–14

Present Invalid State			D Inputs			Next State			
Q_2	Q_1	Q_0	D_2	D_1	D_0	Q_2	Q_1	Q_0	
0	0	0	1	1	1	1	1	1	valid state
0	1	1	0	0	0	0	0	0	
1	0	0	1	1	1	1	1	1	
1	1	0	1	0	1	1	0	1	valid state

$000 \rightarrow 111$
$011 \rightarrow 000 \rightarrow 111$
$100 \rightarrow 111$
$110 \rightarrow 101$

9–5 Three flip-flops, sixteen 3-input AND gates, two 4-input OR gates, four 2-input OR gates, and one inverter

9–6 Five decade counters are required. $10^5 = 100,000$

9–7 $f_{Q0} = 1 \text{ MHz}/[(10)(2)] = 50 \text{ kHz}$

9–8 See Figure 9–84.

9–9 $8AC0_{16}$ would be loaded. $16^4 - 8AC0_{16} = 65,536 - 32,520 = 30,016$
$f_{TC4} = 10 \text{ MHz}/30,016 = 333.2 \text{ Hz}$

9–10 See Figure 9–85.

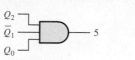

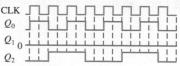

FIGURE 9–84   **FIGURE 9–85**

TRUE/FALSE QUIZ

1. F   **2.** T   **3.** T   **4.** F   **5.** T   **6.** F   **7.** T   **8.** F   **9.** T   **10.** F

SELF-TEST

1. (c)   **2.** (d)   **3.** (a)   **4.** (b)   **5.** (b)   **6.** (c)   **7.** (a)   **8.** (c)

9. (b)   **10.** (c)   **11.** (d)   **12.** (a)   **13.** (c)   **14.** (b)   **15.** (b)   **16.** (d)

Programmable Logic

CHAPTER OUTLINE

CHAPTER OBJECTIVES

- Discuss the types of programmable logic, SPLDs and CPLDs, and explain their basic structure
- Describe the basic architecture of two types of SPLDs—the PAL and the GAL
- Explain the basic structure of a programmable logic array (PLA)
- Discuss the operation of macrocells
- Distinguish between CPLDs and FPGAs
- Explain the basic operation of a look-up table (LUT)
- Define *intellectual property* and *platform FPGA*
- Discuss embedded functions
- Show a basic software design flow for a programmable device
- Explain the design flow elements of design entry, functional simulation, synthesis, implementation, timing simulation, and downloading
- Discuss several methods of testing a programmable logic device, including boundary scan logic

KEY TERMS

Key terms are in order of appearance in the chapter.

- PAL
- GAL
- Macrocell
- Registered
- CPLD
- LAB
- LUT
- FPGA
- CLB
- Intellectual property
- Design flow
- Target device
- Schematic entry
- Text entry
- Functional simulation
- Compiler
- Timing simulation
- Downloading
- Break point
- Boundary scan

VISIT THE WEBSITE

Study aids for this chapter are available at
http://www.pearsonhighered.com/careersresources/

INTRODUCTION

The distinction between hardware and software is hazy. Today, new digital circuits are programmed into hardware using languages like VHDL. The density (number of equivalent gates on a single chip) has increased dramatically over the past few years. The maximum number of gates in an FPGA (a type of PLD known as a field-programmable gate array) is doubling every 18 months, according to Moore's law. At the same time, the price for a PLD is decreasing.

PLDs, such as the FPGA, can be used in conjunction with processors and software in an embedded system, or the FPGA can be the sole component with all the logic functions programmed in. An embedded system is one that is dedicated to a single task or a very limited number of tasks unlike the computer, which is multipurpose and can be programmed to perform just about any task. With PLDs, logic is described with software and then implemented with the internal gates of the PLD.

In this chapter, the basic architecture (internal structure and organization) of SPLDs, CPLDs, and FPGAs is discussed. A discussion of software development tools covers the generic design flow for programming a device, including design entry, functional simulation, synthesis, implementation, timing simulation, and downloading.

10–1 Simple Programmable Logic Devices (SPLDs)

Two major types of simple programmable logic devices (SPLDs) are the PAL and the GAL. *PAL* stands for programmable array logic, and *GAL* stands for generic array logic. Generally, a PAL is one-time programmable (OTP), and a GAL is a type of PAL that is reprogrammable. The term *GAL* is a designation originally used by Lattice Semiconductor and later licensed to other manufacturers. The basic structure of both PALs and GALs is a programmable AND array and a fixed OR array, which is a basic sum-of-products architecture.

After completing this section, you should be able to

- Describe SPLD operation
- Show how a sum-of-products expression is implemented in a PAL or GAL
- Explain simplified PAL/GAL logic diagrams
- Describe a basic PAL/GAL macrocell

SPLD: The PAL

A **PAL** (programmable array logic) consists of a programmable array of AND gates that connects to a fixed array of OR gates. Generally, PALs are implemented with fuse process technology and are, therefore, one-time programmable (OTP).

The PAL structure allows any sum-of-products (SOP) logic expression with a defined number of variables to be implemented. As you have learned, any combinational logic function can be expressed in SOP form. A simple PAL structure is shown in Figure 10–1 for two input variables and one output; most PALs have many inputs and many outputs. As you learned earlier, a programmable array is essentially a grid or matrix of conductors that form rows and columns with a programmable link at each cross point. Each programmable link, which is a fuse in the case of a PAL, is called a *cell*. Each row is connected to the input of an AND gate, and each column is connected to an input variable or its complement. By programming the presence or absence of a fuse connection, any combination of input variables or complements can be applied to an AND gate to form any desired product term. The AND gates are connected to an OR gate, creating a sum-of-products (SOP) output.

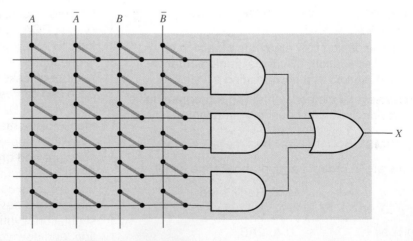

FIGURE 10–1   Basic AND/OR structure of a PAL.

Implementing a Sum-of-Products Expression

An example of a simple PAL is programmed as shown in Figure 10–2 so that the product term AB is produced by the top AND gate, $A\overline{B}$ is produced by the middle AND gate, and $\overline{A}\,\overline{B}$ is produced by the bottom AND gate. As you can see, the fuses are left intact to connect the desired variables or their complements to the appropriate AND gate inputs. The fuses are opened where a variable or its complement is not used in a given product term. The final output from the OR gate is the SOP expression,

$$X = AB + A\overline{B} + \overline{A}\,\overline{B}$$

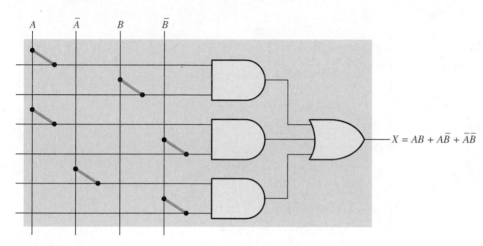

FIGURE 10–2   PAL implementation of a sum-of-products expression.

SPLD: The GAL

The **GAL** is essentially a PAL that can be reprogrammed. It has the same type of AND/OR organization that the PAL does. The basic difference is that a GAL uses a reprogrammable process technology, such as EEPROM (E^2CMOS), instead of fuses, as shown in Figure 10–3.

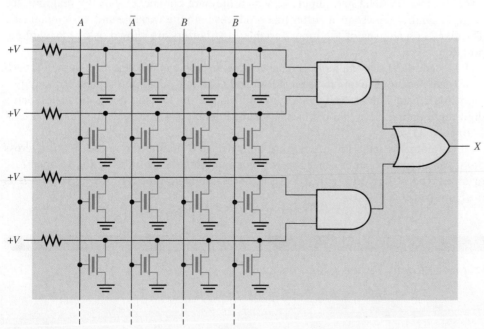

FIGURE 10–3   Simplified GAL array.

Simplified Notation for PAL/GAL Diagrams

Actual PAL and GAL devices have many AND and OR gates in addition to other elements and are capable of handling many variables and their complements. Most PAL and GAL diagrams that you may see on a data sheet use simplified notation, as illustrated in Figure 10–4, to keep the schematic from being too complicated.

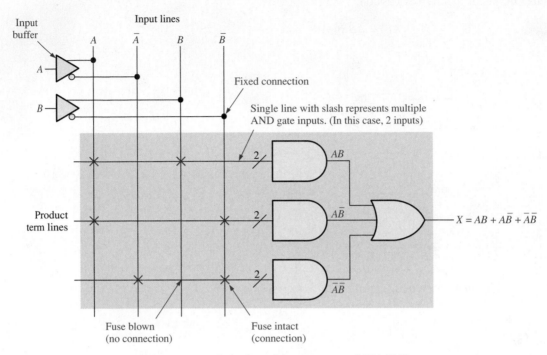

FIGURE 10–4   A portion of a programmed PAL/GAL.

The input variables to a PAL or GAL are usually buffered to prevent loading by a large number of AND gate inputs to which they are connected. On the diagram, the triangle symbol represents a buffer that produces both the variable and its complement. The fixed connections of the input variables and buffers are shown using standard dot notation.

PALs and GALs have a large number of programmable interconnection lines, and each AND gate has multiple inputs. Typical PAL and GAL logic diagrams represent a multiple-input AND gate with an AND gate symbol having a single input line with a slash and a digit representing the actual number of inputs. Figure 10–4 illustrates this for the case of 2-input AND gates.

Programmable links in an array are indicated in a diagram by a red X at the cross point for an intact fuse or other type of link and the absence of an X for an open fuse or other type of link. In Figure 10–4, the 2-variable logic function $AB + A\overline{B} + \overline{A}\,\overline{B}$ is programmed.

EXAMPLE 10–1

Show how a PAL is programmed for the following 3-variable logic function:

$$X = A\overline{B}C + \overline{A}B\overline{C} + \overline{A}\,\overline{B} + AC$$

Solution

The programmed array is shown in Figure 10–5. The intact fusible links are indicated by small red Xs. The absence of an X means that the fuse is open.

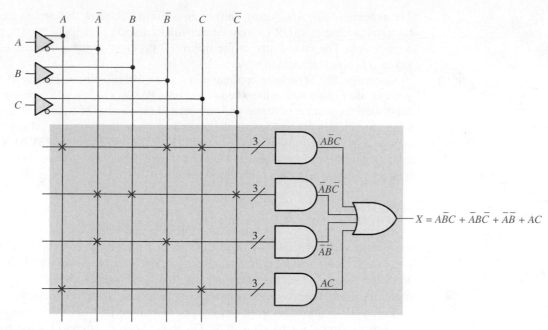

FIGURE 10–5

Related Problem*

Write the expression for the output if the fusible links connecting input A to the top row and to the bottom row in Figure 10–5 are open.

*Answers are at the end of the chapter.

PAL/GAL General Block Diagram

A block diagram of a PAL or GAL is shown in Figure 10–6. Remember, the basic difference is that a GAL has a reprogrammable array and the PAL is one-time programmable.

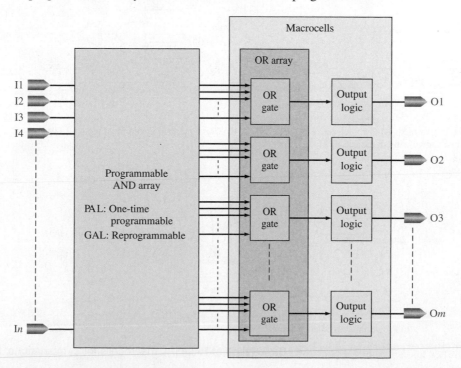

FIGURE 10–6   General block diagram of a PAL or GAL.

The programmable AND array outputs go to fixed OR gates that are connected to additional output logic. An OR gate combined with its associated output logic is typically called a *macrocell*. The complexity of the macrocell depends on the particular device, and in GALs it is often reprogrammable.

Generally, SPLD package configurations range from 20 pins to 28 pins. Two factors that you can use to help determine whether a certain PAL or GAL is adequate for a given logic design are the number of inputs and outputs and the number of equivalent gates or density. Other parameters to consider are the maximum operating frequency, delay times, and dc supply voltage. Two common types of SPLD are the 16V8 and the 22V10. Various SPLD manufacturers may have different ways of defining density, so you have to use the specified number of equivalent gates with this in mind.

Macrocells

A **macrocell** generally consists of one OR gate and some associated output logic. The macrocells vary in complexity, depending on the particular type of PAL or GAL. A macrocell can be configured for combinational logic, registered logic, or a combination of both. **Registered** logic means that there is a flip-flop in the macrocell to provide for sequential logic functions. The registered operation of macrocells is covered in Section 10–3.

Figure 10–7 illustrates three basic types of macrocells with combinational logic. Part (a) shows a simple macrocell with the OR gate and an inverter with a tristate control that can make the inverter like an open circuit to completely disconnect the output. The output of the tristate inverter can be either LOW, HIGH, or disconnected. Part (b) is a macrocell that can be either an input or an output. When it is used as an input, the tristate inverter is disconnected, and the input goes to the buffer that is connected to the AND array. Part (c) is a macrocell that can be programmed to have either an active-HIGH or an active-LOW output, or it can be used as an input. One input to the exclusive-OR (XOR) gate can be programmed to be either HIGH or LOW. When the programmable XOR input is HIGH, the OR gate output is inverted because $0 \oplus 1 = 1$ and $1 \oplus 1 = 0$. Similarly, when the programmable XOR input is LOW, the OR gate output is not inverted because $0 \oplus 0 = 0$ and $1 \oplus 0 = 1$.

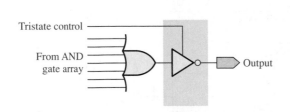

(a) Combinational output (active-LOW). An active-HIGH output would be shown without the bubble on the tristate gate symbol.

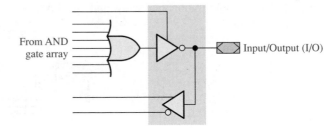

(b) Combinational input/output (active-LOW)

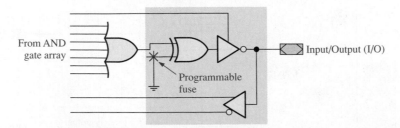

(c) Programmable polarity output

FIGURE 10–7 Basic types of PAL/GAL macrocells for combinational logic.

1. What does PAL stand for?

2. What does GAL stand for?

3. What is the difference between a PAL and a GAL?

4. Basically, what does a macrocell contain?

10–2 Complex Programmable Logic Devices (CPLDs)

The complex programmable logic device (CPLD) is basically a single device containing multiple SPLDs and providing more capacity for larger logic designs. In this section, the focus is the concepts of traditional CPLD architecture, keeping in mind that CPLDs may vary somewhat in architecture and/or in parameters such as density, process technology, power consumption, supply voltage, and speed.

After completing this section, you should be able to

◆ Describe a typical CPLD

◆ Discuss the basic CPLD architecture

◆ Explain how product terms are generated in CPLDs

The CPLD

A **CPLD** (complex programmable logic device) consists basically of multiple SPLD arrays with programmable interconnections. Although the way CPLDs are internally organized varies with the manufacturer, Figure 10–8 illustrates a generic CPLD. We will refer to each

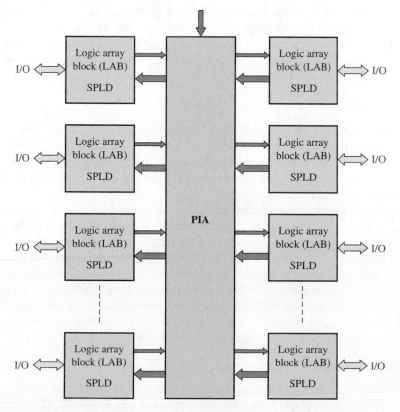

FIGURE 10–8 Basic block diagram of a generic CPLD.

SPLD array in a CPLD as a **LAB** (logic array block). Other designations are sometimes used, such as *function block, logic block,* or *generic block.* The programmable interconnections are generally called the *PIA* (**programmable interconnect array**) although some manufacturers, such as Xilinx, use the term *AIM* (advanced interconnect matrix) or a similar designation. The LABs and the interconnections between LABs are programmed using software. A CPLD can be programmed for complex logic functions based on the SOP structure of the individual LABs (actually SPLDs). Inputs can be connected to any of the LABs, and their outputs can be interconnected to any other LABs via the PIA.

Most programmable logic manufacturers make a series of CPLDs that range in density, process technology, power consumption, supply voltage, and speed. Manufacturers usually specify CPLD density in terms of macrocells or logic array blocks. Densities can range from tens of macrocells to over 1500 macrocells in packages with up to several hundred pins. As PLDs become more complex, maximum densities will increase. Most CPLDs are reprogrammable and use EEPROM or SRAM process technology for the programmable links. Power consumption can range from a few milliwatts to a few hundred milliwatts. DC supply voltages are typically from 2.5 V to 5 V, depending on the specific device.

Several manufacturers, (for example, Altera, Xilinx, Lattice, and Atmel) produce CPLDs. As you will learn, CPLDs and other programmable logic devices are really a combination of hardware and software.

Classic CPLD Architecture

The **architecture** of a CPLD is the way in which the internal elements are organized and arranged. The architecture of specific CPLDs is similar to the block diagram of a generic CPLD (shown in Figure 10–8). It has the classic PAL/GAL structure that produces SOP functions. The density ranges from 2 LABs to 16 LABs, depending on the particular device in the series. Remember, a LAB is roughly equivalent to one SPLD, and package sizes for CPLDs vary from 44 pins to 208 pins. Typically, a series of CPLDs uses the EEPROM-based process technology. In-system programmable (ISP) versions use the JTAG standard interface.

Figure 10–9 shows a general block diagram of a typical CPLD. Four LABs are shown, but there can be up to sixteen, depending on the particular device in a series. Each of the four LABs consists of sixteen macrocells, and multiple LABs are linked together via the PIA, which is a programmable global (goes to all LABs) bus structure to which the general-purpose inputs, the I/Os, and the macrocells are connected.

The Macrocell

A simplified diagram of a typical macrocell is shown in Figure 10–10. The macrocell contains a small programmable AND array with five AND gates, an OR gate, a product-term selection matrix for connecting the AND gate outputs to the OR gate, and associated logic that can be programmed for input, combinational logic output, or registered output. This macrocell is covered in more detail in Section 10–3.

Although based on the same concept, this macrocell differs somewhat from the macrocell discussed in Section 10–1 in relation to SPLDs because it contains a portion of the programmable AND array and a product-term selection matrix. As shown in Figure 10–10, five AND gates feed product terms from the PIA into the product-term selection matrix. The product term from the bottom AND gate can be fed back inverted into the programmable array as a shared expander for use by other macrocells. The parallel expander inputs allow borrowing of unused product terms from other macrocells to expand an SOP expression. The product-term selection matrix is an array of programmable connections that is used to connect selected outputs from the AND array and from the expander inputs to the OR gate.

Shared Expanders

A complemented product term that can be used to increase the number of product terms in an SOP expression is available from each macrocell in a LAB. Figure 10–11 illustrates how a shared expander term from another macrocell can be used to create additional product terms. In this case, each of the five AND gates in a macrocell array is limited to four inputs

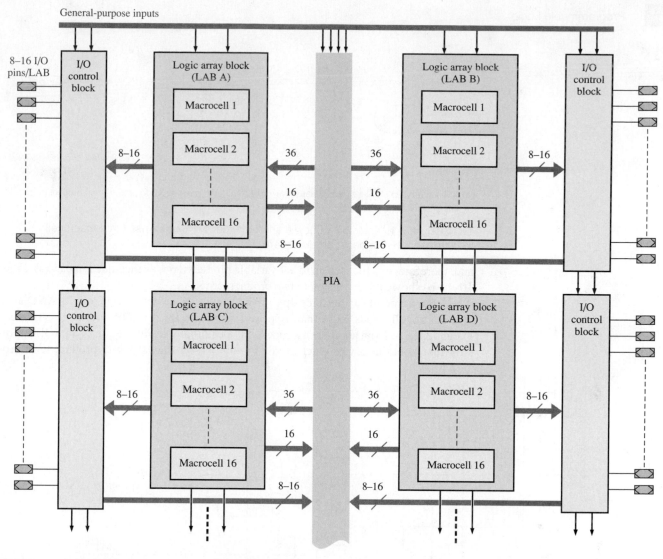

FIGURE 10–9 Basic block diagram of a typical CPLD.

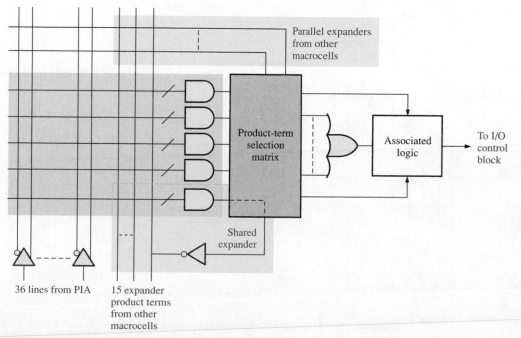

FIGURE 10–10 Simplified diagram of a macrocell in a typical CPLD.

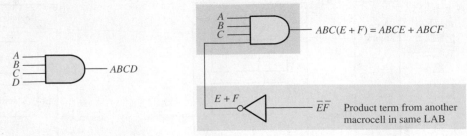

(a) A 4-input AND array gate can produce one 4-variable product term.

(b) AND gate is expanded to produce two product terms.

FIGURE 10–11   Example of how a shared expander can be used in a macrocell to increase the number of product terms.

and, therefore, can produce up to a 4-variable product term, as illustrated in part (a). Figure 10–11(b) shows the expansion to two product terms.

Each macrocell can produce up to five product terms generated from its AND array. If a macrocell needs more than five product terms for its SOP output, it can use an expander term from another macrocell. Suppose that a design requires an SOP expression that contains six product terms. Figure 10–12 shows how a product term from

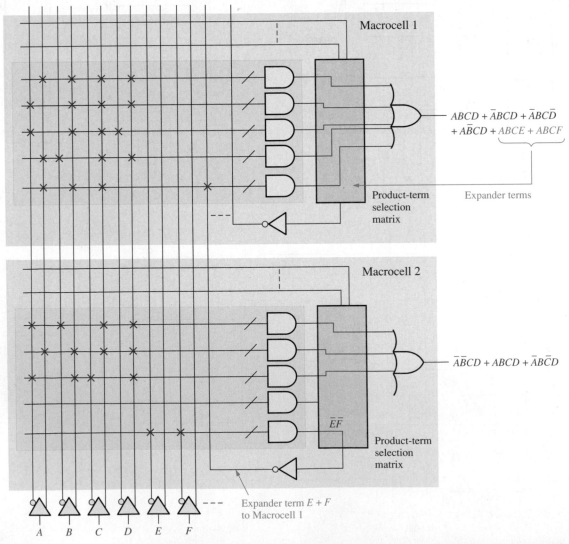

FIGURE 10–12   Simplified illustration of using a shared expander term from another macrocell to increase an SOP expression. The red Xs and lines represent the connections produced in the hardware by the software compiler running the programmed design.

another macrocell can be used to increase an SOP output. Macrocell 2, which is under-utilized, generates a shared expander term $(E + F)$ that connects to the fifth AND gate in macrocell 1 to produce an SOP expression with six product terms.

Parallel Expanders

Another way to increase the number of product terms for a macrocell is by using parallel expanders in which additional product terms are ORed with the terms generated by a macrocell instead of being combined in the AND array, as in the shared expander. A given macrocell can borrow unused product terms from neighboring macrocells. The basic concept is illustrated in Figure 10–13 where a simplified circuit that can produce two product terms borrows three additional product terms.

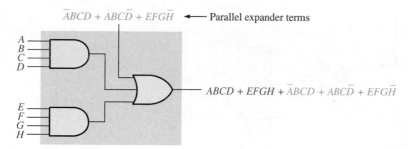

FIGURE 10–13 Basic concept of the parallel expander.

Figure 10–14 shows how one macrocell can borrow parallel expander terms from another macrocell to increase the SOP output. Macrocell 2 uses three product terms from macrocell 1 to produce an eight-term SOP expression.

LUT CPLD Architecture

This architecture differs from the classic CPLD previously discussed. As shown by the block diagram in Figure 10–15, this device contains logic array blocks (LABs) each with multiple **logic elements** (LEs). An LE is the basic logic design unit and is analogous to the macrocell. The programmable interconnects are arranged in a row and column arrangement running between the LABs, and input/output elements (IOEs) are oriented around the perimeter. The architecture of this type of CPLD is similar to that of FPGAs, which we discuss in Section 10–4.

A main difference between this type of CPLD and the classic AND/OR array CPLD previously discussed is the way in which a logic function is developed. Look-up tables (LUTs) are used instead of AND/OR arrays. An **LUT** is basically a type of memory that can be programmed to produce SOP functions (discussed in more detail in Section 10–4). These two approaches are contrasted in Figure 10–16.

As mentioned, the LUT CPLD has a row/column arrangement of interconnects instead of the channel-type interconnects found in most classic CPLDs. These two approaches are contrasted in Figure 10–17 and can be understood by comparing Figure 10–9 and Figure 10–15.

Most CPLDs use a nonvolatile process technology for the programmable links. The LUT CPLD, however, uses a SRAM-based process technology that is **volatile**—all programmed logic is lost when power is turned off. The memory embedded on the chip stores the program data using nonvolatile memory technology and reconfigures the CPLD on power up.

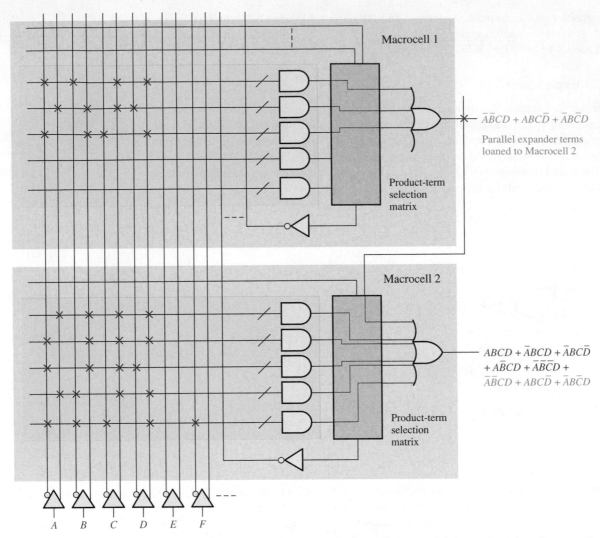

$\overline{A}BCD + ABC\overline{D} + \overline{A}BC\overline{D}$

Parallel expander terms
loaned to Macrocell 2

$ABCD + \overline{A}BCD + \overline{A}BC\overline{D}$
$+ A\overline{B}CD + \overline{A}\overline{B}C\overline{D} +$
$\overline{A}\overline{B}CD + ABC\overline{D} + \overline{A}BC\overline{D}$

FIGURE 10–14 Simplified illustration of using parallel expander terms from another macrocell to increase an SOP expression. The red Xs and lines represent the connections produced in the hardware by the software complier running the programmed design.

PLA (Programmable Logic Array)

As you have learned, the architecture of a CPLD is the way in which the internal elements are organized and arranged. The architecture of some PLDs is based on a PLA (programmable logic array) structure rather than on a PAL (programmable array logic) structure, which we have discussed. Figure 10–18 compares a simple PAL structure with a simple PLA structure. The PAL has a programmable AND array followed by a fixed OR array and produces an SOP expression, as shown by the example in Figure 10–18(a). The **PLA** has a programmable AND array followed by a programmable OR array, as shown by the example in Figure 10–18(b).

Specific CPLD Devices

Several manufacturers produce CPLDs. Table 10–1 lists device families from selected companies. As time passes, a series may become obsolete or a new series may be added. You can check the websites for the most current information.

CPLDs vary greatly in terms of complexity. Table 10–2 lists some of the parameter ranges that are available. Keep in mind that these numbers are subject to change as technology advances.

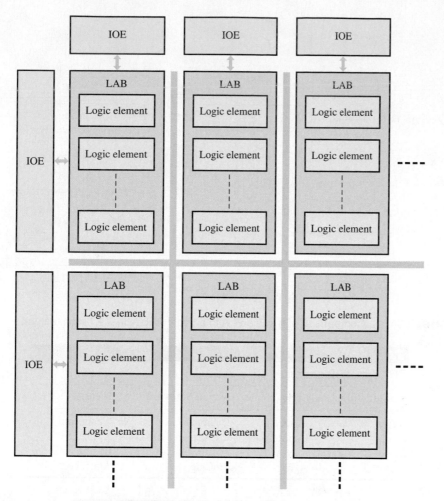

FIGURE 10–15 Simplified block diagram of an LUT CPLD.

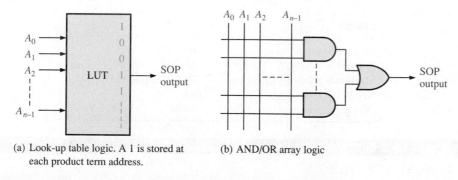

(a) Look-up table logic. A 1 is stored at each product term address.

(b) AND/OR array logic

FIGURE 10–16 Two types of logic function generation in CPLDs.

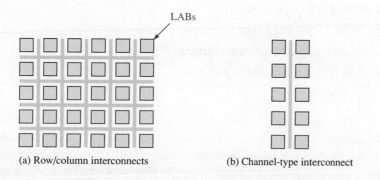

(a) Row/column interconnects

(b) Channel-type interconnect

FIGURE 10–17 LUT CPLDs have row/column interconnects. Classic CPLDs have channel-type interconnects.

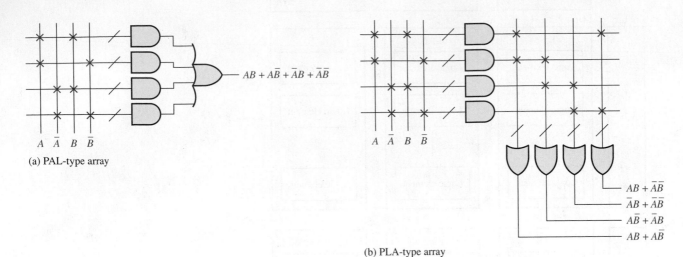

(a) PAL-type array

$$AB + A\bar{B} + \bar{A}B + \bar{A}\bar{B}$$

(b) PLA-type array

FIGURE 10–18   Comparison of a basic PLA to a basic PAL type PLD architecture.

TABLE 10–1

CPLD manufacturers.

Manufacturer	Series Name	Design Software	Website
Altera	MAX	Quartus II	Altera.com
Xilinx	Coolrunner	ISE Design Suite	Xilinx.com
Lattice	ispMACH	ispLEVER classic	Latticesemi.com
Atmel	ATF	ProChip Designer	Atmel.com

TABLE 10–2

Selected CPLD parameters.

Feature	Range
Number of macrocells	10–1700
Number of LABs	10–221
Maximum operating frequency	20.4 MHz–400 MHz
Number of I/Os	10–1156
DC operating voltage	1.8 V, 2.5 V, 3.3 V, 5 V

SECTION 10–2 CHECKUP

1. What is a CPLD?

2. What does LAB stand for?

3. Describe a LAB in a typical CPLD.

4. What is the purpose of a shared expander?

5. What is the purpose of a parallel expander?

6. How does a PLA differ from a PAL?

10–3   Macrocell Modes

CPLD macrocells were introduced previously. A macrocell can be configured for combinational logic or registered logic outputs and inputs by programming. The term *registered* refers to the use of flip-flops. In this section, you will learn about the typical macrocell, including the combinational and the registered modes of operation. Although macrocell architecture varies among different CPLDs, a typical macrocell architecture is used for illustration.

After completing this section, you should be able to

⬥ Describe the operation of a typical CPLD macrocell

Logic diagrams often use the symbol shown in Figure 10–19 to represent a multiplexer. In this case, the multiplexer has two data inputs and a select input that provides for programmable selection; the select input is usually not shown on a logic diagram.

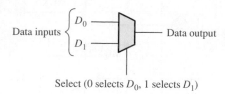

Select (0 selects D_0, 1 selects D_1)

FIGURE 10–19 Commonly used symbol for a multiplexer. It can have any number of inputs.

Figure 10–20 shows a complete macrocell including the flip-flop (register). The XOR gate provides for complementing the SOP function from the OR gate to produce a function in POS form. A 1 on the top input of the XOR gate complements the OR output, and a 0 lets the OR output pass uncomplemented (in SOP form). MUX 1 provides for selection of either the XOR output or an input from the I/O. MUX 2 can be programmed to select either the global clock or a clock signal based on a product term. MUX 3 can be programmed to select either a HIGH (V_{CC}) or a product-term enable for the flip-flop. MUX 4 can select the global clear or a product-term clear. MUX 5 is used to bypass the flip-flop and connect the combinational logic output to the I/O or to connect the registered output to the I/O. The flip-flop can be programmed as a D, T (toggle), or J-K flip-flop.

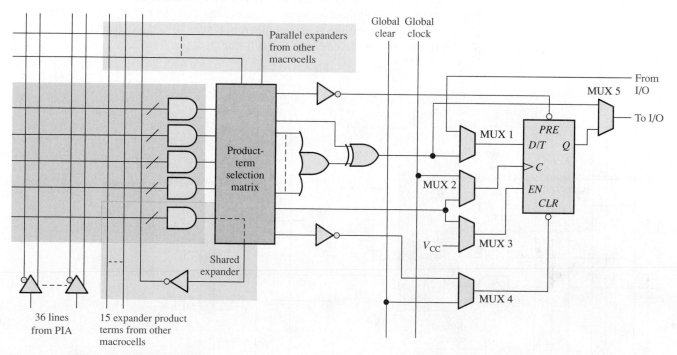

FIGURE 10–20 A CPLD macrocell.

The Combinational Mode

When a macrocell is programmed to produce an SOP combinational logic function, the logic elements in the data path are as shown in red in Figure 10–21. As you can see, only one mux is used and the register (flip-flop) is bypassed.

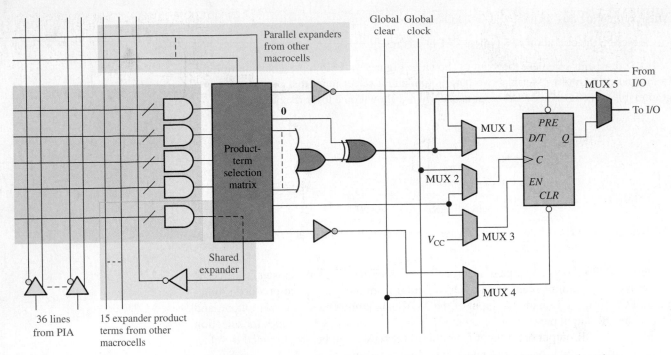

FIGURE 10–21 A macrocell configured for generation of an SOP logic function. Red indicates data path.

The Registered Mode

When a macrocell is programmed for the registered mode with the SOP combinational logic output providing the data input to the register and clocked by the global clock, the elements in the data path are as shown in red in Figure 10–22. As you can see, four multiplexers (mux) are used and the register (flip-flop) is active.

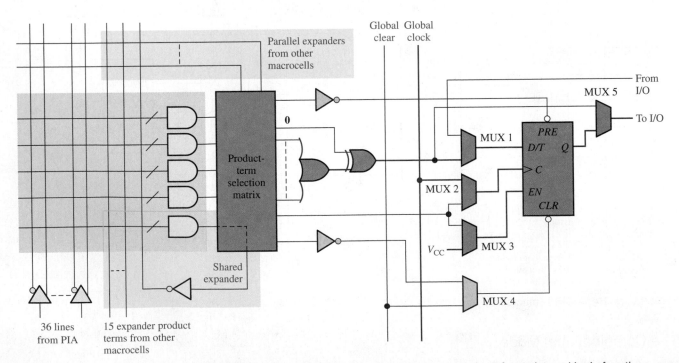

FIGURE 10–22 A macrocell configured for generation of a registered logic function. Red indicates data path.

1. Explain the purpose of the XOR gate in the macrocell.

2. What are the two major modes of a macrocell?

3. What does the term *registered* refer to?

4. Besides the OR gate, XOR gate, and flip-flop, what other logic element is commonly used in a macrocell?

10-4 Field-Programmable Gate Arrays (FPGAs)

As you have learned, the classic CPLD architecture consists of PAL/GAL or PLA-type logic blocks with programmable interconnections. Basically, the FPGA (field-programmable gate array) differs in architecture, does not use PAL/PLA type arrays, and has much greater densities than CPLDs. A typical FPGA has many times more equivalent gates than a typical CPLD. The logic-producing elements in FPGAs are generally much smaller than in CPLDs, and there are many more of them. Also, the programmable interconnections are generally organized in a row and column arrangement in FPGAs.

After completing this section, you should be able to

- Describe the basic structure of a field-programmable gate array (FPGA)

- Compare an FPGA to a CPLD

- Discuss look-up tables (LUTs)

- Discuss the SRAM-based FPGA

- Define the FPGA core

The three basic elements in an **FPGA** are the configurable logic block (CLB), the interconnections, and the input/output (I/O) blocks, as illustrated in Figure 10–23. The configurable logic blocks (CLBs) in an FPGA are not as complex as the LABs or function blocks (FBs) in a CPLD, but generally there are many more of them. When the CLBs are relatively simple, the FPGA architecture is called *fine grained*. When the CLBs are larger and more complex, the architecture is called *coarse grained*. The I/O blocks around the perimeter of the structure provide individually selectable input, output, or bidirectional access to the outside world. The distributed matrix of programmable interconnections provide for interconnection of the CLBs and connection to inputs and outputs. Large FPGAs can have tens of thousands of CLBs in addition to memory and other resources.

Most programmable logic manufacturers make a series of FPGAs that range in density, power consumption, supply voltage, speed, and to some degree vary in architecture. FPGAs are reprogrammable and use SRAM or antifuse process technology for the programmable links. Densities can range from hundreds of logic modules to hundreds of thousands of logic modules in packages with up to over 1,000 pins. DC supply voltages are typically 1.8 V to 5 V, depending on the specific device.

Configurable Logic Blocks

Typically, an FPGA logic block consists of several smaller logic modules that are the basic building units, somewhat analogous to macrocells in a CPLD. Figure 10–24 shows the fundamental configurable logic blocks (CLBs) within the global row/column programmable interconnects that are used to connect logic blocks. Each **CLB** (also known as logic array block, LAB) is made up of multiple smaller logic modules and a local programmable interconnect that is used to connect logic modules within the CLB.

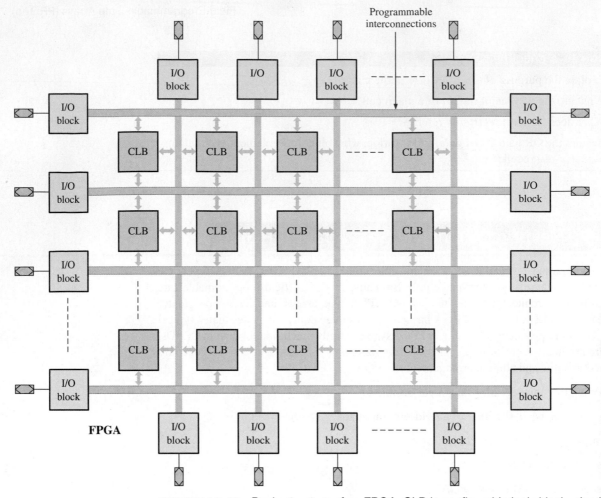

FIGURE 10–23 Basic structure of an FPGA. CLB is configurable logic block, also known as logic array block (LAB).

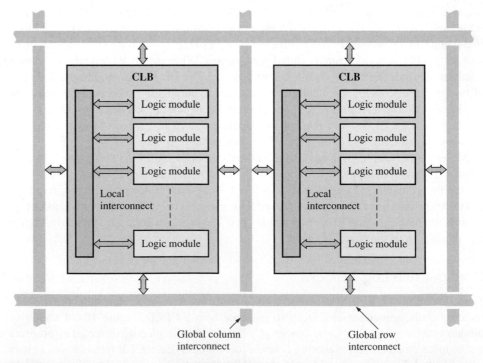

FIGURE 10–24 Basic configurable logic blocks (CLBs) within the global row/column programmable interconnects.

Logic Modules

A logic module in an FPGA logic block can be configured for combinational logic, registered logic, or a combination of both. A flip-flop is part of the associated logic and is used for registered logic. A block diagram of a typical LUT-based logic module is shown in Figure 10–25. As you know, an LUT (look-up table) is a type of memory that is programmable and used to generate SOP combinational logic functions. The LUT essentially does the same job as the PAL or PLA does.

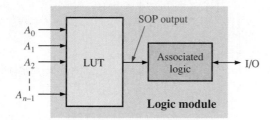

FIGURE 10–25 Basic block diagram of a logic module in an FPGA.

Generally, the organization of an LUT consists of a number of memory cells equal to 2^n, where n is the number of input variables. For example, three inputs can select up to eight memory cells, so an LUT with three input variables can produce an SOP expression with up to eight product terms. A pattern of 1s and 0s can be programmed into the LUT memory cells, as illustrated in Figure 10–26 for a specified SOP function. Each 1 means the associated product term appears in the SOP output, and each 0 means that the associated product term does not appear in the SOP output. The resulting SOP output expression is

$$\overline{A}_2\overline{A}_1\overline{A}_0 + \overline{A}_2 A_1 A_0 + A_2 \overline{A}_1 A_0 + A_2 A_1 A_0$$

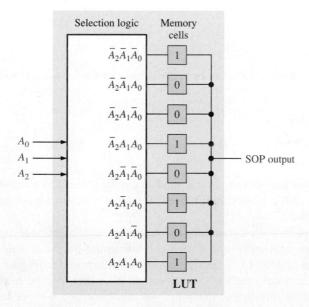

FIGURE 10–26 The basic concept of an LUT programmed for a particular SOP output.

EXAMPLE 10–2

Show a basic 3-variable LUT programmed to produce the following SOP function:

$$A_2 A_1 \overline{A}_0 + A_2 \overline{A}_1 \overline{A}_0 + \overline{A}_2 A_1 A_0 + A_2 \overline{A}_1 A_0 + \overline{A}_2 \overline{A}_1 A_0$$

Solution

A 1 is stored for each product term in the SOP expression, as shown in Figure 10–27.

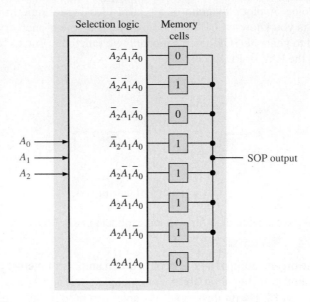

FIGURE 10–27

Related Problem

How many memory cells would be in an LUT with four input variables? What would be the maximum possible number of product terms in the SOP output?

Operating Modes of a Logic Module

Typically, a logic module (LM) can be programmed for the following modes of operation:

- Normal mode
- Extended LUT mode
- Arithmetic mode
- Shared arithmetic mode

In addition to these four modes, a logic module can be utilized as a register chain to create counters and shift registers. In this section, we will discuss the normal mode and the extended LUT mode.

The *normal mode* is used primarily for generating combinational logic functions. A logic module can implement one or two combinational output functions with its two LUTs. Examples of four LUT configurations are illustrated in Figure 10–28. Generally, two SOP functions, each with four variables or less, can be implemented in an LM without sharing inputs. For example, you can have two 4-variable functions, one 4-variable function and one 3-variable function, or two 3-variable functions. By sharing inputs, you can have any combination of a total of eight inputs up to a maximum of six inputs for each LUT. In the normal mode, you are limited to 6-variable SOP functions.

The *extended LUT mode* allows expansion to a 7-variable function, as illustrated in Figure 10–29. The multiplexer formed by the AND-OR circuit with a complemented input is part of the dedicated logic in a logic module.

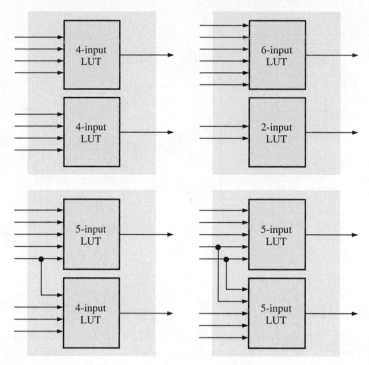

FIGURE 10–28 Examples of possible LUT configurations in a logic module (LM) in the normal mode.

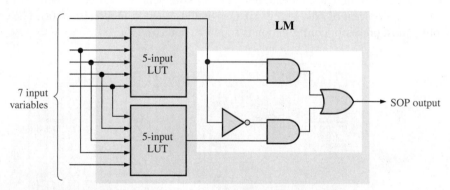

FIGURE 10–29 Expansion of a logic module (LM) to produce a 7-variable SOP function in the extended LUT mode.

EXAMPLE 10–3

A logic module is configured in the extended LUT mode, as shown in Figure 10–30. For the specific LUT outputs shown, determine the final SOP output.

Solution

The SOP output expression is as follows:

$$\overline{A}_5 A_4 A_3 A_2 A_1 A_0 + A_5 \overline{A}_4 A_3 A_2 A_1 A_0 + A_5 A_4 A_3 A_2 A_1 A_0 + A_6 A_5 A_4 A_4 A_3 \overline{A}_2 \overline{A}_0 + A_6 A_5 \overline{A}_4 A_3 A_2 \overline{A}_0 + A_6 A_5 A_4 A_3 A_2 \overline{A}_0$$

Related Problem

Show an LM configured in the normal mode to produce one SOP function with five product terms from one LUT and three product terms from the other LUT.

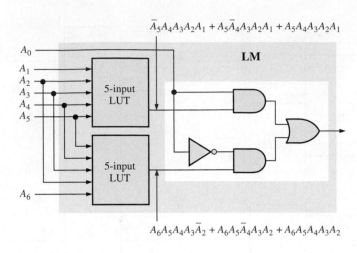

$$\overline{A}_5A_4A_3A_2A_1 + A_5\overline{A}_4A_3A_2A_1 + A_5A_4A_3A_2A_1$$

$$A_6A_5A_4A_3\overline{A}_2 + A_6A_5\overline{A}_4A_3A_2 + A_6A_5A_4A_3A_2$$

FIGURE 10–30

SRAM-Based FPGAs

FPGAs are either nonvolatile because they are based on antifuse technology or they are volatile because they are based on SRAM technology. (The term *volatile* means that all the data programmed into the configurable logic blocks are lost when power is turned off.) Therefore, SRAM-based FPGAs include either a nonvolatile configuration memory embedded on the chip to store the program data and reconfigure the device each time power is turned back on or they use an external memory with data transfer controlled by a host processor. The concept of on-the-chip memory is illustrated in Figure 10–31(a). The concept of the host processor configuration is shown in part (b).

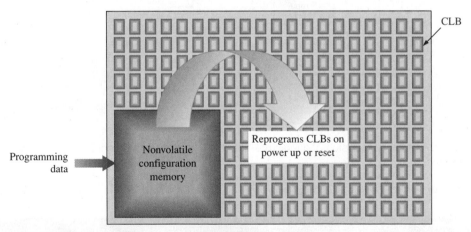

(a) Volatile FPGA with on-the-chip nonvolatile configuration memory

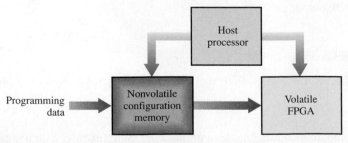

(b) Volatile FPGA with on-board memory and host processor

FIGURE 10–31 Basic concepts of volatile FPGA configurations.

FPGA Cores

FPGAs, as we have discussed, are essentially like "blank slates" that the end user can program for any logic design. FPGAs are available that also contain hard-core logic. A **hard core** is a portion of logic in an FPGA that is put in by the manufacturer to provide a specific function and that cannot be reprogrammed. For example, if a customer needs a small microprocessor as part of a system design, it can be programmed into the FPGA by the customer or it can be provided as hard core by the manufacturer. If the embedded function has some programmable features, it is known as a **soft-core** function. An advantage of the hard-core approach is that the same design can be implemented using much less of the available capacity of the FPGA than if the user programmed it in the field, resulting in less space on the chip ("real estate") and less development time for the user. Also, hard-core functions have been thoroughly tested. The disadvantage of the hard core is that the specifications are fixed during manufacturing and the customer must be able to use the hard-core logic "as is." It cannot be changed later.

Hard cores are generally available for functions that are commonly used in digital systems, such as a microprocessor, standard input/output interfaces, and digital signal processors. More than one hard-core function can be programmed in an FPGA. Figure 10–32 illustrates the concept of a hard core surrounded by configurable logic programmed by the user. This is a basic embedded system because the hard-core function is embedded in the user-programmed logic.

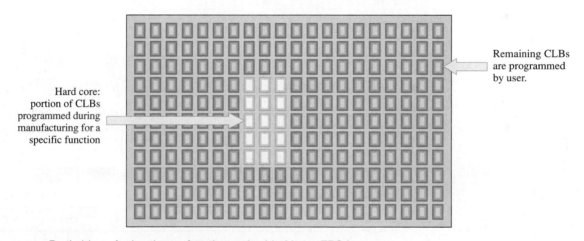

Hard core: portion of CLBs programmed during manufacturing for a specific function

Remaining CLBs are programmed by user.

FIGURE 10–32 Basic idea of a hard-core function embedded in an FPGA.

Hard core designs are generally developed by and are the property of the FPGA manufacturer. Designs owned by the manufacturer are termed **intellectual property (IP)**. A company usually lists the types of intellectual property that are available on its website. Some intellectual properties are a mix of hard core and soft core. A processor that has some flexibility in the selection and adjustment of certain parameters by the user is an example.

Those FPGAs containing either or both hard-core and soft-core embedded processors and other functions are known as **platform FPGAs** because they can be used to implement an entire system without the need for external support devices.

Embedded Functions

A block diagram of a typical FPGA is shown in Figure 10–33. The FPGA contains embedded memory functions as well as digital signal processing (DSP) functions. DSP functions, such as digital filters, are commonly used in many systems. As you can see in the block diagram, the embedded blocks are arranged throughout the FPGA interconnection matrix and input/output elements (IOEs) are placed around the FPGA perimeter.

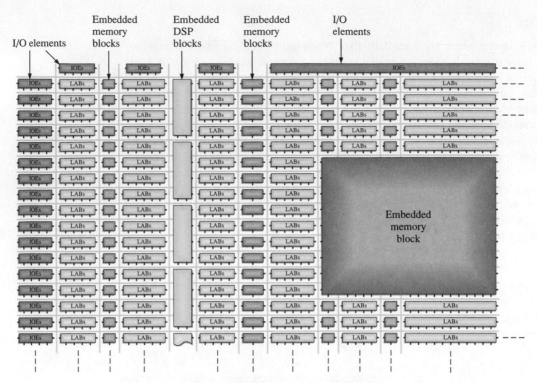

FIGURE 10–33   Example FPGA block diagram.

Specific FPGA Devices

Several manufacturers produce FPGAs as well as CPLDs. Table 10–3 lists device families from selected companies. Check the website for the most current information.

TABLE 10–3

FPGA manufacturers.

Manufacturer	Series Name(s)	Design Software	Website
Altera	Stratix Aria Cyclone	Quartus II	Altera.com
Xilinx	Spartan Artix Kintex Virtex	ISE Design Suite	Xilinx.com
Lattice	iCE40 MachX02 Lattice ECP3 LatticeXP2 LatticeGC/M	Lattice Diamond iCEcube2	Latticesemi.com
Atmel	AT40	IDS	Atmel.com

FPGAs vary greatly in terms of complexity. Table 10–4 lists some of the parameter ranges that are available. Keep in mind that these numbers are subject to change as technology advances.

TABLE 10-4

Selected FPGA parameters.

Feature	Range
Number of LEs	1,500–813,000
Number of CLBs	26–359,000
Embedded memory	26 kb–63 Mb
Number of I/Os	18–1200
DC operating voltage	1.8 V, 2.5 V, 3.3 V, 5 V

SECTION 10-4 CHECKUP

1. How does an FPGA differ from a CPLD?

2. What does CLB stand for?

3. Describe an LUT and discuss its purpose.

4. What is the difference between a local interconnect and a global interconnect in an FPGA?

5. What is an FPGA core?

6. Define the term *intellectual property* in relation to an FPGA manufacturer.

7. What produces combinational logic functions in an LM?

8. Name the two types of embedded functions.

10-5 Programmable Logic Software

In order to be useful, programmable logic must have both hardware and software components combined into a functional unit. All manufacturers of SPLDs, CPLDs, and FPGAs provide software support for each hardware device. These software packages are in a category of software known as computer-aided design (CAD). In this section, programmable logic software is presented in a generic way using the traffic signal controller from Chapters 6 and 7 Applied Logic for illustration. Tutorials for two types of software, Altera Quartus II and Xilinx ISE, are provided on the website.

After completing this section, you should be able to

* Explain the programming process in terms of design flow

* Describe the design entry phase

* Describe the functional simulation phase

* Describe the synthesis phase

* Describe the implementation phase

* Describe the timing simulation phase

* Describe the download phase

The programming process is generally referred to as **design flow**. A basic design flow diagram for implementing a logic design in a programmable device is shown in Figure 10–34. Most specific software packages incorporate these elements in one form or another and process them automatically. The device being programmed is usually referred to as the **target device**.

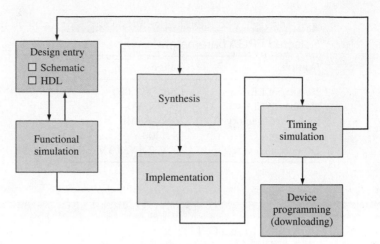

FIGURE 10–34   General design flow diagram for programming a SPLD, CPLD, or FPGA.

You must have four things to get started programming a device: a computer, development software, a programmable logic device (SPLD, CPLD, or FPGA), and a way to connect the device to the computer. These essentials are illustrated in Figure 10–35. Part (a) shows a computer that meets the system requirements for the particular software you are using. Part (b) shows the software acquired either on a CD from the device manufacturer or downloaded from the device manufacturer's website. Most manufacturers provide free software that can be downloaded and used for a limited time (Examples are Altera Quartus II and Xilinx ISE.). Part (c) shows a programmable logic device. Part (d) illustrates two means of physically connecting the device to the computer via cable by using either the

FIGURE 10–35   Essential elements for programming an SPLD, CPLD, or FPGA.
(d) photo courtesy of Digilent, Inc.

programming fixture into which the device is inserted or the development board on which the device is mounted. After the software has been installed on your computer, you must become familiar with the particular software tools before attempting to connect and program a device.

Design Entry

Assume that you have a logic circuit design that you wish to implement in a programmable device. You can enter the design on your computer in either of two basic ways: **schematic entry** or **text entry**. In order to use text entry, you must be familiar with an HDL such as VHDL, Verilog, or AHDL. Most programmable logic manufacturers provide software packages that support VHDL and Verilog because they are standard HDLs. Some also support AHDL, ABEL, or other proprietary HDLs. Schematic entry allows you to place symbols of logic gates and other logic functions from a library on the screen and connect them as required by your design. A knowlege of an HDL is not required for schematic entry.

Building a Logic Design

In addition to programming languages such as VHDL and Verilog, schematic capture can also be used in PLD development. When you enter a complete logic circuit schematic on the screen, it is called a "flat" schematic. Complex logic circuits may be hard to fit onto the screen and difficult to read. You can enter logic circuits in segments, save each segment as a block symbol, and then connect the block symbols graphically to form a complex circuit, as shown in Figure 10–36 for the traffic signal controller (Chapters 6 and 7), which we will use for illustration of the process. This is called a hierarchical approach.

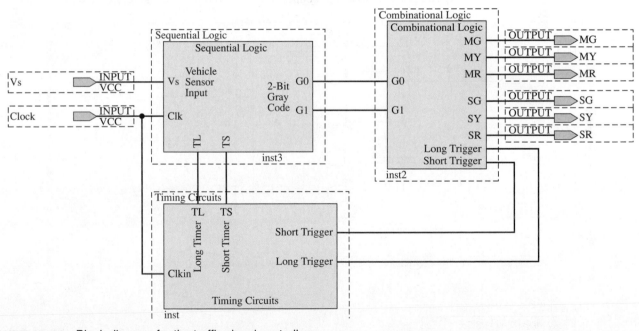

FIGURE 10–36 Block diagram for the traffic signal controller.

The sequential logic section of the traffic signal controller is created using schematic capture and compared to the same application created using VHDL. Figure 10–37 shows the use of VHDL to create the sequential logic component of the system. The sequential logic portion of the traffic light application was developed in Chapter 7. The code for the expressions assigned to D0 and D1 are created straight from the Boolean expressions.

$$D_1 = G_0\overline{T}_\text{S} + G_1T_\text{S}$$
$$D_0 = \overline{G}_1\overline{T}_\text{L}V_\text{s} + \overline{G}_1G_0 + G_0T_\text{L}V_\text{s}$$

```
 1  library ieee;
 2  use ieee.std_logic_1164.all;
 3
 4  entity SequentialLogic is
 5  port(VS, TL, TS, Clk: in std_logic;
 6       G0, G1: inout std_logic);
 7  end entity SequentialLogic;
 8
 9  architecture SequenceBehavior of SequentialLogic is
10  component dff is
11  port(D,Clk: in std_logic;
12       Q: out std_logic);
13  end component dff;
14
15  signal D0, D1: std_logic;
16  begin
17    D1 <= (G0 and not TS) or
18          (G1 and TS);
19
20    D0 <= (not G1 and not TL and VS)  or
21          (not G1 and G0) or
22          (G0 and TL and VS);
23
24  DFF0: dff port map(D => D0, Clk => Clk, Q => G0);
25  DFF1: dff port map(D => D1, Clk => Clk, Q => G1);
26  end architecture SequenceBehavior;
```

FIGURE 10–37   Text entry with VHDL description of the sequential logic for the traffic signal controller.

Figure 10–38(a) shows the sequential logic block created using schematic entry (also known as schematic capture) techniques. Breaking the schematic into separate logic circuits allows for functional compartmentalization and easier development. The Boolean expressions are implemented using separate logic gates with graphical representation of wires and I/O components needed to connect them. The completed and tested module is reduced to a

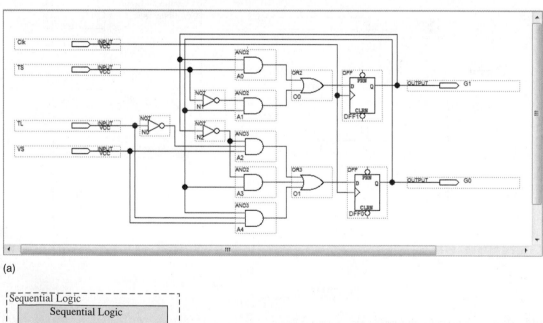

(a)

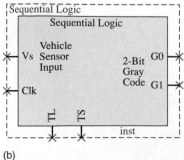

(b)

FIGURE 10–38   The sequential logic using schematic entry.

simple block symbol, as shown in Figure 10–38(b), and can be inserted as a component, as shown in Figure 10–36. A block symbol can also be created using VHDL code.

Functional Simulation

The purpose of the **functional simulation** in the design flow is to make sure that the design you entered works as it should in terms of its logic operation, before synthesizing into a hardware design. Basically, after a logic circuit is compiled, it can then be simulated by applying input waveforms and checking the output for all possible input combinations. Functional simulation is accomplished graphically using a waveform editor or programmatically using a test bench. Graphical waveform editors allow drawing of test stimulus using waveform drawing features and drag and drop techniques.

Graphical Approach

Graphical generation tools allow for the easy creation of drawn stimulus waveforms for simple testing applications. Graphical waveforms are created to provide the input stimulus for the sequential logic component of the traffic signal control system as an example. Inputs Clk, TL, TS, and VS will be created using graphical tools. Output identifiers G0 and G1 require no input stimulus and are simply dragged and dropped into the Wave window. The clock definition is created using the Define Clock feature to drive the system clock Clk and limit the simulation run time. The offset, duty cycle, period, logic values, cancel, and initial edge are provided. Inputs VS, TL, and TS are created using the same graphical techniques. You can view the drawn stimulus waveforms prior to simulation. Typical windows are shown in Figure 10–39.

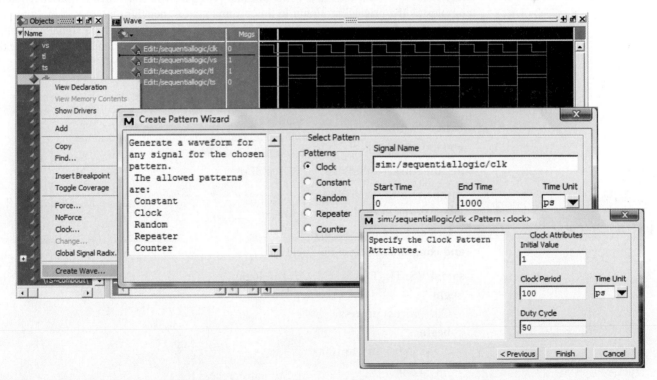

FIGURE 10–39   Functional simulation.

After you have specified the input waveforms, the simulation is ready to run. When the simulation is started, the output waveforms for G0 and G1 will be displayed as shown in Figure 10–40. This allows you to verify that the design is good or that it is working properly. In this case, the output waveform is corrected to the selected input waveforms. An incorrect output waveform would indicate a flaw in the functionality of the logic; you would have to go back, check the original design, and then re-enter a revised design.

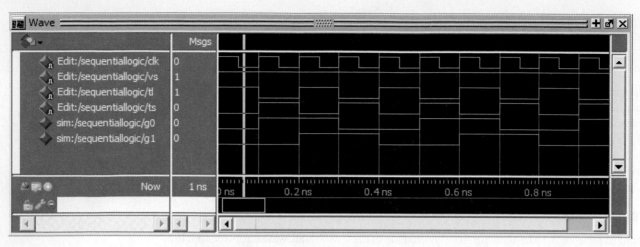

FIGURE 10–40   After the functional simulation is run, the output waveform should indicate that the logic is functioning properly.

Test Bench Approach

A programmatic approach to design simulation is to create an additional program file called a *test bench*. A test bench is similar in construction to the program code and is typically written in the same HDL as the original program. The test bench program can be as complex as the original program. In this example, a test bench program is written to provide the input stimulus for the sequential logic component of the traffic signal controller. The following test bench program is written in VHDL to simulate input waveforms for the sequential logic module.

```
library IEEE;
use IEEE.std_logic_1164.all;

entity TestSL is
end entity TestSL;
```
> Input stimulus for the SequentialLogic unit under test (UUT) is created programmatically within the program so the entity is left blank.

```
architecture TestSLBehavior of TestSL is

component SequentialLogic is
port(VS, TL, TS, CLK: in std_logic;
     G0, G1: inout std_logic);
end component SequentialLogic;

signal VS, TL, TS, Clk, G0, G1: std_logic;
begin
  Clk_process:process
  begin
    for iterate in 1 to 10000
      loop
      CLK<='1';
      wait for 50 us;
      CLK<='0';
      wait for 50 us;
    end loop;
    wait;
  end process;
```
> Stimulus process for CLKin input. A loop structure is used to limit the number clock cycles to 10000.

```
TLS_process: process
begin
        TL <= '0';
        TS <= '1';
        wait for 100 us;                  Stimulus process for
        TL <= '1';                        TL and TS input.
        TS <= '0';
        wait for 100 us;
    end process;
    stim_proc: process
    begin
        VS <= '0';
        wait for 100 us;                  Stimulus process for VSin
        VS <= '1';                        input.
        wait;
    end process;
UUT: SequentialLogic port map
(VS => VS, TL => TL, TS => TS, Clk => Clk, G0 => G0, G1 => G1);
end architecture TestSLBehavior;
```

> Creating a separate stimulus process for inputs CLK, TL, TS, and VS allows for independent control of the input identifiers.

After the test bench simulation is run, the output waveform on the waveform editor screen should indicate that the logic is functioning properly.

Synthesis

Once the design has been entered and functionally simulated to verify that its logical operation is correct, the compiler automatically goes through several phases to prepare the design to be downloaded to the target device. During this synthesis phase of the design flow, the design is optimized in terms of minimizing the number of gates, replacing logic elements with other logic elements that can perform the same function more efficiently, and eliminating any redundant logic. The final output from the synthesis phase is a netlist that describes the optimized version of the logic circuit.

To demonstrate the process of design optimization, the schematic capture version of the sequential logic section of the system is presented with redundant ORGates and NotGates, as shown in Figure 10–41(a). The AND-OR logic that was entered in the design entry phase, shown in Figure 10–41(a), could result in the optimized circuit shown in Figure 10–41(b). In this illustration, the compiler removed two 2-input OR gates and replaced them with a single 3-input OR gate. Also, one of the redundant inverters was eliminated.

Netlist

A **netlist** is a connectivity list that describes components and how they are connected together. Generally, a netlist contains references to descriptions of the components or elements used. Each time a component, such as a logic gate, is used in a netlist, it is called an *instance*. Each instance has a definition that lists the connections that can be made to that kind of component and some basic properties of that component. These connection points are called *ports or pins*. Usually, each instance will have a unique name; for example, if you have two instances of AND gates, one might be "and1" and the other "and2". Aside from their names, they might otherwise be identical. Nets are the "wires" that connect things together in the circuit. Net-based netlists usually describe

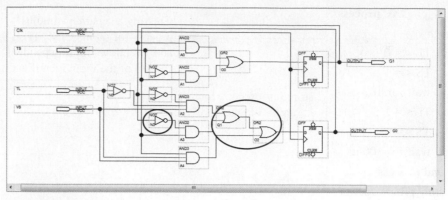

(a) Original logic design

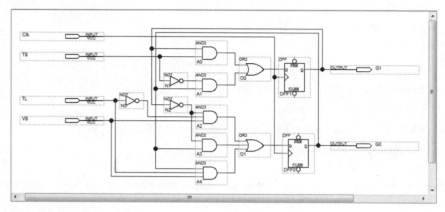

(b) Logic after synthesis

FIGURE 10–41 Example of logic optimization during synthesis. The final version is reduced by eliminating one inverter and combining two 2-input OR gates into a single 3-input OR gate.

all the instances and their attributes, then describe each net, and specify which port they are connected to on each instance. The synthesis software generates a netlist, as shown in Figure 10–42(a). The netlist indicates the type of information that is necessary to describe a circuit. One format used for netlists is **EDIF** (Electronic Design Interchange Format). Using the netlist, the software creates a schematic representation of the net assignments, as shown in Figure 10–42(b).

Implementation (Software)

After the design has been synthesized, the **compiler** implements the design, which is basically a "mapping" of the design so that it will fit in the specific target device based on its architecture and pin configurations. This process is called *fitting* or *place and routing*. To accomplish the implementation phase of the design flow, the software must "know" about the specific device and have detailed pin information. Complete data on all potential target devices are generally stored in the software library.

Timing Simulation

This part of the design flow occurs after the implementation and before downloading to the target device. The **timing simulation** verifies that the circuit works at the design frequency and that there are no timing problems that will affect the overall operation. Since a functional simulation has already been done, the circuit should work properly from a logic

Netlist(SequentialLogic)
Net<name>: instance<name>,<from>,<to>;
Instances: and0,and1,and2,and3,and4,or0,or1,inv0,inv1,inv2,DFF0,DFF1;
Input/outputs:I1,I2,I3,I4,O1,O2
net1:   DFF0, inport2; DFF1, inport2; I1;
net2:   and0, inport2; inv1, outport1; I2;
net3:   inv0, outport1; and4, inport2; I3;
net4:   and2, inport3, and4, inport4; I4;
net5:   and2, inport2;
net6:   DFF1, outport1; and0, inport1; inv2, output1; O0;
net7:   DFF0, outport1; and1, inport2; and3, inport2; and4, inport1; O1;
net8:   and1, inport1;
net9:   and2, inport1; and3, inport1;
net10: or0, inport1;
net11: or0, inport2;
net12: or1, inport1;
net13: or1, inport2;
net14: or1, inport3;
net15: DFF1, inport1;
net16: DFF0, inport1;
end;

(a) Netlist

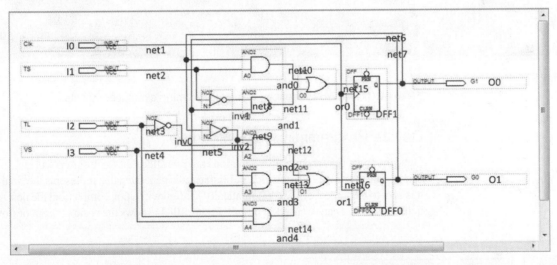

(b) Schematic representation of netlist

FIGURE 10–42   Synthesis produces netlist and schematic for the optimized sequential logic.

point of view. The development software uses information about the specific target device, such as propagation delays of the gates, to perform a timing simulation of the design. For the functional simulation, the specification of the target device was not required; but for the timing simulation, the target device must be chosen. The Waveform Editor can be used to view the result of the timing simulation just as with the functional simulation, as illustrated in Figure 10–43. If there are no problems with the timing, as shown in part (a), the design is ready to download. However, suppose that the timing simulation reveals a "glitch" due to propagation delay, as shown in Figure 10–43(b). A glitch is a very short duration spike in the waveform. In this event, you would need to carefully analyze the design for the cause, then re-enter the modified design, and repeat the design flow process. Remember, you have not committed the design to hardware at this point.

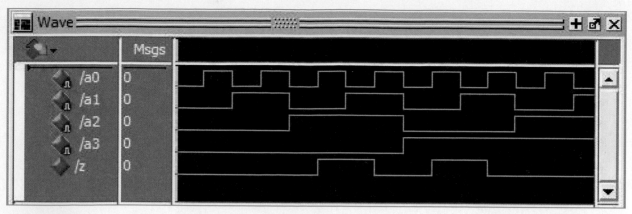

(a) Good result

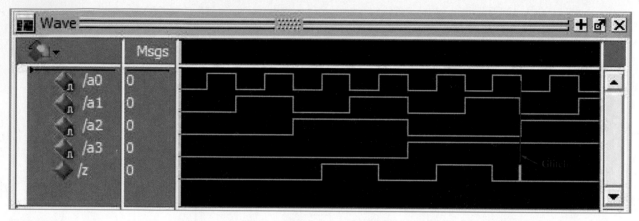

(b) Timing problem

FIGURE 10–43 Hypothetical examples of timing simulation results.

Device Programming (Downloading)

Once the functional and timing simulations have verified that the design is working properly, you can initiate the download sequence. A **bitstream** is generated that represents the final design, and it is sent to the target device to automatically configure it. Upon completion, the design is actually in hardware and can be tested in-circuit. Figure 10–44 shows the basic concept of **downloading**.

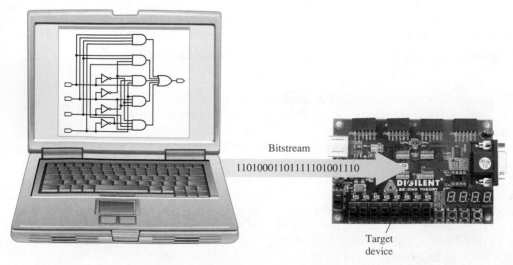

FIGURE 10–44 Downloading a design to the target device.   (Photo courtesy of Digilent, Inc.)

SECTION 10–5 CHECKUP

1. List the phases of the design flow for programmable logic.

2. List the essential elements for programming a CPLD or FPGA.

3. What is the purpose of a netlist?

4. Which comes first in the design flow, the functional simulation or the timing simulation?

10–6 Boundary Scan Logic

Boundary scan is used for both the testing and the programming of the internal logic of a programmable device. The JTAG standard for boundary scan logic is specified by IEEE Std. 1149.1. Most programmable logic devices are JTAG compliant. In this section, the basic architecture of a JTAG IEEE Std. 1149.1 device is introduced and discussed in terms of the details of its boundary scan register and control logic structure.

After completing this section, you should be able to

 ◆ Describe the required elements of a JTAG-compliant device

 ◆ List the mandatory JTAG inputs and outputs

 ◆ State the purpose of the boundary scan register

 ◆ State the purpose of the instruction register

 ◆ Explain what the bypass register is for

IEEE Std. 1149.1 Registers

All programmable logic devices that are compliant with IEEE Std. 1149.1 require the elements shown in the simplified diagram in Figure 10–45. These are the boundary scan register, the bypass register, the instruction register, and the TAP (test access port) logic. Another register, the identification register, is optional and not shown in the figure.

Boundary Scan (BS) Register The interconnected BSCs (boundary scan cells) form the boundary scan register. The serial input to the register is the TDI (test data in), and the serial output is TDO (test data out). Data from the internal logic and the input and output pins of the device can also be parallel shifted into the BS register. The BS register is used to test connections between PLDs and the internal logic that has been programmed into the device.

Bypass (BP) Register This required data register (typically only one flip-flop) optimizes the shifting process by shortening the path between the TDI and the TDO in case the BS register or other data register is not used.

Instruction Register This required register stores instructions for the execution of various boundary scan operations.

Identification (ID) Register An identification register is an optional data register that is not required by IEEE Std. 1149.1. However, it is used in some boundary scan architectures to store a code that identifies the particular programmable device.

IEEE Std. 1149.1 Boundary Scan Instructions

Several standard instructions are used to control the boundary scan logic. In addition to these, other optional instructions are available.

 • **BYPASS** This instruction switches the BP register into the TDI/TDO path.

 • **EXTEST** This instruction switches the BS register into the TDI/TDO path and allows external pin tests and interconnection tests between the output of one programmable logic device and the input of another.

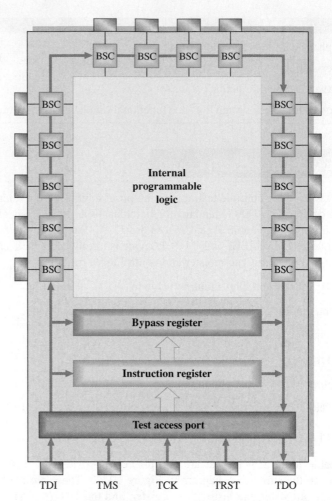

FIGURE 10–45   Greatly simplified diagram of a JTAG compliant (IEEE Std. 1149.1) programmable logic device (CPLD or FPGA). The BSCs (boundary scan cells) form the boundary scan register. Only a small number of BSCs are shown for illustration.

- **INTEST** This instruction switches the BS register into the TDI/TDO path and allows testing of the internal programmed logic.
- **SAMPLE/PRELOAD** This instruction is used to sample data at the device input pins and apply the data to the internal logic. Also, it is used to apply data (preload) from the internal logic to the device output pins.
- **IDCODE** This instruction switches the optional identification register into the TDI/ TDO path so the ID code can be shifted out to the TDO.

IEEE Std. 1149.1 Test Access Port (TAP)

The Test Access Port (TAP) consists of control logic, four mandatory inputs and outputs, and one defined optional input, Test Reset (TRST).

- **Test Data In (TDI)** The TDI provides for serially shifting test and programming data as well as instructions into the boundary scan logic.
- **Test Data Out (TDO)** The TDO provides for serially shifting test and programming data as well as instructions out of the boundary scan logic.
- **Test Mode Select (TMS)** The TMS switches between the states of the TAP controller.
- **Test Clock (TCK)** The TCK provides timing for the TAP controller which generates control signals for the data registers and the instruction register.

A block diagram of the boundary scan logic is shown in Figure 10–46. Both instructions and data are shifted in on the TDI line. The TAP controller directs instructions into the instruction register or data into the appropriate data register. A decoded instruction from the instruction decoder selects which data register is to be accessed via MUX 1 and also if an instruction or data are to be shifted out on the TDO line via MUX 2. Also, a decoded instruction provides for setting up the boundary scan register in one of five basic modes. The boundary scan cell and its modes of operation are described next.

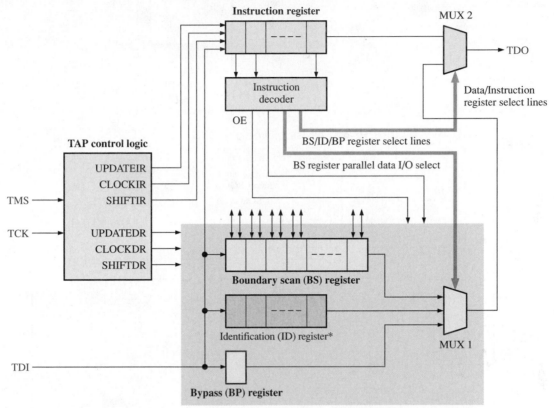

FIGURE 10–46 Boundary scan logic diagram.

The Boundary Scan Cell (BSC)

The boundary scan register is made up of boundary scan cells. A block diagram of a basic bidirectional BSC is shown in Figure 10–47. As indicated, data can be serially shifted in and out of the BSC. Also, data can be shifted into the BSC from the internal programmable logic, from a device input pin, or from the previous BSC. Additionally, data can be shifted out of the BSC to the internal programmable logic, to a device output pin, or to the next BSC.

The architecture of a generic boundary scan cell is shown in Figure 10–48. The cell consists of two identical logic circuits, each containing two flip-flops and two multiplexers. Essentially, one circuit allows data to be shifted from the internal programmable logic or to a device output pin. The other circuit allows data to be shifted from a device input pin or to the internal programmable logic.

There are five modes in which the BSC can operate in terms of data flow. The first BSC mode allows data to flow serially from the previous BSC to the next BSC, as illustrated in Figure 10–49. A 1 on the SHIFT input selects the SDI. The data on the SDI line are clocked into Capture register A on the positive edge of the CLOCK. The data are then clocked into Capture register B on the negative edge of the CLOCK and appear

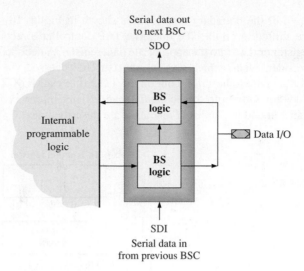

FIGURE 10–47   A basic bidirectional BSC.

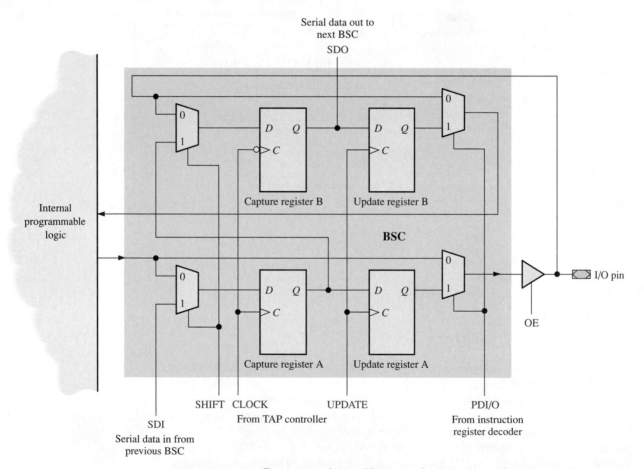

FIGURE 10–48   Representative architecture of a typical boundary scan cell.

on the SDO line. This is equivalent to serially shifting data through the boundary scan register.

The second BSC mode allows data to flow directly from the internal programmable logic to a device output pin, as illustrated in Figure 10–50. The 0 on the PDI/O (parallel data I/O) control line selects the data from the internal programmable logic. The 1 on the OE (output enable) line enables the output buffer.

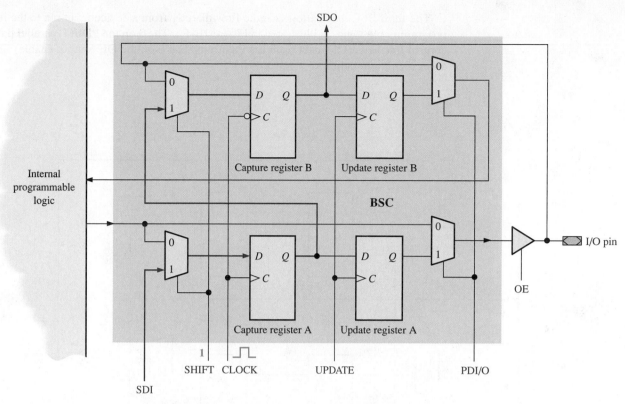

FIGURE 10–49 Data path for serially shifting data from one BSC to the next. There is a 1 on the SHIFT input and a CLOCK pulse is applied. The red lines indicate data flow.

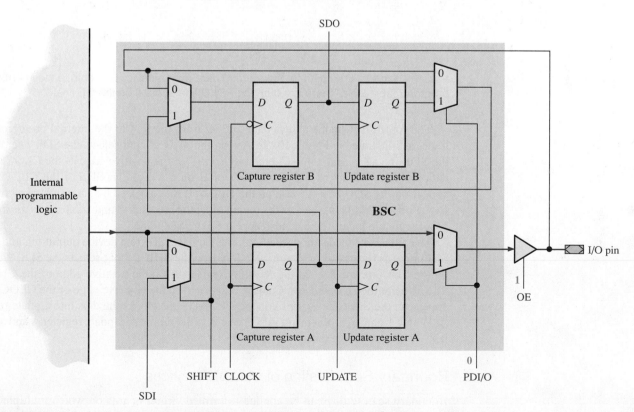

FIGURE 10–50 Data path for transferring data from the internal programmable logic to a device output pin. There is a 0 on the PDI/O line and a 1 on the OE line.

The third BSC mode allows data to flow directly from a device input pin to the internal programmable logic, as illustrated in Figure 10–51. The 0 on the PDI/O (parallel data I/O) control line selects the data from the input pin. The 0 on the OE (output enable) line disables the output buffer.

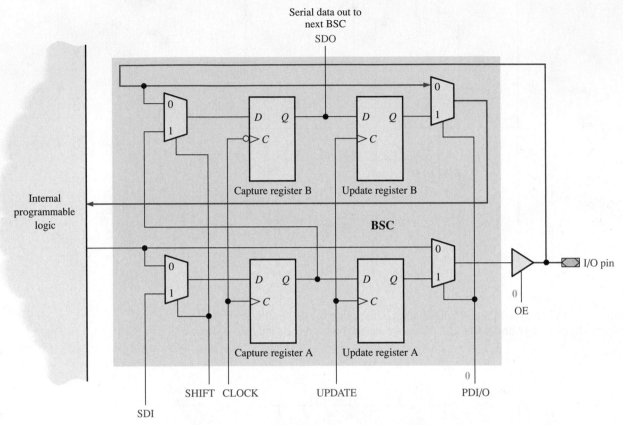

FIGURE 10–51   Data path for transferring data from a device input pin to the internal programmable logic. There is a 0 on the PDI/O line and a 0 on the OE line.

The fourth BSC mode allows data to flow from the SDI to the internal programmable logic, as illustrated in Figure 10–52. A 1 on the SHIFT input selects the SDI. The data on the SDI line are clocked into Capture register A on the positive edge of the CLOCK. The data are then clocked into Capture register B on the negative edge of the CLOCK and appear on the SDO line. A pulse on the UPDATE line clocks the data into Update register B. A 1 on the PDI/O line selects the output of Update register B and applies it to the internal programmable logic. The data also appear on the SDO line.

The fifth BSC mode allows data to flow from the SDI to a device output pin and to the SDO line, as illustrated in Figure 10–53. A 1 on the SHIFT input selects the SDI. The data on the SDI line are clocked into Capture register A on the positive edge of the CLOCK. The data are then clocked into Capture register B on the negative edge of the CLOCK and appear on the SDO line. A pulse on the UPDATE line clocks the data into Update register A. With a 1 on OE, a 1 on the PDI/O line selects the output of Update register A and applies it to the device output pin.

Boundary Scan Testing of Multiple Devices

Boundary scan testing can be applied to printed circuit boards on which multiple JTAG (IEEE Std. 1149.1) devices are mounted to check interconnections as well as internal logic. This concept is illustrated by tracing the path of data shown in red through the boundary scan registers in Figure 10–54.

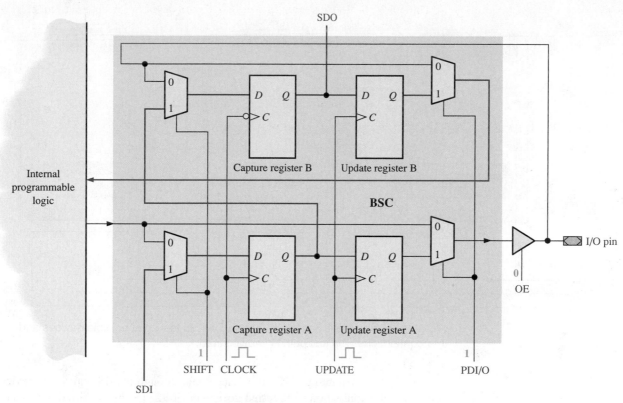

FIGURE 10–52 Data path for transferring data from the SDI to the internal programmable logic and the SDO. There is a 1 on the SHIFT line, a 1 on the PDI/O line, and a 0 on the OE line. A pulse is applied to the CLOCK line followed by a pulse on the UPDATE line.

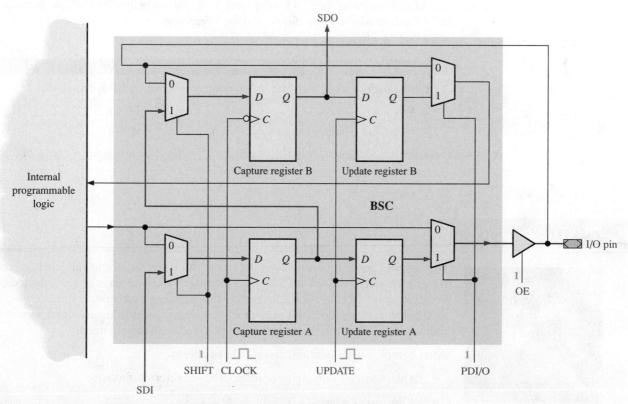

FIGURE 10–53 Data path for transferring data from the SDI to a device output pin and the SDO. There is a 1 on the SHIFT line, a 1 on the PDI/O line and a 1 on the OE line. A pulse is applied to the CLOCK line followed by a pulse on the UPDATE line.

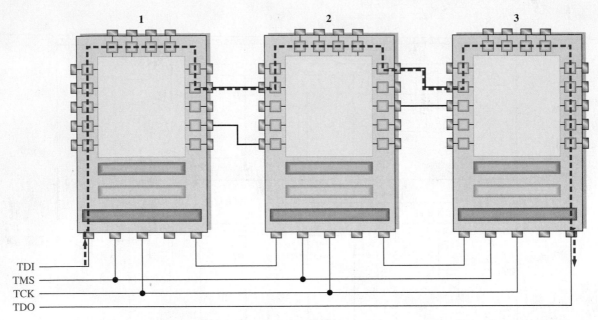

FIGURE 10–54 Basic concept of boundary scan testing of multiple devices and interconnections. The test path is shown in red.

The bit is shifted into the TDI of device 1 and through the BS register of device 1 to a cell where the connection to be tested goes to device 2. The bit is shifted out to the device output pin and through the interconnection to the input pin of device 2. The bit continues through the BS register of device 2 to an output pin and through the interconnection to the input pin of device 3. It is then shifted through the BS register of device 3 to the TDO. If the bit coming out of the TDO is the same as the bit going into the TDI, the boundary scan cells through which it was shifted and the interconnections from device 1 to device 2 and from device 2 to device 3 are good.

SECTION 10–6 CHECKUP

1. List the boundary scan inputs and outputs required by IEEE Std.1149.1.

2. What is the TAP?

3. Name the mandatory registers in boundary scan logic.

4. Describe five modes in which a boundary scan cell can operate in terms of data flow.

10–7 Troubleshooting

During program code development, simulation tools can be used to validate logic modules for proper operation prior to PLD programming. Two basic ways to test a device that has been programmed with a logic design are traditional and automated. Boundary scan is an automated method used in this section. The focus is on simulation prior to device programming and boundary scan testing once the PLD has been programmed.

After completing this section, you should be able to

◆ Explain troubleshooting techniques using waveform simulation

◆ Define *break point*

◆ Discuss boundary scan testing

Troubleshooting with Waveform Simulation

As discussed, simulation waveform stimulus can be accomplished using a test bench program or graphically using a waveform editor. The following illustration demonstrates simulation troubleshooting techniques applied to the SequentialLogic section of the traffic signal controller created in VHDL.

Functional Simulation

Prior to download to the target device, simulation tools are useful to identify unexpected behavior. In the following illustration, the waveform output in Figure 10–55 shows that the sequential logic Gray code output from identifiers g0 and g1 does not respond to the waveform test stimulus as expected. In a timing simulation, the PLD chip libraries are loaded, and testing is conducted against a model of the target device where typically outputs start at a zero state. In the functional simulation, the basic logic is tested. Since functional simulation does not make assumptions about initial states, a circular dependency could exist where the output of one function is used to determine the outcome of a second where neither may be resolved. A break point can be inserted in the program code to determine where undetermined states may exist, so they can be addressed in the program code if needed. A **break point** is a flag placed within the program source code where the application is stopped temporarily, allowing investigation of program identifiers and the status of the I/O.

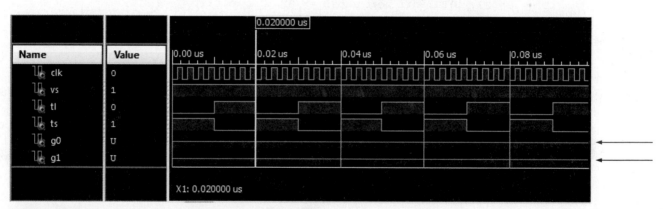

FIGURE 10–55

To investigate this behavior, you can insert a break point into the program code, so you can view the condition of identifiers G0 and G1 as the simulation progresses.

In the sequential logic component of the traffic signal controller, identifiers D0 and D1 are dependent on the output of flip-flops DFF0 and DFF1. Since D0 feeds DFF0, for example, D0 could be in an undetermined state at startup, causing G0 to also be in an undetermined state. The functional simulation would point this out as shown since G0 and G1 are left in an undetermined state. As shown in Figure 10–56, in this case, a break point is set by right-clicking line number 22 and selecting "Set Breakpoint 22". Multiple break points may be defined as needed to investigate the behavior of the program under simulation.

The simulation has stopped at the predefined break point inserted at identifier D0. By examining the condition of the supporting identifiers D0, TL, VS and G1, you determine the problem to be related to the D flip-flop components whose output value G1 is listed as "U" or undefined. D0 is dependent on identifier G1 and the flip-flop. DFF1 is in turn dependent on D1. The output of the flip-flop does not allow resolution of the Boolean expressions assigned to D0 or D1.

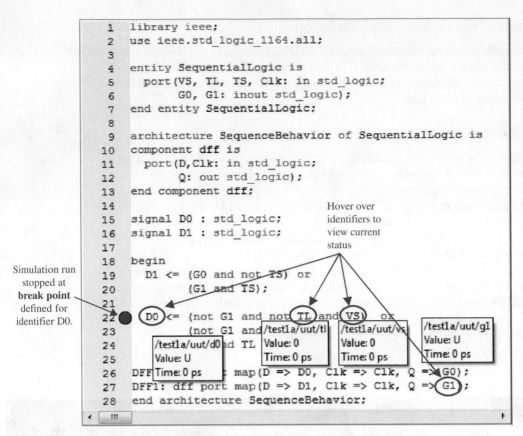

FIGURE 10–56

Examining the D flip-flop definition, you see that the flip-flop simply writes the value of the D input to output Q upon a rising clock edge. Figure 10–57(a) shows that the output Q is not preinitialized, causing the output to start in an uninitialized state. To correct this problem, a new signal QT is created and initialized to 0 in Figure 10–57(b). The value of identifier D is written to signal QT upon a rising clock edge and QT is written to output Q.

```
1  library ieee;
2  use ieee.std_logic_1164.all;
3
4  entity dff is
5    port(D,Clk: in std_logic;
6         Q: out std_logic);
7  end dff;
8
9  architecture dffbehavior of dff is
10 begin
11   process (Clk)
12   begin
13     if (Clk'event and Clk='1') then
14       Q <= D;
15     end if;
16   end process;
17 end dffbehavior;
```

(a)

```
1  library ieee;
2  use ieee.std_logic_1164.all;
3  entity dff is
4    port(D,Clk: in std_logic;
5         Q: out std_logic);
6  end dff;
7
8  architecture dffbehavior of dff is
9  signal QT :std_logic:= '0';
10 begin
11   process (Clk)
12   begin
13     if (Clk'event and Clk='1') then
14       QT <= D;
15     end if;
16   end process;
17   Q <= QT;
18 end dffbehavior;
```

(b)

Signal QT allows for the pre-initialization of a 0 to output Q

FIGURE 10–57

Initializing the D flip-flop output to 0 allows the Boolean expressions for D0 and D1 to resolve to a value of 1 or 0. A second simulation shows that the sequential logic portion of the traffic signal controller is now able to output a valid Gray code, as shown in Figure 10–58.

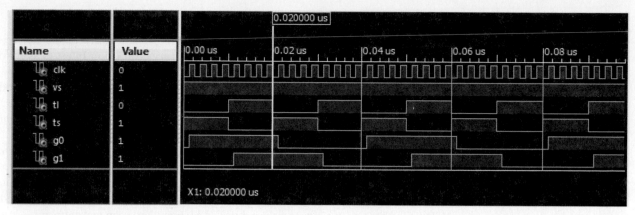

FIGURE 10–58

Boundary Scan Testing

Limited access to test points led to the concept of placing the test points within the integrated circuit devices themselves. Most CPLDs and FPGAs include boundary scan logic as part of their internal structure independent of the functionality of the logic programmed into the device. These devices are JTAG compliant.

A circuit, known as a boundary scan cell, is placed between the programmable logic and each input and output pin of the device, as shown in Figure 10–59. The cells are basically memory cells that store a 1 or a 0. The cells connected to the programmable logic inputs are called input cells, and those connected to the programmable logic outputs are called output cells. **Boundary scan** testing is based on the JTAG standard (IEEE Std. 1149.1). The four JTAG inputs and outputs—TDI (test data in), TDO (test data out), TCK (test clock), and TMS (test mode select)—are known as the test access port (TAP).

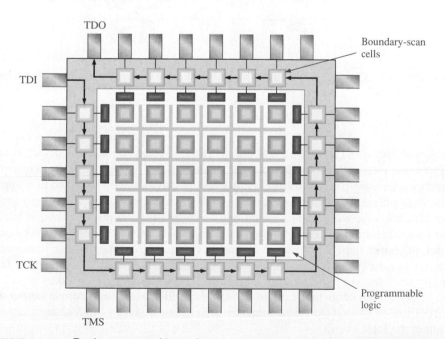

FIGURE 10–59 Basic concept of boundary scan logic in a programmable logic device.

Intest

When boundary scan cells are used to test the internal functionality of the device, the test mode is called Intest. The basic concept of boundary scan using Intest is as follows: A software-driven pattern of 1s and 0s is shifted in via the TDI pin and is placed on the programmable logic inputs. As a result of these applied input bits, the logic will produce output bit(s) in response. The resulting output bit(s) is (are) then shifted out on the TDO pin and checked for errors. An incorrect output, of course, indicates a fault in the programmed logic, I/O cells, or boundary scan cells.

Figure 10–60 shows a boundary scan Intest pattern 1011 for an AND-OR logic circuit that has been programmed into a device. Sixteen combinations of four TDI bits would test the circuit in all possible states according to the list in Table 10–5. The 4-bit combinations are serially shifted into the boundary scan cells, and the corresponding output is shifted out on TDO for checking. This process is controlled by boundary scan test software.

TABLE 10–5

Boundary scan test bit pattern for the programmed device in Figure 10–60.

TDI	TDO
0000	1
0001	1
0010	0
0011	1
0100	1
0101	1
0110	1
0111	1
1000	1
1001	1
1010	0
1011	1
1100	1
1101	1
1110	1
1111	1

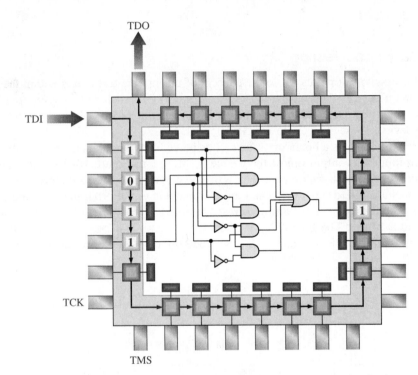

FIGURE 10–60 Example of a bit pattern in the boundary scan Intest for the internal logic.

Extest

When boundary scan cells are used to test the external connections to the device in addition to some internal functionality, the test mode is called Extest. The basic concept of boundary scan using Extest is as follows: A software-driven pattern of 1s and 0s is applied to the input pins of the device and entered into the input cells. As a result of these applied input bits, the logic will produce output bit(s) in response. The resulting output bit(s) is (are) then taken from the output pin of the device and checked for errors. An incorrect output, of course, indicates a fault in the input or output pin connections or interconnections, an incorrect device, or improperly installed device. Obviously, some internal faults can also be detected in the Extest mode. For example, faults in the boundary scan cells, I/O cells or certain faults in the programmed logic will produce an incorrect output. Figure 10–61 shows an example of a boundary scan Extest that tests the four inputs and the output of the logic circuit.

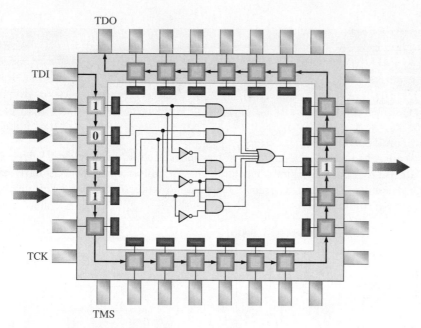

FIGURE 10-61 Example of a bit pattern in the boundary scan Extest for external faults.

If a fault is detected in the Extest mode, it can be either external (a bad pin connection) or internal (a faulty connection, boundary scan cell, or logic element) to the device. Therefore, in order to isolate an Extest detected fault, an Intest should be run following the Extest. If both tests show a fault, then it is internal to the device.

In the Extest mode, it is necessary to probe contacts to the input and output pins of the device. These pins have to be available at a connector to the circuit board or on test pads so they can be checked by the automatic test equipment.

Boundary Scan Description Language (BSDL)

This test software is part of the JTAG standard IEEE 1149.1 and uses VHDL to describe how the boundary scan logic is implemented in a specific device and how it operates. BSDL provides a standard data format for describing how IEEE 1149.1 is implemented in a JTAG-compliant device. When you use boundary scan test software tools that support BSDL, you can usually obtain BSDL from the device manufacturer.

Each device that contains dedicated boundary scan logic is supported by a BSDL file that describes that particular device. Certain things described in the BSDL file include the device type and descriptions of the I/O pins and TAP (test access port) pins. BSDL also provides a mapping of logical signals onto the physical pins and a description of the boundary scan logic architecture contained in the device. A bit test pattern for testing the device can be defined using BSDL.

SECTION 10-7 CHECKUP

1. Describe the purpose of a programmer-defined break point.
2. Explain the basic concept of boundary scan.
3. What are the two modes of boundary scan test?
4. Name four JTAG signals used with boundary scan.
5. What is BSDL?

Applied Logic

Elevator Controller: Part 2

In this section, the elevator controller that was introduced in the Applied Logic in Chapter 9 will be programmed for implementation in a PLD. Refer to Chapter 9 to review the elevator operation. The logic diagram is repeated in Figure 10–62 with labels changed to facilitate programming.

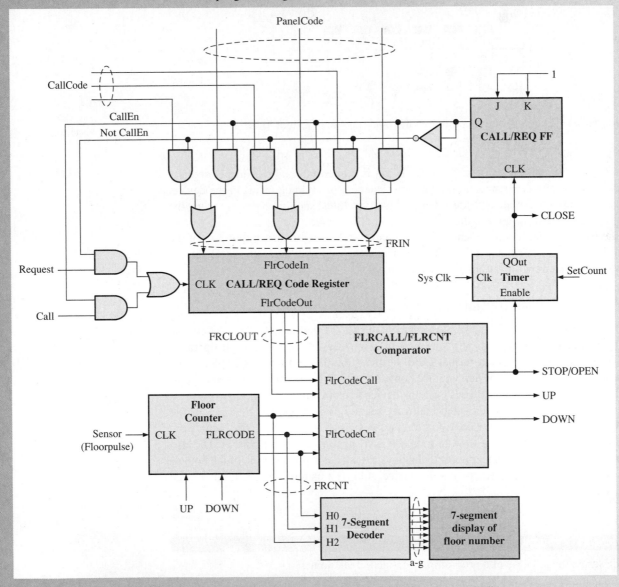

FIGURE 10–62   Programming model of the elevator controller.

The VHDL program code for the elevator controller will include component definitions for the Floor Counter, the FLRCALL/FLRCNT Comparator, the Code Register, the Timer, the Seven-Segment Decoder, and the CALL/REQ Flip-Flop. The VHDL program codes for these six components are as follows. (Blue annotated notes are not part of the program.)

Floor Counter

```vhdl
library ieee;
use ieee.std_logic_1164.all;
use ieee.numeric_std.all;

entity FLOORCOUNTER is
    port (UP, DOWN, Sensor: in std_logic;
            FLRCODE: out std_logic_vector(2 downto 0));
end entity FLOORCOUNTER;

architecture LogicOperation of FLOORCOUNTER is
signal FloorCnt: unsigned(2 downto 0) := "000";
begin
    process(UP, DOWN, Sensor, FloorCnt)
    begin
        FLRCODE <= std_logic_vector(FloorCnt);

        if (Sensor'EVENT and Sensor = '1') then
            if UP = '1' and DOWN = '0' then
                FloorCnt <= FloorCnt + 1;
            elsif Up = '0' and DOWN = '1' then
                FloorCnt <= FloorCnt - 1;
            end if;
        end if;
    end process;
end architecture LogicOperation;
```

ieee.numeric_std_all is included to enable casting of unsigned identifier. Unsigned FloorCnt is converted to std_logic_vector.

UP, DOWN: Floor count direction signals
Sensor: Elevator car floor sensor
FLRCODE: 3-digit floor count

Floor count is initialized to 000.

Numeric unsigned FloorCnt is converted to std_logic_vector data type and sent to std_logic_vector output FLRCODE.

Sensor event high pulse causes the floor count to increment when UP is set high or decrement by one when DOWN is set low.

FLRCALL/FLRCNT Comparator

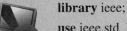

```vhdl
library ieee;
use ieee.std_logic_1164.all;
use ieee.std_logic_arith.all;

entity FLRCALLCOMPARATOR is
    port (FlrCodeCall, FlrCodeCnt: in std_logic_vector(2 downto 0);
        UP, DOWN, STOP: inout std_logic;
end entity FLRCALLCOMPARATOR;

architecture LogicOperation of FLRCALLCOMPARATOR is
begin
    STOP <= '1' when (FlrCodeCall = FlrCodeCnt) else '0';
    UP <= '1' when (FlrCodeCall > FlrCodeCnt) else '0';
    DOWN <= '1' when (FlrCodeCall < FlrCodeCnt) else '0';
end architecture LogicOperation;
```

FlrCodeCall, FlrCodeCnt: Compared values
UP, DOWN, STOP: Output control signals

STOP, UP, and DOWN signals are set or reset based on =, >, and < relational comparisons.

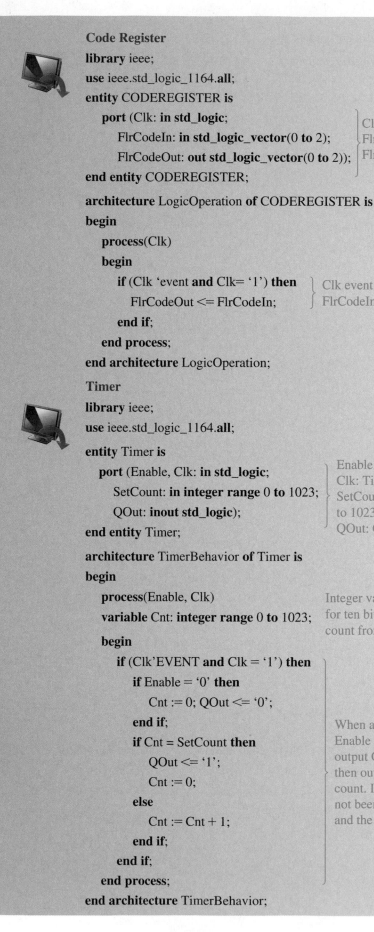

Code Register

```vhdl
library ieee;
use ieee.std_logic_1164.all;
entity CODEREGISTER is
    port (Clk: in std_logic;
        FlrCodeIn: in std_logic_vector(0 to 2);
        FlrCodeOut: out std_logic_vector(0 to 2));
end entity CODEREGISTER;

architecture LogicOperation of CODEREGISTER is
begin
    process(Clk)
    begin
        if (Clk 'event and Clk= '1') then
            FlrCodeOut <= FlrCodeIn;
        end if;
    end process;
end architecture LogicOperation;
```

Clk: Clk Pulse input
FlrCodeIn: 3-digit floor panel input
FlrCodeOut: 3-digit floor panel output

Clk event high pulse sends the FlrCodeIn floor number to FlrCodeOut.

Timer

```vhdl
library ieee;
use ieee.std_logic_1164.all;

entity Timer is
    port (Enable, Clk: in std_logic;
        SetCount: in integer range 0 to 1023;
        QOut: inout std_logic);
end entity Timer;

architecture TimerBehavior of Timer is
begin
    process(Enable, Clk)
    variable Cnt: integer range 0 to 1023;
    begin
        if (Clk'EVENT and Clk = '1') then
            if Enable = '0' then
                Cnt := 0; QOut <= '0';
            end if;
            if Cnt = SetCount then
                QOut <= '1';
                Cnt := 0;
            else
                Cnt := Cnt + 1;
            end if;
        end if;
    end process;
end architecture TimerBehavior;
```

Enable: Enable timer count input
Clk: Timer clock input
SetCount: Counter set input. Limit to 1023 for ten bits.
QOut: Counter output

Integer variable Cnt range limited to 1023 for ten bits used to count from 0 to terminal count from integer port input SetCount.

When a Clk clock event is HIGH, input Enable is checked for a '0' to clear Cnt and output Qout. If Cnt is equal to SetCount, then output QOut is set to '1' ending the count. If the terminal count in SetCount has not been reached, Cnt is incremented by one and the count process continues.

Seven Segment Decoder

a, b, c, d, e, f, g: Seven-segment display element output
H0, H1, H2: Hexadecimal count input

Seven-segment logic operation

```
library ieee;
use ieee.std_logic_1164.all;

entity SevenSegment is
    port (a, b, c, d, e, f, g: out std_logic; H0, H1, H2: inout std_logic);
end entity SevenSegment;

architecture SevenSegmentBehavior of SevenSegment is
begin
    a <= H1 or (H2 and H0) or (not H2 and not H0);
    b <= not H2 or (not H0 and not H1) or (H0 and H1);
    c <= H0 or not H1 or H2;
    d <= (not H0 and not H2) or (not H2 and H1) or
        (H1 and not H0) or (H2 and not H1 and H0);
    e <= (not H0 and not H2) or (H1 and not H0);
    f <= (not H1 and H2) or (not H1 and not H0) or (H2 and not H0);
    g <= (not H2 and H1) or (H1 and not H0) or (H2 and not H1);
end architecture SevenSegmentBehavior;
```

CALL/REQ FF

```
library ieee;
use ieee.std_logic_1164.all;

entity JKFlipFlop is
    port (J,K,Clk: in std_logic; Q: inout std_logic);
end entity JKFlipFlop;

architecture LogicOperation of JKFlipFlop is
signal QNot: std_logic := '1';
begin
    process (J, K, Clk)
    begin
        if (Clk'EVENT and Clk = '1') then
            if J = '1' and K = '0' then
                Q <= '1';
            elsif J = '0' and K = '1' then
                Q <= '0';
            elsif J = '1' and K = '1' then
                Q <= QNot;
            end if;
        end if;
    end process;
    QNot <= not Q;
end architecture LogicOperation;
```

The complete VHDL program code for the elevator controller using the previously defined components is as follows. Comments shown in green preceded by two hyphens are for explanatory purposes and are not recognized by the program for processing purposes.

CallCode: Request number from floor
PanelCode: Request number from car
Call: Request pulse for CallCode
Request: Request pulse for PanelCode
Sensor: Floor level pulse input
Clk: Elevator system clock
UP, DOWN: Direction for elevator car
STOPOPEN: Motor stop and door open command
CLOSE: Door close command

Elevator Controller

```vhdl
library ieee;
use ieee.std_logic_1164.all;

entity ELEVATOR is
    port (CallCode, PanelCode: in std_logic_vector(2 downto 0);
          Call, Request, Sensor, Clk: in std_logic;
          UP, DOWN, STOPOPEN, CLOSE: inout std_logic;
          a, b, c, d, e, f, g: out std_logic);
end entity ELEVATOR;

architecture LogicOperation of ELEVATOR is
component FLOORCOUNTER is
    port (UP, DOWN, Sensor: in std_logic;
          FLRCODE: out std_logic_vector(2 downto 0));
end component FLOORCOUNTER;
```
Component definition for FLOOR COUNTER

```vhdl
component FLRCALLCOMPARATOR is
    port (FlrCodeCall, FlrCodeCnt: in std_logic_vector(2 downto 0);
          UP, DOWN, STOP : inout std_logic);
end component FLRCALLCOMPARATOR;
```
Component definition for FLRCALL/FLRCNT COMPARATOR

```vhdl
component CODEREGISTER
    port (Clk: in std_logic;
          FlrCodeIn: in std_logic_vector(0 to 2);
          FlrCodeOut: out std_logic_vector(0 to 2));
end component CODEREGISTER;
```
Component definition for CODEREGISTER

```vhdl
component Timer is
    port (Enable, Clk: in std_logic;
          SetCount: in integer range 0 to 1023;
          QOut: inout std_logic);
end component Timer;
```
Component defintion for Timer

```vhdl
component SevenSegment is
    Port (a, b, c, d, e, f, g: out std_logic;
          H0, H1, H2: inout std_logic);
end component SevenSegment;
```
Component definition for SevenSegment Decoder

```vhdl
component JKFlipFlop
    port (J, K, Clk: in std_logic;
          Q: out std_logic);
end component JKFlipFlop;
```
Component definition for CALL/REQ flip-flop

```
-- Signal definitions used to interconnect components and output control signals
signal FRCNT, FRCLOUT, FRIN: std_logic_vector(0 to 2);
signal CallEn: std_logic;
begin

    Gnd <= '0';

    process (CallEn, CallCode, PanelCode) -- Select Floor or Panel call code based on
        begin                                CALL/REQ

            if (CallEn = '1') then
                FRIN <= CallCode; -- If CALL Enabled, select code from call buttons from floor
            else
                FRIN <= PanelCode; -- If CALL not Enabled, select code from elevator
            end if;                          panel buttons

        end process;
```

```
-- Component instantiations
CALLREQ: JKFlipFlop port map(J=>'1', K=>'1', Clk=>Close, Q=> CallEn);

CODEREG: CODEREGISTER port map(Call => (Call and CallEn) or (Request and not
CallEn), FlrCodeIn=> FRIN, FlrCodeOut => FRCLOUT);

FLCLCOMP: FLRCALLCOMPARATOR port map(FlrCodeCall=> FRCL
FlrCodeCnt => FRCNT, Up=>UP, Down=>DOWN, Stop=>STOPOPEN);

FLRCNT: FLOORCOUNTER port map(UP=>UP, DOWN=>DOWN, Sensor=>Sensor,
FLRCODE=>FRCNT);

DISPLAY: SevenSegment port map(a=>a,b=>b,c=>c,d=>d,e=>e,f=>f,g=>g,
H0=>FRCNT(2),H1=>FRCNT(1),H2=>FRCNT(0));

TIMER1: Timer port map (Enable=>STOPOPEN, Clk=> Clk,SetCount=>10,
QOut=>Close);

end architecture LogicOperation;
```

The Programming and PLD Implementation Process

The elevator controller is implemented in a PLD using Altera Quartus II and ModelSim software. The Altera Quartus II software package is an integrated development environment (IDE) supplied by Altera for the creation of HDL applications combined with the ModelSim simulation software. A short summary of the programming process and PLD implementation follows. An expanded description of the elevator controller programming process can be found on the website as well as an Altera Quartus II tutorial. Altera Quartus II is available as a free download from Altera.com.

Project Creation   To start the programming process, a project is created. A project allows the IDE to identify a location to store your application and to create self-generated support files needed to organize your application as well as to keep track of project preferences, rules, and definitions.

Project Definition   To complete the project, you will need to respond to general questions defining the location of your project, the PLD device to be used, and the primary language. Additional questions will determine how you will simulate and verify your application.

Completed Project Definition   With the project definitions completed, the VHDL program source code for the previously defined components and Elevator Controller files are added to your project.

Compiling the Application   By compiling the program at this time, part of the input and output identifier information is automatically entered as you are now ready to make pin assignments to your I/O port identifiers. However, the basic design can be simulated before making the pin assignments.

Graphical Waveform Simulation   In order to simulate the elevator controller design, first start the ModelSim application. Graphical waveform generation tools allow for the easy creation of stimulus waveforms. Graphical waveforms are created to provide the input stimulus to test the elevator controller application. Inputs call, request, callcode, panelcode, sensor, and clk will be created using graphical tools. Output identifiers *up, down, stopopen, close,* and seven-segment outputs *a* through *g* require no input stimulus.

Pin Assignments   A pin assignment editor is used to associate an I/O port identifier with an external pin. Many newer pin editors utilize drag-and-drop features to allow the user to select an identifier with the mouse, then drag and drop to a graphic representation of the target device. Pin assignments can also be accomplished using traditional text entry.

Device Programming   With the pins selected and saved, the project is recompiled once again, generating the output file to be loaded on the target device (PLD). The second compiling operation associates the selected pin to the program identifier. In order to program the target device, the project board on which it is mounted must be connected to the programming computer according to the project board manufacturer's instructions. The target device is typically JTAG compliant and connected through a USB port. Other JTAG compliant target boards may use other inputs such as Ethernet, serial, parallel, or FireWire as described by the manufacturer.

Downloading to the PLD   With the simulation, pin assignment, and recompiling complete, it is time to download the application to the development environment (project board with PLD).

Hardware Testing   With the project loaded, the application can be tested against actual hardware.

Putting Your Knowledge to Work

Modify the elevator controller program for a building with ten floors rather than eight.

SUMMARY

- A PAL is a one-time programmable (OTP) SPLD consisting of a programmable array of AND gates that connects to a fixed array of OR gates.
- The PAL structure allows any sum-of-products (SOP) logic expression with a defined number of variables to be implemented.
- The GAL is essentially a PAL that can be reprogrammed.
- In a PAL or GAL, a macrocell generally consists of one OR gate and some associated output logic.
- A CPLD is a complex programmable logic device that consists basically of multiple SPLD arrays with programmable interconnections.
- Each SPLD array in a CPLD is called a logic array block (LAB).
- A macrocell can be configured for either of two modes: the combinational mode or the registered mode.

- An FPGA (field-programmable gate array) differs in architecture, does not use PAL/PLA type arrays, and has much greater densities than typical CPLDs.
- Most FPGAs use either antifuse or SRAM-based process technology.
- Each configurable logic block (CLB) in an FPGA is made up of multiple smaller logic modules and a local programmable interconnect that is used to connect logic modules within the CLB.
- FPGAs are based on LUT architecture.
- LUT stands for *look-up table,* which is a type of memory that is programmable and used to generate SOP combinational logic functions.
- A hard core is a portion of logic embedded in an FPGA that is put in by the manufacturer to provide a specific function and which cannot be reprogrammed.
- A soft core is a portion of logic embedded in an FPGA that has some programmable features.
- Designs owned by the manufacturer are termed *intellectual property* (IP).
- The programming process is generally referred to as design flow.
- The device being programmed is usually referred to as the target device.
- In software packages for programmable logic, the operations are controlled by an application program called the compiler.
- During downloading, a bitstream is generated that represents the final design, and it is sent to the target device to automatically configure it.
- A method of internally testing a programmable device is called boundary scan, which is based on the JTAG standard (IEEE Std. 1149.1).
- The boundary scan logic in a CPLD consists of a boundary scan register, a bypass register, an instruction register, and a test access port (TAP).

KEY TERMS

Key terms and other bold terms in the chapter are defined in the end-of-book glossary.

Boundary scan   A method for internally testing a PLD based on the JTAG standard (IEEE Std. 1149.1).

Break point   A flag placed within a program source code to stop a program for investigation.

CLB   Configurable logic block; a unit of logic in an FPGA that is made up of multiple smaller logic modules and a local programmable interconnect that is used to connect logic modules within the CLB.

Compiler   An application program in development software packages that controls the operation of the software.

CPLD   A complex programmable logic device that consists basically of multiple SPLD arrays with programmable interconnections.

Design flow   The process or sequence of operations carried out to program a target device.

Downloading   The final step in a design flow in which the logic design is implemented in the target device.

FPGA   Field-programmable gate array; a programmable logic device that uses the LUT as the basic logic element and generally employs either antifuse or SRAM-based process technology.

Functional simulation   A software process that tests the logical or functional operation of a design.

GAL   A reprogrammable type of SPLD that is similar to a PAL except that it uses a reprogrammable process technology, such as EEPROM (E^2CMOS), instead of fuses.

Intellectual property (IP)   Designs owned by a manufacturer of programmable logic devices.

LAB   Logic array block; an SPLD array in a CPLD.

LUT   Look-up table; a type of memory that can be programmed to produce SOP functions.

Macrocell   Part of a PAL, GAL, or CPLD that generally consists of one OR gate and some associated output logic.

PAL   A type of one-time programmable SPLD that consists of a programmable array of AND gates that connects to a fixed array of OR gates.

Registered   A macrocell operational mode that uses a flip–flop.

Schematic entry   A method of placing a logic design into software using schematic symbols.

Target device   The programmable logic device that is being programmed.

Text entry   A method of placing a logic design into software using a hardware description language (HDL).

Timing simulation   A software process that uses information on propagation delays and netlist data to test both the logical operation and the worst-case timing of a design.

TRUE/FALSE QUIZ

Answers are at the end of the chapter.

1. A PAL consists of a programmable array of AND gates connected to a fixed array of OR gates.
2. SPLD stands for *synchronous programmable logic device*.
3. Typically, a macrocell consists of an OR gate and its associated output logic.
4. CPLD stands for *complex programmable logic device*.
5. An FPGA is a fast programmable gate array.
6. A typical FPGA has a greater gate density than a CPLD.
7. Logic array blocks are found in CPLDs.
8. The process of programming a PLD is known as design flow.
9. The device being programmed is called a target device.
10. Two types of programmable design entry are schematic and HDL.

SELF-TEST

Answers are at the end of the chapter.

1. Two types of SPLDs are
   (a) CPLD and PAL
   (b) PAL and FPGA
   (c) PAL and GAL
   (d) GAL and SRAM

2. A PAL consists of a
   (a) programmable AND array and a programmable OR array
   (b) programmable AND array and a fixed OR array
   (c) fixed AND array and a programmable OR array
   (d) fixed AND/OR array

3. A macrocell consists of a
   (a) fixed OR gate and other associated logic
   (b) programmable OR array and other associated logic
   (c) fixed AND gate and other associated logic
   (d) fixed AND/OR array with a flip-flop

4. A device that is not a type of PLD is the
   (a) CPLD   (b) GAL   (c) ROM   (d) FPGA

5. The basic AND/OR structure of SPLDs and CPLDs produces types of Boolean expressions known as
   (a) POS
   (b) SOP
   (c) product of complements
   (d) sum of complements

6. The term *LAB* stands for
   (a) logic AND block
   (b) logic array block
   (c) last asserted bit
   (d) logic assembly block

7. Two modes of macrocell operation are
   (a) input and output
   (b) registered and sequential
   (c) combinational and registered
   (d) parallel and shared

8. When a macrocell is configured to produce an SOP function, it is in the
   (a) combinational mode   (b) parallel mode
   (c) registered mode   (d) shared mode

9. A typical macrocell consists of
   (a) gates, multiplexers, and a flip-flop
   (b) gates and a shift register
   (c) a Gray code counter
   (d) a fixed logic array

10. Based on the complexity of its configurable logic blocks (CLBs), an FPGA can be classified
    as either
    (a) volatile or nonvolatile
    (b) programmable or reprogrammable
    (c) fine grained or coarse grained
    (d) platform or embedded

11. Nonvolatile FPGAs are generally based on
    (a) fuse technology   (b) antifuse technology
    (c) EEPROM technology   (d) SRAM technology

12. An FPGA with an embedded logic function that cannot be programmed is said to be
    (a) nonvolatile   (b) platform
    (c) hard core   (d) soft core

13. Hard core designs are generally developed by and are the property of the FPGA manufacturer.
    These designs are called
    (a) intellectual property (b) proprietary logic
    (c) custom designs   (d) IEEE standards

14. For text entry of a logic design,
    (a) logic symbols must be used
    (b) an HDL must be used
    (c) only Boolean algebra is used
    (d) a special code must be used

15. In a functional simulation, the user must specify the
    (a) specific target device (b) output waveform
    (c) input waveforms   (d) HDL

16. The final output of the synthesis phase of a design flow is the
    (a) netlist   (b) bitstream
    (c) timing simulation   (d) device pin numbers

17. EDIF stands for
    (a) electronic device interchange format
    (b) electrical design integrated fixture
    (c) electrically destructive input function
    (d) electronic design interchange format

18. The boundary scan TAP stands for
    (a) test access point   (b) test array port
    (c) test access port   (d) terminal access path

19. A typical boundary scan cell contains
    (a) flip–flops only
    (b) flip–flops and multiplexer logic
    (c) latches and flip–flops
    (d) latches and an encoder

20. The JTAG standard has the following inputs and outputs
    (a) Intest, extest, TDI, TDO (b) TDI, TDO, TCK, TMS
    (c) ENT, CLK, SHF, CLR   (d) TCK, TMS, TMO, TLF

21. The acronym BSDL stands for
    (a) board standard digital logic
    (b) boundary scan down load
    (c) bistable digital latch
    (d) boundary scan description language

PROBLEMS

Answers to odd-numbered problems are at the end of the book.

Section 10–1 Simple Programmable Logic Devices (SPLDs)

1. Determine the Boolean output expression for the simple PAL array shown in Figure 10–63. The Xs represent connected links.

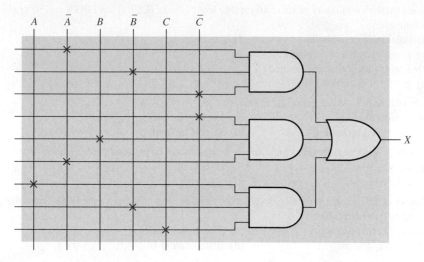

FIGURE 10–63

2. Show how the PAL-type array in Figure 10–64 should be programmed to implement each of the following SOP expressions. Use an X to indicate a connected link.

   (a) $Y = A\bar{B}C + \bar{A}B\bar{C} + ABC$
   (b) $Y = A\bar{B}C + \bar{A}\,\bar{B}C + \bar{A}BC$

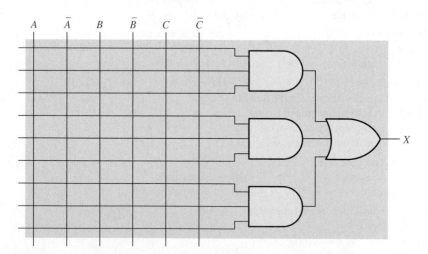

FIGURE 10–64

3. Modify the array in Figure 10–64 for the expression

$$Y = A\bar{B}CD + \bar{A}B\bar{C}D + A\bar{B}C\bar{D} + \bar{A}\,\bar{B}\,\bar{C}\,\bar{D}$$

4. Explain how a programmed polarity output in a PAL works.

Section 10–2 Complex Programmable Logic Devices (CPLDs)

5. Describe how a CPLD differs from an SPLD.

6. Refer to the block diagram in Figure 10–9 and determine the number of

    (a) inputs from the PIA to a LAB
    (b) outputs from a LAB to the PIA
    (c) inputs from an I/O control block to the PIA
    (d) outputs from a LAB to an I/O control block

7. Determine the product term for the AND gate in a CPLD array shown in Figure 10–65(a). If the AND gate is expanded, as shown in Figure 10–65(b), determine the SOP output.

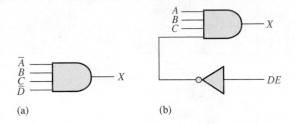

(a)   (b)

FIGURE 10–65

8. Determine the output of the macrocell logic in Figure 10–66 if $AB\overline{C}D + \overline{A}BCD$ is applied to the parallel expander input.

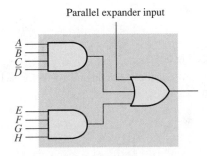

FIGURE 10–66

9. Determine the output of the array in Figure 10–67. The Xs represent connected links.

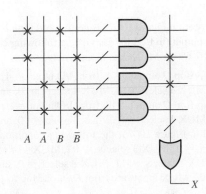

FIGURE 10–67

10. Modify the array in Figure 10–67 to produce an output $X = \overline{A}\,\overline{B}\,\overline{C} + \overline{A}B\overline{C} + ABC + A\overline{B}C$

11. Determine the output expressions for X_1 and X_2 from macrocells 1 and 2 in Figure 10–68.

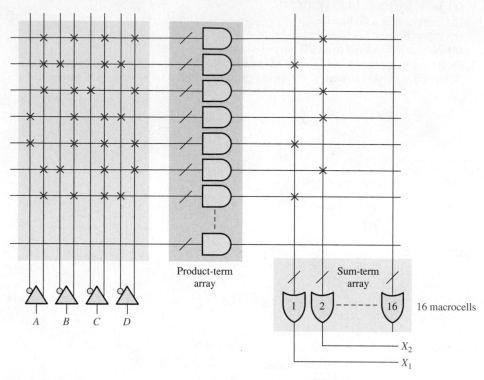

FIGURE 10–68

Section 10–3 Macrocell Modes

12. Determine the data output for the multiplexer in Figure 10–69 for each of the following conditions:

   (a) $D_0 = 1, D_1 = 0$, Select $= 0$
   (b) $D_0 = 1, D_1 = 0$, Select $= 1$

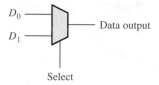

FIGURE 10–69

13. Determine how the macrocell in Figure 10–70 is configured (combinational or registered) and the data bit that is on the output (to I/O) for each of the following conditions. The flip-flop is a D type. Refer to Figure 10–69 for MUX data input arrangement.

   (a) XOR output $= 1$, flip-flop Q output $= 1$, from I/O input $= 1$, MUX 1 select $= 1$, MUX 2 select $= 0$, MUX 3 select $= 0$, MUX 4 select $= 0$, and MUX 5 select $= 0$.
   (b) XOR output $= 0$, flip-flop Q output $= 0$, from I/O input $= 1$, MUX 1 select $= 1$, MUX 2 select $= 0$, MUX 3 select $= 1$, MUX 4 select $= 0$, and MUX 5 select $= 1$.

14. For the CPLD macrocell in Figure 10–71, the following conditions are programmed: MUX 1 select $= 1$, MUX 2 select $= 1$, MUX 3 selects $= 01$, MUX 4 select $= 0$, MUX 5 select $= 1$, MUX 6 selects $= 11$, MUX 7 selects $= 11$, MUX 8 select $= 1$, and the OR output $= 1$. The flip-flop is a D type and the MUX inputs are from D_0 at the top to D_n at the bottom.

   (a) Is the macrocell configured for combinational or registered logic?
   (b) Which clock is applied to the flip-flop?
   (c) What is the data bit on the D input to the flip-flop?
   (d) What is the output of MUX 8?

15. Repeat Problem 14 for MUX 1 select $= 0$.

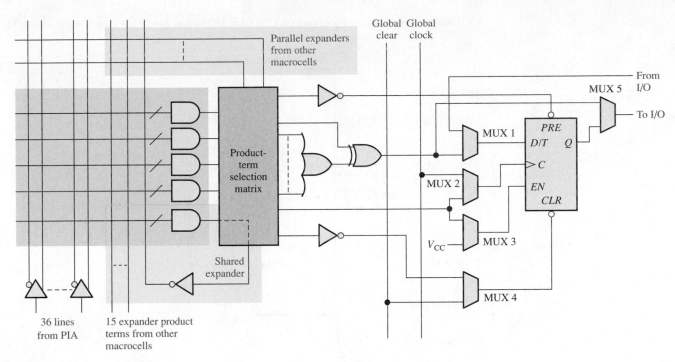

FIGURE 10–70

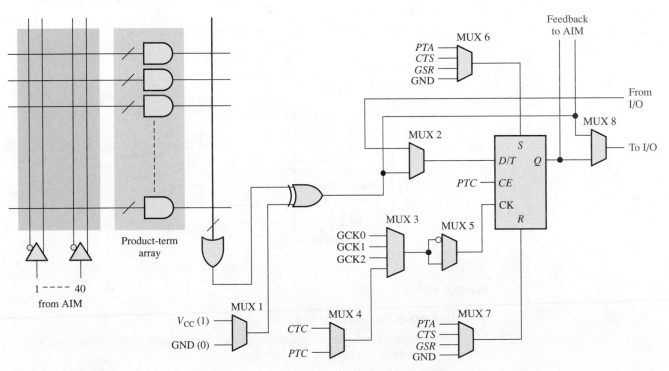

FIGURE 10–71

Section 10–4 Field-Programmable Gate Arrays (FPGAs)

16. Generally, what elements make up a configurable logic block (CLB) in an FPGA? What elements make up a logic module?

17. Determine the output expression of the LUT for the internal conditions shown in Figure 10–72.

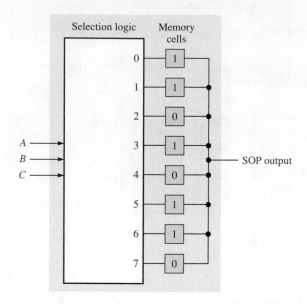

FIGURE 10–72

18. Show how to reprogram the LUT in Figure 10–72 to produce the following SOP output:

$$\overline{AB}\overline{C} + A\overline{B}\,\overline{C} + ABC$$

19. Show a logic module configured in the normal mode to produce one 4-variable SOP function and one 2-variable SOP function.

20. Determine the final SOP output function for the logic module shown in Figure 10–73.

$$A_4A_3\overline{A}_2A_1 + \overline{A}_4\overline{A}_3\overline{A}_2A_1$$

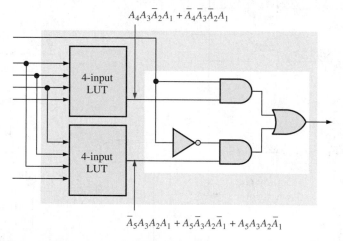

$$\overline{A}_5A_3A_2A_1 + A_5\overline{A}_3A_2\overline{A}_1 + A_5A_3A_2\overline{A}_1$$

FIGURE 10–73

Section 10–5 Programmable Logic Software

21. Show the logic diagram that you would enter in the Graphic Editor for the circuit described by each of the VHDL programs.

(a) **entity** AND_OR **is**
    port (A0, A1, A2, A3: **in** bit; X: **out** bit);
end entity AND_OR;
architecture LogicFunction **of** AND_OR **is**
begin
    X <= (A0 **and** A1) **or** (A2 **and not** A3);
end architecture LogicFunction;

(b) **entity** LogicCircuit **is**
    port (A, B, C, D: **in** bit; X: **out** bit);
  end entity LogicCircuit;
  architecture Function **of** LogicCircuit **is**
  begin
     X <= (A **and** B) **or** (C **and** D) **and**
       (A **and not** B) **and** (**not** C **and not** D);
  end architecture Function;

22. Show the logic circuit that you would enter in the Graphic Editor for the following Boolean expression. Simplify before entering, if possible.

$$X = \overline{A}BCD + A\overline{B}CD + AB\overline{C}D + ABC\overline{D} + ABCD + \overline{A}\,\overline{B}\,\overline{C}D$$

23. The input waveforms for the logic circuit described in Problem 22 are as shown in the Waveform Editor of Figure 10–74. Determine the output waveform that is produced after running a simulation.

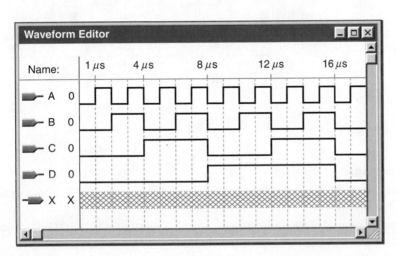

FIGURE 10–74

24. Repeat Problem 23 for the following Boolean expression:

$$X = \overline{A}BC\overline{D} + A\overline{B}\,\overline{C}D + ABCD + A\overline{B}C\overline{D} + \overline{A}B\overline{C}D$$

Section 10–6 Boundary Scan Logic

25. In a given boundary scan cell, assume that data flow serially from the previous BCS to the next BSC. Describe what happens as the data pass through the given BCS.

26. Describe the conditions and what happens in a given BCS when data flow directly from the internal programmable logic to a device output pin.

27. Describe the conditions and what happens in a given BCD when data flow from a device input pin to the internal programmable logic.

28. Describe the data path for transferring data from the SDI to the internal programmable logic.

Section 10–7 Troubleshooting

29. Develop a boundary scan test bit pattern to test the logic that is programmed into the device shown in Figure 10–75 for all possible input combinations.

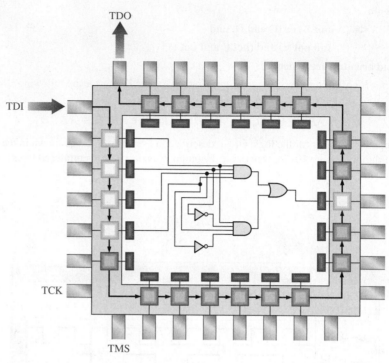

FIGURE 10–75

Applied Logic

30. List the changes to Figure 10–62 required to use the elevator controller for a 16-story building including a basement.

31. Explain the purpose of the AND-OR logic associated with the CALL/REQ Code Register.

32. Modify the VHLD code for the seven-segment decoder in order to add another floor to the eight-floor building.

ANSWERS

SECTION CHECKUPS

Section 10–1 Simple Programmable Logic Devices (SPLDs)

1. PAL: Programmable Array Logic

2. GAL: Generic Array Logic

3. A GAL is reprogrammable. A PAL is one-time programmable.

4. Basically, a macrocell consists of an OR gate and associated output logic including a flip-flop.

Section 10–2 Complex Programmable Logic Devices (CPLDs)

1. CPLD: Complex Programmable Logic Device

2. LAB: Logic Array Block

3. A LAB consists of 16 macrocells in a typical CPLD.

4. A shared expander is used to increase the number of product terms from a macrocell by ANDing additional sum terms (complemented product terms) from other macrocells.

5. A parallel expander is used to increase the number of product terms from a macrocell by ORing unused product terms from other macrocells in a LAB.

6. A PLA has a programmable AND array and a programmable OR array. A PAL has a fixed OR array.

Section 10–3 Macrocell Modes

1. The XOR gate is used as a programmable inverter for the data. It can be programmed to invert or not invert.

2. Combinational and registered

3. Registered refers to the use of a flip-flop.

4. Multiplexer

Section 10–4 Field-Programmable Gate Arrays (FPGAs)

1. Generally, an FPGA is organized with a row/column interconnect structure and uses LUTs rather than AND/OR logic for generating combinational logic functions.

2. CLB: Configurable Logic Block

3. LUT: Look-Up Table. A programmable type of memory that is used to store and generate combinational logic functions.

4. A local interconnect is used to connect logic modules within a CLB. A global interconnect is used to connect a CLB with other CLBs.

5. A core is a portion of logic embedded in an FPGA to provide a specific function.

6. *Intellectual property* refers to the hard-core designs that are developed and owned by the FPGA manufacturer.

7. An LUT produces combinational logic functions in an LM.

8. Memory and DSP (digital signal processing)

Section 10–5 Programmable Logic Software

1. Design entry, functional simulation, synthesis, implementation, timing simulation, downloading

2. Computer running PLD development software, a programming fixture or a development board, and an interface cable

3. A netlist provides information necessary to describe a circuit.

4. The functional simulation comes before the timing simulation.

Section 10–6 Boundary Scan Logic

1. TDI, TMS, TCK, TDO

2. TAP: Test access port

3. Boundary scan register, bypass register, instruction register, and TAP

4. Transfer of data from SDI to SDO, transfer of data from internal programmable logic to device output pin, transfer of data from device input pin to internal programmable logic, transfer of data from SDI to internal programmable logic, and transfer of data from SDI to device output pin and to the SDO line.

Section 10–7 Troubleshooting

1. A break point is a user-defined location in a program where the simulation is stopped temporarily.

2. Boundary scan enables the internal testing and programming of a programmable logic device and testing of interconnections between two or more devices. It is based on the JTAG IEEE Std. 1149.1. Boundary scan uses specific logic internal to the device for testing.

3. Intest and Extest

4. TDI, TDO, TCK, TMS

5. BSDL: Boundary Scan Description Language

RELATED PROBLEMS FOR EXAMPLES

10–1 $X = \overline{B}C + \overline{A}B\overline{C} + \overline{A}\,\overline{B} + C$

10–2 Sixteen; sixteen

10–3 See Figure 10–76.

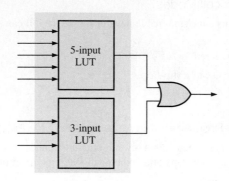

FIGURE 10–76

TRUE/FALSE QUIZ

1. T   **2.** F   **3.** T   **4.** T   **5.** F

6. T   **7.** F   **8.** T   **9.** T   **10.** T

SELF-TEST

1. (c)   **2.** (b)   **3.** (a)   **4.** (c)   **5.** (b)   **6.** (b)   **7.** (c)

8. (a)   **9.** (b)   **10.** (c)   **11.** (b)   **12.** (b)   **13.** (a)   **14.** (a)

15. (c)   **16.** (a)   **17.** (d)   **18.** (c)   **19.** (b)   **20.** (b)   **21.** (d)

Data Storage

CHAPTER OBJECTIVES

- Define the basic memory characteristics
- Explain what a RAM is and how it works
- Explain the difference between static RAMs (SRAMs) and dynamic RAMs (DRAMs)
- Explain what a ROM is and how it works
- Describe the various types of PROMs
- Discuss the characteristics of a flash memory
- Describe the expansion of ROMs and RAMs to increase word length and word capacity
- Discuss special types of memories such as FIFO and LIFO
- Describe the basic organization of magnetic disks and magnetic tapes
- Describe the basic operation of magneto-optical disks and optical disks
- Describe the key elements in a memory hierarchy
- Describe several characteristics of cloud storage
- Describe basic methods for memory testing
- Develop flowcharts for memory testing

KEY TERMS

Key terms are in order of appearance in the chapter.

- Memory
- Byte
- Word
- Cell
- Address
- Capacity
- Write
- Read
- RAM
- ROM
- SRAM
- DRAM
- Bus
- PROM
- EPROM
- Flash memory
- FIFO
- LIFO
- Hard disk
- Blu-ray
- Memory hierarchy
- Cloud storage
- Server

VISIT THE WEBSITE

Study aids for this chapter are available at
http://www.pearsonhighered.com/careersresources/

INTRODUCTION

Chapter 8 covered shift registers, which are a type of storage device. The memory devices covered in this chapter are generally used for longer-term storage of larger amounts of data than registers can provide.

Computers and other types of systems require the permanent or semipermanent storage of large amounts of binary data. Microprocessor-based systems rely on storage devices for their operation because of the necessity for storing programs and for retaining data during processing.

In this chapter semiconductor memories and magnetic and optical storage media are covered. Also, memory hierarchy and cloud storage are discussed.

11–1 Semiconductor Memory Basics

Memory is the portion of a computer or other system that stores binary data. In a computer, memory is accessed millions of times per second, so the requirement for speed and accuracy is paramount. Very fast semiconductor memory is available today in modules with several GB (a gigabyte is one billion bytes) of capacity. These large-memory modules use exactly the same operating principles as smaller units, so we will use smaller ones for illustration in this chapter to simplify the concepts.

After completing this chapter, you should be able to

- Explain how a memory stores binary data
- Discuss the basic organization of a memory
- Describe the write operation
- Describe the read operation
- Describe the addressing operation
- Explain what RAMs and ROMs are

InfoNote

The general definition of *word* is a complete unit of information consisting of a unit of binary data. When applied to computer instructions, a word is more specifically defined as two bytes (16 bits). As an important part of assembly language used in computers, the DW (Define Word) directive means to define data in 16-bit units. This definition is independent of the particular microprocessor or the size of its data bus. Assembly language also allows definitions of bytes (8 bits) with the DB directive, double words (32 bits) with the DD directive, and quad-words (64 bits) with the QD directive.

Units of Binary Data: Bits, Bytes, Nibbles, and Words

As a rule, memories store data in units that have from one to eight bits. The smallest unit of binary data, as you know, is the **bit**. In many applications, data are handled in an 8-bit unit called a **byte** or in multiples of 8-bit units. The byte can be split into two 4-bit units that are called **nibbles**. Bytes can also be grouped into words. The term **word** can have two meanings in computer terminology. In memories, it is defined as a group of bits or bytes that acts as a single entity that can be stored in one memory location. In assembly language, a word is specifically defined as two bytes.

The Basic Memory Array

Each storage element in a memory can retain either a 1 or a 0 and is called a **cell**. Memories are made up of arrays of cells, as illustrated in Figure 11–1 using 64 cells as an example. Each block in the **memory array** represents one storage cell, and its location can be identified by specifying a row and a column.

The 64-cell array can be organized in several ways based on units of data. Figure 11–1(a) shows an 8 × 8 array, which can be viewed as either a 64-bit memory or an 8-byte memory. Part (b) shows a 16 × 4 array, which is a 16-nibble memory, and part (c) shows a 64 × 1

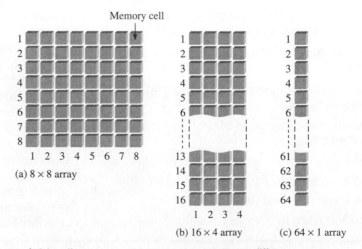

FIGURE 11–1 A 64-cell memory array organized in three different ways.

array, which is a 64-bit memory. A memory is identified by the number of words it can store times the word size. For example, a 16k × 8 memory can store 16,384 words of eight bits each. The inconsistency here is common in memory terminology. The actual number of words is always a power of 2, which, in this case, is $2^{14} = 16,384$. However, it is common practice to state the number to the nearest thousand, in this case, 16k.

Memory Address and Capacity

A representation of a small 8 × 8 memory chip is shown in Figure 11–2(a). The location of a unit of data in a memory array is called its **address**. For example, in part (b), the address of a bit in the 2-dimensional array is specified by the row and column as shown. In part (c), the address of a byte is specified only by the row. So, as you can see, the address depends on how the memory is organized into units of data. Personal computers have random-access memories organized in bytes. This means that the smallest group of bits that can be addressed is eight.

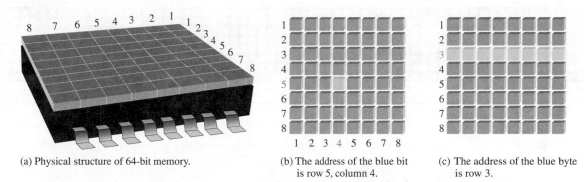

(a) Physical structure of 64-bit memory.

(b) The address of the blue bit is row 5, column 4.

(c) The address of the blue byte is row 3.

FIGURE 11–2 Examples of memory address in a 2-dimensional memory array.

Figure 11–3(a) illustrates the expansion of the 8 × 8 (64-bit) array to a 64-byte memory. The address of a byte in the array is specified by the row and column, as shown. In this case, the smallest group of bits that can be accessed is eight. This can be viewed as a 3-dimensional array, as shown in part (b).

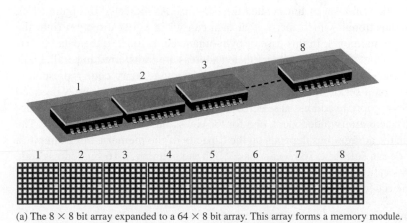

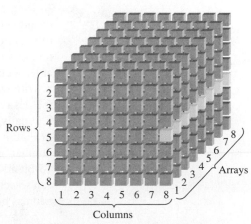

(a) The 8 × 8 bit array expanded to a 64 × 8 bit array. This array forms a memory module.

(b) The address of the blue byte is row 5, column 8.

FIGURE 11–3 Example of memory address in an expanded (multiple) array.

The **capacity** of a memory is the total number of data units that can be stored. For example, in the bit-organized memory array in Figure 11–2(b), the capacity is 64 bits. In the byte-organized memory array in Figure 11–2(c), the capacity is 8 bytes, which is also

64 bits. In Figure 11–3, the capacity is 64 bytes. Computer memories typically have multiple gigabytes of internal memory. Computers usually transfer and store data as 64-bit words, in which case all eight bits of row five in each chip in Figure 11–3(a) would be accessed.

Memory Banks and Ranks

A **bank** is a section of memory within a single memory array (chip). A memory chip may have one or more banks. Memory banks can be used for storing frequently used information. Easier and faster access can be achieved by knowing the section of memory in which the data are stored. A **rank** is a group of chips that make up a memory module that stores data in units such as words or bytes. These terms are illustrated in Figure 11–4.

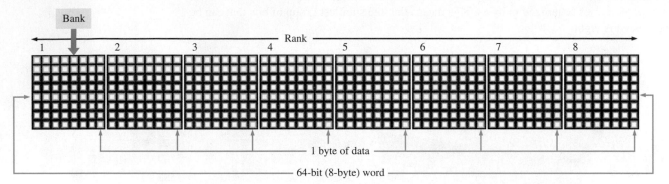

FIGURE 11–4   Simple illustration of memory bank and memory rank.

Basic Memory Operations

Addressing is the process of accessing a specified location in memory. Since a memory stores binary data, data must be put into the memory and data must be copied from the memory when needed. The **write** operation puts data into a specified address in the memory, and the **read** operation copies data out of a specified address in the memory. The addressing operation, which is part of both the write and the read operations, selects the specified memory address.

Data units go into the memory during a write operation and come out of the memory during a read operation on a set of lines called the *data bus*. As indicated in Figure 11–5, the data bus is bidirectional, which means that data can go in either direction (into the memory or out of the memory). In this case of byte-organized memories, the data bus has at least eight lines so that all eight bits in a selected address are transferred in parallel. For a write or a read operation, an address is selected by placing a binary code representing the desired address on a set of lines called the *address bus*. The address code is decoded internally, and the appropriate address is selected. In the case of the multiple-array memory in Figure 11–5(b) there are two decoders, one for the rows and one for the columns. The number of lines in the address bus depends on the capacity of the memory. For example, a 15-bit address code can select 32,768 locations (2^{15}) in the memory, a 16-bit address code can select 65,536 locations (2^{16}) in the memory, and so on. In personal computers a 32-bit address bus can select 4,294,967,296 locations (2^{32}), expressed as 4G.

The Write Operation

A simplified write operation is illustrated in Figure 11–6. To store a byte of data in the memory, a code held in the address register is placed on the address bus. Once the address code is on the bus, the address decoder decodes the address and selects the specified location in the memory. The memory then gets a write command, and the data byte held in the data register is placed on the data bus and stored in the selected memory address, thus completing the write operation. When a new data byte is written into a memory address, the current data byte stored at that address is overwritten (replaced with a new data byte).

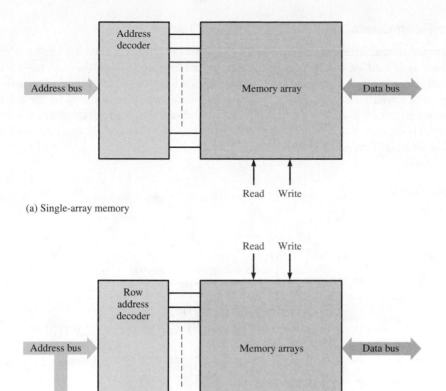

(a) Single-array memory

(b) Multiple-array memory

FIGURE 11–5 Block diagram of a single-array memory and a multiple-array memory showing address bus, address decoder(s), bidirectional data bus, and read/write inputs.

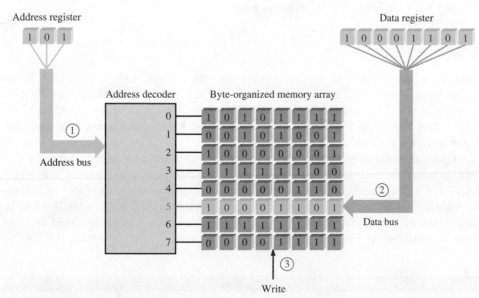

① Address code 101 is placed on the address bus and address 5 is selected.

② Data byte is placed on the data bus.

③ Write command causes the data byte to be stored in address 5, replacing previous data.

FIGURE 11–6 Illustration of the write operation.

The Read Operation

A simplified read operation is illustrated in Figure 11–7. Again, a code held in the address register is placed on the address bus. Once the address code is on the bus, the address decoder decodes the address and selects the specified location in the memory. The memory then gets a read command, and a "copy" of the data byte that is stored in the selected memory address is placed on the data bus and loaded into the data register, thus completing the read operation. When a data byte is read from a memory address, it also remains stored at that address. This is called *nondestructive read.*

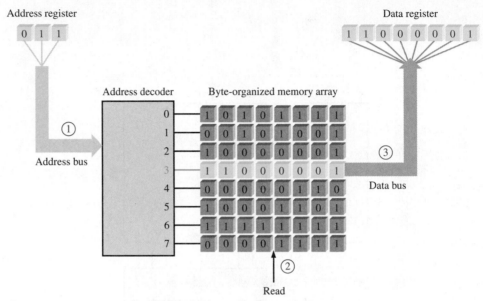

① Address code 011 is placed on the address bus and address 3 is selected.

② Read command is applied.

③ The contents of address 3 is placed on the data bus and shifted into data register. The contents of address 3 is not erased by the read operation.

FIGURE 11–7   Illustration of the read operation.

RAMs and ROMs

The two major categories of semiconductor memories are the RAM and the ROM. **RAM** (random-access memory) is a type of memory in which all addresses are accessible in an equal amount of time and can be selected in any order for a read or write operation. All RAMs have both *read* and *write* capability. Because RAMs lose stored data when the power is turned off, they are **volatile** memories.

ROM (read-only memory) is a type of memory in which data are stored permanently or semipermanently. Data can be read from a ROM, but there is no write operation as in the RAM. The ROM, like the RAM, is a random-access memory but the term *RAM* traditionally means a random-access *read/write* memory. Several types of RAMs and ROMs will be covered in this chapter. Because ROMs retain stored data even if power is turned off, they are **nonvolatile** memories.

SECTION 11–1 CHECKUP

Answers are at the end of the chapter.

1. What is the smallest unit of data that can be stored in a memory?

2. What is the bit capacity of a memory that can store 256 bytes of data?

3. What is a write operation?

4. What is a read operation?

5. How is a given unit of data located in a memory?

6. Describe the difference between a RAM and a ROM.

11–2   The Random-Access Memory (RAM)

A RAM is a read/write memory in which data can be written into or read from any selected address in any sequence. When a data unit is written into a given address in the RAM, the data unit previously stored at that address is replaced by the new data unit. When a data unit is read from a given address in the RAM, the data unit remains stored and is not erased by the read operation. This nondestructive read operation can be viewed as copying the content of an address while leaving the content intact. A RAM is typically used for short-term data storage because it cannot retain stored data when power is turned off.

After completing this section, you should be able to

- Name the two categories of RAM

- Explain what a SRAM is

- Describe the SRAM storage cell

- Explain the difference between an asynchronous SRAM and a synchronous burst SRAM

- Explain the purpose of a cache memory

- Explain what a DRAM is

- Describe the DRAM storage cells

- Discuss the types of DRAM

- Compare the SRAM with the DRAM

The RAM Family

The two major categories of RAM are the *static RAM* (SRAM) and the *dynamic RAM* (DRAM). **SRAMs** generally use latches as storage elements and can therefore store data indefinitely *as long as dc power is applied.* **DRAMs** use capacitors as storage elements and cannot retain data very long without the capacitors being recharged by a process called **refreshing.** Both SRAMs and DRAMs will lose stored data when dc power is removed and, therefore, are classified as volatile memories.

Data can be read much faster from SRAMs than from DRAMs. However, DRAMs can store much more data than SRAMs for a given physical size and cost because the DRAM cell is much simpler and more cells can be crammed into a given chip area than in the SRAM.

The basic types of SRAM are the *asynchronous SRAM* and the *synchronous SRAM* with a burst feature. The basic types of DRAM are the *Fast Page Mode DRAM* (FPM DRAM), the *Extended Data Out DRAM* (EDO DRAM), the *Burst EDO DRAM* (BEDO DRAM), and the *synchronous DRAM* (SDRAM). These are shown in Figure 11–8.

Static RAMs (SRAMs)

Memory Cell

All SRAMs are characterized by latch memory cells. As long as dc power is applied to a **static memory** cell, it can retain a 1 or 0 state indefinitely. If power is removed, the stored data bit is lost.

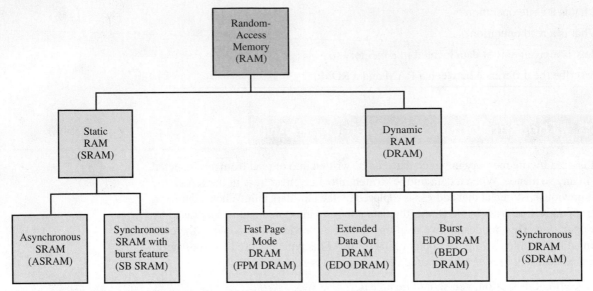

FIGURE 11–8   The RAM family.

Figure 11–9 shows a basic SRAM latch memory cell. The cell is selected by an active level on the Select line and a data bit (1 or 0) is written into the cell by placing it on the Data in line. A data bit is read by taking it off the Data out line.

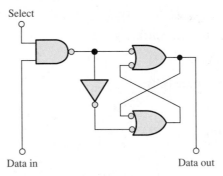

FIGURE 11–9   A typical SRAM latch memory cell.

Static Memory Cell Array

The memory cells in a SRAM are organized in rows and columns, as illustrated in Figure 11–10 for the case of an $n \times 4$ array. All the cells in a row share the same Row Select line. Each set of Data in and Data out lines go to each cell in a given column and are connected to a single data line that serves as both an input and output (Data I/O) through the data input and data output buffers.

To write a data unit, in this case a nibble (4 bits), into a given row of cells in the memory array, the Row Select line is taken to its active state and four data bits are placed on the Data I/O lines. The Write line is then taken to its active state, which causes each data bit to be stored in a selected cell in the associated column. To read a data unit, the Read line is taken to its active state, which causes the four data bits stored in the selected row to appear on the Data I/O lines.

Basic Asynchronous SRAM Organization

An asynchronous SRAM is one in which the operation is not synchronized with a system clock. To illustrate the general organization of a SRAM, a 32k $\times$ 8 bit memory is used. A logic symbol for this memory is shown in Figure 11–11.

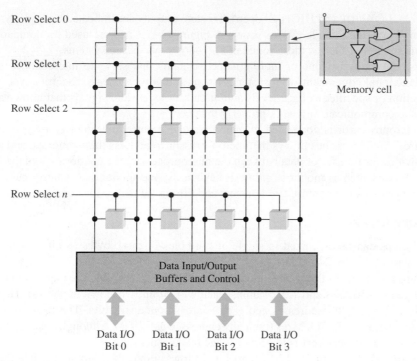

FIGURE 11–10 Basic SRAM array.

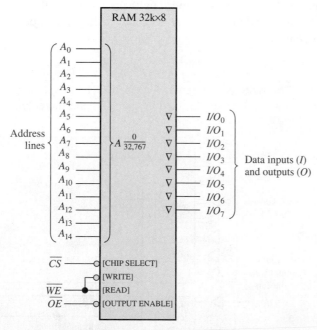

FIGURE 11–11 Logic diagram for an asynchronous 32k × 8 SRAM.

In the READ mode, the eight data bits that are stored in a selected address appear on the data output lines. In the WRITE mode, the eight data bits that are applied to the data input lines are stored at a selected address. The data input and data output lines (I/O_0 through I/O_7) share the same lines. During READ, they act as output lines (O_0 through O_7) and during WRITE they act as input lines (I_0 through I_7).

Tri-state Outputs and Buses

Tri-state buffers in a memory allow the data lines to act as either input or output lines and connect the memory to the data bus in a computer. These buffers have three output states:

HIGH (1), LOW (0), and HIGH-Z (open). Tri-state outputs are indicated on logic symbols by a small inverted triangle (∇), as shown in Figure 11–11, and are used for compatibility with bus structures such as those found in microprocessor-based systems.

Physically, a **bus** is one or more conductive paths that serve to interconnect two or more functional components of a system or several diverse systems. Electrically, a bus is a collection of specified voltage levels and/or current levels and signals that allow various devices to communicate and work properly together.

A microprocessor is connected to memories and input/output devices by certain bus structures. An address bus allows the microprocessor to address the memories, and a data bus provides for transfer of data between the microprocessor, the memories, and the input/output devices such as monitors, printers, keyboards, and modems. A control bus allows the microprocessor to control data transfers and timing for the various components.

Memory Array

SRAM chips can be organized in single bits, nibbles (4 bits), bytes (8 bits), or multiple bytes (words with 16, 24, 32 bits, etc.).

Figure 11–12 shows the organization of a small 32k $\times$ 8 SRAM. The memory cell array is arranged in 256 rows and 128 columns, each with 8 bits, as shown in part (a). There are actually $2^{15} = 32{,}768$ addresses and each address contains 8 bits. The capacity of this example memory is 32,768 bytes (typically expressed as 32 kB). Although small by today's standards, this memory serves to introduce the basic concepts.

The SRAM in Figure 11–12(b) works as follows. First, the chip select, $\overline{CS}$, must be LOW for the memory to operate. (Other terms for chip select are *enable* or *chip enable*.) Eight of the fifteen address lines are decoded by the row decoder to select one of the 256 rows. Seven of the fifteen address lines are decoded by the column decoder to select one of the 128 8-bit columns.

Read

In the READ mode, the write enable input, $\overline{WE}$, is HIGH and the output enable, $\overline{OE}$, is LOW. The input tri-state buffers are disabled by gate G_1, and the column output tri-state

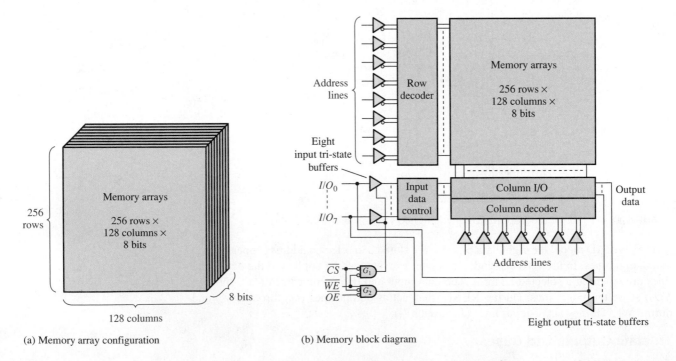

(a) Memory array configuration

(b) Memory block diagram

FIGURE 11–12 Basic organization of an asynchronous 32k $\times$ 8 SRAM.

buffers are enabled by gate G_2. Therefore, the eight data bits from the selected address are routed through the column I/O to the data lines (I/O_0 though I/O_7), which are acting as data output lines.

Write

In the WRITE mode, $\overline{WE}$ is LOW and $\overline{OE}$ is HIGH. The input tri-state buffers are enabled by gate G_1, and the output tri-state buffers are disabled by gate G_2. Therefore, the eight input data bits on the data lines are routed through the input data control and the column I/O to the selected address and stored.

Read and Write Cycles

Figure 11–13 shows typical timing diagrams for a memory read cycle and a write cycle. For the read cycle shown in part (a), a valid address code is applied to the address lines for a specified time interval called the *read cycle time, t_{RC}*. Next, the chip select ($\overline{CS}$) and the output enable ($\overline{OE}$) inputs go LOW. One time interval after the $\overline{OE}$ input goes LOW, a valid data byte from the selected address appears on the data lines. This time interval is called the *output enable access time, t_{GQ}*. Two other access times for the read cycle are the *address access time, t_{AQ}*, measured from the beginning of a valid address to the appearance of valid data on the data lines and the *chip enable access time, t_{EQ}*, measured from the HIGH-to-LOW transition of $\overline{CS}$ to the appearance of valid data on the data lines.

During each read cycle, one unit of data, a byte in this case, is read from the memory.

For the write cycle shown in Figure 11–13(b), a valid address code is applied to the address lines for a specified time interval called the *write cycle time, t_{WC}*. Next, the chip

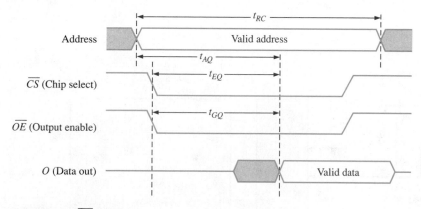

(a) Read cycle ($\overline{WE}$ HIGH)

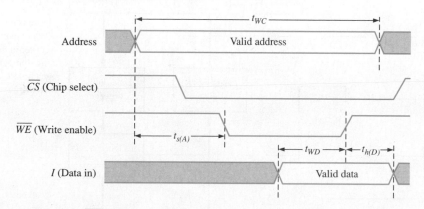

(b) Write cycle ($\overline{WE}$ LOW)

FIGURE 11–13   Timing diagrams for typical read and write cycles for the SRAM in Figure 11–12.

select ($\overline{CS}$) and the write enable ($\overline{WE}$) inputs go LOW. The required time interval from the beginning of a valid address until the $\overline{WE}$ input goes LOW is called the *address setup time*, $t_{s(A)}$. The time that the $\overline{WE}$ input must be LOW is the write pulse width. The time that the input $\overline{WE}$ must remain LOW after valid data are applied to the data inputs is designated t_{WD}; the time that the valid input data must remain on the data lines after the $\overline{WE}$ input goes HIGH is the *data hold time*, $t_{h(D)}$.

During each write cycle, one unit of data is written into the memory.

Synchronous SRAM with Burst Feature

Unlike the asynchronous SRAM, a synchronous SRAM is synchronized with the system clock. For example, in a computer system, the synchronous SRAM operates with the same clock signal that operates the microprocessor so that the microprocessor and memory are synchronized for faster operation.

The fundamental concept of the synchronous feature of a SRAM can be shown with Figure 11–14, which is a simplified block diagram of a 32k × 8 memory for purposes of illustration. The synchronous SRAM is similar to the asynchronous SRAM in terms of the memory array, address decoder, and read/write and enable inputs. The basic difference is that the synchronous SRAM uses clocked registers to synchronize all inputs with the system clock. The address, the read/write input, the chip enable, and the input data are all latched into their respective registers on an active clock pulse edge. Once this information is latched, the memory operation is in sync with the clock.

For the purpose of simplification, a notation for multiple parallel lines or bus lines is introduced in Figure 11–14, as an alternative to drawing each line separately. A set of

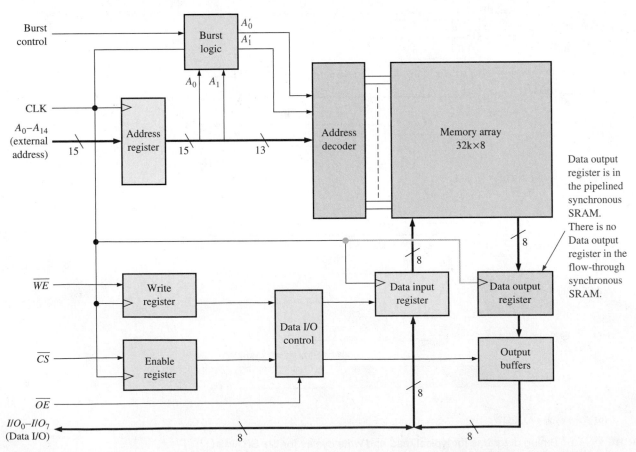

FIGURE 11–14   A basic block diagram of a synchronous SRAM with burst feature.

parallel lines can be indicated by a single heavy line with a slash and the number of separate lines in the set. For example, the following notation represents a set of 8 parallel lines:

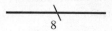

The address bits A_0 through A_{14} are latched into the Address register on the positive edge of a clock pulse. On the same clock pulse, the state of the write enable ($\overline{WE}$) line and chip select ($\overline{CS}$) are latched into the Write register and the Enable register respectively. These are one-bit registers or simply flip-flops. Also, on the same clock pulse the input data are latched into the Data input register for a Write operation, and data in a selected memory address are latched into the Data output register for a Read operation, as determined by the Data I/O control based on inputs from the Write register, Enable register, and the Output enable ($\overline{OE}$).

Two basic types of synchronous SRAM are the *flow-through* and the *pipelined*. The flow-through synchronous SRAM does not have a Data output register, so the output data flow asynchronously to the data I/O lines through the output buffers. The **pipelined** synchronous SRAM has a Data output register, as shown in Figure 11–14, so the output data are synchronously placed on the data I/O lines.

The Burst Feature

As shown in Figure 11–14, synchronous SRAMs normally have an address burst feature, which allows the memory to read or write up to four sequential locations using a single address. When an external address is latched in the address register, the two lowest-order address bits, A_0 and A_1, are applied to the burst logic. This produces a sequence of four internal addresses by adding 00, 01, 10, and 11 to the two lowest-order address bits on successive clock pulses. The sequence always begins with the base address, which is the external address held in the address register.

The address burst logic in a typical synchronous SRAM consists of a binary counter and exclusive-OR gates, as shown in Figure 11–15. For 2-bit burst logic, the internal burst address sequence is formed by the base address bits A_2–A_{14} plus the two burst address bits A_1' and A_0'.

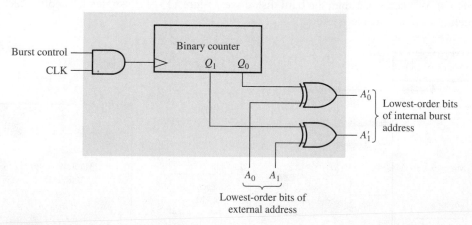

FIGURE 11–15 Address burst logic.

To begin the burst sequence, the counter is in its 00 state and the two lowest-order address bits are applied to the inputs of the XOR gates. Assuming that A_0 and A_1 are both 0, the internal address sequence in terms of its two lowest-order bits is 00, 01, 10, and 11.

Cache Memory

One of the major applications of SRAMs is in cache memories in computers. **Cache memory** is a relatively small, high-speed memory that stores the most recently used instructions or data from the larger but slower main memory. Cache memory can also use dynamic

RAM (DRAM), which is discussed next. Typically, SRAM is several times faster than DRAM. Overall, a cache memory gets stored information to the microprocessor much faster than if only high-capacity DRAM is used. Cache memory is basically a cost-effective method of improving system performance without having to resort to the expense of making all of the memory faster.

The concept of cache memory is based on the idea that computer programs tend to get instructions or data from one area of main memory before moving to another area. Basically, the cache controller "guesses" which area of the slow dynamic memory the CPU (central-processing unit) will need next and moves it to the cache memory so that it is ready when needed. If the cache controller guesses right, the data are immediately available to the microprocessor. If the cache controller guesses wrong, the CPU must go to the main memory and wait much longer for the correct instructions or data. Fortunately, the cache controller is right most of the time.

Cache Analogy

There are many analogies that can be used to describe a cache memory, but comparing it to a home refrigerator is perhaps the most effective. A home refrigerator can be thought of as a "cache" for certain food items while the supermarket is the main memory where all foods are kept. Each time you want something to eat or drink, you can go to the refrigerator (cache) first to see if the item you want is there. If it is, you save a lot of time. If it is not there, then you have to spend extra time to get it from the supermarket (main memory).

L1 and L2 Caches

A first-level cache (L1 cache) is usually integrated into the processor chip and has a very limited storage capacity. L1 cache is also known as *primary cache*. A second-level cache (L2 cache) may also be integrated into the processor or as a separate memory chip or set of chips external to the processor; it usually has a larger storage capacity than an L1 cache. L2 cache is also known as *secondary cache*. Some systems may have higher-level caches (L3, L4, etc.), but L1 and L2 are the most common. Also, some systems use a disk cache to enhance the performance of the hard disk because DRAM, although much slower than SRAM, is much faster than the hard disk drive. Figure 11–16 illustrates L1 and L2 cache memories in a computer system.

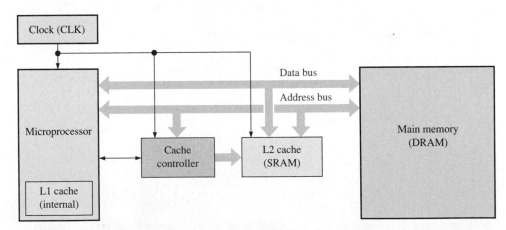

FIGURE 11–16 Block diagram showing L1 and L2 cache memories in a computer system.

Dynamic RAM (DRAM) Memory Cells

Dynamic memory cells store a data bit in a small capacitor rather than in a latch. The advantage of this type of cell is that it is very simple, thus allowing very large memory arrays to be constructed on a chip at a lower cost per bit. The disadvantage is that the storage capacitor cannot hold its charge over an extended period of time and will lose the stored data bit unless its charge is refreshed periodically. To refresh requires additional memory circuitry and complicates the operation of the DRAM. Figure 11–17 shows a typical DRAM cell consisting of a single MOS transistor (MOSFET) and a capacitor.

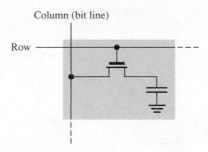

FIGURE 11–17   A MOS DRAM cell.

In this type of cell, the transistor acts as a switch. The basic simplified operation is illustrated in Figure 11–18 and is as follows. A LOW on the $R/\overline{W}$ line (WRITE mode) enables the tri-state input buffer and disables the output buffer. For a 1 to be written into the cell, the D_{IN} line must be HIGH, and the transistor must be turned on by a HIGH on the row line. The transistor acts as a closed switch connecting the capacitor to the bit line. This connection allows the capacitor to charge to a positive voltage, as shown in Figure 11–18(a). When a 0 is to be stored, a LOW is applied to the D_{IN} line. If the capacitor is storing a 0, it remains uncharged, or if it is storing a 1, it discharges as indicated in Figure 11–18(b). When the row line is taken back LOW, the transistor turns off and disconnects the capacitor from the bit line, thus "trapping" the charge (1 or 0) on the capacitor.

To read from the cell, the $R/\overline{W}$ (Read/$\overline{Write}$) line is HIGH, enabling the output buffer and disabling the input buffer. When the row line is taken HIGH, the transistor turns on and connects the capacitor to the bit line and thus to the output buffer (sense amplifier), so the data bit appears on the data-output line (D_{OUT}). This process is illustrated in Figure 11–18(c).

For refreshing the memory cell, the $R/\overline{W}$ line is HIGH, the row line is HIGH, and the refresh line is HIGH. The transistor turns on, connecting the capacitor to the bit line. The output buffer is enabled, and the stored data bit is applied to the input of the refresh buffer, which is enabled by the HIGH on the refresh input. This produces a voltage on the bit line corresponding to the stored bit, thus replenishing the capacitor. This is illustrated in Figure 11–18(d).

DRAM Organization

The major application of DRAMs is in the main memory of computers. The difference between DRAMs and SRAMs is the type of memory cell. As you have seen, the DRAM memory cell consists of one transistor and a capacitor and is much simpler than the SRAM cell. This allows much greater densities in DRAMs and results in greater bit capacities for a given chip area, although much slower access time.

Again, because charge stored in a capacitor will leak off, the DRAM cell requires a frequent refresh operation to preserve the stored data bit. This requirement results in more complex circuitry than in a SRAM. Several features common to most DRAMs are now discussed, using a generic 1M × 1 bit DRAM as an example.

Address Multiplexing

DRAMs use a technique called *address multiplexing* to reduce the number of address lines. Figure 11–19 shows the block diagram of a 1,048,576-bit (1 Mb) DRAM with a 1M × 1

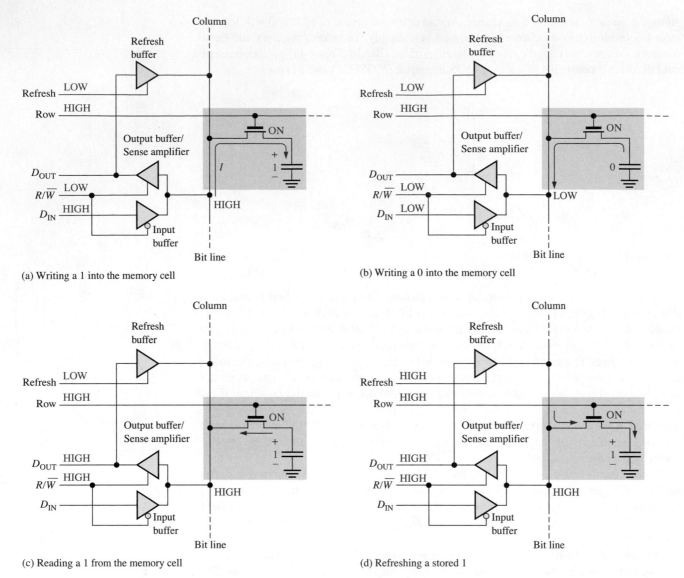

(a) Writing a 1 into the memory cell

(b) Writing a 0 into the memory cell

(c) Reading a 1 from the memory cell

(d) Refreshing a stored 1

FIGURE 11–18 Basic operation of a DRAM cell.

organization. We will focus on the blue blocks to illustrate address multiplexing. The green blocks represent the refresh logic.

The ten address lines are time multiplexed at the beginning of a memory cycle by the row address select ($\overline{RAS}$) and the column address select ($\overline{CAS}$) into two separate 10-bit address fields. First, the 10-bit row address is latched into the row address register. Next, the 10-bit column address is latched into the column address register. The row address and the column address are decoded to select one of the 1,048,576 addresses ($2^{20} = 1,048,576$) in the memory array. The basic timing for the address multiplexing operation is shown in Figure 11–20.

Read and Write Cycles

At the beginning of each read or write memory cycle, $\overline{RAS}$ and $\overline{CAS}$ go active (LOW) to multiplex the row and column addresses into the registers, and decoders. For a read cycle, the $R/\overline{W}$ input is HIGH. For a write cycle, the $R/\overline{W}$ input is LOW. This is illustrated in Figure 11–21.

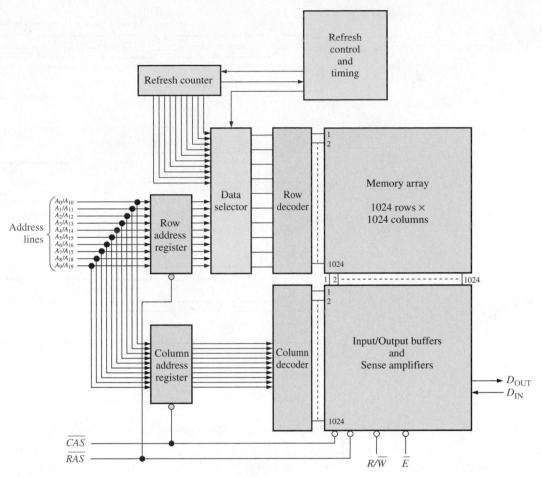

FIGURE 11–19 Simplified block diagram of a 1M × 1 DRAM.

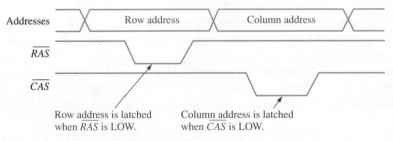

FIGURE 11–20 Basic timing for address multiplexing.

Fast Page Mode

In the normal read or write cycle described previously, the row address for a particular memory location is first loaded by an active-LOW $\overline{RAS}$ and then the column address for that location is loaded by an active-LOW $\overline{CAS}$. The next location is selected by another $\overline{RAS}$ followed by a $\overline{CAS}$, and so on.

A "page" is a section of memory available at a single row address and consists of all the columns in a row. Fast page mode allows fast successive read or write operations at each column address in a selected row. A row address is first loaded by $\overline{RAS}$ going LOW and remaining LOW while $\overline{CAS}$ is toggled between HIGH and LOW. A single row address is selected and remains selected while $\overline{RAS}$ is active. Each successive $\overline{CAS}$ selects another column in the selected row. So, after a fast page mode cycle, all of the addresses in the

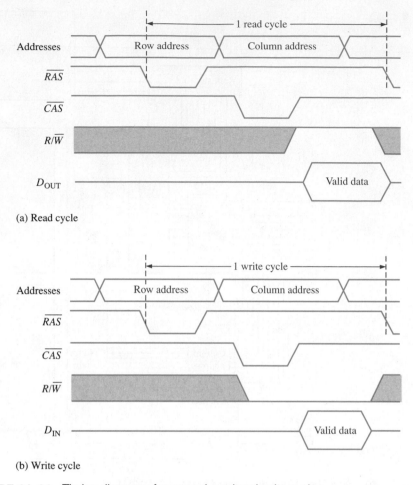

(a) Read cycle

(b) Write cycle

FIGURE 11–21   Timing diagrams for normal read and write cycles.

selected row have been read from or written into, depending on $R/\overline{W}$. For example, a fast page mode cycle for the DRAM in Figure 11–19 requires $\overline{CAS}$ to go active 1024 times for each row selected by $\overline{RAS}$.

Fast page mode operation for read is illustrated by the timing diagram in Figure 11–22. When $\overline{CAS}$ goes to its nonasserted state (HIGH), it disables the data outputs. Therefore, the transition of $\overline{CAS}$ to HIGH must occur only after valid data are latched by the external system.

Refresh Cycles

As you know, DRAMs are based on capacitor charge storage for each bit in the memory array. This charge degrades (leaks off) with time and temperature, so each bit must be periodically refreshed (recharged) to maintain the correct bit state. Typically, a DRAM must be refreshed every several milliseconds, although for some devices the refresh period can be much longer.

A read operation automatically refreshes all the addresses in the selected row. However, in typical applications, you cannot always predict how often there will be a read cycle, and so you cannot depend on a read cycle to occur frequently enough to prevent data loss. Therefore, special refresh cycles must be implemented in DRAM systems.

Burst refresh and *distributed refresh* are the two basic refresh modes for refresh operations. In burst refresh, all rows in the memory array are refreshed consecutively each refresh period. For a memory with a refresh period of 8 ms, a burst refresh of all rows occurs once every 8 ms. The normal read and write operations are suspended during a burst

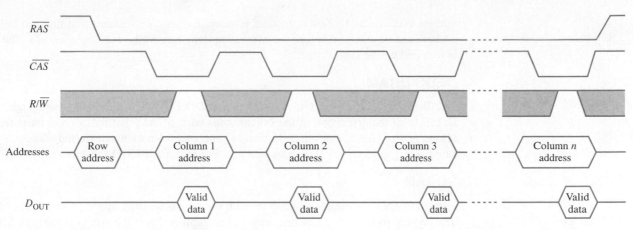

FIGURE 11–22 Fast page mode timing for a read operation.

refresh cycle. In distributed refresh, each row is refreshed at intervals interspersed between normal read or write cycles. For example, the memory in Figure 11–19 has 1024 rows. As an example, for an 8 ms refresh period, each row must be refreshed every 8 ms/1024 = 7.8 μs when distributed refresh is used.

The two types of refresh operations are $\overline{RAS}$ *only refresh* and $\overline{CAS}$ *before* $\overline{RAS}$ *refresh.* $\overline{RAS}$-only refresh consists of a $\overline{RAS}$ transition to the LOW (active) state, which latches the address of the row to be refreshed while $\overline{CAS}$ remains HIGH (inactive) throughout the cycle. An external counter is used to provide the row addresses for this type of operation.

The $\overline{CAS}$ before $\overline{RAS}$ refresh is initiated by $\overline{CAS}$ going LOW before $\overline{RAS}$ goes LOW. This sequence activates an internal refresh counter that generates the row address to be refreshed. This address is switched by the data selector into the row decoder.

Types of DRAMs

Now that you have learned the basic concept of a DRAM, let's briefly look at the major types. These are the *Fast Page Mode (FPM) DRAM,* the *Extended Data Out (EDO) DRAM,* the *Burst Extended Data Out (BEDO) DRAM,* and the *Synchronous (S) DRAM.*

FPM DRAM

Fast page mode operation was described earlier. Recall that a page in memory is all of the column addresses contained within one row address.

The idea of the **FPM DRAM** is based on the probability that the next several memory addresses to be accessed are in the same row (on the same page). Fortunately, this happens a large percentage of the time. FPM saves time over pure random accessing because in FPM the row address is specified only once for access to several successive column addresses whereas for pure random accessing, a row address is specified for each column address.

Recall that in a fast page mode read operation, the $\overline{CAS}$ signal has to wait until the valid data from a given address are accepted (latched) by the external system (CPU) before it can go to its nonasserted state. When $\overline{CAS}$ goes to its nonasserted state, the data outputs are disabled. This means that the next column address cannot occur until after the data from the current column address are transferred to the CPU. This limits the rate at which the columns within a page can be addressed.

EDO DRAM

The Extended Data Out DRAM, sometimes called *hyper page mode DRAM,* is similar to the FPM DRAM. The key difference is that the $\overline{CAS}$ signal in the **EDO DRAM** does not disable the output data when it goes to its nonasserted state because the valid data from the current address can be held until $\overline{CAS}$ is asserted again. This means that the next column address can be accessed before the external system accepts the current valid data. The idea is to speed up the access time.

BEDO DRAM

The Burst Extended Data Out DRAM is an EDO DRAM with address burst capability. Recall from the discussion of the synchronous burst SRAM that the address burst feature allows up to four addresses to be internally generated from a single external address, which saves some access time. This same concept applies to the **BEDO DRAM.**

SDRAM

Faster DRAMs are needed to keep up with the ever-increasing speed of microprocessors. The Synchronous DRAM is one way to accomplish this. Like the synchronous SRAM discussed earlier, the operation of the **SDRAM** is synchronized with the system clock, which also runs the microprocessor in a computer system. The same basic ideas described in relation to the synchronous burst SRAM, also apply to the SDRAM.

This synchronized operation makes the SDRAM totally different from the other asynchronous DRAM types. With asynchronous memories, the microprocessor must wait for the DRAM to complete its internal operations. However, with synchronous operation, the DRAM latches addresses, data, and control information from the processor under control of the system clock. This allows the processor to handle other tasks while the memory read or write operations are in progress, rather than having to wait for the memory to do its thing as is the case in asynchronous systems.

DDR SDRAM

DDR stands for double data rate. A DDR SDRAM is clocked on both edges of a clock pulse, whereas a SDRAM is clocked on only one edge. Because of the double clocking, a DDR SDRAM is theoretically twice as fast as an SDRAM. Sometimes the SDRAM is referred to as an SDR SDRAM (single data rate SDRAM) for contrast with the DDR SDRAM.

SECTION 11–2 CHECKUP

1. List two types of SRAM.
2. What is a cache?
3. Explain how SRAMs and DRAMs differ.
4. Describe the refresh operation in a DRAM.
5. List four types of DRAM.

11–3   The Read-Only Memory (ROM)

A ROM contains permanently or semipermanently stored data, which can be read from the memory but either cannot be changed at all or cannot be changed without specialized equipment. A ROM stores data that are used repeatedly in system applications, such as tables, conversions, or programmed instructions for system initialization and operation. ROMs retain stored data when the power is off and are therefore nonvolatile memories.

After completing this section, you should be able to

- List the types of ROMs
- Describe a basic mask ROM storage cell
- Explain how data are read from a ROM
- Discuss internal organization of a typical ROM

The ROM Family

Figure 11–23 shows how semiconductor ROMs are categorized. The mask ROM is the type in which the data are permanently stored in the memory during the manufacturing process. The **PROM,** or programmable ROM, is the type in which the data are electrically stored by the user with the aid of specialized equipment. Both the mask ROM and the PROM can be of either MOS or bipolar technology. The **EPROM,** or erasable PROM, is strictly a MOS device. The **UV EPROM** is electrically programmable by the user, but the stored data must be erased by exposure to ultraviolet light over a period of several minutes. The electrically erasable PROM (**EEPROM** or E^2PROM) can be erased in a few milliseconds. The UV EPROM has been largely displaced by the EEPROM.

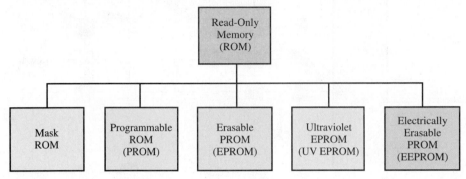

FIGURE 11–23   The ROM family.

The Mask ROM

The mask ROM is usually referred to simply as a ROM. It is permanently programmed during the manufacturing process to provide widely used standard functions, such as popular conversions, or to provide user-specified functions. Once the memory is programmed, it cannot be changed. Most IC ROMs utilize the presence or absence of a transistor connection at a row/column junction to represent a 1 or a 0.

Figure 11–24 shows MOS ROM cells. The presence of a connection from a row line to the gate of a transistor represents a 1 at that location because when the row line is taken HIGH, all transistors with a gate connection to that row line turn on and connect the HIGH (1) to the associated column lines. At row/column junctions where there are no gate connections, the column lines remain LOW (0) when the row is addressed.

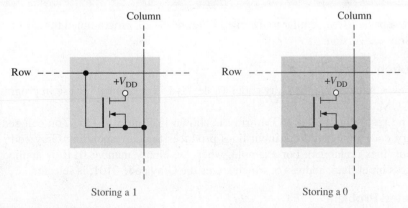

FIGURE 11–24   ROM cells.

To illustrate the ROM concept, Figure 11–25 shows a small, simplified ROM array. The blue squares represent stored 1s, and the gray squares represent stored 0s. The basic read operation is as follows. When a binary address code is applied to the address input lines, the corresponding row line goes HIGH. This HIGH is connected to the column lines through the transistors at each junction (cell) where a 1 is stored. At each cell where a 0 is stored, the column line stays LOW because of the terminating resistor. The column lines form the data output. The eight data bits stored in the selected row appear on the output lines.

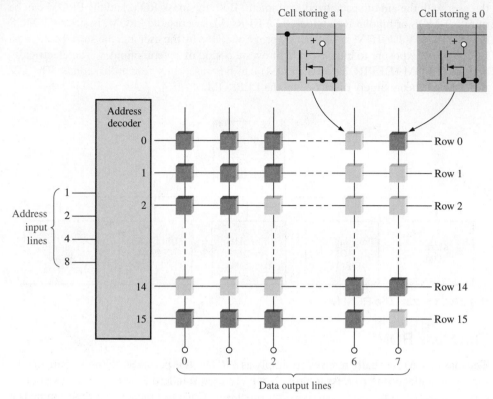

FIGURE 11–25   A representation of a 16 × 8-bit ROM array.

As you can see, the example ROM in Figure 11–25 is organized into 16 addresses, each of which stores 8 data bits. Thus, it is a 16 × 8 (16-by-8) ROM, and its total capacity is 128 bits or 16 bytes. ROMs can be used as look-up tables (LUTs) for code conversions and logic function generation.

EXAMPLE 11–1

Show a basic ROM, similar to the one in Figure 11–25, programmed for a 4-bit binary-to-Gray conversion.

Solution

Review Chapter 2 for the Gray code. Table 11–1 is developed for use in programming the ROM.

The resulting 16 × 4 ROM array is shown in Figure 11–26. You can see that a binary code on the address input lines produces the corresponding Gray code on the output lines (columns). For example, when the binary number 0110 is applied to the address input lines, address 6, which stores the Gray code 0101, is selected.

Related Problem*

Using Figure 11–26, determine the Gray code output when a binary code of 1011 is applied to the address input lines.

---

*Answers are at the end of the chapter.

TABLE 11–1

Binary				Gray			
B_3	B_2	B_1	B_0	G_3	G_2	G_1	G_0
0	0	0	0	0	0	0	0
0	0	0	1	0	0	0	1
0	0	1	0	0	0	1	1
0	0	1	1	0	0	1	0
0	1	0	0	0	1	1	0
0	1	0	1	0	1	1	1
0	1	1	0	0	1	0	1
0	1	1	1	0	1	0	0
1	0	0	0	1	1	0	0
1	0	0	1	1	1	0	1
1	0	1	0	1	1	1	1
1	0	1	1	1	1	1	0
1	1	0	0	1	0	1	0
1	1	0	1	1	0	1	1
1	1	1	0	1	0	0	1
1	1	1	1	1	0	0	0

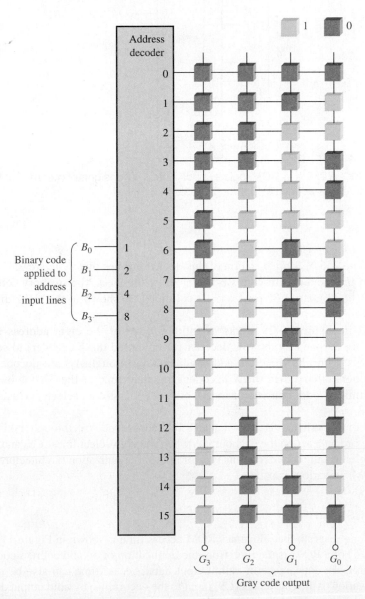

FIGURE 11–26 Representation of a ROM programmed as a binary-to-Gray code converter.

Internal ROM Organization

Most IC ROMs have a more complex internal organization than that in the basic simplified example just presented. To illustrate how an IC ROM is structured, let's use a 1024-bit device with a 256 × 4 organization. The logic symbol is shown in Figure 11–27. When any one of 256 binary codes (eight bits) is applied to the address lines, four data bits appear on the outputs if the chip select inputs are LOW. (256 addresses require eight address lines.)

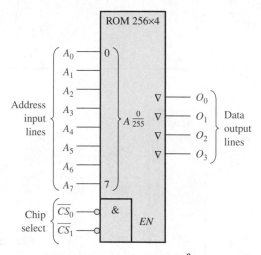

FIGURE 11–27 A 256 × 4 ROM logic symbol. The $A\frac{0}{255}$ designator means that the 8-bit address code selects addresses 0 through 255.

Although the 256 × 4 organization of this device implies that there are 256 rows and 4 columns in the memory array, this is not actually the case. The memory cell array is actually a 32 × 32 matrix (32 rows and 32 columns), as shown in the block diagram in Figure 11–28.

The ROM in Figure 11–28 works as follows. Five of the eight address lines (A_0 through A_4) are decoded by the row decoder (often called the Y decoder) to select one of the 32 rows. Three of the eight address lines (A_5 through A_7) are decoded by the column decoder (often called the X decoder) to select four of the 32 columns. Actually, the column decoder consists of four 1-of-8 decoders (data selectors), as shown in Figure 11–28.

The result of this structure is that when an 8-bit address code (A_0 through A_7) is applied, a 4-bit data word appears on the data outputs when the chip select lines ($\overline{CS_0}$ and $\overline{CS_1}$) are LOW to enable the output buffers. This type of internal organization (architecture) is typical of IC ROMs of various capacities.

ROM Access Time

A typical timing diagram that illustrates ROM access time is shown in Figure 11–29. The **access time,** t_a, of a ROM is the time from the application of a valid address code on the input lines until the appearance of valid output data. Access time can also be measured from the activation of the chip select ($\overline{CS}$) input to the occurrence of valid output data when a valid address is already on the input lines.

InfoNote

ROM is used in a computer to store the BIOS (Basic Input/Output System). These are programs that are used to perform fundamental supervisory and support functions for the computer. For example, BIOS programs stored in the ROM control certain video monitor functions, provide for disk formatting, scan the keyboard for inputs, and control certain printer functions.

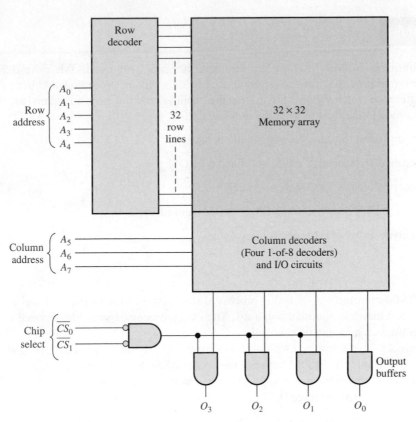

FIGURE 11–28   A 1024-bit ROM with a 256 × 4 organization based on a 32 × 32 array.

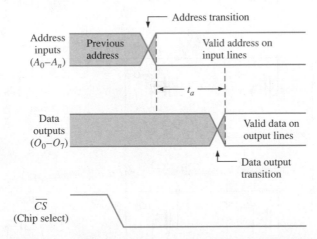

FIGURE 11–29   ROM access time (t_a) from address change to data output with chip select already active.

SECTION 11–3 CHECKUP

1. What is the bit storage capacity of a ROM with a 512 × 8 organization?

2. List the types of read-only memories.

3. How many address bits are required for a 2048-bit memory organized as a 256 × 8 memory?

11–4   Programmable ROMs

Programmable ROMs (PROMs) are basically the same as mask ROMs once they have been programmed. As you have learned, ROMs are a type of programmable logic device. The difference is that PROMs come from the manufacturer unprogrammed and are custom programmed in the field to meet the user's needs.

After completing this section, you should be able to

- ◆ Distinguish between a mask ROM and a PROM
- ◆ Describe a basic PROM memory cell
- ◆ Discuss EPROMs including UV EPROMs and EEPROMs
- ◆ Analyze an EPROM programming cycle

PROMs

A **PROM** uses some type of fusing process to store bits, in which a memory *link* is burned open or left intact to represent a 0 or a 1. The fusing process is irreversible; once a PROM is programmed, it cannot be changed.

Figure 11–30 illustrates a MOS PROM array with fusible links. The fusible links are manufactured into the PROM between the source of each cell's transistor and its column line. In the programming process, a sufficient current is injected through the fusible link to burn it open to create a stored 0. The link is left intact for a stored 1.

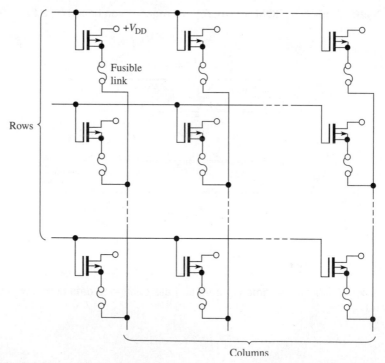

FIGURE 11–30   MOS PROM array with fusible links. (All drains are commonly connected to V_{DD}.)

Three basic fuse technologies used in PROMs are metal links, silicon links, and *pn* junctions. A brief description of each of these follows.

1. Metal links are made of a material such as nichrome. Each bit in the memory array is represented by a separate link. During programming, the link is either "blown" open

or left intact. This is done basically by first addressing a given cell and then forcing a sufficient amount of current through the link to cause it to open.

2. Silicon links are formed by narrow, notched strips of polycrystalline silicon. Programming of these fuses requires melting of the links by passing a sufficient amount of current through them. This amount of current causes a high temperature at the fuse location that oxidizes the silicon and forms an insulation around the now-open link.

3. Shorted junction, or avalanche-induced migration, technology consists basically of two *pn* junctions arranged back-to-back. During programming, one of the diode junctions is avalanched, and the resulting voltage and heat cause aluminum ions to migrate and short the junction. The remaining junction is then used as a forward-biased diode to represent a data bit.

EPROMs

An **EPROM** is an erasable PROM. Unlike an ordinary PROM, an EPROM can be reprogrammed if an existing program in the memory array is erased first.

An EPROM uses an NMOSFET array with an isolated-gate structure. The isolated transistor gate has no electrical connections and can store an electrical charge for indefinite periods of time. The data bits in this type of array are represented by the presence or absence of a stored gate charge. Erasure of a data bit is a process that removes the gate charge.

A typical EPROM is represented in Figure 11–31 by a logic diagram. Its operation is representative of that of other typical EPROMs of various sizes. As the logic symbol shows, this device has 2048 addresses ($2^{11} = 2048$), each with eight bits. Notice that the eight outputs are tri-state (∇).

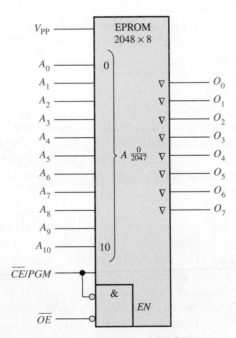

FIGURE 11–31   The logic symbol for a 2048 × 8 EPROM.

To read from the memory, the output enable input ($\overline{OE}$) must be LOW and the power-down/program ($\overline{CE}/PGM$) input LOW.

To program or write to the device, a high dc voltage is applied to V_{PP} and $\overline{OE}$ is HIGH. The eight data bits to be programmed into a given address are applied to the outputs (O_0

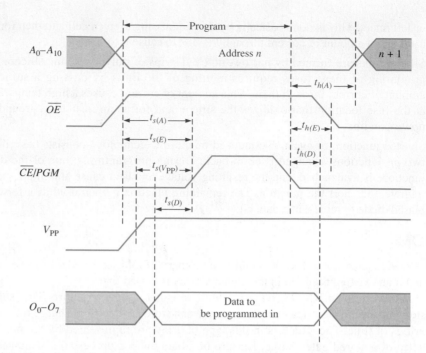

FIGURE 11–32   Timing diagram for a 2048 $\times$ 8 EPROM programming cycle, with critical setup times (t_s) and hold times (t_h) indicated.

through O_7), and the address is selected on inputs A_0 through A_{10}. Next, a HIGH level pulse is applied to the $\overline{CE}/PGM$ input. The addresses can be programmed in any order. A timing diagram for the programming is shown in Figure 11–32. These signals are normally produced by an EPROM programmer.

Two basic types of erasable PROMs are, the electrically erasable PROM (EEPROM) and the ultraviolet erasable PROM (UV EPROM). The UV EPROM is much less used than the EEPROM.

EEPROMs

An electrically erasable PROM can be both erased and programmed with electrical pulses. Since it can be both electrically written into and electrically erased, the EEPROM can be rapidly programmed and erased in-circuit for reprogramming. Two types of EEPROMs are the floating-gate MOS and the metal nitride-oxide silicon (MNOS). The application of a voltage on the control gate in the floating-gate structure permits the storage and removal of charge from the floating gate.

UV EPROMs

You can recognize the UV EPROM device by the UV transparent window on the package. The isolated gate in the **FET** of an ultraviolet EPROM is "floating" within an oxide insulating material. The programming process causes electrons to be removed from the floating gate. Erasure is done by exposure of the memory array chip to high-intensity ultraviolet radiation through the UV window on top of the package. The positive charge stored on the gate is neutralized after several minutes to an hour of exposure time.

SECTION 11–4 CHECKUP

1. How do PROMs differ from ROMs?

2. What represents a data bit in an EPROM?

3. What is the normal mode of operation for a PROM?

11–5   The Flash Memory

The ideal memory has high storage capacity, nonvolatility, in-system read and write capability, comparatively fast operation, and cost effectiveness. The traditional memory technologies such as ROM, PROM, EPROM, EEPROM, SRAM, and DRAM individually exhibit one or more of these characteristics. Flash memory has all of the desired characteristics.

After completing this section, you should be able to

- Discuss the basic characteristics of a flash memory
- Describe the basic operation of a flash memory cell
- Compare flash memories with other types of memories
- Discuss the USB flash drive

Flash memories are high-density read/write memories (high-density translates into large bit storage capacity) that are nonvolatile, which means that data can be stored indefinitely without power. High-density means that a large number of cells can be packed into a given surface area on a chip; that is, the higher the density, the more bits that can be stored on a given size chip. This high density is achieved in flash memories with a storage cell that consists of a single floating-gate MOS transistor. A data bit is stored as charge or the absence of charge on the floating gate depending if a 0 or a 1 is stored.

Flash Memory Cell

A single-transistor cell in a flash memory is represented in Figure 11–33. The stacked gate MOS transistor consists of a control gate and a floating gate in addition to the drain and source. The floating gate stores electrons (charge) as a result of a sufficient voltage applied to the control gate. A *0 is stored when there is more charge* and a *1 is stored when there is less or no charge.* The amount of charge present on the floating gate determines if the transistor will turn on and conduct current from the drain to the source when a control voltage is applied during a read operation.

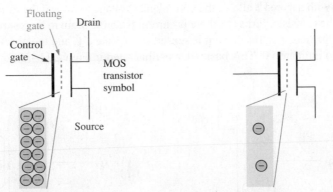

Many electrons = more charge = stored 0.   Few electrons = less charge = stored 1.

FIGURE 11–33 The storage cell in a flash memory.

Basic Flash Memory Operation

There are three major operations in a flash memory: the *programming* operation, the *read* operation, and the *erase* operation.

Programming

Initially, all cells are at the 1 state because charge was removed from each cell in a previous erase operation. The programming operation adds electrons (charge) to the floating gate of those cells that are to store a 0. No charge is added to those cells that are to store a 1. Application of a sufficient positive voltage to the control gate with respect to the source during programming attracts electrons to the floating gate, as indicated in Figure 11–34. Once programmed, a cell can retain the charge for up to 100 years without any external power.

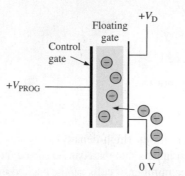
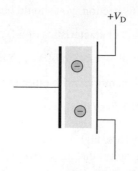

To store a 0, a sufficient positive voltage is applied to the control gate with respect to the source to add charge to the floating gate during programming.

To store a 1, no charge is added and the cell is left in the erased condition.

FIGURE 11–34 Simplified illustration of storing a 0 or a 1 in a flash cell during the programming operation.

Read

During a read operation, a positive voltage is applied to the control gate. The amount of charge present on the floating gate of a cell determines whether or not the voltage applied to the control gate will turn on the transistor. If a 1 is stored, the control gate voltage is sufficient to turn the transistor on. If a 0 is stored, the transistor will not turn on because the control gate voltage is not sufficient to overcome the negative charge stored in the floating gate. Think of the charge on the floating gate as a voltage source that opposes the voltage applied to the control gate during a read operation. So the floating gate charge associated with a stored 0 prevents the control gate voltage from reaching the turn-on threshold, whereas the small or zero charge associated with a stored 1 allows the control gate voltage to exceed the turn-on threshold.

When the transistor turns on, there is current from the drain to the source of the cell transistor. The presence of this current is sensed to indicate a 1, and the absence of this current is sensed to indicate a 0. This basic idea is illustrated in Figure 11–35.

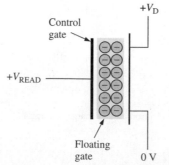
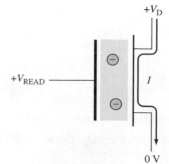

When a 0 is read, the transistor remains off because the charge on the floating gate prevents the read voltage from exceeding the turn-on threshold.

When a 1 is read, the transistor turns on because the absence of charge on the floating gate allows the read voltage to exceed the turn-on threshold.

FIGURE 11–35 The read operation of a flash cell in an array.

Erase

During an erase operation, charge is removed from all the memory cells. A sufficient positive voltage is applied to the transistor source with respect to the control gate. This is opposite in polarity to that used in programming. This voltage attracts electrons from the floating gate and depletes it of charge, as illustrated in Figure 11–36. A flash memory is always erased prior to being reprogrammed.

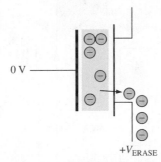

0 V

$+V_{\text{ERASE}}$

To erase a cell, a sufficient positive voltage is applied to the source with respect to the control gate to remove charge from the floating gate during the erase operation.

FIGURE 11–36 Simplified illustration of removing charge from a cell during erase.

Flash Memory Array

A simplified array of flash memory cells is shown in Figure 11–37. Only one row line is accessed at a time. When a cell in a given bit line turns on (stored 1) during a read operation, there is current through the bit line, which produces a voltage drop across the active load. This voltage drop is compared to a reference voltage with a comparator circuit and an output level indicating a 1 is produced. If a 0 is stored, then there is no current or little current in the bit line and an opposite level is produced on the comparator output.

The memory stick is a storage medium that uses flash memory technology in a physical configuration smaller than a stick of chewing gum. Memory sticks are typically available up to 64 GB capacities and as a kit with a PC card adaptor. Because of its compact design, it is ideal for use in small digital electronics products, such as laptop computers and digital cameras.

Comparison of Flash Memories with Other Memories

Let's compare flash memories with other types of memories with which you are already familiar.

Flash vs. ROM, EPROM, and EEPROM

Read-only memories are high-density, nonvolatile devices. However, once programmed the contents of a ROM can never be altered. Also, the initial programming is a time-consuming and costly process. The EEPROM has a more complex cell structure than either the ROM or UV EPROM and so the density is not as high, although it can be reprogrammed without being removed from the system. Because of its lower density, the cost/bit is higher than ROMs or EPROMs. Although the UV EPROM is a high-density, nonvolatile memory, it can be erased only by removing it from the system and using ultraviolet light. It can be reprogrammed only with specialized equipment.

A flash memory can be reprogrammed easily in the system because it is essentially a READ/WRITE device. The density of a flash memory compares with the ROM and EPROM because both have single-transistor cells. A flash memory (like a ROM, EPROM, or EEPROM) is nonvolatile, which allows data to be stored indefinitely with power off.

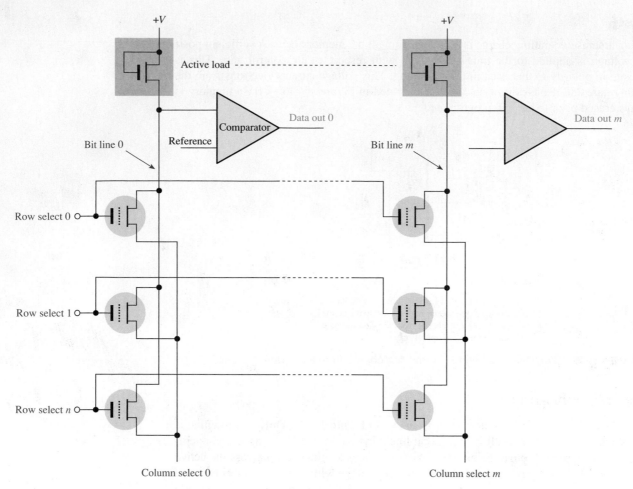

FIGURE 11–37 Basic flash memory array.

Flash vs. SRAM

As you have learned, static random-access memories are volatile READ/WRITE devices. A SRAM requires constant power to retain the stored data. In many applications, a battery backup is used to prevent data loss if the main power source is turned off. However, since battery failure is always a possibility, indefinite retention of the stored data in a SRAM cannot be guaranteed. Because the memory cell in a SRAM is basically a flip-flop consisting of several transistors, the density is relatively low.

A flash memory is also a READ/WRITE memory, but unlike the SRAM it is nonvolatile. Also, a flash memory has a much higher density than a SRAM.

Flash vs. DRAM

Dynamic random-access memories are volatile high-density READ/WRITE devices. DRAMs require not only constant power to retain data but also that the stored data must be refreshed frequently. In many applications, backup storage such as hard disk must be used with a DRAM.

Flash memories exhibit higher densities than DRAMs because a flash memory cell consists of one transistor and does not need refreshing, whereas a DRAM cell is one transistor plus a capacitor that has to be refreshed. Typically, a flash memory consumes much less power than an equivalent DRAM and can be used as a hard disk replacement in many applications.

Table 11–2 provides a comparison of the memory technologies.

TABLE 11–2

Comparison of types of memories.

Memory Type	Nonvolatile	High-Density	One-Transistor Cell	In-System Writability
Flash	Yes	Yes	Yes	Yes
SRAM	No	No	No	Yes
DRAM	No	Yes	Yes	Yes
ROM	Yes	Yes	Yes	No
EEPROM	Yes	No	No	Yes
UV EPROM	Yes	Yes	Yes	No

USB Flash Drive

A USB flash drive consists of a flash memory connected to a standard USB connector housed in a small case about the size of a cigarette lighter. The USB connector can be plugged into a port on a personal computer and obtains power from the computer. These memories are usually rewritable and can have a storage capacity up to 512 GB (a number which is constantly increasing), with most ranging from 2 GB to 64 GB. A typical USB flash drive is shown in Figure 11–38(a), and a basic block diagram is shown in part (b).

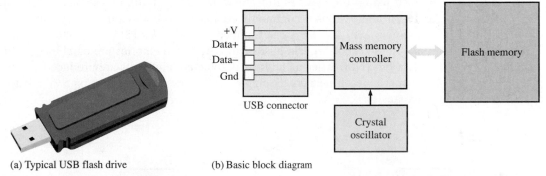

(a) Typical USB flash drive   (b) Basic block diagram

FIGURE 11–38 The USB flash drive.

The USB flash drive uses a standard USB A-type connector for connection to the computer, as shown in Figure 11–39(a). Peripherals such as printers use the USB B-type connector, which has a different shape and physical pin configuration. The USB icon is shown in part (b).

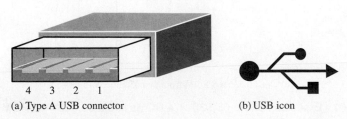

(a) Type A USB connector   (b) USB icon

FIGURE 11–39 Connector and symbol.

SECTION 11–5 CHECKUP

1. What types of memories are nonvolatile?
2. What is a major advantage of a flash memory over a SRAM or DRAM?
3. List the three modes of operation of a flash memory.

11–6   Memory Expansion

Available memory can be expanded to increase the word length (number of bits in each address) or the word capacity (number of different addresses) or both. Memory expansion is accomplished by adding an appropriate number of memory chips to the address, data, and control buses. SIMMs and DIMMs, which are types of memory expansion modules, are introduced.

After completing this section, you should be able to

◆ Define *word-length expansion*

◆ Show how to expand the word length of a memory

◆ Define *word-capacity expansion*

◆ Show how to expand the word capacity of a memory

◆ Discuss types of memory modules

Word-Length Expansion

To increase the **word length** of a memory, the number of bits in the data bus must be increased. For example, an 8-bit word length can be achieved by using two memories, each with 4-bit words as illustrated in Figure 11–40(a). As you can see in part (b), the 16-bit address bus is commonly connected to both memories so that the combination memory still has the same number of addresses ($2^{16} = 65,536$) as each individual memory. The 4-bit data buses from the two memories are combined to form an 8-bit data bus. Now when an address is selected, eight bits are produced on the data bus—four from each memory. Example 11–2 shows the details of $65,536 \times 4$ to $65,536 \times 8$ expansion.

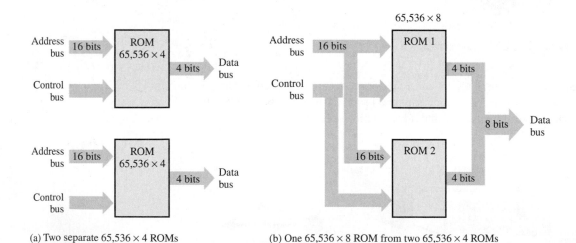

(a) Two separate $65,536 \times 4$ ROMs   (b) One $65,536 \times 8$ ROM from two $65,536 \times 4$ ROMs

FIGURE 11–40   Expansion of two $65,536 \times 4$ ROMs into a $65,536 \times 8$ ROM to illustrate word-length expansion.

EXAMPLE 11–2

Expand the $65,536 \times 4$ ROM ($64k \times 4$) in Figure 11–41 to form a $64k \times 8$ ROM. Note that "64k" is the accepted shorthand for 65,536. Why not "65k"? Maybe it's because 64 is also a power-of-two.

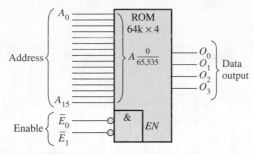

FIGURE 11–41   A 64k × 4 ROM.

Solution

Two 64k × 4 ROMs are connected as shown in Figure 11–42. Notice that a specific address is accessed in ROM 1 and ROM 2 at the same time. The four bits from a selected address in ROM 1 and the four bits from the corresponding address in ROM 2 go out in parallel to form an 8-bit word on the data bus. Also notice that a LOW on the enable line, $\overline{E}$, which forms a simple control bus, enables *both* memories.

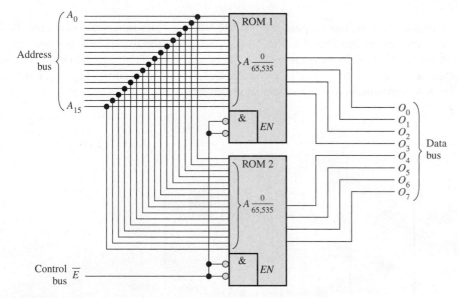

FIGURE 11–42

Related Problem

Describe how you would expand a 64k × 1 ROM to a 64k × 8 ROM.

EXAMPLE 11–3

Use the memories in Example 11–2 to form a 64k × 16 ROM.

Solution

In this case you need a memory that stores 65,536 16-bit words. Four 64k × 4 ROMs are required to do the job, as shown in Figure 11–43.

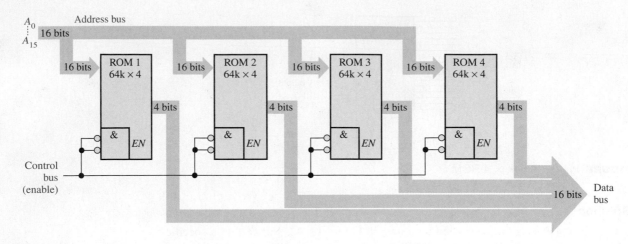

FIGURE 11–43

Related Problem

How many 64k $\times$ 1 ROMs would be required to implement the memory shown in Figure 11–43?

A ROM has only data outputs, but a RAM has both data inputs and data outputs. For word-length expansion in a RAM (SRAM or DRAM), the data inputs *and* data outputs form the data bus. Because the same lines are used for data input and data output, tri-state buffers are required. Most RAMs provide internal tri-state circuitry. Figure 11–44 illustrates RAM expansion to increase the data word length.

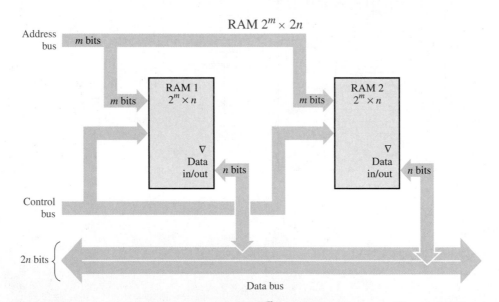

FIGURE 11–44   Illustration of word-length expansion with two $2^m \times n$ RAMs forming a $2^m \times 2n$ RAM.

EXAMPLE 11–4

Use 1M $\times$ 4 SRAMs to create a 1M $\times$ 8 SRAM.

Solution

Two 1M $\times$ 4 SRAMs are connected as shown in the simplified block diagram of Figure 11–45.

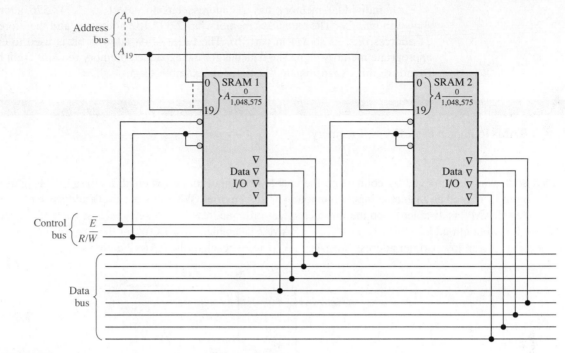

FIGURE 11–45

Related Problem

Use 1M $\times$ 8 SRAMs to create a 1M $\times$ 16 SRAM.

Word-Capacity Expansion

When memories are expanded to increase the **word capacity**, the *number of addresses is increased*. To achieve this increase, the number of address bits must be increased, as illustrated in Figure 11–46, (where two 1M $\times$ 8 RAMs are expanded to form a 2M $\times$ 8 memory).

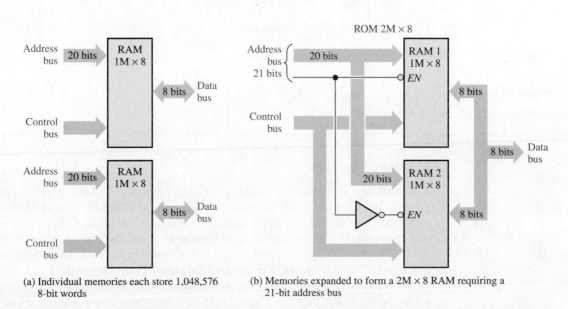

(a) Individual memories each store 1,048,576 8-bit words

(b) Memories expanded to form a 2M $\times$ 8 RAM requiring a 21-bit address bus

FIGURE 11–46 Illustration of word-capacity expansion.

Each individual memory has 20 address bits to select its 1,048,576 addresses, as shown in part (a). The expanded memory has 2,097,152 addresses and therefore requires 21 address bits, as shown in part (b). The twenty-first address bit is used to enable the appropriate memory chip. The data bus for the expanded memory remains eight bits wide. Details of this expansion are illustrated in Example 11–5.

EXAMPLE 11–5

Use 512k $\times$ 4 RAMs to implement a 1M $\times$ 4 memory.

Solution

The expanded addressing is achieved by connecting the enable ($\overline{E}_0$) input to the twentieth address bit (A_{19}), as shown in Figure 11–47. Input $\overline{E}_1$ is used as an enable input common to both memories. When the twentieth address bit (A_{19}) is LOW, RAM 1 is selected (RAM 2 is disabled), and the nineteen lower-order address bits (A_0–A_{18}) access each of the addresses in RAM 1. When the twentieth address bit (A_{19}) is HIGH, RAM 2 is enabled by a LOW on the inverter output (RAM 1 is disabled), and the nineteen lower-order address bits ($A_0 - A_{18}$) access each of the RAM 2 addresses.

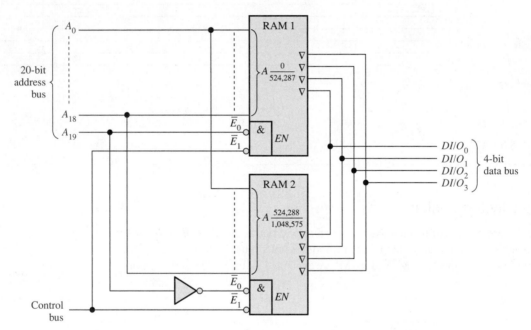

FIGURE 11–47

Related Problem

What are the ranges of addresses in RAM 1 and in RAM 2 in Figure 11–47?

Memory Modules

SDRAMs are available in modules consisting of multiple memory ICs arranged on a printed circuit board (PCB). The most common type of SDRAM memory module is called a **DIMM** (dual in-line memory module). Another version of the DIMM is the SODIMM (small-outline DIMM). A type of memory module, generally found in older equipment and essentially obsolete, is the **SIMM** (single in-line memory module). The SIMM has connection pins on one side of a PCB where the DIMM uses both sides of the board. DIMMs plug into a socket on the system mother board for memory expansion. A generic representation of a memory module is shown in Figure 11–48 with the system board connectors into which the modules are inserted.

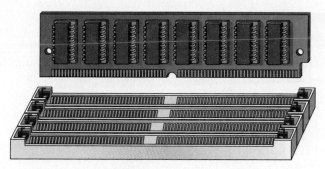

FIGURE 11–48   A memory module with connectors.

DIMMs generally contain DDR SDRAM memory chips. DDR means double data rate, so a DDR SDRAM transfers two blocks of data for each clock cycle rather than one like a standard SDRAM. Three basic types of modules are DDR, DDR2, and DDR3.

- DDR modules have 184 pins and require a 2.5 voltage source.
- DDR2 modules have 240 pins and require a 1.8 voltage source.
- DDR3 modules have 240 pins and require a 1.5 voltage source.

The DDR, DDR2, and DDR3 have transfer data rates of 1600 MB/s, 3200 MB/s, and 6400 MB/s respectively.

Memory components are extremely sensitive to static electricity. Use the following precautions when handling memory chips or modules such as DIMMs:

- Before handling, discharge your body's static charge by touching a grounded surface or wear a grounding wrist strap containing a high-value resistor if available. A convenient, reliable ground is the ac outlet ground.
- Do not remove components from their antistatic bags until you are ready to install them.
- Do not lay parts on the antistatic bags because only the inside is antistatic.
- When handling DIMMs, hold by the edges or the metal mounting bracket. Do not touch components on the boards or the edge connector pins.
- Never slide any part over any type of surface.
- Avoid plastic, vinyl, styrofoam, and nylon in the work area.

When installing DIMMs, follow these steps:

1. Line up the notches on the DIMM board with the notches in the memory socket.
2. Push firmly on the module until it is securely seated in the socket.
3. Generally, the latches on both sides of the socket will snap into place when the module is completely inserted. These latches also release the module, so it can be removed from the socket.

SECTION 11–6 CHECKUP

1. How many 16k $\times$ 1 RAMs are required to achieve a memory with a word capacity of 16k and a word length of eight bits?

2. To expand the 16k $\times$ 8 memory in question 1 to a 32k $\times$ 8 organization, how many more 16k $\times$ 1 RAMs are required?

3. What does DIMM stand for?

11-7   Special Types of Memories

In this section, the first in–first out (FIFO) memory, the last in–first out (LIFO) memory, the memory stack, and the charge-coupled device memory are covered.

After completing this section, you should be able to

- Describe a FIFO memory
- Describe a LIFO memory
- Discuss memory stacks
- Explain how to use a portion of RAM as a memory stack
- Describe a basic CCD memory

First In–First Out (FIFO) Memories

This type of memory is formed by an arrangement of shift registers. The term **FIFO** refers to the basic operation of this type of memory, in which the first data bit written into the memory is the first to be read out.

One important difference between a conventional shift register and a FIFO register is illustrated in Figure 11–49. In a conventional register, a data bit moves through the register only as new data bits are entered; in a FIFO register, a data bit immediately goes through the register to the right-most bit location that is empty.

Conventional Shift Register						FIFO Shift Register					
Input	X	X	X	X	Output	Input	—	—	—	—	Output
0	0	X	X	X	⟶	0	—	—	—	0	⟶
1	1	0	X	X	⟶	1	—	—	1	0	⟶
1	1	1	0	X	⟶	1	—	1	1	0	⟶
0	0	1	1	1	⟶	0	0	1	1	0	⟶

X = unknown data bits.
In a conventional shift register, data stay to the left until "forced" through by additional data

— = empty positions.
In a FIFO shift register, data "fall" through (go right).

FIGURE 11–49   Comparison of conventional and FIFO register operation.

Figure 11–50 is a block diagram of a FIFO serial memory. This particular memory has four serial 64-bit data registers and a 64-bit control register (marker register). When data are entered by a shift-in pulse, they move automatically under control of the marker register to the empty location closest to the output. Data cannot advance into occupied positions. However, when a data bit is shifted out by a shift-out pulse, the data bits remaining in the registers automatically move to the next position toward the output. In an asynchronous FIFO, data are shifted out independent of data entry, with the use of two separate clocks.

FIFO Applications

One important application area for the FIFO register is the case in which two systems of differing data rates must communicate. Data can be entered into a FIFO register at one rate and taken out at another rate. Figure 11–51 illustrates how a FIFO register might be used in these situations.

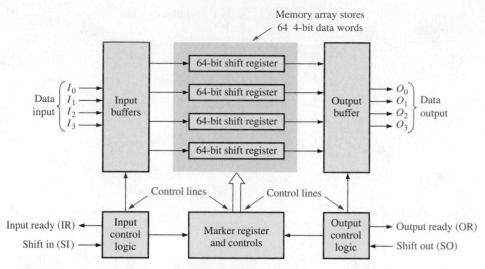

Memory array stores
64 4-bit data words

FIGURE 11–50 Block diagram of a typical FIFO serial memory.

Irregular-rate data → FIFO register → Constant-rate data

(a) Irregular telemetry data can be stored and retransmitted at a constant rate.

Lower-rate data → FIFO register → Higher-rate data

(b) Data input at a slow keyboard rate can be stored and then transferred at a higher rate for processing.

Constant-rate data → FIFO register → Burst data

(c) Data input at a constant rate can be stored and then output in bursts.

Burst data → FIFO register → Constant-rate data

(d) Data in bursts can be stored and reformatted into a constant-rate output.

FIGURE 11–51 Examples of the FIFO register in data-rate buffering applications.

Last In–First Out (LIFO) Memories

The **LIFO** (last in–first out) memory is found in applications involving microprocessors and other computing systems. It allows data to be stored and then recalled in reverse order; that is, the last data byte to be stored is the first data byte to be retrieved.

Register Stacks

A LIFO memory is commonly referred to as a push-down stack. In some systems, it is implemented with a group of registers as shown in Figure 11–52. A stack can consist of any number of registers, but the register at the top is called the *top-of-stack.*

To illustrate the principle, a byte of data is loaded in parallel onto the top of the stack. Each successive byte pushes the previous one down into the next register. This process is illustrated in Figure 11–53. Notice that the new data byte is always loaded into the top register and the previously stored bytes are pushed deeper into the stack. The name *push-down stack* comes from this characteristic.

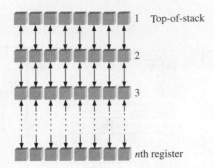

FIGURE 11–52 Register stack.

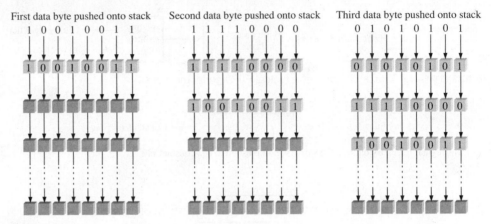

FIGURE 11–53 Simplified illustration of pushing data onto the stack.

Data bytes are retrieved in the reverse order. The last byte entered is always at the top of the stack, so when it is pulled from the stack, the other bytes pop up into the next higher locations. This process is illustrated in Figure 11–54.

RAM Stack

Another approach to LIFO memory used in microprocessor-based systems is the allocation of a section of RAM as the stack rather than the use of a dedicated set of registers. As you have seen, for a register stack the data move up or down from one location to the next. In

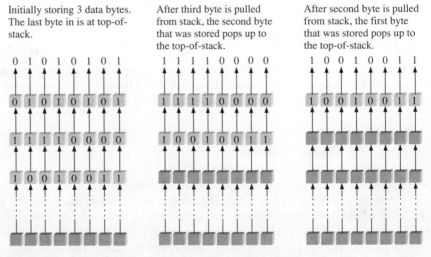

FIGURE 11–54 Simplified illustration of pulling data from the stack.

a RAM stack, the data do not move but the top-of-stack moves under control of a register called the stack pointer.

Consider a random-access memory that is byte organized—that is, one in which each address contains eight bits—as illustrated in Figure 11–55. The binary address 0000000000001111, for example, can be written as 000F in hexadecimal. A 16-bit address can have a *minimum* hexadecimal value of 0000_{16} and a *maximum* value of $FFFF_{16}$. With this notation, a 64 kB memory array can be represented as shown in Figure 11–55. The lowest memory address is 0000_{16} and the highest memory address is $FFFF_{16}$.

Now, consider a section of RAM set aside for use as a stack. A special separate register, the stack pointer, contains the address of the top of the stack, as illustrated in Figure 11–56. A 4-digit hexadecimal representation is used for the binary addresses. In the figure, the addresses are chosen for purposes of illustration.

Now let's see how data are pushed onto the stack. The stack pointer is initially at address $FFEE_{16}$, which is the top of the stack as shown in Figure 11–56(a). The stack pointer is then decremented (decreased) by two to $FFEC_{16}$. This moves the top of the stack to a lower memory address, as shown in Figure 11–56(b). Notice that the top of the stack is not stationary as in the fixed register stack but moves downward (to lower addresses) in the RAM as data words are stored. Figure 11–56(b) shows that two bytes (one data word) are then pushed onto the stack. After the data word is stored, the top of the stack is at $FFEC_{16}$.

Figure 11–57 illustrates the POP operation for the RAM stack. The last data word stored in the stack is read first. The stack pointer that is at FFEC is incremented (increased) by two to address $FFEE_{16}$ and a POP operation is performed as shown in part (b). Keep in mind that RAMs are nondestructive when read, so the data word still remains in the memory after a POP operation. A data word is destroyed only when a new word is written over it.

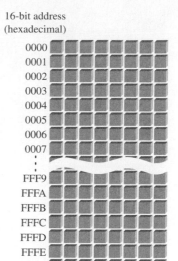

16-bit address (hexadecimal)

0000
0001
0002
0003
0004
0005
0006
0007
⋮
FFF9
FFFA
FFFB
FFFC
FFFD
FFFE
FFFF

FIGURE 11–55 Representation of a 64 kB memory with the 16-bit addresses expressed in hexadecimal.

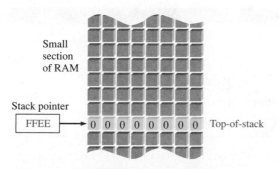
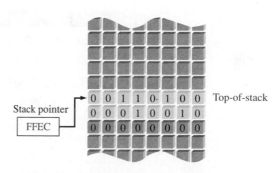

(a) The stack pointer is initially at FFEE before the data word 0001001000110100 (1234) is pushed onto the stack.

(b) The stack pointer is decremented by two and the data word 0001001000110100 is placed in the two locations prior to the original stack pointer location.

FIGURE 11–56 Illustration of the PUSH operation for a RAM stack.

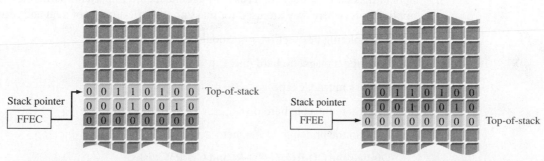

(a) The stack pointer is at FFEC before the data word is copied (popped) from the stack.

(b) The stack pointer is incremented by two and the last data word stored is copied (popped) from the stack.

FIGURE 11–57 Illustration of the POP operation for the RAM stack.

A RAM stack can be of any depth, depending on the number of continuous memory addresses assigned for that purpose.

CCD Memories

The **CCD** (charge-coupled device) memory stores data as charges on capacitors and has the ability to convert optical images to electrical signals. Unlike the DRAM, however, the storage cell does not include a transistor. High density is the main advantage of CCDs, and these devices are widely used in digital imaging.

The CCD memory consists of long rows of semiconductor capacitors, called *channels*. Data are entered into a channel serially by depositing a small charge for a 0 and a large charge for a 1 on the capacitors. These charge packets are then shifted along the channel by clock signals as more data are entered.

As with the DRAM, the charges must be refreshed periodically. This process is done by shifting the charge packets serially through a refresh circuit. Figure 11–58 shows the basic concept of a CCD channel. Because data are shifted serially through the channels, the CCD memory has a relatively long access time. CCD arrays are used in many modern cameras to capture video images in the form of light-induced charge.

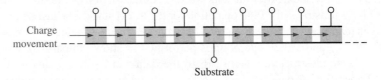

Charge
movement

Substrate

FIGURE 11–58   A CCD (charge-coupled device) channel.

SECTION 11–7 CHECKUP

1. What is a FIFO memory?

2. What is a LIFO memory?

3. Explain the PUSH operation in a memory stack.

4. Explain the POP operation in a memory stack.

5. What does the term *CCD* stand for?

11–8   Magnetic and Optical Storage

In this section, the basics of magnetic disks, magnetic tape, magneto-optical disks, and optical disks are introduced. These storage media are important, particularly in computer applications, where they are used for mass nonvolatile storage of data and programs.

After completing this section, you should be able to

- Describe a magnetic hard disk

- Discuss magnetic tape

- Discuss removable hard disks

- Explain the principle of magneto-optical disks

- Discuss the CD-ROM, CD-R, and CD-RW disks

- Describe the WORM

- Discuss the DVD-ROM

Magnetic Storage

Magnetic Hard Disks

Computers use hard disks as the internal mass storage media. **Hard disks** are rigid "platters" made of aluminum alloy or a mixture of glass and ceramic covered with a magnetic coating. Hard disk drives mainly come in three diameter sizes, 3.5 in., 2.5 in., and 1.8 in. Older formats of 8 in. and 5.25 in. are considered obsolete. A hard disk drive is hermetically sealed to keep the disks dust-free.

Typically, two or more disks are stacked on top of each other on a common shaft or spindle that turns the assembly at several thousand rpm. A separation between each disk allows for a magnetic read/write head that is mounted on the end of an actuator arm, as shown in Figure 11–59. There is a read/write head for both sides of each disk since data are recorded on both sides of the disk surface. The drive actuator arm synchronizes all the read/write heads to keep them in perfect alignment as they "fly" across the disk surface with a separation of only a fraction of a millimeter from the disk. A small dust particle could cause a head to "crash," causing damage to the disk surface.

FIGURE 11–59   A hard disk drive.   FrameAngel/Shutterstock

Basic Read/Write Head Principles

The hard drive is a random-access device because it can retrieve stored data anywhere on the disk in any order. A simplified diagram of the magnetic surface read/write operation is shown in Figure 11–60. The direction or polarization of the magnetic domains on the disk surface is controlled by the direction of the magnetic flux lines (magnetic field) produced

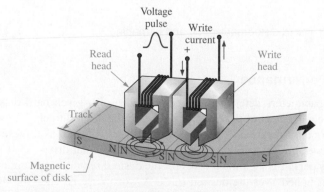

FIGURE 11–60   Simplified read/write head operation.

by the write head according to the direction of a current pulse in the winding. This magnetic flux magnetizes a small spot on the disk surface in the direction of the magnetic field. A magnetized spot of one polarity represents a binary 1, and one of the opposite polarity represents a binary 0. Once a spot on the disk surface is magnetized, it remains until written over with an opposite magnetic field.

When the magnetic surface passes a read head, the magnetized spots produce magnetic fields in the read head, which induce voltage pulses in the winding. The polarity of these pulses depends on the direction of the magnetized spot and indicates whether the stored bit is a 1 or a 0. The read and write heads are usually combined in a single unit.

Hard Disk Format

A hard disk is organized or formatted into tracks and sectors, as shown in Figure 11–61(a). Each track is divided into a number of sectors, and each track and sector has a physical address that is used by the operating system to locate a particular data record. Hard disks typically have from a few hundred to thousands of tracks and are available with storage capacities of up to 1 TB or more. As you can see in the figure, there is a constant number of tracks/sector, with outer sectors using more surface area than the inner sectors. The arrangement of tracks and sectors on a disk is known as the *format*.

A hard disk stack is illustrated in Figure 11–61(b). Hard disk drives differ in the number of disks in a stack, but there is always a minimum of two. All of the same corresponding tracks on each disk are collectively known as a cylinder, as indicated.

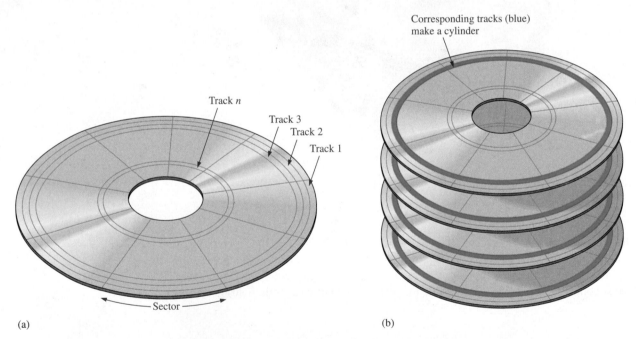

FIGURE 11–61   Hard disk organization and formatting.

Hard Disk Performance

Several basic parameters determine the performance of a given hard disk drive. A *seek* operation is the movement of the read/write head to the desired track. The **seek time** is the average time for this operation to be performed. Typically, hard disk drives have an average seek time of several milliseconds, depending on the particular drive.

The **latency period** is the time it takes for the desired sector to spin under the head once the head is positioned over the desired track. A worst case is when the desired sector is just past the head position and spinning away from it. The sector must rotate almost a full revolution back to the head position. *Average latency period* assumes that the disk must make half of a revolution. Obviously, the latency period depends on the constant rotational speed of the disk. Disk rotation speeds are different for different disk drives but typically are from 4200 rpm to 15,000 rpm.

The sum of the average seek time and the average latency period is the *access time* for the disk drive.

Removable Hard Disk

A removable hard disk drive with a capacity of 1 TB is available. Keep in mind that the technology is changing so rapidly that there most likely will be further advancements at the time you are reading this.

Magnetic Tape

Tape is used for backup data from mass storage devices and typically is slower than disks because data on tape is accessed serially rather than randomly. There are several types that are available, including QIC, 8 mm, and DLT.

QIC is an abbreviation for quarter-inch cartridge and looks much like audio tape cassettes with two reels inside. Various QIC standards have from 28 to 108 tracks that can store from 80 MB to 1.6 GB. More recent innovations under the Travan standard have lengthened the tape and increased its width allowing storage capacities up to 10 GB. QIC tape drives use read/write heads that have a single write head with a read head on each side. This allows the tape drive to verify data just written when the tape is running in either direction. In the record mode, the tape moves past the read/write heads at approximately 100 inches/second, as indicated in Figure 11–62.

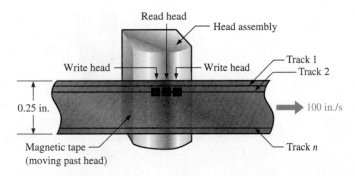

FIGURE 11–62   QIC tape.

8 mm tape was originally designed for the video industry but has been adopted by the computer industry as a reliable way to store large amounts of computer data.

DLT is an abbreviation for digital linear tape. DLT is a half-inch wide tape, which is 60% wider than 8 mm and, of course, twice as wide as standard QIC. Basically, DLT differs in the way the tape-drive mechanism works to minimize tape wear compared to other systems. DLT offers the highest storage capacity of all the tape formats with capacities ranging up to 800 GB.

Magneto-Optical Storage

As the name implies, magneto-optical (MO) storage devices use a combination of magnetic and optical (laser) technologies. A **magneto-optical disk** is formatted into tracks and sectors similar to magnetic disks.

The basic difference between a purely magnetic disk and an MO disk is that the magnetic coating used on the MO disk requires heat to alter the magnetic polarization. Therefore, the MO is extremely stable at ambient temperature, making data unchangeable. To write a data bit, a high-power laser beam is focused on a tiny spot on the disk, and the temperature of that tiny spot is raised above a temperature level called the Curie point (about 200°C). Once heated, the magnetic particles at that spot can easily have their direction (polarization) changed by a magnetic field generated by the write head. Information is read from the disk with a less-powerful laser than used for writing, making use of the Kerr effect where the polarity of the reflected laser light is altered depending on the orientation of the magnetic particles. Magnetic spots of one polarity represent 0s and magnetic spots of the opposite polarity represent 1s. Basic MO operation is shown in Figure 11–63, which represents a small cross-sectional area of a disk.

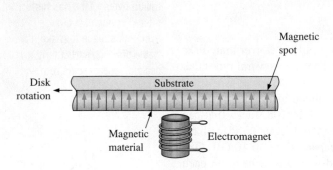

(a) Unrecorded disk

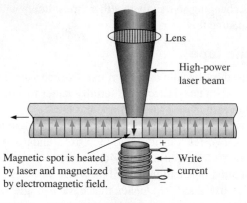

(b) Writing: A high-power laser beam heats the spot, causing the magnetic particles to align with the electromagnetic field.

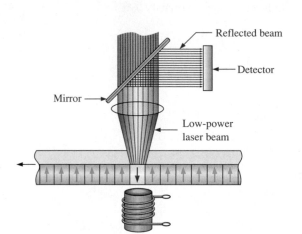

(c) Reading: A low-power laser beam reflects off of the reversed-polarity magnetic particles and its polarization shifts. If the particles are not reversed, the polarization of the reflected beam is unchanged.

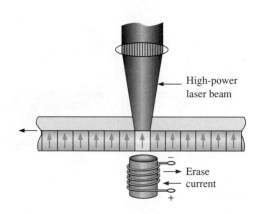

(d) Erasing: The electromagnetic field is reversed as the high-power laser beam heats the spot, causing the magnetic particles to be restored to the original polarity.

FIGURE 11–63 Basic principle of a magneto-optical disk.

Optical Storage

CD-ROM

The most common Compact Disk–Read-Only Memory is a 120 mm diameter disk with a sandwich of three coatings: a polycarbonate plastic on the bottom, a thin aluminum sheet for reflectivity, and a top coating of lacquer for protection. The **CD-ROM** disk is formatted in a single spiral track with sequential 2 kB sectors and has a capacity of 680 MB. Data are prerecorded at the factory in the form of minute indentations called *pits* and the flat area surrounding the pits called *lands*. The pits are stamped into the plastic layer and cannot be erased.

A CD player reads data from the spiral track with a low-power infrared laser, as illustrated in Figure 11–64. The data are in the form of pits and lands as shown. Laser light reflected from a pit is 180° out-of-phase with the light reflected from the lands. As the disk rotates, the narrow laser beam strikes the series of pits and lands of varying lengths, and a photodiode detects the difference in the reflected light. The result is a series of 1s and 0s corresponding to the configuration of pits and lands along the track.

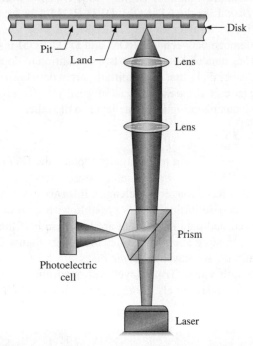

FIGURE 11–64   Basic operation of reading data from a CD-ROM.

WORM

Write Once/Read Many (**WORM**) is a type of optical storage that can be written onto one time after which the data cannot be erased but can be read many times. To write data, a low-power laser is used to burn microscopic pits on the disk surface. 1s and 0s are represented by the burned and nonburned areas.

CD-R

This is essentially a type of WORM. The difference is that the CD-Recordable allows multiple write sessions to different areas of the disk. The **CD-R** disk has a spiral track like the CD-ROM, but instead of mechanically pressing indentations on the disk to represent data, the CD-R uses a laser to burn microscopic spots into an organic dye surface. When heated beyond a critical temperature with a laser during read, the burned spots change color and reflect less light than the nonburned areas. Therefore, 1s and 0s are represented on a CD-R by burned and nonburned areas, whereas on a CD-ROM they are represented by pits and lands. Like the CD-ROM, the data cannot be erased once it is written.

CD-RW

The CD-Rewritable disk can be used to read and write data. Instead of the dye-based recording layer in the CD-R, the **CD-RW** commonly uses a crystalline compound with a special property. When it is heated to a certain temperature, it becomes crystalline when it cools; but if it is heated to a certain higher temperature, it melts and becomes amorphous when it cools. To write data, the focused laser beam heats the material to the melting temperature resulting in an amorphous state. The resulting amorphous areas reflect less light than the crystalline areas, allowing the read operation to detect 1s and 0s. The data can be erased or overwritten by heating the amorphous areas to a temperature above the crystallization temperature but lower than the melting temperature that causes the amorphous material to revert to a crystalline state.

DVD-ROM

Originally DVD was an abbreviation for Digital Video Disk but eventually came to represent *Digital Versatile Disk*. Like the CD-ROM, **DVD-ROM** data are prestored on the disk. However, the pit size is smaller than for the CD-ROM, allowing more data to be stored on a track. The major difference between CD-ROM and DVD-ROM is that the CD is single-sided, while the DVD has data on both sides. Also, in addition to double-sided DVD disks, there are also multiple-layer disks that use semitransparent data layers placed over the main data layers, providing storage capacities of tens of gigabytes. To access all the layers, the laser beam requires refocusing going from one layer to the other.

Blu-Ray

The **Blu-ray** Disc (BD) is designed to eventually replace the DVD. The BD is the same size as DVDs and CDs. The name *Blu-ray* refers to the blue laser used to read the disc. DVDs use a red laser that has a longer wavelength. Information can be stored on a BD at a greater density and video definition than is possible with a DVD. The smaller Blu-ray laser beam can read recorded data in pits that are less than half the size of the pits on a DVD. A Blu-ray Disc can store about five times more data than a DVD. Typical storage capacities for conventional Blu-ray dual-layer discs are 50 GB, which is the industry standard for feature-length video. Triple layer and quadruple layer discs (BD-XL) can store 100 GB and 128 GB, respectively. Storage capacities up to 1 TB are currently under development.

SECTION 11–8 CHECKUP

1. List the major types of magnetic storage.
2. Generally, how is a magnetic disk organized?
3. How are data written on and read from a magneto-optical disk?
4. List the types of optical storage.

11–9   Memory Hierarchy

A memory system performs the data storage function in a computer. The memory system holds data temporarily during processing and also stores data and programs on a long-term basis. A computer has several types of memory, such as registers, cache, main, and hard disk. Other types of storage can also be used, such as magnetic tape, optical disk, and magnetic disk. Memory hierarchy as well as the system processor determines the processing speed of a computer.

After completing this section, you should be able to

- Discuss several types of memory
- Define memory hierarchy
- Describe key elements in a memory hierarchy

Three key characteristics of memory are cost, capacity, and access time. Memory cost is usually specified in cost per bit. The capacity of a memory is measured in the amount of data (bits or bytes) it can store. The access time is the time it takes to acquire a specified unit of data from the memory. The greater the capacity, the smaller the cost and the greater the access time. The smaller the access time, the greater the cost. The goal of using a memory hierarchy is to obtain the shortest possible average access time while minimizing the cost.

The speed with which data can be processed depends both on the processor speed and on the time it takes to access stored data. **Memory hierarchy** is the arrangement of various memory elements within the computer architecture to maximize processing speed and minimize cost. Memory can be classified according to its "distance" from the processor in terms of the number of machine cycles or access time required to get data for processing. Distance is measured in time, not in physical location. Faster memory elements are considered closer to the processor compared to slower types of memory elements. Also, the cost per bit is much greater for the memory close to the processor than for the memory that is further from the processor. Figure 11–65 illustrates the arrangement of elements in a typical memory hierarchy.

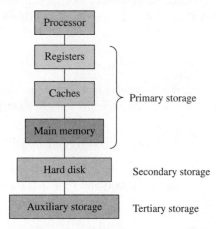

FIGURE 11–65 Typical memory hierarchy.

A primary distinction between the storage elements in Figure 11–65 is the time required for the processor to access data and programs. This access time is known as **memory latency**. The greater the latency, the further from the processor a storage element is considered to be. For example, typical register latency can be up to 1 or 2 ns, cache latency can be up to about 50 ns, main memory latency can be up to about 90 ns, and hard disk latency can be up to about 20 ms. Auxiliary memory latency can range up to several seconds.

Registers

Registers are memory elements that are located within the processor. They have a very small latency as well as a low capacity (number of bits that can be stored). One goal of programming is to keep as much frequently used data in the registers as possible. The number of registers in a processor can vary from the tens to hundreds.

Caches

The next level in the hierarchy is the memory cache, which provides temporary storage. The L1 cache is located in the processor, and the L2 cache is outside of the processor. A programming goal is to keep as much of a program as possible in the cache, especially the parts of a program that are most extensively used. There can be more than two caches in a memory system.

Main Memory

Main memory generally consists of two elements: RAM (random-access memory) and ROM (read-only memory). The RAM is a working memory that temporarily stores less frequently used data and program instructions. The RAM is volatile, which means that the stored contents are lost when the power is turned off. The ROM is for permanent storage of frequently used programs and data; ROM is nonvolatile. Registers, caches, and main memory are considered primary storage.

Hard Disk

The hard disk has a very high latency and is used for mass storage of data and programs on a permanent basis. The hard disk is also used for virtual memory, space allocated for data when the primary memory fills up. In effect, virtual memory simulates primary memory with the disadvantage of high latency. Capacities range up to about 1 terabyte (TB).

$$1 \text{ TB} = 1,000,000,000,000 \text{ B} = 10^{12} \text{ B}$$

In addition to the internal hard disk, secondary storage can also include off-line storage. Off-line storage includes DVDs, CD-ROM, CD-RW, and USB flash drive. Off-line storage is removable storage.

Auxiliary Storage

Auxiliary storage, also called *tertiary storage,* includes magnetic tape libraries and optical jukeboxes. A **tape library** can store immense amounts of data (up to hundreds of petabytes). A petabyte (PB) is

$$1 \text{ PB} = 1,000,000,000,000,000 \text{ B} = 10^{15} \text{ B}$$

An **optical jukebox** is a robotic storage device that automatically loads and unloads optical disks. It may have as many as 2,000 slots for disks and can store hundreds of petabytes.

Relationship of Cost, Capacity, and Access Time

Figure 11–66 shows how capacity (the amount of data a memory can store) and cost per unit of storage varies as the distance from the processor, in terms of access time or latency, increases. The capacity increases and the cost decreases as access time increases.

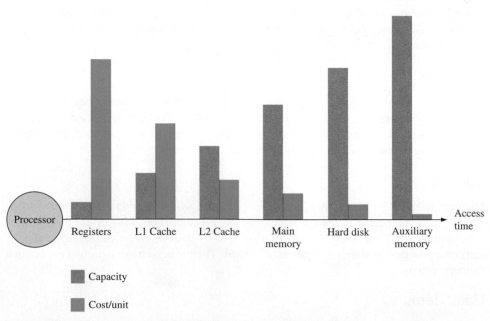

FIGURE 11–66 Changes in memory capacity and cost per unit of data as latency (access time) increases.

Memory Hierarchy Performance

In a computer system, the overall processing speed is usually limited by the memory, not the processor. Programming determines how well a particular memory hierarchy is utilized. The goal is to process data at the fastest rate possible. Two key factors in establishing maximum processor performance are locality and hit rate.

If a block of data is referenced, it will tend to be either referenced again soon or a nearby data block will be referenced soon. Frequent referencing of the same data block is known as *temporal locality*, and the program code should be arranged so that the piece of the data in the cache is reused frequently. Referencing an adjacent data block is known as *spatial locality*, and the program code should be arranged to use consecutive pieces of data on a frequent basis.

A **miss** is a failed attempt by the processor to read or write a block of data in a given level of memory (such as the cache). A miss causes the processor to have to go to a lower level of memory (such as main memory), which has a longer latency. The three types of misses are instruction read miss, data read miss, and data write miss. A successful attempt to read or write a block of data in a given level of memory is called a *hit*. Hits and misses are illustrated in Figure 11–67, where the processor is requesting data from the cache.

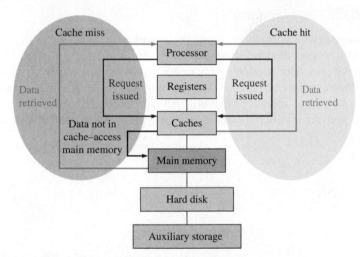

FIGURE 11–67   Illustration of a cache hit and a miss.

The **hit rate** is the percentage of memory accesses that find the requested data in the given level of memory. The *miss rate* is the percentage of memory accesses that fail to find the requested data in the given level of memory and is equal to 1– hit rate. The time required to access the requested information in a given level of memory is called the *hit time*. The higher the hit rate (hit to miss ratio), the more efficient the memory hierarchy is.

SECTION 11–9 CHECKUP

1. State the purpose of memory hierarchy.
2. What is access time?
3. How does memory capacity affect the cost per bit?
4. Does higher level memory generally have lower capacity than lower level memory?
5. What is a hit? A miss?
6. What determines the efficiency of the memory hierarchy?

11–10   Cloud Storage

Cloud storage is a system, usually maintained by a third party, for securely storing data in a remote location that can be conveniently accessed through the Internet. A file on a computer can be stored on secure remote servers and accessed by various user devices such as computers, smart phones, and tablets. Cloud storage eliminates the need for local backup storage such as external hard drives or CDs. When you use cloud storage, you are essentially storing your files or documents on Internet servers instead of or in addition to a computer. The term *cloud* may have originated from the use of a symbol that resembled a cloud on early network diagrams.

After completing this section, you should be able to

◆ Describe cloud storage

◆ Explain what a server is

◆ State the advantages of cloud storage

◆ Describe several properties of cloud storage

The Cloud Storage System

A **cloud storage** system consists of a remote network of servers (also called *nodes*) that are connected to a user device through the Internet, as shown in Figure 11–68. Some cloud storage systems accommodate only certain types of data such as e-mail or digital pictures, while others store all types of data and range in size from small operations with a few servers to very large operations that utilize hundreds of servers. A facility that houses cloud storage systems is called a **data center**. A typical storage cloud system can serve multiple users.

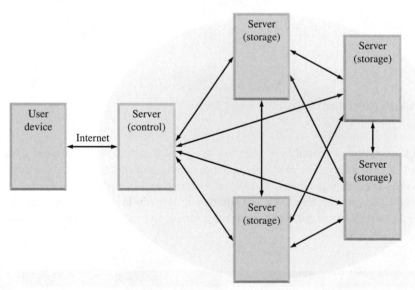

FIGURE 11–68   A typical cloud storage system architecture consists of a master control server and several storage servers that can be accessed by a user device over the Internet.

Servers typically operate within a client-server architecture, where the client is the user that is subscribing to the cloud storage. Theoretically, a **server** is any computerized process that shares a resource with one or more clients. More practically, a storage server is a computer and software with a large memory capacity that responds to requests across a network to provide file storage and access as well as services such as file sharing. The control server

(a) A typical rack of servers   (b) A typical server room in a data center

FIGURE 11–69 Cloud servers.   (a) Jojje/Shutterstock (b) Oleksiy Mark/Shutterstock

coordinates the activities within the storage cloud network among other servers and manages user access. A server rack and data center are shown in Figure 11–69.

At its simplest level, a cloud storage system needs just one storage server connected to the Internet. When copies of a file are sent by a client to the server over the Internet, the data are stored. When the client wishes to retrieve the data, the storage server (node) sends it back through a Web-based interface or allows the client to manipulate the file on the server itself.

Most cloud storage systems have many storage servers (hundreds in some cases) to provide both capacity and redundancy. A grouping of servers is sometimes called a *cluster*. Depending on the system architecture, a given system may have multiple clusters. A simple system with four storage servers illustrating file storage redundancy is shown in Figure 11–70. When a client sends data to the cloud, it is stored in multiple servers. This redundancy guarantees availability of data at any time to the client and makes the system highly reliable. Redundancy is necessary because a server requires periodic maintenance or may break down and need repairs. In addition to storage server redundancy, most cloud storage systems use power supply redundancy so that all servers are not operating from the same power source.

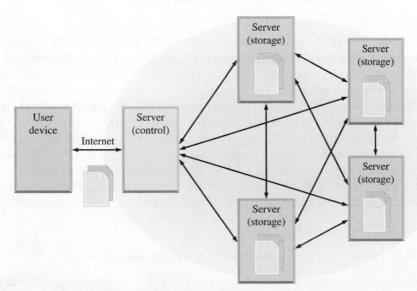

FIGURE 11–70 A simple cloud storage system with storage redundancy. In this case, the data are stored on four different servers.

In addition to reliability that provides assurance that a client's data are accurately stored and can be retrieved at any time, a second major factor for cloud storage is security that the data cannot be compromised. Generally, three methods are used to provide data security:

- *Encryption* or encoding, which prevents the data from being read or interpreted without proper decryption tools
- *Authentication*, which requires a name and password for access
- *Authorization*, which requires a list of only those people who can have access to the data

Cloud storage has certain advantages over traditional data storage in a computer. One advantage is that you can store and retrieve data from any physical location that has Internet access. A second advantage is that you don't have to use the same computer to store and retrieve data or carry a physical storage device for data backup around with you. Also, the user does not have to maintain the storage components. Another advantage of cloud storage is that other people can access your data (data sharing).

Architecture

The term *architecture* relates to how a cloud storage system is structured and organized. The primary purpose of cloud storage architecture is to deliver the service for data storage in a specific way. Architectures vary but generically most consist of a front end, a control, and a back end, as depicted in Figure 11–71.

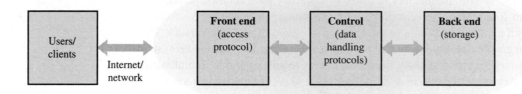

FIGURE 11–71   Generic architecture of a cloud storage system.

A cloud storage system uses various protocols within the architecture that determine how the data are accessed and handled. A **protocol** is a standardized set of software regulations, requirements, and procedures that control and regulate the transmission, processing, and exchange of data among devices. For example, common Internet protocols are HTTP (Hypertext Transfer Protocol), FTP (File Transfer Protocol), TCP/IP (Transfer Control Protocol/Internet Protocol), and SMTP (Simple Mail Transfer Protocol).

An API is an Application Programming Interface, which is essentially a protocol for access and utilization of a cloud storage system. There are many types of APIs. For example, a commonly used one is the REST API. REST stands for Representational State Transfer. An API is a software-to-software interface, not a user interface. With APIs, applications talk to each other "behind the scene" without user knowledge.

Cloud Storage Properties

The following cloud storage properties determine the performance of the system.

- *Latency.* The time between a request for data and the delivery of the data to the user is the **latency** of a system. Delay is due to the time for each component of the cloud storage system to respond to a request and to the time for data to be transferred to the user.

- *Bandwidth.* Bandwidth is a measure of the range of frequencies that can be simultaneously transferred to the cloud and is defined as a range of frequencies that can be handled by the system. Generally, the wider the bandwidth, the shorter the latency and vice versa.

- *Scalability.* The **scalability** property indicates the ability of a cloud storage system to handle increasing amounts of data in a smooth and easy manner; or it is the cloud's ability to improve movement of data through the system (throughput) when additional resources (typically hardware) are added. When the performance of a system improves proportionally to the storage capacity added, the system is said to be scalable. Scaling vertically (scale up) occurs when resources (hardware and memory) are added to a single server (node). Scaling horizontally (scale out) occurs when more servers (nodes) are added to a system.

- *Elasticity.* **Elasticity** is a cloud's ability to deal with variations in the amount of data (load) being transferred in and out of the storage system without service interrupts. There is a subtle difference between scalability and elasticity when describing a system's behavior. Essentially, *scalability* is a static parameter that indicates how much the system can be expanded, and *elasticity* is a dynamic parameter that refers to the implementation of scalability. For example, a storage system may be scalable from one to 100 servers. If the system is currently operating with 20 servers (nodes) and the data load doubles, its elasticity allows 20 more nodes to be added for a total of 40. Likewise, if the data load decreases by half, the elasticity allows 10 nodes to be removed. A server can be added or removed by powering it up or down in a proper manner without disrupting service to the user. Elasticity results in cost efficiency because only the number of servers required for the data load at any given time are consuming power.

- *Multitenancy.* The **multitenancy** property of a cloud storage system allows multiple users to share the same software applications and hardware and the same data storage mechanism but not to see each other's data.

SECTION 11–10 CHECKUP

1. What is a cloud storage system?
2. What is a server?
3. How does a user connect to a cloud storage system?
4. Name three advantages of a cloud system.

11–11 Troubleshooting

Because memories can contain large numbers of storage cells, testing each cell can be a lengthy and frustrating process. Fortunately, memory testing is usually an automated process performed with a programmable test instrument or with the aid of software for in-system testing. Most microprocessor-based systems provide automatic memory testing as part of their system software.

After completing this section, you should be able to

- ◆ Discuss the checksum method of testing ROMs
- ◆ Discuss the checkerboard pattern method of testing RAMs

ROM Testing

Since ROMs contain known data, they can be checked for the correctness of the stored data by reading each data word from the memory and comparing it with a data word that is known to be correct. One way of doing this is illustrated in Figure 11–72. This process requires a reference ROM that contains the same data as the ROM to be tested. A special test instrument is programmed to read each address in both ROMs simultaneously and to compare the contents. A flowchart in Figure 11–73 illustrates the basic sequence.

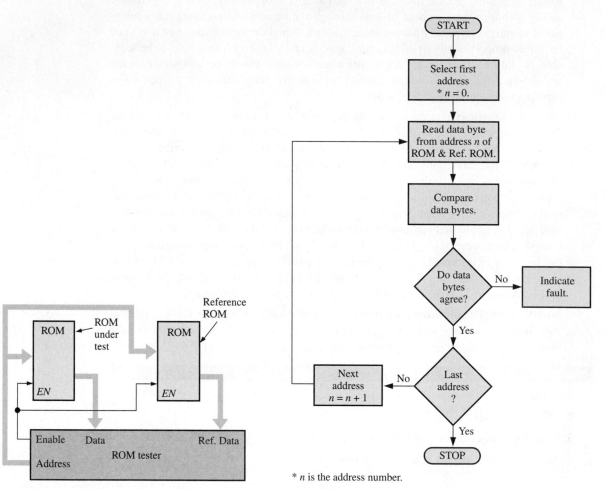

FIGURE 11–72   Block diagram for a complete contents check of a ROM.

FIGURE 11–73   Flowchart for a complete contents check of a ROM.

Checksum Method

Although the previous method checks each ROM address for correct data, it has the disadvantage of requiring a reference ROM for each different ROM to be tested. Also, a failure in the reference ROM can produce a fault indication.

In the checksum method a number, the sum of the contents of all the ROM addresses, is stored in a designated ROM address when the ROM is programmed. To test the ROM, the contents of all the addresses except the checksum are added, and the result is compared with the checksum stored in the ROM. If there is a difference, there is definitely a fault. If the checksums agree, the ROM is most likely good. However, there is a remote possibility that a combination of bad memory cells could cause the checksums to agree.

This process is illustrated in Figure 11–74 with a simple example. The checksum in this case is produced by taking the sum of each column of data bits and discarding the carries. This is actually an XOR sum of each column. The flowchart in Figure 11–75 illustrates the basic checksum test.

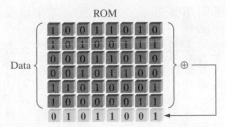

FIGURE 11–74 Simplified illustration of a programmed ROM with the checksum stored at a designated address.

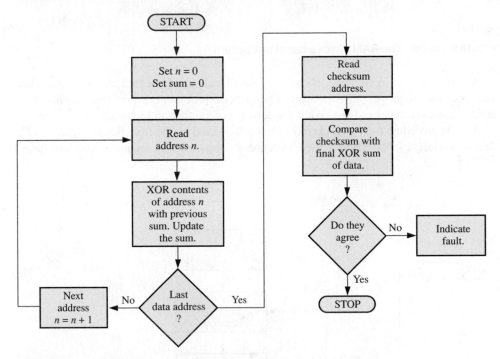

FIGURE 11–75 Flowchart for a basic checksum test.

The checksum test can be implemented with a special test instrument, or it can be incorporated as a test routine in the built-in (system) software or microprocessor-based systems. In that case, the ROM test routine is automatically run on system start-up.

RAM Testing

To test a RAM for its ability to store both 0s and 1s in each cell, first 0s are written into all the cells in each address and then read out and checked. Next, 1s are written into all the cells in each address and then read out and checked. This basic test will detect a cell that is stuck in either a 1 state or a 0 state.

Some memory faults cannot be detected with the all-0s–all-1s test. For example, if two adjacent memory cells are shorted, they will always be in the same state, both 0s or both 1s. Also, the all-0s–all-1s test is ineffective if there are internal noise problems such that the contents of one or more addresses are altered by a change in the contents of another address.

The Checkerboard Pattern Test

One way to more fully test a RAM is by using a checkerboard pattern of 1s and 0s, as illustrated in Figure 11–76. Notice that all adjacent cells have opposite bits. This pattern checks for a short between two adjacent cells; if there is a short, both cells will be in the same state.

After the RAM is checked with the pattern in Figure 11–76(a), the pattern is reversed, as shown in part (b). This reversal checks the ability of all cells to store both 1s and 0s.

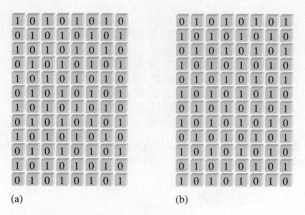

(a)   (b)

FIGURE 11–76   The RAM checkerboard test pattern.

A further test is to alternate the pattern one address at a time and check all the other addresses for the proper pattern. This test will catch a problem in which the contents of an address are dynamically altered when the contents of another address change.

A basic procedure for the checkerboard test is illustrated by the flowchart in Figure 11–77. The procedure can be implemented with the system software in microprocessor-based

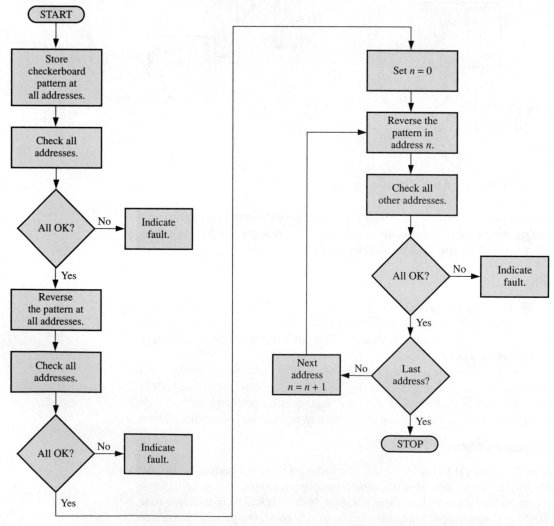

FIGURE 11–77   Flowchart for basic RAM checkerboard test.

systems so that either the tests are automatic when the system is powered up or they can be initiated from the keyboard.

SECTION 11–11 CHECKUP

1. Describe the checksum method of ROM testing.

2. Why can the checksum method not be applied to RAM testing?

3. List the three basic faults that the checkerboard pattern test can detect in a RAM.

SUMMARY

- Types of semiconductor memories:

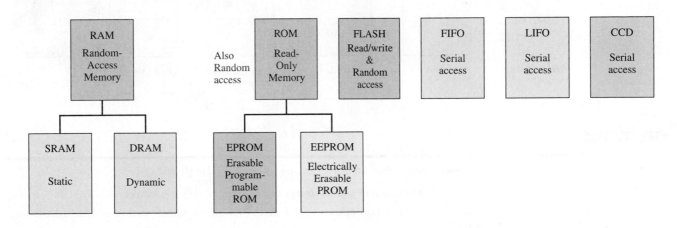

- Types of SRAMs (Static RAMs) and DRAMs (Dynamic RAMs):

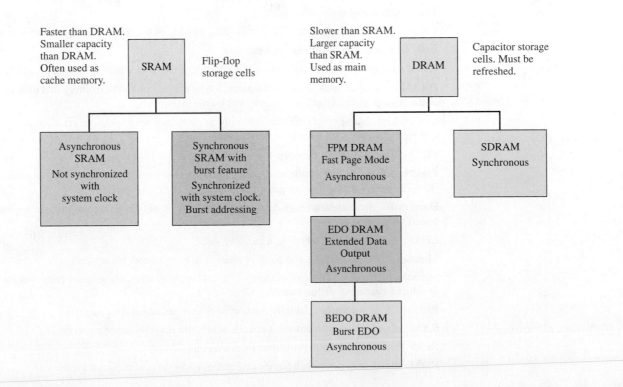

- Types of magnetic storage:

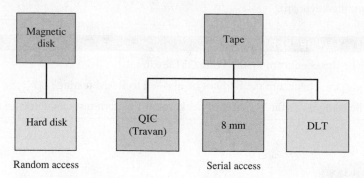

Random access   Serial access

- Types of optical (laser) storage:

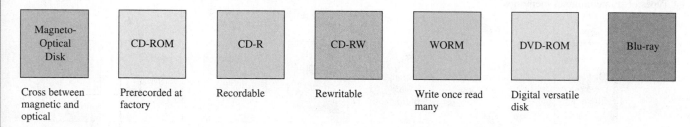

| Cross between magnetic and optical | Prerecorded at factory | Recordable | Rewritable | Write once read many | Digital versatile disk | |

KEY TERMS

Key terms and other bold terms in the chapter are defined in the end-of-book glossary.

Address   The location of a given storage cell or group of cells in a memory.

Blue-ray   A disc storage technology that uses a blue laser to achieve more density and definition than a DVD.

Bus   One or more interconnections that interface one or more devices based on a standardized specification.

Byte   A group of eight bits.

Capacity   The total number of data units (bits, nibbles, bytes, words) that a memory can store.

Cell   A single storage element in a memory.

Cloud storage   A network of servers that is connected to a user device through the Internet.

DRAM   Dynamic random-access memory; a type of semiconductor memory that uses capacitors as the storage elements and is a volatile, read/write memory.

EPROM   Erasable programmable read-only memory; a type of semiconductor memory device that typically uses ultraviolet light to erase data.

FIFO   First in–first out memory.

Flash memory   A nonvolatile read/write random-access semiconductor memory in which data are stored as charge on the floating gate of a certain type of FET.

Hard disk   A magnetic storage device; typically, a stack of two or more rigid disks enclosed in a sealed housing.

LIFO   Last in–first out memory; a memory stack.

Memory   The portion of a computer or other system that stores binary data.

Memory hierarchy   The arrangement of various memory elements within a computer architecture to achieve maximum performance.

PROM   Programmable read-only memory; a type of semiconductor memory.

RAM   Random-access memory; a volatile read/write semiconductor memory.

Read   The process of retrieving data from a memory.

ROM   Read-only memory; a nonvolatile random-access semiconductor memory.

Server Any computerized process that shares a resource with one or more clients. A computer and software with a large memory capacity that responds to requests across a network to provide file storage and access as well as services such as file sharing.

SRAM Static random-access memory; a type of volatile read/write semiconductor memory.

Word A group of bits or bytes that acts as a single entity that can be stored in one memory location; two bytes.

Write The process of storing data in a memory.

TRUE/FALSE QUIZ

Answers are at the end of the chapter.

1. A data byte consists of eight bits.
2. A memory cell can store a byte of data.
3. The write operation stores data in memory.
4. The read operation always erases the data byte.
5. RAM is a *random address memory*.
6. Stored data are lost if power is removed from a static RAM.
7. Cache is a type of memory used for intermediate or temporary storage of data.
8. Dynamic RAMs must be periodically refreshed to retain data.
9. ROM is a *random output memory*.
10. A flash memory uses a flashing beam of light to store data.
11. Registers are at the top of a memory hierarchy.
12. Cloud storage is accessed through the Internet.

SELF-TEST

Answers are at the end of the chapter.

1. The bit capacity of a memory that has 1024 addresses and can store 8 bits at each address is
   (a) 1024   (b) 8192   (c) 8   (d) 4096

2. A 32-bit data word consists of
   (a) 2 bytes   (b) 4 nibbles   (c) 4 bytes   (d) 3 bytes and 1 nibble

3. Data are stored in a random-access memory (RAM) during the
   (a) read operation   (b) enable operation
   (c) write operation   (d) addressing operation

4. Data that are stored at a given address in a random-access memory (RAM) are lost when
   (a) power goes off   (b) the data are read from the address
   (c) new data are written at the address   (d) answers (a) and (c)

5. A ROM is a
   (a) nonvolatile memory   (b) volatile memory
   (c) read/write memory   (d) byte-organized memory

6. A memory with 256 addresses has
   (a) 256 address lines   (b) 6 address lines
   (c) 1 address line   (d) 8 address lines

7. A byte-organized memory has
   (a) 1 data output line   (b) 4 data output lines
   (c) 8 data output lines   (d) 16 data output lines

8. The storage cell in a SRAM is
   (a) a flip-flop   (b) a capacitor   (c) a fuse   (d) a magnetic domain

9. A DRAM must be
   (a) replaced periodically   (b) refreshed periodically
   (c) always enabled   (d) programmed before each use

10. A flash memory is
   (a) volatile   (b) a read-only memory
   (c) a read/write memory   (d) nonvolatile
   (e) answers (a) and (c)   (f) answers (c) and (d)

11. SRAM, DRAM, flash, and EEPROM are all
   (a) magneto-optical storage devices (b) semiconductor storage devices
   (c) magnetic storage devices   (d) optical storage devices

12. Optical storage devices employ
   (a) ultraviolet light   (b) electromagnetic fields
   (c) optical couplers   (d) lasers

13. Memory latency is
   (a) average down time   (b) time to reference a block of data
   (c) processor access time   (d) the hit rate

14. A facility that houses a cloud storage system is called a
   (a) server   (b) data center
   (c) computer center   (d) cloud house

PROBLEMS

Answers to odd-numbered problems are at the end of the book.

Section 11–1 Semiconductor Memory Basics

1. Identify the ROM and the RAM in Figure 11–78.

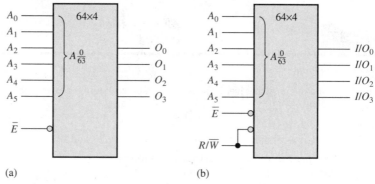

(a)   (b)

FIGURE 11–78

2. Explain why RAMs and ROMs are both random-access memories.

3. Explain the purposes of the address bus and the data bus.

4. What memory address (0 through 256) is represented by each of the following hexadecimal numbers:

   (a) $0A_{16}$ (b) $3F_{16}$ (c) CD_{16}

Section 11–2 The Random-Access Memory (RAM)

5. A static memory array with four rows similar to the one in Figure 11–10 is initially storing all 0s. What is its content after the following conditions? Assume a 1 selects a row.

$$\text{Row } 0 = 1, \text{Data in (Bit 0)} = 1$$

$$\text{Row } 1 = 0, \text{Data in (Bit 1)} = 1$$

$$\text{Row } 2 = 1, \text{Data in (Bit 2)} = 1$$

$$\text{Row } 3 = 0, \text{Data in (Bit 3)} = 0$$

6. Draw a basic logic diagram for a 512×8-bit static RAM, showing all the inputs and outputs.

7. Assuming that a 64k $\times$ 8 SRAM has a structure similar to that of the SRAM in Figure 11–12. determine the number of rows and 8-bit columns in its memory cell array.

8. Redraw the block diagram in Figure 11–12 for a 64k $\times$ 8 memory.

9. Explain the difference between a SRAM and a DRAM.

10. What is the capacity of a DRAM that has twelve address lines?

Section 11–3 The Read-Only Memory (ROM)

11. For the ROM array in Figure 11–79, determine the outputs for all possible input combinations, and summarize them in tabular form (Blue cell is a 1, gray cell is a 0).

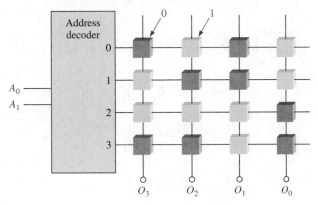

FIGURE 11–79

12. Determine the truth table for the ROM in Figure 11–80.

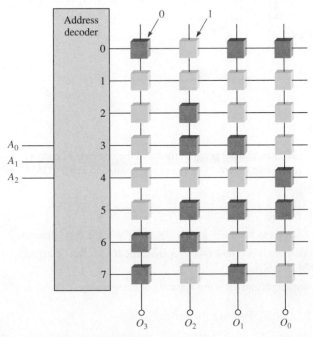

FIGURE 11–80

13. Using a procedure similar to that in Example 11–1, design a ROM for conversion of single-digit BCD to excess-3 code.

14. What is the total *bit* capacity of a ROM that has 14 address lines and 8 data outputs?

Section 11–4 Programmable ROMs

15. Assuming that the PROM matrix in Figure 11–81 is programmed by blowing a fuse link to create a 0, indicate the links to be blown to program an X^3 look-up table, where X is a number from 0 through 7.

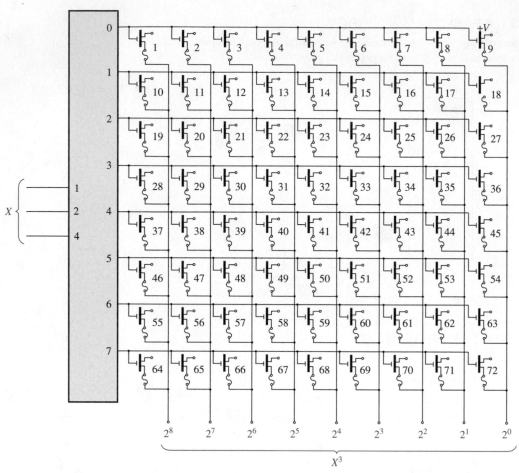

FIGURE 11–81

16. Determine the addresses that are programmed and the contents of each address after the programming sequence in Figure 11–82 has been applied to an EPROM like the one shown in Figure 11–31.

Section 11–6 Memory Expansion

17. Use 16k $\times$ 4 DRAMs to build a 64k $\times$ 8 DRAM. Show the logic diagram.

18. Using a block diagram, show how 64k $\times$ 1 dynamic RAMs can be expanded to build a 256k $\times$ 4 RAM.

19. What is the word length and the word capacity of the memory of Problem 17? Problem 18?

Section 11–7 Special Types of Memories

20. Complete the timing diagram in Figure 11–83 by showing the output waveforms that are initially all LOW for a FIFO serial memory like that shown in Figure 11–50.

21. Consider a 4096 $\times$ 8 RAM in which the last 64 addresses are used as a LIFO stack. If the first address in the RAM is 000_{16}, designate the 64 addresses used for the stack.

22. In the memory of Problem 21, sixteen bytes are pushed into the stack. At what address is the first byte in located? At what address is the last byte in located?

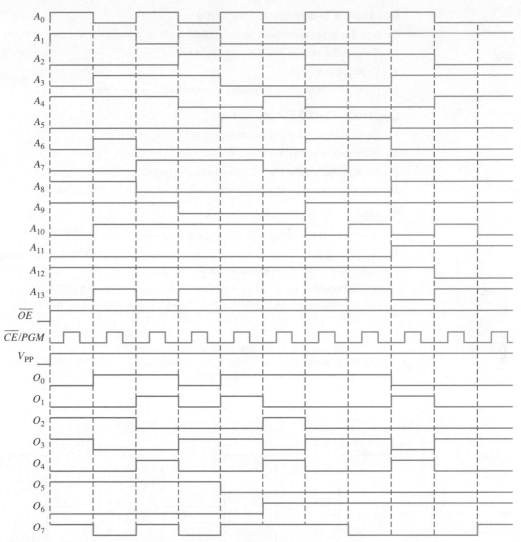

FIGURE 11–82

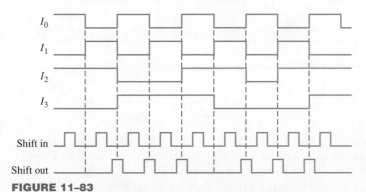

FIGURE 11–83

Section 11–8 Magnetic and Optical Storage

23. Describe the general format of a hard disk.

24. Explain seek time and latency period in a hard disk drive.

25. Why does magnetic tape require a much longer access time than does a disk?

26. Explain the differences in a magneto-optical disk, a CD-ROM, and a WORM.

27. What is the main difference between a Blu-ray Disc and a DVD?

28. Name two advantages of Blu-ray compared to a DVD.

Section 11–9 Memory Hierarchy

29. List five types of memory in order of their distance in time from the processor.

30. Discuss the hit rate in a memory system.

31. Describe miss rate.

32. If the hit rate in a certain memory system is 0.7, what is the miss rate?

Section 11–10 Cloud Storage

33. Draw a diagram of a cloud storage system with six servers.

34. What does a server in a cloud storage system provide?

35. What is the architecture of a cloud storage system?

36. List five properties of a cloud storage system and briefly discuss each.

Section 11–11 Troubleshooting

37. Determine if the contents of the ROM in Figure 11–84 are correct.

38. A 128 × 8 ROM is implemented as shown in Figure 11–85. The decoder decodes the two most significant address bits to enable the ROMs one at a time, depending on the address selected.

   (a) Express the lowest address and the highest address of each ROM as hexadecimal numbers.
   (b) Assume that a single checksum is used for the entire memory and it is stored at the highest address. Develop a flowchart for testing the complete memory system.
   (c) Assume that each ROM has a checksum stored at its highest address. Modify the flowchart developed in part (b) to accommodate this change.
   (d) What is the disadvantage of using a single checksum for the entire memory rather than a checksum for each individual ROM?

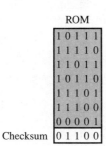

ROM

1	0	1	1	1
1	1	1	1	0
1	1	0	1	1
1	0	1	1	0
1	1	1	0	1
1	1	1	0	0
0	0	0	0	1

Checksum 0 1 1 0 0

FIGURE 11–84

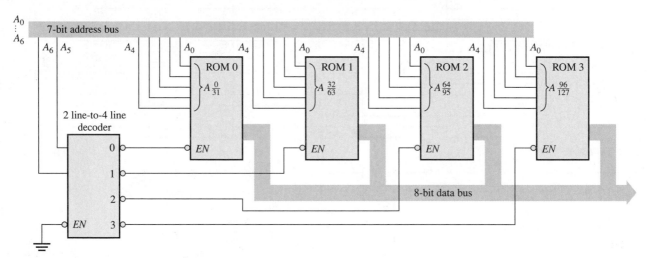

FIGURE 11–85

39. Suppose that a checksum test is run on the memory in Figure 11–85 and each individual ROM has a checksum at its highest address. What IC or ICs will you replace for each of the following error messages that appear on the system's video monitor?

   (a) ADDRESSES 40–5F FAULTY   (b) ADDRESSES 20–3F FAULTY
   (c) ADDRESSES 00–7F FAULTY

ANSWERS

SECTION CHECKUPS

Section 11–1 Semiconductor Memory Basics

1. Bit is the smallest unit of data.

2. 256 bytes is 2048 bits.

3. A write operation stores data in memory.

4. A read operation takes a copy of data from memory.

5. A unit of data is located by its address.

6. A RAM is volatile and has read/write capability. A ROM is nonvolatile and has only read capability.

Section 11–2 The Random-Access Memory (RAM)

1. Asynchronous and synchronous with burst feature

2. A small fast memory between the CPU and main memory

3. SRAMs have latch storage cells that can retain data indefinitely while power is applied. DRAMs have capacitive storage cells that must be periodically refreshed.

4. The refresh operation prevents data from being lost because of capacitive discharge. A stored bit is restored periodically by recharging the capacitor to its nominal level.

5. FPM, EDO, BEDO, Synchronous

Section 11–3 The Read-Only Memory (ROM)

1. 512×8 equals 4096 bits.

2. Mask ROM, PROM, EPROM, UV EPROM, EEPROM

3. Eight bits of address are required for 256 byte locations ($2^8 = 256$).

Section 11–4 Programmable ROMs

1. PROMs are field-programmable; ROMs are not.

2. Presence or absence of stored charge

3. Read is the normal mode of operation for a PROM.

Section 11–5 The Flash Memory

1. Flash, ROM, EPROM, and EEPROM are nonvolatile.

2. Flash is nonvolatile; SRAM and DRAM are volatile.

3. Programming, read, erase

Section 11–6 Memory Expansion

1. Eight RAMs

2. Eight RAMs

3. DIMM: Dual in-line memory module

Section 11–7 Special Types of Memories

1. In a FIFO memory the *first* bit (or word) *in* is the *first* bit (or word) *out*.

2. In a LIFO memory the *last* bit (or word) *in* is the *first* bit (or word) *out*. A stack is a LIFO.

3. The PUSH operation or instruction adds data to the memory stack.

4. The POP operation or instruction removes data from the memory stack.

5. CCD is a charge-coupled device.

Section 11–8 Magnetic and Optical Storage

1. Magnetic storage: hard disk, tape, and magneto-optical disk

2. A magnetic disk is organized in tracks and sectors.

3. A magneto-optical disk uses a laser beam and an electromagnet.

4. Optical storage: CD-ROM, CD-R, CD-RW, DVD-ROM, WORM, Blu-ray Disc (BD)

Section 11–9 Memory Hierarchy

1. The purpose of memory hierarchy is to obtain the fastest access time at the lowest cost.

2. Access time is the time it takes a processor to retrieve (read) or write a block of data stored in the memory.

3. Generally, the higher the capacity the lower the cost per bit.

4. Yes

5. A hit is when the processor finds the requested data at the first place it looks. A miss is when the processor fails to find the requested data and has to go to another level of memory to find it.

6. The hit rate

Section 11–10 Cloud Storage

1. A cloud storage system is a remote network of servers connected to a user device through the Internet.

2. A server is any computerized process that shares a resource with one or more clients. Practically, a storage server is a computer and software with a large memory capacity that responds to requests across a network to provide file storage and access as well as services such as file sharing.

3. A user connects via Internet access.

4. Data storage and retrieval from any physical location with Internet access, any computer can be used and a local physical backup storage device is not necessary, and other users can be permitted to access your data.

Section 11–11 Troubleshooting

1. The contents of the ROM are added and compared with a prestored checksum.

2. Checksum cannot be used because the contents of a RAM are not fixed.

3. (1) a short between adjacent cells; (2) an inability of some cells to store both 1s and 0s; (3) dynamic altering of the contents of one address when the contents of another address change.

RELATED PROBLEMS FOR EXAMPLES

11–1 $G_3G_2G_1G_0 = 1110$

11–2 Connect eight 64k $\times$ 1 ROMs in parallel to form a 64k $\times$ 8 ROM.

11–3 Sixteen 64k $\times$ 1 ROMs

11–4 See Figure 11–86.

11–5 ROM 1: 0 to 524,287; ROM 2: 524,288 to 1,048,575

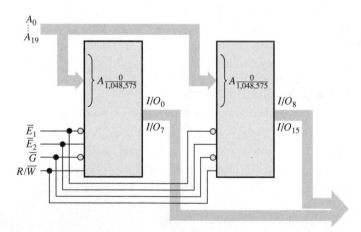

FIGURE 11–86

TRUE/FALSE QUIZ

1. T	**2.** F	**3.** T	**4.** F	**5.** F	**6.** T
7. T	**8.** T	**9.** F	**10.** F	**11.** T	**12.** T

SELF-TEST

1. (b)	**2.** (c)	**3.** (c)	**4.** (d)	**5.** (a)	**6.** (d)	**7.** (c)
8. (a)	**9.** (b)	**10.** (f)	**11.** (b)	**12.** (d)	**13.** (c)	**14.** (b)

CHAPTER 12

Signal Conversion and Processing

CHAPTER OUTLINE

CHAPTER OBJECTIVES

- Explain how analog signals are converted to digital form
- Discuss the purpose of filtering
- Describe the sampling process
- State the purpose of analog-to-digital conversion
- Explain how several types of ADCs operate
- State the purpose of digital-to-analog conversion
- Explain how DACs operate
- List the essential elements in a digital signal processing system
- Explain the basic concepts of a digital signal processor (DSP)
- Describe the basic architecture of a DSP
- Name some of the functions that a DSP performs

KEY TERMS

Key terms are in order of appearance in the chapter.

- Sampling
- Nyquist frequency
- Aliasing
- Analog-to-digital converter (ADC)
- Quantization
- Digital-to-analog converter (DAC)
- Digital signal processor (DSP)
- DSP core
- MIPS
- MFLOPS
- MMACS
- Pipeline
- Fetch
- Decode
- Execute

VISIT THE WEBSITE

Study aids for this chapter are available at
http://www.pearsonhighered.com/careersresources/

INTRODUCTION

This chapter provides an introduction to interfacing digital and analog systems using methods of analog-to-digital and digital-to-analog conversions.

Digital signal processing is a technology that is widely used in many applications, such as automotive, consumer, graphics/imaging, industrial, instrumentation, medical, military, telecommunications, and voice/speech applications. Digital signal processing incorporates mathematics, software programming, and processing hardware to manipulate analog signals.

12–1   Analog-to-Digital Conversion

In order to process signals using digital techniques, the incoming analog signal must be converted into digital form.

After completing this section, you should be able to

- ◆ Explain the basic process of converting an analog signal to digital
- ◆ Describe the purpose of the sample-and-hold function
- ◆ Define the Nyquist frequency
- ◆ Define the reason for *aliasing* and discuss how it is eliminated
- ◆ Describe the purpose of an ADC

Sampling and Filtering

An anti-aliasing filter and a sample-and-hold circuit are two functions typically found in a digital signal processing system. The sample-and-hold function does two operations, the first of which is sampling. **Sampling** is the process of taking a sufficient number of discrete values at points on a waveform that will define the shape of the waveform. The more samples you take, the more accurately you can define a waveform. Sampling converts an analog signal into a series of impulses, each representing the amplitude of the signal at a given instant in time. Figure 12–1 illustrates the process of sampling.

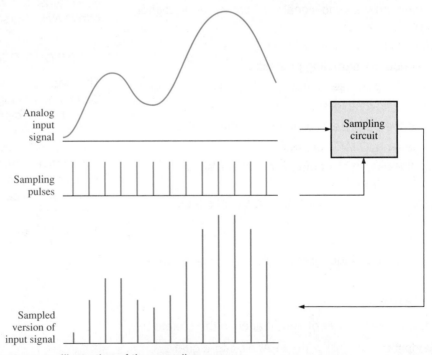

FIGURE 12–1   Illustration of the sampling process.

When an analog signal is to be sampled, there are certain criteria that must be met in order to accurately represent the original signal. All analog signals (except a pure sine wave) contain a spectrum of component frequencies. For a pure sine wave, these frequencies appear in multiples called *harmonics*. The harmonics of an analog signal are sine waves of different frequencies and amplitudes. When the harmonics of a given periodic waveform are added, the result is the original signal. Before a signal can be sampled, it must be passed through a low-pass filter (anti-aliasing filter) to eliminate harmonic frequencies above a certain value as determined by the Nyquist frequency.

The Sampling Theorem

Notice in Figure 12–1 that there are two input waveforms. One is the analog signal and the other is the sampling pulse waveform. The sampling theorem states that, in order to represent an analog signal, the sampling frequency, f_{sample}, must be at least twice the highest frequency component $f_{a(max)}$ of the analog signal. Another way to say this is that the highest analog frequency can be no greater than one-half the sampling frequency. The frequency $f_{a(max)}$ is known as the **Nyquist frequency** and is expressed in Equation 12–1. In practice, the sampling frequency should be more than twice the highest analog frequency.

$$f_{sample} > 2f_{a(max)} \hspace{3cm} \text{Equation 12–1}$$

To intuitively understand the sampling theorem, a simple "bouncing-ball" analogy may be helpful. Although it is not a perfect representation of the sampling of electrical signals, it does serve to illustrate the basic idea. If a ball is photographed (sampled) at one instant during a single bounce, as illustrated in Figure 12–2(a), you cannot tell anything about the path of the ball except that it is off the floor. You can't tell whether it is going up or down or the distance of its bounce. If you take photos at two equally-spaced instants during one bounce, as shown in part (b), you can obtain only a minimum amount of information about its movement and nothing about the distance of the bounce. In this particular case, you know only that the ball has been in the air at the times the two photos were taken and that the maximum height of the bounce is at least equal to the height shown in each photo. If you take four photos, as shown in part (c), then the path that the ball follows during a bounce begins to emerge. The more photos (samples) that you take, the more accurately you can determine the path of the ball as it bounces.

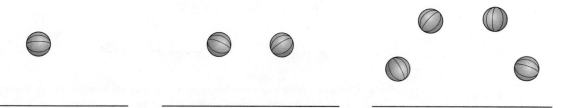

(a) One sample of a ball during a single bounce

(b) Two samples of a ball during a single bounce. This is the absolute minimum required to tell anything about its movement, but generally insufficient to describe its path.

(c) Four samples of a ball during a single bounce form a rough picture of the path of the ball.

FIGURE 12–2 Bouncing ball analogy of sampling theory.

The Need for Filtering

Low-pass filtering is necessary to remove all frequency components (harmonics) of the analog signal that exceed the Nyquist frequency. If there are any frequency components in the analog signal that exceed the Nyquist frequency, an unwanted condition known as **aliasing** will occur. An alias is a signal produced when the sampling frequency is not at least twice the signal frequency. An alias signal has a frequency that is less than the highest frequency in the analog signal being sampled and therefore falls within the spectrum or frequency band of the input analog signal causing distortion. Such a signal is actually "posing" as part of the analog signal when it really isn't, thus the term *alias*.

Another way to view aliasing is by considering that the sampling pulses produce a spectrum of harmonic frequencies above and below the sample frequency, as shown in Figure 12–3. If the analog signal contains frequencies above the Nyquist frequency, these frequencies overlap into the spectrum of the sample waveform as shown and interference occurs. The lower frequency components of the sampling waveform become mixed in with the frequency spectra of the analog waveform, resulting in an aliasing error.

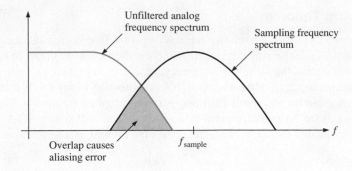

FIGURE 12–3   A basic illustration of the condition $f_{sample} < 2f_{a(max)}$.

A low-pass anti-aliasing filter must be used to limit the frequency spectrum of the analog signal for a given sample frequency. To avoid an aliasing error, the filter must at least eliminate all analog frequencies above the minimum frequency in the sampling spectrum, as illustrated in Figure 12–4. Aliasing can also be avoided by sufficiently increasing the sampling frequency. However, the maximum sampling frequency is usually limited by the performance of the analog-to-digital converter (ADC) that follows it.

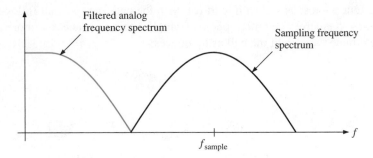

FIGURE 12–4   After low-pass filtering, the frequency spectra of the analog and the sampling signals do not overlap, thus eliminating aliasing error.

An Application

An example of the application of sampling is in digital audio equipment. The sampling rates used are 32 kHz, 44.1 kHz, or 48 kHz (the number of samples per second). The 48 kHz rate is the most common, but the 44.1 kHz rate is used for audio CDs and prerecorded tapes. According to the Nyquist rate, the sampling frequency must be at least twice the audio signal. Therefore, the CD sampling rate of 44.1 kHz captures frequencies up to about 22 kHz, which exceeds the 20 kHz specification that is common for most audio equipment.

Many applications do not require a wide frequency range to obtain reproduced sound that is acceptable. For example, human speech contains some frequencies near 10 kHz and, therefore, requires a sampling rate of at least 20 kHz. However, if only frequencies up to 4 kHz (ideally requiring an 8 kHz minimum sampling rate) are reproduced, voice is very understandable. On the other hand, if a sound signal is not sampled at a high enough rate, the effect of aliasing will become noticeable with background noise and distortion.

Holding the Sampled Value

The holding operation is the second part of the sample-and-hold function. After filtering and sampling, the sampled level must be held constant until the next sample occurs. This is necessary for the ADC to have time to process the sampled value. This sample-and-hold operation results in a "stairstep" waveform that approximates the analog input waveform, as shown in Figure 12–5.

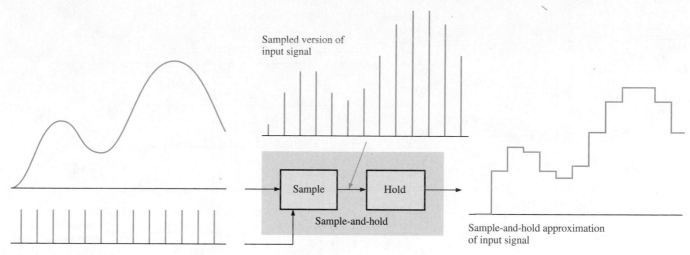

FIGURE 12-5 Illustration of a sample-and-hold operation.

Analog-to-Digital Conversion

Analog-to-digital conversion is the process of converting the output of the sample-and-hold circuit to a series of binary codes that represent the amplitude of the analog input at each of the sample times. The sample-and-hold process keeps the amplitude of the analog input signal constant between sample pulses; therefore, the analog-to-digital conversion can be done using a constant value rather than having the analog signal change during a conversion interval, which is the time between sample pulses. Figure 12–6 illustrates the basic function of an **analog-to-digital converter (ADC)**, which is a circuit that performs analog-to-digital conversion. The sample intervals are indicated by dashed lines.

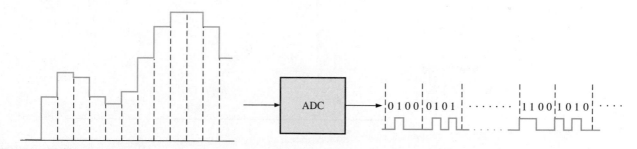

FIGURE 12-6 Basic function of an analog-to-digital converter (ADC) (The binary codes and number of bits are arbitrarily chosen for illustration only). The ADC output waveform that represents the binary codes is also shown.

Quantization

The process of converting an analog value to a code is called **quantization**. During the quantization process, the ADC converts each sampled value of the analog signal to a binary code. The more bits that are used to represent a sampled value, the more accurate is the representation.

To illustrate, let's quantize a reproduction of the analog waveform into four levels (0–3). Two bits are required for four levels. As shown in Figure 12–7, each quantization level is represented by a 2-bit code on the vertical axis, and each sample interval is numbered along the horizontal axis. The sampled data is held for the entire sample period. This data is quantized to the next lower level, as shown in Table 12–1 (for example, compare samples 3 and 4, which are assigned different levels).

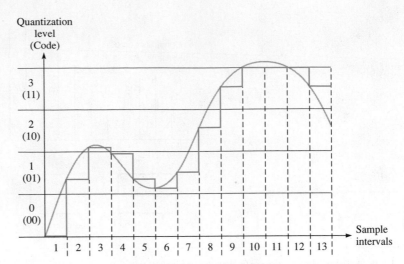

FIGURE 12–7   Sample-and-hold output waveform with four quantization levels. The original analog waveform is shown in light gray for reference.

TABLE 12–1

Two-bit quantization for the waveform in Figure 12–7.

Sample Interval	Quantization Level	Code
1	0	00
2	1	01
3	2	10
4	1	01
5	1	01
6	1	01
7	1	01
8	2	10
9	3	11
10	3	11
11	3	11
12	3	11
13	3	11

If the resulting 2-bit digital codes are used to reconstruct the original waveform, you would get the waveform shown in Figure 12–8. This operation is done by **digital-to-analog converters (DACs)**, which are circuits that perform digital-to-analog conversions. As you can see, quite a bit of accuracy is lost using only two bits to represent the sampled values.

Now, let's see how more bits will improve the accuracy. Figure 12–9 shows the same waveform with sixteen quantization levels (4 bits). The 4-bit quantization process is summarized in Table 12–2.

If the resulting 4-bit digital codes are used to reconstruct the original waveform, you would get the waveform shown in Figure 12–10. As you can see, the result is much more like the original waveform than for the case of four quantization levels in Figure 12–8. This shows that greater accuracy is achieved with more quantization bits. Typical integrated circuit ADCs use from 12 to 24 bits, and the sample-and-hold function is sometimes contained on the ADC chip. Several types of ADCs are introduced in the next section.

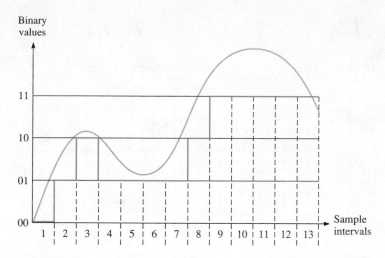

FIGURE 12–8 The reconstructed waveform in Figure 12–7 using four quantization levels (2 bits). The original analog waveform is shown in light gray for reference.

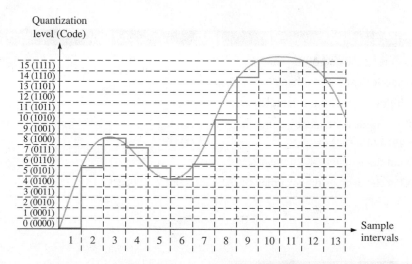

FIGURE 12–9 Sample-and-hold output waveform with sixteen quantization levels. The original analog waveform is shown in light gray for reference.

TABLE 12–2

Four-bit quantization for the waveform in Figure 12–9.

Sample Interval	Quantization Level	Code
1	0	0000
2	5	0101
3	8	1000
4	7	0111
5	5	0101
6	4	0100
7	6	0110
8	10	1010
9	14	1110
10	15	1111
11	15	1111
12	15	1111
13	14	1110

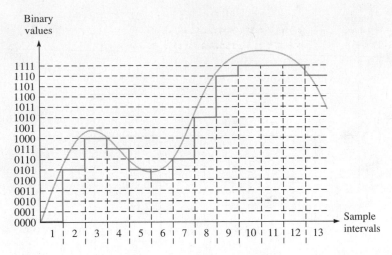

FIGURE 12–10   The reconstructed waveform in Figure 12–9 using sixteen quantization levels (4 bits). The original analog waveform is shown in light gray for reference.

SECTION 12–1 CHECKUP

Answers are at the end of the chapter.

1. What does sampling mean?
2. Why must you hold a sampled value?
3. If the highest frequency component in an analog signal is 20 kHz, what is the minimum sample frequency?
4. What does quantization mean?
5. What determines the accuracy of the quantization process?

12–2   Methods of Analog-to-Digital Conversion

As you have seen, analog-to-digital conversion is the process by which an analog quantity is converted to digital form. It is necessary when measured quantities must be in digital form for processing or for display or storage. Some common types of analog-to-digital converters (ADCs) are now examined. Two important ADC parameters are *resolution,* which is the number of bits, and *throughput,* which is the sampling rate an ADC can handle in units of samples per second (sps).

After completing this section, you should be able to

♦ Explain what an operational amplifier is

♦ Show how the op-amp can be used as an inverting amplifier or a comparator

♦ Explain how a flash ADC works

♦ Discuss dual-slope ADCs

♦ Describe the operation of a successive-approximation ADC

♦ Describe a delta-sigma ADC

♦ Discuss testing ADCs for a missing code, incorrect code and offset

A Quick Look at an Operational Amplifier

Before getting into analog-to-digital converters (ADCs), let's look briefly at an element that is common to most types of ADCs and digital-to-analog converters (DACs). This element is the operational amplifier, or op-amp for short. This is an abbreviated coverage of the op-amp.

An **op-amp** is a linear amplifier that has two inputs (inverting and noninverting) and one output. It has a very high voltage gain and a very high input impedance, as well as a very low output impedance. The op-amp symbol is shown in Figure 12–11(a). When used as an inverting amplifier, the op-amp is configured as shown in part (b). The feedback resistor, R_f, and the input resistor, R_i, control the voltage gain according to the formula in Equation 12–2, where V_{out}/V_{in} is the closed-loop voltage gain (closed loop refers to the feedback from output to input provided by R_f). The negative sign indicates inversion.

$$\frac{V_{out}}{V_{in}} = -\frac{R_f}{R_i}$$
Equation 12–2

In the inverting amplifier configuration, the inverting input of the op-amp is approximately at ground potential (0 V) because feedback and the extremely high open-loop gain make the differential voltage between the two inputs extremely small. Since the noninverting input is grounded, the inverting input is at approximately 0 V, which is called *virtual ground*.

When the op-amp is used as a comparator, as shown in Figure 12–11(c), two voltages are applied to the inputs. When these input voltages differ by a very small amount, the op-amp is driven into one of its two saturated output states, either HIGH or LOW, depending on which input voltage is greater.

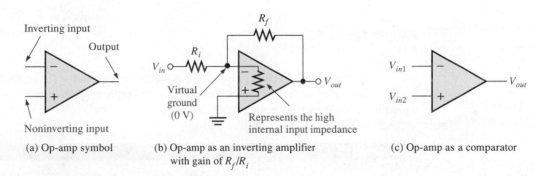

FIGURE 12–11   The operational amplifier (op-amp).

Flash (Simultaneous) Analog-to-Digital Converter

The flash method utilizes special high-speed comparators that compare reference voltages with the analog input voltage. When the input voltage exceeds the reference voltage for a given comparator, a HIGH is generated. Figure 12–12 shows a 3-bit converter that uses seven comparator circuits; a comparator is not needed for the all-0s condition. A 4-bit converter of this type requires fifteen comparators. In general, $2^n - 1$ comparators are required for conversion to an n-bit binary code. The number of bits used in an ADC is its **resolution**. The large number of comparators necessary for a reasonable-sized binary number is one of the disadvantages of the **flash ADC**. Its chief advantage is that it provides a fast conversion time because of a high *throughput*, measured in samples per second (sps).

The reference voltage for each comparator is set by the resistive voltage-divider circuit. The output of each comparator is connected to an input of the priority encoder. The encoder is enabled by a pulse on the *EN* input, and a 3-bit code representing the value of the input appears on the encoder's outputs. The binary code is determined by the highest-order input having a HIGH level.

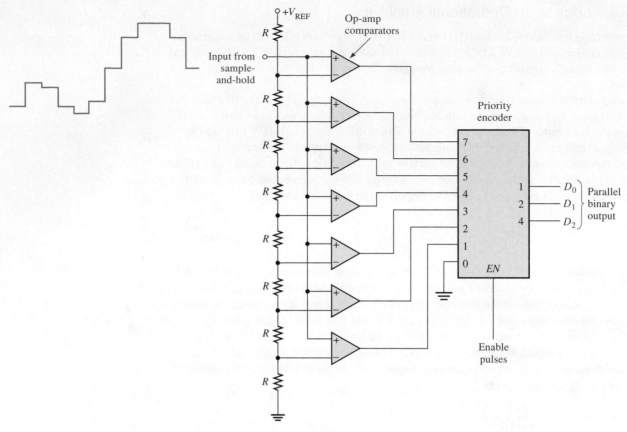

FIGURE 12–12   A 3-bit flash ADC.

The frequency of the enable pulses and the number of bits in the binary code determine the accuracy with which the sequence of binary codes represents the input of the ADC. The signal is sampled each time the enable pulse is active.

EXAMPLE 12–1

Determine the binary code output of the 3-bit flash ADC in Figure 12–12 for the input signal in Figure 12–13 and the encoder enable pulses shown. For this example, $V_{REF} = +8$ V.

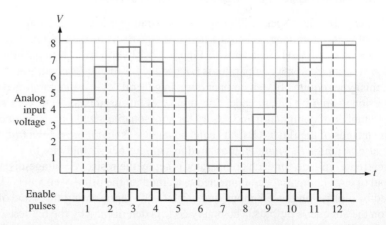

FIGURE 12–13   Sampling of values on a waveform for conversion to binary code.

Solution

The resulting digital output sequence is listed as follows and shown in the waveform diagram of Figure 12–14 in relation to the enable pulses:

100, 110, 111, 110, 100, 010, 000, 001, 011, 101, 110, 111

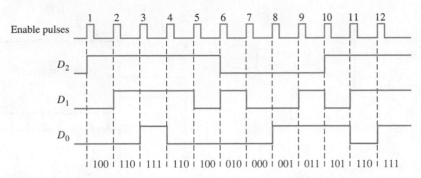

FIGURE 12–14 Resulting digital outputs for sample-and-hold values. Output D_0 is the LSB of the 3-bit binary code.

Related Problem*

If the enable pulse frequency in Figure 12–13 were halved, determine the binary numbers represented by the resulting digital output sequence for 6 pulses. Is any information lost?

*Answers are at the end of the chapter.

Dual-Slope Analog-to-Digital Converter

A dual-slope ADC is common in digital voltmeters and other types of measurement instruments. A ramp generator (integrator) is used to produce the dual-slope characteristic. A block diagram of a dual-slope ADC is shown in Figure 12–15.

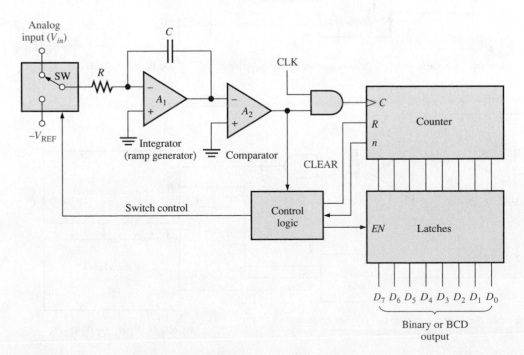

FIGURE 12–15 Basic dual-slope ADC.

Figure 12–16 illustrates dual-slope conversion. Start by assuming that the counter is reset and the output of the integrator is zero. Now assume that a positive input voltage is applied to the input through the switch (SW) as selected by the control logic. Since the

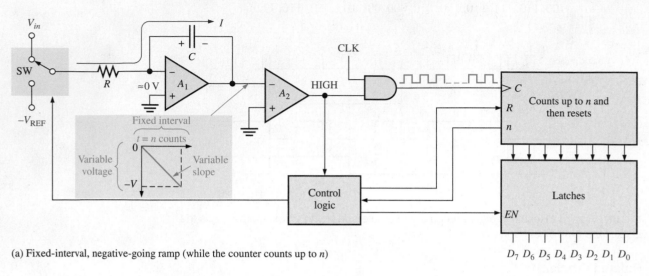

(a) Fixed-interval, negative-going ramp (while the counter counts up to *n*)

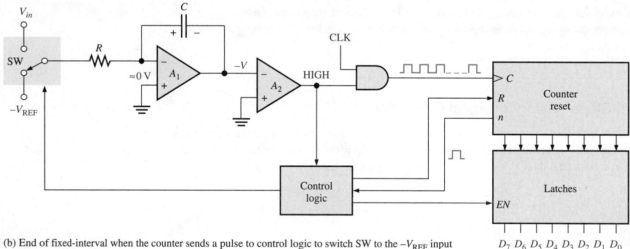

(b) End of fixed-interval when the counter sends a pulse to control logic to switch SW to the $-V_{REF}$ input

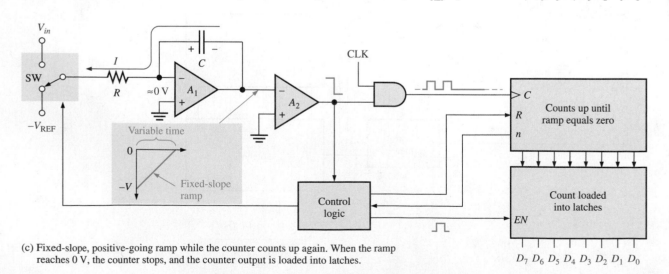

(c) Fixed-slope, positive-going ramp while the counter counts up again. When the ramp reaches 0 V, the counter stops, and the counter output is loaded into latches.

FIGURE 12–16 Illustration of dual-slope conversion.

inverting input of A_1 is at virtual ground, and assuming that V_{in} is constant for a period of time, there will be constant current through the input resistor R and therefore through the capacitor C. Capacitor C will charge linearly because the current is constant, and as a result, there will be a negative-going linear voltage ramp on the output of A_1, as illustrated in Figure 12–16(a).

When the counter reaches a specified count (n), it will be reset (R), and the control logic will switch the negative reference voltage $(-V_{REF})$ to the input of A_1, as shown in Figure 12–16(b). At this point the capacitor is charged to a negative voltage $(-V)$ proportional to the input analog voltage.

Now the capacitor discharges linearly because of the constant current from the $-V_{REF}$, as shown in Figure 12–16(c). This linear discharge produces a positive-going ramp on the A_1 output, starting at $-V$ and having a constant slope that is independent of the charge voltage. As the capacitor discharges, the counter advances from its RESET state. The time it takes the capacitor to discharge to zero depends on the initial voltage $-V$ (proportional to V_{in}) because the discharge rate (slope) is constant. When the integrator (A_1) output voltage reaches zero, the comparator (A_2) switches to the LOW state and disables the clock to the counter. The binary count is latched, thus completing one conversion cycle. The binary count is proportional to V_{in} because the time it takes the capacitor to discharge depends only on $-V$, and the counter records this interval of time.

Successive-Approximation Analog-to-Digital Converter

One of the most widely used methods of analog-to-digital conversion is successive-approximation. It has a much faster conversion time than the dual-slope conversion, but it is slower than the flash method. It also has a fixed conversion time that is the same for any value of the analog input.

Figure 12–17 shows a basic block diagram of a 4-bit successive approximation ADC. It consists of a DAC (DACs are covered in Section 12–3), a successive-approximation register (SAR), and a comparator. The basic operation is as follows: The input bits of the DAC are enabled (made equal to a 1) one at a time, starting with the most significant bit (MSB). As each bit is enabled, the comparator produces an output that indicates whether the input signal voltage is greater or less than the output of the DAC. If the DAC output is greater than the input signal, the comparator's output is LOW, causing the bit in the register to reset. If the output is less than the input signal, the 1 bit is retained in the register. The system does this with the MSB first, then the next most

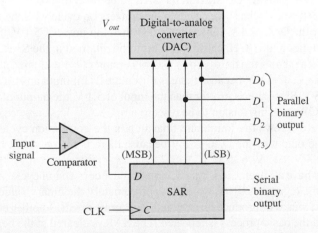

FIGURE 12–17 Successive-approximation ADC.

significant bit, then the next, and so on. After all the bits of the DAC have been tried, the conversion cycle is complete.

In order to better understand the operation of the successive-approximation ADC, let's take a specific example of a 4-bit conversion. Figure 12–18 illustrates the step-by-step conversion of a constant input voltage (5.1 V in this case). Let's assume that the DAC has the following output characteristics: $V_{out} = 8$ V for the 2^3 bit (MSB), $V_{out} = 4$ V for the 2^2 bit, $V_{out} = 2$ V for the 2^1 bit, and $V_{out} = 1$ V for the 2^0 bit (LSB).

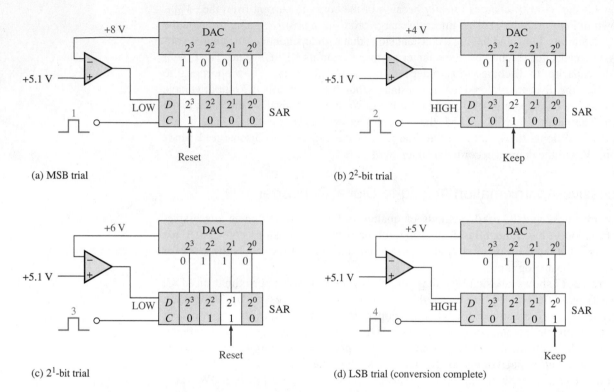

FIGURE 12–18   Illustration of the successive-approximation conversion process.

Figure 12–18(a) shows the first step in the conversion cycle with the MSB = 1. The output of the DAC is 8 V. Since this is greater than the input of 5.1 V, the output of the comparator is LOW, causing the MSB in the SAR to be reset to a 0.

Figure 12–18(b) shows the second step in the conversion cycle with the 2^2 bit equal to a 1. The output of the DAC is 4 V. Since this is less than the input of 5.1 V, the output of the comparator switches to a HIGH, causing this bit to be retained in the SAR.

Figure 12–18(c) shows the third step in the conversion cycle with the 2^1 bit equal to a 1. The output of the DAC is 6 V because there is a 1 on the 2^2 bit input and on the 2^1 bit input; 4 V + 2 V = 6 V. Since this is greater than the input of 5.1 V, the output of the comparator switches to a LOW, causing this bit to be reset to a 0.

Figure 12–18(d) shows the fourth and final step in the conversion cycle with the 2^0 bit equal to a 1. The output of the DAC is 5 V because there is a 1 on the 2^2 bit input and on the 2^0 bit input; 4 V + 1 V = 5 V.

The four bits have all been tried, thus completing the conversion cycle. At this point the binary code in the register is 0101, which is approximately the binary value of the input of 5.1 V. Additional bits will produce an even more accurate result. Another conversion cycle now begins, and the basic process is repeated. The SAR is cleared at the beginning of each cycle.

IMPLEMENTATION: ANALOG-TO-DIGITAL CONVERTER

The ADC0804 is an example of a successive-approximation ADC. A block diagram is shown in Figure 12–19. This device operates from a +5 V supply and has a resolution of eight bits with a conversion time of 100 μs. Also, it has an on-chip clock generator. Optionally, an external clock can be used. The data outputs are tri-state, so they can be interfaced with a microprocessor bus system.

FIGURE 12–19 The ADC0804 analog-to-digital converter.

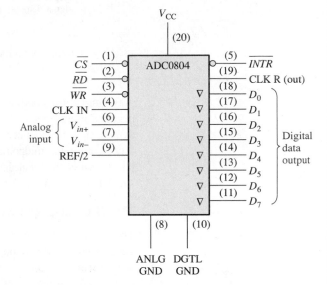

The basic operation of the device is as follows: The ADC0804 contains the equivalent of a 256-resistor DAC network. The successive-approximation logic sequences the network to match the analog differential input voltage ($V_{in+} - V_{in-}$) with an output from the resistive network. The MSB is tested first. After eight comparisons (sixty-four clock periods), an 8-bit binary code is transferred to output latches, and the interrupt ($\overline{INTR}$) output goes LOW. The device can be operated in a free-running mode by connecting the $\overline{INTR}$ output to the write ($\overline{WR}$) input and holding the conversion start ($\overline{CS}$) LOW. To ensure startup under all conditions, a LOW $\overline{WR}$ input is required during the power-up cycle. Taking $\overline{CS}$ low anytime after that will interrupt the conversion process.

When the $\overline{WR}$ input goes LOW, the internal successive-approximation register (SAR) and the 8-bit shift register are reset. As long as both $\overline{CS}$ and $\overline{WR}$ remain LOW, the ADC remains in a RESET state. Conversion starts one to eight clock periods after $\overline{CS}$ or $\overline{WR}$ makes a LOW-to-HIGH transition.

When a LOW is at both the $\overline{CS}$ and $\overline{RD}$ inputs, the tri-state output latch is enabled and the output code is applied to the D_0–D_7 lines. When either the $\overline{CS}$ or the $\overline{RD}$ input returns to a HIGH, the D_0–D_7 outputs are disabled.

Sigma-Delta Analog-to-Digital Converter

Sigma-delta is a widely used method of analog-to-digital conversion, particularly in telecommunications using audio signals. The method is based on **delta modulation** where the difference between two successive samples (increase or decrease) is quantized; other ADC methods were based on the absolute value of a sample. Delta modulation is a 1-bit quantization method.

The output of a delta modulator is a single-bit data stream where the relative number of 1s and 0s indicates the level or amplitude of the input signal. The number of 1s over a given number of clock cycles establishes the signal amplitude during that interval. A maximum number of 1s corresponds to the maximum positive input voltage. A number of 1s equal to one-half the

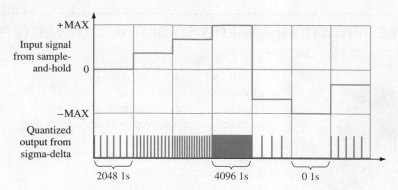

FIGURE 12–20   A simplified illustration of sigma-delta analog-to-digital conversion.

maximum corresponds to an input voltage of zero. No 1s (all 0s) corresponds to the maximum negative input voltage. This is illustrated in a simplified way in Figure 12–20. For example, assume that 4096 1s occur during the interval when the input signal is a positive maximum. Since zero is the midpoint of the dynamic range of the input signal, 2048 1s occur during the interval when the input signal is zero. There are no 1s during the interval when the input signal is a negative maximum. For signal levels in between, the number of 1s is proportional to the level.

The Sigma-Delta ADC Functional Block Diagram

The basic block diagram in Figure 12–21 accomplishes the conversion illustrated in Figure 12–20. The analog input signal and the analog signal from the converted quantized bit stream from the DAC in the feedback loop are applied to the summation (Σ) point. The difference (Δ) signal out of the Σ is integrated, and the 1-bit ADC increases or decreases the number of 1s depending on the difference signal. This action attempts to keep the quantized signal that is fed back equal to the incoming analog signal. The 1-bit quantizer is essentially a comparator followed by a latch.

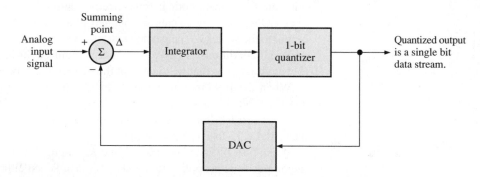

FIGURE 12–21   Partial functional block diagram of a sigma-delta ADC.

To complete the sigma-delta conversion process using one particular approach, the single bit data stream is converted to a series of binary codes, as shown in Figure 12–22. The counter counts the 1s in the quantized data stream for successive intervals. The code in the counter then represents the amplitude of the analog input signal for each interval. These codes are shifted out into the latch for temporary storage. What comes out of the latch is a series of n-bit codes, which completely represent the analog signal.

Testing Analog-to-Digital Converters

One method for testing ADCs is shown in Figure 12–23. A DAC is used as part of the test setup to convert the ADC output back to analog form for comparison with the test input.

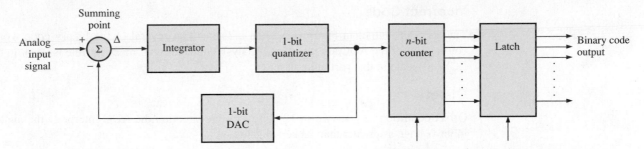

FIGURE 12–22   One type of sigma-delta ADC.

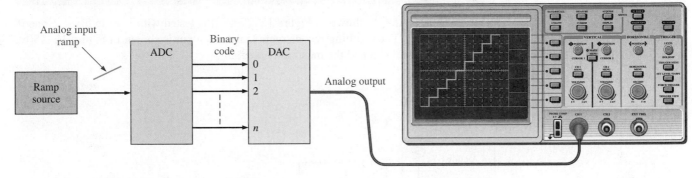

FIGURE 12–23   A method for testing ADCs.

A test input in the form of a linear ramp is applied to the input of the ADC. The resulting binary output sequence is then applied to the DAC test unit and converted to a stairstep ramp. The input and output ramps are compared for any deviation.

Analog-to-Digital Conversion Errors

Again, a 4-bit conversion is used to illustrate the principles. Let's assume that the test input is an ideal linear ramp.

Missing Code

The stairstep output in Figure 12–24(a) indicates that the binary code 1001 does not appear on the output of the ADC. Notice that the 1000 value stays for two intervals and then the output jumps to the 1010 value.

In a flash ADC, for example, a failure of one of the op-amp comparators can cause a missing-code error.

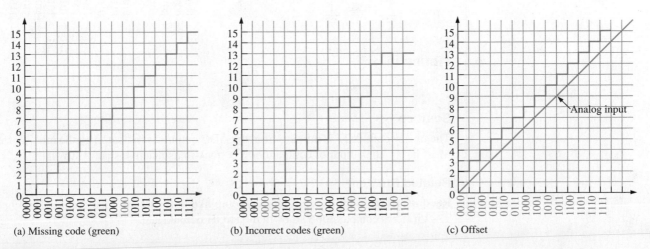

(a) Missing code (green)   (b) Incorrect codes (green)   (c) Offset

FIGURE 12–24   Illustrations of analog-to-digital conversion errors.

Incorrect Code

The stairstep output in Figure 12–24(b) indicates that several of the binary code words coming out of the ADC are incorrect. Analysis indicates that the 2^1-bit line is stuck in the LOW (0) state in this particular case.

Offset

Offset conditions are shown in 12–24(c). In this situation the ADC interprets the analog input voltage as greater than its actual value.

EXAMPLE 12–2

A 4-bit flash ADC is shown in Figure 12–25(a). It is tested with a setup like the one in Figure 12–23. The resulting reconstructed analog output is shown in Figure 12–25(b). Identify the problem and the most probable fault.

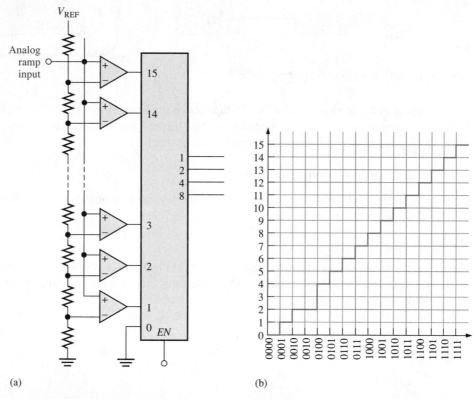

(a)   (b)

FIGURE 12–25

Solution

The binary code 0011 is missing from the ADC output, as indicated by the missing step. Most likely, the output of comparator 3 is stuck in its inactive state (LOW).

Related Problem

Reconstruct the analog output in a test setup like in Figure 12–23 if the ADC in Figure 12–25(a) has comparator 8 stuck in the HIGH output state.

1. What is the fastest method of analog-to-digital conversion?

2. Which analog-to-digital conversion method produces a single-bit data stream?

3. Does the successive-approximation converter have a fixed conversion time?

4. Name two types of output errors in an ADC.

12–3   Methods of Digital-to-Analog Conversion

Digital-to-analog conversion is an important part of a digital processing system. Once the digital data has been processed, it is converted back to analog form. In this section, we will examine the theory of operation of two basic types of digital-to-analog converters (DACs) and learn about their performance characteristics.

After completing this section, you should be able to

◆ Explain the operation of a binary-weighted-input DAC

◆ Explain the operation of an $R/2R$ ladder DAC

◆ Discuss resolution, accuracy, linearity, monotonicity, and settling time in a DAC

◆ Discuss the testing of DACs for nonmonotonicity, differential nonlinearity, low or high gain, and offset error

Binary-Weighted-Input Digital-to-Analog Converter

One method of digital-to-analog conversion uses a resistor network with resistance values that represent the binary weights of the input bits of the digital code. Figure 12–26 shows a 4-bit DAC of this type. Each of the input resistors will either have current or have no current, depending on the input voltage level. If the input voltage is zero (binary 0), the current is also zero. If the input voltage is HIGH (binary 1), the amount of current depends on the input resistor value and is different for each input resistor, as indicated in the figure.

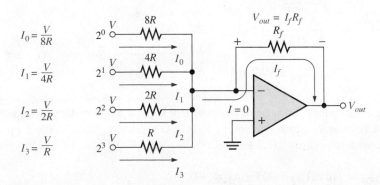

$$I_0 = \frac{V}{8R}$$

$$I_1 = \frac{V}{4R}$$

$$I_2 = \frac{V}{2R}$$

$$I_3 = \frac{V}{R}$$

$$V_{out} = I_f R_f$$

FIGURE 12–26   A 4-bit DAC with binary-weighted inputs.

Since there is practically no current into the op-amp inverting $(-)$ input, all of the input currents sum together and go through R_f. Since the inverting input is at $0\,V$ (virtual ground), the drop across R_f is equal to the output voltage, so $V_{out} = I_f R_f$.

The values of the input resistors are chosen to be inversely proportional to the binary weights of the corresponding input bits. The lowest-value resistor (R) corresponds to the highest binary-weighted input (2^3). The other resistors are multiples of R (that is, $2R$, $4R$, and $8R$) and correspond to the binary weights 2^2, 2^1, and 2^0, respectively. The input currents are also proportional to the binary weights. Thus, the output voltage is proportional to the sum of the binary weights because the sum of the input currents is through R_f.

Disadvantages of this type of DAC are the number of different resistor values and the fact that the voltage levels must be exactly the same for all inputs. For example, an 8-bit converter requires eight resistors, ranging from some value of R to $128R$ in binary-weighted steps. This range of resistors requires tolerances of one part in 255 (less than 0.5%) to accurately convert the input, making this type of DAC very difficult to mass-produce.

EXAMPLE 12–3

Determine the output of the DAC in Figure 12–27(a) if the waveforms representing a sequence of 4-bit numbers in Figure 12–27(b) are applied to the inputs. Input D_0 is the least significant bit (LSB).

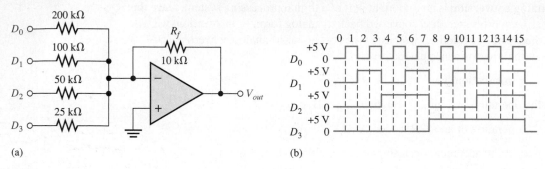

(a)   (b)

FIGURE 12–27

Solution

First, determine the current for each of the weighted inputs. Since the inverting ($-$) input of the op-amp is at 0 V (virtual ground) and a binary 1 corresponds to $+5$ V, the current through any of the input resistors is 5 V divided by the resistance value.

$$I_0 = \frac{5\text{ V}}{200\text{ k}\Omega} = 0.025\text{ mA}$$

$$I_1 = \frac{5\text{ V}}{100\text{ k}\Omega} = 0.05\text{ mA}$$

$$I_2 = \frac{5\text{ V}}{50\text{ k}\Omega} = 0.1\text{ mA}$$

$$I_3 = \frac{5\text{ V}}{25\text{ k}\Omega} = 0.2\text{ mA}$$

Almost no current goes into the inverting op-amp input because of its extremely high impedance. Therefore, assume that all of the current goes through the feedback resistor R_f. Since one end of R_f is at 0 V (virtual ground), the drop across R_f equals the output voltage, which is negative with respect to virtual ground.

$$V_{out(D0)} = (10\text{ k}\Omega)(-0.025\text{ mA}) = -0.25\text{ V}$$

$$V_{out(D1)} = (10\text{ k}\Omega)(-0.05\text{ mA}) = -0.5\text{ V}$$

$$V_{out(D2)} = (10\text{ k}\Omega)(-0.1\text{ mA}) = -1\text{ V}$$

$$V_{out(D3)} = (10\text{ k}\Omega)(-0.2\text{ mA}) = -2\text{ V}$$

From Figure 12–27(b), the first binary input code is 0000, which produces an output voltage of 0 V. The next input code is 0001, which produces an output voltage of -0.25 V. The next code is 0010, which produces an output voltage of -0.5 V. The next code is 0011, which produces an output voltage of -0.25 V $+ -0.5$ V $= -0.75$ V. Each successive binary code increases the output voltage by -0.25 V, so for this particular straight binary sequence on the inputs, the output is a stairstep waveform going from 0 V to -3.75 V in -0.25 V steps. This is shown in Figure 12–28.

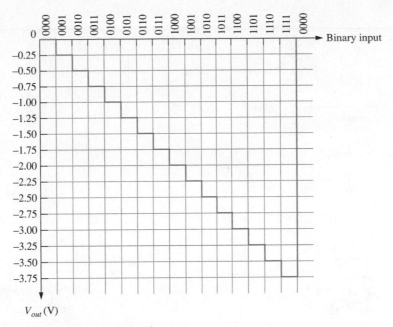

FIGURE 12–28 Output of the DAC in Figure 12–27.

Related Problem

Reverse the input waveforms to the DAC in Figure 12–27 (D_3 to D_0, D_2 to D_1, D_1 to D_2, D_0 to D_3) and determine the output.

The $R/2R$ Ladder Digital-to-Analog Converter

Another method of digital-to-analog conversion is the $R/2R$ ladder, as shown in Figure 12–29 for four bits. It overcomes one of the problems in the binary-weighted-input DAC in that it requires only two resistor values.

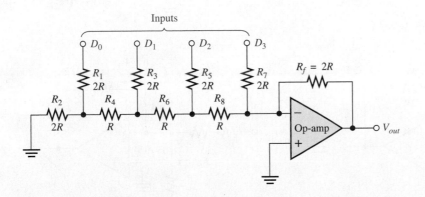

fg12_02900

FIGURE 12–29 An $R/2R$ ladder DAC.

Start by assuming that the D_3 input is HIGH (+5 V) and the others are LOW (ground, 0 V). This condition represents the binary number 1000. A circuit analysis will show that this reduces to the equivalent form shown in Figure 12–30(a). Essentially no current goes

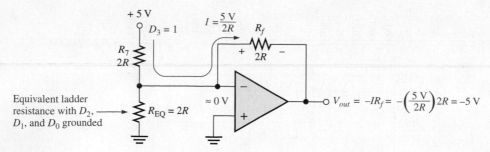

(a) Equivalent circuit for $D_3 = 1, D_2 = 0, D_1 = 0, D_0 = 0$

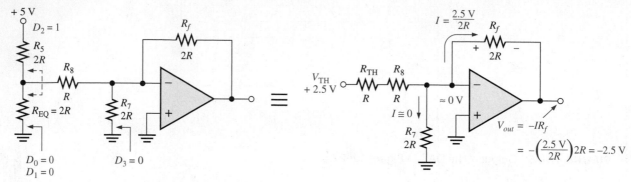

(b) Equivalent circuit for $D_3 = 0, D_2 = 1, D_1 = 0, D_0 = 0$

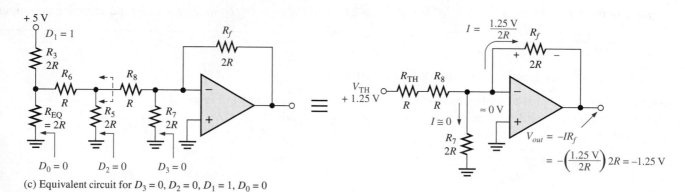

(c) Equivalent circuit for $D_3 = 0, D_2 = 0, D_1 = 1, D_0 = 0$

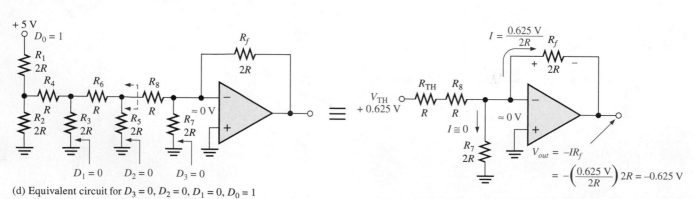

(d) Equivalent circuit for $D_3 = 0, D_2 = 0, D_1 = 0, D_0 = 1$

FIGURE 12–30   Analysis of the $R/2R$ ladder DAC.

through the $2R$ equivalent resistance because the inverting input is at virtual ground. Thus, all of the current ($I = 5$ V/$2R$) through R_7 also goes through R_f, and the output voltage is -5 V. The operational amplifier keeps the inverting ($-$) input near zero volts (≈ 0 V) because of negative feedback. Therefore, all current goes through R_f rather than into the inverting input.

Figure 12–30(b) shows the equivalent circuit when the D_2 input is at +5 V and the others are at ground. This condition represents 0100. If we thevenize* looking from R_8, we get 2.5 V in series with R, as shown. This results in a current through R_f of $I = 2.5$ V/2R, which gives an output voltage of -2.5 V. Keep in mind that there is no current into the op-amp inverting input and that there is no current through the equivalent resistance to ground because it has 0 V across it, due to the virtual ground.

Figure 12–30(c) shows the equivalent circuit when the D_1 input is at +5 V and the others are at ground. This condition represents 0010. Again thevenizing looking from R_8, you get 1.25 V in series with R as shown. This results in a current through R_f of $I = 1.25$ V/2R, which gives an output voltage of -1.25 V.

In part (d) of Figure 12–30, the equivalent circuit representing the case where D_0 is at +5 V and the other inputs are at ground is shown. This condition represents 0001. Thevenizing from R_8 gives an equivalent of 0.625 V in series with R as shown. The resulting current through R_f is $I = 0.625$ V/2R, which gives an output voltage of -0.625 V.

Notice that each successively lower-weighted input produces an output voltage that is halved, so that the output voltage is proportional to the binary weight of the input bits.

Performance Characteristics of Digital-to-Analog Converters

The performance characteristics of a DAC include resolution, accuracy, linearity, monotonicity, and settling time, each of which is discussed in the following list:

- *Resolution.* The resolution of a DAC is the reciprocal of the number of discrete steps in the output. This, of course, is dependent on the number of input bits. For example, a 4-bit DAC has a resolution of one part in $2^4 - 1$ (one part in fifteen). Expressed as a percentage, this is (1/15)100 = 6.67%. The total number of discrete steps equals $2^n - 1$, where n is the number of bits. Resolution can also be expressed as the number of bits that are converted.

- *Accuracy.* Accuracy is derived from a comparison of the actual output of a DAC with the expected output. It is expressed as a percentage of a full-scale, or maximum, output voltage. For example, if a converter has a full-scale output of 10 V and the accuracy is ±0.1%, then the maximum error for any output voltage is (10 V)(0.001) = 10 mV. Ideally, the accuracy should be no worse than ±1/2 of a least significant bit. For an 8-bit converter, the least significant bit is 0.39% of full scale. The accuracy should be approximately ±0.2%.

- *Linearity.* A linear error is a deviation from the ideal straight-line output of a DAC. A special case is an offset error, which is the amount of output voltage when the input bits are all zeros.

- *Monotonicity.* A DAC is **monotonic** if it does not take any reverse steps when it is sequenced over its entire range of input bits.

- *Settling time.* Settling time is normally defined as the time it takes a DAC to settle within ±1/2 LSB of its final value when a change occurs in the input code.

EXAMPLE 12–4

Determine the resolution, expressed as a percentage, of the following:

(a) an 8-bit DAC

(b) a 12-bit DAC

---

*Thevenin's theorem states that any circuit can be reduced to an equivalent voltage source in series with an equivalent resistance.

Solution

(a) For the 8-bit converter,

$$\frac{1}{2^8 - 1} \times 100 = \frac{1}{255} \times 100 = \mathbf{0.392\%}$$

(b) For the 12-bit converter,

$$\frac{1}{2^{12} - 1} \times 100 = \frac{1}{4095} \times 100 = \mathbf{0.0244\%}$$

Related Problem

Calculate the resolution for a 16-bit DAC.

Testing Digital-to-Analog Converters

The concept of DAC testing is illustrated in Figure 12–31. In this basic method, a sequence of binary codes is applied to the inputs, and the resulting output is observed. The binary code sequence extends over the full range of values from 0 to $2^n - 1$ in ascending order, where n is the number of bits.

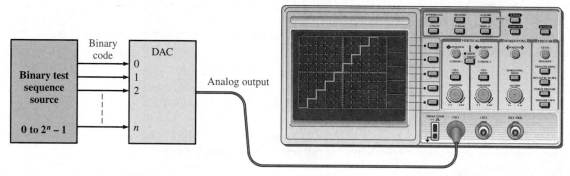

FIGURE 12–31   Basic test setup for a DAC.

The ideal output is a straight-line stairstep as indicated. As the number of bits in the binary code is increased, the resolution is improved. That is, the number of discrete steps increases, and the output approaches a straight-line linear ramp.

Digital-to-Analog Conversion Errors

Several digital-to-analog conversion errors to be checked for are shown in Figure 12–32, which uses a 4-bit conversion for illustration purposes. A 4-bit conversion produces fifteen discrete steps. Each graph in the figure includes an ideal stairstep ramp for comparison with the faulty outputs.

Nonmonotonicity

The step reversals in Figure 12–32(a) indicate nonmonotonic performance, which is a form of nonlinearity. In this particular case, the error occurs because the 2^1 bit in the binary code is interpreted as a constant 0. That is, a short is causing the bit input line to be stuck LOW.

Differential Nonlinearity

Figure 12–32(b) illustrates differential nonlinearity in which the step amplitude is less than it should be for certain input codes. This particular output could be caused by the 2^2 bit having an insufficient weight, perhaps because of a faulty input resistor. We could also see steps with amplitudes greater than normal if a particular binary weight were greater than it should be.

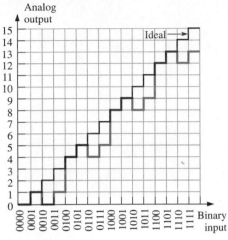

(a) Nonmonotonic output (green)

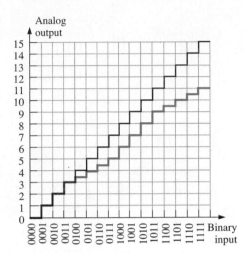

(b) Differential nonlinearity (green)

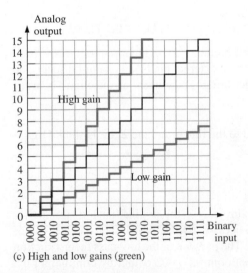

(c) High and low gains (green)

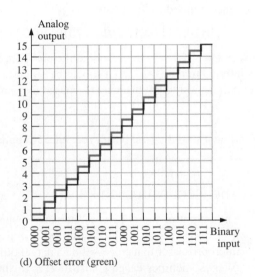

(d) Offset error (green)

FIGURE 12–32 Illustrations of several digital-to-analog conversion errors.

Low or High Gain

Output errors caused by low or high gain are illustrated in Figure 12–32(c). In the case of low gain, all of the step amplitudes are less than ideal. In the case of high gain, all of the step amplitudes are greater than ideal. This situation may be caused by a faulty feedback resistor in the op-amp circuit.

Offset Error

An offset error is illustrated in Figure 12–32(d). Notice that when the binary input is 0000, the output voltage is nonzero and that this amount of offset is the same for all steps in the conversion. A faulty op-amp may be the culprit in this situation.

EXAMPLE 12–5

The DAC output in Figure 12–33 is observed when a straight 4-bit binary sequence is applied to the inputs. Identify the type of error, and suggest an approach to isolate the fault.

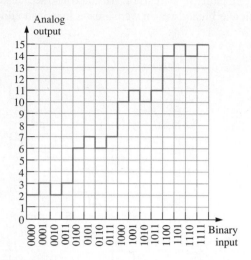

FIGURE 12–33

Solution

The DAC in this case is nonmonotonic. Analysis of the output reveals that the device is converting the following sequence, rather than the actual binary sequence applied to the inputs.

0010, 0011, 0010, 0011, 0110, 0111, 0110, 0111, 1010, 1011, 1010, 1011, 1110, 1111, 1110, 1111

Apparently, the 2^1 bit is stuck in the HIGH (1) state. To find the problem, first monitor the bit input pin to the device. If it is changing states, the fault is internal to the DAC and it should be replaced. If the external pin is not changing states and is always HIGH, check for an external short to $+V$ that may be caused by a solder bridge somewhere on the circuit board.

Related Problem

Determine the output of a DAC when a straight 4-bit binary sequence is applied to the inputs and the 2^0 bit is stuck HIGH.

The Reconstruction Filter

The output of the DAC is a "stairstep" approximation of the original analog signal after it has been processed by the **digital signal processor (DSP)**, which is a special type of microprosessor that processes data in real time. The purpose of the low-pass reconstruction filter (sometimes called a postfilter) is to smooth out the DAC output by eliminating the higher frequency content that results from the fast transitions of the "stairsteps," as roughly illustrated in Figure 12–34.

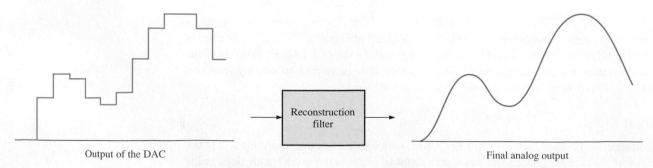

Output of the DAC

Reconstruction filter

Final analog output

FIGURE 12–34   The reconstruction filter smooths the output of the DAC.

SECTION 12–3 CHECKUP

1. What is the disadvantage of the DAC with binary-weighted inputs?

2. What is the resolution of a 4-bit DAC?

3. How do you detect nonmonotonic behavior in a DAC?

4. What effect does low gain have on a DAC output?

12–4   Digital Signal Processing

Digital signal processing converts signals that naturally occur in analog form, such as sound, video, and information from sensors, to digital form and uses digital techniques to enhance and modify analog signal data for various applications.

After completing this section, you should be able to

◆ Discuss digital signal processing

◆ Draw a basic block diagram of a digital signal processing system

A digital signal processing system first translates a continuously varying analog signal into a series of discrete levels. This series of levels follows the variations of the analog signal and resembles a staircase, as illustrated for the case of a sine wave in Figure 12–35. The process of changing the original analog signal to a "stairstep" approximation is accomplished by a sample-and-hold circuit.

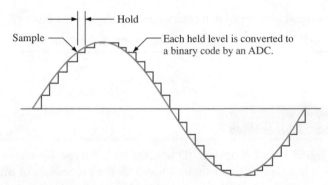

FIGURE 12–35   An original analog signal (sine wave) and its "stairstep" approximation.

Next, the "stairstep" approximation is quantized into binary codes that represent each discrete step on the "stairsteps" by a process called analog-to-digital (A/D) conversion. The circuit that performs A/D conversion is an analog-to-digital converter (ADC).

Once the analog signal has been converted to a binary coded form, it is applied to a DSP (digital signal processor). The DSP can perform various operations on the incoming data, such as removing unwanted interference, increasing the amplitude of some signal frequencies and reducing others, encoding the data for secure transmissions, and detecting and correcting errors in transmitted codes. DSPs make possible, among many other things, the cleanup of sound recordings, the removal of echos from communications lines, the enhancement of images from CT scans for better medical diagnosis, and the scrambling of cellular phone conversations for privacy.

After a DSP processes a signal, the signal can be converted back to an enhanced version of the original analog signal. This is accomplished by a digital-to-analog converter (DAC). Figure 12–36 shows a basic block diagram of a typical digital signal processing system.

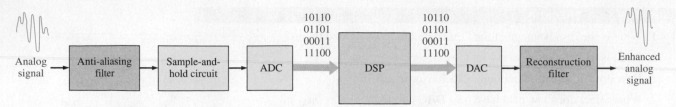

FIGURE 12–36 Basic block diagram of a typical digital signal processing system.

DSPs are actually a specialized type of microprocessor but are different from general-purpose microprocessors in a couple of significant ways. Typically, microprocessors are designed for general-purpose functions and operate with large software packages. DSPs are used for special-purpose applications; they are very fast number crunchers that must work in real time by processing information as it happens using specialized algorithms (programs). The analog-to-digital converter (ADC) in a system must take samples of the incoming analog data often enough to catch all the relevant fluctuations in the signal amplitude, and the DSP must keep pace with the sampling rate of the ADC by doing its calculations as fast as the sampled data are received. Once the digital data are processed by the DSP, they go to the digital-to-analog converter (DAC) and reconstruction filter for conversion back to analog form.

SECTION 12–4 CHECKUP

1. What does DSP stand for?

2. What does ADC stand for?

3. What does DAC stand for?

4. An analog signal is changed to a binary coded form by what circuit?

5. A binary coded signal is changed to analog form by what circuit?

12–5 The Digital Signal Processor (DSP)

Essentially, a digital signal processor (DSP) is a special type of microprocessor that processes data in real time. Its applications focus on the processing of digital data that represents analog signals. A DSP, like a microprocessor, has a central processing unit (CPU) and memory units in addition to many interfacing functions. Every time you use your cellular telephone, you are using a DSP, and this is only one example of its many applications.

After completing this chapter, you should be able to

- Explain the basic concepts of a DSP

- List some of the applications of DSPs

- Describe the basic functions of a DSP in a cell phone

- Discuss the TMS320C6000 series DSP

The digital signal processor (DSP) is the heart of a digital signal processing system. It takes its input from an ADC and produces an output that goes to a DAC, as shown in Figure 12–37. As you have learned, the ADC changes an analog waveform into data in the form of a series of binary codes that are then applied to the DSP for processing. After being processed by the DSP, the data go to a DAC for conversion back to analog form.

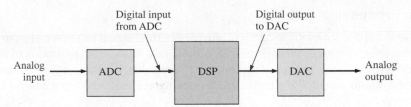

FIGURE 12-37   The DSP has a digital input and produces a digital output.

DSP Programming

DSPs are typically programmed in either assembly language or in C. Because programs written in assembly language can usually execute faster and because speed is critical in most DSP applications, assembly language is used much more in DSPs than in general-purpose microprocessors. Also, DSP programs are usually much shorter than traditional microprocessor programs because of their very specialized applications where much redundancy is used. In general, the instruction sets for DSPs tend to be smaller than for microprocessors.

DSP Applications

The DSP, unlike the general-purpose microprocessor, must typically process data in *real time;* that is, as it happens. Many applications in which DSPs are used cannot tolerate any noticeable delays, requiring the DSP to be extremely fast. In addition to cell phones, digital signal processors (DSPs) are used in multimedia computers, video recorders, CD players, hard disk drives, digital radio modems, and other applications to improve the signal quality. DSPs are also used in television applications. For example, television converters use DSP to provide compatibility with various television standards.

An important application of DSPs is in signal compression and decompression. In CD systems, for example, the music on the CD is in a compressed form so that it doesn't use as much storage space. It must be decompressed in order to be reproduced. Also signal compression is used in cell phones to allow a greater number of calls to be handled simultaneously in a local cell.

Telecommunications

The field of telecommunications involves transferring all types of information from one location to another, including telephone conversations, television signals, and digital data. Among other functions, the DSP facilitates multiplexing many signals onto one transmission channel because information in digital form is relatively easy to multiplex and demultiplex.

At the transmitting end of a telecommunications system, DSPs are used to compress digitized voice signals for conservation of bandwidth. Compression is the process of reducing the data rate. Generally, a voice signal is converted to digital form at 8000 samples per second (sps), based on a Nyquist frequency of 4 kHz. If 8 bits are used to encode each sample, the data rate is 64 kbps. In general, reducing (compressing) the data rate from 64 kbps to 32 kbps results in no loss of sound quality. When the data are compressed to 8 kbps, the sound quality is reduced noticeably. When compressed to the minimum of 2 kbps, the sound is greatly distorted but still usable for some applications where only word recognition and not quality is important. At the receiving end of a telecommunications system, the DSP decompresses the data to restore the signal to its original form.

Echoes, a problem in many long distance telephone connections, occur when a portion of a voice signal is returned with a delay. For shorter distances, this delay is barely noticeable; but as the distance between the transmitter and the receiver increases, so does the delay time of the echo. DSPs are used to effectively cancel the annoying echo, which results in a clear, undisturbed voice signal.

InfoNote

Sound cards used in computers use an ADC to convert sound from a microphone, audio CD player, or other source into a digital signal. The ADC sends the digital signal to a digital signal processor (DSP). Based on instructions from a ROM, one function of the DSP is to compress the digital signal so it uses less storage space. The DSP then sends the compressed data to the computer's processor which, in turn, sends the data to a hard drive or CD ROM for storage. To play a recorded sound, the stored data is retrieved by the processor and sent to the DSP where it is decompressed and sent to a DAC. The output of the DAC, which is a reproduction of the original sound signal, is applied to the speakers.

Music Processing

The DSP is used in the music industry to provide filtering, signal addition and subtraction, and signal editing in music preparation and recording. Also, another application of the DSP is to add artificial echo and reverberation, which are usually minimized by the acoustics of a sound studio, in order to simulate ideal listening environments from concert halls to small rooms.

Speech Generation and Recognition

DSPs are used in speech generation and recognition to enhance the quality of man/machine communication. The most common method used to produce computer-generated speech is digital recording. In digital recording, the human voice is digitized and stored, usually in a compressed form. During playback the stored voice data are uncompressed and converted back into the original analog form. Approximately an hour of speech can be stored using about 3 MB of memory.

Speech recognition is much more difficult to accomplish than speech generation. The DSP is used to isolate and analyze each word in the incoming voice signal. Certain parameters are identified in each word and compared with previous examples of the spoken word to create the closest match. Most systems are limited to a few hundred words at best. Also, significant pauses between words are usually required and the system must be "trained" for a given individual's voice. Speech recognition is an area of tremendous research effort and will eventually be applied in many commercial applications.

Radar

In *ra*dio *d*etection *a*nd *r*anging (radar) applications, DSPs provide more accurate determination of distance using data compression techniques, decrease noise using filtering techniques, thereby increasing the range, and optimize the ability of the radar system to identify specific types of targets. DSPs are also used in similar ways in sonar systems.

Image Processing

The DSP is used in image-processing applications such as the computed tomography (CT) and magnetic resonance imaging (MRI), which are widely used in the medical field for looking inside the human body. In CT, X-rays are passed through a section of the body from many directions. The resulting signals are converted to digital form and stored. This stored information is used to produce calculated images that appear to be slices through the human body that show great detail and permit better diagnosis.

Instead of X-rays, MRI uses magnetic fields in conjunction with radio waves to probe inside the human body. MRI produces images, just as CT, and provides excellent discrimination between different types of tissue as well as information such as blood flow through arteries. MRI depends entirely on digital signal processing methods.

In applications such as video telephones, digital television, and other media that provide moving pictures, the DSP uses image compression to reduce the number of bits needed, making these systems commercially feasible.

Filtering

DSPs are commonly used to implement digital filters for the purposes of separating signals that have been combined with other signals or with interference and noise and for restoring signals that are distorted. Although analog filters are quite adequate for some applications, the digital filter is generally much superior in terms of the performance that can be achieved. One drawback to digital filters is that the execute time required produces a delay from the time the analog signal is applied until the time the output appears. Analog filters present no delay problems because as soon as the input occurs, the response appears on the output. Analog filters are also less expensive than digital filters. Regardless of this, the overall performance of the digital filter is far superior in many applications.

The DSP in a Cellular Telephone

The digital cellular telephone is an example of how a DSP can be used. Figure 12–38 shows a simplified block diagram of a digital cell phone. The voice **codec** (codec is the abbreviation for coder/decoder) contains, among other functions, the ADC and DAC necessary to convert between the analog voice signal and a digital voice format. Sigma-delta conversion is typically used in most cell phone applications. For transmission, the voice signal from the microphone is converted to digital form by the ADC in the codec and then it goes to the DSP for processing. From the DSP, the digital signal goes to the rf (radio frequency) section where it is modulated and changed to the radio frequency for transmission. An incoming rf signal containing voice data is picked up by the antenna, demodulated, and changed to a digital signal. It is then applied to the DSP for processing, after which the digital signal goes to the codec for conversion back to the original voice signal by the DAC. It is then amplified and applied to the speaker.

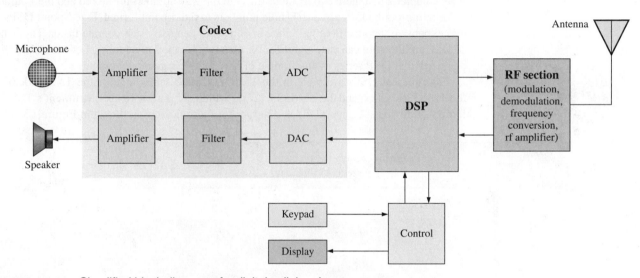

FIGURE 12–38   Simplified block diagram of a digital cellular phone.

Functions Performed by the DSP

In a cellular phone application, the DSP performs many functions to improve and facilitate the reception and transmission of a voice signal. Some of these DSP functions are as follows:

- *Speech compression.* The rate of the digital voice signal is reduced significantly for transmission in order to meet the bandwidth requirements.

- *Speech decompression.* The rate of the received digital voice signal is returned to its original rate in order to properly reproduce the analog voice signal.

- *Protocol handling.* The cell phone communicates with the nearest base in order to establish the location of the cell phone, allocates time and frequency slots, and arranges handover to another base station as the phone moves into another cell.

- *Error detection and correction.* During transmission, error detection and correction codes are generated and, during reception, detect and correct errors induced in the rf channel by noise or interference.

- *Encryption.* Converts the digital voice signal to a form for secure transmission and converts it back to original form during reception.

Basic DSP Architecture

As mentioned before, a DSP is basically a specialized microprocessor optimized for speed in order to process data in real time. Many DSPs are based on what is known as the *Harvard architecture,* which consists of a central processing unit (CPU) and two memories, one for data and the other for the program, as shown by the block diagram in Figure 12–39.

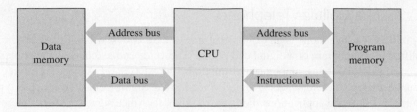

FIGURE 12–39   Many DSPs use the Harvard architecture (two memories).

A Specific DSP

DSPs are manufactured by several companies including Texas Instruments, Motorola, and Analog Devices. DSPs are available for both fixed-point and floating-point processing. Recall from Chapter 2 that these two methods differ in the way numbers are stored and manipulated. All floating-point DSPs can also handle numbers in fixed-point format. Fixed-point DSPs are less expensive than the floating-point versions and, generally, can operate faster. The details of DSP architecture can vary significantly, even within the same family. Let's look briefly at one particular DSP series as an example of how a DSP is generally organized.

Examples of DSPs available in the TMS320C6000 series include the TMS320C62xx, the TMS320C64xx, and the TMS320C67xx, which are part of Texas Instrument's TMS320 family of devices. A general block diagram for these devices is shown in Figure 12–40.

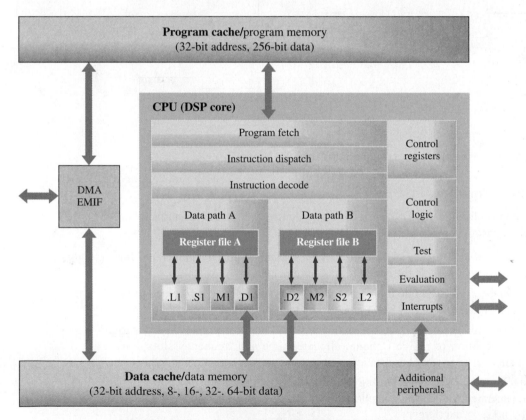

FIGURE 12–40   General block diagram of the TMS320C6000 series DSP.

The DSPs have a central processing unit (CPU), also known as the **DSP core,** that contains 64 general-purpose 32-bit registers in the C64xx and 32 general-purpose 32-bit registers in the C62xx and the C67xx. The C67xx can handle floating-point operations, whereas the C62xx and C64xx are fixed-point devices.

Each DSP has eight functional units that contain two 16-bit multipliers and six arithmetic logic units (ALUs). The performance of the three DSPs in the C6000 series in terms of

TABLE 12–3

TMS320C6000 series DSP data processing performance.

DSP	Type	Application	Processing Speed	Multiply/ Accumulate Speed
C62xx	Fixed-point	General-purpose	1200–2400 MIPS	300–600 MMACS
C64xx	Fixed-point	Special-purpose	3200–4800 MIPS	1600–2400 MMACS
C67xx	Floating-point	General-purpose	600–1000 MFLOPS	200–333 MMACS

MIPS (Million Instructions Per Second), **MFLOPS** (Million Floating-point Operations Per Second), and **MMACS** (Million Multiply/Accumulates per Second) is shown in Table 12–3.

Data Paths in the CPU

In the CPU, the program fetch, instruction dispatch, and instruction decode sections can provide eight 32-bit instructions to the functional units during every clock cycle. The CPU is split into two data paths, and instruction processing occurs in both data paths A and B. Each data path contains half of the general-purpose registers (16 in the C62xx and C67xx or 32 in the C64xx) and four functional units. The control register and logic are used to configure and control the various processor operations.

Functional Units

Each data path has four functional units. The M units (labeled .M1 and .M2 in Figure 12–40) are dedicated multipliers. The L units (labeled .L1 and .L2) perform arithmetic, logic, and miscellaneous operations. The S units (labeled .S1 and .S2) perform compare, shift, and miscellaneous arithmetic operations. The D units (labeled .D1 and .D2) perform load, store, and miscellaneous operations.

Pipeline

A **pipeline** allows multiple instructions to be processed simultaneously. A pipeline operation consists of three stages through which all instructions flow: *fetch, decode, execute.* Eight instructions at a time are first fetched from the program memory; they are then decoded, and finally they are executed.

During **fetch,** the eight instructions (called a packet) are taken from memory in four *phases,* as shown in Figure 12–41.

- *Program address generate (PG).* The program address is generated by the CPU.
- *Program address send (PS).* The program address is sent to the memory.
- *Program access ready wait (PW).* A memory read operation occurs.
- *Program fetch packet receive (PR).* The CPU receives the packet of instructions.

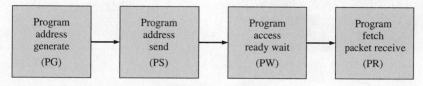

FIGURE 12–41 The four fetch phases of the pipeline operation.

Two phases make up the instruction **decode** stage of pipeline operation, as shown in Figure 12–42. The instruction dispatch (DP) phase is where the instruction packets are split into execute packets and assigned to the appropriate functional units. The instruction decode (DC) phase is where the instructions are decoded.

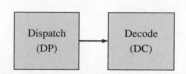

FIGURE 12–42 The two decode phases of the pipeline operation.

The **execute** stage of the pipeline operation is where the instructions from the decode stage are carried out. The execute stage has a maximum of five phases (E1 through E5), as shown in Figure 12–43. All instructions do not use all five phases. The number of phases used during execution depends on the type of instruction. Part of the execution of an instruction requires getting data from the data memory.

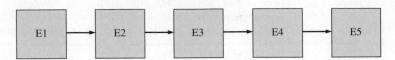

FIGURE 12–43   The five execute phases of pipeline operation.

Internal DSP Memory and Interfaces

As you can see in Figure 12–40, there are two internal memories, one for data and one for program. The program memory is organized in 256 bit packets (eight 32-bit instructions) and there are 64 kB of capacity. The data memory also has a capacity of 64 kB and can be accessed in 8-, 16-, 32-, or 64-bit word lengths, depending on the specific device in the series. Both internal memories are accessed with a 32-bit address. The DMA (Direct Memory Access) is used to transfer data without going through the CPU. The EMIF (External Memory Interface) is used to support external memories when required in an application. Additional interface is provided for serial I/O ports and other external devices.

Timers

There are two general-purpose timers in the DSP that can be used for timed events, counting, pulse generation, CPU interrupts, and more.

Packaging

The TMS 3206000 series processors are available in 352-pin ball grid array (BGA) packages, as shown in Figure 12–44, and are implemented with CMOS technology.

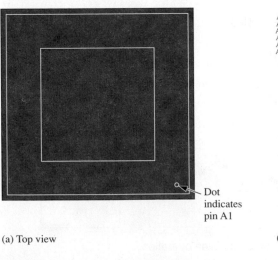

(a) Top view

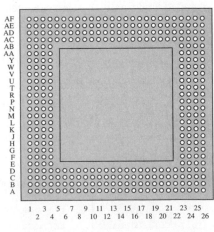

(b) Bottom view

(c) Side view

FIGURE 12–44   A 352-pin BGA package.

1. What is meant by the Harvard architecture?

2. What is a DSP core?

3. Name two categories of DSPs according to the type of numbers handled.

4. What are the two types of internal memory?

5. Define (a) MIPS (b) MFLOPS (c) MMACS.

6. Basically, what does pipelining accomplish?

7. Name the three stages of pipeline operation.

8. What happens during the fetch phase?

SUMMARY

- Sampling converts an analog signal into a series of impulses, each representing the signal amplitude at a given instant in time.

- The sampling theorem states that the sampling frequency must be at least twice the highest sampled frequency (Nyquist frequency).

- Analog-to-digital conversion changes an analog signal into a series of digital codes.

- Four types of analog-to-digital converters (ADCs) are flash (simultaneous), dual-slope, successive-approximation, and sigma-delta.

- Digital-to-analog conversion changes a series of digital codes that represent an analog signal back into the analog signal.

- Two types of digital-to-analog converters (DACs) are binary-weighted input and $R/2R$ ladder.

- Digital signal processing is the digital processing of analog signals, usually in real-time, for the purpose of modifying or enhancing the signal in some way.

- In general, a digital signal processing system consists of an anti-aliasing filter, a sample-and-hold circuit, an analog-to-digital converter, a DSP (digital signal processor), a digital-to-analog converter, and a reconstruction filter.

- A DSP is a specialized microprocessor optimized for speed in order to process data as it occurs (real-time).

- Most DSPs are based on the Harvard architecture, which means that there is a data memory and a program memory.

- A pipeline operation consists of fetch, decode, and execute stages.

KEY TERMS

Key terms and other bold terms in the chapter are defined in the end-of-book glossary.

Aliasing   The effect created when a signal is sampled at less than twice the signal frequency. Aliasing creates unwanted frequencies that interfere with the signal frequency when the signal is recovered.

Analog-to-digital converter (ADC)   A circuit used to convert an analog signal to digital form.

Decode   A stage of the DSP pipeline operation in which instructions are assigned to functional units and are decoded.

Digital signal processor (DSP)   A special type of microprocessor that processes data in real time.

Digital-to-analog converter (DAC)   A circuit used to convert the digital representation of an analog signal back to the analog signal.

DSP core   The central processing unit of a DSP.

Execute   A stage of the DSP pipeline operation in which the decoded instructions are carried out.

Fetch   A stage of the DSP pipeline operation in which an instruction is obtained from the program memory.

MFLOPS   Million floating-point operations per second.

MIPS   Million instructions per second.

MMACS   Million multiply/accumulates per second.

Nyquist frequency   The highest signal frequency that can be sampled at a specified sampling frequency; a frequency equal to or less than half the sampling frequency.

Pipeline   Part of the DSP architecture that allows multiple instructions to be processed simultaneously.

Quantization   The process whereby a binary code is assigned to each sampled value during analog-to-digital conversion.

Sampling   The process of taking a sufficient number of discrete values at points on a waveform that will define the shape of the waveform.

TRUE/FALSE QUIZ

Answers are at the end of the chapter.

1. An analog signal is converted to a digital signal by an ADC.
2. A DAC is a digital approximation computer.
3. The Nyquist frequency is twice the sampling frequency.
4. A higher sampling rate is more accurate than a lower sampling rate for a given analog signal.
5. Resolution is the number of bits used by an analog-to-digital converter.
6. Successful approximation is an analog-to-digital conversion method.
7. Delta modulation is based on the difference of two successive samples.
8. Two types of DAC are the binary-weighted input and the $R/2R$ ladder.
9. The process of converting an analog value to a code is called *quantization*.
10. A flash ADC differs from a simultaneous ADC.

SELF-TEST

Answers are at the end of the chapter.

1. An ADC is an
   (a) alphanumeric data code
   (b) analog-to-digital converter
   (c) analog device carrier
   (d) analog-to-digital comparator

2. A DAC is a
   (a) digital-to-analog computer
   (b) digital analysis calculator
   (c) data accumulation converter
   (d) digital-to-analog converter

3. Sampling of an analog signal produces
   (a) a series of impulses that are proportional to the amplitude of the signal
   (b) a series of impulses that are proportional to the frequency of the signal
   (c) digital codes that represent the analog signal amplitude
   (d) digital codes that represent the time of each sample

4. According to the sampling theorem, the sampling frequency should be
   (a) less than half the highest signal frequency
   (b) greater than twice the highest signal frequency
   (c) less than half the lowest signal frequency
   (d) greater than the lowest signal frequency

5. A hold action occurs
   (a) before each sample   **(b)** during each sample
   (c) after the analog-to-digital conversion   **(d)** immediately after a sample

6. The quantization process
   (a) converts the sample-and-hold output to binary code
   (b) converts a sample impulse to a level
   (c) converts a sequence of binary codes to a reconstructed analog signal
   (d) filters out unwanted frequencies before sampling takes place

7. Generally, an analog signal can be reconstructed more accurately with
   (a) more quantization levels   **(b)** fewer quantization levels
   (c) a higher sampling frequency   **(d)** a lower sampling frequency
   (e) either answer (a) or (c)

8. A flash ADC uses
   (a) counters   **(b)** op-amps
   (c) an integrator   **(d)** flip-flops
   (e) answers (a) and (c)

9. A dual-slope ADC uses
   (a) a counter   **(b)** op-amps
   (c) an integrator   **(d)** a differentiator
   (e) answers (a) and (c)

10. The output of a sigma-delta ADC is
   (a) parallel binary codes   **(b)** multiple-bit data
   (c) single-bit data   **(d)** a difference voltage

11. In a binary-weighted DAC, the resistors on the inputs
   (a) determine the amplitude of the analog signal
   (b) determine the weights of the digital inputs
   (c) limit the power consumption
   (d) prevent loading on the source

12. In an *R/2R* DAC, there are
   (a) four values of resistors
   (b) one resistor value
   (c) two resistor values
   (d) a number of resistor values equal to the number of inputs

13. A digital signal processing system usually operates in
   (a) real time   **(b)** imaginary time
   (c) compressed time   **(d)** computer time

14. The term *Harvard architecture* means
   (a) a CPU and a main memory
   (b) a CPU and two data memories
   (c) a CPU, a program memory, and a data memory
   (d) a CPU and two register files

15. The minimum number of general-purpose registers in the TMS320C6000 series DSPs is
   (a) 32   **(b)** 64
   (c) 16   **(d)** 8

16. The two internal memories in the TMS320C6000 series each have a capacity of
   (a) 1 MB   **(b)** 512 kB
   (c) 64 kB   **(d)** 32 kB

17. In the TMS320C6000 series pipeline operation, the number of instructions processed simultaneously is
   (a) eight   **(b)** four
   (c) two   **(d)** one

18. The stage of the pipeline operation in which instructions are retrieved from the memory is called
   (a) execute   **(b)** accumulate
   (c) decode   **(d)** fetch

PROBLEMS

Answers to odd-numbered problems are at the end of the book.

Section 12–1 Analog-to-Digital Conversion

1. The waveform shown in Figure 12–45 is applied to a sampling circuit and is sampled every 3 ms. Show the output of the sampling circuit. Assume a one-to-one voltage correspondence between the input and output.

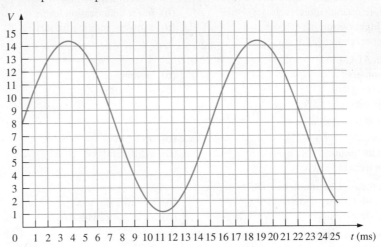

FIGURE 12–45

2. The output of the sampling circuit in Problem 1 is applied to a hold circuit. Show the output of the hold circuit.

3. If the output of the hold circuit in Problem 2 is quantized using two bits, what is the resulting sequence of binary codes?

4. Repeat Problem 3 using 4-bit quantization.

5. (a) Reconstruct the analog signal from the 2-bit quantization in Problem 3.

   (b) Reconstruct the analog signal from the 4-bit quantization in Problem 4.

6. Graph the analog function represented by the following sequence of binary numbers:
   1111, 1110, 1101, 1100, 1010, 1001, 1000, 0111, 0110, 0101, 0100, 0101, 0110, 0111, 1000, 1001, 1010, 1011, 1100, 1100, 1100, 1011, 1010, 1001.

Section 12–2 Methods of Analog-to-Digital Conversion

7. The input voltage to a certain op-amp inverting amplifier is 10 mV, and the output is 2 V. What is the closed-loop voltage gain?

8. To achieve a closed-loop voltage gain of 330 with an inverting amplifier, what value of feedback resistor do you use if $R_i = 1.0\,k\Omega$?

9. What is the gain of an inverting amplifier that uses a $47\,k\Omega$ feedback resistor if the input resistor is $2.2\,k\Omega$?

10. How many comparators are required to form an 8-bit flash converter?

11. Determine the binary output code of a 3-bit flash ADC for the analog input signal in Figure 12–46.

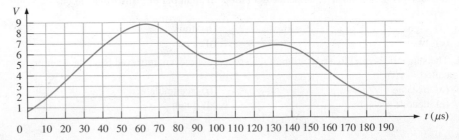

FIGURE 12–46

12. Repeat Problem 11 for the analog waveform in Figure 12–47.

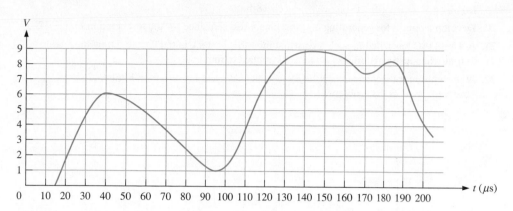

FIGURE 12–47

13. For a certain 2-bit successive-approximation ADC, the maximum ladder output is +8 V. If a constant +6 V is applied to the analog input, determine the sequence of binary states for the SAR.

14. Repeat Problem 13 for a 4-bit successive-approximation ADC.

15. An ADC produces the following sequence of binary numbers when an analog signal is applied to its input: 0000, 0001, 0010, 0011, 0100, 0101, 0110, 0111, 0110, 0101, 0100, 0011, 0010, 0001, 0000.

(a) Reconstruct the input digitally.

(b) If the ADC failed so that the code 0111 were missing, what would the reconstructed output look like?

Section 12–3 Methods of Digital-to-Analog Conversion

16. In the 4-bit DAC in Figure 12–26, the lowest-weighted resistor has a value of 10 kΩ. What should the values of the other input resistors be?

17. Determine the output of the DAC in Figure 12–48(a) if the sequence of 4-bit numbers in part (b) is applied to the inputs. The data inputs have a low value of 0 V and a high value of +5 V.

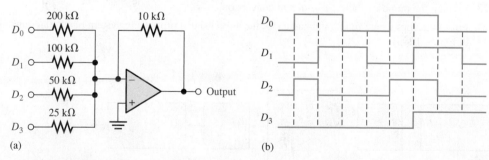

(a)   (b)

FIGURE 12–48

18. Repeat Problem 17 for the inputs in Figure 12–49.

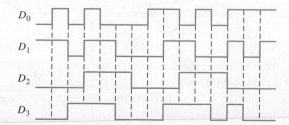

FIGURE 12–49

19. Determine the resolution expressed as a percentage, for each of the following DACs:

    (a) 3-bit   (b) 10-bit   (c) 18-bit

20. Develop a circuit for generating an 8-bit binary test sequence for the test setup in Figure 12–31.

21. A 4-bit DAC has failed in such a way that the MSB is stuck in the 0 state. Draw the analog output when a straight binary sequence is applied to the inputs.

22. A straight binary sequence is applied to a 4-bit DAC, and the output in Figure 12–50 is observed. What is the problem?

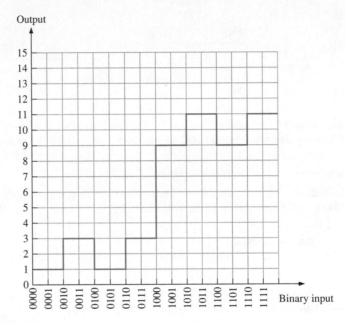

FIGURE 12–50

Section 12–4 Digital Signal Processing

23. Explain the purpose of analog-to-digital conversion.

24. Fill in the appropriate functional names for the digital signal processing system block diagram in Figure 12–51.

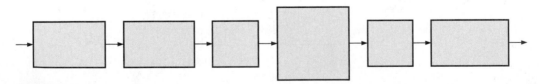

FIGURE 12–51

25. Explain the purpose of digital-to-analog conversion.

Section 12–5 The Digital Signal Processor (DSP)

26. A TMS320C62xx DSP has 32-bit instructions and is operating at 2000 MIPS. How many bytes per second is the DSP processing?

27. If the clock rate of a TMS320C64xx DSP is 400 MHz, how many instructions can it provide to the CPU functional units in one second?

28. How many floating-point operations can a DSP do in one second if it is specified at 1000 MFLOPS?

29. List and describe the four phases of the fetch operation in a TMS320C6000 series DSP.

30. List and describe the two phases of the decode operation in a TMS320C6000 series DSP.

ANSWERS

SECTION CHECKUPS

Section 12–1 Analog-to-Digital Conversion

1. Sampling is the process of converting an analog signal into a series of impulses, each representing the amplitude of the analog signal.

2. A sampled value is held to allow time to convert the value to a binary code.

3. The minimum sampling frequency is 40 kHz.

4. Quantization is the process of converting a sampled level to a binary code.

5. The number of bits determine quantization accuracy.

Section 12–2 Methods of Analog-to-Digital Conversion

1. The simultaneous (flash) method is fastest.

2. The sigma-delta method produces a single-bit data stream.

3. Yes, successive approximation has a fixed conversion time.

4. Missing code, incorrect code, and offset are types of ADC output errors.

Section 12–3 Methods of Digital-to-Analog Conversion

1. In a binary-weighted DAC, each resistor has a different value.

2. $(1/(2^4 - 1))100\% = 6.67\%$

3. A step reversal indicates nonmonotonic behavior in a DAC.

4. Step amplitudes in a DAC are less than ideal with low gain.

Section 12–4 Digital Signal Processing

1. DSP stands for digital signal processor.

2. ADC stands for analog-to-digital converter.

3. DAC stands for digital-to-analog converter.

4. The ADC changes an analog signal to binary coded form.

5. The DAC changes a binary coded signal to analog form.

Section 12–5 The Digital Signal Processor (DSP)

1. Harvard architecture means that there is a CPU and two memories, one for data and one for programs.

2. The DSP core is the CPU.

3. DSPs can be fixed-point or floating-point.

4. Internal memory types are data and program.

5. **(a)** MIPS—million instructions per second

   (b) MFLOPS—million floating-point operations per second

   (c) MMACS—million multiply/accumulates per second

6. Pipelining provides for the processing of multiple instructions simultaneously.

7. The stages of pipeline operation are fetch, decode, and execute.

8. During fetch, instructions are retrieved from the program memory.

RELATED PROBLEMS FOR EXAMPLES

12–1 100, 111, 100, 000, 011, 110. Yes, information is lost.

12–2 See Figure 12–52.

12–3 See Figure 12–53.

12–4 $(1/(2^{16} - 1))100\% = 0.00153\%$

12–5 See Figure 12–54.

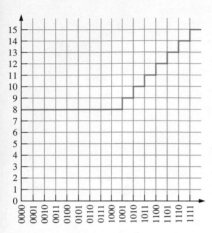

FIGURE 12–52

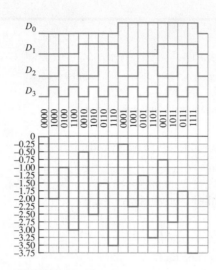

FIGURE 12–53

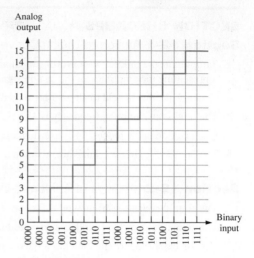

FIGURE 12–54

TRUE/FALSE QUIZ

1. T   **2.** F   **3.** F   **4.** T   **5.** T   **6.** F   **7.** T   **8.** T   **9.** T   **10.** F

SELF-TEST

1. (b)   **2.** (d)   **3.** (a)   **4.** (b)   **5.** (d)   **6.** (a)

7. (e)   **8.** (b)   **9.** (e)   **10.** (c)   **11.** (b)   **12.** (c)

13. (a)   **14.** (c)   **15.** (a)   **16.** (c)   **17.** (a)   **18.** (d)

Data Transmission

CHAPTER OUTLINE

CHAPTER OBJECTIVES

- Discuss various types of data transmission media
- Describe the methods for data transmission
- Explain data transmission modes
- Define *modulation*
- Describe the types of modulation for transmission of digital data
- Explain how digital signals are used to transmit analog information
- Define *multiplexing* and *demultiplexing*
- Discuss the types of multiplexing and demultiplexing
- Discuss the types of buses
- List bus characteristics
- Explain bus protocols
- Explain the multiplexed bus and tri-state outputs
- Discuss the PCI series, AGP, ISA, IEEE-488, and SCSI types of parallel buses
- Describe the Universal Serial Bus (USB)
- Discuss the RS-232/422/423/485, SPI, I²C, CAN, Firewire, and SCSI types of serial buses

KEY TERMS

Key terms are in order of appearance in the chapter.

- Coaxial cable
- EMI
- Optical fiber
- Electromagnetic waves
- Bit rate
- Baud
- RZ
- NRZ
- Manchester encoding
- Simplex
- Half-duplex
- Full-duplex
- Modulation
- PAM
- Bus
- Bus protocol
- Handshake
- GPIB
- SCSI
- Universal serial bus (USB)
- RS-232 bus
- Tri-state buffers

VISIT THE WEBSITE

Study aids for this chapter are available at
www.pearsonhighered.com/careersresources/

INTRODUCTION

Data transmission is the transfer of data over a medium or channel from one point to one or more points. Channels can be wire, cable, optical fibers, or wireless. Information (data) transmission falls into two basic categories, analog and digital. As you know, analog information is transmitted as a continuously varying signal. Digital data is transmitted either as a sequence of pulses, called *baseband* transmission, or by modulating an analog signal, called *broadband* transmission. Both types are generally considered digital transmission. A modem is used in broadband modulation and demodulation. Buses are an important part of many data transmission systems and are covered in this chapter.

13–1   Data Transmission Media

All data transmission systems in their most basic form have a data source (sending device) at one end and a receiving device at the other. The two devices are connected by a transmission medium, which can be wire, coaxial cable, twisted pair cable, optical fiber cable, or space (wireless). A digital signal is a changing electrical or electromagnetic quantity that carries information through the medium. When data are sent without modulation, usually over wires or cables, it is called *baseband transmission*. When data are modulated and sent through a wireless medium, it is called *broadband transmission*.

After completing this section, you should be able to

- Define a data transmission system

- Describe wire and cable transmission media

- Describe optical fiber transmission media

- Describe wireless transmission

InfoNote

Early fundamental work in data transmission and information theory was done by Harry Nyquist, Ralph Hartley, Claude Shannon, and others.

The basic block diagram in Figure 13–1 illustrates the essential elements in a data transmission system.

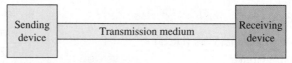

FIGURE 13–1   Basic data transmission system.

Wire Connections

The simplest connection between sending and receiving devices is a wire or a conductive trace on a printed circuit board (PCB). This type of connection is typically limited to internal data transmission over very short distances within the same system or between nearby systems, such as a computer and/or peripherals. Data buses or conductive traces connect one element to another on a PCB and between PCBs in close proximity or between parts of a system, as illustrated in Figure 13–2.

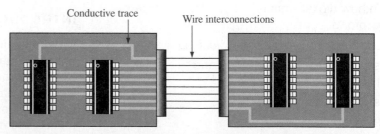

FIGURE 13–2   Conductive traces on PCBs and wire interconnections between boards.

Coaxial Cable

Coaxial cable (coax) consists of a center conductor within an insulating dielectric material. A copper braided or foil shield surrounds the dielectric to protect the conductor against electromagnetic interference (**EMI**). The shield is encased in a protective insulating jacket, as shown in Figure 13–3. BNC (Bayonet Neill-Concelman) connectors are typically used for coaxial connections. Coax is used in data transmission applications with data rates up to about 1 GHz. Two common applications for coax are cable TV and Internet connections.

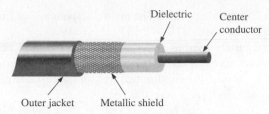

FIGURE 13–3   Construction view of a coaxial cable.

Twisted Pair Cable

Unshielded twisted pair (**UTP**) cable is used extensively for indoor telephone application as well as some outdoor uses. It is found in many computer networks and video applications, such as security cameras, and also in the USB (universal serial bus) cable. UTP is color-coded according to a standard 25-pair color code. Most cables use a subset of these standard colors.

Cross talk, a type of distortion, is minimized when twisted pairs are bundled together. The two wires in each pair are twisted so that they cross each other at nearly 90°, ideally cancelling any electromagnetic fields generated by the signals in the wires. UTP cables are limited to use in low-noise environments and to lower signal frequencies than coax, such as audio and other signals up to about 1 MHz. UTP cables use standard RJ-45 connectors. A common four-pair UTP cable is shown in Figure 13–4(a), and an RJ-45 connector is shown in part (b). Shielded twisted pair (**STP**) cable encased in a metal sleeve or conduit is also available and provides more protection from EMI.

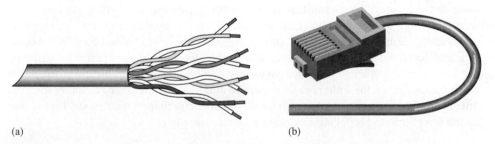

(a)   (b)

FIGURE 13–4   Example of an unshielded twisted pair (UTP) cable and connector.

Optical Fiber Cable

The structure of a single **optical fiber** is shown in Figure 13–5(a). An optical fiber can be as small as a human hair, so many single fibers can be bundled into a cable, as shown in Figure 13–5(b).

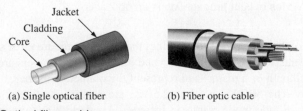

(a) Single optical fiber   (b) Fiber optic cable

FIGURE 13–5   Optical fiber cables.

Instead of using electrical pulses to transmit information through copper lines, fiber optics uses light pulses transmitted through optical fibers. Fiber-optic systems have several advantages over electrical transmission media. Advantages include faster data rates, higher signal capacity (more signals at a time), and better transmission over longer distances;

optical fibers are not susceptible to EMI. The main disadvantage of fiber optics is the cost, which is higher than that of coax, UTP, and STP.

Optical fiber is commonly used as a medium for telecommunication and networking. Because light propagates through the fiber with little attenuation compared to electrical cables, optical fiber is useful for long-distance transmission. Data rates from 10 GHz to 40 GHz are common, although rates over 100 GHz are used.

When light is introduced at one end of an optical fiber called the *core*, it "bounces" along until it emerges from the other end, as shown in Figure 13–6. The fiber is typically made of pure glass, plastic, or other material that is surrounded by a highly reflective cladding that effectively acts as a mirrored surface, using a phenomenon called *total internal reflection* to produce an almost lossless reflection. This allows the light to move around bends in the fiber.

FIGURE 13–6   Light propagating through an optical fiber while reflecting off the internal surface.

Modes of Light Propagation

Two basic modes of light propagation in optical fibers are multimode and single-mode, as illustrated in Figure 13–7. In **multimode**, the light entering the fiber will tend to propagate through the core in multiple rays (modes), basically due to varying angles as each light ray moves along. Some of the rays will go straight down the core, while others will bounce back and forth. Still others will scatter due to the sharp angle at which they strike the cladding, resulting in attenuation in light energy. Multimode also exhibits time dispersion, which means that all the light rays do not arrive at the end of the fiber at exactly the same time. In **single-mode**, the core is much smaller in diameter than in multimode. Light entering the fiber tends to propagate in a straight line as a single ray.

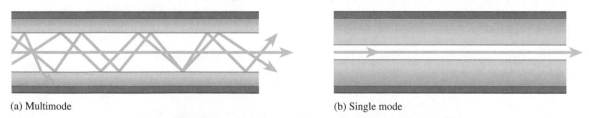

(a) Multimode   (b) Single mode

FIGURE 13–7   Modes of light propagation in an optical fiber.

The diameter of the optical fiber determines the mode. There are three sizes most widely used in data transmission: 50/125, 62.5/125 and 8.3/125. The numbers are in microns (one micron is one millionth of a meter) and represent the diameters of the fiber core and cladding, respectively. The 50/125 and the 62.5/125 are multimode fibers. The 8.3/125 is a single-mode fiber. Single mode results in increased bandwidth and distance for transmission, but the costs are higher than for multimode.

A Fiber-Optic Data Communications Link

A simplified block diagram of a fiber-optic communications link is shown in Figure 13–8. The source provides the electrical signal that is to be transmitted. The electrical signal is

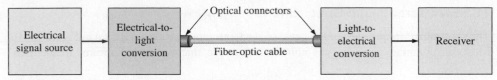

FIGURE 13–8   Basic block diagram of a fiber-optic communications link.

converted to a light signal and coupled to the fiber-optic cable. At the receiving end, the light signal is coupled out of the cable into the receiver, which converts it back to an electrical signal. The signal is then processed and sent to the end user. The electrical signal modulates the light intensity and produces a light signal that carries the same information as the electrical signal. Special connectors are used to connect the fiber-optic cable to various equipment.

Various types of connectors are used in fiber-optic systems. Some of these are described as follows and are shown in Figure 13–9:

- **ST** An AT&T trademark and is one of the most widely used for multimode networks
- **SC** A snap-in type multimode connector
- **FC** A popular single-mode connector type
- **LC** A single-mode connector
- **LX-5** Similar to an LC connector except it has a shutter over the end of the fiber
- **MT** A 12-fiber connector used for ribbon cables
- **FDDI** All duplex, meaning the connector can accommodate two optical fibers for two-way communication

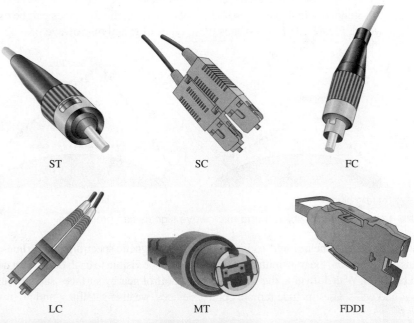

FIGURE 13–9   Typical types of optical fiber connectors.

Wireless Transmission

The transmission of data through air and space via electromagnetic waves without the use of physical connections between sending and receiving systems is known as *wireless transmission*. Wireless transmission generally can be categorized by the type of signal in terms of application, frequency, or how the data are configured. Another medium of wireless communications is water where sonar is used. Sonar produces low-frequency sound waves that do not fall into the electromagnetic spectrum.

The Electromagnetic Spectrum

The spectrum of frequencies for the range of **electromagnetic waves** is shown in Figure 13–10. Most data communications occur within the radio wave, microwave, and infrared frequencies.

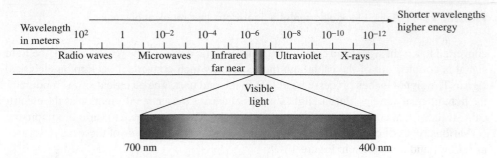

FIGURE 13–10   The electromagnetic spectrum.

Three ways in which radio wave (rf) and microwave signals propagate through Earth's atmosphere (air) are ground wave, ionospheric, and line-of-sight. In ground wave propagation, the radio waves follow the curvature of Earth and can be up to about 2 MHz in frequency; the standard AM broadcast band is an example. Radio frequencies in the 30 MHz to 85 MHz range bounce off of the ionosphere. These signals can change with time of day and weather conditions. Most ham radio bands are examples of where signals bounce off the ionosphere. In line-of-sight (LOS) propagation, the receiver must be in view of the transmitter. The distance is limited to about 100 km (horizon to horizon) from a ground-based transmitter to a ground-based receiver. Long distances are achieved by placing a series of repeater towers so that each tower is within the line-of-sight of the previous tower. In the case of satellites, which use line-of-sight propagation, the distances can be extended around the world. Figure 13–11 illustrates these types of rf and microwave propagation.

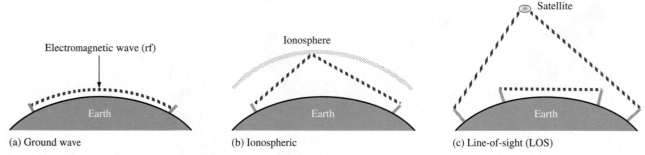

FIGURE 13–11   Ways in which rf and microwave signals can propagate.

Communication in the infrared region of the electromagnetic spectrum can be line-of-sight or diffused. With LOS, the transmitter and receiver must be visible to each other with no obstacles in between. With diffusion, the IR waves reflect off of nearby surfaces such as buildings, ceilings, and walls. Uses include remote control devices, weather satellites, and night vision.

SECTION 13–1 CHECKUP

Answers are at the end of the chapter.

1. List the types of data transmission media.
2. What is the purpose of a coax shield?
3. Name three ways in which radio waves are propagated in wireless transmission.
4. Which type of electromagnetic radiation has the highest frequencies?
5. Generally, what is the difference between baseband and broadband?

13–2 Methods and Modes of Data Transmission

Data transmission over a communications channel can be configured in several ways. A communication channel is the pathway over which data are sent and can be in the form of any of the media discussed in Section 13–1. The methods by which data are transmitted can be one bit at a time (serial) or several bits at a time (parallel), and the data can be either synchronized or unsynchronized. The modes describe the direction of the data.

After completing this section, you should be able to

- Distinguish between serial and parallel data
- Distinguish between synchronous and asynchronous data
- Describe the three modes of data transmission

Serial and Parallel Data

Serial data transmission is when data are transmitted one bit at a time in a bit stream, as illustrated in Figure 13–12(a). Parallel data transmission is when data are transmitted several bits at a time, as shown in part (b). In general, a given number of bits can be transmitted faster in parallel than in series, resulting in higher data rates. However, when several bits are sent simultaneously on separate lines in parallel, slight differences in the properties of the lines can cause skewing in the data, making the data more susceptible to error, so the data rate may need to be reduced to prevent errors. Error detection and correction methods can be used in these cases.

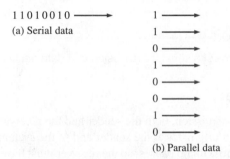

FIGURE 13–12

Generally, data are processed in parallel by computers but are transmitted serially to outside systems. For example, data from a computer to a printer are typically sent over a USB, which is serial. Data that are sent over long distances via one of the transmission media are typically in a serial format. In some cases, data can be sent in parallel over a single channel by using different frequencies for each bit, so the bits can be transmitted at the same time.

Serial-to-parallel and parallel-to-serial conversions are used in most data transmission systems. The basic concept is shown in Figure 13–13. A simple parallel-to-serial converter is a parallel in/serial out shift register. The parallel bits are loaded into the register and then shifted out in serial format at a specific clock frequency that determines the data rate in bits

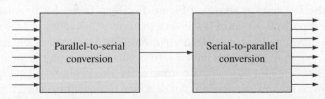

FIGURE 13–13 Digital data conversions.

per second (bps). Likewise, a serial in/parallel out shift register can be used as a serial-to-parallel converter. The receiving device must be able to recognize each valid group of data bits that it is receiving serially. Two types of data transmission in terms of how a receiving device knows what a sending device is transmitting are asynchronous and synchronous.

Asynchronous Data

Data are sent in short "bursts" known as packets in asynchronous transmission. A data **packet** is one complete piece of information of a longer message. Typically, many packets make up the entire message. A data packet consists of data bits representing alphabetic or numeric characters, a parity bit, and start/stop bits. There is a pause between data packets so that the receiver recognizes the start bit that precedes each packet. At the end of the data packet, there are one or more stop bits that tell the receiver the packet is complete.

In **asynchronous** systems, the sending and receiving devices operate with separate oscillators having the same clock frequency. Because the separate clock frequencies may drift over time, they are typically re-synchronized on each data packet with the start bit. Most commonly, data are sent in small packets of perhaps 10 or 11 bits. Eight of these bits carry the information. Between packets, when the channel is idle, there is a continuous logic level. A data packet always begins with a start bit with the opposite logic level as the idle period to alert the receiver that a data packet is starting. A parity bit follows the eight data bits, and a stop bit signals the end of the packet. This is illustrated in Figure 13–14.

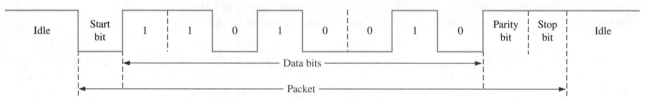

FIGURE 13–14   Example of a serial transmission of a data packet for a given data code.

Synchronous Data

In **synchronous** data transmission, both the sender and the receiver derive timing from the same clock signal, which originates at the sender end of the system. The bits are transmitted in a continuous stream with no pauses, so the receiver must have some way to recognize where a data block starts and ends. In order for the receiver to know when to read information bits from the channel, it must determine exactly when the data begin and the time between bits. When this timing information is determined, the receiver is synchronized with the transmitter. Unlike asynchronous transmission, the data blocks usually contain more than one character of information. Synchronous transmission is generally faster than asynchronous transmission.

One method of synchronization is by using separate channels to transmit the data and the timing information (synchronization and clock pulses). Because the transmitter originates both the data and the timing pulses, the receiver will read the data channel only when told to do so by the transmitter (via the timing channel), and synchronization is achieved. The disadvantage of this method is that it requires two physical lines.

Two data formats that require separate data and timing are **RZ** (return to zero) and **NRZ** (nonreturn to zero). In the RZ format, a single pulse during a bit time represents a 1 and the absence of a pulse is a zero, as shown in Figure 13–15(a). In the NRZ format, a high level during a bit time represents a 1 and a low level represents a 0. A series of 1s is represented by a continuous high level, and a series of 0s is represented by a continuous low level. The waveform does not return to the low level until a zero occurs after a string of 1s and does not go back to the high level until a 1 occurs after a string of 0s. This is illustrated in Figure 13–15(b). Another more commonly used method of data synchronization, called biphase or **Manchester encoding**, is to embed the timing signal in the data at the transmitter so that only

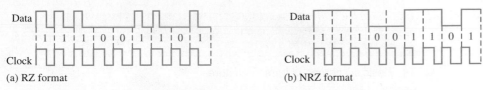

FIGURE 13-15   Data formats that require separate timing for synchronization.

one channel is required. The receiver extracts the embedded timing signal and uses it to synchronize to the transmitter.

Figure 13–16 illustrates Manchester encoding. A rising edge in the biphase code is a 1 and a falling edge is a 0, as indicated by the up and down arrowheads. The edges occur at the middle of the bit time. The biphase code is sent to the receiver, and the clock is extracted from the data with a phase-locked loop. Sometimes a series of all 1s or all 0s are included in the transmission to allow the receiver to synchronize.

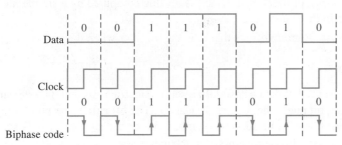

FIGURE 13-16   Example of Manchester encoded data and timing.

EXAMPLE 13-1

Determine the biphase (Manchester) code for the data and clock shown in Figure 13–17(a).

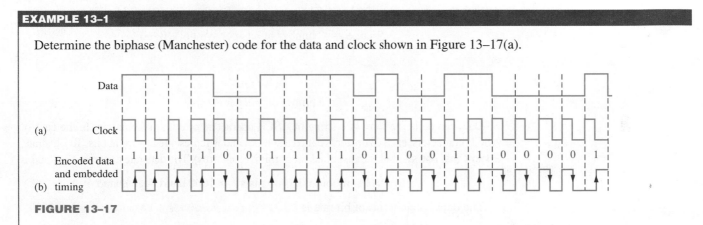

FIGURE 13-17

Solution

The encoded data and embedded timing are shown in Figure 13–17(b). As the arrowheads indicate, the rising edges are 1s and the falling edges are 0s that occur in the middle of each bit time (period of the clock).

Related Problem*

If the data were all 1s, what would the Manchester code look like?

*Answers are at the end of the chapter.

Synchronous Frames

Synchronous data are sent in frames that include other bits, as shown by the generic protocol in Figure 13–18. (Frame formats vary because there are numerous standards in use.)

- **Preamble**   A group of bits at the beginning of a frame that is used to alert the receiver that a new frame has arrived and to synchronize the receiver's clock with the transmitted clock

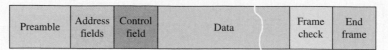

FIGURE 13–18 Basic synchronous frame structure.

- **Address fields** A group of bits containing the address(s) of the sender and the receiver. One or both addresses may be present in a given protocol.
- **Control field** This group of bits identifies the type of data being sent, such as handshaking (establishes a connection), file transfers, and the size of the data.
- **Data field** This sequence is the actual information being sent and can be of a fixed length or a variable length. If it is a fixed-length field, a group of bits called a *pad* is used to fill in if the actual data field is less than the fixed field.
- **Frame check** This field contains an error check such as parity, CRC (cyclic redundancy check), or checksum, which is a value computed by a simple algorithm of the data bits in the frame.
- **End frame** A group of bits that tells the receiver when the end of the frame occurs.

Data Rate

Data rate is the speed of data transfer. In a serial data transmission the rate can be stated as bit rate or baud; bit rate is the preferred measure. The **bit rate** is the number of bits (1s and 0s) per second (bps); the **baud** is the symbol rate or the number of data symbols (sometimes known as transitions or events) per second.

A symbol (transition) can consist of one or more bits. Therefore, bit rate is always greater than or equal to the baud. The relationship between bit rate and baud is

$$\text{Bit rate} = (\text{Number of bits per symbol})(\text{Baud})$$

or

$$\text{Baud} = \frac{\text{Bit rate}}{\text{Number of bits per symbol}}$$

Data can be in the form of a string of ASCII characters or other information. In the case of ASCII characters, each character is called a symbol and is represented by eight bits. To illustrate, assume that one symbol is transmitted every millisecond (ms). The data rate expressed as baud is

$$\text{Baud} = (1 \text{ symbol/ms})(1000 \text{ ms/s}) = 1000 \text{ baud} = 1 \text{ kbaud}$$

The data rate in terms of bit rate is

$$\text{Bit rate} = (8 \text{ bits/symbol})(1000 \text{ symbols/s}) = 8000 \text{ bps} = 8 \text{ kbps}$$

EXAMPLE 13–2

A certain analog waveform is represented by sixteen-voltage levels that are being transmitted. Each level (symbol) is represented by a 4-bit code. Assuming that eight symbols are transmitted in 1 μs, express the data rate as bit rate and as baud.

Solution

$$\text{Bit rate} = (4 \text{ bits/symbol})(8 \text{ symbols/}\mu\text{s}) = 32 \text{ bits/}\mu\text{s} = \textbf{32 Mbps}$$

$$\text{Baud} = \frac{32 \text{ Mbps}}{4 \text{ bits per symbol}} = \textbf{8 Mbaud}$$

Related Problem

Determine the bit rate if a symbol is represented by 8 bits and the baud is 5000 symbols/s.

Transmission Efficiency

The efficiency of a data transmission channel is the ratio of data bits to total bits in a packet. For example, in Figure 13–14 there are eight data bits, a start bit, a parity bit, and a stop bit. The nondata bits are considered overhead bits. There are eleven total bits transmitted in a packet so the efficiency of the transmission is

$$\text{Efficiency} = \frac{\text{Data bits}}{\text{Total bits}} = \frac{8 \text{ bits}}{11 \text{ bits}} = 0.727 \text{ or } 72.7\%$$

EXAMPLE 13–3

A certain system transmits a block of information containing ten packets each with eight data bits, a start bit, and a stop bit. Additional "overhead" bits include a 4-bit synchronization code at the beginning of the block and a parity bit at the end of the block. Determine the transmission efficiency.

Solution

$$\text{Data bits} = (8 \text{ data bits})(10 \text{ packets}) = 80 \text{ bits}$$

$$\text{Overhead bits} = (1)(10 \text{ start bits}) + (1)(10 \text{ stop bits}) + 4 \text{ synchronization bits} + 1 \text{ parity bit} = 25 \text{ bits}$$

$$\text{Total bits} = \text{Data bits} + \text{Overhead bits} = 80 + 25 = 105$$

$$\text{Efficiency} = \frac{\text{Data bits}}{\text{Total bits}} = \frac{80}{105} = \mathbf{0.762 \text{ or } 76.2\%}$$

Related Problem

Determine the efficiency if each packet has 12 data bits and the same number of overhead bits as stated in the example.

Transmission Modes

Three modes that characterize data channel (media) connections are simplex, half-duplex, and full-duplex. In the **simplex** mode, data flow in only one direction from the sender (transmitter) to the receiver. In a computer, for example, data flow one way from the computer to the printer. In the **half-duplex** mode, the data flow both ways but not at the same time in the same channel. For example, a sender transmits information to the receiver and the receiver responds back to the sender after it has received the information. In the **full-duplex** mode, the data flow both ways simultaneously in the same channel. The bandwidth of the channel is divided between the two directions. Figure 13–19 illustrates these three modes.

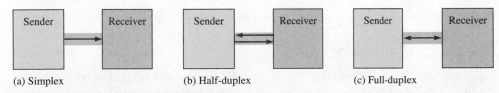

(a) Simplex   (b) Half-duplex   (c) Full-duplex

FIGURE 13–19   Data transmission modes.

SECTION 13–2 CHECKUP

1. Explain the difference between serial and parallel data.

2. What is the purpose of synchronization in a data transmission system?

3. Name three types of data formats.

4. List the modes of data transmission?

13–3 Modulation of Analog Signals with Digital Data

Three major classes of modulation techniques for wireless transmission of digital data are amplitude-shift keying (ASK), frequency-shift keying (FSK), and phase-shift keying (PSK). Another common modulation method is basically a combination of ASK and FSK and is known as quadrature amplitude modulation (QAM). Several variations of these main modulation techniques are also in use. All of these techniques convey information by changing a property of an rf carrier signal, usually a sine wave for the purpose of conveying digital information over a wireless medium.

After completing this section, you should be able to

- Describe ASK
- Describe FSK
- Describe PSK
- Describe QAM

Amplitude-Shift Keying

Amplitude-shift keying (**ASK**) is a form of **modulation** in which a digital signal varies the amplitude of a higher frequency sine wave (carrier). In its simplest form, a sinusoidal carrier signal is turned on and off by the data signal and, therefore, this method is also known as on-off keying (OOK). When the carrier is *on*, a binary 1 is represented, and when the carrier is *off*, a binary 0 is represented. ASK is very susceptible to noise interference and is not typically used for wireless data transmission. ASK is most commonly used in fiber optics where the presence of light represents a binary 1 and the absence of light represents a binary 0. Figure 13–20 illustrates the concept of ASK. The presence of the sine wave carrier is a 1 and the absence is a 0. When modulated by digital data (1s and 0s), this method is sometimes known as binary amplitude-shift keying (BASK).

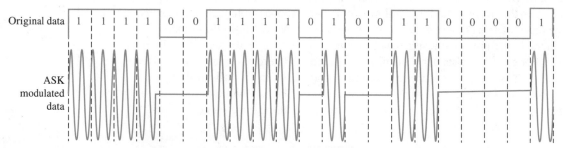

FIGURE 13–20   Illustration of amplitude-shift keying (ASK).

Frequency-Shift Keying

Frequency-shift keying (**FSK**) is a form of modulation in which a digital signal modulates the frequency of a higher frequency sine wave (carrier). A carrier signal with a lower frequency generally represents a binary 0, and a carrier signal with a higher frequency represents a binary 1. When modulated by digital data (1s and 0s), this method is sometimes known as binary frequency-shift keying (BFSK). Figure 13–21 illustrates FSK.

Phase-Shift Keying

Phase-shift keying (**PSK**) is a form of modulation in which a digital signal modulates the phase of a higher frequency sine wave. A carrier signal of one phase generally represents a binary 1, and a carrier signal that is 180° out-of-phase represents a binary 0. When

FIGURE 13–21   Illustration of frequency-shift keying (FSK).

modulated by digital data (1s and 0s), this method is sometimes known as binary phase-shift keying (BPSK). In one of its many variations, PSK applications include wireless LAN and bluetooth. Figure 13–22 illustrates PSK.

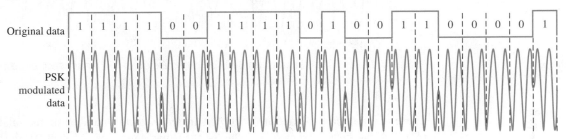

FIGURE 13–22   Illustration of phase-shift keying (PSK).

Quadrature Amplitude Modulation

Quadrature amplitude modulation **(QAM)** is widely used in telecommunications and in digital cable TV. Digital QAM uses a combination of PSK and ASK to send information. Quadrature refers to a 90° phase difference. Each combination of phase and amplitude is called a *modulation state* or *symbol* and represents a combination of two or more bits. To illustrate the basic concept of QAM, let's use what is known as 8-QAM where each of the eight modulation states (2^3) represents a unique three-bit combination. As shown in Figure 13–23, in 8-QAM,

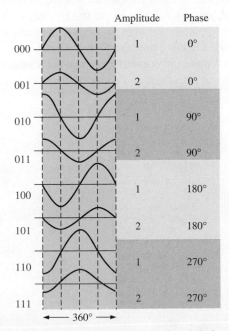

FIGURE 13–23   Eight amplitude/phase combinations (modulation states) represent one of the eight 3-bit groups. Only one cycle of each modulation state is shown.

there are two different amplitudes with a quadrature phase difference between each pair. Since there are four quadratures (90°) in 360°, there are a total of eight different amplitude/phase combinations or modulation states.

A digital 8-QAM data transmission can represent any combination of three bits in any sequence. Figure 13–24 illustrates an 8-QAM transmission of the binary sequence of numbers 0 through 7. Depending on the carrier frequency and the time specified for each bit group, multiple cycles of each modulation state can represent each bit group. For simplicity, only two cycles per bit group are shown.

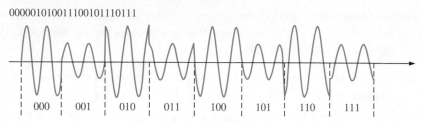

FIGURE 13–24   Illustration of an 8-QAM transmission of the binary sequence shown.

M-QAM

There are numerous variations in QAM in terms of the number of modulation states (M) that can be represented. For the 8-QAM just illustrated, $M = 8$. Higher M values of QAM, such as 16-QAM, 64-QAM, and 256-QAM, are also commonly used. These higher M values are achieved by using more amplitude levels and/or phases. For example, a 64-QAM can have 16 amplitude levels and four phases and can represent 6-bit binary groups. A 256-QAM can have 32 amplitude levels and eight phases and can represent 8-bit binary groups.

Constellation Maps

Modulated transmission of digital data can be represented by a constellation map, which is a vector representation that graphically shows the symbol values and corresponding phases being transmitted by a system. As you have seen, when data are transmitted, a pattern is modulated into the signal, such as in PSK, where the bit pattern is represented by various phase shifts. A constellation map is useful in the design and analysis of a data transmission system and in visually understanding how the system works.

Figure 13–25 shows a 4-quadrant constellation map for a 3-bit PSK transmission. Each green dot represents the ideal amplitude and phase of the modulated signal. The amplitude

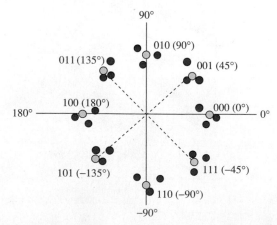

FIGURE 13–25   Constellation map for a 3-bit PSK transmission. The phases are 0°, 45°, 90°, 135°, 180°, −45°, −90°, and −135°, as indicated.

constant is represented by the distance of the dots from the origin. Eight different phases or symbols represent the 3-bit binary combinations.

In an actual transmission, the medium can affect both amplitude and phase shift. In the figure, the cluster of red dots around each green dot represents nonideal signal values. When the signal is received, these nonideal values can be adjusted to the ideal value (nearest green dot) as long as there is adequate separation of the clusters; there should be no confusion as to which ideal value the signal belongs. If there is any overlap, errors can occur.

EXAMPLE 13–4

Develop an ideal constellation map for the 8-QAM transmission represented in Figures 13–23 and 13–24.

Solution

There are four phases and four amplitudes that represent a 3-bit code in this system. The ideal constellation map is shown in Figure 13–26. There are two amplitudes for each phase represented by the distance from the origin.

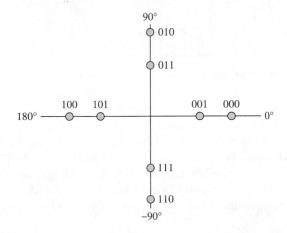

FIGURE 13–26

Related Problem

How many phases and amplitudes could be used to represent a 4-bit code?

SECTION 13–3 CHECKUP

1. List four types of modulation techniques.

2. What parameter is changed in ASK?

3. What parameter is changed in FSK?

4. What is QAM?

5. What parameter is changed in PSK?

13–4 Modulation of Digital Signals with Analog Data

As you learned in the last section, analog signals are commonly used to carry digital data. In this section, you will see that digital signals can be used to carry analog information. These techniques are usually referred to as pulse modulation. A pulse parameter such as amplitude or pulse width is varied to represent an analog quantity.

After completing this section, you should be able to

- ◆ Describe pulse amplitude modulation
- ◆ Describe pulse position modulation
- ◆ Discuss pulse code modulation
- ◆ Explain delta modulation

Pulse Amplitude Modulation

In pulse amplitude modulation (**PAM**), the heights or amplitudes of the pulses are varied according to the modulating analog signal; each pulse represents a value of the analog signal. PAM is the simplest, but least used, type of pulse modulation although it is used in the Ethernet communications standard. A simple PAM sequence is shown in Figure 13–27.

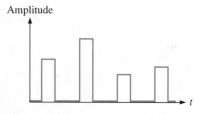

FIGURE 13–27   A simple PAM signal.

A basic method of producing a PAM representation of an analog signal is to use a constant-amplitude pulse source to sample the analog wave that has a frequency lower than the pulses, as shown in Figure 13–28 for a sine wave input; any form of analog signal can be converted to a PAM output. The pulses turn the switch on (closed) and off (open) to sample the waveform. When there is a pulse, the sample switch is closed; the amplitude of the sine wave at that point goes to the hold element that maintains the initial analog value occurring at the beginning of each pulse for the duration of the pulse. The output goes to zero between pulses.

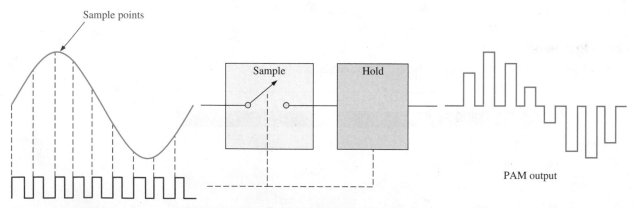

FIGURE 13–28   Basic method of pulse amplitude modulation.

Pulse Width Modulation

In pulse width modulation (**PWM**), the width or duration of the pulses and duty cycle are varied according to the modulating analog signal; each pulse represents a value of the analog signal. PWM (also known as *pulse duration modulation*, PDM), is commonly used in control applications. Braking systems, motor speed control, and renewable energy systems are just three examples.

Figure 13–29 illustrates one method of PWM generation, called the *intersective method*, which uses a sawtooth waveform. A triangular waveform can also be used. Again, a sine wave input is used, but the input can be any type of analog waveform. The sawtooth inter-

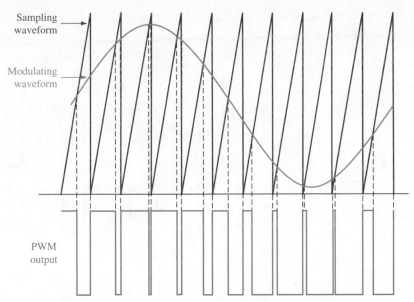

FIGURE 13–29   Illustration of PWM.

sects the sinusoidal modulating signal twice during each cycle. The sawtooth is either increasing above the sine wave or decreasing below the sine wave. When the sawtooth is increasing above the sine wave, a low level is generated; when it is decreasing below the sine wave, a high level is generated. The resulting output is a series of pulses with widths proportional to the amplitude of the sine wave.

An intersective PWM system can be implemented simply with a sawtooth or triangular wave generator and a comparator, as shown in Figure 13–30.

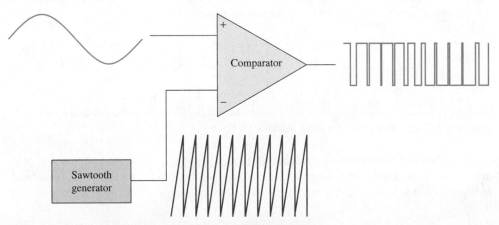

FIGURE 13–30   A basic method of pulse width modulation.

EXAMPLE 13–5

Determine the PAM signal and the PWM signal for the modulating signal in Figure 13–31. Assume ten cycles of the sampling pulse waveform or sawtooth waveform during the portion of the modulating signal shown.

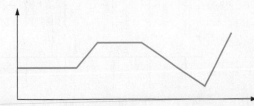

FIGURE 13–31   Modulating waveform.

Solution

Figure 13–32 shows the PAM and PWM results.

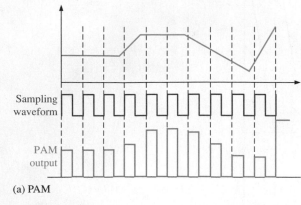

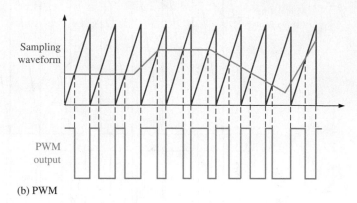

(a) PAM   (b) PWM

FIGURE 13–32

Related Problem

How would the outputs in Figure 13–32 be affected by an increase in frequency of the pulse and sawtooth waveforms?

Pulse Position Modulation

In pulse position modulation (**PPM**), also known as pulse phase modulation, the position of each pulse relative to a reference or timing signal is varied proportional to the modulating signal waveform. The amplitude and width of the pulses in a PPM system are kept constant. An example of a PPM signal is shown in Figure 13–33 where the PPM pulses are shifted relative to the leading edges of the timing waveform.

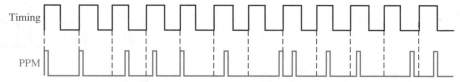

FIGURE 13–33   Example of a PPM signal with timing.

As with other types of pulse modulation, there is generally more than one way to produce a modulated waveform. One method is to derive the PPM from PWM, as illustrated in Figure 13–34. Notice that the leading (positive-going) edges of the PWM signal occur at

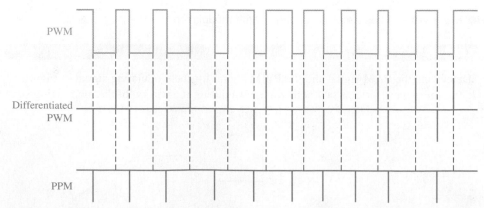

FIGURE 13–34   A method of generating PPM.

fixed intervals, while the trailing (negative-going) edges vary relative to the leading edges. When the PWM signal is passed through a differentiator, short positive pulses (spikes) are generated on the leading edges and short negative pulses occur on the trailing edges. The differentiated signal is rectified to remove the positive pulses and generate the PPM waveform shown, which can be inverted to produce positive pulses. A simplified block diagram of a PPM system is shown in Figure 13–35.

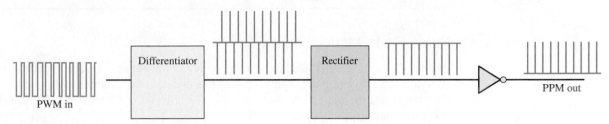

FIGURE 13–35   PPM system block diagram.

PPM is widely used in optical communications, such as fiber optics and in certain types of rf systems, such as radio control for model planes, boats, and cars. It is less sensitive to channel interference than PAM or PWM because noise can alter pulse amplitude and width but not so much the position.

PPM Encoding

A certain number of data bits (D) are encoded by a single pulse in one of 2^D possible positions during a specified fixed time period (T). The data rate is D/T bits per second (bps). Figure 13–36 illustrates the case of four time periods and two data bits per time period. There are $2^D = 2^2 = 4$ possible positions in each time period. As you can see, each position represents a 2-bit binary number. In the first time period, the pulse position represents 00, in the second time period the pulse position represents 10, et cetera. Any pulse could be in any of the four positions within each period, depending on the data being encoded. The code for this particular set of pulse positions is 00100111; it is shown in NRZ format in the figure.

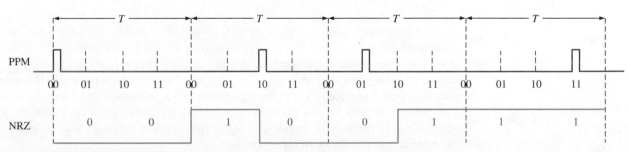

FIGURE 13–36   Encoding of a PPM signal.

EXAMPLE 13–6

For a PPM system with four data bits and a time period of 1 μs, determine the data rate. How many possible pulse positions are there in each time period?

Solution

The data rate is

$$\frac{D}{T} = \frac{4}{1\ \mu s} = \textbf{4 Mbps}$$

The number of possible pulse positions in each period is

$$2^D = 2^4 = 16$$

Related Problem

For eight data bits, what is the data rate ($T = 1 \ \mu s$), and what is the number of possible pulse positions in each period?

Pulse Code Modulation

Pulse code modulation (**PCM**) involves sampling of an analog signal amplitude at regular intervals and converting the sampled values to a digital code. (Sampling was mentioned in Chapter 1, and the sample-and-hold process was covered in Chapter 12.) The concept of PCM is demonstrated in Figure 13–37.

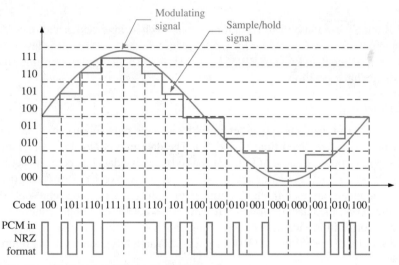

FIGURE 13–37   Concept of PCM with eight levels.

The modulating signal is a sine wave in this illustration, and its amplitude is divided into eight levels as shown. Each level is represented by a 3-bit binary number ($2^3 = 8$). The sine wave is sampled at fixed intervals; and the sampled value is held until the next sample, resulting in the green stair-step waveform that approximates the sine wave. The value at each sample is converted to a 3-bit binary number, and a pulse sequence is generated where each pulse represents a 1 and the absence of a pulse represents a 0. In this case, the PCM waveform is shown in NRZ format. The higher the sampling rate and the more levels used, the more accurate is the PCM representation.

PCM is used for digital audio in computers, Blu-ray, CD, and DVD formats. Also, it is used in digital telephone systems. A simplified block diagram of the PCM process is shown in Figure 13–38.

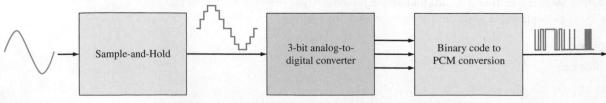

FIGURE 13–38   Block diagram of a PCM system.

EXAMPLE 13–7

In a PCM system with 32 levels, determine the number of code bits for each sample of an analog signal.

Solution

$$\text{Code bits} = 32 = 2^5$$

Five bits represent each sample.

Related Problem

What is the number of PCM code bits for each sample if the system has 64 levels?

Digital Data Systems

All digital data systems have certain common components and variations that depend on the type of data format. The three main data transmission combinations are digital-to-analog, analog-to-digital, and digital-to-digital. Figure 13–39 shows a general functional block diagram of a data transmission system. Each block is not used in all cases, depending on the type of data format and the type of communications channel.

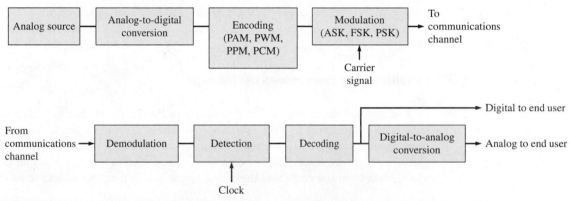

FIGURE 13–39   General function block diagram of a data transmission system.

SECTION 13–4 CHECKUP

1. List four types of pulse modulation methods.

2. What parameter is used in PAM to represent the value of the modulating signal?

3. What parameter is used in PWM to represent the value of the modulating signal?

4. What parameter is used in PPM to represent the value of the modulating signal?

5. What is used in PCM to represent the value of the modulating signal?

13–5   Multiplexing and Demultiplexing

This topic was introduced briefly in Chapter 1. Multiplexing (also known as muxing) is a method used to transmit digital data from multiple sources over a single communication channel. Multiplexing is widely used in telecommunications and computer networks. Demultiplexing (demuxing) is the process of separating data from a single channel to multiple channels. Muxing is used on the sending end of a data communication system, and demuxing is used on the receiving end. Two major types of multiplexing are time-division and frequency-division. Time-division multiplexing is a type of baseband communications. Baseband is where digital or analog signals are sent using the entire channel bandwidth. Frequency-division multiplexing is a type of broadband communications where analog signals of different frequencies are transmitted simultaneously.

After completing this section, you should be able to

*   Describe time-division multiplexing

*   Discuss frequency-division multiplexing

Time-Division Multiplexing

Time-division multiplexing (**TDM**) is a technique in which data from several sources are interleaved on a time basis and sent on a single communication channel or data link. Let's say that there are three sources of digital data to be transmitted. Certain time slots are allotted for each channel so that an element of data (bits or bytes) from source 1 is sent during time slot 1, an element of data from source 2 is sent during time slot 2, and an element of data from source 3 is sent during time slot 3. This is repeated for time slots that follow until all the data have been sent. Figure 13–40 illustrates the TDM concept.

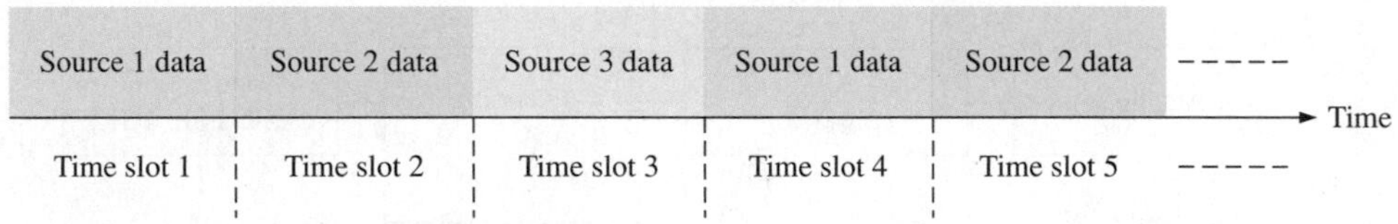

Source 1 data	Source 2 data	Source 3 data	Source 1 data	Source 2 data	- - - - -
Time slot 1	Time slot 2	Time slot 3	Time slot 4	Time slot 5	- - - - -

→ Time

FIGURE 13–40   Basic concept of TDM.

A simplified illustration of TDM is shown in Figure 13–41. Multiple data sources are switched (multiplexed) in a time sequence (t_1 through t_4) onto a single line (communications channel), and the single stream of data is switched back onto multiple lines in a synchronized time sequence. That is, data from source 1 go to the data 1 output during time slot t_1, data from source 2 go to the data 2 output during time slot t_2, and so on.

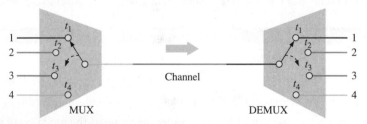

FIGURE 13–41   Simple illustration of TDM.

Bit-Interleaved TDM

In this method, a single data bit from a source is transmitted on the channel, followed by a data bit from another source, and so on. A time slot is reserved on the channel for each input source. These time slots are synchronized with the sender and receiver so that the receiver knows to which output the data bit in each time slot should go.

Bit-interleaving is demonstrated in Figure 13–42. In this case, the TDM channel data are transmitted at a rate four times greater than the data rate of the individual sources. Samples are sequentially taken of each data source (four in this case) during a bit time slot to determine if the bit is a 1 or a 0. The resulting values are sequentially placed onto the channel in 1, 2, 3, 4 order, as shown. This process is repeated for each of the bit times that follow.

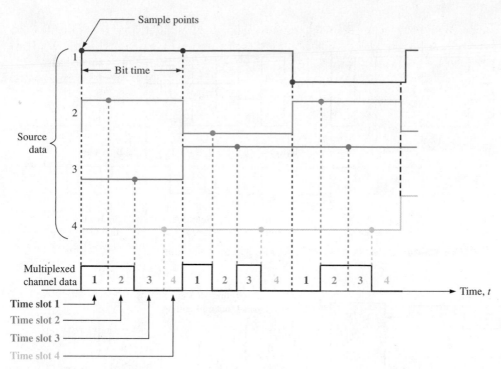

FIGURE 13–42 Illustration of TDM bit interleaving.

Byte-Interleaved TDM

In this method, bytes for each input source are sequentially placed onto the data channel. As in bit-interleaving, synchronization between the mux and demux at each end of the communications channel is required. The basic concept is shown in Figure 13–43.

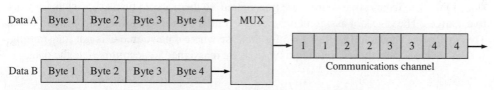

FIGURE 13–43 Basic idea of byte-interleaved TDM.

The byte-interleaved data are from two sources (in this case) and are sent at twice the rate as either source, as illustrated in Figure 13–44. As you can see in the figure, it is necessary to delay the data before multiplexing until an entire byte is complete, using a process called *buffering*. Notice that a byte of Data A and a byte of Data B occur during the Byte n interval. These two bytes are interleaved during the Byte $n + 1$ interval, so the multiplexed data are one byte delayed from the input data (A and B). This continues for each successive data byte.

Synchronous TDM

When the time slots allotted to each source are fixed, each time slot is transmitted whether or not the source has data to send. This results in an inefficient use of the communications channel because sometimes some of the time slots are empty, as illustrated in Figure 13–45. Here, data source C is not transmitting data, so its assigned time slots are blank.

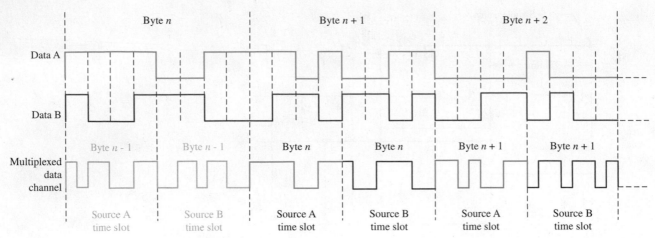

FIGURE 13–44   Byte-interleaved TDM with two data sources.

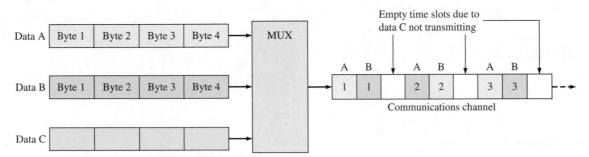

FIGURE 13–45   Example of a 3-source synchronous TDM with one data source inactive.

Statistical TDM

The statistical TDM approach improves channel efficiency by making use of all the time slots. Only data from active sources are transmitted, so there are no blank time slots for inactive sources. The time slot assignment is variable rather than fixed, as in synchronous TDM. This method is shown in Figure 13–46 for the case where data source C is not transmitting. If data source C becomes active, the time slots are reassigned to accommodate the data.

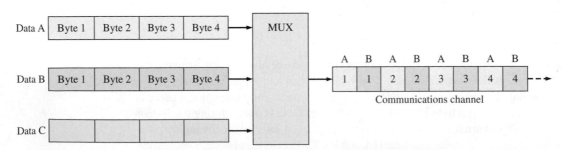

FIGURE 13–46   Example of a 3-source statistical TDM with one source inactive.

TDM is used by the telephone company in North America for nearly all voice traffic with what is known as the T1 system. A T1 line can carry 24 digitized telephone conversations and is capable of transmitting data at a rate of 1.544 Mbps. A voice signal is sampled 8,000 times per second, and each sample is converted to a byte of digital data. A voice signal requires a transmission rate of

$$\text{Voice transmission rate} = (8000 \text{ samples/s})(8 \text{ bits/sample}) = 64 \text{ kbps}$$

The number of digitized voice signals that can be multiplexed on a T1 line is

$$\text{Voice signals} = \frac{1.544 \text{ Mbps}}{64 \text{ kbps}} = 24$$

A T1 transmission over the channel consists of sequential 193-bit frames, as shown in Figure 13–47. Each frame is made up of twenty-four 8-bit slots plus one signaling bit.

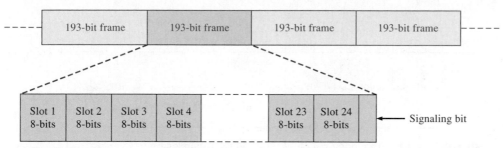

FIGURE 13–47   T1 channel transmission.

Frequency-Division Multiplexing

Frequency-division multiplexing (**FDM**) is a broadband technique in which the total bandwidth available to a system is divided into frequency sub-bands and information is sent in analog form. Each sub-band is assigned to a given source. The sources can transmit at the same time but at different frequencies. At the receiving end, the signals are demuxed using band-pass (BP) filtering. Figure 13–48 illustrates the concept of FDM.

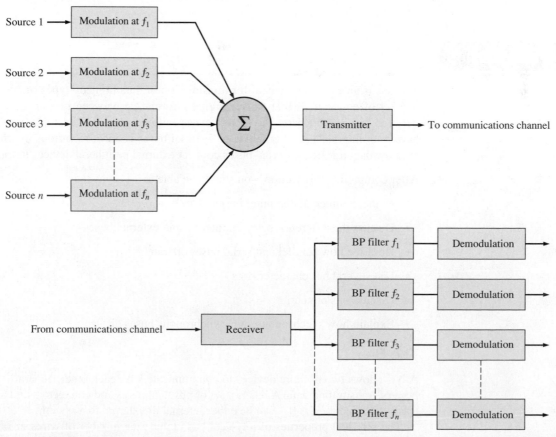

FIGURE 13–48   Basic FDM system. Σ stands for summation.

As mentioned, all sources transmit at the same time but at different frequencies. The general spectrum of a composite FDM transmission is shown in Figure 13–49. The bandwidth (BW) of each source is centered around the carrier frequency for that source and is separated from the adjacent bandwidths by the guard band.

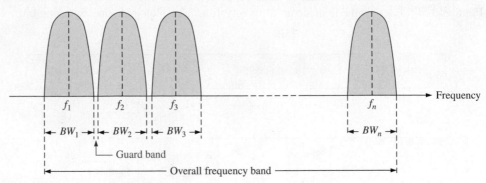

FIGURE 13–49   Frequency spectrum of an FDM transmission.

SECTION 13–5 CHECKUP

1. Discuss the reason that multiplexing is used in data communications.
2. What is TDM?
3. What is FDM?
4. Which has the higher efficiency, synchronous TDM or statistical TDM?
5. What is a guard band?

13–6   Bus Basics

The bus is an essential element for transmission of data in many types of systems. A **bus** is a set of connections that carries digital information between two or more systems or between two or more parts of a system in a specified format. For example, most computers have both internal and external buses. Internal buses connect to various internal elements to allow data transfer. External buses connect external peripheral devices to the computer.

After completing this section, you should be able to

- Name the internal computer buses
- Discuss the difference between internal and external buses
- Describe how parallel and serial buses operate
- List several bus characteristics
- Discuss bus protocol
- Explain how synchronous and asynchronous buses differ

The Bus

A bus allows two or more devices to communicate with each other, generally for the purpose of transmitting data. A bus is a set of physical wires and connectors and a set of electrical specifications. A bus can be either internal or external to a system.

The physical properties of a typical bus include the number of wires or PCB conductors (width), the configuration and length of the wires or conductors, and the types and configurations of the connectors. The electrical properties of a typical bus include but are not limited to some or all of the following: signal format, signal voltage levels, clock frequency, data transfer speed, bandwidth, data frame format, data rate, handshaking protocol, error detection, impedances, and line termination. Each device connected to a bus must be compatible with the bus specifications in order to communicate. A sending device can also be a receiving device, and a receiving device can also be a sending device. Figure 13–50 illustrates the concept of a typical bus.

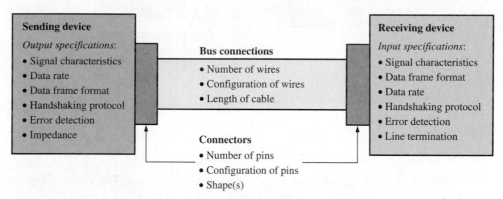

FIGURE 13–50   Physical and electrical definition of a typical bus.

Parallel and Serial Buses

Buses can be either parallel or serial. A **parallel bus** carries data bits simultaneously, and a **serial bus** carries data bits sequentially one at a time. Figure 13–51 is a simple comparison of parallel and serial buses showing eight bits being transmitted.

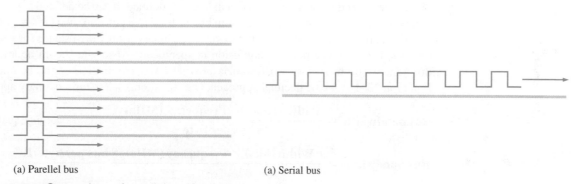

(a) Parellel bus   (a) Serial bus

FIGURE 13–51   Comparison of parallel and serial buses.

It would seem that a parallel bus would transmit data faster than a serial bus because multiple data bits can be sent simultaneously. However, this is not always the case. As data rates increase, things like crosstalk across parallel bus lines, timing skew between bus lines, and EMI (electromagnetic interference) become problems that limit the speed. Serial buses are not limited by those factors and can actually transmit data at higher rates than parallel buses in many situations.

Internal and External Buses

Internal buses carry information within a system, that is, from one part of the system to another part of the same system. External buses (also known as *expansion buses*) are used to connect one system to another separate system. For example, a computer connects to peripheral units such as a monitor, keyboard, mouse, and printer through external buses, as illustrated in Figure 13–52.

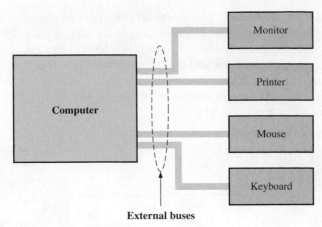

External buses

FIGURE 13–52 Example of external bus application.

General Bus Characteristics

A bus is typically described by the following parameters:

- **Width** The number of bits that a bus can transmit at one time. The width of typical buses can vary from 1 bit for a serial bus up to 64 bits for a parallel bus.
- **Frequency** The clock frequency at which a bus can operate
- **Transfer speed** The number of bytes per clock cycle
- **Bandwidth** The number of bytes per clock cycle times the number of clock cycles per second; that is, transfer speed times frequency. Bus bandwidth is sometimes called *throughput*.

Bus bandwidth can be specified in two ways, which result in slightly different values. The difference depends on how the prefix M in MBps is defined. It can be defined in decimal form as a power of ten ($10^6 = 1,000,000$) or in binary form as a power of two ($2^{20} = 1,048,576$). In the decimal form, the M stands for *mega*; in the binary form, the M stands for *mebi* (mega-binary). This can be a point of confusion in specifications, so you should be aware of the difference. The following two formulas provide for determination of the bus bandwidth. Equation 13–1 is for the decimal approach, and Equation 13–2 is for the binary approach.

$$\text{Bus bandwidth} = \frac{\text{Width (bits)} \times \text{Frequency (MHz)}}{8 \text{ bits per byte}}$$

Equation 13–1

$$\text{Bus bandwidth} = \frac{((\text{Width (bits)} \times \text{Frequency (MHz)})/8 \text{ bits per byte})10^6}{2^{20}}$$

Equation 13–2

EXAMPLE 13–8

A certain bus is specified with a width of 32 bits and a frequency of 66 MHz. Determine the bus bandwidth expressed as two different values, according to the decimal and binary definitions of M. Note that Bps is bytes per second.

Solution

Using the decimal definition of M (10^6) in the unit of MBps,

$$\text{Bandwidth} = \frac{32 \text{ bits} \times 66 \text{ MHz}}{8 \text{ bits per byte}} = \textbf{264 MBps}$$

Using the binary definition of M (2^{20}),

$$\text{Bandwidth} = \frac{((32 \text{ bits} \times 66 \text{ MHz})/8 \text{ bits per byte}) \, 10^6}{2^{20}} = \textbf{252 MBps}$$

Related Problem

What is the frequency for a bus that has a width of 64 bits and bandwidth of 125 MBps as specified in decimal form?

Table 13–1 lists some typical buses and their characteristics.

TABLE 13–1

Bus	Width (bits)	Frequency (MHz)	Transfer Speed (bytes/cycle)	Bandwidth (MBps)	
				Decimal	Binary
ISA (16 bit)	16	8.3	2	16.6	15.9
PCI	32	33	4	132	125.9
PCI-X	32	66	4	264	251.8
AGP	32	66	4	264	251.8

Bus Protocol

Bus protocol is a set of rules that allow two or more devices to communicate through a bus. Buses provide for data transfer, address selection, and control. Each device connected to a bus has an address assigned to it for identification and command signals as well as control signals to implement the protocol. These signals allow the devices to work properly together by identifying each other and communicating back and forth. One device can send a request to another device and get an acknowledgement or reply.

Handshaking

The **handshake** is a routine by which two devices initiate and complete a bus transfer. Figure 13–53 shows a simple handshake process, including a timing diagram, in which

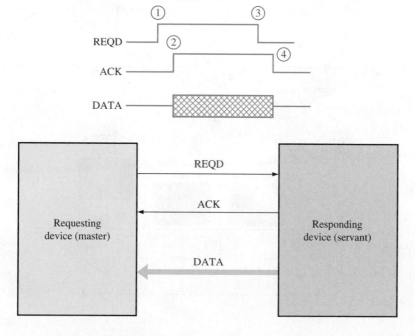

① Master sends request for data (REQD) to servant.

② Servant sends an acknowledgement (ACK) and places data on bus.

③ Master receives data and removes request.

④ Servant removes acknowledge and is ready for next request.

FIGURE 13–53 Simple example of handshake and data transfer.

a requesting device (sometimes called the *master*) and a responding device (sometimes called the *servant*) initiate and complete a data transfer.

Synchronous and Asynchronous Buses

A synchronized bus includes a clock in the control lines and has a fixed protocol that is relative to the clock. A synchronous bus is fast but has the disadvantage that every device connected to it must operate at the same clock frequency. Also, the physical length of the bus may be limited because of having to carry a high-frequency clock signal.

An asynchronous bus is not clocked, so it can serve various devices with different clock rates. The asynchronous bus uses a handshake protocol to establish communication, as previously described.

Single-Ended vs. Differential Buses

Data communications between devices can be classified as either single-ended or differential in terms of the physical bus configuration. In general, single-ended operation is limited in both data rate and distance (cable length). Differential operation provides much higher data rates and longer transmission distances. **Single-ended operation** uses one wire for data and one wire for ground, where the signal voltage on the wire is with respect to ground. **Differential operation** uses two wires for data and one wire for ground. The data signal is sent on one wire in the twisted pair and its complement (inversion) is sent on the other wire. The difference between the two data wires is the differential signal. Figure 13–54 shows both single-ended and differential operation.

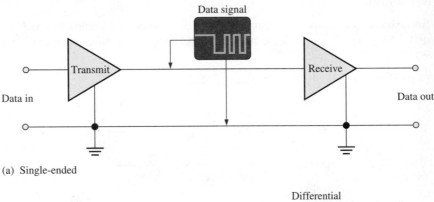

(a) Single-ended

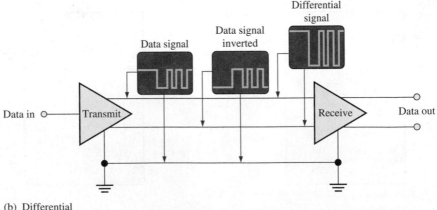

(b) Differential

FIGURE 13–54 A comparison of single-ended and differential bus operation.

A single-ended transmission is simpler and lower in cost compared to a differential transmission. Differential operation is much less sensitive to noise because of common-mode rejection. This means that when a common noise signal appears on each line, the two noise signals cancel due to the differential operation where the difference between them is zero. Since the data signals are the same but opposite in phase, they are effectively added and the data signal is reinforced while the noise signal is cancelled. Due to the twisted pair, crosstalk is reduced at higher frequencies, thus allowing longer cables.

SECTION 13–6 CHECKUP

1. List two factors that may limit the speed of a parallel bus.

2. What is the basic difference between an internal and an external bus?

3. Name four bus characteristics.

4. What is bus protocol?

5. Discuss the difference between single-ended and differential bus operation.

13–7   Parallel Buses

A bus is not only a set of physical connections (PCB traces or cables) but it is also a set of signals and operating parameters that are defined in the bus specification. Any devices connected to a given bus must be compatible with the bus specifications. In this section, we briefly look at several important parallel bus standards. Further details and information on each bus can be found on the Internet.

After completing this section, you should be able to

◆ Discuss the PCI, PCI-X, and PCI-Express buses

◆ Explain the basics of the IEEE-488 bus

◆ Discuss the parallel SCSI bus

The PCI Bus

The **PCI** (peripheral component interconnect) **bus** is an internal synchronous bus for interconnecting chips, expansion boards, and processor/memory subsystems. The original PCI bus had a width of 32 bits and a frequency of 33 MHz. Another version has a width of 64 bits and a frequency of 66 MHz. Still later versions enable 64-bit data transfers using up to a 133 MHz clock to enable bandwidths of up to 1066 MBps.

The original PCI standard required 5 V power and signal levels. As the standard evolved, the option for 3.3 V was added. The latest standard provides for 3.3 V only. The 32-bit PCI connector has 62 pins and 124 contacts (62 per side). Thirty-two of the contacts are used for both a 32-address and 32 bits of data, which are multiplexed. The remaining pins are used for command and control signals, power, ground, etc. A 64-bit PCI connector has an additional 32 pins for a total of 94. The 32-bit PCI connector is shown in Figure 13–55. It has 64 pins of which 62 are used.

The PCI-X Bus

The **PCI-X bus** is a high-performance enhancement of the PCI and is backward compatible with the PCI bus, although it is a faster bus and has some additional features. A 64-bit bus, the PCI-X runs at a frequency of 133 MHz. The PCI-X 2.0 revision supports frequencies of 266 MHz and 533 MHz. Some additional features increase system reliability by minimizing errors at high transfer rates. Servers are the major application for the PCI-X.

PIN	SIDE A	SIDE B	
1	−12 V	TRST#	
2	TCK	+12 V	JTAG port pins
3	Gnd	TMD	
4	TDO	TDI	
5	+5 V	+5 V	
6	+5 V	INTA#	
7	INTB#	INTC#	Interrupt lines
8	INTD#	+5 V	
9	PRSNT1#	Reserved	Indicates 7.5 or 24 W power
10	Reserved	IOPWR	+5 V or +3.3 V
11	PRSNT1#	Reserved	Indicates 7.5 or 24 V power
12	Gnd	Gnd	
13	Gnd	Gnd	Key notch for 3.3 V cards
14	Reserved	3.3 Vaux	Standby power
15	Gnd	RST#	Bus reset
16	CLK	IOPWR	33 MHz/66 MHz clock
17	Gnd	GNT#	Bus grant motherboard to card
18	REQ#	Gnd	Bus request card to motherboard
19	IOPWR	PME#	Power management event/3.3 V
20	AD[31]	AD[30]	
21	AD[29]	+3.3 V	
22	Gnd	AD[28]	
23	AD[27]	AD[26]	Address/data bus
24	AD[25]	Gnd	
25	+3.3 V	AD[24]	
26	C/BE[3]#	IDSEL	
27	AD[23]	+3.3 V	
28	Gnd	AD[22]	
29	AD[21]	AD[20]	
30	AD[19]	Gnd	
31	+3.3 V	AD[18]	
32	AD[17]	AD[16]	

PIN	SIDE A	SIDE B	
33	C/BE[2]#	+3.3 V	
34	Gnd	FRAME#	Bus transfer in progress
35	IRDY#	Gnd	Initiator ready
36	+3.3 V	TRDY#	Target ready
37	DEVSEL#	Gnd	Target selected
38	Gnd	STOP#	Target halt request
39	LOCK#	+3.3 V	Locked transaction
40	PERR#	SMBCLK	Parity error/SMBus clock
41	+3.3 V	SMBDAT	SMBus data
42	SERR#	Gnd	System error
43	+3.3 V	PAR	Even parity over AD bus
44	C/BE[1]#	AD[15]	
45	AD[14]	+3.3 V	
46	Gnd	AD[13]	Address/data bus
47	AD[12]	AD[11]	
48	AD[10]	Gnd	
49	MGGEN/Gen	AD[09]	
50	Gnd	Gnd	Key notch for +5 V cards
51	Gnd	Gnd	
52	AD[08]	C/BE[0]#	
53	AD[07]	+3.3 V	
54	+3.3 V	AD[06]	
55	AD[05]	AD[04]	
56	AD[03]	Gnd	Address/data bus
57	Gnd	AD[02]	
58	AD[01]	AD[00]	
59	IOPWR	IOPWR	
60	ACK64#	REQ64#	For 64-bit expansion
61	+5 V	+5 V	
62	+5 V	+5 V	
63			
64			

FIGURE 13–55   Pin layout and functional designation for a 32-bit PCI connector.

The PCI-Express Bus

The **PCI-Express** is also designated as PCIe or PCI-E. This bus differs from the PCI and PCI-X buses in that it does not use a **shared bus**. Both PCI and PCI-X use a shared bus configuration, as shown in Figure 13–56(a). Each PCIe device has a dedicated path, called a *lane*, to a single chip known as a *switch*, as shown in part (b). More lanes result in a faster data transfer. High speed makes PCI-Express ideal for video and graphics applications.

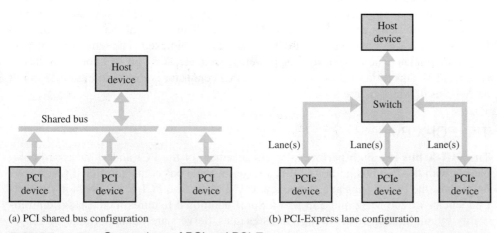

(a) PCI shared bus configuration   (b) PCI-Express lane configuration

FIGURE 13–56   Comparison of PCI and PCI-Express.

A single PCI-Express lane contains two pairs of conductors. One pair of conductors from a given device receives data and the other pair sends data in a serial format. A single lane configuration is known as x1 and is illustrated in Figure 13–57(a). A x2 configuration is shown in part (b), and a x16 configuration is shown in part (c). Other possible configurations are x4, x8, and x32.

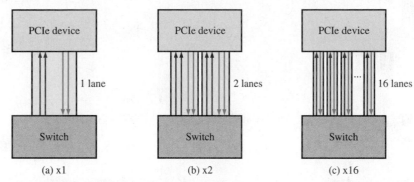

FIGURE 13–57   PCI-Express lane configurations.

Bidirectional data are transferred serially on each lane. For multiple-lane configurations, the serial data on each lane are in parallel with the serial data on all the other lanes. The data are transferred at one bit per cycle on a x1 connection, two bits per cycle on a x2, and sixteen bits per cycle on a x16. The PCI-Express has a bandwidth of 4 Gbps per lane. PCI-Express supports hot swapping in which expansion cards can be added or removed without turning off the system. PCI and PCIe are software compatible but not hardware compatible.

The IEEE-488 Bus

This bus standard has been around a long time and is also known as the General-Purpose Interface Bus (**GPIB**). Widely used in test and measurement applications, it was developed by Hewlett-Packard in the 1960s. The IEEE 488 specifies 24 lines that are used to transfer eight parallel data bits at a time and provide eight control signals that include three handshake lines and five bus-management lines. Also included are eight ground lines used for shielding and ground returns. The maximum data transfer rate for the IEEE 488 standard is 1 Mbps. A superset of this standard, called the HS488, has a maximum data rate of 8 Mbps.

To connect test equipment to a computer using the IEEE-488 bus, an interface card is installed in the computer, which turns the computer into a system controller. In a typical GPIB setup, up to 14 controlled devices (test and measurement instruments) can be connected to the system controller. When the system controller issues a command for a controlled device to perform a specified operation, such as a frequency measurement, it is said that the controller "talks" and the controlled device "listens."

A **listener** is an instrument capable of receiving data over the GPIB when it is addressed by the system controller (computer). Examples of listeners are printers, monitors, programmable power supplies, and programmable signal generators. A **talker** is an instrument capable of sending data over the GPIB. Examples are DMMs and frequency counters that can output bus-compatible data. Some instruments can send and receive data and are called talker/listeners; examples are computers, modems, and certain measurement instruments. The system controller can specify each of the other instruments on the bus as either a talker or a listener for the purpose of data transfer. The controller is usually a talker/listener.

A typical GPIB arrangement is shown in Figure 13–58 as an example. The three basic bus signal groupings are shown as the *data bus, data transfer control bus,* and *interface management bus.*

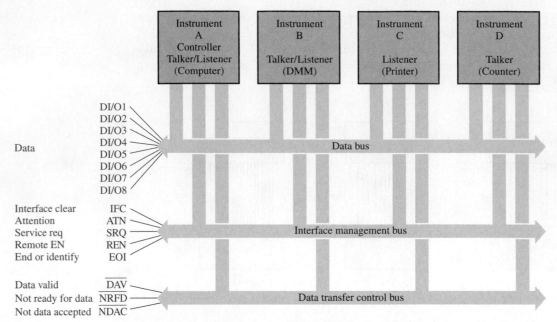

FIGURE 13–58   A typical IEEE 488 (GPIB) connection.

The parallel data lines are designated DI/O1 through DI/O8 (data input/output). One byte of data is transferred on this bidirectional part of the bus. Every byte that is transferred undergoes a handshaking operation via the data transfer control. The three active-LOW handshaking lines indicate if data are valid ($\overline{\text{DAV}}$), if the addressed instrument is not ready for data ($\overline{\text{NRFD}}$), or if the data are not accepted ($\overline{\text{NDAC}}$). More than one instrument can accept data at the same time, and the slowest instrument sets the rate of transfer. Figure 13–59 shows the timing diagram for the GPIB handshaking sequence, and Table 13–2 describes the handshaking signals.

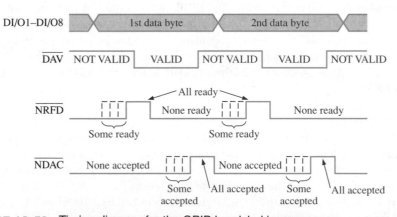

FIGURE 13–59   Timing diagram for the GPIB handshaking sequence.

The five signals of the interface management bus control the orderly flow of data. The ATN (attention) line is monitored by all instruments on the bus. When ATN is active, the system controller selects the specific interface operation, designates the talkers and the listeners, and provides specific addressing for the listeners. Each GPIB instrument has a specific identifying address that is used by the system controller. Table 13–3 describes the GPIB interface management lines and their functions.

TABLE 13-2

The GPIB handshaking signals.

Name	Description
$\overline{\text{DAV}}$	**Data Valid:** After the talker detects a HIGH on the $\overline{\text{NRFD}}$ line, a LOW is placed on this line by the talker when the data on its I/O are settled and valid.
$\overline{\text{NRFD}}$	**Not Ready for Data:** The listener places a LOW on this line to indicate that it is not ready for data. A HIGH indicates that it is ready. The $\overline{\text{NRFD}}$ line will not go HIGH until all addressed listeners are ready to accept data.
$\overline{\text{NDAC}}$	**Not Data Accepted:** The listener places a LOW on this line to indicate that it has not accepted data. When it accepts data from its I/O, it releases its $\overline{\text{NDAC}}$ line. The $\overline{\text{NDAC}}$ line to the talker does not go HIGH until the last listener has accepted data.

TABLE 13-3

The GPIB management lines.

Name	Description
ATN	**Attention:** Causes all the devices on the bus to interpret data, as a controller command or address and activates the handshaking function.
IFC	**Interface Clear:** Initializes the bus.
SRQ	**Service Request:** Alerts the controller that a device needs to communicate.
REN	**Remote Enable:** Enables devices to respond to remote program control.
EOI	**End or Identify:** Indicates the last byte of data to be transferred.

The GPIB is limited to a maximum cable length of 15 meters, and there can be no more than one instrument per meter with a maximum capacitive loading of 50 pF each. The cable length limitation can be overcome by the use of bus extenders and modems. A bus extender provides for cable-interfacing of instruments that are separated by a distance greater than allowed by the GPIB specifications or for communicating over greater distances via modem-interfaced telephone lines.

The IEEE-488 connector and pin configuration are shown in Figure 13–60.

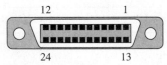

Pin	Designation	Description	Pin	Designation	Description
1	DIO1	Data input/output bit.	13	DIO5	Data input/output bit.
2	DIO2	Data input/output bit.	14	DIO6	Data input/output bit.
3	DIO3	Data input/output bit.	15	DIO7	Data input/output bit.
4	DIO4	Data input/output bit.	16	DIO8	Data input/output bit.
5	EOI	End-or-identify.	17	REN	Remote enable.
6	DAV	Data valid.	18	GND	(wire twisted with DAV)
7	NRFD	Not ready for data.	19	GND	(wire twisted with NRFD)
8	NDAC	Not data accepted.	20	GND	(wire twisted with NDAC)
9	IFC	Interface clear.	21	GND	(wire twisted with IFC)
10	SRQ	Service request.	22	GND	(wire twisted with SRQ)
11	ATN	Attention.	23	GND	(wire twisted with ATN)
12	SHIELD		24	Logic ground	

FIGURE 13-60 The IEEE-488 (GPIB) bus connector and pin assignments.

The Parallel SCSI Bus

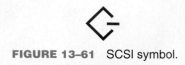

FIGURE 13–61   SCSI symbol.

The **SCSI** (small computer system interface) bus, generally pronounced "skuzy", is a parallel I/O bus with a width of either 8, 16, or 32 bits, depending on the version. For many years SCSI has been one of the most widely used buses in storage servers and data centers. SCSI is also used for the purpose of transferring data between a computer and peripheral devices, such as hard disks, tape drives, scanners, and CD drives. Figure 13–61 shows the SCSI symbol.

The original version of the SCSI parallel bus standard was introduced in 1986 and designated SCSI-1. The current SCSI standard is known as SCSI-5, which was preceded by SCSI-1, SCSI-2, SCSI-3, and SCSI-4. Later versions are backward compatible with earlier versions. There are many variations of the SCSI standard version with designations such as *asynchronous, synchronous, fast, ultra,* and *wide,* which have different speeds, widths, and number of devices that can be connected, as shown in Table 13–4.

TABLE 13–4

Evolution of the parallel SCSI standard.

Version	Variations	Maximum Target Devices Connected	Bus Width	Data Transfer Rates
SCSI-1	Asynchronous/Synchronous	7	8 bits	4 MBps/5 MBps
SCSI-2	Wide, Fast, Fast/Wide	7/15	8/16 bits	10 MBps/20 MBps
SCSI-3	Ultra, Ultra/Wide, Ultra2, Ultra2/Wide, Ultra160	7/15	8/16 bits	20 MBps/40 MBps/ 80 MBps/160 MBps
SCSI-4	Ultra320	15	16 bits	320 MBps
SCSI-5	Ultra640	15	8/16/32 bits	640 MBps

SCSI Signals

A parallel SCSI bus contains nine control signals in addition to data, dc voltages, and ground. These signals are listed in Table 13–5.

TABLE 13–5

SCSI parallel bus signals.

Signal	Description
BSY	Busy, Bus in use
SEL	Select
RST	Reset
C/D	Control/Data
MSG	Message
REQ	Request
ACK	Acknowledge a request
ATN	Attention
I/O	Input or output

Up to eight or sixteen devices, including the host, can be connected to a SCSI bus, but only two devices can communicate at any given time. Communication begins when an initiating device sends a request and the target device acknowledges and performs the request. Single-ended (SE) and differential (LVD or HVD) are the three electrical specifications. Single-ended operation is limited to a cable length of 6 meters, and differential operation allows up to 25 meters. LVD is low-voltage differential and HVD is high-voltage differential. SCSI devices can operate either asynchronously or synchronously. The serial SCSI bus is introduced in Section 13-9.

SECTION 13–7 CHECKUP

1. What does PCI stand for?

2. List two alternate designations for the PCI-Express bus.

3. What is a *lane*?

4. What is the IEEE designation for the GPIB?

5. What does SCSI stand for?

13–8 The Universal Serial Bus (USB)

Although there are several serial bus standards available, the USB is one of the most widely used. Recall that a serial bus transfers data one bit at a time. As with the parallel bus, a serial bus is not only a set of physical connections but it is also a set of signals and operating parameters that are defined in the bus specification. As with other buses, only the basics are introduced here. More details and information can be found on the Internet.

After completing this section, you should be able to

♦ Discuss the USB

♦ Identify and describe USB cables and connectors

♦ Discuss USB signals

The **universal serial bus (USB)** is a widely used standard serial bus for connecting peripherals to a computer. There are typically two or more USB ports on computers and, with USB hubs, up to 127 devices can be connected. USB allows the devices to be connected or disconnected while the computer is running (hot swapping). Figure 13–62 shows the symbol for USB.

FIGURE 13–62   USB symbol.

The original USB standard was 1.0, which was followed by 1.1. USB 2.0 replaced the two original versions and more recently USB 3.0 was introduced. The earlier versions are still in use, especially 2.0. In terms of data rate, USB has four classifications: low-speed, full-speed, high-speed, and super-speed. Table 13–6 shows how the data rate classifications apply to each of the USB versions and Table 13–7 shows the data rate values.

TABLE 13–6

	Low-Speed	Full-Speed	High-Speed	Super-Speed
USB 1.0	•	•		
USB 1.1	•	•		
USB 2.0	•	•	•	
USB 3.0	•	•	•	•

TABLE 13–7

Data Rate	Maximum Value
Low-speed	0.1875 MBps
Full-speed	1.5 MBps
High-speed	60 MBps
Super-speed	625 MBps

Cable length is an important specification for buses. Table 13–8 lists maximum cable lengths for three USB versions and maximum total lengths when multiple cables are strung together using USB hubs. A hub is a common connection device with multiple ports.

TABLE 13–8

	USB 1.1	USB 2.0	USB 3.0
Max cable length	9.8 ft. (3.0 m)	16.4 ft. (5.0 m)	9.8 ft. (3.0 m)
Maximum total length	49.2 ft. (15 m)	82.0 ft. (25 m)	49.2 ft. (15 m)

USB Cable and Connectors

USB versions up to and including 2.0 have a four-wire cable that includes a twisted pair to reduce or eliminate noise for data transmission, a +5 V wire, and a ground wire color-coded, as shown in Figure 13–63(a). The standard type A and type B connectors are shown in parts (b) and (c) with pin designations. USB hosts (computer) and devices (peripherals) have sockets, and all USB cables have a type A plug at one end and a type B plug at the other. The sockets on a host are Type A, and the sockets on peripheral devices are Type B. Hubs have both Type A and Type B. The USB standard also specifies smaller connectors designated mini and micro.

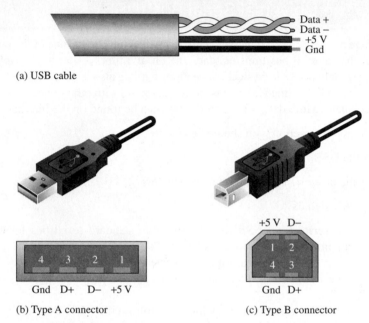

(a) USB cable

(b) Type A connector   (c) Type B connector

FIGURE 13–63   USB cable and connectors for USB standards through 2.0.

USB Data Format

Serial data are transmitted on the twisted pair (D+ and D−) using half-duplex differential mode to minimize EMI and improve the signal-to-noise ratio. Data are sent in packets using NRZI (non-return-to-zero invert) encoding format with a 3.3 V level (differentially, there are 6.6 V between the two data lines). A packet format can contain the following fields:

Sync field   All packets start with a sync (synchronization) field. The sync field consists of 8 bits for low and 32 bits (full speed) for high speed and is used to synchronize the receiver clock with that of the transmitter.

PID field   The packet identification field is used to identify the type of packet that is being transmitted. There are four bits in the PID; however, to ensure it is received correctly, the four bits are complemented and repeated, making an 8-bit PID code.

ADDR field   The address field specifies to which device the packet is sent. The seven bits in this field allow for 127 devices to be supported. Address 0 is invalid.

Data field   The data field contains up to 1024 bytes of data.

ENDP field   The endpoint field is made up of four bits, allowing 16 possible endpoints. An endpoint is a data source or load. Low-speed devices, however, can only have two additional endpoints on top of the default pipe (four endpoints max). Endpoints can be described as sources or sinks of data.

CRC field   Cyclic redundancy checks are performed on the data within the packet using from 5–16 bits, depending on the type of packet.

EOP field   This packet field signals the end of a packet.

Four types of USB packets are token, data, handshake, and start-of-frame, as shown in Figure 13–64 with the packet format for each type. Each field is labeled and the number of bits shown. The token packet indicates the type of transaction, the data packet contains the actual data, the handshake packet acknowledges a transaction, and the start-of-frame packet begins a new frame. The token packet, data packet, handshake packet, and start-of-frame packet each have a different packet format as specified by the PID field.

Sync 8/32	PID 8	ADDR 7	ENDP 4	CRC 5	EOP 3

(a) Token packet

Sync 8/32	PID 8	Data 0-8192	CRC 16	EOP 3

(b) Data packet

Sync 8/32	PID 8	EOP 3

(c) Handshake packet

Sync 8/32	PID 8	Frame number 11	CRC 5	EOP 3

(d) Start-of-frame packet

FIGURE 13–64   Types of USB packets.

USB 3.0

USB 3.0 is a recent version of the USB standard. The 3.0 version, known as SuperSpeed USB, is ten times faster than USB 2.0 at 4.8 Gbps. The 3.0 connectors (types A and B) are different then version 2.0 because they now have nine contacts instead of four. The 3.0 type A connector looks about the same as that for 2.0 except that the extra five pins are further inside and make contact only with a 3.0 mating connector. The connector is compatible with a 2.0 device where the front four pins are accessible. Type A and type B connectors for USB 3.0 along with the standard symbol are shown in Figure 13–65. There is also a micro-B connector available.

(a) Type A   (b) Type B   (c) Symbol

FIGURE 13–65   USB 3.0 connectors and symbol.

USB 3.0 is, for the most part, backward compatible with USB 2.0, but the speed is limited to the 2.0 specification. The USB 3.0 cable consists of two additional twisted pairs for data and an additional ground, for a total of nine wires. Unlike the previous versions, version 3.0 is full-duplex, meaning that data can be sent and received simultaneously. One twisted pair is for receiving data, and two additional twisted pairs are for sending high-speed data.

The recent USB 3.1 specification increases the data rate to 10 Gbps, twice that of the 3.0. USB 3.1 is backward compatible with 3.0 and 2.0 and a new connector, type C, is included in the specification.

Figure 13–66 shows the USB in a typical computer system. The computer acts as a host and uses Type A connectors. The hub functions as both a host and a device.

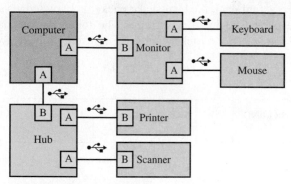

FIGURE 13–66   Example of USB applications.

SECTION 13–8 CHECKUP

1. What does USB stand for?
2. What are the functions of the four pins in a USB 2.0 connector?
3. Why are twisted pairs used in USB cables?
4. Describe the basic differences between USB 2.0 and USB 3.0.

13–9   Other Serial Buses

Although the USB is widely used, many other serial buses are available. A few of the these important bus standards are introduced in this section. For more detailed information on any bus standard, search the Internet.

After completing this section, you should be able to

* Discuss the RS-232, RS-422, RS-423, and RS-485 bus standards
* Describe the SPI bus
* Discuss the I^2C bus
* Explain the CAN bus
* Describe the Firewire (IEEE 1394) bus
* Discuss the serial SCSI bus

The RS-232/422/423/485 Buses

RS-232 Bus

Also known as **EIA-232**, the **RS-232 bus** was once standard on computers for connection to peripheral devices. The standard provides for single-ended data transmission in either synchronous or asynchronous formats. It has been replaced by the USB because of its limited speed, relatively large voltage requirements, and large connector size. However, RS-232 devices are still used in industrial and telecommunication applications as well as scientific instrumentation. The devices connected by the RS-232 are classified as DTE (data terminal equipment) or DCE (data communication equipment). Since newer computers have no RS-232 ports, USB-to-RS-232 converters can be used to connect to older RS-232 compatible peripherals, if necessary. The standard is designed for one transmitting device and one receiving device with a maximum cable length of 50 feet between them.

The maximum RS-232 data rate is 20 kbps. The data format typically consists of seven or eight bits of data, a start bit, a parity bit in some cases depending on the protocol, and a stop bit. A transmitted signal level between $+5$ V and $+15$ V represents a binary 0 and between -5 V and -15 V represents a binary 1. The data is transmitted in NRZ format, as Figure 13–67 shows.

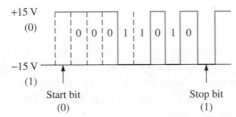

FIGURE 13–67 Example of RS-232 transmitted data format. A parity bit is not included.

The standard 25-pin connector for RS-232 is shown in Figure 13–68(a). A smaller 9-pin connector, is shown in part (b).

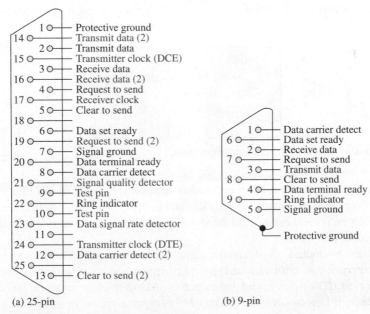

(a) 25-pin   (b) 9-pin

FIGURE 13–68 Standard RS-232 connectors.

RS-422/423/485

The **RS-422 bus** provides for differential transmission for greater distances (longer cable length) and has higher data rates than the RS-232 bus. Also, the standard defines the number of receiving devices as ten for a line with one driver (transmitting device) compared to one receiving device for the RS-232. The **RS-423 bus** is similar to the RS-232 in that it is single-ended, but it has a higher data rate and longer cable length. The RS-485 bus is a differential standard and can accommodate multiple drivers and receivers. Table 13–9 summarizes some of the features of the RS (EIA) buses.

TABLE 13–9				
Specifications	**RS-232**	**RS-423**	**RS-422**	**RS-485**
Operation	Single-ended	Single-ended	Differential	Differential
Drivers/Receivers	1/1	1/10	1/10	32/32
Cable length	50 ft	4000 ft	4000 ft	4000 ft
Max data rate	20 kbps	100 kbps	10 Mbps	10 Mbps
Driver output signal level (+/− min/max)	5 V/15 V	3.6 V/6 V	2 V/6 V	1.5 V/6 V

The SPI Bus

The **serial-to-peripheral interface (SPI) bus** is a synchronous serial communications bus that uses four wires for communication between a "master" device and a "slave" device. This standard was developed by Motorola; it operates in full-duplex mode up to a data rate of 10 Mbps and can accommodate multiple slaves. The four signal wires are

1. MOSI (master out slave in) is initiated by the master and received by the slave.
2. MISO (master in slave out) is initiated by the slave and received by the master.
3. SCLK (serial clock) is generated by the master for synchronizing data transfers.
4. SS (slave select) is generated by the master to select an individual slave.

Other names are sometimes assigned to these signals such as SDI (serial data in) for MOSI and SDO (serial data out) for MISO. Figure 13–69 shows a master with a single slave.

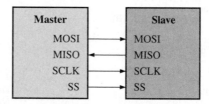

FIGURE 13–69   SPI master/slave configuration.

The SPI bus is typically used in embedded systems and on PCBs for communication between microprocessors or microcontrollers and peripheral IC chips or between two processors. Much PCB "real estate" can be saved compared to using an internal parallel bus with many more connections.

SPI applications include digital audio, signal processing, and telecommunications. SPI is used to communicate with various types of peripherals such as sensors, camera lenses, flash memory, LCD displays, and video games. Many microprocessors and microcontrollers include SPI controllers that can be used as either a master or a slave.

The I²C Bus

I²C bus (pronounced I squared C) is also stated at I2C (I two C) and stands for *inter-integrated circuit*. It is an internal serial bus primarily for connecting ICs on a PCB. A main advantage is that it requires only two lines (plus ground) and therefore saves considerable board space compared to a parallel bus. Two signals (SDA and SCL) are used to communicate between compatible devices. Data are sent serially on the SDA line, and a clock is sent on the SCL line. Like SPI, I²C is based on the master/slave concept where the master device drives the clock line and the slaves respond to the master. Only the master can initiate a transfer over the bus, but slaves can transfer data under control of the master using clock rates up to 100 kHz in the standard mode. Two other modes, enhanced and high-speed, allow 400 kbps and 3.4 Mbps, respectively.

When transferring data from master to slave, the master device sends a start bit, followed by a slave address, and a write bit. The master waits for an acknowledge (ACK) signal from the slave and then sends the data and waits for an acknowledge before sending a stop bit, as illustrated in Figure 13–70(a). The yellow segments are from the master and the gray elements are from the slave. Similarly, when the master requires data from the slave, it sends a start bit followed by the address and a read bit. The slave returns an acknowledge followed by the data. When the master receives the data, it issues an acknowledge and a stop bit, as shown in Figure 13–70(b).

Start	Address	Write	ACK	Data	ACK	Stop

(a) Data transfer from master to slave

Start	Address	Read	ACK	Data	ACK	Stop

(b) Data transfer from slave to master

FIGURE 13–70 I²C data transfers. Yellow is from master. Gray is from slave.

The CAN Bus

The **controller area network (CAN) bus**, a differential serial bus, was developed for automotive applications and is also commonly used in aerospace systems, as well as other applications. The bus consists of a terminated twisted pair of signal lines, called CAN H and CAN L, plus ground. Vehicles sold in the United States are required by the SAE (Society of Automotive Engineers) to use the CAN bus protocol. The European Union has similar requirements.

Devices, called *nodes*, can be connected to the bus but are not assigned specific addresses as in the I²C bus. Two CAN specifications are in use. The standard or basic CAN 2.0A has 11-bit message identifiers and can operate up to 250 kbps, and the full CAN has 29-bit message identifiers and can be used up to 1 Mbps. The message identifier is a label for the contents of a message and goes to each node on the bus. Each receiving node performs a test on the identifier to determine if it is relevant to that node and is used to arbitrate the bus to determine if the message is of highest priority. All of the nodes on the bus can transmit and receive messages. The bus is available to a node with a message with the highest priority (dominant) and can override a message with lower priority (recessive). When the dominant message has been processed, the recessive message is retransmitted.

The standard CAN data frame is shown in Figure 13–71. Data are transmitted in NRZ format. The frame begins with a start-of-frame (SOF) bit followed by an arbitration field and a control field. The arbitration field contains the message identifier and a remote transmission request (RTR) bit. The control field has two reserve bits and a data length code (DLC) that specifies the length of the data field that follows and can contain up to 8 bytes. The cyclic redundancy check (CRC) field provides for error detection. The acknowledge (ACK) verifies the receipt of correct data, and the frame ends with the end-of-frame field (EOF).

SOF	Arbitration field	Control field	Data field	CRC field	ACK	EOF
(1 bit)	Identifier (11 bits) RTR (1 bit)	Reserve (2 bits) DLC (4 bits)	(0–8 bytes)	(16 bits)	(2 bits)	(7 bits)

FIGURE 13–71 Standard CAN data frame format.

An Application

An automobile typically has many control units (usually several dozen) for various subsystems, including the engine control unit and other control units for transmission; ABS; cruise control; power steering; audio system; window, door, and mirror controls; airbags; and others. Figure 13–72 is a block diagram of a partial automotive control system using two CAN buses, one low-speed and one high-speed to control various functions throughout the vehicle.

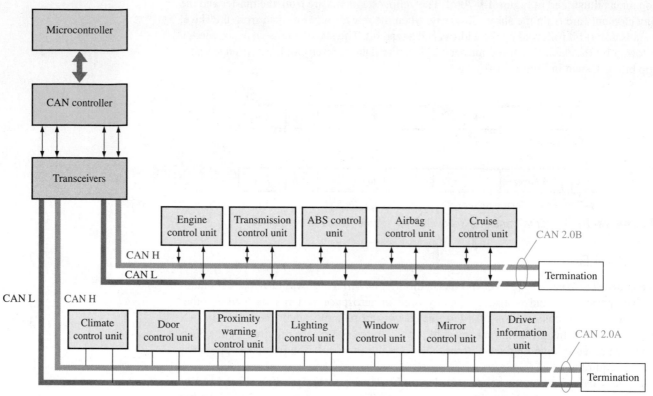

FIGURE 13–72   The CAN bus in an automotive control system.

Each unit connected to the bus contains sensors and other functions that allow it to carry out its unique purpose. For example, the ABS (antilock braking system) can receive a message from sensors in each wheel indicating that the brake is about to lock up. A sudden and rapid deceleration in the wheel indicates an imminent lock-up condition. The ABS unit then sends a message that causes the valve in the brake line to release pressure to allow acceleration. Then, when acceleration is sensed, the unit causes a pump to restore the pressure. A rapid release and restore cycle occurs until the brakes are brought under control. A pulsing of the brake pedal can be felt when the operation occurs.

As another example, part of the engine control unit's operation is to sense parameters such as engine temperature, oil pressure, fuel consumption, and rpm, and send messages to the driver unit. All of the units on the bus operate as a system to keep the vehicle running as smoothly and as safely as possible, while providing a comfortable environment for the driver and passengers.

The Firewire Bus

Firewire, also known as IEEE-1394 and iLink, is a high-speed external serial bus developed by Apple Inc. Firewire is used in high-speed communications and real-time data transfer. It is used in professional audio and video equipment, camcorders, DVD players, external hard drives, and in computers used for audio and video editing, as well as in some auto and aircraft applications. It is similar to the USB except that it has a higher data rate and can handle more data.

Three types of connectors are used in the Firewire standard: a 4-pin connector, a 6-pin connector, and a 9-pin connector. The cable for the 4-pin connector consists of two twisted pairs that carry data. The cable for the 6-pin connector has the two twisted pairs for data plus a power line and a ground line. The cable for the 9-pin connector has the same wires as the 6-pin configuration plus two wires that provide for a grounded shield and one wire that is currently unused. The Firewire symbol is shown in Figure 13–73(a). End views of the three connector types are shown in part (b), and the pin designations are shown in part (c).

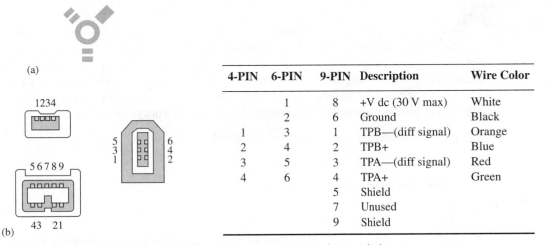

4-PIN	6-PIN	9-PIN	Description	Wire Color
	1	8	+V dc (30 V max)	White
	2	6	Ground	Black
1	3	1	TPB—(diff signal)	Orange
2	4	2	TPB+	Blue
3	5	3	TPA—(diff signal)	Red
4	6	4	TPA+	Green
		5	Shield	
		7	Unused	
		9	Shield	

FIGURE 13–73 Firewire symbol with cable and connector wires and pins.

The Firewire bus address has a total of 64 bits. Ten are for bus ID, six are for node ID, and 48 are for individual addresses. This allows up to 1023 buses, each having up to 63 nodes. The six transfer modes in the IEEE-1394 standard and its revisions are S100, S200, S400, S800, S1600, and S3200. The S100 is the base rate of 98.304 Mbps. The S200 is twice the base rate at 196.608 Mbps, and the S400 is four times the base rate at 393.216 Mbps. The S800 is 786.432 Mbps while the S1600 and S3200 are 16 and 32 times the base rate, respectively (1.6 Gbps and 3.2 Gbps). Firewire cable length cannot exceed 15 ft (4.572 m). To increase this length, up to 16 cables can be connected together.

Firewire versus USB

In general, any capable node can control the bus in a Firewire system, but a single host is used to control the bus in USB. USB networks use a tiered-star topology and Firewire uses a tree topology. A Firewire device can communicate with any node at any time if the conditions allow, but a USB 2.0 device cannot communicate with the host device unless requested by the host. However, USB 3.0 allows Firewire-like communications between devices. USB provides 5 V power while Firewire provides up to 30 V. As a result, Firewire can supply more power to a device than USB. As mentioned before, Firewire is faster than USB.

Serial SCSI

Serial Attached SCSI (**SAS**) is a data-transfer technology for transmitting data to and from storage devices. It has become a replacement for parallel SCSI bus technology that is commonly used in data centers, workstations, and servers. The serial SCSI overcomes some of the limitations of the parallel SCSI. The SAS supports up to a 12 Gbps data rate and allows up to 65,535 devices to be connected using expanders compared to 15 devices for parallel SCSI.

SECTION 13–9 CHECKUP

1. List all the buses introduced in this section.
2. What does SPI stand for?
3. What does I2C stand for?
4. What does CAN stand for?
5. What is another designation for Firewire?

13–10 Bus Interfacing

All the components in a computer or other systems are interconnected by buses, which serve as communication paths. Physically, a bus is a set of conductive paths that serves to interconnect two or more functional components of a system or several diverse systems. Electrically, a bus is a collection of specified voltage levels and/or current levels and signals that allow the various devices connected to the bus to work properly together.

After completing this section, you should be able to

- ◆ Discuss the concept of a multiplexed bus
- ◆ Explain the reason for tri-state outputs

Basic Multiplexed Buses

As you have learned, in computers the microprocessor controls and communicates with the memories and the input/output (I/O) devices via the *internal bus structure,* as indicated in Figure 13–74. A bus is multiplexed so that any of the devices connected to it can either send or receive data to or from one of the other devices. A sending device is often called a master or **source**, and a receiving device is often called a servant or **acceptor**. At any given time, there is only one source active. For example, the RAM may be sending data to the input/output (I/O) interface under control of the microprocessor.

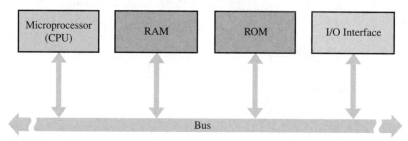

FIGURE 13–74 The interconnection of microprocessor-based system components by a bidirectional, multiplexed bus.

Bus Signals

With synchronous bus control, the microprocessor (CPU) usually originates all control and timing signals. The other devices then synchronize their operations to those control and timing signals. With asynchronous bus control, the control and timing signals are generated jointly by a source and a receiver using a handshaking routine. A typical handshaking sequence is given in Figure 13–75. Handshaking routines may differ from one system to another, as you can see by comparing this sequence with the one shown in Figure 13–53.

An important control function is called **bus arbitration**. Arbitration prevents two sources from trying to use the bus at the same time.

FIGURE 13–75   An example of a handshaking sequence.

Connecting Devices to a Bus

Tri-state buffers are normally used to interface the outputs of a source device to a bus. Usually more than one source is connected to a bus, but only one can have access at any given time. All the other sources must be disconnected from the bus to prevent **bus contention**.

Tri-state circuits are used to connect a source to a bus or disconnect it from a bus, as illustrated in Figure 13–76(a) for the case of two sources. The select input is used to connect

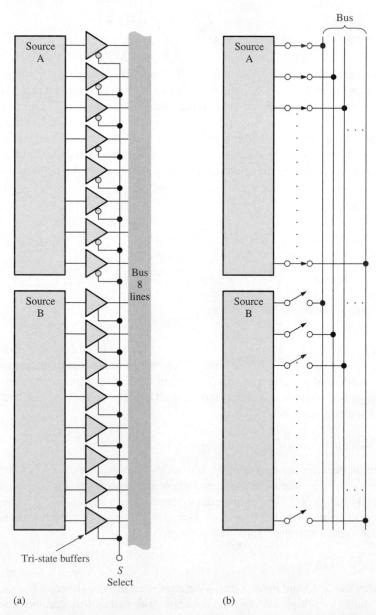

FIGURE 13–76   Tri-state buffer interface to a bus.

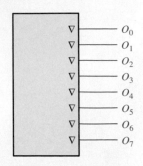

FIGURE 13–77 Method of indicating tri-state outputs on an IC device.

either source A or source B but not both at the same time to the bus. When the select input is LOW, source A is connected and source B is disconnected. When the select input is HIGH, source B is connected and source A is disconnected. A switch equivalent of this action is shown in part (b) of the figure.

When the enable input of a tri-state circuit is not active, the device is in a high-impedance (**high-Z**) state and acts like an open switch. Many digital ICs provide internal tri-state buffers for the output lines. A tri-state output is indicated by a ∇ symbol as shown in Figure 13–77.

Tri-State Buffer Operation

Figure 13–78(a) shows the logic symbol for a noninverting tri-state buffer with an active-HIGH enable. Part (b) of the figure shows one with an active-LOW enable.

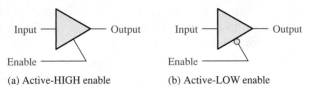

(a) Active-HIGH enable   (b) Active-LOW enable

FIGURE 13–78 Tri-state buffer symbols.

The basic operation of a tri-state buffer can be understood in terms of switching action as illustrated in Figure 13–79. When the enable input is active, the gate operates as a normal noninverting circuit. That is, the output is HIGH when the input is HIGH and LOW when the input is LOW, as shown in parts (a) and (b) respectively. The HIGH and LOW levels represent two of the states. The buffer operates in its third state when the enable input is not active. In this state, the circuit acts as an open switch, and the output is completely disconnected from the input, as shown in part (c). This is sometimes called the *high-impedance* or *high-Z* state.

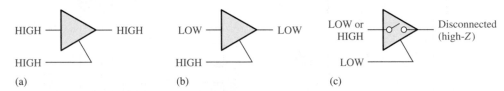

(a)   (b)   (c)

FIGURE 13–79 Tri-state buffer operation.

Many microprocessors, memories, and other integrated circuit functions have tri-state buffers that serve to interface with the buses. Such buffers are necessary when two or more devices are connected to a common bus. To prevent the devices from interfering with each other, the tri-state buffers are used to disconnect all devices except the ones that are communicating at any given time.

Bus Contention

Bus contention occurs when two or more devices try to output opposite logic levels on the same common bus line. The most common form of bus contention is when one device has not completely turned off before another device connected to the bus line is turned on. This generally occurs in memory systems when switching from the READ mode to the WRITE mode or vice versa and is the result of a timing problem.

Multiplexed I/Os

Some devices that send and receive data have combined input and output lines, called I/O ports, that must be multiplexed onto the data bus. Bidirectional tri-state buffers interface this type of device with the bus, as illustrated in Figure 13–80(a).

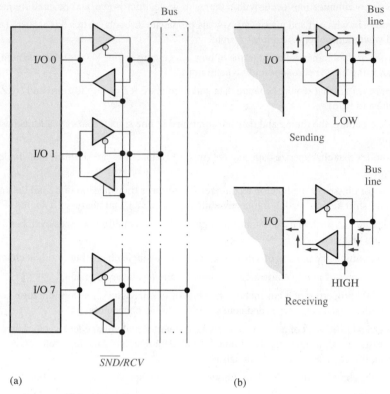

FIGURE 13–80 Multiplexed I/O operation.

Each I/O port has a pair of tri-state buffers. When the $\overline{SND}/RCV$ ($\overline{Send}$/Receive) line is LOW, the upper tri-state buffer in each pair is enabled and the lower one disabled. In this state, the device is acting as a source and sending data to the bus. When the $\overline{SND}/RCV$ line is HIGH, the lower tri-state buffer in each pair is enabled so that the device is acting as an acceptor and receiving data from the bus. This operation is illustrated in Figure 13–80(b). Some devices provide for multiplexed I/O operation with internal circuitry.

SECTION 13–10 CHECKUP

1. Why are tri-state buffers required to interface digital devices to a bus?
2. What is the purpose of a bus system?

SUMMARY

- Three essential elements in a data transmission system are sending device, transmission media, and receiving device.
- The simplest connection between sending and receiving devices is a wire or a conductive trace on a printed circuit board (PCB).
- A coaxial cable consists of a center conductor within an insulating dielectric material surrounded by a copper braided or foil shield encased in a protective jacket.
- BNC (Bayonet Neill-Concelman) connectors are typically used for coaxial connections.
- A twisted pair minimizes crosstalk when bundled together with other twisted pairs into cables.
- Instead of using electrical pulses to transmit information through copper lines, fiber optics uses light pulses transmitted through optical fibers.
- The transmission of data through air and space without the use of physical connections between sending and receiving systems is known as wireless transmission.

- Most data communications occur within the radio wave, microwave, and infrared frequencies.
- Three ways in which rf and microwave signals propagate through Earth's atmosphere (air) are ground wave, ionospheric, and line-of-sight.
- Two types of data transmission in terms of how a receiving device knows what a sending device is transmitting are synchronous and asynchronous.
- Two data formats that require separate data and timing are RZ (return to zero) and NRZ (nonreturn to zero).
- In biphase coding, the timing and data are combined in one signal; also called Manchester encoding.
- Three modes that characterize data channel (media) connections are simplex, half-duplex, and full-duplex.
- Three major classes of modulation techniques for wireless transmission of digital data are amplitude-shift keying (ASK), frequency-shift keying (FSK), and phase-shift keying (PSK).
- Multiplexing is a method used to transmit digital data from multiple sources over a single communication channel.
- Demultiplexing is the process of separating data from a single channel to multiple channels.
- Two major types of multiplexing are time-division and frequency-division.
- The physical properties of a bus include the number of conductors, the configuration and length of the conductors, and the types and configurations of the connectors.
- The electrical properties of a bus include signal format, signal voltage levels, clock frequency, data transfer speed, bandwidth, data frame format, data rate, handshaking protocol, error detection, impedances, and line termination.
- General bus characteristics are width, frequency, transfer speed, and bandwidth.
- A synchronous bus contains a clock, and an asynchronous bus is unclocked.
- Single-ended operation uses one wire for data and one wire for ground, where the signal voltage on the wire is with respect to ground.
- Differential operation uses two wires for data and one wire for ground.
- The PCI (peripheral component interconnect) bus is an internal synchronous bus for interconnecting chips, expansion boards, and processor/memory subsystems.
- PCI-Express uses a lane configuration via a switching device.
- In addition to PCI buses, other important parallel buses are ISA, IEEE-488, and SCSI.
- USB is a widely used standard serial bus for connecting peripherals to a computer.
- A USB packet can include the following fields: synchronization, packet identification, address, data, end point, CRC, and end of packet.
- Four types of USB packets are token, data, handshake, and start-of-frame.
- In addition to USB, some important serial buses are RS-232/422/423/485, SPI, I^2C, CAN, Firewire, and serial SCSI.
- Tri-state devices are used to interface circuitry to a bus.

KEY TERMS

Key terms and other bold terms in the chapter are defined in the end-of-book glossary.

Baud   The number of symbols per second in a data transmission.

Bit rate   The number of bits per second in a data transmission.

Bus   A set of connections and specifications for the transfer of data among two or more devices.

Bus protocol   A set of rules that allow two or more devices to communicate through a bus.

Coaxial cable   A type of data transmission media in which a shielded conductor is used to minimize EMI.

Electromagnetic waves   Related to the electromagnetic spectrum which includes radio waves, microwaves, infrared, visible, ultraviolet, X-rays, and gamma rays.

EMI   Electromagnetic interference.

Full-duplex   A connection in which the data flow both ways simultaneously in the same channel.

GPIB   General-purpose interface bus based on the IEEE-488 standard.

Half-duplex   A connection in which the data flow both ways but not at the same time in the same channel.

Handshake   A routine by which two devices initiate and complete a bus transfer.

Manchester encoding   A method of encoding called biphase in which a 1 is represented by a positive-going transition and a 0 is represented by a negative-going transition.

Modulation   The process of altering a parameter of a higher frequency signal proportional to the amplitude of a lower frequency information-carrying signal.

NRZ   Nonreturn to zero. A type of data format in which the signal level remains at one (1) for successive 1s.

Optical fiber   A type of data transmission media used for transmitting light signals.

PAM   Pulse amplitude modulation. A method of modulation in which the height or amplitude of the pulses are varied according to the modulating analog signal, and each pulse represents a value of amplitude of the analog signal.

RS-232   A bus standard, also known as EIA-232, used in industrial and telecommunication applications as well as scientific instrumentation, but largely replaced by USB in computer applications.

RZ   Return to zero. A type of data format in which the signal level goes to or remains at zero after each data bit.

SCSI   Small computer system interface bus.

Simplex   A connection in which data flow in only one direction from the sender (transmitter) to the receiver.

Tri-state buffer   A circuit used to interface one device to another to prevent loading.

USB   Universal serial bus. A widely used standard serial bus for connecting peripherals to a computer.

TRUE/FALSE QUIZ

Answers are at the end of the chapter.

1. The simplest connection for sending and receiving devices in a data transmission system is a coaxial cable.

2. BNC is a type of connector used for coax.

3. The purpose of a twisted pair is to minimize crosstalk.

4. Shielded twisted pair cable in a conduit provides more protection from EMI than UTP.

5. Fiber-optic cables are used to transmit electrical pulses through thin optical fibers.

6. ST is a type of fiber-optic connector.

7. Most data communications occur within the visible spectrum.

8. Types of rf and microwave signals that propagate through Earth's atmosphere are ground wave, ionospheric, and line-of-sight.

9. In general, a given number of bits can be transmitted faster in parallel than in series.

10. In asynchronous systems, the sending and receiving devices operate with separate oscillators having different clock frequencies.

11. Data rate is the speed of data transfer.

12. Bit rate is always less than or equal to the baud.

13. One method of synchronization is by using separate channels to transmit the data and the timing information.

14. Manchester encoding is a commonly used method that embeds the timing signal in the data so that only one channel is required.

15. Biphase and Manchester are two different types of code.

16. In a simplex connection, data flow in both directions between the sender (transmitter) and the receiver.

17. BASK stands for binary amplitude-shift keying.

18. PSK applications include wireless LAN and bluetooth.

19. In quadrature amplitude modulation there are eight phase quadrants.

20. QAM is widely used in telecommunications and in digital cable TV.

21. PWM and PDM are the same.

22. PPM is less sensitive to channel interference than PAM or PWM.

23. Pulse code modulation involves sampling of an analog signal amplitude at regular intervals.

24. The three main data transmission combinations are digital-to-analog, analog-to-digital, and digital-to-digital.

25. TDM stands for transmitted data multiplexing.

26. FDM is a baseband technique in which the total bandwidth available to a system is divided into frequency sub-bands and information is sent in analog form.

27. A bus connects two or more devices to allow them to communicate.

28. A bus is only defined by the wires and connectors.

29. A parallel bus is always faster than a serial bus.

30. Bus width is the width of each conductor in a parallel bus.

31. Handshaking is part of bus protocol.

32. A single-ended transmission is simpler and lower in cost compared to a differential transmission.

33. A tri-state driver has a HIGH state, a LOW state, and a shorted state.

34. PCI stands for *peripheral computer interface*.

35. Two types of PCI buses are PCI-X and PCI-E.

36. IEEE-488 is known as the GPIB.

37. SCSI stands for *serial computer system interface*.

38. USB is a widely used standard serial bus for connecting peripherals to a computer.

39. The SPI bus is typically used in embedded systems and on PCBs for communications between microprocessors or microcontrollers and peripheral IC chips or between two processors.

40. An internal serial bus primarily for connecting ICs on a PCB is the I^2C bus.

41. Most automotive systems use the Firewire bus.

42. CAN stands for *computer area node*.

SELF-TEST

Answers are at the end of the chapter.

1. The main purpose of the shield in a coaxial cable is to
   (a) make the cable stronger   (b) dissipate heat
   (c) prevent EMI   (d) prevent distortion

2. UTP is color-coded according to
   (a) a 10-pair color code   (b) a 25-pair color code
   (c) the resistor color code   (d) the primary colors

3. Advantages that fiber-optic systems have over electrical transmission media are
   (a) higher data rate, less susceptible to noise, and longer transmission distance
   (b) lower cost, higher data rate, and simplicity
   (c) higher data rate, higher EMI, less distortion
   (d) higher baud, availability, and reliability

4. The modes of light propagation in optical fibers are
   (a) simplex and duplex   (b) multimode and single-mode
   (c) synchronous and asynchronous   (d) scatter and direct

5. The electromagnetic spectrum includes
   (a) radio waves, microwaves, and audio waves
   (b) radio waves, microwaves, and sonic waves
   (c) visible light, infrared, and alpha waves
   (d) radio waves, microwave, and infrared

6. Satellites use
   (a) ground propagation   (b) line-of-sight propagation
   (c) ionospheric propagation   (d) triangulation

7. In asynchronous transmission, data are sent in short bursts known as
   (a) packets   (b) frames
   (c) bundles   (d) quanta

8. For a given bit rate, the baud compared to the bit rate is
   (a) always greater   (b) always less
   (c) equal or greater   (d) equal or less

9. The efficiency of a data transmission system is
   (a) the ratio of baud to bit rate
   (b) the ratio of actual data rate to ideal data rate
   (c) the ratio of data bits to total bits
   (d) the ratio of parity bits to data bits

10. The Manchester code format is
   (a) NRZ   (b) biphase
   (c) RZ   (d) FDM

11. A synchronous data frame does not contain a(n)
   (a) preamble   (b) data field
   (c) address   (d) vector field

12. Three types of data channel connections in terms of data flow are
   (a) input, output, neutral
   (b) simplex, half-duplex, full-duplex
   (c) simplex, duplex, triplex
   (d) uniplex, diplex, biplex

13. In FSK modulation,
   (a) the frequency of a carrier signal is varied by a digital signal.
   (b) the frequency of a digital signal is varied by a carrier signal.
   (c) the phase of a carrier signal is varied by a digital signal.
   (d) the amplitude of a carrier signal is varied by a digital signal.

14. Types of modulation in which a parameter of a sine-wave carrier signal is varied by a digital signal are
   (a) PAM, PWM, PPM   (b) QAM, PAM, ASK
   (c) FSK, PSK, PPM   (d) FSK, ASK, PSK

15. QAM stands for
   (a) quadrature analysis method
   (b) quadrature amplitude modulation
   (c) quasi-amplitude modulation
   (d) quadratic amplitude modulation

16. In QAM, the parameters that are varied are
   (a) amplitude and frequency   (b) phase and frequency
   (c) amplitude and phase   (d) pulse width and position

17. Three methods of modulating a digital signal with analog data are
   (a) PAM, PWM, PPM   (b) PAM, ASK, PPM
   (c) FSK, QAM, PAM   (d) QAM, PAM, PWM

18. The most likely type of modulation to be used in motor speed control is
   (a) PAM   (b) PPM
   (c) PWM   (d) QAM

19. A method in which an analog signal is sampled and converted to a digital code is
   (a) PCM   (b) PDM
   (c) PDC   (d) PPM

20. TDM stands for
    (a) time duration modulation
    (b) time division modulation
    (c) time division multiplexing
    (d) time division method

21. Two methods of combining data in TDM are
    (a) fast and slow
    (b) bit-interleaved and byte-interleaved
    (c) time-domain and frequency-domain
    (d) simplex and duplex

22. A method by which data from multiple sources are sent simultaneously is
    (a) TDM
    (b) FCM
    (c) FSK
    (d) FDM

23. Properties that define a bus include
    (a) type of connectors
    (b) length and type of cable or connection
    (c) data rate and encoding
    (d) all of these

24. The speed of a parallel bus can be limited by
    (a) crosstalk
    (b) EMI
    (c) skew
    (d) all of these

25. The method by which two devices initiate and complete a bus transfer is
    (a) handshaking
    (b) saluting
    (c) give and take
    (d) multiple access protocol

26. PCI is the acronym for
    (a) peripheral controller interface
    (b) peripheral computer interface
    (c) protocol compatible interface
    (d) peripheral component interconnect

27. In a PCI system, the individual paths from switch to peripherals are called
    (a) pipes
    (b) lanes
    (c) highways
    (d) channels

28. The following is not a classification of USB:
    (a) low-speed
    (b) full-speed
    (c) high-speed
    (d) intermediate speed

29. A 3.0 USB cable contains
    (a) two twisted pairs
    (b) one twisted pair
    (c) three twisted pairs
    (d) two straight wires

30. Four types of USB packets are
    (a) token, data, handshake, and start-of-frame
    (b) token, data, handshake, and control
    (c) identification, address, synchronization, and data
    (d) none of these

31. The RS-232 encoding method is
    (a) RZ
    (b) Manchester
    (c) biphase
    (d) NRZ

32. The bus typically used to connect systems in an automobile is the
    (a) SPI
    (b) CAN
    (c) I^2C
    (d) PCI

PROBLEMS

Answers to odd-numbered problems are at the end of the book.

Section 13–1 Data Transmission Media

1. List four parts of a coaxial cable.
2. What do the acronyms UTP and STP mean?
3. Name four advantages of fiber optic media over electrical transmission media.
4. List three parts of an optical fiber.
5. Describe the multimode in an optical fiber.

6. Which of the optical fiber sizes (50/125, 62.5/125, 8.3/125) operates in multimode and which operate(s) in single mode?

7. Draw a basic block diagram of a fiber optics communications link.

8. A frequency of 100 MHz falls into what part of the electromagnetic spectrum?

9. In what frequency range does visible light fall?

Section 13–2 Methods and Modes of Data Transmission

10. If data bits are transmitted serially at a 1 MHz rate, how many bits can be transmitted in 1 ms?

11. If a byte of data is transmitted in parallel in 1 μs, what is the data rate in bits per second?

12. Eight voltage levels are being transmitted by a system where each level (symbol) represents a 3-bit code. Assuming that 12 symbols are transmitted in 0.5 μs, determine the bit rate and the baud.

13. Assume a 5-bit code is used for each symbol transmitted. If the bit rate is 25 MHz, what is the baud?

14. A certain data packet contains a total of 20 bits of which 16 are data bits. Determine the efficiency.

15. Show the data waveform for the bit sequence 101011100011 in NRZ and in RZ formats.

16. For the bit sequence in Problem 15, show the Manchester code.

17. Determine the bit sequence represented by the Manchester code in Figure 13–81.

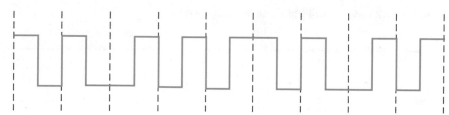

FIGURE 13–81

18. List and describe each part of a typical synchronous data frame structure.

Section 13–3 Modulation of Analog Signals with Digital Data

19. Determine the binary code represented by the ASK signal in Figure 13–82. Presence of a signal is a 1 and absence of a signal is 0.

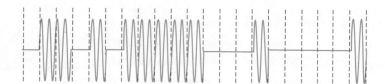

FIGURE 13–82

20. Show how you would respresent four successive bits (1010) using FSK.

21. Repeat Problem 20 for PSK.

22. Refer to Figure 13–23 and determine the sequence of bits represented by the QAM signal in Figure 13–83.

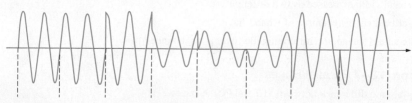

FIGURE 13–83

23. Sketch a constellation map for a 4-bit PSK system.

24. Repeat Problem 23 for a 4-bit QAM system.

Section 13–4 Modulation of Digital Signals with Analog Data

25. Describe the PWM intersective method.

26. Develop the PAM and PWM signals for the waveform in Figure 13–84.

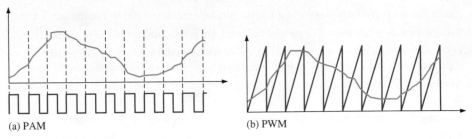

(a) PAM

(b) PWM

FIGURE 13–84

27. In a certain PPM system there are 2^3 positions in each period, T. If $T = 10$ ms, determine the data rate.

28. Show the NRZ code for the PPM signal in Figure 13–85.

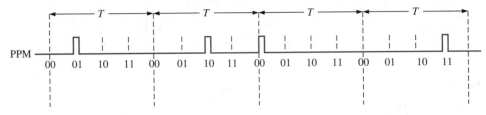

FIGURE 13–85

29. In a PCM code, how many bits are required to represent 16 voltage levels of a modulating signal?

30. Show the 4-bit PCM code in NRZ format for four successive samples of an analog waveform. The sampled values are 1, 3, 5, and 7.

Section 13–5 Multiplexing and Demultiplexing

31. Explain the difference between bit-interleaved and byte-interleaved TDM.

32. Describe the difference between synchronous and statistical TDM.

33. What type of filters are used in an FDM system and what is their purpose?

34. What is the frequency separation called between each source in an FDM system?

Section 13–6 Bus Basics

35. List six physical characteristics and five electrical and performance characteristics of a bus.

36. Explain the difference between bus width and bus bandwidth.

37. A certain bus is specified with a width of 16 bits and a frequency of 100 MHz. Determine the bus bandwidth expressed as two different values.

38. Describe a simple example of a handshake.

39. State an advantage of a differential bus over a single-ended bus.

Section 13–7 Parallel Buses

40. Explain the difference between PCI and PCI-X buses.

41. Explain how the PCI-Express differs from PCI and PCI-X.

42. What does x2 mean in a PCI-Express bus?

43. The terms *listener* and *talker* are associated with which bus standard?

44. Provide the description of each of the SCSI signals: BSY, SEL, RST, C/D, REQ, ACK, ATN, and MSG.

45. Consider the GPIB interface between a talker and a listener as shown in Figure 13–86(a). From the handshaking timing diagram in part (b), determine how many data bytes are actually transferred to the listening device.

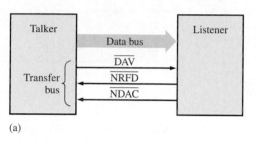
(a)

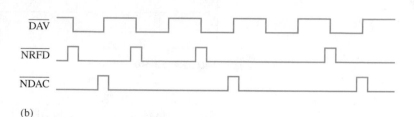
(b)

FIGURE 13–86

46. Describe the operations depicted in the GPIB timing diagram of Figure 13–87. Develop a basic block diagram of the system involved in this operation.

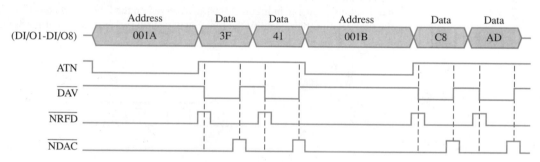

FIGURE 13–87

Section 13–8 The Universal Serial Bus (USB)

47. Identify each of the symbols in Figure 13–88.

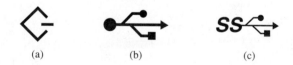
(a)   (b)   (c)

FIGURE 13–88

48. List the four types of USB packets.

49. Describe each of the fields in the USB data packet in Figure 13–89.

Sync 8/32	PID 8	Data 0-8192	CRC 16	EOP 3

FIGURE 13–89

50. What type of data encoding is used in USB 3.0?

51. Determine the maximum number of bytes in a USB data field.

52. What is the maximum separation of two USB 2.0 devices?

Section 13–9 Other Serial Buses

53. Describe how RS-232 and RS-422 differ.

54. List the four signals in an SPI bus and describe each one.

55. Describe the main use for the I²C bus.

56. Fill in the field descriptions for the blank CAN bus data format in Figure 13–90.

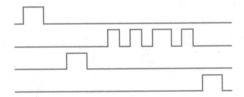

FIGURE 13–90

57. Refer to Figure 13–72 and list additional units that may appear on a CAN automotive system.

58. What is the data rate for the IEEE-1394 bus standard in the S100 mode? What is the data rate in the S1600 mode?

Section 13–10 Bus Interfacing

59. In a simple serial transfer of eight data bits from a sending device to an receiving device, the handshaking sequence in Figure 13–91 is observed on the four generic bus lines. By analyzing the time relationships, identify the function of each signal and indicate if it originates at the sender or at the receiver.

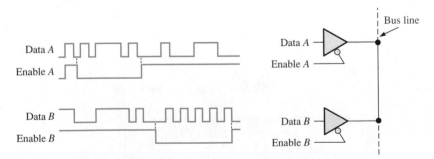

FIGURE 13–91

60. Determine the signal on the bus line in Figure 13–92 for the data input and enable waveforms shown.

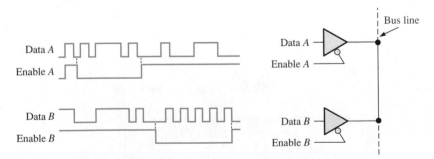

FIGURE 13–92

61. In Figure 13–93(a), data from the two sources are being placed on the data bus under control of the select line. The select waveform is shown in Figure 13–93(b). Determine the data bus waveforms for the device output codes indicated.

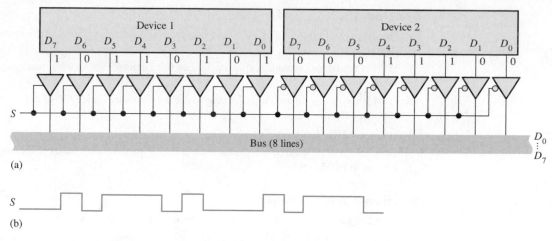

FIGURE 13–93

ANSWERS

SECTION CHECKUPS

Section 13–1 Data Transmission Media

1. Wire, coaxial cable, twisted pair cable, optical fiber cable, and wireless
2. The shield protects against EMI.
3. Ground wave, ionospheric, line-of-sight
4. Gamma radiation has the highest frequencies
5. Baseband uses digital modulation (a series of pulses). Broadband uses a digitally modulated analog signal.

Section 13–2 Methods and Modes of Data Transmission

1. Serial data are one bit at a time in sequence. Parallel data are simultaneous multiple bits at a time.
2. Synchronization allows the receiver to recognize the beginning and end of a data transmission.
3. RZ, NRZ, biphase are three types of data format.
4. Simplex, half-duplex, and full-duplex are three modes of data transmission.

Section 13–3 Modulation of Analog Signals with Digital Data

1. Modulation techniques are ASK, FSK, PSK, and QAM.
2. The amplitude of the analog signal is changed in ASK.
3. The frequency of the analog signal is changed in FSK.
4. QAM is quadrature amplitude modulation.
5. The phase of the analog signal is changed in PSK.

Section 13–4 Modulation of Digital Signals with Analog Data

1. Pulse modulation methods are PAM, PWM, PPM, and PCM
2. Pulse amplitude
3. Pulse width
4. Pulse position
5. A binary code

Section 13–5 Multiplexing and Demultiplexing

1. The purpose of multiplexing is to send data from several sources on a single communication channel.

2. TDM is time division multiplexing where data from multiple sources are interleaved on a time basis.

3. FDM is frequency division multiplexing where data from multiple sources are sent simultaneously at different frequencies.

4. Statistical TDM has the higher efficiency.

5. The guard band is the frequency separation between the frequency bands of the multiple sources in FDM.

Section 13–6 Bus Basics

1. Speed of a parallel bus can be limited by crosstalk, EMI, and clock skew.

2. An internal bus connects parts of a single system. An external bus connects one system to another separate system.

3. Bus characteristics include width, frequency, transfer rate, and bandwidth.

4. Bus protocol is a set of rules used by two or more devices to establish and maintain communication.

5. A single-ended system uses one wire for data and one wire for ground, where the signal voltage on the wire is with respect to ground. In a differential system, two wires are used for data and one wire for ground. The data signal is sent on one wire and its complement (inversion) is sent on the other wire. The difference between the two data wires is the differential signal.

Section 13–7 Parallel Buses

1. PCI is peripheral component interconnect.

2. PCI-Express is also designated PCIe and PCI-E.

3. A lane is a dedicated path to a single chip known as a switch.

4. GPIB is IEEE-488.

5. SCSI is small computer system interface.

Section 13–8 The Universal Serial Bus (USB)

1. USB is universal serial bus, a widely used standard bus.

2. USB pins are D+, D−, +5 V, ground.

3. The twisted pair reduces or eliminates noise.

4. USB 3.0 can run at higher speeds than USB 2.0. USB 3.0 has shorter cable lengths than USB 2.0.

Section 13–9 Other Serial Buses

1. RS-232, RS-422, RS-423, RS-495, SPI, I^2C, CAN, Firewire, and serial SCSI

2. SPI is serial-to-peripheral interface.

3. I^2C is inter-integrated circuit.

4. CAN is controller area network.

5. Firewire is IEEE-1394.

Section 13–10 Bus Interfacing

1. Tri-state buffers allow devices to be completely disconnected from the bus when not in use, thus preventing interference with other devices.

2. A bus interconnects all the devices in a system and makes communication between devices possible.

RELATED PROBLEMS FOR EXAMPLES

13–1 See Figure 13–94.

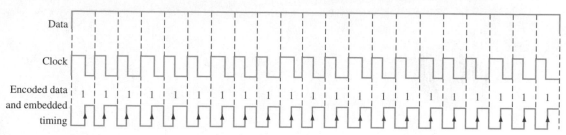

FIGURE 13–94

13–2 Bit rate $= 40$ kbps

13–3 Efficiency $= 0.828$ (82.8%)

13–4 Eight amplitudes and eight phases can be used to represent a 4-bit code.

13–5 There would be more pulses closer together in both cases providing a more accurate representation of the analog signal.

13–6 Data rate $= 8$ Mbps; 256 pulse positions

13–7 Six PCM code bits to represent 64 levels

13–8 15.625 MHz

TRUE/FALSE QUIZ

1. F	**2.** T	**3.** T	**4.** T	**5.** F	**6.** T	**7.** F	**8.** T	**9.** T
10. F	**11.** T	**12.** F	**13.** T	**14.** T	**15.** F	**16.** F	**17.** T	**18.** T
19. F	**20.** T	**21.** T	**22.** T	**23.** T	**24.** T	**25.** F	**26.** F	**27.** T
28. F	**29.** F	**30.** F	**31.** T	**32.** T	**33.** F	**34.** F	**35.** T	**36.** T
37. F	**38.** T	**39.** T	**40.** T	**41.** F	**42.** F			

SELF-TEST

1. (c)	**2.** (b)	**3.** (a)	**4.** (b)	**5.** (d)	**6.** (b)	**7.** (a)	**8.** (d)
9. (c)	**10.** (b)	**11.** (d)	**12.** (b)	**13.** (a)	**14.** (d)	**15.** (b)	**16.** (c)
17. (a)	**18.** (c)	**19.** (a)	**20.** (c)	**21.** (b)	**22.** (d)	**23.** (d)	**24.** (d)
25. (a)	**26.** (d)	**27.** (b)	**28.** (d)	**29.** (c)	**30.** (a)	**31.** (d)	**32.** (b)

Data Processing and Control

CHAPTER OBJECTIVES

- Name the basic units of a computer
- Name the computer buses and how they are used
- Discuss the considerations for a practical computer system
- Describe the purpose of buffers, decoders, and wait-state generators in a computer system
- Define and explain the advantage of DMA
- Name the basic elements of a microprocessor
- Describe the basic architecture of a microprocessor
- Explain basic microprocessor (CPU) operation
- List and describe some microprocessor addressing modes
- Define and describe microprocessor polling, interrupts, exceptions, and bus requests
- Discuss the operating system of a computer
- Explain pipelining, multitasking, and multiprocessing
- Describe a simple assembly language program
- List some typical microprocessor instructions
- Distinguish between assembly language and machine language
- Describe the architecture of a microcontroller and explain how it differs from a microprocessor
- Discuss embedded systems
- Discuss some microcontroller applications
- Describe a system on chip (SoC)

KEY TERMS

Key terms are in order of appearance in the chapter.

- CPU
- Microprocessor
- Main memory
- Caching
- BIOS
- System bus
- Signal loading
- Buffer
- Wait state
- Pipelining
- ALU
- Program
- Op-code
- Operand
- Interrupt
- Exception
- Interrupt vector table
- Bus master
- DMA
- Hardware
- Software
- Operating system
- Multitasking
- Multiprocessing
- Machine language
- Assembly language
- High-level language
- Microcontroller
- System on chip

VISIT THE WEBSITE

Study aids for this chapter are available at
http://www.pearsonhighered.com/careersresources/

INTRODUCTION

This chapter provides a basic introduction to computers, microprocessors, and microcontrollers. It gives you a fundamental coverage of basic concepts related to data processing and control. For the most part, a generic approach is used to present basic concepts of the topics. The total computer system with practical considerations is covered. Various aspects of a microprocessor and

its role as the CPU in computer systems are presented and programming is briefly discussed. Microcontrollers and system on chip (SoC) are also introduced, and some applications are described.

14–1   The Computer System

General-purpose computers, with which most are familiar, and special-purpose computers are used to control various functions or perform specific tasks in areas such as automotive, consumer appliances, manufacturing processes, and navigation. The general-purpose computer system, which can be programmed to do many different things, is the focus in this section.

After completing this section, you should be able to

♦ Describe the basic elements of a general-purpose computer

♦ Discuss each part of a computer

♦ Explain a peripheral device

All computer systems work with information, or data, to produce a desired result. To accomplish this, computer systems must perform the following tasks:

• Acquire information from data sources, including human operators, sensors, memory and storage devices, communication networks, and other computer systems

• Process information by interpreting, evaluating, manipulating, converting, formatting, or otherwise working with acquired data in some intended fashion as directed by a step-by-step set of instructions called a *program*

• Provide information in a meaningful form to data recipients, including human operators, actuators, memory and storage devices, communication networks, and other computer systems

Specific sections and components in computer systems accomplish each of these tasks. Information processing is performed by the central processing unit, or CPU, which is the brain of the computer system. The CPU acquires information through the input section of the computer system, provides information through the output section, and uses the system memory and storage to store and retrieve information as needed. The CPU transfers information to and from other sections of the computer system over special groups of signal lines called *buses*. Figure 14–1 shows a block diagram of a general-purpose computer system. Each block will be discussed in terms of its purpose and function.

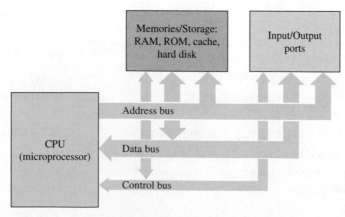

FIGURE 14–1   Basic computer block diagram.

The Central Processing Unit

The central processing unit (**CPU**) performs much of what is associated with the term *computer*. It executes the instruction sequences (called *programs*) in the computer system, directly processes much of the data that pass through the computer system, and controls and coordinates the various sections that make up the computer system. To play such a large role in the computer system, the CPU consists of four separate units: the arithmetic logic unit (ALU), the instruction decoder, the timing and control unit, and the register set. The CPU is basically a **microprocessor** (or simply processor). A single IC package can contain two or more processors, forming a **multicore processor**.

Memory and Storage

Computer systems must have some means of storing and retrieving the information with which they work and use two types of devices—memory devices and storage devices—to do so. Although the usage and meanings of the terms can overlap somewhat, they primarily differ in the construction of the devices and the information they contain. Memory devices typically are semiconductor devices that store information electronically, interface with the computer system through the system buses, and contain dynamic information, such as programs and program variables, that is frequently accessed or modified. Storage devices typically store information on some physical medium, interface with the system through a peripheral interface, and contain primarily static information, such as program and data files, that is accessed or modified relatively infrequently. Memory devices are faster than storage devices; however, memory devices have lower storage capacities and higher cost per bit than storage devices.

Memory in computer systems can be classified both by the type of memory and the function it performs. The different types and characteristics of memory were discussed in Chapter 11. Here we examine the functional requirements of memory in computer systems.

Main Memory

The **main memory** is the computer system memory that contains programs and data associated with them, such as program variables, the program stack, and information the operating system requires to execute the program. The earliest 8-bit processors (for example, the Intel 8080, Motorola 6800, and InMOS 6502) had 16-bit address buses that could access $2^{16} = 65,536$ bytes (64 kilobytes or 64 kB) of memory. However, the main memory in 8-bit PCs was actually less than this because other devices in the system used part of the address space. The 16-bit computers that followed had 20-bit address buses that could access $2^{20} = 1,048,756$ bytes (1 megabyte or 1 MB) of memory. Modern computers require gigabytes of main memory to support the requirements of their graphical user interface (GUI) operating systems and application programs. Main memory must meet the requirements of a large storage capacity at an economical price and also allow the computer system to modify data within it. Because of these requirements, computer systems typically use some form of dynamic RAM (DRAM) for main memory that features large capacity, low cost per bit, and read/write capability.

Cache Memory

Cache memory is memory that computer systems use to overcome the relatively slow speed of main memory DRAM. **Caching** is a process that copies frequently accessed instructions or data from slow main memory into faster cache memory to reduce access time and improve system performance. Because of these requirements, computer systems use some form of static RAM (SRAM) for cache memory.

Basic Input/Output System (BIOS) Memory

The design of every computer system differs to some extent from other systems. The basic input/output system (**BIOS**) memory contains system-specific low-level code that runs the power on self-test (POST), installs specialized software called drivers to configure and provide access to the computer system hardware, and loads the operating system. The BIOS memory must retain its contents when power is removed so that the BIOS code is ready to run when the computer first powers up. This requires computer systems to use some form of nonvolatile memory for BIOS.

The earliest personal computers used read-only memory (ROM) for BIOS, so any change to the BIOS required the user to replace the ROM chip (which was often socketed) itself. Later computers used a low-power CMOS device with a back-up battery to preserve the contents when the system power was shut off. This allowed users to change and save BIOS settings when they made changes to system hardware configuration. Most recently, computers have used EEPROM and flash devices so that users can easily upgrade the BIOS firmware to the latest revision. **Firmware** is software programs or data that have been written into ROM.

Content-Addressable Memory

Computers often use specialized types of memory in addition to those types mentioned previously. One specialized type of memory is the content-addressable (or associative) memory, whose operation differs from that of conventional memory. Conventional memory returns the data stored at a specified address. Content-addressable memory returns the address that contains a specified data value. Computers use content-addressable memory for special data tables that support caching and paging operations.

FIFO

Another specialized type of memory is the FIFO (first-in, first-out) memory. Conventional memory, such as SRAM and DRAM, allow computers to store data and to retrieve data from any memory location in any order. FIFO memory returns data only in the order in which the data were stored. As the acronym *FIFO* indicates, the first data stored in memory must be the first data taken out of memory. Computers use FIFO memory for special data structures called **queues**. Queues temporarily store data for which the sequence of data must be preserved, such as program instructions.

Input/Output Ports

Input/output (or I/O) ports are interfaces that allow computers to transfer data to and from external entities such as users, peripherals (such as mice, keyboards, video monitors, scanners, printers, modems, and network adapters), and other computers. I/O ports vary greatly in complexity and capability. An I/O port can be serial or parallel, operate as an input, output, or both, and transfer several thousand to several billion bits per second. Many I/O ports, such as RS-232, USB 3.0, SCSI-5, Firewire, and Ethernet ports, conform to official or de facto standards to simplify computer system connections. These standards are usually developed by international organizations and typically specify not only the type of connectors but also the pin assignments, electrical signal levels, signal timing, data transmission rates, and communication protocols (i.e., the format, organization, and meaning of data patterns). EIA 802, for example, is the international standard for Ethernet communications and IEEE 1394 is the standard for Firewire. These standards ensure that all devices that comply with the standard will be able to communicate with each other.

Processors support I/O ports and operations in one of two ways. One way is memory-mapped I/O, in which the processor treats I/O ports as memory locations and external circuitry converts standard read and write operations into I/O port accesses. The second way is direct I/O, in which specific processor pins and instructions are exclusively dedicated to data input and output operations. In either case, general-purpose processors require additional circuitry and program code to implement specific communications standards and protocols. Specialized microcontrollers like the Motorola MC68360 and NXP LPC2292 improve on this by incorporating additional circuitry and embedded firmware to support

UART, I2C, Ethernet, CAN, SPI and other popular communication standards on their I/O ports with a minimum of driver coding and external interface circuitry.

System Bus

As you have learned, computers acquire, process, and provide information. Computers must be able (a) to specify where to acquire and return information, (b) to transfer the information from its source to its destination, and (c) to coordinate the movement of data within the computer system. The mechanism by which the computer accomplishes this is the **system bus**, which consists of three component buses: the address bus, the data bus, and the control bus.

The Address Bus

The **address bus** is the means by which a processor specifies the system location from which data are to be read or to which data are to be written. For example, the processor sends an address code to the memory specifying where certain data are stored. If the address bus is 32 bits wide, 2^{32} or 4,294,967,296 memory locations can be accessed.

The Data Bus

The **data bus** consists of signal lines over which the computer system transfers information from one device to another. Because the processor can both read data from and write data to system devices, each data line is bidirectional. The number of data lines determines the width of the data bus, which is a factor in how quickly the processor can process data. The earliest microprocessors had 4-bit and 8-bit data buses, but modern processors have 64-bit data buses.

The Control Bus

The **control bus** is the collection of signals that controls the transfer of data within the system and coordinates the operation of system hardware. Unlike the address and data buses, which consist of functionally identical signals that function as a group, the individual signals lines that make up the control bus vary in characteristics, nature, and function. Control signals can be unidirectional or bidirectional, can function individually or with other control signals, can be active-HIGH or active-LOW, can operate synchronously or asynchronously, and can be edge-oriented or level-oriented. Despite this individual diversity, computer systems and processor operations are similar enough that the signals that make up the control bus—read, write, interrupt, and others—are also similar.

A Typical Computer System

The block diagram in Figure 14–2 shows the main elements in a typical computer system and how they are interconnected. Notice that the computer itself is connected with several peripheral units. For the computer to accomplish a given task, it must communicate with the "outside world" by interfacing with people, sensing devices, or devices to be controlled through input and output ports.

Computer Software

In addition to the hardware, a major part of a computer system is the software. The software makes the hardware perform. The two major categories of software used in computers are the system software and the application software.

The system software is called the operating system (OS) and allows the user to interface with the computer. The most common operating systems are Windows and Mac OS. Many other operating systems are used in special-purpose and mainframe computers.

System software performs two basic functions. It manages all the hardware and software in a computer. For example, the operating system manages and allots space on the hard disk. System software also provides a consistent interface between applications software

InfoNote

Grace Hopper, a mathematician and pioneer programmer, developed considerable troubleshooting skills as a naval officer working with the Harvard Mark I computer in the 1940s. She found and documented in the Mark I's log the first real computer bug. It was a moth that had been trapped in one of the electromechanical relays inside the machine, causing the computer to malfunction. From then on, when asked if anything was being accomplished, those working on the computer would reply that they were "debugging" the system. The term stuck, and finding problems in a computer (or other electronic system), particularly the software, would always be known as debugging.

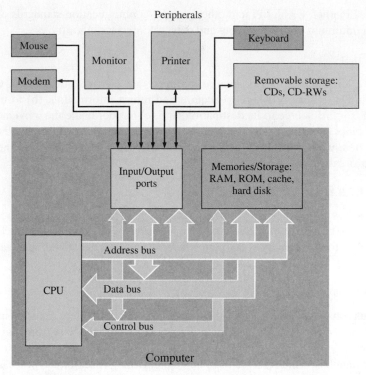

FIGURE 14–2 Basic block diagram of a typical computer system including common peripherals. The computer itself is shown in the gray block.

and hardware. This allows an applications program to work on various computers that may differ in hardware details. The operating system on your computer allows you to have several programs running at the same time. This is called *multitasking*.

Application software is used to accomplish a specific job or task, such as word processing, accounting, tax preparation, circuit simulation, graphic design, to name only a very few.

SECTION 14–1 CHECKUP

Answers are at the end of the chapter.

1. What are the major functional blocks in a computer?
2. What are peripherals?
3. What is the difference between computer hardware and computer software?
4. How does content-addressable memory differ from conventional memory?
5. Compare and contrast the characteristics of the address, data, and control buses in a computer system.

14–2 Practical Computer System Considerations

Practical computer designs incorporate special circuitry that resolves four issues that exist in real-world systems: shared signal lines, signal loading, device selection, and system timing.

After completing this section, you should be able to

- Identify design considerations for practical computer systems
- Explain the role and operation of buffers, decoders, and wait-state generators in practical computer systems

Figure 14–3 shows a block diagram of a practical computer system, based on the consideration for shared signal lines, signal loading, device selection, and system timing.

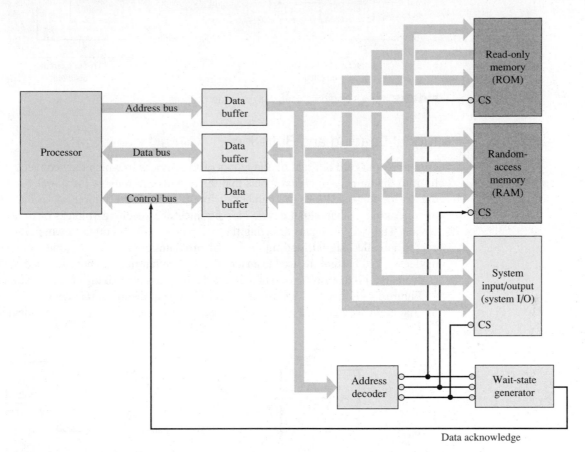

FIGURE 14–3   Block diagram of a practical computer system.

Shared Signal Lines

When the outputs of two or more devices connect to the same signal line, the potential for bus contention exists. Bus contention occurs when device outputs attempt to drive a signal line to different voltage levels. This causes high current to flow from one output into the other, which can damage the devices. Typically, bus contention occurs when device outputs are at different logic levels. However, even when devices are at the same logic level, the variation for different devices will cause some device output voltages to be higher than others so that bus contention will occur. Two special types of output, the tri-state output and open collector output, allow devices to share signal lines, while avoiding bus contention.

The term *tri-state* is a registered trademark of National Semiconductor but is often used interchangeably with the generic terms *three-state* or *3-state*. As the name suggests, the tri-state output adds a third output state, called the high-impedance or high-Z state, to the usual logic LOW and HIGH states. The tri-state switch is effectively a switch that disconnects the output of the tri-state device from the signal line so that it does not interfere with other devices from driving the line. When a tri-state device is enabled, it outputs a logic LOW or HIGH as other digital devices. When a tri-state device is disabled, the output assumes the high-Z state and the output is said to be tri-stated. When tri-state outputs share a signal line, only one output at a time must be enabled to ensure that bus contention will not occur. Figure 14–4 shows the operation of tri-state outputs.

Devices that are designed to connect to processor buses, such as memory and interface devices, typically have tri-state outputs built into them. Devices that do not have tri-state outputs or open-collector outputs must use tri-state buffers to connect to buses.

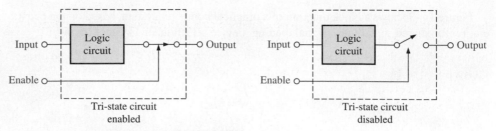

FIGURE 14-4   Logic devices with tri-state outputs.

Signal Loading and Buffering

Digital outputs are affected by the inputs of the devices to which they connect. There is a limit to the number of digital inputs that the outputs can reliably drive; this limit is called the device *fan-out*. When the number of inputs exceeds the fan-out of an output device, the operation of that output device may not meet the specified voltages or timing for that device. The issue of inputs affecting the performance of an output to which they are connected is called **signal loading**. To avoid problems with signal loading, special digital devices called buffers are used to ensure that device fan-outs are not exceeded. A **buffer** is a special circuit that isolates the output of a device from the loading effects of other devices.

Figure 14–5 illustrates the use of buffers to prevent the nine input devices from exceeding the eight-load fan-out of the output device. Note that up to seven input devices could

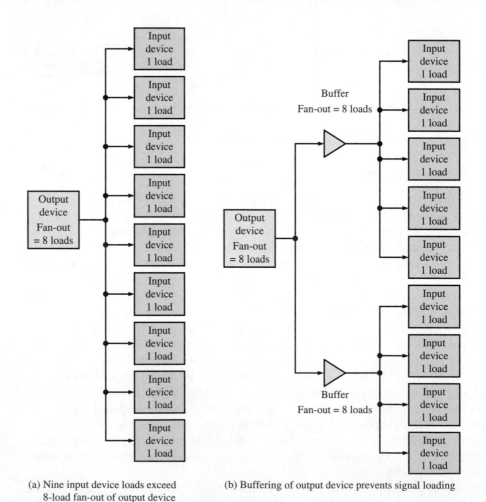

(a) Nine input device loads exceed   (b) Buffering of output device prevents signal loading
8-load fan-out of output device

FIGURE 14-5   Buffers are used to prevent overloading of driving device.

have been connected directly to the output device and a single buffer used to connect the remaining input devices. This would have reduced the parts count, but one characteristic of buffers is that each buffer will increase the propagation delay. If a single buffer were used, the response of the input devices connected to the buffer would be slower than that of the input devices connected directly to the output device. Using two buffers as shown helps match the propagation delay to all the input devices.

The buffers shown in Figure 14–5 are simple noninverting buffers, which means that the buffer output signal is identical to the buffer input signal. There are other types of buffers to ensure that devices will not degrade the performance of a device to which they are connected. These buffers include tri-state buffers like those mentioned previously, inverting buffers that invert the input signal, bidirectional buffers that can pass information through the buffer in both directions as on the data bus, and Schmitt triggers. A Schmitt trigger is a special device that helps prevent logic devices from acting erratically due to system noise affecting slowly changing inputs.

Device Selection

The processor uses the address bus to access ROM, RAM, hardware I/O ports, and other system devices. A question that naturally arises is how a device knows when the processor is attempting to access it rather than some other system device. The answer is that these devices have a special input, usually called a chip select (CS) or chip enable (CE), that enables the device. When the processor must access a specific device, it must assert the select line of the intended device.

While in theory processors could provide separate control lines to select system devices, this is not practical for general-purpose computers because there is no way for the system designers to know what devices a system will contain. Instead, system designers use PLDs or dedicated hardware decoders, similar to that in Figure 14–6, to decode processor addresses and generate the device select lines. For this example, the processor uses a 16-bit address bus where the upper (most significant) four bits are used to generate device select outputs.

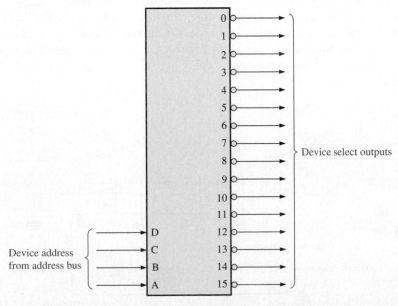

FIGURE 14–6   Address decoding for the purpose of device selection.

System Timing

A final issue with practical computer systems is system timing. In a computer system, the processor signals must meet the setup and hold times for each peripheral so that data are properly stored and accessed. As you have seen, decoding logic or buffers in the system can slow the processor signals. In some cases, the processor runs much faster than the peripherals that are available; in other cases, fast peripherals are available but their cost prohibits designers from using them. In addition, some peripherals, such as SRAM, are inherently faster than others, such as DRAM, so the signal timing that meets the setup and hold times for some devices will not meet the setup and hold times for others. To resolve this issue, three different types of system buses can be used: synchronous, asynchronous, and semisynchronous.

Synchronous buses include a synchronizing clock to ensure that signals from the processor meet the setup and hold times of the peripheral. Synchronous buses are faster than asynchronous or semisynchronous buses.

Asynchronous buses will automatically insert wait states in a bus cycle until a signal indicates that the bus cycle can finish. A **wait state** holds the state of the bus signals for one processor clock cycle so that the read or write operation is "frozen" for one clock period when the processor is accessing memory or other devices that are slow to respond. Several wait states may be necessary. Computer CPUs run at very high speeds, while memory technology does not seem to be able to catch up. Typical processors like the Intel Core 2 and the AMD Athlon 64 X2 run with a clock of several GHz, while the main memory clock generally is in the several hundred to over 1000 MHz range. Even some second-level CPU caches run slower than the processor core. In order to minimize the use of wait states, which slow the computer down, techniques such as CPU caches, instruction pipelines, instruction prefetch, and simultaneous multithreading are used.

Semisynchronous buses are similar to asynchronous buses except that a semisynchronous bus will complete the bus cycle unless a signal indicates that the processor should not complete bus cycle. Until the processor can complete the cycle, it will insert wait states.

Memory and other peripheral devices do not, as a rule, have signals indicating when data are ready. The signals that instruct the processor to insert wait states must be generated by an additional logic circuit, called a *wait-state generator*, which can be basically a programmable timer or shift register. The wait-state generator is clocked by the same clock as the processor and enabled by the device select line for a specific memory or other device. After the wait-state generator is enabled by the device select line, it will generate a ready signal after a specific number of clock cycles. Figure 14–7 shows an 8-bit parallel-in/serial-out shift register circuit that can insert up to six wait states for an asynchronous processor by delaying the ready signal to the processor by up to six clock cycles.

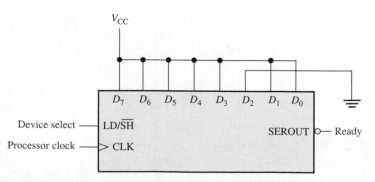

FIGURE 14–7 A wait-state generator programmed for one wait state.

The circuit of Figure 14–7, which inserts wait states for a single device, can be expanded to support more than one device. If two or more devices have the same number of wait states, their device select lines can be ANDed together (assuming the select lines are active-LOW).

EXAMPLE 14–1

For the wait-state generator in Figure 14–8, how many waits states will be generated when the device is selected?

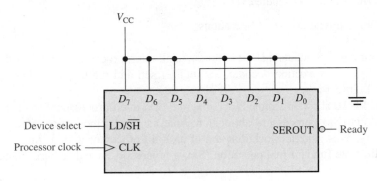

FIGURE 14–8

Solution

The initial pattern loaded into the shift register is 11101111_2. This shifted pattern for each clock and the corresponding number of wait states are

Clock 1 (0 wait states): 11110111_2

Clock 2 (1 wait state): 11111011_2

Clock 3 (2 wait states): 11111101_2

Clock 4 (3 wait states): 11111110_2

On the fourth clock after Device select goes LOW, the most significant bit of the SEROUT line for the shift register goes LOW. This causes the Ready output to go LOW, terminating the bus cycle. Therefore, the wait-state generator inserts **three** wait states.

Related Problem*

Which data input line of the shift register must be tied LOW for the wait-state generator in Figure 14–8 to insert five wait states?

---

*Answers are at the end of the chapter.

SECTION 14–2 CHECKUP

1. Define *bus contention* and discuss types of devices used to prevent it.

2. How does a processor enable various devices?

3. Define *wait state* and give its purpose.

4. What is the purpose of a buffer?

14–3   The Processor: Basic Operation

As you have learned, a microprocessor forms the CPU of a computer system. A microprocessor is a single integrated circuit that consists of several units, each designed for a specific job. The specific units, their design and organization, are called the *architecture* (do not confuse the term with the VHDL element). The architecture determines the instruction set and the process for executing those instructions.

After completing this section, you should be able to

♦ Name the four basic elements of a microprocessor

♦ Describe the fetch/execute cycle

♦ Explain the read and write operations

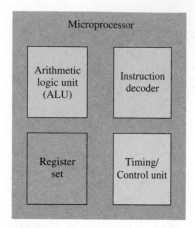

FIGURE 14–9   Elements of a microprocessor (CPU).

The four basic elements that are common to all microprocessors are the arithmetic logic unit (ALU), the instruction decoder, the register set, and the timing and control unit, as shown in Figure 14–9.

Figure 14–10 shows a simple block diagram of a microprocessor. The elements shown are common to most processors, although the internal arrangement or architecture and complexity vary. This generic block diagram of an 8-bit processor with a small register set is used to illustrate fundamental operation. Today, processors have data buses that are 64 bits.

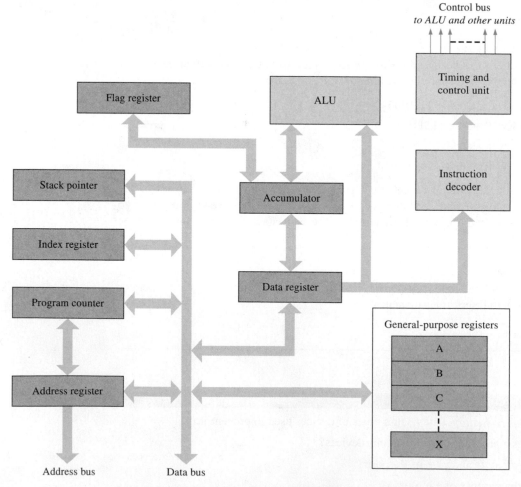

FIGURE 14–10   Basic model of a simplified processor.

The Fetch/Execute Cycle

When a program is being run, the processor goes through a repetitive cycle consisting of two fundamental phases, as shown in Figure 14–11. One phase is called *fetch* and the other is called *execute*. During the **fetch** phase, an instruction is read from the memory and decoded by the instruction decoder. During the **execute** phase, the processor carries out the sequence of operations called for by the instruction. As soon as one instruction has been executed, the processor returns to the fetch phase to get the next instruction from the memory.

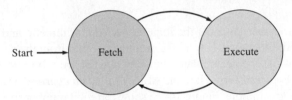

FIGURE 14–11   The fetch/execute cycle of a processor.

Pipelining

A technique where the microprocessor begins executing the next instruction in a program before the previous instruction has been completed is called **pipelining**. That is, several instructions are in the pipeline simultaneously, each at a different processing stage.

Typically, a pipeline is divided into stages or segments, and each stage can execute its operation concurrently with the other stages. When a segment completes an operation, it passes the result to the next segment in the pipeline and fetches the next operation from the preceding segment. The final results of each instruction emerge at the end of the pipeline in rapid succession. Figure 14–12 is a simplified illustration of nonpipelined processing compared to pipelined processing using three stages of execution.

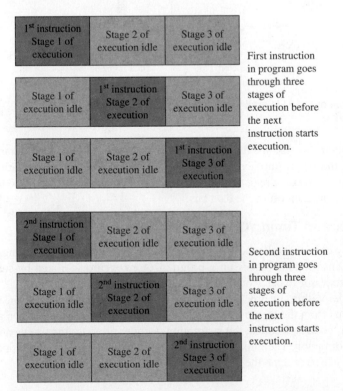

(a) Nonpipelined execution of a program showing three stages of execution

(b) Pipelined execution of a program showing three stages

FIGURE 14–12   Illustration of pipelining.

As shown in the figure, in nonpipelined processing of a program, one instruction at a time is executed through all of its stages before the next instruction begins execution. As you can see in part (a), all the stages of execution are idle (gray) except the one that is active (red). In pipelined processing, as soon as one instruction has finished an execution stage, the next instruction begins that stage. Pipelining results in much shorter overall execution times. Once the pipeline is "full," there are no idle processing stages.

Processor Elements

ALU

This part of the processor contains the logic to perform arithmetic and logic operations. Data are transferred into the **ALU** from the accumulator and from the data register. For the model in Figure 14–10, the accumulator and data register are 8-bit registers that hold one byte of data. Each byte transferred into the ALU is called an *operand* because it is operated on by the ALU. As an example, Figure 14–13 shows an 8-bit number from the accumulator being added to an 8-bit number from the data register. The result of this addition operation (sum) is put back into the accumulator and replaces the original operand that was stored there. When the ALU performs an operation on two operands, the result always goes into the accumulator to replace the previous operand.

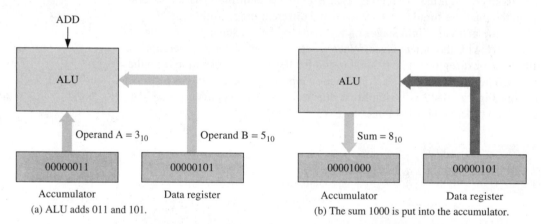

(a) ALU adds 011 and 101. (b) The sum 1000 is put into the accumulator.

FIGURE 14–13   Example of the ALU adding two operands.

As demonstrated in Figure 14–13, one function of the accumulator is to store an operand prior to an operation by the ALU. Another function is to store the result of the operation after it has been performed. The data register temporarily stores data that is to be put onto the data bus or that has been taken off of the data bus.

Instruction Decoder and Timing/Control Unit

An instruction is a binary code that tells the processor what it is to do. An orderly arrangement of many different instructions makes up a program. A **program** is a step-by-step procedure used by the processor to carry out a specified task.

The instruction decoder within the processor decodes an instruction code that has been transferred on the data bus from the memory. The instruction code is commonly known as an **op-code**. When the op-code is decoded, the instruction decoder provides the timing and control unit with this information. The timing and control unit can then produce the proper signals and timing sequence to execute the instruction.

Register Set

Processors typically have two categories of registers for temporary storage of data: general-purpose registers and special-purpose registers. General-purpose registers are used to store any type data that may be required by a program. Special-purpose registers are dedicated to a specific function. Some typical special-purpose registers are described as follows.

Flag register This register is sometimes called a condition code register or status register. It indicates the status of the contents of the accumulator or certain other conditions within the processor. For example, it can indicate a zero result, a negative result, the occurrence of a carry, or the occurrence of an overflow from the accumulator.

Program counter This counter produces the sequence of memory addresses from which the program instructions are taken. The content of the program counter is always the memory address from which the next byte is to be taken. In some processors, the program counter is known as the *instruction pointer*.

Address register This register temporarily stores an address from the program counter in order to place it on the address bus. As soon as the program counter loads an address into the address register, it is incremented (increased by 1) to the address of the next instruction.

Stack pointer The stack pointer is a register that is mainly used during program subroutines and interrupts. It is used in conjunction with the memory stack.

Index register The index register is used as one means of addressing the memory in a mode of addressing called *indexed addressing*.

The Processor and the Memory

The processor is connected to a memory with the address bus and data bus. Also, there are certain control signals that must be sent between the processor and the memory, such as the read and write controls. The address bus is unidirectional so the address bits go only one way, from the processor to the memory. The data bus is bidirectional, so data bits are transferred between the processor and memory in either direction. This is illustrated in Figure 14–14.

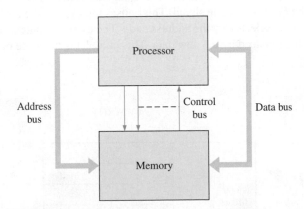

FIGURE 14–14 A processor and memory.

The Read Operation

To transfer data from the memory to the processor, a **read** operation must be performed, as shown in Figure 14–15, using an 8-bit data bus and a 16-bit address bus for illustration. To start, the program counter contains the address of the data to be read from the memory. This address is loaded into the address register and placed onto the address bus. The program counter is then incremented (advanced by one) to the next address and waits. Once the address code is on the bus, the processor timing and control unit sends a read signal to the memory. At the memory, the address bits are decoded and the desired memory location is selected. The read signal causes the contents of the selected address to be placed on the

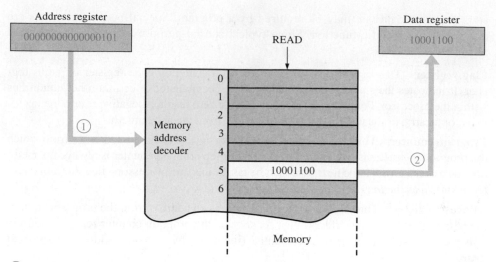

① Address 5_{10} is placed on address bus and followed by the read signal.

② Contents of address 5_{10} in memory is placed on data bus and stored in data register.

FIGURE 14–15   Illustration of the read operation.

data bus. The data are then loaded into the data register to be used by the processor, completing the read operation. In this illustration, each memory location contains one byte of data. When a byte is read from memory, it is not destroyed but remains in the memory. This process of "copying" the contents of a memory location without destroying the contents is called *nondestructive read*.

The Write Operation

To transfer data from the processor to the memory, a **write** operation is required, as illustrated in Figure 14–16. A data byte held in the data register is placed on the data bus, and the processor sends the memory a write signal. This causes the byte on the data bus to be stored at the memory location selected by the address code. The existing contents of that particular memory location are replaced by the new data. This completes the write operation.

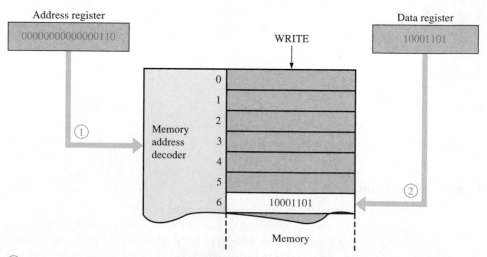

① Address code for address 6_{10} is placed on address bus.

② Data are placed on data bus and followed by the write signal. Data are stored at address 6_{10} in memory.

FIGURE 14–16   Illustration of the write operation.

Roles of the CPU

The CPU has three major roles in a computer system. The first role of the CPU is to control the system hardware. Specifically, the CPU determines how data move through the computer system, which devices are active, and when specific operations and data transactions occur. In computers, some of this control is decentralized by assigning some tasks (such as peripheral access and communications and graphics processing) to devices that can perform those tasks more quickly and efficiently than the CPU itself. Even so, the CPU still coordinates the operation of the computer system as a whole.

The second role of the CPU is to provide hardware support for the operating system software. The first computers were large mainframes that were too expensive to devote to a single user or program. The operating systems allowed these computers to support multiple users and programs, but they required special hardware to ensure that users and programs would not accidentally or deliberately interfere with each other. As the operating systems in personal computers evolved from single-user single-application platforms to multitasking and multiprocessing systems, the microprocessors have incorporated the features required to support them.

The third role of the CPU is to execute application programs. The CPU accesses the system hardware and controls the flow of data through the system largely because some application program requires that it do so. This role greatly influenced the development of many early complex instruction set computing (CISC) microprocessors. Reduced instruction set computing (RISC) processors emphasize smaller and more efficient instruction sets than those in CISC processors and place the burden of high-level programming support on the **compilers**, which are programs that convert the source code written by programmers to executable code that is executed by the processor.

SECTION 14–3 CHECKUP

1. Describe the fetch/execute cycle.

2. Name the four elements in a microprocessor.

3. What is the ALU and its purpose?

4. What happens during a read operation?

5. What happens during a write operation?

14–4 The Processor: Addressing Modes

A processor must address the memory to obtain data or store data. There are several ways in which the processor can generate an address when it is executing an instruction. These ways are called addressing modes and they provide for wide programming flexibility. Each instruction in a processor's instruction set generally has a certain addressing mode associated with it. The type and number of addressing modes vary from one processor to another. In this section, five common addressing modes are discussed, and generic instructions are used for illustration.

After completing this section, you should be able to

- Explain inherent addressing
- Explain immediate addressing
- Explain direct addressing

◆ Explain indexed addressing

◆ Explain relative addressing

Inherent Addressing

Inherent addressing is sometimes known as *implied addressing*. The one-byte instructions using this mode generally require no operand, or the operand is implied by the op-code, which is a mnemonic form of an instruction. An **operand** is the object to be manipulated by the instruction. For example, an instruction used to clear the accumulator (CLRA) has an implied operand of all zeros. The implied all-zeros operand ends up in the accumulator after the instruction is executed. Another example is a halt or wait instruction (WAI), which requires no operand because it simply tells the processor to stop all operations. The sequence that the processor goes through in handling an instruction with inherent addressing is illustrated in Figure 14–17. The op-codes used for illustration are similar to the op-codes of a typical processor.

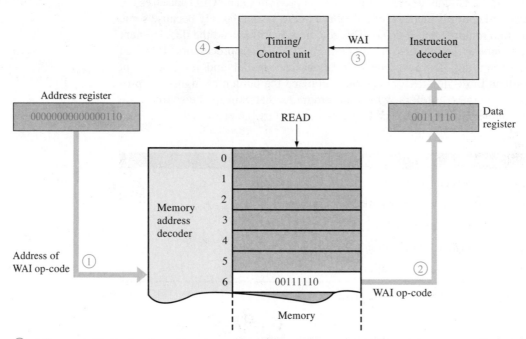

① Address code (6_{10}) is placed on address bus.

② Data are placed on data bus and stored in data register by the read signal.

③ Instruction is decoded.

④ Timing/Control unit stops processor operation.

FIGURE 14–17 Fetch/execute cycle for the wait (WAI) instruction. This illustrates inherent addressing.

Immediate Addressing

Immediate addressing is used in conjunction with two-byte instructions where the first byte is the op-code and the second byte is the operand. The load accumulator (LDA) and the add to accumulator (ADDA) instructions are two examples that use immediate addressing.

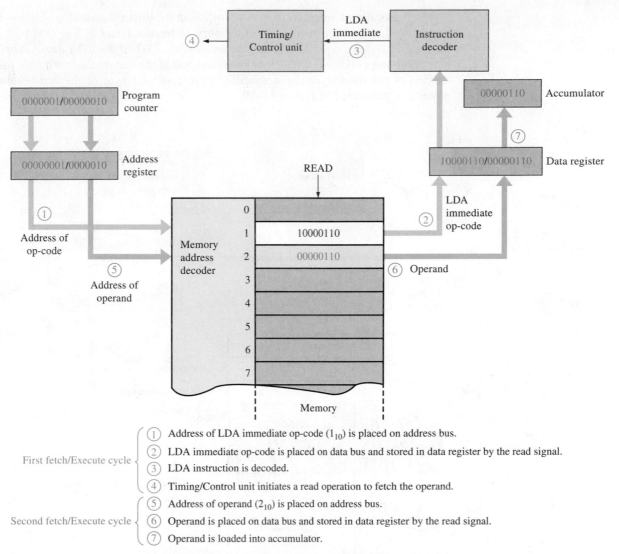

First fetch/Execute cycle
- ① Address of LDA immediate op-code (1_{10}) is placed on address bus.
- ② LDA immediate op-code is placed on data bus and stored in data register by the read signal.
- ③ LDA instruction is decoded.
- ④ Timing/Control unit initiates a read operation to fetch the operand.

Second fetch/Execute cycle
- ⑤ Address of operand (2_{10}) is placed on address bus.
- ⑥ Operand is placed on data bus and stored in data register by the read signal.
- ⑦ Operand is loaded into accumulator.

FIGURE 14–18 Illustration of immediate addressing. The process steps are numbered in sequence, and the cycle operations are color-coded.

The LDA immediate op-code is stored in one memory address, and the operand is stored in the address immediately following the op-code. That is, the op-code and operand are at consecutive memory addresses. When the LDA immediate instruction is fetched and executed, it tells the processor to get the contents of the next memory location (operand) and load it into the accumulator, as illustrated in Figure 14–18.

Direct Addressing

For an instruction using direct addressing, the first part is the op-code and the second part is the address of the operand, not the operand itself. For example, the LDA instruction uses direct addressing as well as immediate addressing. LDA direct has a different op-code than LDA immediate. Let's assume each part is one byte for simplicity.

The LDA direct instruction is used to illustrate direct addressing. Figure 14–19 shows the LDA direct instruction in memory addresses 1 and 2. The first byte is the op-code, and the second byte is the operand address. When the LDA direct instruction is fetched and executed, it tells the processor to load the accumulator with the operand located at the memory address specified by the second byte of the instruction. The process is illustrated in Figure 14–19.

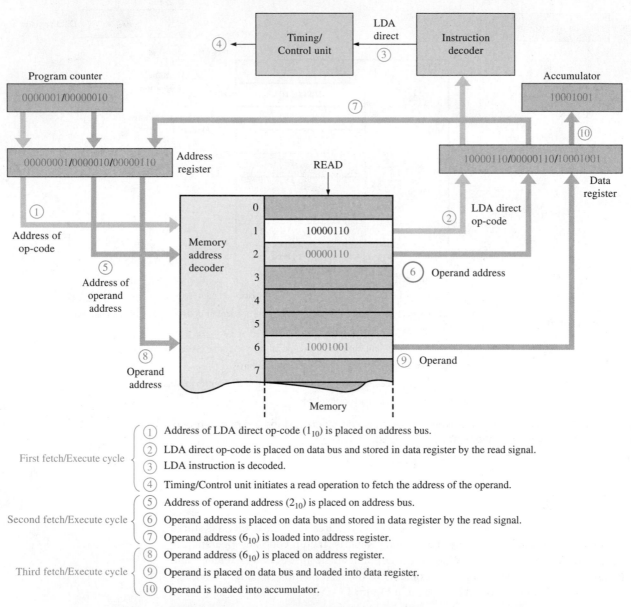

First fetch/Execute cycle
1. Address of LDA direct op-code (1_{10}) is placed on address bus.
2. LDA direct op-code is placed on data bus and stored in data register by the read signal.
3. LDA instruction is decoded.
4. Timing/Control unit initiates a read operation to fetch the address of the operand.

Second fetch/Execute cycle
5. Address of operand address (2_{10}) is placed on address bus.
6. Operand address is placed on data bus and stored in data register by the read signal.
7. Operand address (6_{10}) is loaded into address register.

Third fetch/Execute cycle
8. Operand address (6_{10}) is placed on address register.
9. Operand is placed on data bus and loaded into data register.
10. Operand is loaded into accumulator.

FIGURE 14–19 Illustration of direct addressing.

Indexed Addressing

Indexed addressing is used in conjunction with the index register. An instruction using indexed addressing consists of the op-code and the offset address. When an indexed instruction is executed, the offset address is added to the contents of the index register to produce

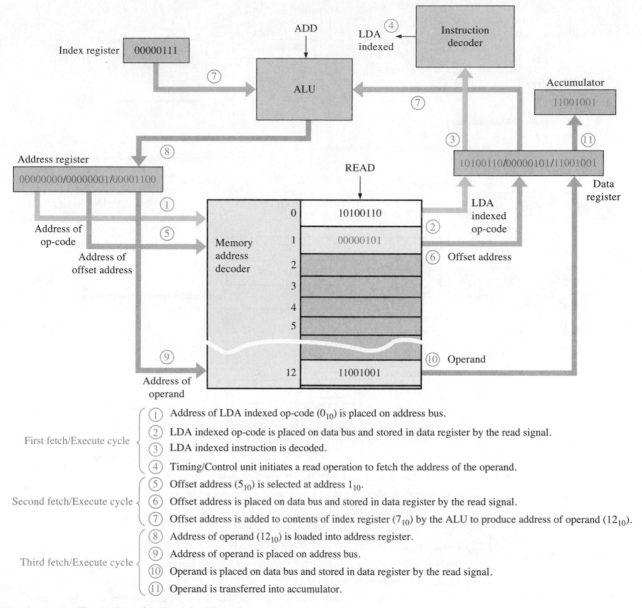

FIGURE 14–20 Illustration of indexed addressing.

an operand address. In Figure 14–20 the LDA (load accumulator) instruction is again used to illustrate indexed addressing.

Relative Addressing

Relative addressing is used by a class of instructions known as *branch instructions*. Basically, a branch instruction allows the CPU to go back or skip ahead for a specified number of addresses in a program instead of going to the next address in sequence. Branching instructions are used to form program loops. For a relative addressing instruction (branch instruction), the first byte is the op-code and the second byte is the relative address. When a branch instruction is executed, the relative address is added to the contents of the program counter to form the address to which the program is

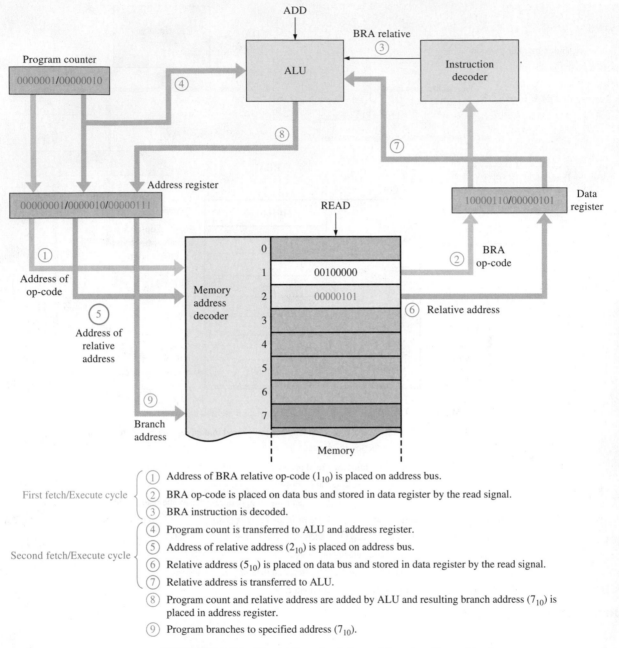

FIGURE 14–21 Illustration of relative addressing (branching).

to branch. Figure 14–21 illustrates relative addressing using a branch relative always (BRA) instruction that can branch both forward or backward. Forward branching is shown.

SECTION 14–4 CHECKUP

1. List five types of addressing.

2. What is an op-code?

3. What is an operand?

4. Explain branching.

14–5   The Processor: Special Operations

During normal operation the CPU fetches instructions from system memory, and these instructions are decoded by the instruction decoder. Each decoded instruction affects the operation of the timing and control unit, which in turn synchronizes the operation of the CPU, system buses, and system components to execute the instruction. In this section, specific CPU operations (polling, interrupts, exceptions, and bus requests) that occur when special circumstances or events arise that preempt normal processor operation are discussed.

After completing this section, you should be able to

- ◆ Define *polling*
- ◆ Define the terms *interrupt* and *exception*
- ◆ Describe the process by which a processor responds to and services an interrupt
- ◆ Explain how an interrupt service routine differs from a subroutine
- ◆ Explain why computer systems use bus requests

A computer runs programs that limit what the computer is permitted to do and how it will respond to situations that arise. Some situations are predictable and others are not. Even when a situation is predictable, just when it will occur may not be. As an example, every word processor program must respond to input from a keyboard, but the program cannot predict just when someone will press a key.

Polling

One technique to deal with unpredictable events is to have the CPU poll, or repeatedly check, the keyboard. The same occurs for other peripheral devices that may require attention from the CPU. Each time the CPU polls a device, it must stop the program that it is currently processing, go through the polling sequence, provide service if needed, and then return to the point where it left off in its current program. This process is inefficient and is suitable only for devices that can be serviced at regular and predictable intervals. Figure 14–22 illustrates **polling**, where the CPU sequentially selects each peripheral device via the multiplexer to see if it needs service.

Interrupts and Exceptions

A more efficient approach than polling is to have the processor perform its normal operations and deviate from them only when some special event requires the processor to take special action to handle it. Some sources use the term *exception* for any event that requires special handling by the processor. Other sources use *exception*, *software interrupt* (*SWI*), or *trap* for an event due to software and *interrupt* or *hardware interrupt* (*HWI*) for an event due to hardware. We will use **interrupt** to refer to a hardware event and **exception** for a software event that require the CPU to deviate from its normal operation.

When the processor receives an interrupt or an exception, it finishes executing the current instruction and then runs a special sequence of instructions called an *interrupt service routine* (*ISR*) or *exception handler*. An ISR similar to calling a standard program subroutine but with three important differences. Because the processor cannot know when an interrupt will occur, it automatically saves on the register stack status information about the program that is executing at the time the interrupt or exception occurs. The information includes the contents of the condition code register as well as the address of the next instruction to be executed when the ISR is finished. Sometimes the accumulator and condition code register, which make up the program status word, are both saved. The ISR must save on the stack any other

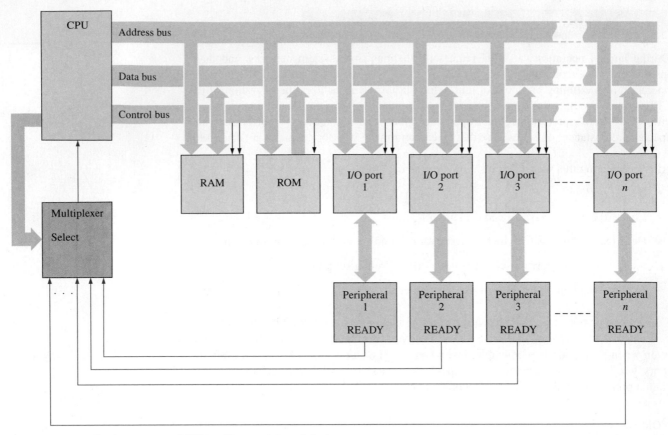

FIGURE 14–22   Basic concept of CPU polling peripheral devices.

registers it may use to ensure that the interrupted process will not be affected when it resumes executing.

Secondly, the processor obtains the address of the ISR based on the specific interrupt or exception that occurs. In some systems, a programmable interrupt controller (*PIC*) provides the address of the ISR over the data bus when the processor acknowledges an interrupt request. Other systems use autovectored interrupts that obtain the address for each interrupt from entries in an **interrupt vector table** stored in memory. Each vector, or ISR address, in the table specifies the starting address of an ISR. The programmer must write the ISRs and place the starting address for each in the correct location of the interrupt vector table. If no ISR exists for an entry in the vector interrupt table, or if the interrupt vector table is not properly initialized, interrupts and exceptions can cause the processor to behave erratically, "hang" (stop responding), or "crash" (abort and restart).

A third difference is that the ISR uses a special *return from interrupt (RTI)* instruction, which restores the additional status information as well as the address of the next instruction. RTI is used rather than a standard return from subroutine (RET) instruction, which restores only the address of the next instruction, to exit and return processor control to the interrupted process. Before executing the RTI instruction, the ISR must restore any registers it saved on the stack.

Specific interrupts and exceptions vary with each processor, but the following list describes some typical ones.

Reset   This is sometimes called a cold boot. A cold boot completely restarts the system so that the processor runs the power on self-test (POST), initializes the hardware, loads the hardware drivers and operating system, and performs all other tasks necessary to prepare the system for operation.

Software reset This is a software exception and is sometimes called a warm boot. This also restarts the system but bypasses many of the hardware initialization tasks performed by a cold boot.

Divide by zero This is a software exception and occurs when the processor attempts to divide a number by zero.

System timer This is a hardware interrupt and occurs when a special timer asserts a signal indicating that a specified time interval or "time tick" (such as 1/60th of a second) has elapsed since the last occurrence.

Unrecognized instruction This is a hardware interrupt that occurs when the instruction decoder determines that the value it contains is not a valid instruction.

As the above list shows, ISRs must perform a variety of tasks. Just what the ISR does can be as complex or simple as the programmer desires. Figure 14–23 shows the basic concept of interrupts where a device called a programmable interrupt controller (**PIC**) is used to monitor peripheral devices for interrupt requests and send the appropriate address to the CPU so it can take the required action.

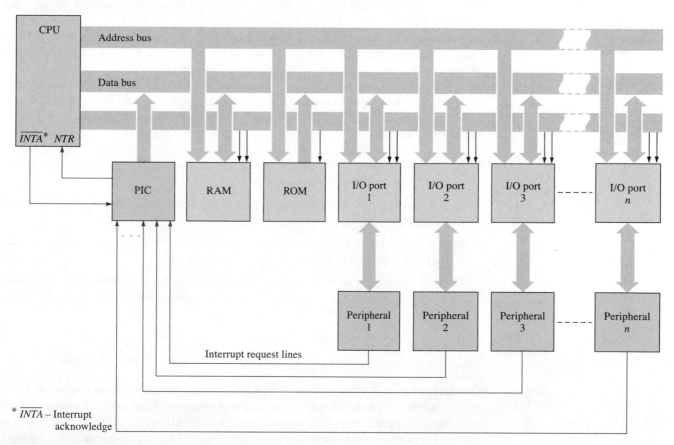

FIGURE 14–23 Basic concept of interrupt control.

Bus Request Operations

The device in a computer that drives the address bus and the bus control signals is called the **bus master**. In a simple computer architecture, only the CPU can be bus master, which means that all communications between I/O devices must involve the CPU. More complex architectures allow other devices (or multiple CPUs) to take turns at controlling the bus. For example, a network controller card can be used to access a disk controller directly while the CPU performs other tasks that do not require the bus. Any device can place data on the data bus when the CPU reads from that device, but only the bus master drives the address bus and control signals.

Although processors operate at high speeds, they are not always efficient at transferring data. When a processor transfers data from one device to another, it must use a bus cycle to read in the data from the source device and use another bus cycle to write the data back out to the destination device. The overhead in reading data into the processor and writing it out again greatly slows data transfers. The bus request operation allows other bus masters to take control of the system buses and rapidly transfer data between system devices.

Bus request operations are similar to interrupts and exceptions but differ in three important ways. Bus request operations do not complete the current instruction cycle before proceeding. Instructions can take hundreds or even thousands of clock cycles, and the circumstances that generated the bus request may be too urgent to be delayed. For example, a CD drive may be on the verge of a buffer under run and require data immediately to refill the buffer, or a memory controller may need to immediately refresh the system DRAM to prevent data from being lost. Interrupts and exceptions allow the processor to complete the current instruction cycle before processing the interrupt or exception.

Secondly, in a bus request operation, the processor passes control of the system buses to the requesting device, which then handles all bus operations. The processor continues to execute instructions in the ISR or exception handler during interrupts.

A third difference is that once the processor grants the bus request and relinquishes the system buses, the processor cannot regain control of the system until the requesting device relinquishes control or the processor is reset. The sequence of events during a bus request operation is as follows:

1. The bus master requesting control of the system buses submits a request by asserting the processor's bus request (BR) line.

2. The processor tri-states the system buses and signals that it has released control of the buses by asserting the bus grant (BG) line.

3. The requesting bus master uses the system address, data, and control lines to transfer data between system devices.

4. After completing the data transfers, the requesting bus master tri-states the system buses and signals the end of the bus request operation by asserting the bus grant acknowledge (BGACK) line.

Direct Memory Access (DMA)

One important class of bus master is the **DMA** (direct memory access) controller. These devices are designed specifically to transfer large amounts of data between system devices in a fraction of the time that the system processor would require. To utilize a DMA controller, the processor first writes the starting source address, starting destination address, and number of bytes to transfer to registers within the DMA controller. The processor next enables the transfer by writing to a control register within the controller, which then initiates the bus request operation. Computer systems typically use DMA controllers to transfer data between memory and hardware peripherals, such as when loading a program or data file from a hard drive to memory or when transferring a message from system memory to the transmit buffer of an Ethernet controller. DMA controllers can also move data between memory devices, for example, when moving data from main memory to cache memory.

DMA speeds up data transfers between RAM and certain peripheral devices. Basically, DMA bypasses the CPU for certain types of data transfers, thus eliminating the time consumed by normal fetch and execute cycles required for each CPU read or write operation. Transfers between the disk drive and RAM are particularly suited for DMA because of the large amount of data and the serial nature of the transfers. Generally, the DMA controller can handle data transfers several times faster than the CPU. Figure 14–24 shows a comparison of a data transfer handled by the CPU (part a) and one handled by the DMA (part b).

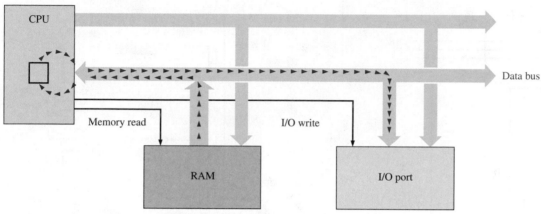

(a) Data transfer handled by the CPU

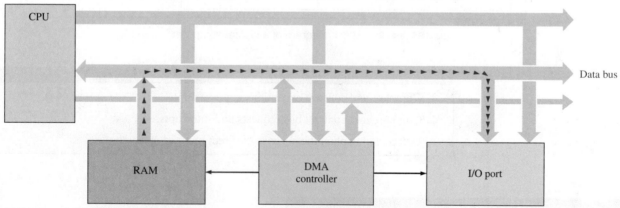

(b) Data transfer handled by the DMA controller

FIGURE 14–24   Illustration of DMA vs CPU data transfer.

Bus masters other than DMA controllers also use bus request operations. Processors in multiprocessor systems use bus request operations to access shared memory and other system resources. Memory controllers use bus request operations to perform background memory operations, such as refreshing DRAM and ensuring that the data in main memory and cache memory are consistent.

Figure 14–25 shows a computer system block diagram with a DMA controller and a PIC.

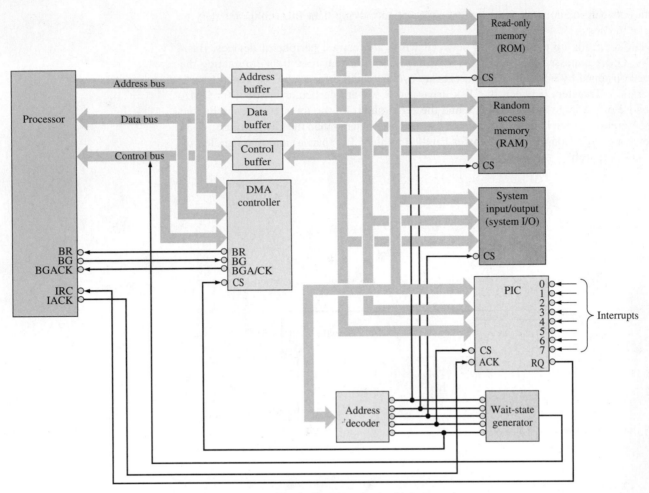

FIGURE 14–25   Block diagram of a typical computer.

SECTION 14–5 CHECKUP

1. Compare and contrast exceptions and interrupts.

2. Compare and contrast bus requests and interrupts.

3. Define and explain the purpose of direct memory accesses.

14–6   Operating Systems and Hardware

Each computer system consists of two main components. The microprocessor, memories, interface circuits, peripherals, power supplies, and other electronic components make up what is collectively referred to as computer **hardware**. The programs that the microprocessor executes and that control the computer system are collectively referred to as computer **software**. One general rule is anything in a computer system that you can physically touch is hardware, and anything that you can't physically touch is software.

After completing this section, you should be able to

◆ Explain the three basic duties of an operating system

◆ Discuss how an operating system functions in a computer system

- Compare and contrast the difference between multitasking and nonmultitasking operating systems

- Differentiate between multitasking and multiprocessing

- Identify and discuss the issues presented by multitasking

Operating System Basics

The **operating system** (OS) of a computer is a special program that establishes the environment in which application programs operate. The operating system provides the functional interface between application programs in the system, called **processes**, and the computer hardware. Because the operating system must work closely with the computer hardware, it is often written in assembly language or programming language with low-level hardware support, such as C++.

An operating system increases the overall complexity of a computer system, but using an operating system offers a number of advantages over running stand-alone application programs. The operating system tests and initalizes hardware in the computer system, eliminating the need for each application to duplicate these functions. Operating systems also provide a standard computing environment so that applications can execute consistently. Finally, operating systems provide system services that allow applications access to commonly used system resources (such as the real-time clock, I/O ports, and data files), which simplify the code for applications programs. A drawback of operating systems is that processes may execute more slowly; accessing system resources through an operating system can take longer than a program accessing them directly. An operating system has three basic duties.

1. To schedule and allocate system resources (CPU time, memory, access to system peripherals)

2. To protect system processes and resources (preventing accidental or deliberate corruption of process code and data, unauthorized access to hardware and memory)

3. To provide system services (messaging between processes, low-level hardware drivers)

Multiple Processes

Computers can run multiple processes in two basic ways. The first way, called **multitasking**, shares a single-core processor among multiple processes. The processor runs more than one process but switches between them so that each process uses only part of the processor's available time. Multitasking systems use different techniques to decide when to switch between processes. One technique allows a process to run until it must wait for some event, such as a keypress, before it can continue and switches to another process that is ready to run. Another technique, called *preemptive multitasking*, allows each process to run for a specific amount of time before the operating system switches to another process. A third technique, called *non-preemptive multitasking*, allows a process to run until it voluntarily relinquishes the processor to another process. Figure 14–26 illustrates how a single-core processor multitasks.

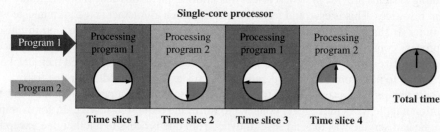

FIGURE 14–26 Simplified model of processor multitasking.

The second way for a computer system to run multiple processes, called **multiprocessing**, uses multiple processors, each of which can either multitask or run a single process. Figure 14–27 illustrates the concept of multitasked multiprocessing.

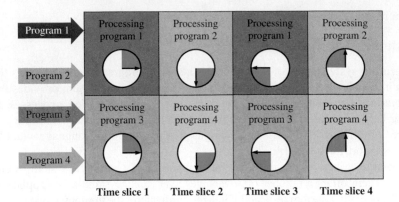

FIGURE 14–27 Multitasked multiprocessing in a multicore processor.

Supervisor and User States

It is difficult for multiple users or processes to coexist in a computer system if processes have unrestricted access to system resources. Once a process takes control of the processor and is running, it can modify or disable any software or hardware in the system that exists to control it. The solution to this is to restrict what the process can access. Some processors use the user/supervisor state bit so only trusted code, like the operating system, can run under certain circumstances. For multiprocess or multiuser systems, the processor executes in supervisor state when it first powers up, while the operating system is running, and when the processor responds to an interrupt. When the operating system loads and transfers control to an application program, it first clears the user/supervisor state bit. This places the process in user state and prevents it from accessing restricted parts of the computer system's hardware or software.

Memory Management Unit

One device in the computer system that has not yet been discussed is the memory management unit, or **MMU**. Memory management units are very sophisticated logic devices that handle many details associated with accessing memory in computer systems, including memory protection, wait-state generation, address translation for handling virtual memory, and cache control. As an example, consider a simplified MMU that simply provides memory protection. The processor can program the MMU with the start and end addresses of a memory range. The MMU then acts as a comparator. If the MMU detects a value on the address bus that is less than the programmed start address or greater than the programmed end address, it will generate a hardware interrupt to the processor.

System Services

Operating systems provide system services that allow applications access to commonly used system resources. This is essential for allowing processes to interact and communicate with each other to share information, coordinate operations, and otherwise function in unison. Interprocess communication uses software interrupts (also called *traps*). When one process wishes to utilize a system service, it loads specific registers with values and then invokes a specific trap to pass control to the operating system's exception handler for that trap.

When the process executes the trap, the processor enters supervisor mode; and the exception handler uses the register contents to fulfill the requested service. If, for example, the requested service was to send several bytes from one process to another, the exception handler would use the starting address of the data and the number of data bytes contained in the processor registers to copy the data from the user memory of the source process to the user memory of the destination process. It would then load a condition code indicating that the service had been completed successfully (or failed) in one of the processor registers and would return processor control to the requesting process.

When processes are meant to interact with other processes, they each must be carefully designed to ensure that messages are passed at the right time and in the right order and that the processes can recover from communication errors. Otherwise, one process may believe that it has sent out a valid message and await a response, while the intended destination process is waiting for the first process to send a message to which it can respond. The result is that neither process can proceed.

SECTION 14–6 CHECKUP

1. What are the three basic duties of an operating system?

2. Compare and contrast multitasking and multiprocessing.

3. Describe how a memory management unit prevents one process from accessing the memory space of another process.

4. Explain how an operating system permits two processes to exchange information.

14–7   Programming

Assembly language is a way to express machine language in English-like terms, so there is a one-to-one correspondence. Assembly language has limited applications and is not portable from one processor to another, so most computer programs are written in high-level languages such as C++, JAVA and BASIC. High-level languages are portable and therefore can be used in different computers. High-level languages must be converted to the machine language for a specific microprocessor by a process called *compiling*.

After completing this section, you should be able to

- ◆ Describe some programming concepts
- ◆ Discuss the levels of programming languages

Levels of Programming Languages

A hierarchy diagram of computer programming languages relative to the computer hardware is shown in Figure 14–28. At the lowest level is the computer hardware (CPU, memory, disk drive, input/output). Next is the **machine language** that the hardware understands because it is written with 1s and 0s (remember, a logic gate can recognize only a LOW (0) or a HIGH (1). The level above machine language is **assembly language** where the 1s and 0s are represented by English-like words. Assembly languages are considered low-level because they are closely related to machine language and are machine dependent, which means a given assembly language can only be used on a specific microprocessor.

The level above assembly language is **high-level language**, which is closer to human language and further from machine language. An advantage of high-level language over assembly language is that it is portable, which means that a program can run on a variety of computers. Also, high-level language is easier to read, write, and maintain than assembly language. Most system software (e.g., Windows), and applications software (e.g., word processors and spreadsheets) are written with high-level languages.

Assembly Language

To avoid having to write out long strings of 1s and 0s to represent microprocessor instructions, English-like terms called mnemonics or op-codes are used. Each type of microprocessor has its own set of mnemonic instructions that represent binary codes for the

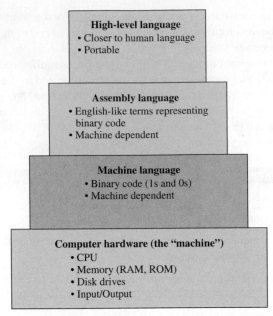

FIGURE 14–28   Hierarchy of programming languages relative to computer hardware.

instructions. All of the mnemonic instructions for a given microprocessor are called the instruction set. Assembly language uses the instruction set to create programs for the microprocessor; and because an assembly language is directly related to the machine language (binary code instructions), it is classified as a low-level language. Assembly language is one step removed from machine language.

Assembly language and the corresponding machine language that it represents is specific to the type of microprocessor or microprocessor family. Assembly language is not portable; that is, you cannot directly run an assembly language program written for one type of microprocessor on another type of microprocessor. For example, an assembly program for the Motorola processors will not work on the Intel processors. Even within a given family different microprocessors may have different instruction sets.

An **assembler** is a program that converts an assembly language program to machine language that is recognized by the microprocessor. Also, programs called **cross-assemblers** translate an assembly language program for one type of microprocessor to an assembly language for another type of microprocessor.

Assembly language is rarely used to create large application programs. However, assembly language is often used in a subroutine (a small program within a larger program) that can be called from a high-level language program. Assembly language is useful in subroutine applications because it usually runs faster and has none of the restrictions of a high-level language. Assembly language is also used in machine control, such as for industrial processes. Another area for assembly language is in video game programming.

Conversion of a Program to Machine Language

All programs written in either an assembly language or a high-level language must be converted into machine language in order for a particular computer to recognize the program instructions.

Assemblers

An assembler translates and converts a program written in assembly language into machine code, as indicated in Figure 14–29. The term **source program** is often used to refer to a program written in either assembly or high-level language. The term **object program** refers to a machine language translation of a source program.

FIGURE 14–29   Assembly to machine conversion using an assembler.

Compilers

A *compiler* is a program that compiles or translates a program written in a high-level language and converts it into machine code, as shown in Figure 14–30. The compiler examines the entire source program and collects and reorganizes the instructions. Every high-level language comes with a specific compiler for a specific computer, making the high-level language independent of the computer on which it is used. Some high-level languages are translated using what is called an *interpreter* that translates each line of program code to machine language.

FIGURE 14–30   High-level to machine conversion with a compiler.

All high-level languages, such as C++, will run on any computer. A given high-level language is valid for any computer, but the compiler that goes with it is specific to a particular type of CPU. This is illustrated in Figure 14–31, where the same high-level language program (written in C++ in this case) is converted by different machine-specific compilers.

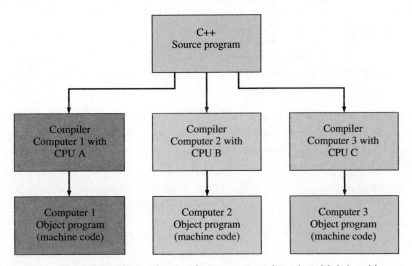

FIGURE 14–31   Machine independence of a program written in a high-level language.

Example of an Assembly Language Program

For a simple assembly language program, let's say that we want the computer to add a list of numbers from the memory and place the sum of the numbers back into the memory. A zero is used as the last number in the list to indicate the end of the list of numbers. The steps required to accomplish this task are as follows:

1. Clear a register (in the microprocessor) for the total or sum of the numbers.
2. Point to the first number in the memory (RAM).
3. Check to see if the number is zero. If it is zero, all the numbers have been added.
4. If the number is not zero, add the number in the memory to the total in the register.
5. Point to the next number in the memory.
6. Repeat steps 3, 4, and 5.

A flowchart is often used to diagram the sequence of steps in a computer program. Figure 14–32 shows the flowchart for the program represented by the six steps.

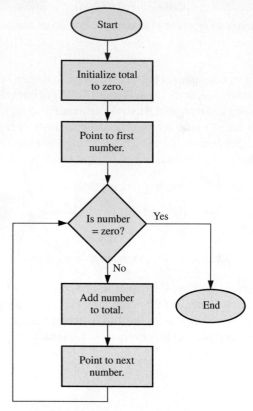

FIGURE 14–32   Flowchart for adding a list of numbers.

The working portion of the assembly language program implements the addition problem shown in the flowchart in Figure 14–32. Two of the registers in the microprocessor are named eax and ebx. The comments preceded by a semicolon are not recognized by the computer; they are for explanation only.

```
        mov eax,0                   ;Replaces the contents of the eax
                                     register with zero.

                                    ;Register eax will store the total of
                                     the addition.

        mov ebx, OFFSET NumArray    ;Places memory address of NumArray
                                     into the ebx register.

next:   cmp dword ptr [ebx],0       ;Compares the number stored in the ebx
                                     register to zero.

        jz done                     ;If the number in the ebx register is
                                     zero, jump to "done".

        add eax,[ebx]               ;Add the number in the ebx register to
                                     the eax register.

        add ebx, 4
        jmp next

done:   mov [ebx],eax
        call WriteInteger           ;WriteInteger utility by Floyd to view
                                     integer values

        exitProg                    ;exitProg utility provided by Floyd
                                     utility to end the executable
```

Depending on the assembler, most programs in assembly language will have a number of assembler directives that are used by the assembler to do a variety of tasks. These tasks include setting up segments, using the appropriate instruction set, describing data sizes, and performing many other "housekeeping" functions. To simplify the explanation, only two directives were shown in the preceding program. The directives were **word ptr**, which is used to indicate the size of the data pointed to by the ebx register, and OFFSET.

EXAMPLE 14-2

Write the instructions for an assembly language program that will find the largest unsigned number in the data and place it in the last position. Assume the last data point is signaled with a zero.

Solution

The flowchart is shown in Figure 14–33.

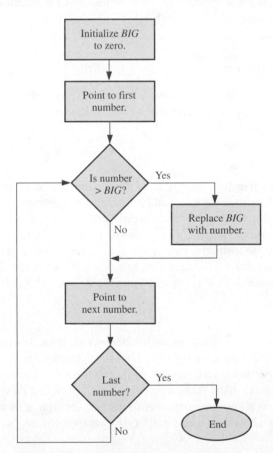

FIGURE 14–33 Flowchart. The variable *BIG* represents the largest value.

The data are assumed to be the same as before. The program listing (with comments) is as follows:

```
mov eax,0              ;initial value of BIG is in the eax
                        register
mov ebx,OFFSET NumArray ;point to the location in memory where
                        the data are stored
```

```
next:  cmp dword ptr [ebx],eax   ;is the data point larger than BIG?
       jbe check                 ;if the data point is smaller, go
                                  to "check"
       mov eax, [ebx]            ;otherwise, put the new largest data
                                  point in eax
check: add ebx,4                 ;point to the next number in memory
                                  (four bytes per word)
       cmp dword ptr [ebx], 0    ;test for the last data point
       jnz next                  ;continue if the data point is not
                                  a zero
       mov [ebx], eax            ;save BIG in memory
       call WriteInteger         ;WriteInteger utility by Floyd to
                                  view integer values
       exitProg                  ;exitProg utility provided by Floyd
                                  utility to end the executable
```

Related Problem

Explain how you could change the flowchart to find the smallest number in the list instead of the largest.

Types of Instructions

The programs in this section only show a few of the hundreds of variations of instructions available to programmers. Generally, an instruction set can be divided into categories, which are described here.

Data Transfer

The most basic data transfer instruction MOV was introduced in the example programs. The MOV instruction, for example, can be used in several ways to copy a byte, a word (16 bits), or a double word (32 bits) between various sources and destinations such as registers, memory, and I/O ports. (A better mnemonic for MOV might have been "COPY" because this is what the instruction actually does.) Other data transfer instructions include IN (get data from a port), OUT (send data to a port), PUSH (copy data onto the stack, a separate area of memory), POP (copy data from the stack), and XCHG (exchange).

Arithmetic

There are a number of instructions and variations of these instructions for addition, subtraction, multiplication, and division. The ADD instruction was used in both example programs. Other arithmetic instructions include INC (increment), DEC (decrement), CMP (compare), SUB (subtract), MUL (multiply), and DIV (divide). Variations of these instructions allow for carry operations and for signed or unsigned arithmetic. These instructions allow for specification of operands located in memory, registers, and I/O ports.

Bit Manipulation

This group of instructions includes those used for three classes of operations: logical (Boolean) operations, shifts, and rotations. The logical instructions are NOT, AND, OR, XOR, and TEST. An example of a shift instruction is SAR (shift arithmetic right). An example of a rotate instruction is ROL (rotate left). When bits are shifted out of an operand, they are lost; but when bits are rotated out of an operand, they are looped back into the other end. These logical, shift, and rotate instructions can operate on bytes or words in registers or memory.

Loops and Jumps

These instructions are designed to alter the normal (one after the other) sequence of instructions. Most of these instructions test the processor's flags to determine which instruction should be processed next. In Example 14–2, the instructions JBE and JNZ were used to alter the path. Other instructions in this group include JMP (unconditional jump), JA (jump above), JO (jump overflow), LOOP (decrement the CX register and repeat if not zero) and many others.

Strings

A **string** is a **contiguous** (one after the other) sequence of bytes or words. Strings are common in computer programs. A simple example is a sentence that the programmer wishes to display on the screen. There are five basic string instructions that are designed to copy, load, store, compare, or scan a string—either as a byte at a time or a word at a time. Examples of string instructions are MOVSB (copy a string, one byte at a time) and MOVSW (copy a string, one word at a time).

Subroutine and Interrupts

A **subroutine** is a miniprogram that can be used repeatedly but programmed only once. For example, if a programmer needs to convert ASCII numbers from a keyboard to a BCD format, a simple programming structure is to make the required instructions a separate process and "call" the process whenever necessary. Instructions in this group include CALL (begin the subroutine) and RET (return to the main program).

Processor Control

This is a small group of instructions that allow direct control of some of the processor's flags and other miscellaneous tasks. An example is the STC (set carry flag) instruction.

High-Level Programming

The basic steps to take when you write a high-level computer program, regardless of the particular programming language that you use, are as follows:

1. Determine and specify the problem that is to be solved or task that is to be done.
2. Create an algorithm; that is, develop a series of steps to accomplish the task.
3. Express the steps using a particular programming language and enter them on the software text editor.
4. Compile (or assemble) and run the program.

A simple program will show an example of high-level programming. The following C++ program implements the same addition problem defined by the flowchart in Figure 14–32 and implemented using assembly language.

```
int total = 0;          //Initialize the total to zero.
int *number = NumArray; //Initialize a pointer to the array of integers.
while (*number != 0x00) //Loop while the value is not found. The
                        //asterisk preceding the pointer identifier
                        //number says the contents of the
                        //memory location pointed to by the
                        //Identifier number are being evaluated.
{
  total = total + *number; //Accumulate summation of total
  number++;               //Increment pointer to next number in memory
}
cout << total;          //C++ cout statement used to view integer value
```

This C++ program is equivalent to the assembly program that adds a series of numbers and produces a total value.

```c++
int total = 0;
int *number = NumArray;
while (*number != 0x00)

{
    total = total + *number;
    number++;
}

cout << total;
```

[C++]

Equivalent →

```asm
        mov eax, 0
        mov ebx, OFFSET NumArray
next:   cmp DWORD PTR [ebx], 0
        jz done

        add eax, [ebx]
        add ebx, 4

        jmp next
        mov [ebx], eax
done:   mov [ebx], eax
        call WriteInteger
```

Assembly

SECTION 14–7 CHECKUP

1. Define *program.*
2. What is an op-code?
3. What is a string?

14–8   Microcontrollers and Embedded Systems

Although a general-purpose microprocessor can interface with a variety of devices over its system buses, its ability to interface with the real world is limited. Most general-purpose microprocessors must use analog-to-digital converters (ADCs), digital-to-analog converters (DACs), universal asynchronous receivers and transmitters (UARTs) and other communication controllers, peripheral interface adapters (PIAs), external timers, and other specialized peripherals to process real-world information. Microcontrollers are used in microprocessor-controlled applications called **embedded systems** that perform a specific set of tasks and incorporate both the hardware and firmware required to perform them. Embedded systems include personal electronic devices such as cell phones, MP3 players, and calculators as well as consumer and industrial products as microwave ovens, automated assembly systems, and robots.

After completing this section, you should be able to

- Describe the general architecture of microcontrollers

- Discuss the types of peripherals found in common microcontrollers

- Describe how microcontroller peripherals are configured

- Describe how microcontrollers are used in various embedded systems

Microcontroller Basics

A special type of processor, called a **microcontroller**, sometimes abbreviated as μC or MCU, combines a microprocessor core, memory, and common peripherals in a single package. Microcontrollers can range in complexity from simple devices with a few dozen pins to very complex devices with hundreds of pins. A common aspect of all these processors is that the design of each seeks to incorporate all the elements of a microprocessor system into a single package. A microcontroller will typically include the following functional units:

- A microprocessor (called the processor core)
- Nonvolatile memory for program code, device configuration data, and similar data that must be preserved when power is removed
- RAM for program data, internal registers, peripheral device buffers, and other data storage
- Peripheral devices such as timers, ADCs, DACs, communication controllers, and I/O ports
- Internal buses to connect the processor core to internal memory
- Internal buses to connect the processor core and internal memory to peripheral devices
- Interface circuitry to connect the microcontroller with external devices

In addition to the above list of microcontroller features, more sophisticated microcontrollers can also include the following:

- A phase-locked loop (PLL) to multiply a low-frequency external clock to a higher internal frequency, increasing the speed of microcontroller operation
- DMA controllers to improve data transfer between internal memory and peripheral devices
- Programmable logic resources, or "fabric", to implement custom functions
- A JTAG interface to support device testing and programming
- Special power modes for low-power and standby operation

Figure 14–34 shows a simplified block diagram of a typical microcontroller.

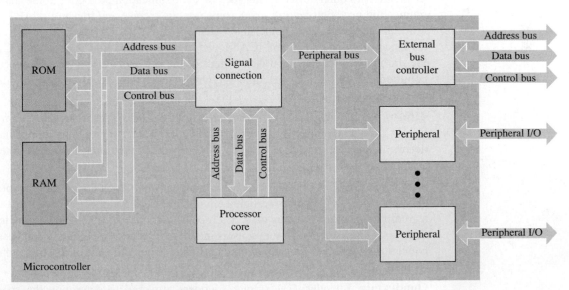

FIGURE 14–34   Simplified microcontroller block diagram.

Microcontroller Peripherals

Microcontrollers feature a wide variety of peripherals. Manufacturers select the type and number of peripherals, as well as the types and amounts of internal memory, to meet the requirements of specific applications, such as communication, automotive, and motion-control products. For example, microcontrollers that target communication applications will include a wide variety and number of communication controllers (such as Ethernet, I^2C, USB, and UART) to support the transmission and reception of data using multiple protocols. In contrast, microcontrollers meant for motion-control applications, such as robotic assemblies, will include ADCs, DACs, encoders, and pulse width modulators (PWMs) for position sensing and motor feedback and control.

Many pins on microcontrollers are multifunctional. This not only helps to reduce the total pin count and cost of the device but also limits the functions that can be used at the same time. The data sheet may state that a communications microcontroller has four USB controllers, two UART controllers, an Ethernet controller, an external memory interface, and 80 general-purpose I/O (GPIO) lines, but it is unlikely that a design can use all of these. A pin on a communications microcontroller, for example, might serve as a transmit line for USB communications, a clear-to-signal line for UART communications, a transmit line for Ethernet communications, an address line for the external bus controller, or a general-purpose I/O (GPIO) pin; but it can be configured for only one function at a time. Since applications rarely can change pin functions "on the fly," the circuit design permanently assigns the function of each microcontroller pin. If the circuit designer must use a set of pins for an external memory interface but also needs other functions that those pins provide, she either must find those functions on other pins (which is why microcontrollers offer more than one instance of a type of peripheral) or use external circuits to provide those functions.

The following describes some of the more common types of peripherals on microcontrollers.

General-Purpose I/O (GPIO) General-purpose I/O pins are typically the default function for many microcontroller pins. As the name suggests, these pins can be configured as input or outputs to read or write data, either as individual bits (for serial transfer of data) or groups of bits (for parraler transfer of data). Typical applications for GPIO lines are to read individual switches, to drive LED indicators, or to select or enable latches or buffers.

Communication Controllers Communication controllers allow microcontrollers to communicate with other devices using specific communication protocols. Some standard communication protocols are universal asynchronous receive and transmit (UART), Ethernet, universal serial bus (USB), inter-integrated circuit (IIC or I2C), serial peripheral interface (SPI), controller area network (CAN), and high-level data link control (HDLC). Because the timing, flow control, and data format of these protocols vary so widely, configuring communication controller functions is typically much more involved than for other peripheral functions.

Timers Microcontroller timers can have multiple uses. These include setting the frequency for a communication controller, indicating when a preset time interval has elapsed, determining the elapsed time between two events, and providing a periodic time tick for a system real-time clock.

ADCs and DACs ADCs and DACs are the means by which digital circuits interact with an analog world. As you know, digital circuits must use a limited set of values to represent a continuous range of analog data. Microcontrollers use ADCs to convert analog voltage and current measurements from sensors into digital values for processing and use DACs to convert digital values into analog voltages and currents to control electric and electronic circuits.

Quadrature Encoders Quadrature encoders are used to determine the speed, direction, and position of a moving object, such as a computer mouse or a stepper motor. A quadrature encoder represents the present position of a tracked object with a Gray code sequence. When the object moves, the Gray code value changes. Each change will increment or decrement a counter to represent a positive or negative change in position. For example, a system could use the sequence $00 \rightarrow 01 \rightarrow 11 \rightarrow 10 \rightarrow 00$ to represent a positive change so that the sequence $00 \rightarrow 10 \rightarrow 11 \rightarrow 01 \rightarrow 00$ would represent a negative change. The counter value represents the position of the tracked object in the physical system relative to the starting point or origin of the tracked object; how fast the counter value changes reflects the speed of the tracked object. Quadrature encoders typically use 32-bit or larger counters to prevent an underflow or overflow condition that would make it seem that the tracked device suddenly changed from a maximum or minimum position or vice versa. Tracked objects require one encoder for each dimension of movement. An object that moves in one dimension, such as a sliding door, needs only one encoder to track movement along the line of travel. An object that moves in two dimensions, such as a computer mouse, requires two encoders to track movement in the plane of travel. Objects that move in three dimensions, such as some robotic assemblies, require three encoders to track movement in the space of travel.

Pulse Width Modulators As you know, a pulse width modulator modulates, or varies, the pulse width of a repetitive digital signal to change the signal's duty cycle (i.e., the ratio of the time that a signal is HIGH to the period of the signal). Pulse width modulators are often used in motor control. Although motor controller circuits can use the amplitude of winding current to set the speed of some motors, a more typical approach is to keep the amplitude of the applied winding current constant and vary the duty cycle to control the speed. Microcontrollers can precisely control the duty cycle to accurately set the motor's running speed. Also, microcontrollers can change the duty cycle very quickly in response to the effects of motor speed due to line or load variations to maintain a constant running speed.

External Memory Controllers Although most microcontrollers contain internal ROM, RAM, EEPROM, flash, and other memory for code, data, and other program information, some applications require more memory than a microcontroller contains. Consequently, many microcontrollers feature external memory controllers that permit interfacing the microcontroller to external memory devices. Some microcontrollers do not contain any internal memory, so external memory must be used. External memory controllers often feature decoded chip select lines that allow programmers to configure the size of the memory range, the port size (8-, 16-, or 32-bits), and the number of wait states for each select line; they can contain memory management units that provide memory protection for multitasking applications. External memory devices are typically limited to SRAM, SDRAM, flash, and other memory types that do not require special bus operations, as do DRAM and EEPROM.

Configuring Peripherals

Microcontroller peripherals must be configured so that they operate the way an application requires them to do. Configuring means loading specific registers associated with the peripheral with values that control the function and operation of the peripheral. The register and values vary with each peripheral, but the registers fall into the general categories described next. Depending upon the number of bits required to configure some aspect of a peripheral, some categories may share one register, while others may require multiple registers to contain the necessary information.

Control Registers Control registers determine how the peripheral will function. For some peripherals a control register may select the specific peripheral as well as the characteristics for that peripheral. For example, the control registers for a communication controller could specify the specific communication protocol and the data rate, data packet size, error detection method, and operating mode (interrupt-driven or polled).

Status Registers Status registers contain information about how the peripheral is operating and conditions associated with peripheral operation. Applications use status registers to detect errors, determine when the peripheral has completed some task, and decide when conditions require some special handling. The microcontroller may automatically clear some status bits when firmware corrects a detected condition, while in other cases firmware may need to manually clear some status bits. For example, if an ADC sets the end-of-conversion status bit to indicate it has completed converting an analog value, reading the converted value from the ADC data register may automatically clear the bit; firmware may need to specifically clear the status bit to allow the ADC to perform another conversion.

Data Registers Data registers contain information that the peripheral processes in some way. The value in a data register can be data for the peripheral to process, data processed by the peripheral, or data currently being processed. The contents of data registers might not change unless firmware changes them, or operation of the peripheral may automatically update them. For example, the initialization register for a timer contains the initial count value that is loaded into the timer and may not change unless firmware writes a new value into the register. In contrast, the timer's count register holds the actual value of the timer and may update each time the counter is clocked. Some peripherals have only a few configuration registers.

The GPIO pins typically have only two registers: a control register that determines whether a pin is an input or output and a data register that contains the signal level of the pin. Other peripherals can have many more registers. A communication controller, for example, can have a control register to specify the communication parameters, a status register to monitor the operation of the controller, a transmit buffer descriptor register to specify the memory locations of data to be transmitted, a transmit length register to specify the number of bytes to transmit, a receive buffer descriptor register to specify the memory locations at which received data are to be stored, a receive length register to indicate the number of bytes received, an interrupt status register to signal communication events during reception and transmission, and an interrupt mask register to prevent or allow recognition of communication events. When configuring microcontroller peripherals, the programmer must carefully read the user manual and understand not only the operation of each peripheral he intends to use but also which configuration registers must be programmed and the configuration values to use.

As the number of products using microcontrollers has grown, manufacturers and third-party vendors have visual development and evaluation tools to simplify the process of programming microcontrollers. Many tools now allow programmers to use drop-down lists, check boxes, and other visual controls to generate C or C++ initialization code by specifying the peripherals they wish to use and how the peripherals should operate. While this is convenient and shortens development time, errors are still possible. Programmers should always review the code to verify it matches what they expected.

Microcontrollers in Embedded Systems

Personal Handheld Systems

Smart phones, digital media devices, calculators, and portable GPS units are only a few examples of portable handheld electronic devices that are microcontroller-based embedded systems. Microcontrollers are widely used in these products because they can easily interface with the input and output hardware, rapidly process data, and consume relatively little power. Some of the most popular microcontrollers for portable handheld devices are those based on the ARM (Advanced RISC Machine) processor.

A block diagram for a microcontroller-based programmable calculator is shown in Figure 14–35. The calculator incorporates a USB communication port. The ROM contains the embedded code that implements the calculator functions and processes while the RAM provides storage for the system stack, system data, user data, and programs. The USB controller transmits and receives data per the USB communications protocol and interfaces to the hardware that makes up the physical USB port. The calculator keypad connects to a parallel port formed by multiple GPIO lines, and the calculator LCD display interfaces with an LCD driver peripheral in the microcontroller to create the human machine interface, or HMI. A timer inside the microcontroller powers down the calculator after it has been active for a preset amount of time to save energy. Other timers in the microcontroller, which are not shown, set the communications rate for the USB controller, provide a real-time clock, and allow the user to set time-of-day alarms.

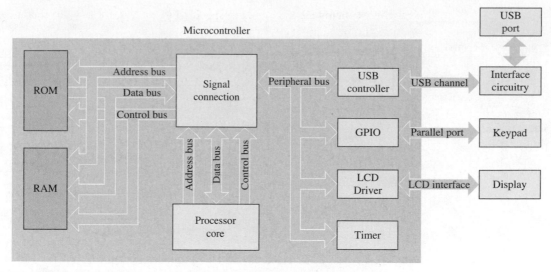

FIGURE 14–35 Microcontroller block diagram for programmable calculator.

Consumer Appliances

Virtually every electronic product today is a "smart" product that can make decisions, perform a preprogrammed sequence of events, or be manually programmed to do so. A short list of these products includes microwave ovens, coffee makers, washers and dryers, refrigerators, ovens, home entertainment components and systems, video game systems, and robotic vacuum cleaners.

Automobile Systems

Automobiles use microcontrollers in a number of embedded systems. Embedded systems in modern automobiles monitor vehicle operation and control the engine, fuel system, brakes, air bags, door locks, environmental system, instrument display, vehicle navigation, and virtually every aspect of vehicle operation. One specific factor that can affect microcontrollers in automotive applications is the operating environment. Microcontrollers must be able to operate properly when exposed to the vehicle's temperature, humidity, vibration, and electrical noise that they will encounter when the vehicle is operating.

Automated Systems

Two large areas of embedded applications are robotics and automation. Robotic and automated assemblies by nature must operate independently, perform repetitive tasks, process real-world data, and respond to circumstances that arise during operation. Embedded microcontroller systems can perform these tasks very well. One particular aspect of automated systems with which the microcontroller must deal is motion control. Microcontrollers must use feedback from the mechanical system to properly control the speed and acceleration of the system to ensure that it operates properly.

Figure 14–36 shows the block diagram for a basic robotics system. Although the block diagram is for a system that operates along a single axis, it can be extended to three axes for three-dimensional movement by using three microcontroller systems.

The ROM contains the embedded code that implements the robotic functions and processes; the RAM provides storage for the system stack and system data. The quadrature encoder receives encoded information from a motor position indicator and increments or decrements a counter depending upon how the encoded pattern changes. The pulse width modulator supplies a pulse train to a motor driver that in turn applies the voltages to the motor windings to turn the motor. GPIO lines detect when optical, magnetic, or other sensors indicate that the mechanical assembly has reached its maximum or minimum position.

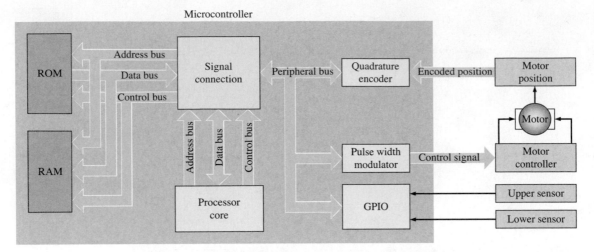

FIGURE 14–36 Basic block diagram for a robotics system.

When the system first powers up, the microcontroller uses the quadrature encoder and pulse width modulator to move the mechanical assembly to its minimum, or home, position and clears the counter so that zero corresponds to this home position and initializes the system. Once the system is initialized, the microcontroller then moves the mechanical assembly as programmed by driving the pulse width modulator to move the motor forward or backward and monitor the counter to determine how far and fast the mechanical assembly has moved. In most robotic systems, the microcontroller performs a complex series of calculations while monitoring the motor position and driving the motor to ensure that the mechanical assembly starts, stops, and operates smoothly.

SECTION 14–8 CHECKUP

1. How does a microcontroller differ from a microprocessor?
2. What are some common functional units found in a typical microcontroller?
3. Discuss an advantage and disadvantage of multifunctional pins on a microcontroller.
4. Which peripherals allow a microcontroller to interact with the real world?
5. How does an embedded system differ from a personal computer system?
6. Identify some types of embedded systems in which microcontrollers are found.

14–9 System on Chip (SoC)

The system on chip (SoC) is a major step up in complexity from the microcontroller and is what makes smaller and more powerful mobile devices possible. A SoC contains a variety of functional blocks integrated on a single semiconductor chip to meet specific application requirements. A SoC generally includes data processing, both digital and analog signal processing, data conversion, memory, interfacing, and other functions. The SoC is found in many devices such as smart phones, tablet computers, and digital cameras. Two important advantages of the SoP are small size and reduced power consumption, which make it ideal for small mobile devices.

After completing this section, you should be able to

♦ Describe a typical SoC

♦ List the functional elements of a SoC

A **system on chip (SoC)** is an integrated circuit that combines all components of a computer or other electronic system on a single chip. The SoC offers reduced manufacturing costs and smaller system configurations; Package sizes can be smaller than a dime, as shown in Figure 14–37.

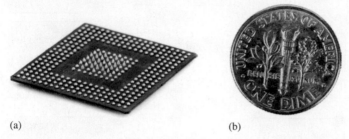

(a)   (b)

FIGURE 14–37   A typical SoC ball-grid package. The bottom of the package with the BG contacts is shown.   (a) Boris Sosnovyy/Shutterstock (b) Eldad Carin/Shutterstock.

A typical SoC consists of the following functional elements, depending on the application:

- A single or multiple-processor (CPU) core
- A digital signal processor (DSP)
- A graphics processor (GPU)
- Memory (ROM, RAM, EEPROM, flash)
- Analog functions such as ADC and DAC
- I/O interfaces such as USB, Firewire, I^{2}C, USART
- Timing sources such as oscillators and phase-locked loops (PLL)
- Voltage regulators and other power management functions
- Bus bridges
- Various peripherals
- Programmable logic and application specific logic

In a system using a microprocessor as the CPU, a variety of other chips must be included to achieve full system capability. The same is true for systems using a microcontroller, although a smaller chip set may be required because the microcontroller typically has memory and some peripherals on a single chip. Actually, the microcontroller often is considered a SoC with limited functionality. The SoC provides all functions necessary for a given system application, such as a computer on a single chip. Figure 14–38 illustrates a simplified generic SoC block diagram. Actual SoCs feature a number of functions that vary from one manufacturer to another.

The CPU (central processing unit) in a typical SoC may feature one or more microprocessors (MPUs) as well as a graphics processor (GPU). Generally, SoCs use processors based on ARM architecture. The ARM processors, developed by Advanced RISC Machines, Ltd. in the 1980s, were very simple in terms of transistor count and instruction set. They used reduced instruction set computer (RISC) architecture which allowed them to have high performance and low energy consumption. The GPU (graphics processing unit) handles complex games and other video requirements that are found on smart phones, tablets, and other devices.

SoCs include various types of memory such as ROM, SRAM, DRAM, and cache as well as the accompanying control functions. A DSP (digital signal processor) is also a feature on many SoCs along with analog functions such as ADC (analog-to-digital conversion) and DAC (digital-to-analog conversion) elements. Of course, interfacing is a crucial part of any system and all SoCs provide a varying number of standard bus and other I/Os. These interfacing elements may include USB, SPI, CAN, I^{2}C, AGP, UART, Bluetooth,

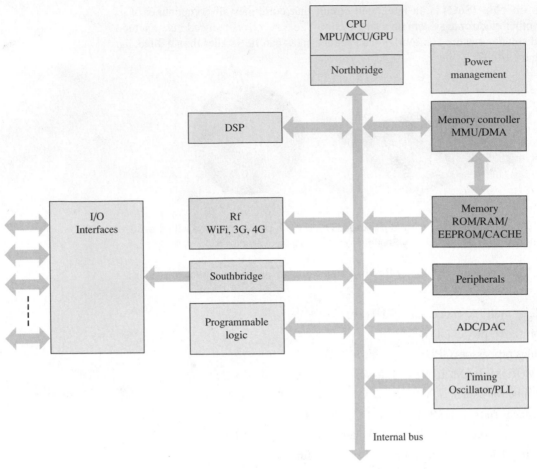

FIGURE 14–38   Generic block diagram of a typical SoC.

Wi-Fi, Ethernet, audio, rf, as well as others. The northbridge is a circuit that connect the CPU to the memory, and to the PCI internal bus. The southbridge is a circuit that controls connections to the I/Os.

SECTION 14–9 CHECKUP

1. What is a SoC?

2. List two advantages of a SoC.

3. Name at least five functional elements of a SoC.

SUMMARY

- The basic functional components of a CPU are the ALU, register set, and timing and control unit.
- A microprocessor implements the functional components of a CPU on a single IC.
- The three basic computer buses are the address bus, data bus, and the control bus.
- Examples of peripheral computer devices include the keyboard, external disk drives, mouse, printer, modem, and scanner.
- Computer systems use I/O ports to access peripherals such as keyboards, mice, video monitors, modems, scanners, and disk drives.
- The three tasks of a computer system are to acquire, process, and return data.
- Computer systems use different types of memory for specialized functions, such as caching and queuing data.

- The functions of a CPU are to control system hardware, to provide hardware support to the system operating system, and to execute programs.

- A multicore processor has two or more microprocessors (cores), each with its own memory cache on a single chip.

- Pipelining, multitasking, and multiprocessing are techniques for decreasing the processing time.

- Buffers are used to prevent output loading but introduce propagation delays.

- Wait states are used to compensate for delays in bus signals due to bus buffers, address decoders, and mismatched timing specifications between hardware.

- Two methods for a processor to execute concurrent multiple processes are multitasking and multiprocessing.

- The three basic duties of a multitasking operating system are to allocate resources, to protect process and system resources, and to provide system services.

- The basic "language" of a computer is called machine code in which instructions are given as a series of binary codes.

- In assembly language, machine instructions are replaced with a short alphabetic English mnemonic that has a one-to-one correspondence to machine code. Assembly language also uses directives to allow the programmer to specify other parameters that are not translated directly into machine code.

- Microcontrollers integrate a microprocessor core with hardware peripherals and are well-suited for embedded applications.

- Embedded systems primarily process real-world signals and operate in real time rather than manipulate application program data.

- A system on chip (SoC) is an integrated circuit that combines all components of a computer or other electronic system on a single chip.

KEY TERMS

Key terms and other bold terms in the chapter are defined in the end-of-book glossary.

ALU   Arithmetic logic unit.

Assembly language   A programming language that uses English-like words and has a one-to-one correspondence to machine language.

BIOS   The set of low-level routines that initialize the computer system hardware and allow high-level programs to interact with the system hardware.

Buffer   A device that prevents loading of an output.

Bus master   Any device that can control and manage the system buses in a computer system.

Caching   The process of copying frequently accessed program instructions from main memory into faster memory to increase processing speed.

CPU   Central processing unit; the "brain" of a computer that processes the program instructions.

DMA   Direct memory access.

Exception   Any software event that requires special handling by the processor.

Hardware   The circuitry and physical components of a computer system, as opposed to the instructions (called software).

High-level language   A type of computer language closest to human language that is a level above assembly language.

Interrupt   Any hardware event that requires special handling by the processor. An event that causes the current process to be temporarily stopped while a service routine is run.

Interrupt vector table   A data structure in memory that contains the addresses of interrupt service routines for the processor.

Machine language   Computer instructions written in binary code that are understood by a computer; the lowest level of programming language.

Main memory   Memory used by computer systems to hold the bulk of programs and associated data.

Microcontroller   An semiconductor device that combines a microprocessor, memory, and various hardware peripherals on a single IC and generally used for special or limited applications.

Microprocessor   A large-scale digital integrated circuit that can be programmed to perform arithmetic, logic, or other operations; the CPU of a computer.

Multiprocessing   The use of multiple processors to multitask or run multiple programs.

Multitasking   A technique by which a single processor runs multiple programs concurrently.

Op-code   The mnemonic representation of a computer instruction.

Operand   The object to be manipulated by the instruction.

Operating system   The software that controls the computer system and oversees the execution of application software.

Pipelining   A technique where the microprocessor begins executing the next instruction in a program before the previous instruction has been completed.

Program   A sequential set of computer instructions designed to accomplish a given task(s).

Signal Loading   The effect of the multiple inputs degrading the voltage or timing specifications of an output.

Software   Computer programs; programs that instruct a computer what to do in order to carry out a given set of tasks.

System bus   The interconnecting paths in a computer system including the address bus, data bus and control bus.

System on Chip (SoC)   An integrated circuit that combines all components of a computer or other electronic system on a single chip.

Wait state   A system bus delay equal to one processor clock cycle.

TRUE/FALSE QUIZ

Answers are at the end of the chapter.

1. The three basic components of a CPU are input, memory, and output.
2. CPU stands for *computer processing unit*.
3. Printers, scanners, and keyboards are examples of computer peripherals.
4. The operating system of a computer is made of hardware.
5. The ALU is a key element in a microprocessor.
6. Microprocessors generally have three types of buses: address, data, and control.
7. Wait states allow other bus masters to share the system buses.
8. A multicore processor has one processor and more than one memory.
9. Pipelining allows a computer to execute a program faster.
10. Three levels of computer programming languages are machine, assembly, and high level.
11. DMA stands for digital memory access.
12. The signals associated with a DMA operation are bus request, bus grant, and bus grant acknowledge.
13. Microprocessor and microcontroller are different names for the same thing.
14. Some examples of microcontroller peripherals are GPIOs, ADCs, and quadrature encoders.
15. SoC stands for sequential output computer.

SELF-TEST

Answers are at the end of the chapter.

1. A basic computer does not include
   - (a) an arithmetic logic unit
   - (b) a control unit
   - (c) peripheral units
   - (d) a memory unit

2. The operation of the timing and control unit in a microprocessor is affected by
   (a) the instruction decoder
   (b) the accumulator
   (c) the arithmetic logic unit
   (d) the register array

3. A 20-bit address bus supports
   (a) 100,000 memory addresses
   (b) 1,048,576 memory addresses
   (c) 2,097,152 memory addresses
   (d) 20,000 memory addresses

4. A bus that is used to transfer information both to and from the microprocessor is the
   (a) address bus
   (b) data bus
   (c) both of the above
   (d) none of the above

5. An example of a peripheral unit is
   (a) the address register
   (b) the CPU
   (c) the video monitor
   (d) the interface adapter

6. In indexed addressing the contents of
   (a) the accumulator are added to the contents of the index register
   (b) the data register are added to the contents of the index register
   (c) the index register are added to the contents of the program counter
   (d) the index register are multiplied by the contents of the program counter

7. Polling is a method used for
   (a) determining the state of the microprocessor
   (b) establishing communications between the CPU and a peripheral
   (c) establishing a priority for communication with several peripherals
   (d) determining the next instruction

8. Polling peripherals
   (a) is less efficient than using interrupts
   (b) requires the CPU to periodically check each polled peripheral
   (c) often has the CPU checking devices that do not require service
   (d) all of the above

9. DMA stands for
   (a) digital microprocessor address
   (b) direct memory access
   (c) data multiplexed access
   (d) direct memory addressing

10. A computer program is a list of
   (a) memory addresses that contain data to be used in an operation
   (b) addresses that contain instructions to be used in an operation
   (c) instructions arranged to achieve a specific result

11. A type of assembly language instruction that alters the course of the program is called a
   (a) loop
   (b) jump
   (c) both of the above
   (d) none of the above

12. A wait state
   (a) terminates a bus cycle
   (b) halts the processor clock for one period
   (c) places the microprocessor in a low-power mode
   (d) delays completion of a bus cycle by one processor clock

13. A bus request differs from an interrupt in that
   (a) an interrupt requires an external bus master to complete the operation
   (b) a bus request will interrupt a current instruction cycle
   (c) an interrupt cannot be masked
   (d) a bus request does not involve a response from the processor

14. Multitasking
   (a) allows a single processor to execute more than one active process concurrently
   (b) allows a processor to share its buses with another bus master
   (c) is the same as running a terminate and stay resident (TSR) application
   (d) requires no hardware support from the CPU

15. A typical microcontroller contains
   (a) nonvolatile memory
   (b) RAM
   (c) a processor core
   (d) all of the above

PROBLEMS

Answers to odd-numbered problems are at the end of the book.

Section 14–1 The Computer System

1. Name the basic elements of a computer.
2. Name the functional units of a CPU.
3. What is a bus?
4. Explain how the control bus signals differ from those of the address and data buses.

Section 14–2 Practical Computer System Considerations

5. Name two special types of output that allow devices to share signal lines while avoiding bus contention.
6. What causes signal loading?
7. In Figure 14–5, determine the number of loads that the output is driving in parts (a) and (b).
8. In a computer system, how does the CPU select a device such as memory or Input/Output?
9. Explain the purpose of a wait state.

Section 14–3 The Processor: Basic Operation

10. Name the basic elements of a microprocessor.
11. List three operations that a microprocessor performs.
12. List the three microprocessor buses.
13. Explain what happens during a fetch/execute cycle.
14. Explain how pipelining works.

Section 14–4 The Processor: Addressing Modes

15. List the sequence of events for the inherent addressing mode.
16. List the sequence of events for the direct addressing mode.
17. List the sequence of events for the indexed addressing mode.
18. A processor is using the relative addressing mode to execute a branch instruction. If the binary number for 125_{10} is in the program counter and the number 55_{10} is in the data register, to what memory address will the processor branch?

Section 14–5 The Processor: Special Operations

19. Describe the purpose of an interrupt vector table.
20. How does an interrupt service routine differ from a normal subroutine?
21. Describe the sequence of events in a bus request operation.
22. Define DMA and describe its purpose in a computer system.

Section 14–6 Operating Systems and Hardware

23. What are the two groups of software that execute in a computer system and what do they include?
24. List the three basic duties of a multitasking operating system.
25. Identify the two ways that computers execute more than one process concurrently and describe how they differ.
26. Describe four difficulties that running multiple concurrent processes can create.
27. What are the functions of a memory management unit?

Section 14–7 Programming

28. What is an assembler?
29. Draw a flowchart for a program that adds the numbers from one to 10 and saves the result in a memory location named TOTAL.

30. Draw a flowchart showing how you could count the number of bytes in a string and place the count in a location in memory called COUNT. Assume the string starts at a location named START and has a 20H (hexadecimal ASCII code for a space) to signal the end. You should not count the space character.

31. Explain what happens when the instruction **mov ax,[bx]** is executed.

32. What is a compiler?

Section 14–8 Microcontrollers and Embedded Systems

33. What is a microcontroller?

34. Identify the functional components of the microcontroller block diagram shown in Figure 14–39.

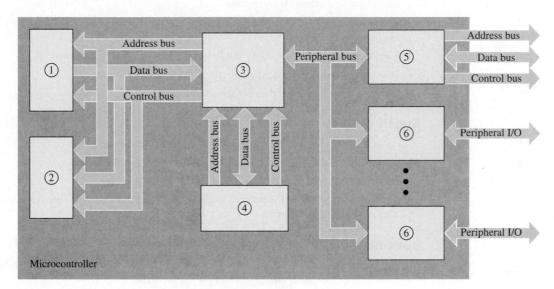

FIGURE 14–39

35. Why are microcontrollers so widely used in embedded applications?

36. How does a microcontroller differ from a microprocessor?

Section 14–9 System on Chip (SoC)

37. How does a SoC differ from a microcontroller?

38. The processors used in SoCs are generally based on what type of architecture?

ANSWERS

SECTION CHECKUPS

Section 14–1 The Computer System

1. The major functional blocks in a computer are CPU, memory/storage, input/output ports.

2. Peripherals are devices external to the computer.

3. Hardware is the microprocessor, memory, hard disk, etc. Software is the program that runs the computer.

4. Conventional memory returns the data stored at a specified address. Content-addressable memory returns the address that contains a specified data value.

5. The address bus is unidirectional. The data bus is bidirectional. The control bus consists of signals that are bidirectional and operate independently rather than as a single functional group.

Section 14-2 Practical Computer System Considerations

1. Bus contention is the condition that results when two outputs are connected to the same signal line attempt to drive it to different voltage levels. Two types of devices intended to prevent bus contention on shared signal lines are tri-state devices and open collector (or open drain) devices.

2. A processor uses an address decoder to select and enable various devices.

3. Wait states in a computer system are additional clock cycles inserted in a processor bus cycle to satisfy the timing specifications of system devices.

4. A buffer prevents excess loading of a device and allows multiple devices to share the same bus.

Section 14-3 The Processor: Basic Operation

1. During a fetch/execute cycle, the CPU retrieves (fetches) an instruction from memory and carries out (executes) the instruction.

2. The four elements in a microprocessor are ALU, timing and control unit, register set, and instruction decoder.

3. The ALU is the arithmetic logic unit, which carries out decoded instructions.

4. During a read operation, data are acquired from memory.

5. During a write operation, data are stored in memory.

Section 14-4 The Processor: Addressing Modes

1. Addressing modes are inherent, direct, immediate, indexed, and relative.

2. An op-code is the representation of an computer program instruction.

3. An operand is an object that is manipulated by an instruction.

4. Branching occurs when the processor leaves the normal sequence in a program and branches to another place in the program.

Section 14-5 The Processor: Special Operations

1. Although the terms are sometimes used interchangeably, interrupts are generated by hardware sources and exceptions are generated by software sources.

2. A bus request can interrupt an instruction cycle, cannot be masked, and allows the external device to take control of the system buses. An interrupt must permit an instruction cycle to complete, can be masked, and is serviced by the processor which retains control of the system buses.

3. A direct memory access is a data transfer operation for which a special bus master called a DMA controller rather than the microprocessor controls the system buses. Direct memory accesses allow data transfers to occur much more rapidly than with the microprocessor because the controller does not attempt to process the data for each transfer as would the microprocessor.

Section 14-6 Operating Systems and Hardware

1. The three basic duties of an operating system are to allocate system resources, to protect processes and system resources, and to provide system services.

2. Multitasking allows a single processor to execute multiple processes concurrently by allocating processor time to each process. Multiprocessing allows a system to execute multiple processes by operating more than one processor.

3. A memory management unit prevents a process from accessing the memory space of another by comparing the contents of the address bus against a permitted range and generating an interrupt when a violation occurs so that the operating system can take appropriate action.

4. An operating system allows processes that are restricted to their own memory spaces to exchange information by providing a system service, activated by a software interrupt, that has the operating system pass the information between the processes.

Section 14-7 Programming

1. A program is a sequence of computer instructions designed to perform a specified task.

2. An op-code is an instruction expressed in mnemonic form.

3. A string is a contiguous sequence of bytes or words.

Section 14–8 Microcontrollers and Embedded Systems

1. A microcontroller basically includes many elements of a computer system where a microprocessor is only the CPU.

2. Common functional units found in a typical microcontroller are the processor core, ROM, RAM, one or more hardware peripherals, a signal connection block between the microcontroller functional blocks, and an external memory controller.

3. An advantage of multifunctional pins on a microcontroller is that it reduces the number of pins and physical size of the device. A disadvantage is that it is not always possible to have access to all the microcontroller features that an application may require so that the design requires additional external hardware.

4. Peripherals that allow microcontrollers to interact with the real world include timers, ADCs, DACs, communication controllers, GPIOs, quadrature encoders, and PWMs.

5. An embedded system is designed to interact directly with the real world and perform a specific function. A personal computer system processes data and can be configured with application software to perform a number of tasks.

6. Some embedded systems in which microcontrollers are found include personal electronics, consumer electronics, automotive systems, and communication devices.

Section 14–9 System on Chip (SoC)

1. A SoC is a system chip, a complete computer on a single silicon chip.

2. Small size and reduced power consumption.

3. Functional elements of a SoC: one or more processors; DSP; GPU; memory; ADC and DAC; I/O interfaces; timing sources; power; bus bridges; peripherals; and programmable logic.

RELATED PROBLEMS FOR EXAMPLES

14–1 D_6 must be LOW.

14–2 Change first block (initialization block) to "BIG = FFFF"; this is the largest possible unsigned number. Change first question to "Is number < BIG?"

TRUE/FALSE QUIZ

1. F   **2.** F   **3.** T   **4.** F   **5.** T   **6.** T   **7.** F   **8.** F   **9.** T   **10.** T   **11.** F
12. F   **13.** F   **14.** T   **15.** F

SELF-TEST

1. (c)   **2.** (a)   **3.** (b)   **4.** (b)   **5.** (c)   **6.** (b)   **7.** (b)   **8.** (d)   **9.** (b)   **10.** (c)   **11.** (c)
12. (d)   **13.** (c)   **14.** (a)   **15.** (d)

Chapter 1

1. Digital can be transmitted and stored more efficiently and reliably.

3. Clock
Thermometer
Speedometer

5. (a) 11010001   (b) 000101010

7. (a) 550 ns   (b) 600 ns   (c) 2.7 μs   (d) 10 V

9. 250 Hz

11. 50%

13. 8 μs; 1 μs

15. $L_{on} = SW1 + SW2 + SW1 \cdot SW2$

17. AND gate

19. (a) adder   (b) multiplier

     (c) multiplexer   (d) comparator

21. 01010000

23. SPLD: Simple Programmable Logic Device
CPLD: Complex Programmable Logic Device
HDL: Hardware Description Language
FPGA: Field-Programmable Gate Array
GAL: Generic Array Logic

25. Place-and-route or fitting is the process where the logic structures described by the netlist are mapped into the actual structure of the specific target device. This results in an output called a bitstream.

27. DIP pins go through holes in a circuit board. SMT pins connect to surface pads.

29. 7 V

31. 125 Hz

33. Troubleshooting is the process of recognizing, isolating, and correcting a fault or failure in a system.

35. In the signal-tracing method, a signal is tracked as it progresses through a system until a point is found where the signal disappears or is incorrect.

37. When a failure is reported, determine when and how the failure occurred and what are the symptoms.

39. An incorrect output can be caused by an incorrect dc supply voltage, improper ground, incorrect component value, or a faulty component.

41. To isolate a fault in a system, apply half-splitting or signal tracing.

43. When a fault has been isolated to a particular circuit board, the options are to repair the board or replace the board with a known good board.

Chapter 2

1. (a) 1   (b) 100   (c) 100,000

3. (a) 400; 70; 1   (b) 9000; 300; 50; 6

     (c) 100,000; 20,000; 5000; 0; 0; 0

5. (a) 3   (b) 4   (c) 7   (d) 8   (e) 9

     (f) 12   (g) 11   (h) (15)

7. (a) 51.75   (b) 42.25   (c) 65.875

     (d) 120.625   (e) 92.65625   (f) 113.0625

     (g) 90.625   (h) 127.96875

9. (a) 5 bits   (b) 6 bits   (c) 6 bits

     (d) 7 bits   (e) 7 bits   (f) 7 bits

     (g) 8 bits   (h) 8 bits

11. (a) 1010   (b) 10001   (c) 11000

     (d) 110000   (e) 111101   (f) 1011101

     (g) 1111101   (h) 10111010

13. (a) 1111   (b) 10101   (c) 11100

     (d) 100010   (e) 101000   (f) 111011

     (g) 1000001   (h) 1001001

15. (a) 100   (b) 100   (c) 1000

     (d) 1101   (e) 1110   (f) 11000

17. (a) 1001   (b) 1000   (c) 100011

     (d) 110110   (e) 10101001   (f) 10110110

19. all 0s or all 1s

21. (a) 010   (b) 001   (c) 0101

     (d) 00101000   (e) 0001010   (f) 11110

23. (a) 00011101   (b) 11010101

     (c) 01100100   (d) 11111011

25. (a) 00001100   (b) 10111100

     (c) 01100101   (d) 10000011

27. (a) −102   (b) +116   (c) −64

29. (a) 0 10001101 11110000101011000000000

     (b) 1 10001010 11000001100000000000000

31. (a) 00110000   (b) 00011101

     (c) 11101011   (d) 100111110

33. (a) 11000101   (b) 11000000

35. 100111001010

37. (a) 00111000   (b) 01011001   (c) 101000010100

     (d) 010111001000   (e) 0100000100000000

     (f) 1111101100010111   (g) 1000101010011101

39. (a) 35   (b) 146   (c) 26   (d) 141

     (e) 243   (f) 235   (g) 1474   (h) 1792

41. (a) 60_{16} (b) $10B_{16}$ (c) $1BA_{16}$

43. (a) 10   (b) 23   (c) 46   (d) 52   (e) 67

     (f) 367   (g) 115   (h) 532   (i) 4085

45. (a) 001011   (b) 101111   (c) 001000001

     (d) 011010001   (e) 101100000

     (f) 100110101011   (g) 001011010111001

     (h) 100101110000000   (i) 0010000000010001011

47. (a) 00010000   (b) 00010011

     (c) 00011000   (d) 00100001

(e) 00100101   **(f)** 00110110

(g) 01000100   **(h)** 01010111

(i) 01101001   **(j)** 10011000

(k) 000100100101   **(l)** 000101010110

49. (a) 000100000100   **(b)** 000100101000

(c) 000100110010   **(d)** 000101010000

(e) 000110000110   **(f)** 001000010000

(g) 001101011001   **(h)** 010101000111

(i) 0001000001010001

51. (a) 80   **(b)** 237   **(c)** 346   **(d)** 421   **(e)** 754

(f) 800   **(g)** 978   **(h)** 1683   **(i)** 9018   **(j)** 6667

53. (a) 00010100   **(b)** 00010010

(c) 00010111   **(d)** 00010110

(e) 01010010   **(f)** 000100001001

(g) 000110010101   **(h)** 0001001001101001

55. The Gray code makes only one bit change at a time when going from one number in the sequence to the next.

57. (a) 1100   **(b)** 00011   **(c)** 10000011110

59. (a) CAN   **(b)** J   **(c)** =

(d) #   **(e)** >   **(f)** B

61. 48 65 6C 6C 6F 2E 20 48 6F 77 20 61 72 65 20 79 6F 75 3F

63. (b) is incorrect.

65. (a) 110100100   **(b)** 000001001   **(c)** 111111110

67. In each case, you get the other number.

69. The remainder is 0100, indicating an error.

Chapter 3

1. See Figure P–1.

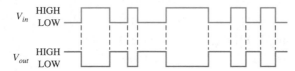

FIGURE P–1

3. See Figure P–2.

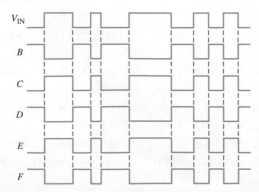

FIGURE P–2

5. See Figure P–3.

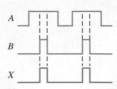

FIGURE P–3

7. See Figure P–4.

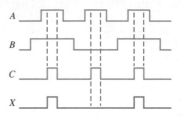

FIGURE P–4

9. See Figure P–5.

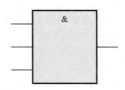

FIGURE P–5

11. See Figure P–6.

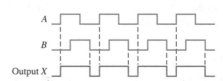

FIGURE P–6

13. See Figure P–7.

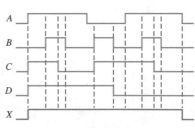

FIGURE P–7

15. See Figure P–8.

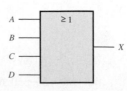

FIGURE P–8

17. See Figure P–9.

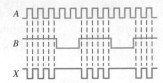

FIGURE P–9

19. See Figure P–10.

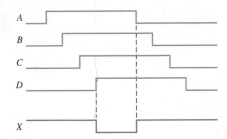

FIGURE P–10

21. See Figure P–11.

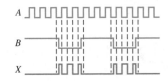

FIGURE P–11

23. See Figure P–12.

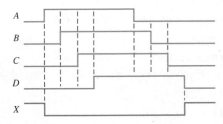

FIGURE P–12

25. $XOR = A\overline{B} + \overline{A}B$; $OR = A + B$

27. See Figure P–13.

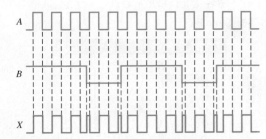

FIGURE P–13

29. $X_1 = \overline{A}B, X_2 = \overline{A}\,\overline{B}, X_3 = A\overline{B}.$

31. **entity** 4InputAND **is**

    port (A, B, C, D: **in** bit; X: **out** bit);

  end entity 4Input AND;

  architecture Function **of** 4InputAND **is**

  begin

    X <= A **and** B **and** C **and** D;

  end architecture Function;

33. CMOS

35. $t_{PLH} = 4.3$ ns; $t_{PHL} = 10.5$ ns

37. 20 mW

39. The gates in parts (b), (c), (e) are faulty.

41. (a) defective output (stuck LOW or open)

    (b) Pin 4 input or pin 6 output internally open.

43. The seat belt input to the AND gate is open.

45. Add an inverter to the enable input line of the AND gate.

47. See Figure P–14.

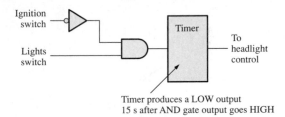

FIGURE P–14

49. The inputs are now active-LOW. Change the OR gates to NAND gates (negative-OR) and add two inverters.

51. See Figure P–15.

FIGURE P–15

53. Input B of NAND gate shorted to VCC.

55. Input B of XOR gate shorted to VCC.

Chapter 4

1. $X = A + B + C + D$

3. $X = \overline{A} + \overline{B} + \overline{C}$

5. (a) $AB = 1$ when $A = 1, B = 1$

    (b) $A\overline{B}C = 1$ when $A = 1, B = 0, C = 1$

    (c) $A + B = 0$ when $A = 0, B = 0$

    (d) $\overline{A} + B + \overline{C} = 0$ when $A = 1, B = 0, C = 1$

    (e) $\overline{A} + \overline{B} + C = 0$ when $A = 1, B = 1, C = 0$

    (f) $\overline{A} + B = 0$ when $A = 1, B = 0$

    (g) $A\overline{B}\,\overline{C} = 1$ when $A = 1, B = 0, C = 0$

7. (a) Commutative

    (b) Commutative

    (c) Distributive

9. (a) $\overline{A}B$ (b) $A + \overline{B}$

(c) $\overline{A}\,\overline{B}C$ (d) $\overline{A} + \overline{B} + \overline{C}$

(e) $\overline{A} + \overline{B}\,\overline{C}$ (f) $\overline{A} + \overline{B} + \overline{C} + \overline{D}$

(g) $(\overline{A} + \overline{B})(\overline{C} + \overline{D})$ (h) $\overline{A}B + C\overline{D}$

11. (a) $(\overline{A} + \overline{B} + \overline{C})(\overline{E} + \overline{F} + \overline{G})(\overline{H} + \overline{I} + \overline{J})(\overline{K} + \overline{L} + \overline{M})$

(b) $\overline{A}B\overline{C} + BC$

(c) $\overline{A}\,\overline{B}\,\overline{C}\,\overline{D}\,\overline{E}\,\overline{F}\,\overline{G}\,\overline{H}$

13. (a) $X = ABCD$ (b) $X = AB + C$

(c) $X = \overline{\overline{A}B}$ (d) $X = (A + B)C$

15. See Figure P–16.

17. (a) See Table P–1.

TABLE P–1

Inputs			Output
$\overline{VCR}$	$\overline{CAMI}$	$\overline{RDY}$	$\overline{RECORD}$
0	0	0	0
0	0	1	1
0	1	0	0
0	1	1	1
1	0	0	0
1	0	1	1
1	1	0	1
1	1	1	1

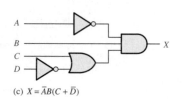

(a) $X = A\overline{B} + \overline{A}B$

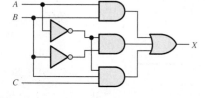

(b) $X = AB + \overline{A}\overline{B} + \overline{A}BC$

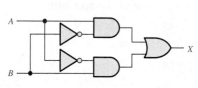

(c) $X = \overline{A}B(C + \overline{D})$

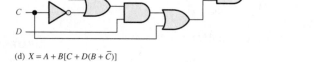

(d) $X = A + B[C + D(B + \overline{C})]$

FIGURE P–16

17. (b) See Table P–2.

TABLE P–2

Inputs			Output
$\overline{RTS}$	$\overline{ENABLE}$	$\overline{BUSY}$	$\overline{SEND}$
0	0	0	1
0	0	1	1
0	1	0	1
0	1	1	1
1	0	0	1
1	0	1	1
1	1	0	0
1	1	1	1

19. (a) A (b) AB (c) C

(d) A (e) $\overline{A}C + \overline{B}C$

21. (a) $BD + BE + \overline{D}F$ (b) $\overline{A}\overline{B}C + \overline{A}\overline{B}D$

(c) B (d) $AB + CD$

(e) ABC

23. (a) $A\overline{B} + AC + BC$ (b) $AC + \overline{B}C$

(c) $AB + AC$

25. (a) Domain: A, B, C

Standard SOP: $\overline{A}B\overline{C} + A\overline{B}\overline{C} + ABC + \overline{A}BC$

(b) Domain: A, B, C

Standard SOP: $ABC + A\overline{B}C + \overline{A}\overline{B}C$

(c) Domain: A, B, C

Standard SOP: $ABC + AB\overline{C} + A\overline{B}C$

27. (a) $101 + 100 + 111 + 011$

(b) $111 + 101 + 001$

(c) $111 + 110 + 101$

29. (a) $(A + B + C)(A + B + \overline{C})(\overline{A} + \overline{B} + C)(\overline{A} + \overline{B} + C)$

(b) $(A + B + C)(A + \overline{B} + C)(A + \overline{B} + \overline{C})$
$(\overline{A} + B + C) + (\overline{A} + \overline{B} + C)$

(c) $(A + B + C)(A + B + \overline{C})(A + \overline{B} + C)$
$(A + \overline{B} + \overline{C})(\overline{A} + B + C)$

31. (a) See Table P–3.

TABLE P–3

A	B	C	X
0	0	0	0
0	0	1	0
0	1	0	1
0	1	1	0
1	0	0	0
1	0	1	1
1	1	0	0
1	1	1	1

31. (b) See Table P–4.

TABLE P–4

X	Y	Z	Q
0	0	0	1
0	0	1	1
0	1	0	0
0	1	1	1
1	0	0	0
1	0	1	1
1	1	0	1
1	1	1	0

33. (a) See Table P–5.

TABLE P–5

A	B	C	X
0	0	0	1
0	0	1	0
0	1	0	1
0	1	1	1
1	0	0	0
1	0	1	1
1	1	0	1
1	1	1	0

33. (b) See Table P–6.

TABLE P–6

W	X	Y	Z	Q
0	0	0	0	1
0	0	0	1	1
0	0	1	0	1
0	0	1	1	1
0	1	0	0	0
0	1	0	1	1
0	1	1	0	1
0	1	1	1	0
1	0	0	0	1
1	0	0	1	1
1	0	1	0	1
1	0	1	1	1
1	1	0	0	0
1	1	0	1	1
1	1	1	0	1
1	1	1	1	1

35. (a) See Table P–7.

TABLE P–7

A	B	C	X
0	0	0	0
0	0	1	0
0	1	0	0
0	1	1	1
1	0	0	1
1	0	1	1
1	1	0	1
1	1	1	1

35. (b) See Table P–8.

TABLE P–8

A	B	C	D	X
0	0	0	0	1
0	0	0	1	0
0	0	1	0	1
0	0	1	1	1
0	1	0	0	0
0	1	0	1	0
0	1	1	0	0
0	1	1	1	0
1	0	0	0	1
1	0	0	1	0
1	0	1	0	0
1	0	1	1	1
1	1	0	0	1
1	1	0	1	1
1	1	1	0	1
1	1	1	1	1

37. See Figure P–17.

AB\C	0	1
00	000	001
01	010	011
11	110	111
10	100	101

FIGURE P–17

39. See Figure P–18.

AB\C	0	1
00	$\bar{A}\bar{B}\bar{C}$	$\bar{A}\bar{B}C$
01	$\bar{A}B\bar{C}$	$\bar{A}BC$
11	$AB\bar{C}$	ABC
10	$A\bar{B}\bar{C}$	$A\bar{B}C$

FIGURE P–18

41. (a) No simplification   (b) AC

(c) $\overline{DF} + E\overline{F}$

43. (a) $AB + AC$

(b) $A + BC$

(c) $B\overline{C}D + A\overline{C}D + BC\overline{D} + AC\overline{D}$

(d) $A\overline{B} + CD$

45. $\overline{B} + C$

47. $\overline{A}\overline{B}\overline{C}D + C\overline{D} + BC + A\overline{D}$

49. (a) No reduction

(b) $(W + X)(W + \overline{Z})(X + \overline{Y})(\overline{W} + \overline{X} + \overline{Y} + \overline{Z})$

51. $(\overline{A} + B + \overline{D})(A + C + D)(A + \overline{B} + C)$
$(B + \overline{C} + \overline{D})(A + \overline{B} + \overline{C} + D)$

53. Minterms: 1, 3, 5, 6, 7

55. See Table P–9.

TABLE P–9

Number of 1s	Minterm	ABCD
0	m_0	0000
1	m_1	0001
2	m_5	0101
	m_6	0110
	m_9	1001
	m_{12}	1100

57. See Table P–10.

TABLE P–10

First Level	Number of 1s in First Level	Second Level
(m_0, m_1) 000x	0	$(m_0 m_1)$ 000x
(m_1, m_5) 0x01 (m_1, m_9) x001	1	(m_1, m_5, m_9) xx01

59. $X = \overline{C}D + \overline{A}\overline{B}C + \overline{A}BC\overline{D} + AB\overline{C}\overline{D}$

61. The VHDL program:

entity SOP **is**

   port(A, B, C: **in** bit; X: **out** bit);

end entity SOP;

architecture Logic **of** SOP **is**

begin

   Y <= (A **and not** B **and** C) **or** (**not** A **and not** B **and** C) **or** (A **and not** B **and not** C) **or** (**not** A **and** B **and** C);

end architecture Logic;

63. The purpose of the invalid code detector is to detect the codes 1010, 1011, 1100, 1101, 1110, and 1111 to activate the display for letters.

65. **Segment d:** The minimum expression requires one 2-input AND gate, one 3-input AND gate, one 2-input OR gate, and 2 inverters.

Segment e: The minimum expression requires one 3-input AND gate.

Segment f: The minimum expression requires one 2-input AND gate.

Segment g: The minimum expression requires one 2-input AND gate, one 3-input AND gate, one 2 input OR gate, and 2 inverters.

67. See Figure P–19.

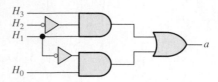

FIGURE P–19

69. The invalid code detector must disable the display when any numerical input (0–9) occurs. A HIGH enables the display and a LOW disables it. A circuit that detects the numeric codes and produces a LOW is shown in Figure P–20.

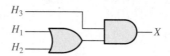

FIGURE P–20

71. Bottom input of U7 is open.

Chapter 5

1. See Figure P–21.

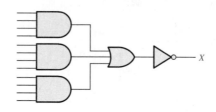 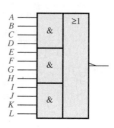

FIGURE P–21

3. (a) $X = ABB$ (b) $X = AB + B$

(c) $X = \overline{A} + B$ (d) $X = (A + B) + AB$

(e) $X = \overline{\overline{AB}C}$ (f) $X = (A + B)(\overline{B} + C)$

5. (a)

A	B	X
0	0	0
0	1	0
1	0	0
1	1	1

5. (b)

A	B	X
0	0	0
0	1	1
1	0	0
1	1	1

5. (c)

A	B	X
0	0	1
0	1	1
1	0	0
1	1	1

5. (d)

A	B	X
0	0	0
0	1	1
1	0	1
1	1	1

5. (e)

A	B	C	X
0	0	0	1
0	0	1	1
0	1	0	1
0	1	1	0
1	0	0	1
1	0	1	1
1	1	0	1
1	1	1	1

5. (f)

A	B	C	X
0	0	0	0
0	0	1	0
0	1	0	0
0	1	1	1
1	0	0	1
1	0	1	1
1	1	0	0
1	1	1	1

7. $X = \overline{A\overline{B} + \overline{A}B} = (\overline{A} + B)(A + \overline{B})$

9. $\overline{ABCD} + \overline{EFGH}$

11. See Figure P–22.

13. See Figure P–23.

15. $X = AB$

17. (a) No simplification

(b) No simplification

(c) $X = A$

(d) $X = \overline{A} + \overline{B} + \overline{C} + EF + \overline{G}$

(e) $X = ABC$

(f) $X = BC\overline{D}E + \overline{A}B\overline{E}FG + BC\overline{E}FG$

19. (a) $X = AC + AD + BC + BD$

(b) $X = \overline{A}CD + \overline{B}CD$

(c) $X = ABD + CD + E$

(d) $X = \overline{A} + B + D$

(e) $X = ABD + \overline{C}D + \overline{E}$

(f) $X = \overline{A}\,\overline{C} + \overline{A}\,\overline{D} + \overline{B}\,\overline{C} + \overline{B}\,\overline{D} + \overline{E}G + \overline{E}H + \overline{F}G + \overline{F}H$

21. See Figure P–24.

23. See Figure P–25.

25. See Figure P–26.

27. See Figure P–27.

29. $X = A + \overline{B}$; see Figure P–28.

31. $X = AB\overline{C}$ see Figure P–29.

33. The output pulse width is greater than the specified minimum.

35. $X \Leftarrow A$ **and** B **and** C

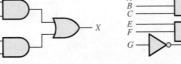

(a) $X = AB + \overline{B}C$

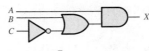

(b) $X = A(B + \overline{C})$

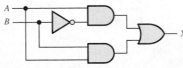

(c) $X = A\overline{B} + AB$

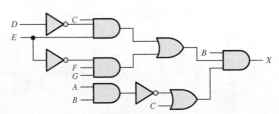

(d) $X = \overline{ABC} + B(EF + \overline{G})$

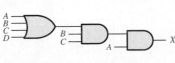

(e) $X = A[BC(A + B + C + D)]$

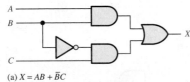

(f) $X = B(C\overline{D}E + \overline{E}FG)\,(\overline{AB} + C)$

FIGURE P–22

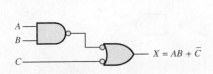

FIGURE P–23

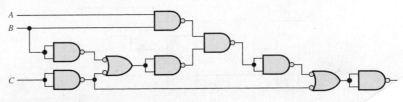

FIGURE P–24

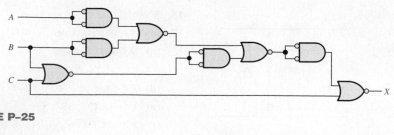

FIGURE P–25

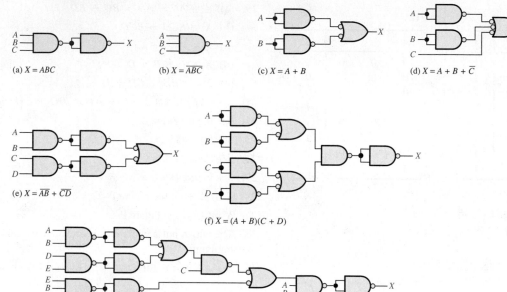

(a) $X = ABC$

(b) $X = \overline{ABC}$

(c) $X = A + B$

(d) $X = A + B + \overline{C}$

(e) $X = \overline{AB} + \overline{CD}$

(f) $X = (A + B)(C + D)$

(g) $X = AB[C(\overline{DE} + \overline{AB}) + \overline{BCE}]$

FIGURE P–26

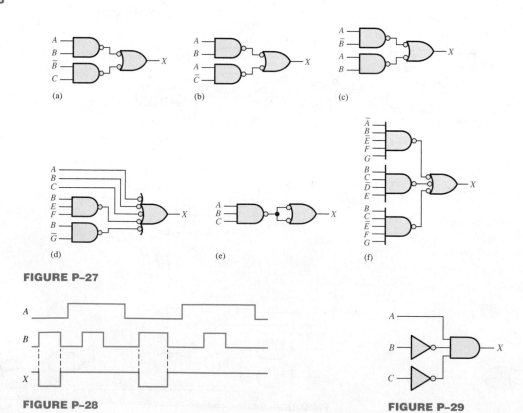

(a)

(b)

(c)

(d)

(e)

(f)

FIGURE P–27

FIGURE P–28

FIGURE P–29

37. (e) entity Circuit5_55e **is**

    port(A, B, C: **in** bit; X: **out** bit);

    end entity Circuit5_55e;

    architecture LogicFunction **of** Circuit5_55e **is**

    begin

        X <= (**not** A **and** B) **or** B **or** (B **and not** C) **or**

        (**not** A **and not** C) **or** (B **and not** C) **or not** C;

    end architecture LogicFunction;

(f) entity Circuit5_55f **is**

    port(A, B, C: **in** bit; X: **out** bit);

    end entity Circuit5_55f;

    architecture LogicFunction **of** Circuit5_55f **is**

    begin

        X <= (A **or** B) **and** (**not** B **or** C);

    end architecture LogicFunction;

39. Number gates from top to bottom and left to right G1, G2, G3, etc. Relabel inputs IN1, IN2, IN3, etc, and output OUT.

    entity Circuit5_56f **is**

    port(IN1, IN2, IN3, IN4, IN5, IN6, IN7, IN8: **in** bit; OUT: **out** bit);

    end entity Circuit5_56f;

    architecture LogicFunction **of** Circuit5_56f **is**

    component NAND_gate **is**

    port(A, B: **in** bit; X: **out** bit);

    end component NAND_gate;

    signal G1OUT, G2OUT, G3OUT, G4OUT, G5OUT, G6OUT: bit;

    begin

    G1: NAND_gate **port map** (A => IN1, B => IN2, X => G1OUT);

    G2: NAND_gate **port map** (A => IN3, B => IN4, X => G2OUT);

    G3: NAND_gate **port map** (A => IN5, B => IN6, X => G3OUT);

    G4: NAND_gate **port map** (A => IN7, B => IN8, X => G4OUT);

    G5: NAND_gate **port map** (A => G1OUT, B => G2OUT, X => G5OUT);

    G6: NAND_gate **port map** (A => G3OUT, B => G4OUT, X => G6OUT);

    G7: NAND_gate **port map** (A => G5OUT, B => G6OUT, X => OUT);

    end architecture LogicFunction;

41. –Data flow approach

    entity Fig5_65 **is**

    port (A, B, C, D, E: **in** bit; X: **out bit**);

    end entity Fig5_65;

    architecture DataFlow **of** Fig5_65 **is**

    begin

        X <= (A **and** B **and** C) **or** (D **and not** E);

    end architecture DataFlow;

–Structural approach

    entity Fig5_65 **is**

    port(IN1, IN2, IN3, IN4, IN5: **in** bit; OUT: **out** bit);

    end entity Fig5_65;

    architecture Structure **of** Fig5_65 **is**

    component AND_gate **is**

    port(A, B: **in** bit; X: **out** bit);

    end component AND_gate;

    component OR_gate **is**

    port(A, B: **in** bit; X: **out** bit);

    end component OR_gate;

    component Inverter **is**

    port(A: **in** bit; X: **out** bit);

    end component Inverter;

    signal G1OUT, G2OUT, G3OUT INVOUT: bit;

    begin

    G1: AND_gate **port map** (A => IN1, B => IN2, X => G1OUT);

    G2: AND_gate **port map** (A => G1OUT, B => IN3, X => G2OUT);

    INV: Inverter **port map** (A => IN5, X => INVOUT);

    G3: AND_gate **port map** (A => IN4, B => INVOUT, X => G3OUT);

    G4: OR_gate **port map** (A => G2OUT, B => G3OUT, X => OUT);

    end architecture Structure;

43. See Table P–11.

TABLE P–11

Inputs				Output
A	B	C	D	X
0	0	0	0	0
1	0	0	0	0
0	1	0	0	0
1	1	0	0	0
0	0	1	0	0
1	0	1	0	0
0	1	1	0	0
1	1	1	0	0
0	0	0	1	0
1	0	0	1	0
0	1	0	1	0
1	1	0	1	1
0	0	1	1	0
1	0	1	1	1
0	1	1	1	1
1	1	1	1	1

45. The AND gates are numbered top to bottom G1, G2, G3, G4. The OR gate is G5 and the inverters are, top to bottom. G6 and G7. Change A_1, A_2, B_1, B_2 to IN1, IN2, IN3, IN4 respectively. Change X to OUT.

entity Circuit5_67 **is**

    port (IN1, IN2, IN3, IN4: **in** bit; OUT: **out** bit);

end entity Circuit5_67;

architecture Logic **of** Circuit5_67 **is**

component AND_gate **is**

    port (A, B: **in** bit; X: **out** bit);

end component AND_gate;

component OR_gate **is**

    port (A, B, C, D: **in** bit; X: **out** bit);

end component OR_gate;

component Inverter **is**

    port (A: **in** bit; X: **out** bit);

end component Inverter;

    signal G1OUT, G2OUT, G3OUT, G4OUT, G5OUT, G6OUT, G7OUT: bit;

begin

    G1: AND_gate **port map** (A => IN1, B => IN2, X => G1OUT);

    G2: AND_gate **port map** (A => IN2, B => G6OUT, X => G2OUT);

    G3: AND_gate **port map** (A => G6OUT, B => G7OUT, X => G3OUT);

    G4: AND_gate **port map** (A => G7OUT, B => IN1, X => G4OUT);

    G5: OR_gate **port map** (A => G1OUT, B => G2OUT, C => G3OUT, D => G4OUT, X => OUT);

    G6: Inverter **port map** (A => IN3, X => G6OUT);

    G7: Inverter **port map** (A => IN4, X => G7OUT);

end architecture Logic;

47. $X = ABC + D\overline{E}$. Since X is the same as the G_3 output, either G_1 or G_2 has failed, with its output stick LOW.

49. See Figure P–30.

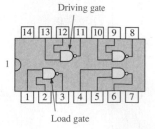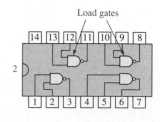

Driving gate   Load gates

Load gate

FIGURE P–30

51. **(a)** See Figure P–31.   **(b)** $X = E$ **(c)** $X = E$

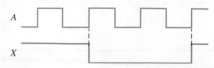

FIGURE P–31

53. The flow sensor measures the solution into the tank. The temperature transducer measures the temperature of the solution. The level sensors indicate when the solution is at the minimum or maximum level.

55. See Figure P–32.

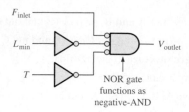

FIGURE P–32

57. See Figure P–33.

L_{min}
T ─── $V_{additive}$

FIGURE P–33

59. **(a)** X = lamp on, A = front door switch on, B = back door switch on. See Figure P–34.

A
B ─── X

FIGURE P–34

    (b) **entity** LampCircuit **is**

        port (A, B: **in** bit; X: **out** bit);

    end entity LampCircuit;

    architecture Function **of** LampCircuit **is**

    begin

        X <= A **xor** B;

    end architecture Function;

61. Output of U3A shorted to ground.

63. Output of U2A is always HIGH (shorted to VCC).

Chapter 6

1. **(a)** $A \oplus B = 0$, $\Sigma = 1$, $(A \oplus B)C_{in} = 0$, $AB = 1$, $C_{out} = 1$

    (b) $A \oplus B = 1$, $\Sigma = 0$, $(A \oplus B)C_{in} = 1$, $AB = 0$, $C_{out} = 1$

    (c) $A \oplus B = 1$, $\Sigma = 1$, $(A \oplus B)C_{in} = 0$, $AB = 0$, $C_{out} = 0$

3. **(a)** $\Sigma = 1$, $C_{out} = 0$;

    (b) $\Sigma = 1$, $C_{out} = 0$;

    (c) $\Sigma = 0$, $C_{out} = 1$;

    (d) $\Sigma = 1$, $C_{out} = 1$

5. 11100

7. $\Sigma_3\Sigma_2\Sigma_1\Sigma_0 = 1101$

9. $\Sigma_1 = 0110$; $\Sigma_2 = 1011$; $\Sigma_3 = 0110$; $\Sigma_4 = 0001$; $\Sigma_5 = 1000$

11. 225 ns

13. $A = B$ is HIGH when $A_0 = B_0$ and $A_1 = B_1$; see Figure P–35.

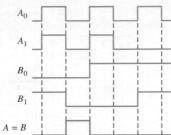

FIGURE P–35

15. (a) $A > B = 1; A = B = 0; A < B = 0$

　(b) $A < B = 1; A = B = 0; A > B = 0$

　(c) $A = B = 1; A < B = 0; A > B = 0$

17. See Figure P–36.

19. $X = A_3 A_2 \overline{A}_1 \overline{A}_0 + \overline{A}_3 \overline{A}_2 \overline{A}_1 A_0 + A_3 \overline{A}_2 A_1$

21. See Figure P–37.

23. $A_3 A_2 A_1 A_0 = 1011$, invalid BCD

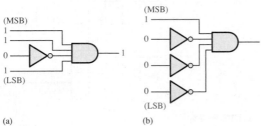

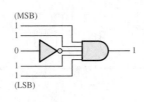

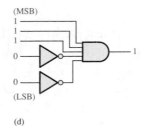

(a)　(b)　(c)　(d)

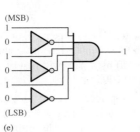

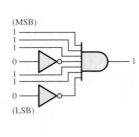

(e)　(f)　(g)　(h)

FIGURE P–36

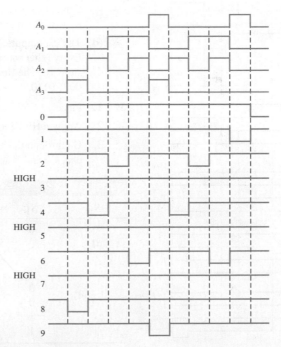

FIGURE P–37

25. (a) $2 = 0010 = 0010_2$

(b) $8 = 1000 = 1000_2$

(c) $13 = 00010011 = 1101_2$

(d) $26 = 00100110 = 11010_2$

(e) $33 = 00110011 = 100001_2$

27. (a) 1010000000 Gray $\rightarrow 1100000000$ binary

(b) 0011001100 Gray $\rightarrow 0010001000$ binary

(c) 1111000111 Gray $\rightarrow 1010000101$ binary

(d) 0000000001 Gray $\rightarrow 0000000001$ binary

See Figure P–38.

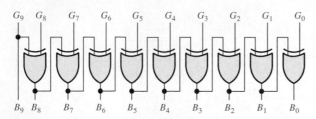

FIGURE P–38

29. See Figure P–39.

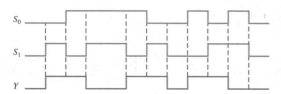

FIGURE P–39

31. See Figure P–40.

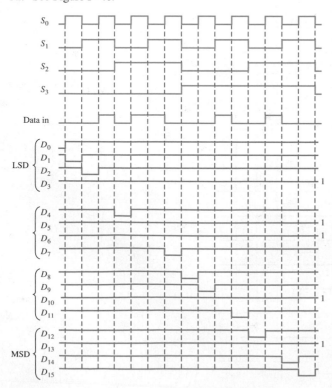

FIGURE P–40

33. See Figure P–41.

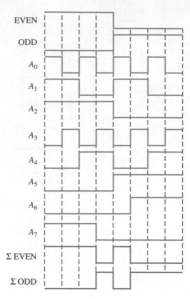

FIGURE P–41

35. (a) OK

(b) segment g burned out; output G open

(c) segment b output stuck LOW

37. (a) The A_1 input of the top adder is open: All binary values corresponding to a BCD number having a value of 0, 1, 4, 5, 8, or 9 will be off by 2. This will first be seen for a BCD value of 0000 0000.

(b) The carry out of the top adder is open: All values not normally involving an output carry will be off by 32. This will first be seen for a BCD value of 0000 0000.

(c) The Σ_4 output of the top adder is shorted to ground: Same binary values above 15 will be short by 16. The first BCD value to indicate this will be 0001 1000.

(d) The Σ_3 output of the bottom adder is shorted to ground: Every other set of 16 values starting with 16 will be short 16. The first BCD value to indicate this will be 0001 0110.

39. 1. Place a LOW pin 7 (Enable).

2. Apply a HIGH to D_0 and a LOW to D_1 through D_7.

3. Go through the binary sequence on the select inputs and check Y and $\overline{Y}$ according to Table P–12.

TABLE P–12

S_2	S_1	S_0	Y	$\overline{Y}$
0	0	0	1	0
0	0	1	0	1
0	1	0	0	1
0	1	1	0	1
1	0	0	0	1
1	0	1	0	1
1	1	0	0	1
1	1	1	0	1

TABLE P–13

D_0	D_1	D_2	D_3	D_4	D_5	D_6	D_7	Y	$\overline{Y}$	S_2	S_1	S_0
L	H	L	L	L	L	L	L	1	0	0	0	1
L	L	H	L	L	L	L	L	1	0	0	1	0
L	L	L	H	L	L	L	L	1	0	0	1	1
L	L	L	L	H	L	L	L	1	0	1	0	0
L	L	L	L	L	H	L	L	1	0	1	0	1
L	L	L	L	L	L	H	L	1	0	1	1	0
L	L	L	L	L	L	L	H	1	0	1	1	1

4. Repeat the binary sequence of select inputs for each set of data inputs listed in Table P–13. A HIGH on the Y output should occur only for the corresponding combinations of select inputs shown.

41. Apply a HIGH in turn to each Data input, D_0 through D_7 with LOWs on all the other inputs. For each HIGH applied to a data input, sequence through all eight binary combinations of select inputs ($S_2 S_1 S_0$) and check for HIGH on the corresponding data output and LOWs on all the other data outputs.

43. See Figure P–42.

45. $\Sigma = \overline{A}\,\overline{B}C_{in} + \overline{A}B\overline{C}_{in} + A\overline{B}\,\overline{C}_{in} + ABC_{in}$
    $C_{out} = \overline{A}BC_{in} + A\overline{B}C_{in} + AB\overline{C}_{in} + ABC_{in}$

See Figure P–43.

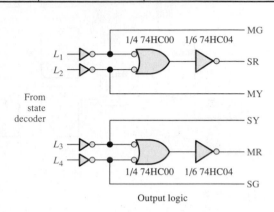

FIGURE P–42

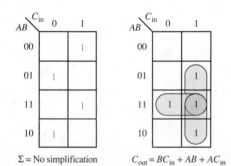

Σ = No simplification $C_{out} = BC_{in} + AB + AC_{in}$

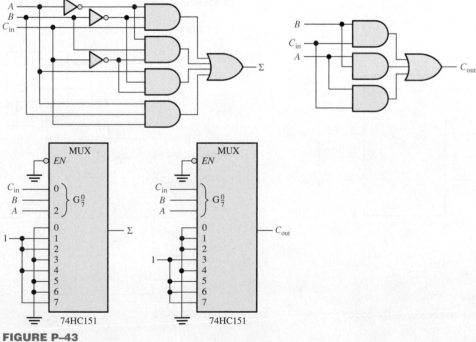

FIGURE P–43

47. See the block diagram in Figure P–44.

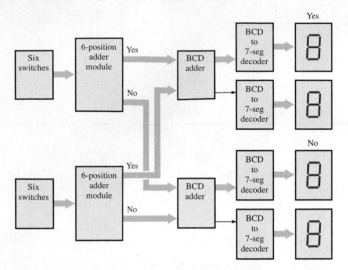

FIGURE P–44

49. See Figure P–45.

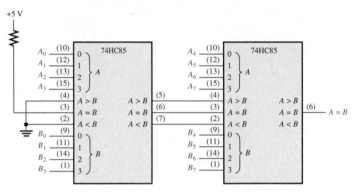

FIGURE P–45

51. See Figure P–46.

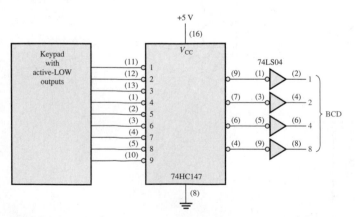

FIGURE P–46

53. Cin of U1 is shorted to VCC.

55. Input C of 4-to-16 line decoder is shorted to ground.

Chapter 7

1. See Figure P–47.

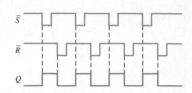

FIGURE P–47

3. See Figure P–48.

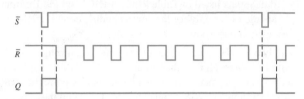

FIGURE P–48

5. See Figure P–49.

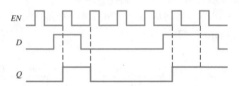

FIGURE P–49

7. See Figure P–50.

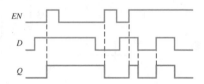

FIGURE P–50

9. See Figure 51.

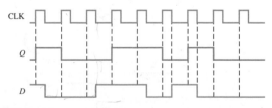

FIGURE P–51

11. See Figure P–52.

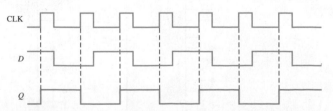

FIGURE P–52

13. See Figure P-53.

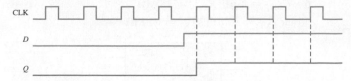

FIGURE P-53

15. See Figure P-54.

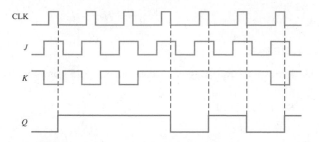

FIGURE P-54

17. See Figure P-55.

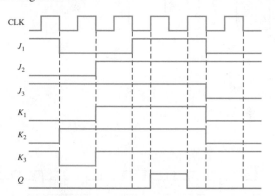

FIGURE P-55

19. Direct current and dc supply voltage

21. 14.9 MHz

23. 150 mA, 750 mW

25. divide-by-2; see Figure P-56.

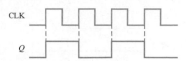

FIGURE P-56

27. 4.62 μs

29. $C_1 = 1\ \mu F$, $R_1 = 227\ k\Omega$ (use 220 kΩ). See Figure P-57.

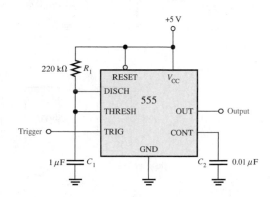

FIGURE P-57

31. $R_1 = 18\ k\Omega$, $R_2 = 9.1\ k\Omega$.

33. The wire from pin 6 to pin 10 and the ground wire are reversed on the protoboard.

35. $\overline{CLR}$ shorted to ground.

37. See Figure P-58. Delays not shown.

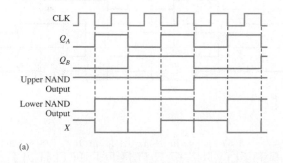

(a)

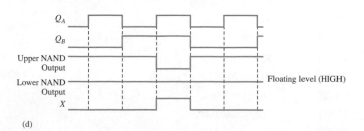

(d)

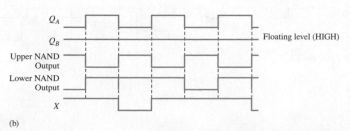

(b)

(c) $X = $ LOW if $Q_B = 1$; $X = \bar{Q}_A$ if $Q_B = 0$

FIGURE P-58

39. See Figure P–59.

6 s: $C_1 = 1\ \mu\text{F}, R_1 = 5.5\ \text{M}\Omega$ (use 5.6 MΩ)

40 s: $C_1 = 2.2\ \mu\text{F}, R_1 = 16.5\ \text{M}\Omega$ (use 15 MΩ)

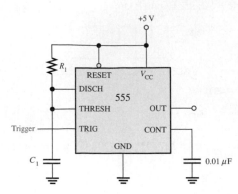

FIGURE P–59

41. See Figure P–60.

6 s: $C_{\text{EXT}} = 1\ \mu\text{F}; R_{\text{EXT}} = 18\ \text{M}\Omega$

40 s: $C_{\text{EXT}} = 10\ \mu\text{F}; R_{\text{EXT}} = 12\ \text{M}\Omega$

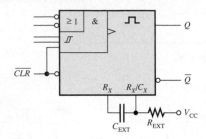

FIGURE P–60

43. Increase the $R_{\text{EXT}}C_{\text{EXT}}$ time constant of the 25 s one-shot by 2.4 times.

45. See Figure P–61.

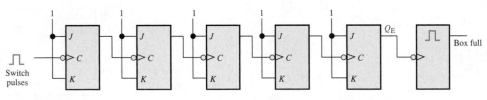

FIGURE P–61

47. R input of U1 is shorted to VCC.

49. The clock input is shorted to VCC or ground.

51. The D input of U2 is shorted to ground.

Chapter 8

1. Shift registers store binary data.

3. Shift data and store data.

5. See Figure P–62.

7.

Initially:	101001111000
CLK1:	010100111100
CLK2:	001010011110
CLK3:	000101001111
CLK4:	000010100111
CLK5:	100001010011
CLK6:	110000101001
CLK7:	111000010100
CLK8:	011100001010
CLK9:	001110000101
CLK10:	000111000010
CLK11:	100011100001
CLK12:	110001110000

9. See Figure P–63.

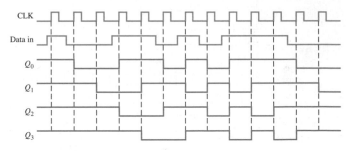

FIGURE P–62

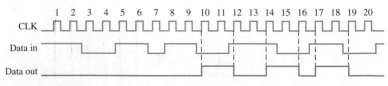

FIGURE P–63

11. See Figure P–64.

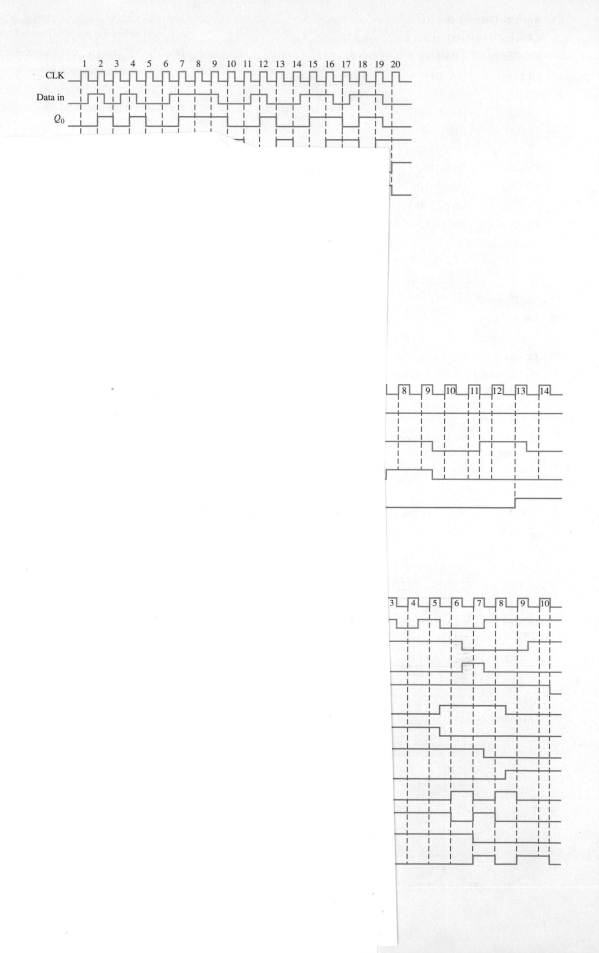

21. Initially (76): 01001100

CLK1:	10011000	left
CLK2:	01001100	right
CLK3:	00100110	right
CLK4:	00010011	right
CLK5:	00100110	left
CLK6:	01001100	left
CLK7:	00100110	right
CLK8:	01001100	left
CLK9:	00100110	right
CLK10:	01001100	left
CLK11:	10011000	left

23. See Figure P–69.

25. (a) 3   (b) 5

     (c) 7   (d) 8

27. See Figure P–70.

29. See Figure P–71.

31. An incorrect code may be produced.

33. D_3 input open

35. (a) No clock at switch closure because of faulty NAND (negative-OR) gate or one-shot; open clock (C) input to key code register; open $SH/\overline{LD}$ input to key code register

     (b) Diode in third row open: Q_2 output of ring counter open

(c) The NAND (negative-OR) gate input connected to the first column is open or shorted.

(d) The "2" input to the column encoder is open.

37. (a) Contents of data output register remain constant.

     (b) Contents of both registers do not change.

     (c) Third stage output of data output register remains HIGH.

     (d) Clock generator is disabled after each pulse by the flip-flop being continuously SET and then RESET.

39. shift register A: 1001

     shift register C: 00000100

41. Control flip-flop: 74HC76

     Clock generator: 555

     Counter: 74HC163

     Data input register: 74HC164

     Data output register: 74HC199

     One-shot: 74121

43. See Figure P–72.

45. See Figure P–73.

47. The D input of FF1 is shorted to ground.

49. The U3 Q' output of the Johnson counter is connected to the U2 D input.

51. The connection between the Q output of U3 and D input of U4 is open.

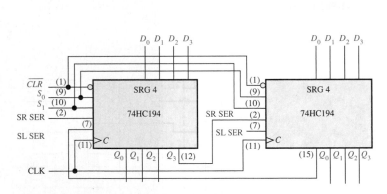

FIGURE P–69

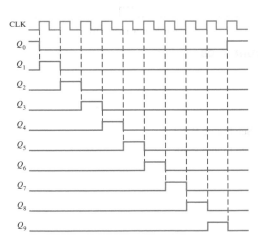

FIGURE P–70

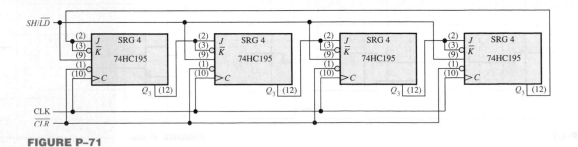

FIGURE P–71

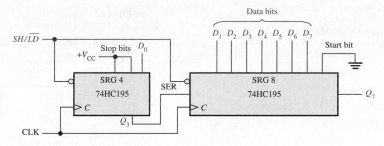

FIGURE P-72

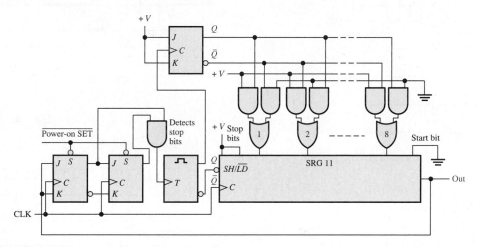

FIGURE P-73

Chapter 9

1. See Figure P-74.

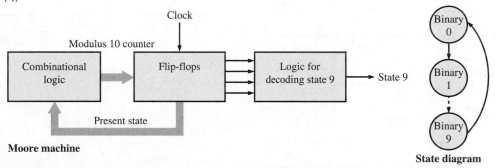

FIGURE P-74

3. See Figure P-75.

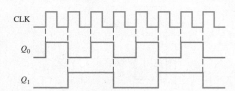

FIGURE P-75

5. Worst-case delay is 24 ns; it occurs when all flip-flops change state from 011 to 100 or from 111 to 000.

7. 8 ns

9. Initially, each flip-flop is reset.

At CLK1:

$J_0 = K_0 = 1$ Therefore Q_0 goes to a 1.

$J_1 = K_1 = 0$ Therefore Q_1 remains a 0.

$J_2 = K_2 = 0$ Therefore Q_2 remains a 0.

$J_3 = K_3 = 0$ Therefore Q_3 remains a 0.

At CLK2:

$J_0 = K_0 = 1$ Therefore Q_0 goes to a 0.

$J_1 = K_1 = 1$ Therefore Q_1 goes to a 1.

$J_2 = K_2 = 0$ Therefore Q_2 remains a 0.

$J_3 = K_3 = 0$ Therefore Q_3 remains a 0.

At CLK3:

$J_0 = K_0 = 1$ Therefore Q_0 goes to a 1.

$J_1 = K_1 = 0$ Therefore Q_1 remains a 1.

$J_2 = K_2 = 0$ Therefore Q_2 remains a 0.

$J_3 = K_3 = 0$ Therefore Q_3 remains a 0.

A continuation of this procedure for the next seven clock pulses will show that the counter progresses through the BCD sequence.

11. See Figure P–76.

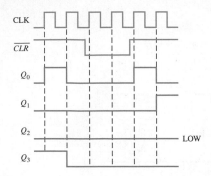

FIGURE P–76

13. See Figure P–77.

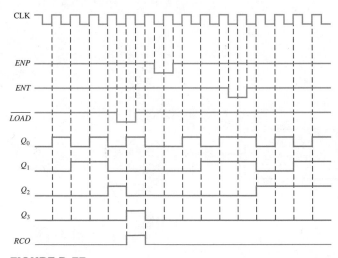

FIGURE P–77

15. See Figure P–78.

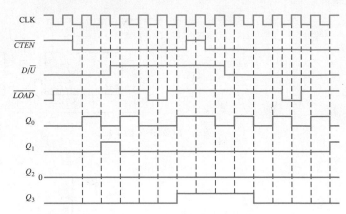

FIGURE P–78

17. See Figure P–79.

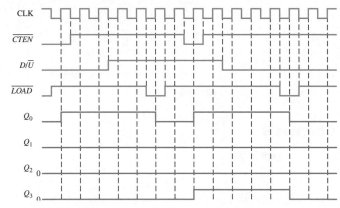

FIGURE P–79

19. The sequence is 0000, 1111, 1110, 1101, 1010, 0101. The counter "locks up" in the 1010 and 0101 states and alternates between them.

21. See Figure P–80.

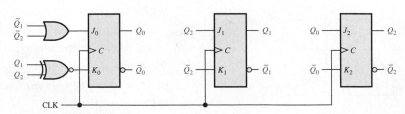

FIGURE P–80

23. See Figure P–81.

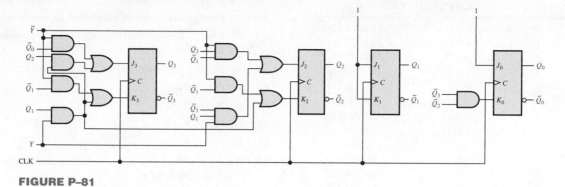

FIGURE P–81

25. See Figure P–82 for divide-by-10,000. Add one more DIV 10 counter to create a divide-by-100,000.

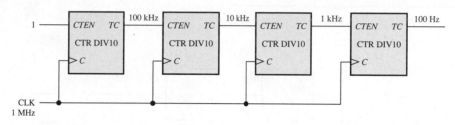

FIGURE P–82

27. See Figure P–83.

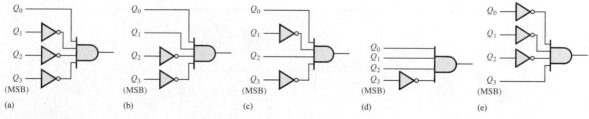

FIGURE P–83

29. CLK2, output 0; CLK4, outputs 2, 0; CLK6, output 4; CLK8, outputs 6, 4, 0; CLK10, output 8; CLK12, outputs 10, 8; CLK14, output 12; CLK16, outputs 14, 12, 8

31. A glitch of the AND gate output occurs on the 111 to 000 transition. Eliminate by ANDing $\overline{CLK}$ with counter outputs (strobe) or use Gray code.

33. Hours tens: 0001

Hours units: 0010

Minutes tens: 0000

Minutes units: 0001

Seconds tens: 0000

Seconds units: 0010

35. 68

37. (a) Q_0 and Q_1 will not change from their initial state.

(b) Normal operation. Q_0 and Q_1 toggle.

(c) Q_0 toggles and Q_1 remains in initial state.

(d) Q_0 goes HIGH and remains HIGH. Q_1 does not change.

(e) Q_0 toggles and Q_1 remains LOW.

39. The D input to FF1 is open acting as a HIGH.

41. Q_0 input to AND gate open and acting as a HIGH

43. See Table P–14.

45. The decode 6 gate interprets count 4 as a 6 (0110) and clears the counter back to 0 (actually 0010 since Q_1 is open). The apparent sequence of the tens portion of the counter is 0010, 0011, 0010, 0011, 0110.

TABLE P–14

Stage	Open	Loaded Count	f_{OUT}
1	0	63C1	250.006 Hz
1	1	63C2	250.012 Hz
1	2	63C4	250.025 Hz
1	3	63C8	250.050 Hz
2	0	63D0	250.100 Hz
2	1	63E0	250.200 Hz
2	2	63C0	250 Hz
2	3	63C0	250 Hz
3	0	63C0	250 Hz
3	1	63C0	250 Hz
3	2	67C0	256.568 Hz
3	3	6BC0	263.491 Hz
4	0	73C0	278.520 Hz
4	1	63C0	250 Hz
4	2	63C0	250 Hz
4	3	E3C0	1.383 kHz

47. See Figure P–84. The floor code is hardwired and is unique to each floor. The fifth floor logic is shown to illustrate.

49. (a) Change floor counter to two bits.

   (b) Change the Call/Req code register and associated logic to two bits.

   (c) Modify the 7-segment decoder for a 2-bit code.

51. See Figure P–85.

53. See Figure P–86.

55. See Figure P–87.

57. See Figure P–88.

59. The input of the U5 AND gate that connects to the Q output of U2 shorted to VCC.

61. Line to LOAD' input always LOW.

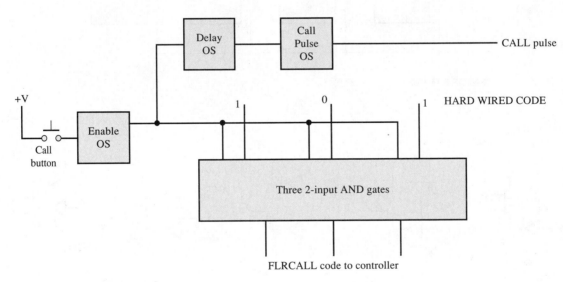

FIGURE P–84

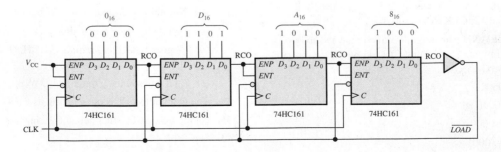

FIGURE P–85

FIGURE P–86

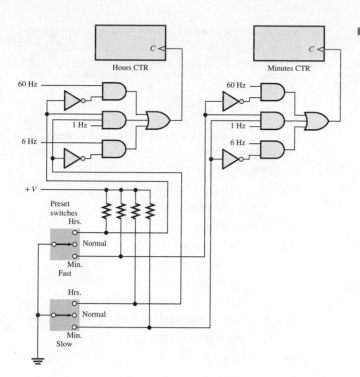

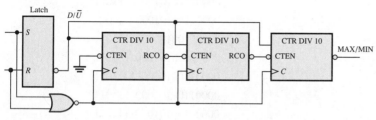

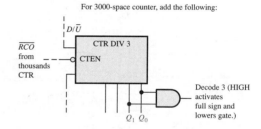

For 3000-space counter, add the following:

FIGURE P–87

FIGURE P–88

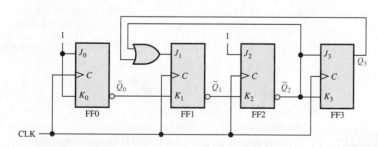

Chapter 10

1. $X = \overline{A}\overline{B}\overline{C} + \overline{A}B\overline{C} + A\overline{B}C$

3. See Figure P–89.

5. A CPLD basically consists of multiple SPLDs that can be connected with a programmable interconnect array.

7. (a) $\overline{A}BC\overline{D}$

   (b) $ABC(\overline{D} + \overline{E}) = ABC\overline{D} + ABC\overline{E}$

9. $X = A\overline{B} + \overline{A}B$

11. $X_1 = A\overline{B}C\overline{D} + \overline{A}BCD + ABC\overline{D}$;
    $X_2 = ABCD + AB\overline{C}D + \overline{A}BC\overline{D} + A\overline{B}CD$

13. (a) Combinational; 1

    (b) Registered; 0

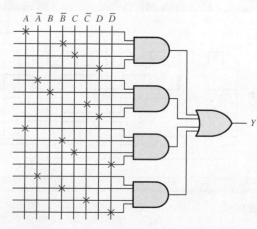

FIGURE P–89

15. (a) Registered **(b)** GCK1

(c) 0 **(d)** 0

17. SOP output $= \overline{A}\,\overline{B}\,\overline{C} + \overline{A}\,\overline{B}C + \overline{A}BC + A\overline{B}C + AB\overline{C}$

19. See Figure P–90.

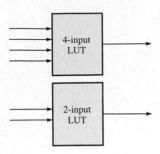

FIGURE P–90

21. See Figure P–91.

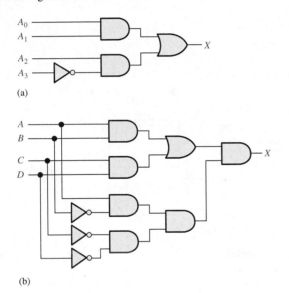

FIGURE P–91

23. See Figure P–92.

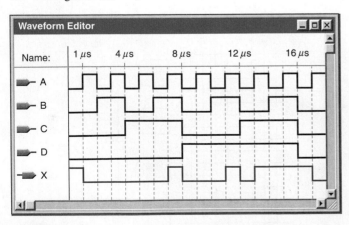

FIGURE P–92

25. Shift input $= 1$, data are applied to SDI, go through the MUX, and are clocked into Capture register A on the leading edge of the clock pulse. From the output of Capture register A, the data go through the upper MUX and are clocked into Capture register B on the trailing edge of the clock pulse.

27. PDI/O $= 0$ and OE $= 0$. The data are applied to the input pin and go through the selected MUX to the internal programmable logic.

29. 000011001010001111011

0	000011001010001111011
1	000011001010001111011
3	000011001010001111011
6	000011001010001111011
12	000011001010001111011
9	000011001010001111011
2	000011001010001111011
5	000011001010001111011
10	000011001010001111011
4	000011001010001111011
8	000011001010001111011
1	000011001010001111011
3	000011001010001111011
7	000011001010001111011
15	000011001010001111011
14	000011001010001111011
13	000011001010001111011
11	000011001010001111011

31. The AND-OR logic switches either the Call code from the floor panel or the Request code from the elevator panel and the associated clock into the register based on the state of the flip-flop.

Chapter 11

1. (a) ROM **(b)** RAM

3. *Address bus* provides for transfer of address code to memory for accessing any memory location in any order for a read or write operation. *Data bus* provides for transfer of data between the microprocessor and the memory or I/O.

5.

	Bit 0	Bit 1	Bit 2	Bit 3
Row 0	1	0	0	0
Row 1	0	0	0	0
Row 2	0	0	1	0
Row 3	0	0	0	0

7. 512 row $\times$ 128 8-bit columns

9. A SRAM stores bits in flip-flops indefinitely as long as power is applied. A DRAM stores bits in capacitors that must be refreshed periodically to retain the data.

11. See Table P–15.

TABLE P–15

Inputs		Outputs			
A_1	A_0	O_3	O_2	O_1	O_0
0	0	0	1	0	1
0	1	1	0	0	1
1	0	1	1	1	0
1	1	0	0	1	0

13. See Figure P–93.

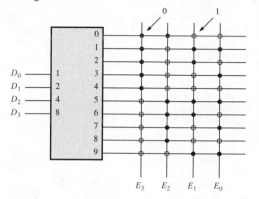

FIGURE P–93

15. Blown links: 1–17, 19–23, 25–31, 34, 37, 38, 40–47, 53, 55, 58, 61, 62, 63, 65, 67, 69

17. Use eight 16k × 4 DRAMs with sixteen address lines. Two of the address lines are decoded to enable the selected memory chips. Four data lines go to each chip.

19. 8 bits, 64k words; 4 bits, 256k words

21. lowest address: $FC0_{16}$

highest address: FFF_{16}

23. A hard disk is formatted into tracks and sectors. Each track is divided into a number of sectors with each sector of a track having a physical address. Hard disks typically have from a few hundred to a few thousand tracks.

25. Magnetic tape has a longer access time than disk because data must be accessed sequentially rather than randomly.

27. Blu-ray uses a blue laser and DVD uses a red laser. A Blu-ray disc has a higher definition and storage density than a DVD.

29. Beginning with closest memory to the processor in terms of access time: Registers, caches, main memory, hard disk, auxiliary storage.

31. The miss rate is the percentage of memory accesses that fail to find the requested data in the given level of memory and is equal to $1 - \text{hit rate}$.

33. See Figure P–94.

35. The architecture is the way in which a cloud storage system is structured and organized. Generically, a cloud storage system consists of a front end that uses access protocols, a control that uses data handling protocols, and a back end that provides storage.

37. Checksum content is in error.

39. (a) ROM 2

(b) ROM 1

(c) All ROMs

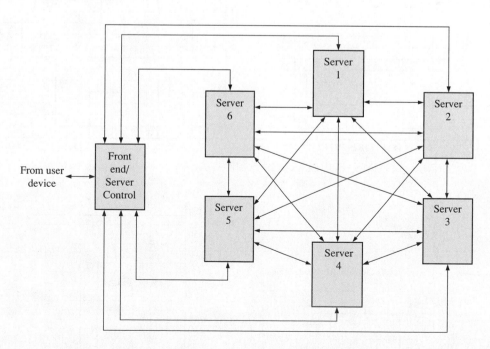

FIGURE P–94

Chapter 12

1. See Figure P–95.

3. 11, 11, 11, 11, 01, 11, 11, 11, 11

5. See Figure P–96.

7. 200

9. −21.4

11. 001, 010, 011, 101, 110, 111, 111, 111, 111, 110, 101, 101, 110, 110, 110, 101, 100, 011, 010, 001

13. 11, 11, 11

15. See Figure P–97.

17. See Figure P–98.

19. (a) 14.3%   (b) 0.098%

    (c) 0.00038%

21. See Figure P–99.

23. An analog-to-digital converter converts an analog signal to a digital code.

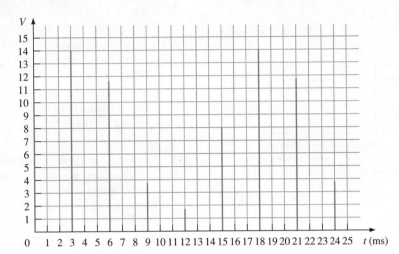

FIGURE P–95

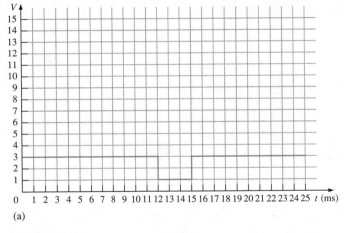

(a)

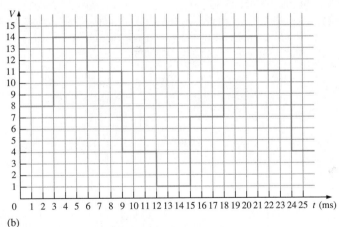

(b)

FIGURE P–96

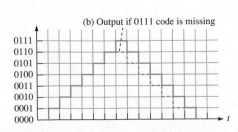

(b) Output if 0111 code is missing

FIGURE P–97

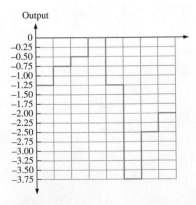

Output

FIGURE P–98

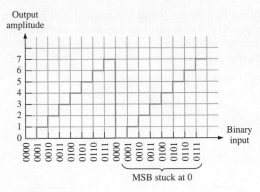

FIGURE P–99

25. A digital-to-analog converter changes a digital code to the corresponding analog signal.

27. $\dfrac{400 \text{ Mb/s}}{32 \text{ bits/instruction}} = 12.5$ million instructions/s

29. **1.** Program address generate (PG). The program address is generated by the CPU.

   2. Program address send (PS). The program address is sent to the memory.

   3. Program access ready wait (PW). A memory read operation occurs.

   4. Program fetch packet receive (PR). The CPU receives the packet of instructions.

Chapter 13

1. A coaxial cable consists of an outer jacket, metallic shield, dielectric, and center conductor.

3. Advantages of fiber optics over electrical transmission media are faster data rates, higher signal capacity (more signals at a time), transmission over longer distances, and not susceptible to EMI.

5. In multimode, the light entering the optical fiber will tend to propagate through the core in multiple rays (modes), basically due to varying angles as each light ray moves along.

7. See Figure P–100.

9. Visible light is in the frequency range 4×10^{14} Hz to 7.5×10^{14} Hz. Figure 13–10 in the textbook shows the range in units of wavelength (nm).

11. 8 Mbps

13. 5 Mbaud

15. See Figure P–101.

17. 001110011

19. 11010111110001000001

21. See Figure P–102.

23. See Figure P–103.

25. In the PWM intersective method, the sawtooth intersects the sinusoidal modulating signal twice during each cycle. The sawtooth is either increasing above the sine wave or

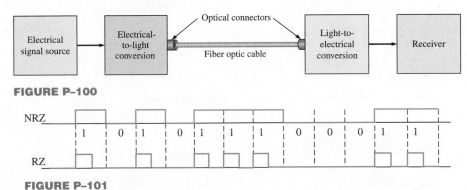

FIGURE P–100

FIGURE P–101

FIGURE P–102

FIGURE P–103

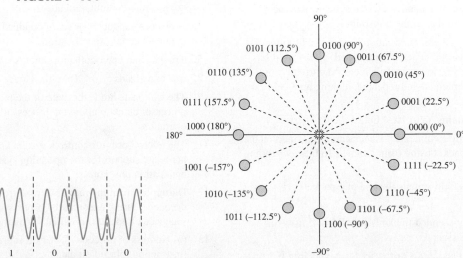

decreasing below the sine wave. When the sawtooth is increasing above the sine wave, a low level is generated, and when it is decreasing below the sine wave, a high level is generated. The resulting output is a series of pulses with widths proportional to the amplitude of the sine wave.

27. Data rate = 300 bps

29. Four bits

31. Bit interleaved: A single data bit from a source is transmitted on the channel, followed by a data bit from another source, and so on.

Byte interleaved: A byte of data from a source is transmitted on the channel, followed by a byte from another source, and so on.

33. In FDM, band-pass filters are used on the receiving end to separate the transmitted signals.

35. *Physical characteristics of a bus:* Number of conductors, length, configuration (serial or parallel), type of connector, number of connector pins, pin configuration.

Electrical/performance characteristics of a bus: Signal format, signal voltage, clock frequency, transfer speed, bandwidth, data frame format, handshaking protocol, error detection, impedances.

37. 200 MBps and 191 MBps

39. A differential bus provides much higher data rates and longer transmission distances than does a single-ended bus.

41. The PCI-Express bus does not use a shared bus as PCI and PCI-X do.

43. The terms *talker* and *listener* are associated with the IEE-488 bus (GPIB).

45. Three data bytes are transferred because the NDAC line goes HIGH three times, each time indicating that a data byte is accepted.

47. **(a)** SCSI

(b) USB

(c) Super speed USB (V 3.0)

49. *Sync Field:* All packets start with a sync (synchronization) field. The sync field consists of 8 bits for low and full speed or 32 bits for high speed and is used to synchronize the receiver clock with that of the transmitter.

PID Field: The packet identification field is used to identify the type of packet that is being transmitted. There are 4 bits in the PID; however, to ensure it is received correctly, the 4 bits are complemented and repeated, making an 8-bit PID code.

Data Field: Contains up to 1024 bytes of data.

CRC Field: Cyclic Redundancy Checks are performed on the data within the packet using from 5 bits to 16 bits, depending on the type of packet.

EOP Field: This field signals the end of a packet.

51. 1024 bytes

53. RS-232 uses single-ended transmission. RS-422 uses differential transmission.

55. I^2C is an internal serial bus primarily for connecting ICs on a PC board.

57. Other possible units on an automotive CAN system include wiper control unit, parking control unit, entertainment system unit, tire pressure monitor, seat position unit, heads-up display unit.

59. See Figure P–104.

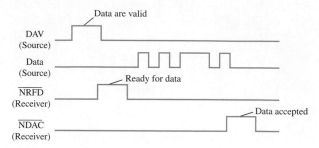

FIGURE P–104

61. See Figure P–105.

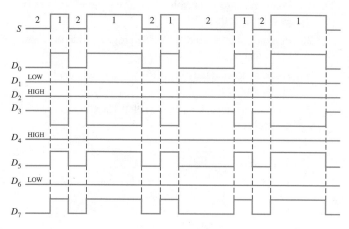

FIGURE P–105

Chapter 14

1. The basic elements of a computer are the CPU, memory/storage, input/output, and buses.

3. A bus is a conductor or set of conductors for transferring data that meet certain specifications.

5. Tri-state and open-collector outputs

7. **(a)** Nine loads   **(b)** Two loads

9. The wait-state holds the state of the bus signals for one processor clock to allow the processor to complete an access operation.

11. The microprocessor controls system hardware, provides hardware support for the operating system, and executes application programs.

13. During fetch, an instruction is read from memory and decoded. During execute, the processor carries out the sequence of operations called for by the instruction.

15. (1) Address of op-code placed on address bus; (2) Op-code (instruction) placed on data bus and stored in data register; (3) Instruction decoded; (4) Instruction carried out.

17. *First fetch/execute cycle:* (1) Address of indexed op-code placed on address bus; (2) Indexed op-code placed on data bus and stored in data register; (3) Indexed instruction decoded; (4) Address of operand fetched.

Second fetch/execute cycle: (5) Offset address selected; (6) Offset address placed on data bus and stored in data register; (7) Offset address added to contents of index register to produce address of operand.

Third fetch/execute cycle: (8) Address of operand transferred to address register; (9) Address of operand placed on address bus; (10) Operand address placed on data bus and stored in data register; (11) Operand loaded into accumulator.

19. The interrupt vector table is used in auto-vectored interrupts to obtain the starting address for an interrupt service routine (ISR).

21. The sequence of events during a bus request operation is as follows:

   1. The bus master requesting control of the system buses submits a request by asserting the processor's bus request (BR) line.

   2. The processor tri-states the system buses and signals that it has released control of the buses by asserting the bus grant (BG) line.

   3. The requesting bus master uses the system address, data, and control lines to transfer data between system devices.

   4. After completing the data transfers, the requesting bus master tri-states the system buses and signals the end of the bus request operation by asserting the bus grant acknowledge (BGACK) line.

23. The first group consists of application software, which includes word processors, spreadsheets, computer games, and other programs, written to accomplish some specific task. The second group consists of system software, a major portion of which is the operating system. The operating system manages the system hardware, supervises the running of applications software, provides a standard operating environment for programs in which they can run and interacts with the computer hardware.

25. Two ways in which computers execute more than one process are multitasking and multiprocessing.

27. MMUs handle memory accessing including memory protection, wait state generation, address translation for virtual memory, and cache control.

29. One possible flow chart is shown in Figure P–106.

31. Move contents of bx register into ax register.

33. A microcontroller is a device that combines a microprocessor with common peripheral units.

35. Microcontrollers are widely used in embedded applications because they provide the interface and processing resources required by embedded systems.

37. An SoC is a system on a chip and has all the components and functions to implement a complete system such as a computer. A microcontroller is similar to a SoC but generally more limited in available functions.

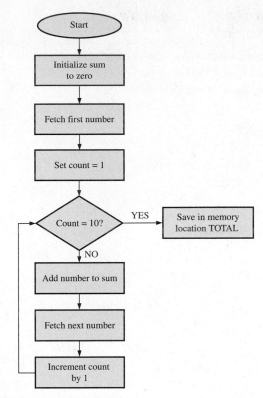

FIGURE P–106

Chapter 15

(Chapter 15 is on the website.)

  1. No; $V_{OH(min)} < V_{IH(min)}$

  3. 0.15 V in HIGH state; 0.25 V in LOW state.

  5. Gate C

  7. 12 ns

  9. Gate C

11. Yes, G_2

13. **(a)** on   **(b)** off

     (c) off   **(d)** on

15. See Figure P–107 for one possible circuit.

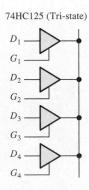

74HC125 (Tri-state)

FIGURE P–107

17. **(a)** HIGH   **(b)** Floating

     (c) HIGH   **(d)** High-Z

19. (a) LOW   (b) LOW
    (c) LOW
21. See Figure P–108.
23. (a) $R_p = 198\ \Omega$
    (b) $R_p = 198\ \Omega$
    (c) $R_p = 198\ \Omega$
25. ALVC
27. (a) A, B to X: 9.9 ns
        C, D to X: 6.6 ns

    (b) A to X_1, X_2, X_3: 14 ns
        B to X_1: 7 ns
        C to X_2: 7 ns
        D to X_3: 7 ns
    (c) A to X: 11.1 ns
        B to X: 11.1 ns
        C to X: 7.4 ns
        D to X: 7.4 ns
29. ECL operates with nonsaturated BJTs.

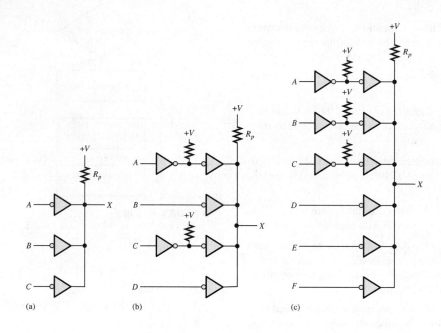

FIGURE P–108

acceptor A receiving device on a bus.

access time The time from the application of a valid memory address to the appearance of valid output data.

addend In addition, the number that is added to another number called the augend.

adder A logic circuit used to add two binary numbers.

address The location of a given storage cell or group of cells in a memory; a unique memory location containing one byte.

address bus A one-way group of conductors from the microprocessor to a memory, or other external device, on which the address code is sent.

adjacency Characteristic of cells in a Karnaugh map in which there is a single-variable change from one cell to another cell next to it on any of its four sides.

aliasing The effect created when a signal is sampled at less than twice the signal frequency. Aliasing creates unwanted frequencies that interfere with the signal frequency.

alphanumeric Consisting of numerals, letters, and other characters.

ALU Arithmetic logic unit; the key processing element of a microprocessor that performs arithmetic and logic operations.

amplitude In a pulse waveform, the height or maximum value of the pulse as measured from its low level.

analog Being continuous or having continuous values, as opposed to having a set of discrete values.

analog-to-digital (A/D) conversion The process of converting an analog signal to digital form.

analog-to-digital converter (ADC) A device used to convert an analog signal to a sequence of digital codes.

AND A basic logic operation in which a true (HIGH) output occurs only when all the input conditions are true (HIGH).

AND array An array of AND gates consisting of a matrix of programmable interconnections.

AND gate A logic gate that produces a HIGH output only when all of the inputs are HIGH.

ANSI American National Standards Institute.

antifuse A type of PLD nonvolatile programmable link that can be left open or can be shorted once as directed by the program.

architecture The VHDL unit that describes the internal operation of a logic function; the internal functional arrangement of the elements that give a device its particular operating characteristics.

array In a PLD, a matrix formed by rows of product-term lines and columns of input lines with a programmable cell at each junction. In VHDL, an array is an ordered set of individual items called elements with a single identifier name.

ASCII American Standard Code for Information Interchange; the most widely used alphanumeric code.

ASK Amplitude shift keying; a form of modulation in which a digital signal modulates the amplitude of a higher frequency sine wave.

assembler A program that converts English-like mnemonics into machine code.

assembly language A programming language that uses English-like words and has a one-to-one correspondence to machine language.

associative law For addition (ORing) and multiplication (ANDing) of three or more variables, the order in which the variables are grouped makes no difference.

astable Having no stable state. An astable multivibrator oscillates between two quasi-stable states.

asynchronous Having no fixed time relationship; not occurring at the same time.

asynchronous counter A type of counter in which each stage is clocked from the output of the preceding stage.

augend In addition, the number to which the addend is added.

bandwidth The frequency at which a sinusoidal input signal is attenuated to 70.7 percent of its original amplitude.

bank A section of memory within a single memory array (chip).

base One of the three regions in a bipolar junction transistor.

base address The beginning address of a segment of memory.

baud The number of symbols per second in a data transmission.

BCD Binary coded decimal; a digital code in which each of the decimal digits, 0 through 9, is represented by a group of four bits.

BEDO DRAM Burst extended data output dynamic random-access memory.

BiCMOS A family of logic circuits that combines CMOS and bipolar logic.

bidirectional Having two directions. In a bidirectional shift register, the stored data can be shifted right or left.

binary Having two values or states; describes a number system that has a base of two and utilizes 1 and 0 as its digits.

BIOS Basic input/output system; a set of programs in ROM that interfaces the I/O devices in a computer system.

bipolar A class of integrated logic circuits implemented with bipolar transistors; also known as TTL.

bistable Having two stable states. Flip-flops and latches are bistable multivibrators.

bit A binary digit, which can be either a 1 or 0.

bit rate The number of bits per second in a data transmission.

bitstream A series of bits describing a final design that is sent to the target device during programming.

bit time The interval of time occupied by a single bit in a sequence of bits; the period of the clock.

BJT Bipolar junction transistor; a semiconductor device used for switching or amplification. A BJT has two junctions, the base-emitter junction and the base-collector junction.

Blue-ray A disc storage technology that uses a blue laser to achieve more density and definition than a DVD.

Boolean addition In Boolean algebra, the OR operation.

Boolean algebra The mathematics of logic circuits.

Boolean expression A formulation of variables and operators used to express the operation of a logic circuit.

Boolean multiplication In Boolean algebra, the AND operation.

boundary scan A method for internally testing a PLD based on the JTAG standard (IEEE Std. 1149.1).

break point A flag placed within a program source code to stop a program for investigation.

buffer A circuit that prevents loading of an input or output.

bus A set of connections and specifications for the transfer of data among two or more devices.

bus arbitration The process that prevents two sources from using a bus at the same time.

bus contention An adverse condition that could occur if two or more devices try to communicate at the same time on a bus.

bus master Any device that can control and manage the system buses in a computer system.

bus protocol A set of rules that allow two or more devices to communicate through a bus.

byte A group of eight bits.

cache memory A relatively small, high-speed memory that stores the most recently used instructions or data from the larger but slower main memory.

caching The process of copying frequently accessed program instructions from main memory into faster memory to increase processing speed.

capacity The total number of data units (bits, nibbles, bytes, words) that a memory can store.

carry The digit generated when the sum of two binary digits exceeds 1.

carry generation The process of producing an output carry in a full-adder when both input bits are 1s.

carry propagation The process of rippling an input carry to become the output carry in a full-adder when either or both of the input bits are 1s and the input carry is a 1.

cascade To connect "end-to-end" as when several counters are connected from the terminal count output of one counter to the enable input of the next counter.

cascading Connecting two or more similar devices in a manner that expands the capability of one device.

CCD Charge-coupled device; a type of semiconductor memory that stores data in the form of charge packets and is serially accessed.

CD-R CD-Recordable; an optical disk storage device on which data can be stored once.

CD-ROM An optical disk storage device on which data are prestored and can only be read.

CD-RW CD-Rewritable; an optical disk storage on which data can be written and overwritten many times.

cell An area on a Karnaugh map that represents a unique combination of variables in product form; a single storage element in a memory; a fused cross point of a row and column in a PLD.

character A symbol, letter, or numeral.

circuit An arrangement of electrical and/or electronic components interconnected in such a way as to perform a specified function.

CLB Configurable logic block; a unit of logic in an FPGA that is made up of multiple smaller logic modules and a local programmable interconntect that is used to connect logic modules within the CLB.

clear An asynchronous input used to reset a flip-flop (make the Q output 0); to place a register or counter in the state in which it contains all 0s.

clock The basic timing signal in a digital system; a periodic waveform used to synchronize operation.

cloud storage A remote network of servers that is connected to a user device through the Internet.

CMOS Complementary metal oxide semiconductor; a class of integrated logic circuits that is implemented with a type of field-effect transistor.

coaxial cable A type of data transmission media in which a shielded conductor is used to minimize EMI.

code A set of bits arranged in a unique pattern and used to represent such information as numbers, letters, and other symbols; in VHDL, program statements.

codec A combined coder and decoder.

collector One of the three regions in a bipolar transistor.

combinational logic A combination of logic gates interconnected to produce a specified Boolean function with no storage or memory capability; sometimes called *combinatorial logic*.

commutative law In addition (ORing) and multiplication (ANDing) of two variables, the order in which the variables are ORed or ANDed makes no difference.

comparator A digital circuit that compares the magnitudes of two quantities and produces an output indicating the relationship of the quantities.

compiler An application program in development software packages that controls the design flow process and translates source code into object code in a format that can be logically tested or downloaded to a target device.

complement The inverse or opposite of a number; in Boolean algebra, the inverse function, expressed with a bar over the variable. The complement of a 1 is a 0, and vice versa.

component A VHDL feature that can be used to predefine the logic function for multiple use throughout a program or programs.

contiguous Joined together.

control bus A set of conductive paths that connects the CPU to other parts of the computer to coordinate its operations and to communicate with external devices.

controller An instrument that can specify each of the other instruments on the bus as either a talker or a listener for the purpose of data transfer.

control unit The portion within the microprocessor that provides the timing and control signals for getting data into and out of the microprocessor and for synchronizing the execution of instructions.

counter A digital circuit capable of counting electronic events, such as pulses, by progressing through a sequence of binary states.

CPLD A complex programmable logic device that consists basically of multiple SPLD arrays with programmable interconnections.

CPU Central processing unit; the main part of a computer responsible for control and processing of data; the core of a DSP that processes the program instructions.

cross-assembler A program that translates an assembly language program for one type of microprocessor to an assembly language for another type of microprocessor.

crosstalk The presence of an unwanted signal via an accidental coupling.

current sinking The action of a circuit in which it accepts current into its output from a load.

current sourcing The action of a circuit in which it sends current out of its output and into a load.

cyclic redundancy check (CRC) A type of error detection code.

data Information in numeric, alphabetic, or other form.

data bus A bidirectional set of conductive paths on which data or instruction codes are transferred into a microprocessor or on which the result of an operation is sent out from the microprocessor.

data center A facility that houses a cloud storage system.

data selector A circuit that selects data from several inputs one at a time in a sequence and places them on the output; also called a multiplexer.

data sheet A document that specifies parameter values and operating conditions for an integrated circuit or other device.

DCE Data communications equipment.

DDR Double data rate.

DDR SDRAM Double data rate, synchronous dynamic random-access memory.

decade Characterized by ten states or values.

decade counter A digital counter having ten states.

decimal Describes a number system with a base of ten.

decode The process of interpreting coded information; changing data in a coded form into a more common form; a stage of the DSP pipeline operation in which instructions are assigned to functional units and are decoded.

decoder A digital circuit (device) that converts coded information into another (familiar) or noncoded form.

decrement To decrease the binary state of a counter by one.

delta modulation A method of analog-to-digital conversion using a 1-bit quantization process.

design flow The process or sequence of operations carried out to program a target device.

D flip-flop A type of bistable multivibrator in which the output assumes the state of the D input on the triggering edge of a clock pulse.

demultiplexer (demux) A circuit (digital device) that switches digital data from one input line to several output lines in a specified time sequence.

dependency notation A notational system for logic symbols that specifies input and output relationships, thus fully defining a given function; an integral part of ANSI/IEEE Std. 91-1984.

difference The result of a subtraction.

differential operation A bus operation that uses two wires for data (one for data and one for the complement of the data) and one wire for ground.

digit A symbol used to express a quantity.

digital Related to digits or discrete quantities; having a set of discrete values as opposed to continuous values.

digital signal processor (DSP) A special type of microprocessor that processes data in real time.

digital-to-analog (D/A) conversion The process of converting a sequence of digital codes to an analog form.

digital-to-analog converter (DAC) A device in which information in digital form is converted to analog form.

DIMM Dual in-line memory module.

diode A semiconductor device that conducts current in only one direction.

DIP Dual in-line package; a type of IC package whose leads must pass through holes to the other side of a PC board.

distributive law The law that states that ORing several variables and then ANDing the result with a single variable is equivalent to ANDing the single variable with each of the several variables and then ORing the product.

dividend In a division operation, the quantity that is being divided.

divisor In a division operation, the quantity that is divided into the dividend.

DLT Digital linear tape; a type of magnetic tape format.

DMA Direct memory access; a method to directly interface a peripheral device to memory without using the CPU for control.

domain All of the variables in a Boolean expression.

"Don't care" A combination of input literals that cannot occur and can be used as a 1 or a 0 on a Karnaugh map for simplification.

downloading A design flow process in which the logic design is transferred from software to hardware.

drain One of the terminals of a field-effect transistor.

DRAM Dynamic random-access memory; a type of semiconductor memory that uses capacitors as the storage elements and is a volatile, read/write memory.

DSP core The central processing unit of a digital system processor.

DTE Data terminal equipment.

duty cycle The ratio of pulse width to period expressed as a percentage.

DVD-ROM Digital versatile disk-ROM; also known as digital video disk-ROM; a type of optical storage device on which data is prestored with a much higher capacity than a CD-ROM.

dynamic memory A type of semiconductor memory having capacitive storage cells that lose stored data over a period of time and, therefore, must be refreshed.

ECL Emitter-coupled logic; a class of integrated logic circuits that are implemented with nonsaturating bipolar junction transistors.

E²CMOS Electrically erasable CMOS (EECMOS); the circuit technology used for the reprogrammable cells in a PLD.

edge-triggered flip-flop A type of flip-flop in which the data are entered and appear on the output on the same clock edge.

EDIF Electronic design interchange format; a standard form of netlist.

EDO DRAM Extended data output dynamic random-access memory.

EEPROM Electrically erasable programmable read-only memory; a type of nonvolatile PLD reprogrammable link based on electrically-erasable programmable read-only memory cells and can be turned on or off repeatedly by programming.

8 mm A type of magnetic tape format.

elasticity The ability of a cloud storage system to deal with variations in the amount of data being transferred without service interruptions.

electromagnetic waves Related to the electromagnetic spectrum, which includes radio waves, microwaves, infrared, visible, ultraviolet, X-rays, and gamma rays.

embedded system Generally, a single-purpose system, such as a processor, built into a larger system for the purpose of controlling the system.

EMI Electromagnetic interference.

emitter One of the three regions in a bipolar junction transistor.

enable To activate or put into an operational mode; an input on a logic circuit that permits its operation.

encoder A digital circuit (device) that converts information to a coded form.

entity The VHDL unit that describes the inputs and outputs of a logic function.

EPROM Erasable programmable read-only memory; A type of PLD nonvolatile programmable link based on electrically programmable read-only memory cells and can be turned either on or off once with programming.

error detection The process of detecting bit errors in a digital code.

even parity The condition of having an even number of 1s in every group of bits.

exception Any software event that requires special handling by the processor.

exclusive-NOR (XNOR)gate A logic gate that produces a LOW only when the two inputs are at opposite levels.

exclusive-OR (XOR) A basic logic operation in which a HIGH occurs when the two inputs are at opposite levels.

exclusive-OR (XOR) gate A logic gate that produces a HIGH only when the two inputs are at opposite levels.

execute A CPU process in which an instruction is carried out; a stage of the DSP pipeline operation in which the decoded instructions are carried out.

exponent The part of a floating-point number that represents the number of places that the decimal point (or binary point) is to be moved.

fall time The time interval between the 90% point and the 10% point on the negative-going edge of a pulse.

fan-out The number of equivalent gate inputs of the same family series that a logic gate can drive.

FDM Frequency division multiplexing; a broadband technique in which the total bandwidth available to a system is divided into frequency sub-bands and information is sent in analog form.

feedback The output voltage or a portion of it that is connected back to the input of a circuit.

FET Field-effect transistor.

fetch A CPU process in which an instruction is obtained from the memory; a stage of the DSP pipeline operation in which an instruction is obtained from the program memory.

FIFO First in–first out memory.

Firewire A high-speed external serial bus standard developed by Apple Inc. and used in high-speed communications and real-time data transfer, also known as IEEE-1394.

firmware Small fixed programs and/or data structures that internally control various electronic devices; usually stored in ROM.

fixed-function logic A category of digital integrated circuits having functions that cannot be altered.

flag A bit that indicates the result of an arithmetic or logic operation or is used to alter an operation.

flash A type of PLD nonvolatile reprogrammable link technology based on a single transistor cell.

flash ADC A simultaneous analog-to-digital converter.

flash memory A nonvolatile read/write random-access semiconductor memory in which data is stored as charge on the floating gate of a certain FET.

flip-flop A basic storage circuit that can store only one bit at a time; a synchronous bistable device.

floating-point number A number representation based on scientific notation in which the number consists of an exponent and a mantissa.

forward bias A voltage polarity condition that allows a semiconductor *pn* junction in a transistor or diode to conduct current.

FPGA Field-programmable gate array; a programmable logic device that uses the LUT as the basic logic elements and generally employs either antifuse or SRAM-based process technology.

FPM DRAM Fast page mode dynamic random-access memory.

frequency (f) The number of pulses in one second for a periodic waveform. The unit of frequency is the hertz.

FSK Frequency shift keying; a form of modulation in which a digital signal modulates the frequency of a higher frequency sine wave.

full-adder A digital circuit that adds two bits and an input carry to produce a sum and an output carry.

full-duplex A connection in which the data flow both ways simultaneously in the same channel.

functional simulation A software process that tests the logical or functional operation of a design.

fuse A type of PLD nonvolatile programmable link that can be left shorted or can be opened once as directed by the program; also called a fusible link.

GAL Generic array logic; a reprogrammable type of SPLD that is similar to a PAL except that it uses a reprogrammable process technology, such as EEPROM (E^2 CMOS), instead of fuses.

gate A logic circuit that performs a basic logic operation, such as AND or OR; one of the three terminals of a field-effect transistor.

glitch A voltage or current spike of short duration, usually unintentionally produced and unwanted.

graphic (schematic) entry A method of entering a logic design into software by graphically creating a logic diagram (schematic) on a design screen.

GPIB General-purpose interface bus based on the IEEE 488 standard.

Gray code An unweighted digital code characterized by a single bit change between adjacent code numbers in a sequence.

half-adder A digital circuit that adds two bits and produces a sum and an output carry. It cannot handle input carries.

half-duplex A connection in which the data flow both ways but not at the same time in the same channel.

Hamming code An error detection and correction code used in data transmission.

handshaking The process of signal interchange by which two digital devices or systems jointly establish communication.

hard core A fixed portion of logic in an FPGA that is put in by the manufacturer to provide a specific function.

hard disk A magnetic disk storage device; typically, a stack of two or more rigid disks enclosed in a sealed housing.

hardware The circuitry and physical components of a computer system (as opposed to the instructions called software).

HDL Hardware description language; a language used for describing a logic design using software.

hexadecimal Describes a number system with a base of 16.

high-level language A type of computer language closest to human language that is a level above assembly language.

high-Z The high-impedance state of a tri-state circuit in which the output is effectively disconnected from the rest of the circuit.

hit rate The percentage of memory accesses that find the requested data in the given level of memory.

hold time The time interval required for the control levels to remain on the inputs to a flip-flop after the triggering edge of the clock in order to reliably activate the device.

HPIB Hewlett-Packard interface bus; same as GPIB (general-purpose interface bus).

hysteresis A characteristic of a threshold-triggered circuit, such as the Schmitt trigger, where the device turns on and off at different input levels.

IEEE Institute of Electrical and Electronics Engineers.

IEEE 488 bus Same as GPIB (general-purpose interface bus); a standard parallel bus used widely for test and measurement interfacing.

IEEE 1394 A serial bus for high-speed data transfer; also known as FireWire.

I^2L Integrated injection logic; an IC technology.

implementation The software process where the logic structures described by the netlist are mapped into the structure of the target device; the physical realization of a conceptual design.

increment To increase the binary state of a counter by one.

inhibit To prevent the passage of a signal from one point to another.

input The signal or line going into a circuit; a signal that controls the operation of a circuit.

input/output (I/O) A terminal of a device that can be used as either an input or as an output.

instruction One step in a computer program; a unit of information that tells the CPU what to do.

in-system programming (ISP) A method for programming SPLDs after they are installed on a printed circuit board and operating in a system.

integer A whole number.

integrated circuit (IC) A type of circuit in which all of the components are integrated on a single chip of semiconductive material of very small size.

intellectual property (IP) Designs owned by the manufacturer of programmable logic devices or other products.

interfacing The process of making two or more electronic devices or systems operationally compatible with each other so that they function properly together.

interrupt Any hardware event that requires special handling by the processor, an event that causes the current process to be temporarily stopped while a service routine is run.

inversion The conversion of a HIGH level to a LOW level or vice versa; also called complementation.

inverter A NOT circuit; a circuit that changes a HIGH to a LOW or vice versa.

I/O port Input/output port; the interface between an internal bus and a peripheral.

IP Instruction pointer; a special register within the CPU that holds the offset address of the next instruction to be executed.

ISA bus Industry standard architecture bus; an internal parallel bus standard.

J-K flip-flop A type of flip-flop that can operate in the SET, RESET, no-change, and toggle modes.

Johnson counter A type of register in which a specific prestored pattern of 1s and 0s is shifted through the stages, creating a unique sequence of bit patterns.

JTAG Joint test action group; the IEEE Std. 1149.1 standard interface for in-system programming.

junction The boundary between an n region and a p region in a BJT.

Karnaugh map An arrangement of cells representing the combinations of literals in a Boolean expression and used for a systematic simplification of the expression.

LAB Logic array block; an SPLD array in a CPLD.

latch A bistable digital circuit used for storing a bit.

latency The time between the request for data and the delivery of the data to the user.

latency period The time it takes for the desired sector to spin under the head once the head is positioned over the desired track of a magnetic hard disk.

LCC Leadless ceramic chip; an SMT package that has metallic contacts molded into its body.

LCD Liquid crystal display.

leading edge The first transition of a pulse.

least significant bit (LSB) Generally, the right-most bit in a binary whole number or code.

LED Light-emitting diode.

LIFO Last in–first out memory, memory stack.

listener An instrument capable of receiving data on a GPIB (general-purpose interface bus) when it is addressed by the computer.

literal A variable or the complement of a variable.

load To enter data into a shift register.

loading The effect of the multiple inputs degrading the voltage or timing specifications of an output.

local bus An internal bus that connects the microprocessor to the cache memory, the main memory, the coprocessor, and the PCI bus controller.

local interconnect A set of lines that allows interconnections among the eight logic elements in a logic array block without using the row and column interconnects.

logic In digital electronics, the decision-making capability of gate circuits, in which a HIGH represents a true statement and a LOW represents a false one.

logic array block (LAB) A group of macrocells that can be interconnected with other LABs or to other I/Os using a programmable interconnect array; also called a function block.

logic element The smallest section of logic in an FPGA that typically contains an LUT, associated logic, and a flip-flop.

look-ahead carry A method of binary addition whereby carries from preceding adder stages are anticipated, thus eliminating carry propagation delays.

LSI Large-scale integration; a level of fixed-function IC complexity in which there are from more than 100 to 10,000 equivalent gates per chip.

LUT Look-up table; a type of memory that can be programmed to produce SOP functions.

machine code The basic binary instructions understood by the processor.

machine language Computer instructions written in binary code that are understood by a computer; the lowest level of programming language.

macrocell An SOP logic array with combinational and registered outputs; part of a PAL or GAL that generally consists of one OR gate and some associated output logic. Multiple interconnected macrocells form a CPLD.

magneto-optical disk A storage device that uses electro-magnetism and a laser beam to read and write data.

magnitude The size or value of a quantity.

main memory Memory used by computer systems to store the bulk of programs and associated data.

Manchester encoding A method of encoding called biphase in which a 1 is represented by a positive-going transition and a 0 is represented by a negative-going transition.

mantissa The magnitude of a floating-point number.

Mealy state machine A state machine in which the outputs depend on both the internal present state and on the inputs.

mechatronics Interdisciplinary field that comprises both mechanical and electronic components.

memory The portion of a computer or other system that stores binary data.

memory array An array of memory cells arranged in rows and columns.

memory hierarchy The arrangement of various memory elements within a computer architecture to achieve maximum performance.

memory latency The time required to access a memory.

MFLOPS Million floating-point operations per second.

microcontroller A semiconductor device that combines a microprocessor, memory, and various hardware peripherals on a single IC.

microprocessor A large-scale digital integrated circuit device that can be programmed with a series of instructions to perform specified functions on data.

minimization The process that results in an SOP or POS Boolean expression that contains the fewest possible terms with the fewest possible literals per term.

minterm A product of literals in which each input variable appears exactly once.

minuend The number from which another number is subtracted.

MIPS Million instructions per second.

miss A failed attempt by the processor to read or write a block of data in a given level of memory.

MMACS Million multiply/accumulates per second.

MMU Memory management unit; a device responsible for handling accesses to memory requested by the CPU.

mnemonic An English-like instruction that is converted by an assembler into a machine code for use by a processor.

modem A modulator/demodulator for interfacing digital devices to analog transmission systems such as telephone lines.

modulation The process of altering a parameter of a higher frequency signal proportional to the amplitude of a lower frequency information-carrying signal.

modulus The number of unique states through which a counter will sequence.

monostable Having only one stable state. A monostable multivibrator, commonly called a one-shot, produces a single pulse in response to a triggering input.

monotonic The characteristic of a DAC defined by the absence of any incorrect step reversals; one type of digital-to-analog linearity.

Moore state machine A state machine in which the outputs depend only on the internal present state.

MOS Metal-oxide semiconductor; a type of transistor technology.

MOSFET Metal-oxide semiconductor field-effect transistor.

most significant bit (MSB) The left-most bit in a binary whole number or code.

MSI Medium-scale integration; a level of fixed-function IC complexity in which there are from 10 to 100 equivalent gates per chip.

multicore processor A microprocessor chip with more than one processor.

multimode The characteristic of an optical fiber in which the light is propagated in multiple rays.

multiplexer (mux) A circuit (digital device) that switches digital data from several input lines onto a single output line in a specified time sequence.

multiplicand The number that is being multiplied by another number.

multiplier The number that multiplies the multiplicand.

multiprocessing A data-processing technique that uses multiple processors to multitask or run multiple programs.

multitasking A technique by which a processor runs multiple programs concurrently.

multitenancy The property of a cloud storage system that allows multiple users to share the same software applications, hardware, and data storage mechanism without seeing each other's data.

multithreading The process of executing different parts of a program, called threads, simultaneously.

multivibrator A class of digital circuits in which the output is connected back to the input (an arrangement called feedback) to produce either two stable states, one stable state, or no stable states, depending on the configuration.

NAND gate A logic circuit in which a LOW output occurs only if all the inputs are HIGH.

negative-AND An equivalent NOR gate operation in which the HIGH is the active input when all inputs are LOW.

negative-OR An equivalent NAND gate operation in which the HIGH is the active input when one or more of the inputs are LOW.

netlist A detailed listing of information necessary to describe a circuit, such as types of elements, inputs, and outputs, and all interconnections.

nibble A group of four bits.

NMOS An n-channel metal-oxide semiconductor.

node A common connection point in a circuit in which a gate output is connected to one or more gate inputs.

noise immunity The ability of a circuit to reject unwanted signals.

noise margin The difference between the maximum LOW output of a gate and the maximum acceptable LOW input of an equivalent gate; the difference between the minimum HIGH output of a gate and the minimum HIGH input of an equivalent gate; the amount by which the actual signal level exceeds the minimum acceptable level for an error-free transmission.

nonvolatile A term that describes a memory that can retain stored data when the power is removed.

NOR gate A logic gate in which the output is LOW when any or all of the inputs are HIGH.

NOT A basic logic operation that performs inversions.

NRZ Nonreturn to zero; a type of data format in which the signal level remains at one (1) for successive 1s.

numeric Related to numbers.

Nyquist frequency The highest signal frequency that can be sampled at a specified sampling frequency; a frequency equal to or less than half the sampling frequency.

object program A machine language translation of a high-level source program.

octal Describes a number system with a base of eight.

odd parity The condition of having an odd number of 1s in every group of bits.

offset address The distance in number of bytes of a physical address from the base address.

OLMC Output logic macrocell; the part of a GAL that can be programmed for either combinational or registered outputs; a block of logic in a GAL that contains a fixed OR gate and other logic for handling inputs and/or outputs.

one-shot A monostable multivibrator.

op-code Operation code; the code representing a particular microprocessor instruction; a mnemonic.

open-collector A type of output in a logic circuit in which the collector of the output transistor is left disconnected from any internal circuitry and is available for external connection; normally used for driving higher-current or higher-voltage loads.

operand The object to be manipulated by the instruction.

operating system The software that controls the computer system and oversees the execution of application software.

operational amplifier (op-amp) A device with two differential inputs that has very high gain, very high input impedance, and very low output impedance.

optical fiber A type of data transmission media used for transmitting light signals.

optical jukebox A type of auxiliary storage for very large amounts of data.

OR A basic logic operation in which a true (HIGH) output occurs when one or more of the input conditions are true (HIGH).

OR gate A logic gate that produces a HIGH output when one or more inputs are HIGH.

oscillator An electronic circuit that is based on the principle of regenerative feedback and produces a repetitive output waveform; a signal source.

OTP One-time programmable.

output The signal or line coming out of a circuit.

overflow The condition that occurs when the number of bits in a sum exceeds the number of bits in each of the numbers added.

packet A formatted block of digital data.

PAL Programmable array logic; a type of one-programmable SPLD that consists of a programmable array of AND gates that connects to a fixed array of OR gates.

PAM Pulse amplitude modulation; a method of modulation in which the height or amplitude of the pulses are varied according to the modulating analog signal, and each pulse represents a value of amplitude of the analog signal.

parallel In digital systems, data occurring simultaneously on several lines; the transfer or processing of several bits simultaneously.

parallel bus A bus that consists of multiple conductors and carries several data bits simultaneously, one on each conductor.

parity In relation to binary codes, the condition of evenness or oddness of the number of 1s in a code group.

parity bit A bit attached to each group of information bits to make the total number of 1s odd or even for every group of bits.

PCI bus An internal synchronous bus for interconnecting chips, expansion boards, and processor/memory subsystems.

PCI-Express Also designated as PCIe or PCI-E. This bus differs from the PCI and PCI-X buses in that it does not use a shared bus.

PCI-X A high-performance enhancement of the PCI bus that is backward compatible with PCI.

PCM Pulse code modulation; A method of modulation that involves sampling of an analog signal amplitude at regular intervals and converting the sampled values to a digital code.

period (T) The time required for a periodic waveform to repeat itself.

periodic Describes a waveform that repeats itself at a fixed interval.

peripheral A device or instrument that provides communication with a computer or provides auxiliary services or functions for the computer.

physical address The actual location of a data unit in memory.

PIC Programable interrupt controller; handles the interrupts on a priority basis.

pipeline As applied to memories, an implementation that allows a read or write operation to be initiated before the previous operation is completed; part of the DSP architecture that allows multiple instructions to be processed simultaneously.

pipelining A technique where the processor begins executing the next instruction before the previous instruction has been completed.

PLA Programmable logic array; an SPLD with programmable AND and OR arrays.

platform FPGA An FPGA that contains either or both hard core and soft core embedded processors and other functions.

PLCC Plastic leaded chip carrier; an SMT package whose leads are turned up under its body in a J-type shape.

PLD Programmable logic device; an integrated circuit that can be programmed with any specified logic function.

PMOS A p-channel metal-oxide semiconductor.

pointer The contents of a register (or registers) that contain an address.

polling The process of checking a series of peripheral devices to determine if any require service from the CPU.

port A physical interface on a computer through which data are passed to or from peripherals.

positive logic The system of representing a binary 1 with a HIGH and a binary 0 with a LOW.

power dissipation The product of the dc supply voltage and the dc supply current in an electronic circuit; the amount of power required by a circuit.

PPM Pulse position modulation; a method of modulation in which the position of each pulse relative to a reference or timing signal is varied proportional to the amplitude of the modulating signal waveform.

prefetching The process of executing instructions at the same time as other instructions are "fetched," eliminating idle time; also called pipelining.

preset An asynchronous input used to set a flip-flop (make the Q output 1).

priority encoder An encoder in which only the highest value input digit is encoded and any other active input is ignored.

probe An accessory used to connect a voltage to the input of an oscilloscope or other instrument.

processes Instances of a computer program that are being executed.

product The result of a multiplication.

product-of-sums (POS) A form of Boolean expression that is basically the ANDing of ORed terms.

product term The Boolean product of two or more literals equivalent to an AND operation.

program A list of computer instructions arranged to achieve a specific result; software.

programmable interconnect array (PIA) An array consisting of conductors that run throughout the CPLD chip and to which connections from the macrocells in each LAB can be made.

programmable logic A category of digital integrated circuits capable of being programmed to perform specified functions.

PROM Programmable read-only semiconductor memory; an SPLD with a fixed AND array and programmable OR array; used as a memory device and normally not as a logic circuit device.

propagation delay time The time interval between the occurrence of an input transition and the occurrence of the corresponding output transition in a logic circuit.

protocol A standardized set of software regulations, requirements, and procedures that control and regulate the transmission, processing, and exchange of data among devices.

pseudo-operation An instruction to the assembler (as opposed to a processor).

PSK Phase shift keying; a form of modulation in which a digital signal modulates the phase of a higher frequency sine wave.

pull-up resistor A resistor with one end connected to the dc supply voltage used to keep a given point in a circuit HIGH when in the inactive state.

pulse A sudden change from one level to another, followed after a time, called the pulse width, by a sudden change back to the original level.

pulse width (t_w) The time interval between the 50% points of the leading and trailing edges of the pulse; the duration of the pulse.

PWM Pulse width modulation; a method of modulation in which the width or duration of the pulses and duty cycle are varied according to the modulating analog signal, and each pulse width represents an amplitude value of the analog signal.

QAM Quadrature amplitude modulation; a form of modulation that uses a combination of PSK and amplitude modulation to send information.

QIC Quarter-inch cassette; a type of magnetic tape.

quantization The process whereby a binary code is assigned to each sampled value during analog-to-digital conversion.

queue A high-speed memory that stores instructions or data.

quotient The result of a division.

race A condition in a logic network in which the difference in propagation times through two or more signal paths in the network can produce an erroneous output.

RAM Random-access memory; a volatile read/write semi-conductor memory.

rank A group of chips that make up a memory module that stores data in units such as words or bytes.

read The process of retrieving data from a memory.

real mode Operation of an Intel processor in a manner to emulate the 8086's 1 MB of memory.

record length The number of samples (data points) that an oscilloscope can capture and store.

recycle To undergo transition (as in a counter) from the final or terminal state back to the initial state.

refresh To renew the contents of a dynamic memory by recharging the capacitor storage cells.

register A digital circuit capable of storing and shifting binary information; typically used as a temporary storage device.

register array A set of temporary storage locations within the microprocessor for keeping data and addresses that need to be accessed quickly by the program.

registered A CPLD macrocell output configuration where the output comes from a flip-flop.

relocatable code A program that can be moved anywhere within the memory space without changing the basic code.

remainder The amount left over after a division.

RESET The state of a flip-flop or latch when the output is 0; the action of producing a RESET state.

resolution The number of bits used to digitally represent a sampled value.

reverse bias A voltage polarity condition that prevents a *pn* junction of a transistor or diode from conducting current.

ring counter A register in which a certain pattern of 1s and 0s is continuously recirculated.

ripple carry A method of binary addition in which the output carry from each adder becomes the input carry of the next higher-order adder.

ripple counter An asynchronous counter.

rise time The time required for the positive-going edge of a pulse to go from 10% of its full value to 90% of its full value.

ROM Read-only semiconductor memory, accessed randomly; also referred to as mask-ROM.

RS-232 A bus standard, also known as EIA-232, used in industrial and telecommunication applications as well as scientific instrumentation, but largely replaced by USB in computer applications.

RS-422 A bus standard for differential data transmission.

RS-423 A bus standard for single-ended data transmission.

RS-485 A bus standard for differential data transmission.

RZ Return to zero; a type of data format in which the signal level goes to or remains at zero after each data bit.

sampling The process of taking a sufficient number of discrete values at points on a waveform that will define the shape of the waveform.

sampling rate The rate at which the analog-to-digital converter (ADC) in an oscilloscope is clocked to digitize the incoming signal.

SAS Serial attached SCSI.

scalability The ability of a cloud storage system to handle increasing amounts of data in a smooth manner. The ability of a cloud storage system to improve the movement of data when additional resources are added.

schematic (graphic) entry A method of placing a logic design into software using schematic symbols.

Schottky A specific type of transistor-transistor logic circuit technology.

SCSI Small computer system interface.

SDRAM Synchronous dynamic random-access memory.

seek time The time for the read/write head in a hard drive to position itself over the desired track for a read operation.

segment A 64k block of memory.

sequential circuit A digital circuit whose logic states follow a specified time sequence.

serial Having one element following another, as in a serial transfer of bits; occurring, as pulses, in sequence rather than simultaneously.

serial bus A bus that carries data bits sequentially one at a time on a single conductor.

server Any computerized process that shares a resource with one or more clients. A computer and software with a large memory capacity that responds to requests across a network to provide file storage and access as well as services such as file sharing.

SET The state of a flip-flop or latch when the output is 1; the action of producing a SET state.

set-up time The time interval required for the control levels to be on the inputs to a digital circuit, such as a flip-flop, prior to the triggering edge of clock pulse.

shared bus A bus, such as PCI, that is shared by multiple devices.

signal A type of VHDL object that holds data.

signal-to-noise ratio (SNR) A measure of the signal strength relative to background noise, usually expressed in decibels (dB).

signal tracing A troubleshooting technique in which waveforms are observed in a step-by-step manner beginning at the input and working toward the output or vice versa. At each point the observed waveform is compared with the correct signal for that point.

sign bit The left-most bit of a binary number that designates whether the number is positive (0) or negative (1).

SIMM Single-in-line memory module.

simplex A connection in which data flows in only one direction from the sender (transmitter) to the receiver.

single-ended operation A bus operation that uses one wire for data and one wire for ground.

single mode The characteristic of an optical fiber in which the light tends to propagate in a single beam or ray.

SMT Surface-mount technology; an IC package technique in which the packages are smaller than DIPs and are mounted on the printed surface of the PC board.

soft core A portion of logic in an FPGA; similar to hard core except it has some programmable features.

software Computer programs; programs that instruct a computer what to do in order to carry out a given set of tasks.

software interrupt An instruction that invokes an interrupt service routine.

SOIC Small-outline integrated circuit; an SMT package that resembles a small DIP but has its leads bent out in a "gull-wing" shape.

source A sending device of a bus; one of the terminals of a field-effect transistor.

source program A program written in either assembly or high-level language.

speed-power product A performance parameter that is the product of the propagation delay time and the power dissipation in a digital circuit.

SPI Serial-to-peripheral interface bus; a synchronous serial communications bus that uses four wires for communication between a "master" device and a "slave" device.

SPLD Simple programmable logic device; an array of AND gates and OR gates that can be programmed to achieve specified logic functions. Four types are PROM, PLA, PAL, and GAL.

SRAM Static random-access memory; a type of PLD volatile reprogrammable link based on static random-access memory cells and can be turned on or off repeatedly with programming.

SSI Small-scale integration; a level of fixed-function IC complexity in which there are up to 10 equivalent gates per chip.

SSOP Shrink small-outline package.

stage One storage element (flip-flop) in a register.

state diagram A graphic depiction of a sequence of states or values.

state machine A logic system or circuit exhibiting a sequence of states conditioned by internal logic and external inputs; any sequential circuit exhibiting a specified sequence of states. Two types of state machine are Moore and Mealy.

static memory A volatile semiconductor memory that uses flip-flops as the storage cells and is capable of retaining data without refreshing.

storage The capability of a digital device to retain bits; the process of retaining digital data for later use.

STP Shielded twisted pair; a type of data transmission medium.

string A contiguous sequence of bytes or words.

strobing A process of using a pulse to sample the occurrence of an event at a specified time in relation to the event.

subroutine A series of instructions that can be assembled together and used repeatedly by a program but programmed only once.

subtracter A logic circuit used to subtract two binary numbers.

subtrahend The number that is being subtracted from the minuend.

sum The result when two or more numbers are added together.

sum-of-products (SOP) A form of Boolean expression that is basically the ORing of ANDed terms.

sum term The Boolean sum of two or more literals equivalent to an OR operation.

synchronous A condition that describes signals or systems that are aligned or synchronized with each other in terms of timed events, two or more systems that have the same timing signal.

synchronous counter A type of counter in which each stage is clocked by the same pulse.

synthesis The software process where the design is translated into a netlist.

system bus The interconnecting paths in a computer system including the address bus, data bus and control bus.

talker An instrument capable of transmitting data on a GPIB (general-purpose interface bus).

tape library A type of auxiliary storage for very large amounts of data.

target device A PLD mounted on a programming fixture or development board into which a software logic design is to be downloaded; the programmable logic device that is being programmed.

TDM Time division multiplexing; a technique in which data from several sources are interleaved on a time basis and sent on a single communication channel or data link.

terminal count The final state in a counter's sequence.

text entry A method of entering a logic design into software using a hardware description language (HDL).

throughput The average speed with which a program is executed.

timer A circuit that can be used as a one-shot or as an oscillator; a circuit that produces a fixed time interval output.

timing diagram A graph of digital waveforms showing the proper time relationship of two or more waveforms and how each waveform changes in relation to the others.

timing simulation A software process that uses information on propagation delays and netlist data to test both the logical operation and the worst-case timing of a design.

toggle The action of a flip-flop when it changes state on each clock pulse.

totem-pole A type of output in TTL circuits.

trailing edge The second transition of a pulse.

transistor A semiconductor device exhibiting current and/or voltage gain. When used as a switching device, it approximates an open or closed switch.

trigger A pulse used to initiate a change in the state of a logic circuit.

tri-state A type of output in logic circuits that exhibits three states: HIGH, LOW, and high-Z; also known as 3-state.

tri-state buffer A circuit used to interface one device to another to prevent loading.

troubleshooting The technique of systematically identifying, isolating, and correcting a fault in a circuit or system.

truth table A table showing the inputs and corresponding output level of a logic circuit.

TTL Transistor-transistor logic; a class of integrated logic circuit that uses bipolar junction transistors. Also called *bipolar*.

ULSI Ultra large-scale integration; a level of IC complexity in which there are more than 100,000 equivalent gates per chip.

unit load A measure of fan-out. One gate input represents a unit load to the output of a gate within the same IC family.

universal gate Either a NAND gate or a NOR gate. The term *universal* refers to the property of a gate that permits any logic function to be implemented by that gate or by a combination of gates of that kind.

universal shift register A register that has both serial and parallel input and output capability.

up/down counter A counter that can progress in either direction through a certain sequence.

USB Universal serial bus; an external serial bus standard.

UTP Unshielded twisted pair; a type of data transmission medium.

UV EPROM Ultraviolet erasable programmable ROM.

variable symbol used to represent an action, a condition, or data that can have a value of 1 or 0, usually designated by an italic letter or word.

VHDL A standard hardware description language; IEEE Std. 1076-1993.

VLSI Very large-scale integration; a level of IC complexity in which there are from more than 10,000 to 100,000 equivalent gates per chip.

volatile The characteristic of a programmable logic device that loses programmed data when power is turned off.

wait state A system bus delay equal to one processor clock cycle. Wait states are used to ensure that the system bus timing satisfies the address, data, and control timing specifications of a system.

weight The value of a digit in a number based on its position in the number.

word A group of bits or bytes that acts as a single entity that can be stored in one memory location; two bytes.

word capacity The number of words that a memory can store.

word length The number of bits in a word.

WORM Write once-read many; a type of optical storage device.

write The process of storing data in a memory.

zero suppression The process of blanking out leading or trailing zeros in a digital display.